Advances in Intelligent Systems and Computing

Volume 1159

The series "Advances in Intelligent Systems and Computing" contains publications on theory, applications, and design methods of Intelligent Systems and Intelligent Computing. Virtually all disciplines such as engineering, natural sciences, computer and information science, ICT, economics, business, e-commerce, environment, healthcare, life science are covered. The list of topics spans all the areas of modern intelligent systems and computing such as: computational intelligence, soft computing including neural networks, fuzzy systems, evolutionary computing and the fusion of these paradigms, social intelligence, ambient intelligence, computational neuroscience, artificial life, virtual worlds and society, cognitive science and systems, Perception and Vision, DNA and immune based systems, self-organizing and adaptive systems, e-Learning and teaching, human-centered and human-centric computing, recommender systems, intelligent control, robotics and mechatronics including human-machine teaming, knowledge-based paradigms, learning paradigms, machine ethics, intelligent data analysis, knowledge management, intelligent agents, intelligent decision making and support, intelligent network security, trust management, interactive entertainment, Web intelligence and multimedia.

The publications within "Advances in Intelligent Systems and Computing" are primarily proceedings of important conferences, symposia and congresses. They cover significant recent developments in the field, both of a foundational and applicable character. An important characteristic feature of the series is the short publication time and world-wide distribution. This permits a rapid and broad dissemination of research results.

**** Indexing: The books of this series are submitted to ISI Proceedings, EI-Compendex, DBLP, SCOPUS, Google Scholar and Springerlink ****

More information about this series at http://www.springer.com/series/11156

Álvaro Rocha · Hojjat Adeli ·
Luís Paulo Reis · Sandra Costanzo ·
Irena Orovic · Fernando Moreira
Editors

Trends and Innovations in Information Systems and Technologies

Volume 1

Springer

Editors
Álvaro Rocha
Departamento de Engenharia Informática
Universidade de Coimbra
Coimbra, Portugal

Luís Paulo Reis
FEUP
Universidade do Porto
Porto, Portugal

Irena Orovic
Faculty of Electrical Engineering
University of Montenegro
Podgorica, Montenegro

Hojjat Adeli
College of Engineering
The Ohio State University
Columbus, OH, USA

Sandra Costanzo
DIMES
Università della Calabria
Arcavacata, Italy

Fernando Moreira
Universidade Portucalense
Porto, Portugal

ISSN 2194-5357 ISSN 2194-5365 (electronic)
Advances in Intelligent Systems and Computing
ISBN 978-3-030-45687-0 ISBN 978-3-030-45688-7 (eBook)
https://doi.org/10.1007/978-3-030-45688-7

This Springer imprint is published by the registered company Springer Nature Switzerland AG
The registered company address is: Gewerbestrasse 11, 6330 Cham, Switzerland

Preface

This book contains a selection of papers accepted for presentation and discussion at the 2020 World Conference on Information Systems and Technologies (WorldCIST'20). This conference had the support of the IEEE Systems, Man, and Cybernetics Society (IEEE SMC), Iberian Association for Information Systems and Technologies/Associação Ibérica de Sistemas e Tecnologias de Informação (AISTI), Global Institute for IT Management (GIIM), University of Montengero, Mediterranean University and Faculty for Business in Tourism of Budva. It took place at Budva, Montenegro, during 7–10 April 2020.

The World Conference on Information Systems and Technologies (WorldCIST) is a global forum for researchers and practitioners to present and discuss recent results and innovations, current trends, professional experiences and challenges of modern information systems and technologies research, technological development and applications. One of its main aims is to strengthen the drive towards a holistic symbiosis between academy, society and industry. WorldCIST'20 built on the successes of WorldCIST'13 held at Olhão, Algarve, Portugal; WorldCIST'14 held at Funchal, Madeira, Portugal; WorldCIST'15 held at São Miguel, Azores, Portugal; WorldCIST'16 held at Recife, Pernambuco, Brazil; WorldCIST'17 held at Porto Santo, Madeira, Portugal; WorldCIST'18 held at Naples, Italy and WorldCIST'19 which took place at La Toja, Spain.

The program committee of WorldCIST'20 was composed of a multidisciplinary group of almost 300 experts and those who are intimately concerned with information systems and technologies. They have had the responsibility for evaluating, in a 'blind review' process, the papers received for each of the main themes proposed for the conference: (A) Information and Knowledge Management; (B) Organizational Models and Information Systems; (C) Software and Systems Modelling; (D) Software Systems, Architectures, Applications and Tools; (E) Multimedia Systems and Applications; (F) Computer Networks, Mobility and Pervasive Systems; (G) Intelligent and Decision Support Systems; (H) Big Data Analytics and Applications; (I) Human–Computer Interaction; (J) Ethics, Computers and Security; (K) Health Informatics; (L) Information Technologies in

Education; (M) Information Technologies in Radiocommunications; (N) Technologies for Biomedical Applications.

The conference also included workshop sessions taking place in parallel with the conference ones. Workshop sessions covered themes such as (i) Innovative Technologies Applied to Rural; (ii) Network Modelling, Learning and Analysis; (iii) Intelligent Systems and Machines; (iv) Healthcare Information Systems Interoperability, Security and Efficiency; (v) Applied Statistics and Data Analysis using Computer Science; (vi) Cybersecurity for Smart Cities Development; (vii) Education through ICT; (viii) Unlocking the Artificial Intelligence Interplay with Business Innovation (ix) and Pervasive Information Systems.

WorldCIST'20 received about 400 contributions from 57 countries around the world. The papers accepted for presentation and discussion at the conference are published by Springer (this book) in three volumes and will be submitted for indexing by ISI, EI-Compendex, SCOPUS, DBLP and/or Google Scholar, among others. Extended versions of selected best papers will be published in special or regular issues of relevant journals, mainly SCI/SSCI and Scopus/EI-Compendex indexed journals.

We acknowledge all of those that contributed to the staging of WorldCIST'20 (authors, committees, workshop organizers and sponsors). We deeply appreciate their involvement and support that were crucial for the success of WorldCIST'20.

April 2020

Álvaro Rocha
Hojjat Adeli
Luís Paulo Reis
Sandra Costanzo
Irena Orovic
Fernando Moreira

Organization

Conference

General Chair

Álvaro Rocha	University of Coimbra, Portugal

Co-chairs

Hojjat Adeli	The Ohio State University, USA
Luis Paulo Reis	University of Porto, Portugal
Sandra Costanzo	University of Calabria, Italy

Local Organizing Committee

Irena Orovic (Chair)	University of Montenegro, Montenegro
Milos Dakovic	University of Montenegro, Montenegro
Andjela Draganic	University of Montenegro, Montenegro
Milos Brajovic	University of Montenegro, Montenegro
Snezana Scepanvic	Mediterranean University, Montenegro
Rade Ratkovic	Faculty of Business and Tourism, Montenegro

Advisory Committee

Ana Maria Correia (Chair)	University of Sheffield, UK
Benjamin Lev	Drexel University, USA
Chatura Ranaweera	Wilfrid Laurier University, Canada
Chris Kimble	KEDGE Business School and MRM, UM2, Montpellier, France
Erik Bohlin	Chalmers University of Technology, Sweden
Eva Onaindia	Polytechnical University of Valencia, Spain
Gintautas Dzemyda	Vilnius University, Lithuania

Janusz Kacprzyk	Polish Academy of Sciences, Poland
Jason Whalley	Northumbria University, UK
João Tavares	University of Porto, Portugal
Jon Hall	The Open University, UK
Justin Zhang	University of North Florida, USA
Karl Stroetmann	Empirica Communication and Technology Research, Germany
Kathleen Carley	Carnegie Mellon University, USA
Keng Siau	Missouri University of Science and Technology, USA
Manlio Del Giudice	University of Rome Link Campus, Italy
Michael Koenig	Long Island University, USA
Miguel-Angel Sicilia	University of Alcalá, Spain
Reza Langari	Texas A&M University, USA
Vedat Verter	McGill University, Canada
Vishanth Weerakkody	Bradford University, UK
Wim Van Grembergen	University of Antwerp, Belgium

Program Committee

Abdul Rauf	RISE SICS, Sweden
Adnan Mahmood	Waterford Institute of Technology, Ireland
Adriana Peña Pérez Negrón	Universidad de Guadalajara, Mexico
Adriani Besimi	South East European University, Macedonia
Agostinho Sousa Pinto	Polytechnic of Porto, Portugal
Ahmed El Oualkadi	Abdelmalek Essaadi University, Morocco
Ahmed Rafea	American University in Cairo, Egypt
Alberto Freitas	FMUP, University of Porto, Portugal
Aleksandra Labus	University of Belgrade, Serbia
Alexandru Vulpe	University Politehnica of Bucharest, Romania
Ali Idri	ENSIAS, University Mohammed V, Morocco
Amélia Badica	Universti of Craiova, Romania
Amélia Cristina Ferreira Silva	Polytechnic of Porto, Portugal
Almir Souza Silva Neto	IFMA, Brazil
Amit Shelef	Sapir Academic College, Israel
Ana Isabel Martins	University of Aveiro, Portugal
Ana Luis	University of Coimbra, Portugal
Anabela Tereso	University of Minho, Portugal
Anacleto Correia	CINAV, Portugal
Anca Alexandra Purcarea	University Politehnica of Bucharest, Romania
Andjela Draganic	University of Montenegro, Montenegro
Aneta Polewko-Klim	University of Białystok, Institute of Informatics, Poland
Aneta Poniszewska-Maranda	Lodz University of Technology, Poland
Angeles Quezada	Instituto Tecnologico de Tijuana, Mexico

Anis Tissaoui	University of Jendouba, Tunisia
Ankur Singh Bist	KIET, India
Ann Svensson	University West, Sweden
Antoni Oliver	University of the Balearic Islands, Spain
Antonio Jiménez-Martín	Universidad Politécnica de Madrid, Spain
Antonio Pereira	Polytechnic of Leiria, Portugal
Armando Toda	University of São Paulo, Brazil
Arslan Enikeev	Kazan Federal University, Russia
Benedita Malheiro	Polytechnic of Porto, ISEP, Portugal
Boris Shishkov	ULSIT/IMI-BAS/IICREST, Bulgaria
Borja Bordel	Universidad Politécnica de Madrid, Spain
Branko Perisic	Faculty of Technical Sciences, Serbia
Bruno Veloso	INESC TEC, Portugal
Carla Pinto	Polytechnic of Porto, ISEP, Portugal
Carla Santos Pereira	Universidade Portucalense, Portugal
Catarina Reis	Polytechnic of Leiria, Portugal
Cengiz Acarturk	Middle East Technical University, Turkey
Cesar Collazos	Universidad del Cauca, Colombia
Christophe Feltus	LIST, Luxembourg
Christophe Soares	University Fernando Pessoa, Portugal
Christos Bouras	University of Patras, Greece
Christos Chrysoulas	London South Bank University, UK
Christos Troussas	University of Piraeus, Greece
Ciro Martins	University of Aveiro, Portugal
Claudio Sapateiro	Polytechnic of Setúbal, Portugal
Costin Badica	University of Craiova, Romania
Cristian García Bauza	PLADEMA-UNICEN-CONICET, Argentina
Cristian Mateos	ISISTAN-CONICET, UNICEN, Argentina
Daria Bylieva	Peter the Great St.Petersburg Polytechnic University, Russia
Dante Carrizo	Universidad de Atacama, Chile
Dayana Spagnuelo	Vrije Universiteit Amsterdam, Netherlands
Dušan Barać	University of Belgrade, Serbia
Edita Butrime	Lithuanian University of Health Sciences, Lithuania
Edna Dias Canedo	University of Brasilia, Brazil
Eduardo Santos	Pontifical Catholic University of Paraná, Brazil
Egils Ginters	Riga Technical University, Latvia
Ekaterina Isaeva	Perm State University, Russia
Elena Mikhailova	ITMO University, Russia
Eliana Leite	University of Minho, Portugal
Erik Fernando Mendez Garcea	Autonomous Regional University of the Andes, Ecuador
Eriks Sneiders	Stockholm University, Sweden
Esteban Castellanos	ESPE, Ecuador

Imen Ben Said	Université de Sfax, Tunisia
Inês Domingues	University of Coimbra, Portugal
Isabel Lopes	Polytechnic of Bragança, Portugal
Isabel Pedrosa	Coimbra Business School ISCAC, Portugal
Isaías Martins	University of Leon, Spain
Issam Moghrabi	Gulf University for Science and Technology, Kuwait
Ivan Dunđer	University of Zabreb, Croatia
Ivan Lukovic	University of Novi Sad, Serbia
Jaime Diaz	University of La Frontera, Chile
Jan Kubicek	Technical University of Ostrava, Czech Republic
Jean Robert Kala Kamdjoug	Catholic University of Central Africa, Cameroon
Jesús Gallardo Casero	University of Zaragoza, Spain
Jezreel Mejia	CIMAT, Unidad Zacatecas, Mexico
Jikai Li	The College of New Jersey, USA
Jinzhi Lu	KTH Royal Institute of Technology, Sweden
Joao Carlos Silva	IPCA, Portugal
João Manuel R. S. Tavares	University of Porto, FEUP, Portugal
João Paulo Pereira	Polytechnic of Bragança, Portugal
João Reis	University of Aveiro, Portugal
João Reis	University of Lisbon, Portugal
João Rodrigues	University of the Algarve, Portugal
João Vidal Carvalho	Polytechnic of Coimbra, Portugal
Joaquin Nicolas Ros	University of Murcia, Spain
Jorge Barbosa	Polytechnic of Coimbra, Portugal
Jorge Buele	Technical University of Ambato, Ecuador
Jorge Esparteiro Garcia	Polytechnic Institute of Viana do Castelo, Portugal
Jorge Gomes	University of Lisbon, Portugal
Jorge Oliveira e Sá	University of Minho, Portugal
José Álvarez-García	University of Extremadura, Spain
José Braga de Vasconcelos	Universidade New Atlântica, Portugal
Jose Luis Herrero Agustin	University of Extremadura, Spain
José Luís Reis	ISMAI, Portugal
Jose Luis Sierra	Complutense University of Madrid, Spain
Jose M. Parente de Oliveira	Aeronautics Institute of Technology, Brazil
José Machado	University of Minho, Portugal
José Paulo Lousado	Polytechnic of Viseu, Portugal
Jose Torres	Universidty Fernando Pessoa, Portugal
José-Luís Pereira	Universidade do Minho, Portugal
Juan M. Santos	University of Vigo, Spain
Juan Manuel Carrillo de Gea	University of Murcia, Spain
Juan Pablo Damato	UNCPBA-CONICET, Argentina
Juncal Gutiérrez-Artacho	University of Granada, Spain
Kalinka Kaloyanova	Sofia University, Bulgaria

Kamel Rouibah	Kuwait University, Kuwait
Khalid Benali	LORIA University of Lorraine, France
Korhan Gunel	Adnan Menderes University, Turkey
Krzysztof Wolk	Polish-Japanese Academy of Information Technology, Poland
Kuan Yew Wong	Universiti Teknologi Malaysia (UTM), Malaysia
Laila Cheikhi	University Mohammed V, Rabat, Morocco
Laura Varela-Candamio	Universidade da Coruña, Spain
Laurentiu Boicescu	E.T.T.I. U.P.B., Romania
Leonardo Botega	University Centre Eurípides of Marília (UNIVEM), Brazil
Leonid Leonidovich Khoroshko	Moscow Aviation Institute (National Research University), Russia
Lia-Anca Hangan	Technical University of Cluj-Napoca, Romania
Lila Rao-Graham	University of the West Indies, Jamaica
Łukasz Tomczyk	Pedagogical University of Cracow, Poland
Luis Alvarez Sabucedo	University of Vigo, Spain
Luis Cavique	University Aberta, Portugal
Luis Gouveia	University Fernando Pessoa, Portugal
Luis Mendes Gomes	University of the Azores, Portugal
Luis Silva Rodrigues	Polythencic of Porto, Portugal
Luiz Rafael Andrade	Tiradentes University, Brazil
Luz Sussy Bayona Oré	Universidad Nacional Mayor de San Marcos, Peru
Maksim Goman	JKU, Austria
Manal el Bajta	ENSIAS, Morocco
Manuel Antonio Fernández-Villacañas Marín	Technical University of Madrid, Spain
Manuel Silva	Polytechnic of Porto and INESC TEC, Portugal
Manuel Tupia	Pontifical Catholic University of Peru, Peru
Manuel Au-Yong-Oliveira	University of Aveiro, Portugal
Marciele Bernardes	University of Minho, Brazil
Marco Bernardo	Polytechnic of Viseu, Portugal
Marco Ronchetti	Universita' di Trento, Italy
Mareca María PIlar	Universidad Politécnica de Madrid, Spain
Marek Kvet	Zilinska Univerzita v Ziline, Slovakia
María de la Cruz del Río-Rama	University of Vigo, Spain
Maria João Ferreira	Universidade Portucalense, Portugal
Maria João Varanda Pereira	Polytechnic of Bragança, Portugal
Maria José Angélico	Polytechnic of Porto, Portugal
Maria José Sousa	University of Coimbra, Portugal
María Teresa García-Álvarez	University of A Coruna, Spain
Mariam Bachiri	ENSIAS, Morocco

Marijana Despotovic-Zrakic	Faculty Organizational Science, Serbia
Mário Antunes	Polytechnic of Leiria and CRACS INESC TEC, Portugal
Marisa Maximiano	Polytechnic Institute of Leiria, Portugal
Marisol Garcia-Valls	Polytechnic University of Valencia, Spain
Maristela Holanda	University of Brasilia, Brazil
Marius Vochin	E.T.T.I. U.P.B., Romania
Marlene Goncalves da Silva	Universidad Simón Bolívar, Venezuela
Maroi Agrebi	University of Polytechnique Hauts-de-France, France
Martin Henkel	Stockholm University, Sweden
Martín López Nores	University of Vigo, Spain
Martin Zelm	INTEROP-VLab, Belgium
Mawloud Mosbah	University 20 Août 1955 of Skikda, Algeria
Michal Adamczak	Poznan School of Logistics, Poland
Michal Kvet	University of Zilina, Slovakia
Miguel António Sovierzoski	Federal University of Technology - Paraná, Brazil
Mihai Lungu	University of Craiova, Romania
Mircea Georgescu	Al. I. Cuza University of Iasi, Romania
Mirna Muñoz	Centro de Investigación en Matemáticas A.C., Mexico
Mohamed Hosni	ENSIAS, Morocco
Monica Leba	University of Petrosani, Romania
Mu-Song Chen	Da-Yeh University, China
Natalia Grafeeva	Saint Petersburg University, Russia
Natalia Miloslavskaya	National Research Nuclear University MEPhI, Russia
Naveed Ahmed	University of Sharjah, United Arab Emirates
Neeraj Gupta	KIET Group of Institutions Ghaziabad, India
Nelson Rocha	University of Aveiro, Portugal
Nikolai Prokopyev	Kazan Federal University, Russia
Niranjan S. K.	JSS Science and Technology University, India
Noemi Emanuela Cazzaniga	Politecnico di Milano, Italy
Noureddine Kerzazi	Polytechnique Montréal, Canada
Nuno Melão	Polytechnic of Viseu, Portugal
Nuno Octávio Fernandes	Polytechnic of Castelo Branco, Portugal
Olimpiu Stoicuta	University of Petrosani, Romania
Patricia Zachman	Universidad Nacional del Chaco Austral, Argentina
Patrick C.-H. Soh	Multimedia University, Malaysia
Paula Alexandra Rego	Polytechnic of Viana do Castelo and LIACC, Portugal
Paulo Maio	Polytechnic of Porto, ISEP, Portugal
Paulo Novais	University of Minho, Portugal

Silviu Vert	Politehnica University of Timisoara, Romania
Simona Mirela Riurean	University of Petrosani, Romania
Slawomir Zolkiewski	Silesian University of Technology, Poland
Solange N. Alves-Souza	University of São Paulo, Brazil
Solange Rito Lima	University of Minho, Portugal
Sonia Sobral	Portucalense University, Portugal
Sorin Zoican	Polytechnic University of Bucharest, Romania
Souraya Hamida	Batna 2 University, Algeria
Sümeyya Ilkin	Kocaeli University, Turkey
Syed Nasirin	Universiti Malaysia Sabah, Malaysia
Taoufik Rachad	University Mohamed V, Morocco
Tatiana Antipova	Institute of Certified Specialists, Russia
Teresa Guarda	University Estatal Peninsula de Santa Elena, Ecuador
Tero Kokkonen	JAMK University of Applied Sciences, Finland
The Thanh Van	HCMC University of Food Industry, Vietnam
Thomas Weber	EPFL, Switzerland
Timothy Asiedu	TIM Technology Services Ltd., Ghana
Tom Sander	New College of Humanities, Germany
Tomaž Klobučar	Jozef Stefan Institute, Slovenia
Toshihiko Kato	University of Electro-Communications, Japan
Tzung-Pei Hong	National University of Kaohsiung, Taiwan
Valentina Colla	Scuola Superiore Sant'Anna, Italy
Veronica Segarra Faggioni	Private Technical University of Loja, Ecuador
Victor Alves	University of Minho, Portugal
Victor Georgiev	Kazan Federal University, Russia
Victor Kaptelinin	Umeå University, Sweden
Vincenza Carchiolo	University of Catania, Italy
Vitalyi Igorevich Talanin	Zaporozhye Institute of Economics and Information Technologies, Ukraine
Wafa Mefteh	Tunisia
Wolf Zimmermann	Martin Luther University Halle-Wittenberg, Germany
Yadira Quiñonez	Autonomous University of Sinaloa, Mexico
Yair Wiseman	Bar-Ilan University, Israel
Yuhua Li	Cardiff University, UK
Yuwei Lin	University of Roehampton, UK
Yves Rybarczyk	Dalarna University, Sweden
Zorica Bogdanovic	University of Belgrade, Serbia

Contents

Information and Knowledge Management

Open Innovation at University: A Systematic Literature Review

Marcelo Juca-Aulestia[1], Milton Labanda-Jaramillo[1,3,4], Jose Guaman-Quinche[2], Edison Coronel-Romero[2,3], Luis Chamba-Eras[2,3](✉), and Luis-Roberto Jácome-Galarza[5]

[1] Carrera de Informática Educativa, Loja, Ecuador
{jose.juca,miltonlab}@unl.edu.ec
[2] Carrera de Ingeniería en Sistemas, Loja, Ecuador
{jose.o.guaman,edisoncor,lachamba}@unl.edu.ec
[3] Grupo de Investigación en Tecnologías de la Información y Comunicación (GITIC), Loja, Ecuador
[4] Grupo de Investigación en Tecnología Educativa (GITED), Universidad Nacional de Loja, Loja, Ecuador
[5] Centro de I+D de Sistemas Computacionales Escuela Superior Politécnica del Litoral, Guayaquil, Ecuador
lrjacome@espol.edu.ec

Abstract. The aim of this paper is to describe the role of open innovation at universities around the world through a systematic literature review (SLR). The research was a methodology for SLR applied to engineer and education. The SLR selected 61 documents in the Scopus database. The obtained results allow us to identify why universities use open collaborative networks to link industry and academics through projects, link the triple helix model in their practices (spin-off), and through policies and strategies, organizations develop open innovation, while universities develop curricular strategies. Finally, university-enterprise financing is considered important for the development of products and services, preserving intellectual property.

Keywords: Knowledge networks · Transfer of knowledge · R&D

1 Introduction

Open innovation (OI) is a much studied topic in the management of organizations and industry, as well as, in small and medium enterprises, but, from the point of view of universities, it is a little explored area. For this reason, it has been decided to analyze the studies within this context, characterizing through fields such as interaction, cooperation, models, internal organization and collaboration.

In the context of Industry 4.0, it is currently demanded that the university be supported by models and platforms that allow it to respond adequately to the vertiginous requirements of the knowledge society in terms of the training of human talent. For this

Á. Rocha et al. (Eds.): WorldCIST 2020, AISC 1159, pp. 3–14, 2020.
https://doi.org/10.1007/978-3-030-45688-7_1

reason, a new substantive function is evident in Higher Education Institutions, which must be configured as an innovation hub that crystallizes the real contributions that universities must make towards society.

In Ecuador, there is a regulatory framework that defines elements that universities must consider in order to innovate and/or reinvent themselves in these times, for example, in the "Código Orgánico de la Economía Social de los Conocimientos", OI is defined as "*...the collaborative contribution of one or several people to find a solution to a problem presented by a third party with whom an employment relationship is not necessarily maintained*" [63], on the other hand, the same code promulgates that "*the Secretariat of Higher Education, Science, Technology and Innovation will facilitate social access to knowledge, in a public and open manner, so as to facilitate and promote open innovation processes*" [63]. Furthermore, in the context of technological innovation, in Ecuador's public institutions, it is necessary to execute and plan the migration of their digital technologies to free digital technologies.

In this systematic literature review (SLR), the methodology adapted by Torres-Carrion [56] was used, which is composed of three phases: planning, execution and reporting of results. 61 studies related to open innovation in universities were found. From this information, four research questions were defined that include the interaction of universities in collaborative networks, open innovation practices of universities, characteristics of environments where universities develop open innovation and the generation and transfer of knowledge from universities.

2 Systematic Literature Review

We use the methodology for SLR applied to engineering and education proposed in [56], which is based on three phases: planning, conducting and reporting.

2.1 Planning

Current State of Open Innovation

Research in the context of OI will be an added value in the development of universities, when transferring knowledge and technology to society, promoting a collaborative and innovative environment in the development of products and services. OI has been implemented in different universities around the world, with the aim of exchanging ideas between university researchers and businessmen from small and large companies, forming true knowledge networks.

Research Questions

Four research questions are defined, which will make it possible to identify the current state of OI at universities:

- RQ1: How do universities interact in collaborative networks?
- RQ2: What OI practices do universities implement?
- RQ3: What are the characteristics of the environment in which universities develop OI?
- RQ4: How is knowledge generated and transferred from universities?

Conceptual Mindset

The function of the conceptual mindset is to guide the SLR in OI at university. On the right are the concepts that do not belong to research and can be discarded, such papers of our research. On the left observe the characteristics that are linked to the central concept and that will be the keywords to execute the search in the different databases.

Semantic Search Structure

A script (search in Scopus databases) was generated with six layers for the search process. The first one involves OI within the university, the second one refers to the networks that exist for collaboration, the third one involves OI in the universities, the fourth and fifth ones refer to the environment and factors that affect innovation and, finally, the sixth one, the research questions that guide the semantic search.

Related Systematic Literature Reviews

There are SLR in the field of OI (see Table 2), which will support the present review. The databases Scopus, the Web of Science platform and the Google Scholar were searched, using the semantic search (see Table 1) and thus addressing the research questions.

Table 1. Layers for the support the semantic search.

L1	University	((“open innovation” and (universit* OR “higher education”))
L2	Networks	(collaboration OR cooperation OR “collaborative skills” OR determinants OR network OR integration)
L3	Practices	(model * OR spin-off OR framework * OR “ideas management” OR entrepreneurship OR enterprise)
L4	Environment	(challenge OR environment OR policy)
L5	Knowledge	(technology OR knowledge OR management)
L6		Q1: (Collaborative networks); Q2: (Open innovation practices); Q3: (Characteristics of the environment); Q4: (Generation and transfer of knowledge)

Selection of Journals and Databases

The selection of the journal was organized according to Scopus databases and 61 papers reviewed, and the list of journals in which the papers are indexed is also presented in the Table 3.

2.2 Conducting the Review

Definition of Inclusion and Exclusion Criteria

In this SLR, general and specific criteria have been defined for the selection to papers of journals, which will allow research questions to be answered, as well as exclusion criteria:

Table 2. Three SLR have been identified.

Study	Analysis	Papers reviewed
[46]	Factors affecting the participation of researchers in knowledge transfer in the context of OI, applies to the professional profile of 382 researchers	63
[24]	The most influential papers, authors and journals in OI are presented. Geographical locations are identified, and frequently used keywords are listed	293
[23]	The role of the main practices is identified in the management of human resources in organizations, where the relationship between these practices and OI has not been studied, and possible research based on human resources management and its role in OI is identified	79

Table 3. Relevant journals where they have been published according to the SJR scientometrics indicator.

Journals	#Papers	SJR		h5 Google
		IF	Cuartil	
Canadian Review of American Studies	1	0,1	Q4	5
Environmental Quality Management	1	0,15	Q4	7
Beijing HangkongHangtianDaxueXuebao/Journal of Beijing University of Aeronautics and Astronautics	1	0,23	Q3	–
Communications in Computer and Information Science	1	0,17	Q3	–
Innovations in Education and Teaching International	1	0,66	Q2	29
Journal of Visual Languages and Computing	1	0,23	Q2	–
Journal of the Canadian Academy of Child and Adolescent Psychiatry	1	0,52	Q2	18
Wireless Networks	1	0,4	Q2	34
Interaction Design and Architecture(s)	1	0,19	Q2	10
International Journal of e-Collaboration	1	0,51	Q2	–
IEEE Pervasive Computing	1	0,47	Q2	31
Multimedia Tools and Applications	1	0,34	Q1	52
International Journal of Human Computer Studies	1	0,69	Q1	39
IEEE Transactions on Image Processing	1	1,81	Q1	102
Production and Operations Management	1	3,28	Q1	48
American Journal of Occupational Therapy	1	0,67	Q1	31
IEEE Transactions on Visualization and Computer Graphics	1	0,96	Q1	65

- General: studies involving OI at universities and published in the last 7 years, between 2012 and 2019.
- Specific: studies that mention the characteristics of the environment in which universities develop OI, the generation of knowledge, the transfer of knowledge and collaborative networks in OI at universities.
- Exclusion: industry, organizations, enterprises, government.

Definition of Analysis Categories
The categories that have been defined are based on the research questions and their different variables:

- RQ1: cooperation, collaboration, collaborative skills, integration and partnerships.
- RQ2: entrepreneurship, spin-off, organizations, models, framework, idea management, integration and practices.
- RQ3: strategies, determining factors, factors, challenges, policy and curriculum.
- RQ4: knowledge transfer, technology transfer and R&D.

Preparing a Data Extraction Form
The Mendeley has been used for the extraction of information according Fig. 1.

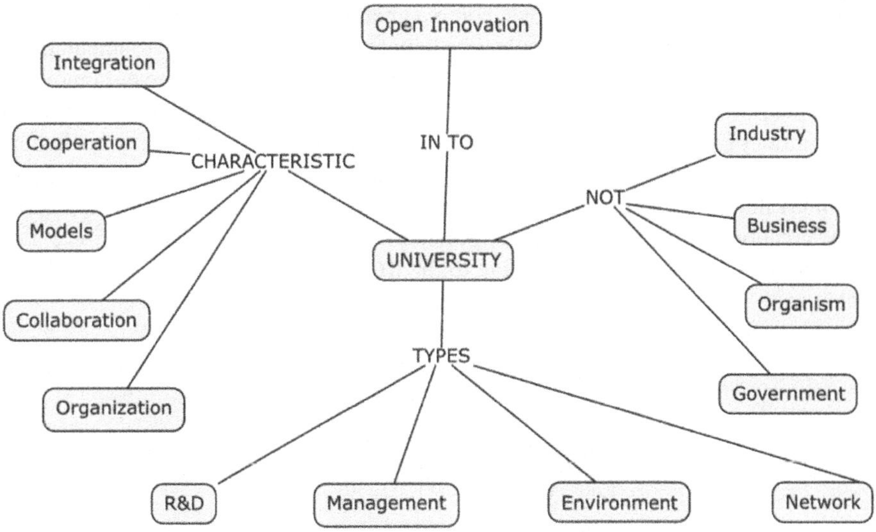

Fig. 1. Conceptual mindset to guide the literature search according to [56].

2.3 Reporting the Review

Table 4 shows the papers according to the RQ1: How do universities interact in collaborative networks?

Table 4. Number of papers to RQ1.

Characteristics	Papers	*f*
Networking	[6, 17, 26, 40, 62]	5
Cooperation	[39, 52, 61]	3
Collaboration	[2, 11, 18, 25, 28, 29, 34, 43, 48, 49, 53, 60]	12
Collaborative skills	–	0
Integration	[12]	1
Partnerships	–	0

Knowledge is the key to innovation in collaborative networks through projects between enterprises and universities, in some cases, a balanced scorecard is used to track, measure and improve the impact on the implementation of activities in projects. In other cases, the collaboration of the university differs greatly from the nature, location and performance of the company, so it is important to define the role to be played by the university, whose main actors are academics and students, so there should be a strong drive towards funding resources at university.

University-enterprise collaboration allows the development of new products and the commercialization of technology, through the management of partnerships in projects for the generation of new knowledge, being a great opportunity for the enterprise to be endorsed by a university, and thus obtain a better reputation in the environment, developing links with radical innovations, for which there should be government innovation policies and institutional mechanisms to promote innovation in small enterprises. The communities of practice as well as the innovation laboratories are a space for collaboration between the university and the industry, being the channel of knowledge exchange with motivational activities for researchers and entrepreneurs, arising with this, the need for the creation of intellectual property policies on the resulting inventions.

Table 5 shows the papers according to the RQ2: What OI practices do universities implement?

Table 5. Number of papers to RQ2.

Characteristics	Papers	*f*
Entrepreneurship	[1, 22, 59]	3
Spin-off	[54, 57]	2
Organizations	[8, 31, 38]	3
Models	[15, 21, 33, 35]	4
Framework	[3, 9, 21, 37]	4
Idea management	[7]	1
Integration	[20]	1
Practices	[10, 23, 44]	3

Models have been implemented based on research and innovation projects between the university and the enterprises in the area of technology evaluation, led by students and researchers, some models indicate technology transfer processes based on trust to identify drivers towards OI directed as a source of knowledge and technology for the enterprise through continuous dialogue; on the other hand, evaluation models have been implemented for university-industry collaboration to work with a group of researchers for decision making in collaborative projects between the university and the industry. There are other models based on university-industry-government called triple helix, for the division of direction in innovation for technology transfer, thus obtaining new perspectives on the same information with external and internal ideas of the university improving synergy and innovation, in addition, benefiting from educational activities, consequence of collaboration, emphasizing the capacities for the implementation of processes.

Table 6 shows the papers according to the RQ3: What are the characteristics of the environment in which universities develop OI?

Table 6. Number of papers to RQ3.

Characteristics	Papers	*f*
Strategies	–	0
Determining factors	–	0
Factors	[5, 16, 29, 46]	4
Challenges	[50]	1
Policy	[19, 27, 55, 58]	4
Curriculum	[30, 32, 37, 45]	4

The factors of OI at universities directly affect the participation of their researchers in different processes, one of those means were social networks, which have contributed to communication, participation and collaboration, in addition, it should be taken into account, the personal and professional profile that exists at university and in enterprises, which are open to collaborate in all decisions, without the presence of conflict of interest and intellectual property problems. It is demonstrated that the policy environment allows for a better scope in the improvement of programs for the funding of universities related to OI in small and medium-sized enterprises, focusing on science and engineering, designing incentives for interaction between researchers and entrepreneurs, opinions being valid to boost innovation.

OI proposes the curricular development between the university and the enterprise, and that satisfy the demands of services and products, it is for that reason, the university has implemented new methodologies for the improvement of the creative education and trains the students for the real world, developing the critical thought, behaviors and abilities, exchanging knowledge, fomenting the cooperation, basing on the confidence and the team work for the accomplishment of projects. The implementation of creative education programs is the basis of the development of ideas for students to generate results and profits to enterprises, and thereby accumulate experience with the implementation of new ideas.

Table 7 shows the papers according to the RQ4: How is knowledge generated and transferred from universities?

Table 7. Number of papers to RQ4.

Characteristics	Papers	*f*
Knowledge transfer	[10, 41, 47]	3
Technology transfer	[13, 42, 59]	3
R&D	[4, 11, 14, 36, 51, 59]	6

Funding is an essential point in R&D, one way of achieving it is through connections from the university with enterprises (small and micro), focusing more on research projects for the development of small enterprises that at the time arose just as social networks, renewable energy, etc. These projects have arisen from the work between universities in cooperation with enterprises, generating new knowledge for the development of technologies and products; on the other hand, technology transfer is important, the university contributes to private enterprise innovation to commercialize and disseminate potential inventions, having as a challenge the investment in the administration of intellectual capital, which is often suppressed by technology.

OI research allows the transfer of knowledge from universities, emphasizing intellectual property, by contract, to enterprises, to ensure development and cooperation; on the other hand, to have the balance between enterprise and university, collaborative innovation is necessary for the fair distribution of income, allowing the transfer of knowledge between university and industry, being the main factor for decision-making processes in organizations.

3 Conclusions and Future Work

We identified 61 research studies of acceptable rigor, credibility and relevance. The papers studied the answers to four research questions about OI at university: (RQ1) two mechanisms have been identified that universities use in collaborative networks in OI: the first, R&D projects that involve industry and academics, taking different roles according to the form of participation; the second, conformation of communities of practice that allow the participation nexus to be less formal, but effective, supported by the use of social networks; (RQ2) OI practices in universities are implemented and consolidated mainly through models and frameworks. The most generalized model is the triple helix model, which involves the university-industry-government. One of the not so generalized practices, but of great projection, constitutes the generation of spin-offs as mechanisms of generation of enterprise, based on research within the universities; (RQ3) the factors for developing OI in universities consist of policies and strategies at the organizational level, while, at the university level, the creation of curricular content related to creativity and the generation of ideas in open environments has acquired great importance; and finally (RQ4) the R&D process, universities are committed to research in association with medium and small companies, and with it finance developments in collaborative environments, the results of which are reflected in ideas and creations in advantage of the university-enterprise. For future work, the new SLR in the context of intellectual protection at university and management of human talent would be useful. Other works would be to design a model based in this paper and to study of the methodologies in base of the triple helix in OI.

Acknowledgement. This work is part of the research project "Modelo de innovación tecnológica abierta y colaborativa en ambientes universitarios públicos en Ecuador" sponsored by the "Universidad Nacional de Loja".

References

1. Alexander, A.T., Miller, K., Fielding, S.: Open for business: universities, entrepreneurial academics and open innovation. In: Exploiting Intellectual Property to Promote Innovation and Create Value (2017)
2. Alexander, A.T., Miller, K., Fielding, S.: Open for business: universities, entrepreneurial academics and open innovation. Int. J. Innov. Manage. **19**(06), 1540013 (2015)
3. Alexandra, K., Maximilian, R.: Towards an integrated capability framework. Bus. Process Manage. J. **23**, 87–107 (2017)
4. Bao, X., Dong, Y., Wang, Y.: R & D cost allocation and income sharing for industry-university cooperation in open innovation context. NTUT J. Intellect. Property Law Manage. **5**(1), 11–32 (2016)
5. Baron, M.: Open innovation capacity of the polish universities. J. Knowl. Econ. 1–23 (2017)
6. Becker, B.A., Eube, C.: Open innovation concept: integrating universities and business in digital age. J. Open Innov.: Technol.Mark. Complex. **4**(1), 12 (2018)
7. Boas, K.M.V., Zambalde, A.L., de Souza Bermejo, P.H., Furtado, T.B., de Freitas Santos, R., Grützmann, A.: Platforms of ideas management and open innovation: the crowdstorm approach applied to public university in Brazil. In: International Conference on Electronic Government and the Information Systems Perspective, pp. 205–217 (2018)
8. Buganza, T., Verganti, R.: Open innovation process to inbound knowledge: collaboration with universities in four leading firms. Eur. J. Innov. Manage. **12**(3), 306–325 (2009)
9. Cadiou, J.-C., Chene, E.: Strengthening partnerships between universities and SMEs within the open innovation framework. Innov. Interdisc. Solutions Underserved Areas **204**, 45–55 (2018)
10. Chaston, I.: Knowledge management systems and open innovation in second tier uk universities. Aust. J. Adult Learn. **52**(1), 153–172 (2012)
11. Chinta, R., Culpan, R.: The role of open innovation in business-university R&D collaborations. In: Open Innovation Through Strategic Alliances: Approaches for Product, Technology, and Business Model Creation (2014)
12. Coras, E.L., Tanţău, A.D.: Towards convergence in European higher education through open innovation. In: Economics: Concepts, Methodologies, Tools, and Applications, pp. 2–3 (2015)
13. Da Matta, V., Vasconcellos, E., Diogenes, D., Polo, E.: Managerial competencies model for promoting open innovation between companies and universities. In: IAMOT 2016 - 25th International Association for Management of Technology Conference, Proceedings: Technology - Future Thinking, pp. 516–535 (2016)

14. Diaz, P.A., Ramirez, Y., Cinthya, S.Q., Zarate, A.M.: Performance factors of university-industry R+D+I cooperations: determinants of an open innovation organizational strategy. In: 2017 Congreso Internacional de Innovación y Tendencias En Ingeniería, CONIITI 2017 – Conference Proceedings, pp. 1–6 (2018)
15. Draghici, A., Baban, C. F., Ivascu, L., Gaureanu, A.: A proposed business model to reinforce universities-industry collaboration in open innovation. In: Proceedings of the 27th International Business Information Management Association Conference - Innovation Management and Education Excellence Vision 2020: From Regional Development Sustainability to Global Economic Growth, pp. 2600–2613 (2016)
16. Fernández López, S., Pérez Astray, B., Rodeiro-Pazos, D., Calvo, N.: Are firms interested in collaborating with universities? An open-innovation perspective in countries of the South West European Space. Serv. Bus. **9**(4), 637–662 (2015)
17. Figaredo, D.D., Álvarez, J.F.Á.: Redes sociales y espacios universitarios. Conocimiento e innovación abierta en el espacio iberoamericano del conocimiento. RUSC Univ. Knowl. Soc. J. **9**(1), 245–257 (2012)
18. Flores, M., Al-Ashaab, A., Magyar, A.: A balanced scorecard for open innovation: measuring the impact of industry-university collaboration. IFIP Adv. Inf. Commun. Technol. **307**, 23–32 (2009)
19. Fujiwara, Y., Yamamoto, H., Fukushima, C., Komori, T., Tanaka, Y.: Medical open innovation initiative in Nagasaki university. Folia Pharmacol. Jpn. **146**(6), 327–331 (2015)
20. Georgy, U.: Open innovation - involving universities in the innovation process of libraries and information institutions. Inf.-Wiss. Praxis **63**(1), 37–44 (2012)
21. Hassanin, M.: A dynamic open innovation framework to accelerate research and regional development in the Egyptian open university. In: Proceedings of the International Conference on E-Learning, ICEL, pp. 125–131 (2012)
22. Hassanin, M.E.M.: Barriers to applying open innovation at universities. In: Innovation and Knowledge Management: A Global Competitive Advantage - Proceedings of the 16th International Business Information Management Association Conference, vol. 2, pp. 992–1001 (2011)
23. Hong, J.F.L., Zhao, X., Stanley Snell, R.: Collaborative-based HRM practices and open innovation a conceptual review. Int. J. Hum. Resour. Manage. **30**(1), 31–62 (2019)
24. Hossain, M., Islam, K.M.Z., Sayeed, M.A., Kauranen, I.: A comprehensive review of open innovation literature. J. Sci. Technol. Pol. Manage. **7**(1), 2–25 (2016)
25. Howells, J., Ramlogan, R., Cheng, S.L.: Universities in an open innovation system: a UK perspective. Int. J. Entrep. Behav. Res. **18**(4), 440–456 (2012)
26. Huggins, R., Prokop, D., Thompson, P.: Universities and open innovation: the determinants of network centrality. J. Technol. Transfer 1–40 (2019)
27. Hughes, A.: Open innovation, the Haldane principle and the new production of knowledge: science policy and university-industry links in the UK after the financial crisis. Prometheus (U. K.) **29**(4), 411–442 (2011)
28. Iskanius, P.: Open innovation in university-industry collaboration: communities of practice. In: Open Innovation: A Multifaceted Perspective, vol. 1, pp. 443–474 (2016)
29. Janeiro, P., Proença, I., da Gonçalves, V.C.: Open innovation: factors explaining universities as service firm innovation sources. J. Bus. Res. **66**(10), 2017–2023 (2013)
30. Jiravansirikul, T., Dheandhanoo, T., Chantamas, M.: University-industry collaboration for game curriculum: the open innovation model. In: 2017 9th International Conference on Information Technology and Electrical Engineering, pp. 1–4 (2018)
31. Jonsson, L., Baraldi, E., Larsson, L.-E., Forsberg, P., Severinsson, K.: Targeting academic engagement in open innovation: tools, effects and challenges for university management. J. Knowl. Econ. **6**(3), 522–550 (2015)

32. Kim, S.A., Ryoo, H.Y., Ahn, H.J.: Student customized creative education model based on open innovation. J. Open Innov. Technol. Mark. Complex. **3**(1), 6 (2017)
33. Laine, K., Leino, M., Pulkkinen, P.: Open innovation between higher education and industry. J. Knowl. Econ. **6**(3), 589–610 (2015)
34. Lam, J.C.K., Hills, P., Ng, C.K.W.: Open innovation: a study of industry-university collaboration in environmental R&D in Hong Kong. Int. J. Technol. Knowl. Soc. **8**(6), 83–102 (2013)
35. Leydesdorff, L., Ivanova, I.: "Open innovation" and "triple helix" models of innovation: can synergy in innovation systems be measured? J. Open Innov.: Technol. Mark. Complex. **2**(1), 11 (2016)
36. Lopes, A.P.V.B.V., Ferrarese, A., De Carvalho, M.M.: Open innovation in the research and development process: an analysis of the cooperation between automotive companies and universities. Gestao e Producao **24**(4), 653–666 (2017)
37. Lucia, Ó., Burdio, J.M., Acero, J., Barragán, L.A., Garcia, J.R.: Educational opportunities based on the university-industry synergies in an open innovation framework. Eur. J. Eng. Educ. **37**(1), 15–28 (2012)
38. Manjarres, C.C.A., Tarazona, M.F.G., Estupinan, L.M.A., Solano, L.J.C., Caballero, W.V.H.: An approach to the open innovation strategy in Higher Education Institutions in the metropolitan area of Bucaramanga, CongresoInternacional de Innovación y TendenciasEnIngeniería, pp. 1–8 (2018)
39. Matulová, P., Štemberková, R., Zdřálek, P., Marešová, P., Kuča, K.: Open innovation as a effective way of transfer technology between universities and the sector of commerce. In: IDIMT 2015: Information Technology and Society - Interaction and Interdependence - 23rd Interdisciplinary Information Management Talks, pp. 459–466 (2015)
40. Meissner, D., Carayannis, E.G.: Value generation from industry-science linkages in light of targeted open innovation. J. Knowl. Manage. **21**(2), 295–307 (2017)
41. Miller, K., Mcadam, R., Moffett, S., Alexander, A., Puthusserry, P.: Knowledge transfer in university quadruple helix ecosystems: an absorptive capacity perspective. R and D Manage. **46**(2), 383–399 (2016)
42. Minshall, T., Seldon, S., Probert, D.: Commercializing a disruptive technology based upon university IP through open innovation. Int. J. Innov. Technol. Manage. **4**(3), 225–239 (2007)
43. Molina Gutiérrez, A., Miranda, J., Chavarría, D., Noguez, J., Ramírez, M., Macías, M.E., Ramírez, J.: Open innovation laboratory for rapid realization of sensing, smart and sustainable products: motives, concepts and uses in higher education. IFIP Adv. Inf. Commun. Technol. **534**, 156–163 (2018)
44. Moretti, F.: Open lab studying the implementation of open innovation practices in a university laboratory. Int. J. Innov. Technol. Manage. **16**(1), (2019)
45. Oganisjana, K.: Promotion of university students' collaborative skills in open innovation environment. J. Open Innov.: Technol. Mark. Complex. **1**(2), 18 (2015)
46. Padilla-Meléndez, A., Garrido-Moreno, A.: Open innovation in universities: What motivates researchers to engage in knowledge transfer exchanges? Int. J. Entrep. Behav. Res. **18**(4), 417–439 (2012)
47. Perkmann, M., Walsh, K.: University-industry relationships and open innovation: towards a research agenda. Int. J. Manage. Rev. **9**(4), 259–280 (2007)
48. Pourdeyhimi, B.: Industry-university partnerships: Managing an "open" innovation network. Textile World **2017**(JUNE), (2017)
49. Ramli, N., Zainol, A.Z.: Malaysian university intellectual property policies in an era of open innovation: time for some uniformity? J. Intell. Prop. Law Pract. **12**(9), 761–767 (2017)
50. Rhéaume, L., Gardoni, M.: The challenges facing corporate universities in dealing with open innovation. J. Workplace Learn. **27**(4), 315–328 (2015)

51. Roper, S., Hewitt-Dundas, N.: Catalysing open innovation through publicly-funded R&D: a comparison of university and company-based research centres. Int. Small Bus. J. **31**(3), 275–295 (2013)
52. Rostoka, Z., Locovs, J., Gaile-Sarkane, E.: Open innovation of new emerging small economies based on university-construction industry cooperation. J. Open Innov.: Technol. Mark. Complex. **5**(1), 10 (2019)
53. Shekar, A.: Open innovation: a framework for collaborative product development between industry and universities. In: Open Innovation: New Product Development Essentials from the PDMA (2014)
54. Shutyak, Y.: Open innovation practice: a case study of university spin-offs. J. Entrep. Manage. Innov. **12**(1), 75–90 (2016)
55. Striukova, L., Rayna, T.: University-industry knowledge exchange. Eur. J. Innov. Manage. **18**(4), 471–492 (2015)
56. Torres-Carrion, P.V., Gonzalez-Gonzalez, C.S., Aciar, S., Rodriguez-Morales, G.: Methodology for systematic literature review applied to engineering and education. In: IEEE Global Engineering Education Conference, EDUCON, pp. 1364–1373 (2018)
57. Van Geenhuizen, M., Soetanto, D.P.: Open innovation among university spin-off firms: what is in it for them, and what can cities do? Innovation **25**(2), 191–207 (2012)
58. Vega, A., Brown, D., Chiasson, M.: Open innovation and SMEs: exploring policy and the scope for improvements in university-based public programmes through a multidisciplinary lens. Int. J. Entrep. Behav. Res. **18**(4), 457–476 (2012)
59. Villasalero, M.: University knowledge, open innovation and technological capital in Spanish science parks: research revealing or technology selling? J. Intell. Cap. **15**(4), 479–496 (2014)
60. Walden, R.: Progressing a university-industry collaboration (UIC) model for open and sustainable innovation. Cumulus Hong Kong 2016: Open Design for E-Very-Thing (2017)
61. Xu, G., Zhou, Y., Xu, L., Li, S.: Effects of control in open innovation: An empirical study of university-industry cooperation in China. Int. J. Technol. Policy Manage. **14**(4), 346–363 (2014)
62. Yuqiu, W.: Study on the university innovation system and its network mechanism in the high-tech context. Q. Educ. Teenagers 120–126 (2016)
63. Registro Oficial: Gobierno del Ecuador, Homepage, https://www.wipo.int/edocs/lexdocs/laws/es/ec/ec075es.pdf. Accessed 28 Dec 2019

A Dynamic Approach for Template and Content Extraction in Websites

Nicolae Cristian-Catalin(✉) and Mihaita Dragan

Faculty of Mathematics and Computer Science, University of Bucharest, Bucharest, Romania
nicolaecristiancatalin@gmail.com, mihaita.dragan@fmi.unibuc.ro

Abstract. Web scraping is a technique used to extract data from websites and it is the pillar of information retrieval in a world wide web that is ever growing. There are two main ways of extracting data from a website: static and dynamic scraping. Static scraping requires input beyond the target website because the user needs to inspect the HTML content of the target and find certain patterns in the templates that are then used to extract data. Static scraping is also very vulnerable to changes in the template of the web page. Dynamic scraping is a very broad topic and it has been tackled from many different angles: tree-based, natural language processing (NLP), computer vision or machine learning techniques. For most websites, the problem can be broken in two big steps: finding the template for the pages we want to extract data from and then removing irrelevant text such as ads, text from controls or JavaScript code.

This paper proposes a solution for dynamic scraping that uses AngleSharp for HTML retrieval and involves a slightly modified approach of the graph technique mentioned in for template finding. Once we find a number of pages then several heuristics can be applied for content extraction and noise filtering. Such heuristics can include: text and hyperlink density, but also removing common content between multiple pages (usually text from controls, static JavaScript) and then of final layer of NLP techniques (breaking the content into sentences, tokenization and part-of-speech tagging).

Keywords: Web scraping · Web graph topology · NLP · DOM content filtering

1 Introduction

The development of the World Wide Web has followed an exponential growth and the amount of information that is available is staggering. Humans can no longer parse all this information in their limited time and this problem gave birth to web scraping. The easier approach, from an algorithmic point of view, is static scraping. This kind of scraping requires a good amount of work in inspecting the targeted web pages and finding the template that enables us to retrieve information. One example would be noticing that relevant information is kept in <DIV> tags that have a specific class-name. Often times, with modern frameworks, these templates will have a much higher degree of complexity that requires more than just specific HTML patterns, but also specific code that is written for a certain web page, thus increasing the demands of static scraping.

Á. Rocha et al. (Eds.): WorldCIST 2020, AISC 1159, pp. 15–20, 2020.
https://doi.org/10.1007/978-3-030-45688-7_2

Dynamic web scraping can be approached from many angles and several branches of computer science can be used in this regard. The first important step is obtaining the (Document Object Model) DOM. Many programming languages offer libraries that can parse and instantiate the DOM. In all browsers, when an HTML document is loaded into a web browser, the DOM is created. Then, every node is usable and ready to be filtered for the purpose of extracting information [5]. Early studies focused on the DOM-tree representation of Web pages and learn a template that wrap data records in HTML tags, such as [6, 7]. The most popular method is to use the structured, tree-like nature of the DOM [3] and to use known algorithms to extract patterns and data. Regular expressions are also a common tool for the task since they can define a clear set of rules for content extraction. Computer vision can also be used to allow information retrieval in a similar way to humans. This approach involves rendering the DOM and then using several heuristics like pixel distance to the top of the page or font size to identify the title, splitting the page into different zones and narrowing the search to the central area and many others. Machine learning is another excellent approach that wields great results since manually labeled web pages can be used as a training set in a supervised learning task [3]. The most successful solutions often use a combination of the above.

The task of dynamic scraping can be broken into two big steps. The first step is finding several relevant pages that fit a certain template while starting from a main page. The second step is the task of extracting information from a HTML document with a minimum of noise (client-side JavaScript code, text from the controls in page such as "Follow us"). The first step is usually based on the oriented-graph nature of websites. The main page of a site can be considered the starting node and then the links to other pages can be considered edges. If a certain site contains a link that returns to our starting page, we have a bidirectional edge. Almost every modern website has a top header menu that is ever present on every other page. Pages that are linked by the main page and then link back to it will form a connected graph. Find the maximal connected graph (or a connected graph of a specified dimension) will give links to similar web pages from which we can start extracting a template. This is the approach proposed in [1] which does seem to give good results. For content extraction, a node level solution such as the one mentioned in [2] seems to offer a reduced amount of complexity with no sacrifice of correctness.

In Sect. 2, we present a few problems we've encountered with the algorithm in [1], used to find a complete subdigraph in a website topology. In Sect. 3 we discuss the methods that are used to extract the template from the complete subdigraph. In Sect. 4 we introduce the node-based filtering system that uses several heuristics, some simply based on node characteristic while others based on NLP processing. In Sect. 5 we summarize our approach and results and suggest further research such as NLP techniques that aggregate data in powerful ways.

2 Retrieving Relevant Web Pages

A good way of finding similar web pages is presented in [1] with a main page that is given as input. It relies on the strong heuristic that web pages are linked with the main page of a site through a menu. Together, these form a connected graph (because of the implied bidirectional connections). The algorithm for finding a complete subdigraph in a website topology is explained in section 2 of [1]. The algorithms wields good results, but it does have some limitations:

- On starting pages that have a lot of outgoing links (large websites like BBC can have over 250) there is a very high complexity, admitted by the authors themselves and proposed as further work.
- Somewhat linked to the first issue, large websites use more than one template for their pages. For example, the BBC news site uses a template for articles, but it also uses a template for categories (e.g. BBC Sport, BBC Travel, BBC Future). There is a high chance of encountering a subdigraph containing pages from another template than the one we expect (the one we can extract data from). A simple, yet effective solution to this issue is sorting the outgoing links by the length of their URLs in a descending manner. This is based on the fact that article links are always longer than most other links on a website, since they almost always contain the title of the article in the URL.

3 Extracting the Template

Once a subdigraph is found, the authors in [1] propose a tree based equal top-down approach of extracting the template. This approach is very precise, but can be somewhat rigid for most websites since templates don't always match node by node. The alternative approach that we have used is to a frequency array for every type of HTML element in the respective DOM. We can pad accordingly with the missing HTML elements between them so all frequency arrays have the same size and elements, just different values. We can then calculate the standard deviation for each HTML element (with a dataset of four, in our example) and then average all these value into one single value that tell us how strongly related the web pages in the subdigraph really are (Fig. 1).

```
// Compute the frequency dictionary for every DOM.
var pagesFrequencyDictionaryList = webDocuments
    .Select(dom => dom.All.GroupBy(element => element.GetType().ToString())
    .ToDictionary(x => x.Key, x => x.Count())).ToList();
// <...> <Pad every frequency array with
// missing tags from other arrays> <...>
var tagsArray = allHTMLTags.ToArray();
for (int iTagIdx = 0; iTagIdx < allHTMLTags.Count; iTagIdx++)
{
    // Get all values for this particular HTML tag.
    List<int> tagValues = new List<int>();
    foreach (var pageDictionary in pagesFrequencyDictionaryList)
        tagValues.Add(pageDictionary[tagsArray[iTagIdx]]);
    // Compute median and standard deviation for the list of values.
    templateFrequency[tagsArray[iTagIdx]] = tagValues.GetMedian();
    templateStandardDeviation[tagsArray[iTagIdx]] =
        tagValues.GetStandardDeviation();
}
double averageStandardDeviation = templateStandardDeviation.Values.Sum() /
   templateStandardDeviation.Values.Count;
```

Fig. 1. First compute a list of dictionaries (frequency arrays), pad them with missing values, get the list of values for each individual tag, compute standard deviation for all tags and average this set of values for a good index of how suitable this graph is for extracting a template.

We can set a threshold for this value so we can reject web pages that do form a complete subdigraph, but their content is largely unrelated. Once a subdigraph is validated by this threshold value, like we used the standard deviation of every HTML element in the frequency array, we can compute the median count for each element (the median is more resilient to outliers than the average and we want to be forgiving to certain elements being present in one instance of the template and not another). This array of median values will be the final template. One disadvantage of this approach is the fact that is losing the connections between the nodes present in the DOM. The advantage is that the approach is forgiving to variations between instances of the template and it is fast to compute.

4 Filtering the Content of the Web Page

Even after parsing the DOM, there is still a lot of work to be done in order to have the main content from the page. The approach in [2] computes two values for every node: text density and hyperlink density. A first filter would only keep all HTML that are known to store information: divs, paragraphs, table cells and so on.

Then, one heuristic that is used is that a content node has higher text density and lower hyperlink density than the noisy node. The second heuristic used is the fact that nodes are grouped so nodes that are adjacent will be of a similar type. This approach does remove a good amount of noisy nodes, but in our experiments it is not powerful enough to certain cases:

- Nodes that contain JavaScript code can have a high text density and a low hyperlink density, so the first heuristic can have such false negatives.
- We can have nodes that sit at the border between a group of content nodes and a group of noisy nodes, which can pass through the second heuristic.
- Text from the controls and labels in the GUI tend to pass both heuristics (text such as "Follow us on Instagram", "Subscribe on Youtube").

One problem we've encountered is that fixed thresholds for text density and hyperlink density can be very prohibitive and will easily erase all relevant content. This is caused by the fact that not every website holds its content scattered among many HTML elements, so we will encounter a very large node with a low text density, but this node can actually be very relevant. A good method is to compute the thresholds at runtime, using the average value for text density and hyperlink density in the current document.

After this step, one method that can easily get rid of text from controls is taking the intersection of nodes with the same content between the pages in the subdigraph. This is, of course, based on the observation that the same buttons will be present on each page and they will figure as common content, but the main content will differ and we won't be removing parts of it. Even after this step, JavaScript nodes that contain hashes and code that varies will remain. The final step is to use NLP techniques [4]. We can break all content we have so far into sentences, tokenize the sentences into words and then tag the words into parts of speech. For NLP, we are using the C# port of OpenNLP. After these three steps, we can use two heuristics:

- Every content sentence must contain at least one verb. The part of speech tagger in OpenNLP has a best effort approach to tagging so we will always have tags, even on nonsensical sentences generated by JavaScript nodes. Nodes that do not have at least one verb can be removed.
- JavaScript nodes usually contain long strings such as hashes. We can simply remove every node that contains a word that is longer than a certain threshold.

The NLP approach is quite powerful and has the advantage of computing tokenized, tagged sentences that can be further used in information retrieval. A stronger, but very costly approach would be trying every word against a dictionary of the respective language. While this would almost all noisy nodes, the computational cost of checking every word against a dictionary would be enormous. A small disadvantage of adding NLP to this problem is that articles we can extract information from are now reduced to being written in languages we are trained to process.

5 Testing the Results

A good method for testing the solution is Levenshtein distance, also know as edit-distance, the number of edits required to go from one string to another. We can regard these edits as errors or missing information.

This will be accomplished by comparing the output content of the algorithm with a hand-gathered, perfectly scraped output. To remove several irrelevant edits, we can ignore newlines and whites paces. A good metric could also be the Levenshtein distance divided by the length of the perfectly scraped article since it shows how much of the article is missing or how much noise is left.

	Perfectly scraped output size	Solution output size	Levenshtein distance	Distance over length factor
HackerNews article 1	5600	5792	924	0.165
HackerNews article 2	3272	3639	663	0.202
HackerNews article 3	3861	4283	608	0.157
HackerNews article 4	5011	5102	1090	0.217
The Guardian article 1	6107	6902	1586	0.259
The Guardian article 2	5410	5940	850	0.155
Blog article 1	5948	6192	1251	0.21
Blog article 2	4500	4689	1064	0.23

As it can be seen, the algorithm performs within reasonable parameters and while the node content filtering could be a bit more severe in order to reduce the noise even more, doing that would risk removing relevant content, which is not preferred.

6 Conclusion

The topic of web scraping is already very developed and current technologies are very powerful. The proposed solution combines a few branches of computer science and mathematics: graph theory, data science and NLP. The solution performs well as long as it gets the template for news article right, this being the problematic part of most scraping solutions since large websites have so many outgoing links it's very easy to get side-tracked. The template extraction method is a bit more flexible than tree-based approaches and is computed in a very fast manner using simple frequency arrays measured in terms of standard deviation and average. Node filtering is performing adequately. Severe filtering would lead to a lot of content being removed so false positives (as in, remaining JavaScript or other noise) is preferred to false negatives (content being tagged as noise), but only in reasonable terms.

Further work could go towards improving the algorithm on large websites where it still gets off-topic or creating a system that can deduce reasonable thresholds on the go (some websites can yield zero data since every node is removed in the node filtering process). Another direction for further work, since we're using NLP techniques and the data from the last step of node filtering, we can maybe break the content into a system of knowledge representation and create a chat bot that can process questions in natural language, query the knowledge base and answer in natural language.

References

1. Alarte, J., Insa, D., Silva, J., Tamarit, S.: Web template extraction based on hyperlink analysis. In: Escobar, S. (ed.): XIV Jornadas sobre Programación Y Lenguajes, PROLE 2014, Revised Selected Papers EPTCS, vol. 173, pp. 16–26 (2015)
2. Liu, Q., Shao, M., Wu, L., Zhao, G., Fan, G.: Main content extraction from web pages based on node characteristics. J. Comput. Sci. Eng. **11**(2), 39–48 (2017)
3. Ferrara, E., De Meob, P., Fiumarac, G., Baumgartnerd, R.F.: Web Data Extraction, Applications and Techniques: A Survey. arXiv:1207.0246v4 [cs.IR], 10 June 2014
4. OpenNLP Documentation. https://opennlp.apache.org/docs/. Accessed 2 Nov 2019
5. Uzun, E., Doruk, A., Nusret Buluş, H., Özhan, E.: Evaluation of HAP, AngleSharp and HTML document in web content extraction. In: International Scientific Conference, Gabrovo, 18 November 2017 (2017)
6. Soderland, S.: Learning information extraction rules for semi-structured and free text. Mach. Learn. **34**(1–3), 233–272 (1999)
7. Muslea, I., Minton, S., Knoblock, C.: Hierarchical wrapper induction for semistructured information sources. Auton. Agents Multi-Agent Syst. 93–114 (2001). https://doi.org/10.1023/A%3A1010022931168

Analysis of Factors Affecting Backers' Fundraising on Reward-Based Crowdfunding

Bih-Huang Jin[1](✉), Yung-Ming Li[2], and Cui-Ping Chen[1]

[1] Department of Business Administration, Tunghai University, Taichung City, Taiwan
bihuang.jin@gmail.com

[2] Institute of Information Management, National Chiao Tung University, Hsinchu, Taiwan
yml@mail.nctu.edu.tw

Abstract. This study use data from Kickstarter to find what factors are correlated with the numbers of backers of all projects and projects of different categories by using multiple linear regression model. The result shows that variables including projects goal, description information such as FAQs, updates and comments as well as the rewards characteristics are significantly affecting the number of projects backers. Besides, variables have different effect on projects backers of different categories, several factors show significantly effect on some projects but totally different results of others. It can help entrepreneurs to better understand backers' funding decision and attract more backers in order to increase the success rate of crowdfunding projects.

Keywords: FinTech · Crowdfunding · Kickstarter · Multiple linear regression analysis

1 Introduction

The dramatic development of network and information technology has brought breakthroughs to all walks of life, especially in finance industry. According to CB Insights report, 2017 was a record year reaching to $16.6B in global fintech companies, where it was just $3.8B in 2013 (Insights 2018). According to the EY FinTech Report (EY 2017), the global utilization of FinTech products has reached 33% comparing to 16% two years ago and over 50% users adopt FinTech money transfer and payments services in the study.

Crowdfunding, as an emerging aspect of FinTech, is a novel method that combining the concept like micro-finance, crowdsourcing and network technology, which allows entrepreneurs to raise funds much easier directly from the crowd instead of traditional intermediaries, such as banks and business investors (Mollick 2014). According to Statista Market Forecast report (Statista 2018), the global transaction amounts of crowdfunding markets will rush to $9.37 billion in 2018 and it is expected to increase to the total amount of over $25 billion in 2022. It uses platforms to connect creators and investors so that they can easily find each other. The innovative projects created by innovators are posted on crowdfunding platforms, and investors select which projects

Á. Rocha et al. (Eds.): WorldCIST 2020, AISC 1159, pp. 21–34, 2020.
https://doi.org/10.1007/978-3-030-45688-7_3

to invest. Though each investor raises small amount money to the projects, it has the possibility to sum up to over thousands or millions of dollars with the power of the crowd instead of traditional intermediaries (Riedl 2013).

Kickstarter, one of the most popular reward-based crowdfunding platforms all over the world, was founded in America in 2009. It has been successfully raised over $3 million to more than 140,000 funded projects on Kickstarter (2018a). However, it is less than 50% projects have successfully reached their goal funds at the end of the funding cycle, with the lowest average success rate of projects belonged to technology category, which is lower than 20%.

As the success rate of projects is still low on crowdfunding platforms, it is important to find out the reasons why some projects would fail. There is some research focusing on the key factors associated with the outcomes of successful projects. Numerous studies (Chien 2016; Jin et al. 2018; Lin 2015) manifest that the increasing backers are determinants of successful funding results even on different crowdfunding platforms. According to Kuppuswamy and Bayus (2017), the dynamics of backers' support over the funding cycle follows the U-shape pattern that it is more likely for backers to invest projects at the early and final stage rather than the middle period, the same as change of pledged funds across the whole funding cycle (Jin et al. 2018). Besides, it is proved that the successful projects on Kickstarter due to the high quality such as the full preparedness (Mollick 2014) and detailed project description (Zhang and Liu 2012). Rewards are also crucial that affecting backers investment on the reward-based crowdfunding platform, and the analysis shows that projects which provide more rewards are more likely to succeed (Lin et al. 2016). However, substantial studies have been performed the key factors that affect the pledged funds of projects, those of backers' funding decision are still lacking. According to the study conducted by Barba-Sánchez and Atienza-Sahuquillo (2017), the dramatic development of network, social media and fast-growing globalization facilitate a market based on network that creators need to provide products fitting consumers demands. It is important to understand what information is the key factors affecting backers' funding decision when they are interested in a project. Although Kuppuswamy and Bayus (2017) concluded that the pattern of backers' support was dynamic during different phases across the funding cycle, it was not clear what factors significantly related to the number of backers in general. What factors will affect the backers' funding decision? It is primary for project creators to figure out what factors will affect backers' funding decision so that they can make strategies to attract more backers and maintain their backers. In order to achieve this goal, we list out the research questions as below:

(1) What factors are associated with the number of backers of all projects?

We choose Kickstarter, one of most popular crowdfunding site, as our object in this paper. There are lot of information investors can see on the projects page, such as funding goals, duration, updates, comments, and different kinds of rewards. Creators will update information of the projects, and investors can leave comments only after they become projects' backers. Most creators will provide rewards, product or non-product, to investors once the project succeeds to life. According to the research by Kuppuswamy and Bayus (2017), there is no obvious evidence show that the backers support increasing at the early and final stage of project funding phase due to the visibility by sorting on special zoom where users can easily find them, such as "Recently Launched" or "Ending Soon". They were focus on the identifying the dynamic of backers support are related to factors explored by other researchers, including the role of friends and family and the effects coming from social influence (Kuppuswamy and Bayus 2018), but not the information originated from projects such as FAQs and comments.

Based on the above elements of the projects page, we will find out whether they are the factors affecting the number of backers of all projects.

(2) What factors are associated with the number of backers for different categories projects?

As we known, projects are classified into 15 categories on Kickstarter, including art, comics, crafts, dance, design, film & videos, fashion, food, games, journalism, music, photography, publishing, technology and theater. According to the official data from Kickstarter (2018b), the success rate of projects of different categories are different, with the highest success rate over 62% of dance projects compared to the one less than 20% of technology projects. Therefore, we will find whether backers are affected by different factors of different categories projects.

2 Related Literature on Crowdfunding Factors and Hypothesis

2.1 Project Description

Projects creators are required to produce detail information on project's webpage. According to Kickstarter creators rules, projects creators should set the goal of funds, duration and rewards before launching the projects (Kickstarter 2018a). According to Mollick (2014), setting appropriate goal of funds and duration of a project would be more likely to improve success rate of projects. Since people raise small amount money to support a project, it seems like high project goal will reduce the chance of success. Longer duration time can be a sign of lack of confidence of both creators and backers. Once launching the project, the funding goal and expired date cannot be modified. Creators can post updates and replay backers' comments as an important interaction with backers. It strengths the relationship between creators and backers since backers can get involving in the process of the projects. Kuppuswamy and Bayus (2018) unveiled that updates were basic of projects success since it was like a signal for backers to follow the process of the project. Both updates and comments are important information for creators to communicate with the community, especially with backers, that are more likely to contribute to increase backers support.

2.2 Rewards

Numerous studies find out that rewards are one of the determinants of project success in reward-based crowdfunding platforms (Lin et al. 2016). Most backers are seeking tangible or intangible rewards, which can be thought as a consumer purchasing behavior (Gerber and Hui 2013). According to the study by Gerber and Hui (2013), they concluded that projects with more rewards were more likely to reach projects campaigns' goals.

According to the study by Lin et al. (2016), backers not only choose to back a project they are interested in but only consider whether worth paying for the projects depending on the price of a reward. They find out that reward sell-out rate is negatively related to high pricing range both of funded and unfunded projects.

2.3 Research Hypothesis

Basic on above related literatures, it is expected that factors, such as projects goal, number of updates, FAQs and comments, amount of rewards and price of reward, are associated with backers' support. Therefore, the hypothesizes are summarized as following.

H1a: Projects goal, the desired funding amount creators want to raise from the backers during the funding period, is related to the number of projects backers.

H1b: The number of FAQs, the frequently asked questions of project campaign, the section provided to all users where they can ask questions about projects, are related to the number of projects backers.

H1c: The number of updates, the information about projects' progress which are posted by creators on the page home pages, are related to the number of projects backers.

H1d: The number of comments, the section where creators usually answer questions from backers during funding period, are related to the number of projects backers.

H1e: Duration time, the fixed days of project's funding period decided by creators, is related to the number of projects backers.

H1f: Reward's characteristics, such as the number of rewards and mean of price of each project's reward, are related to the number of projects backers.

Another research question is to find out whether there are different factors affecting projects' backers of different categories, therefore, we can generate the following hypothesis:

H2: There are different factors related to the number of projects backers of different categories.

3 Research Methodologies

In order to examine the relationship between each variable and backers, we are going to conduct the necessary analysis and multiple regression model. The research questions are shown in Fig. 1.

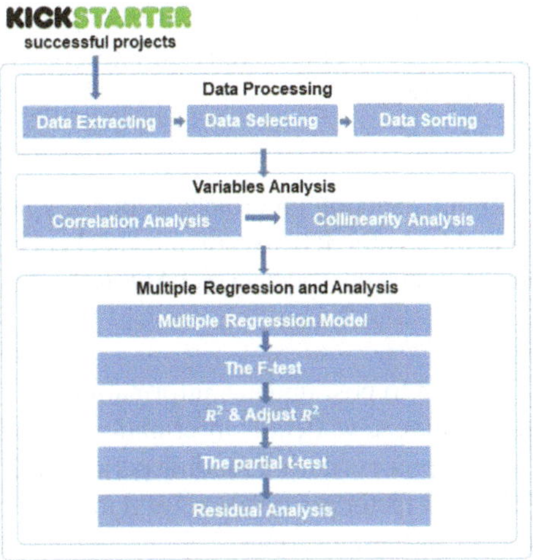

Fig. 1. The research process

We choose Kickstarter, one of the most popular crowdfunding platforms as our study object. According to their website, 14 million backers investing total $3.6 billion since the platform launching in 2009 funded over 140,000 projects successfully. Kickstarter is a global community for innovators to share their ideas and concepts as projects, which are arranged to 15 different categories: Art, Comics, Crafts, Dance, Design, Fashion, Film & videos, Food, Games, Journalism, Music, Photography, Publishing, Technology and Theater. The data came from Kickstarter could be publicly available from projects webpages. We used Python to get information from each project home page, such as project title, pledged funds, the goal of funding, duration time, number of total backers, FAQs, updates, comments, numbers of rewards and each price of reward that each backer can choose to invest. We collected data of successful projects which launching time were from January 1 to December 31 in 2017. All data were classified and stored into Excel documents. After selecting and eliminating missing and useless data, total 16,662 successful projects with complete information were available for the following analysis process. Then we put the data into a statistics software SPSS, which stands for Statistical Product and Service Solutions, in order to conduct multiple regression analysis.

3.1 Variables

Total Backers (Y)
The purpose of this research is to find out what factors affecting the backers funding decision all projects as well as the one in different categories, so we defined the number of projects backers as dependent variable.

Project Goal (X1)
Project goal is the funding target that creators wanted to raise for the project. Projects campaigns on Kickstarter follow the "All-Or-Nothing" rule rather than "Keep-It-All",

which means creators would not receive any amount from investors unless the projects succeed funded ultimately. It can be considered that backers have less risk to support the projects that they are interested.

FAQs (X2)
FAQs, is one part of community section on project webpage where users ask questions about the project directly to the creator. It helps potential backers get further acknowledge about the project they are interested in. The data on FAQs is the number of these postings.

Updates (X3)
Project creators are encouraged to post updates information during as well as after the funding period. Backers would receive details information through email once creators post any updates of the project process during funding period. As Mollick (2014) pointed out, updates are one of the vital factor positively related to the success of a project. The data on update is the number of these postings.

Comments (X4)
According to definition on Kickstarter website, backers are users who have already backed a project. Creators and backers can communicate with each other about projects on this comment section. The data on comment is the number of the total postings.

Duration (X5)
According to the statement from Kickstarter (2018a), a project funding duration time is required no more than 60 days. Once the expired time is decided, it could not be modified during the funding period of project campaign.

Rewards (X6, X7)
Many studies examined that rewards were one of the determinants of project success in reward-based crowdfunding platforms (Lin et al. 2016). Projects with many reward choices tend to attract more funders to invest them (Kuppuswamy and Bayus 2017).

According to the research produced by Lin (2015), we choose "reward count" (X6), and "reward mean price" (X7), which stands for the number of rewards and the mean price of rewards respectively, of a project as the variable to measure the efficiency of reward characteristics. Different reward options with the same price are seem as different types of rewards.

4 Data Analysis and Results

In this section, we show the statistics description of variables and then conduct the multiple regression model and analysis.

4.1 Variables Statistics Descriptions

We extracted data of projects whose launching time were from January 1 to December 31 in 2017 and already completed before we collected it. The statistics description of all successful projects is shown as Table 1.

Table 1. Descriptive statistics of all projects

Variables	All projects (N = 16662)			
	Mean	Std. deviation	Min.	Max.
Backers	314.87	1175.02	1.00	43733.00
Goal	9257.05	19073.50	0.54	750000.00
FAQs	1.19	3.63	0.00	121.00
Updates	10.08	10.23	0.00	173.00
Comments	109.01	1370.96	0.00	112078.00
Duration	29.57	10.03	1.00	60.00
Reward count	9.45	6.62	1.00	131.00
Reward mean price	256.75	395.10	1.00	7353.25

In Table 1, the average projects goal is \$9257.05 invested by 314.87 backers in average of all projects. The mean of FAQs, updates and comments is 1.19, 10.08 and 109.01, respectively. The duration time is near 30 days in average. With regard to the rewards, the average number of projects rewards is 9.45 and the reward mean price is \$256.75 in average. There are many large standard deviations in most variables, such as standard deviation of backers is 1175.02 and the one of project goal is 19073.50.

According to descriptive statistics tables, standard deviation manifests that there is huge variance of most variables, such as the one of backers, projects goal, updates and comments not only of all projects but also of projects in different categories, which will cause the bias of the residual analysis. According to several research Lin (2015) and Mollick (2014), it is necessary to proceed the variables transformation in order to avoid the Heteroskedasticity when huge deviation exists. We will conduct the necessary log transformation of variables. Kuppuswamy and Bayus (2017) used the common logarithm with base of 10 to proceed the log transformation of variables, such as the goal of the project, pledged fund, backers and funding duration. According to the research studied by Lin (2015), it used the log of the value which was equal to an observed number plus 1 since some original value of variables was 0, such as the number of Facebook sharing, updates and comments.

The multiple regression model after variables log transformation is shown as below:

$$\begin{aligned} \lg(Backers_i) = & \beta_0 + \beta_1 \lg(ProjectGoal_i) + \beta_2 \lg(FAQ_i + 1) \\ & + \beta_3 \lg(Update_i + 1) + \beta_4 \lg(Comment_i + 1) \\ & + \beta_5 DurationTime_i + \beta_6 RewardCount_i \\ & + \beta_7 RewardMeanPrice_i + \varepsilon_i \end{aligned} \quad (1)$$

Where the dependent variable (Y) transfers to $\lg(Backers_i)$, several independent variables such as projects goal (X1) changes to $\lg(ProjectGoal_i)$, the number of FAQs (X2) changed to $\lg(FAQ_i + 1)$, the number of updates (X3) changes to $\lg(Update_i + 1)$ and the number of comments (X4) changes to $\lg(Comment_i + 1)$.

We conduct the Pearson correlation analysis to see whether variables highly correlate to each other. The results of all projects is shown in Table 2. According to the study by Buda (2011) and Cohen (2013), it is no strong correlations between variables if the Pearson coefficient falls into the range from −0.60 to 0.60 so that we can continue following multiple regression analysis.

Table 2. Pearson correlation matrix of all projects

Variables	Y	X1	X2	X3	X4	X5	X6	X7
lg(backers) (Y)	1							
lg(goal) (X1)	0.683*	1						
lg(FAQs+1) (X2)	0.433*	0.314*	1					
lg(updates+1) (X3)	0.576*	0.355*	0.337*	1				
lg(comments+1) (X4)	0.754*	0.412*	0.477*	0.590*	1			
Duration (X5)	0.128*	0.237*	0.086*	0.052*	0.056	1		
Reward count (X6)	0.317*	0.347*	0.118*	0.271*	0.161*	0.126*	1	
Reward mean price (X7)	0.170*	0.429*	0.100*	0.043*	0.034	0.119*	0.212*	1

$*\ p < 0.05$

However, the results manifest that some variables correlate with others which could be the sequences of the collinearity. We then utilize the variance inflation factor (VIF) to verify whether strong collinearity exists or not. According to Neter et al. (1989), it is no serious collinearity problem in the regression model if the value of VIF is less than 10.

4.2 Multiple Regression Model

We use a method of ordinary least squares (OLS) to set up the multiple regression model with the desired variables and conducted analysis by using a statistics software called SPSS. According to above section, we calculated the Pearson correlation efficiencies between variables.

To verify the relationship between independent variables, we find that the value VIF of variables is less than 10, which means that there is no serious collinearity effect of the multiple regression model so that we can continue the analysis.

Table 3 is the result of multiple regression model of all projects. The F-value of multiple regression model of all projects is 7292.554 with the significant is less than 0.05 so that we reject the H_0 and the R^2 is 0.754 which shows that the regression model significantly fits the original data set from the real situation.

According to the Table 3, the standard coefficient of independent variables X1 (the log of projects goal) and X4 (the log of the value of comments plus 1) is 0.483 and 0.438, representing that they play most important roles affecting the number of projects backers. Following by variables X3 (the log of the value of updates plus 1), X6 (rewards quantity) and X2 (the log of value of FAQ plus 1), with the standard coefficient is 0.123,

Table 3. Results of regression model of all projects

Variables	Standard coefficient	t	VIF
lg(Goal) (X1)	0.438	87.787*	1.686
lg(FAQs+1) (X2)	0.022	4.888*	1.331
lg(updates+1) (X3)	0.123	24.998*	1.629
lg(comments+1) (X4)	0.483	91.169*	1.903
Duration (X5)	−0.013	−3.316*	1.066
Reward count (X6)	0.063	15.095*	1.194
Reward mean price (X7)	−0.049	−11.389*	1.261
Constant		11.336*	
F-value		7292.554*	
R^2	07854		
Adjusted R^2	0.752		

$*\ p < 0.05$

0.063 and 0.022, respectively. After conducting partial t-test of variables, we can find that variables including X1 (the log of projects goal), X2 (the log of value of FAQ plus 1), X3 (the log of the value of updates plus 1), X4 (the log of the value of comments plus 1) and X6 (rewards quantity) are positively affect the number of backers. While the X5 (duration time) and X7 (reward mean price) are negatively related to the number of backers.

Overall, projects goal is significantly affecting the number of backers that it can be considered as a signal to attract backers' attentions. However, creators should set an appropriate funding goal of project since too much high amount of money is difficult to reach in a short time (Mollick 2014). FAQs, updates and comments are also significantly positive affected backers. The more information creators provided on project home page, the more details potential backers can know about the projects. Longer funding period can be a sign of lack of confidence so that creators should think over the appropriate duration time of project. Backers can get more acknowledge through updates information and communicate with creators at comments zone, which can be considered as a contribution of projects success. Besides, projects with more rewards options seems to be attracted by backers who look for the products as consumers. They are seeking for rewards as products but with less money. Therefore, projects creators need to think more about projects before launching it and improve high quality of the projects such as posting updates information and communicate with backers frequently to keep interacting and manage the relationship with backers in order to increase their confidence and attract more potential backers.

Table 4. Results of regression model of Arts, Comics, Crafts, Dance and Design

Variables	(1)		(2)		(3)		(4)		(5)	
	Arts (N = 1569)		Comics (N = 1203)		Crafts (N = 484)		Dance (N = 164)		Design (N = 1919)	
	Coef.	VIF	Coef.	VIF	Coef.	VIF	Coef.	VIF	Coef.	VIF
lg(Goal) (X1)	33.213* (0.555)	1.492	27.355* (0.548)	1.871	17.766* (0.537)	1.909	10.091* (0.627)	1.849	16.774* (0.284)	1.854
lg(FAQs+1) (X2)	1.243 (0.018)	1.144	2.580* (0.039)	1.083	−2.212* (−0.59)	1.472	−0.091* (−0.005)	1.353	5.924* (0.082)	1.238
lg(updates+1) (X3)	11.162* (0.180)	1.388	4.799* (0.083)	1.386	5.222* (0.138)	1.467	3.023* (0.167)	1.458	5.772* (0.093)	1.684
lg(comments+1) (X4)	22.578* (0.365)	1.394	21.659* (0.396)	1.559	13.518* (0.406)	1.888	3.732* (0.216)	1.606	33.655* (0.560)	1.794
Duration (X5)	−3.961* (−0.056)	1.077	−4.586* (−0.071)	1.115	−0.796* (−0.019)	1.149	0.298 (0.014)	1.031	1.681 (0.022)	1.142
Reward count (X6)	3.531* (0.054)	1.260	−0.350 (−0.006)	1.374	3.957* (0.093)	1.152	0.972 (0.058)	1.725	3.473* (0.046)	1.150
Reward mean price (X7)	−1.041 (−0.016)	1.253	−1.828 (−0.028)	1.097	−2.850* (−0.070)	1.255	−0.960 (−0.060)	1.840	−8.870* (−0.124)	1.274
Constant	0.510*		7.759*		−1.414		−2.839*		5.670*	
F-value	539.005*		495.357*		230.495*		46.104*		654.061*	
R^2	0.707		0.744		0.772		0.674		0.706	
Adjusted R^2	0.706		0.742		0.769		0.660		0.704	

$* \ p < 0.05$

Table 4 shows the results of regression model of Arts, Comics, Crafts, Dance and Design, respectively.

Table 5 shows the results of regression model of Fashion, Film & videos, Food, Games.

Table 6 shows the results of regression model of Music, Photography, Publishing.

Due to the different success rate of projects in 15 different categories, we conducted calculation and found out those variables such as projects goal and comments are significant in all categories. Updates are significant result in 12 categories. FAQs are significant result in 10 categories. Reward quantity of projects are significant result in 10 categories, and reward mean price are significant result in only 5 categories. These factors could be considered as most important factors affecting backers funding decision even though in different kinds of categories, which can be useful for creators when create projects in various categories.

Table 5. Results of regression model of Fashion, Film & videos, Food, Games and Journalism

Variables	(6) Fashion (N = 991)		(7) Film & videos (N = 1626)		(8) Food (N = 624)		(9) Games (N = 2667)		(10) Journalism (N = 69)	
	Coef.	VIF	Coef.	VIF	Coef.	VIF	Coef.	VIF	Coef.	VIF
lg(Goal) (X1)	18.949* (0.448)	1.867	34.438* (0.560)	1.939	20.186* (0.534)	1.721	24.875* (0.313)	1.799	9.265* (0.724)	2.396
lg(FAQs+1) (X2)	2.776* (0.054)	1.256	2.656* (0.032)	1.084	0.743 (0.017)	1.243	6.606* (0.071)	1.304	0.924 (0.054)	1.316
lg(updates+1) (X3)	8.165* (0.162)	1.316	7.770* (0.110)	1.471	7.030* (0.163)	1.327	0.166 (0.002)	1.810	0.264 (0.017)	1.650
lg(comments+1) (X4)	20.678* (0.433)	1.461	21.928* (0.316)	1.518	16.315* (0.397)	1.458	42.632* (0.617)	2.378	5.600* (0.356)	1.586
Duration (X5)	−0.019 (−0.001)	1.133	−0.429 (−0.005)	1.027	−0.158 (−0.003)	1.142	1.909 (0.019)	1.129	−1.747 (−0.098)	1.237
Reward count (X6)	2.795* (0.054)	1.235	4.610* (0.067)	1.545	3.170* (0.077)	1.435	1.057 (0.011)	1.167	0.170 (0.010)	1.400
Reward mean price (X7)	−3.545* (−0.073)	1.398	1.312 (0.018)	1.424	−1.913 (−0.044)	1.305	0.371 (0.004)	1.164	−0.850 (−0.055)	1.615
Constant	3.598*		−1.240		−0.929		11.426*		−1.599	
F-value	336.176*		814.662*		263.569*		1239.699*		42.277*	
R^2	0.705		0.779		0.750		0.765		0.844	
Adjusted R^2	0.703		0.778		0.747		0.765		0.827	

∗ $p < 0.05$

4.3 Residual Analysis

Result of residual fit analysis of all projects is shown in Fig. 2. According to the definition, residual is used as the predicted error term ε_i of regression model. In order to identify the hypothesis that the regression residuals follow the normal distribution which the mean value is 0 and the variance is a constant, we present the histograms, normal probability plots of residuals as well as the P-P plot to check whether the random errors are distributed normally or not. From both histograms and normal probability plots we can clearly see that the residuals in statistical process present the normal distribution. According to the P-P Plots, we can find that there is no obvious trend effect and most items fall into the range between −2 to 2. In summary, the residuals follow the normal distribution which means the errors items are distributed randomly.

Table 6. Results of regression model of Music, Photography, Publishing, Technology and Theater

Variables	(11) Music (N = 1988)		(12) Photography (N = 340)		(13) Publishing (N = 1687)		(14) Technology (N = 851)		(15) Theater (N = 480)	
	Coef.	VIF	Coef.	VIF	Coef.	VIF	Coef.	VIF	Coef.	VIF
lg(Goal) (X1)	29.934* (0.461)	1.533	16.978* (0.587)	1.885	33.691* (0.521)	1.504	12.545* (0.299)	2.134	20.398* (0.664)	1.707
lg(FAQs+1) (X2)	1.316 (0.017)	1.052	−0.329 (−0.009)	1.201	1.780 (0.024)	1.100	3.606* (0.074)	1.583	−0.247 (−0.006)	1.077
lg(updates+1) (X3)	10.172* (0.141)	1.248	3.758* (0.114)	1.446	6.479* (0.096)	1.384	3.570* (0.080)	1.901	4.852* (0.137)	1.283
lg(comments+1) (X4)	24.726* (0.360)	1.370	10.540* (0.349)	1.730	27.967* (0.429)	1.482	21.828* (0.555)	2.425	6.324* (0.176)	1.252
Duration (X5)	0.465 (0.006)	1.015	−0.787 (−0.020)	1.060	−4.990* (−0.064)	1.033	1.602 (0.027)	1.074	−1.469 (−0.037)	1.022
Reward count (X6)	8.115* (0.116)	1.328	1.250 (0.038)	1.418	3.775* (0.052)	1.188	3.009* (0.054)	1.197	3.100* (0.088)	1.291
Reward mean price (X7)	0.698 (0.010)	1.275	−0.990 (−0.029)	1.360	−3.157* −(0.044)	1.203	−8.872* (−0.157)	1.197	−0.851 (−0.025)	1.360
Constant	7.072*		−0.890		2.155*		2.149*		−0.890	
F-value	641.751*		177.615*		660.045*		415.956*		162.556*	
R^2	0.694		0.789		0.733		0.775		0.707	
Adjusted R^2	0.693		0.785		0.732		0.774		0.702	

∗ $p < 0.05$

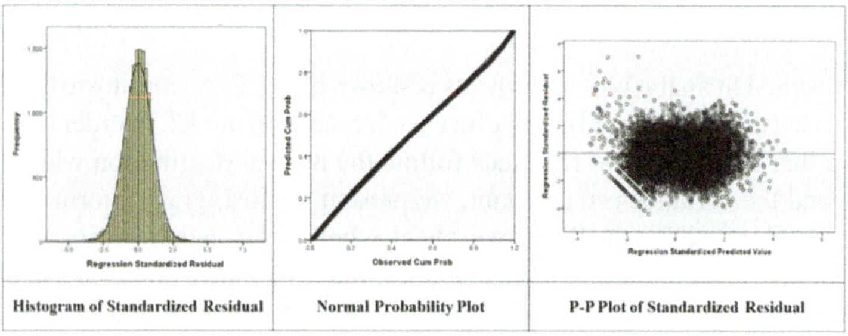

Fig. 2. Residual fit analysis of all projects (N = 16662)

5 Conclusion and Suggestions

The purpose of this research is to find out effective factors affecting the number of projects backers both of all projects and projects of different categories on reward-based crowdfunding platforms. We extracted data of successful funded projects which launching time were from January 1 to December 31 in 2017 on Kickstarter. Total 16662 projects were selected as our objective. We conducted multiple regression analysis and succeeded in finding that independent variables including projects goals, FAQs, updates, comments and reward quantity are significantly affecting projects backers, while other factors, duration time and reward mean price had a significantly negative effect towards the number of backers in general. Projects backers will get more familiar with projects through FAQs, updates and communicate with creators so that creators should pay more effort to manage and enhance relationship with backers in order to attract more potential backers.

Although numerous study discussed the determinants of projects success, rarely research was focus on what factors would affect the projects backers. We used factors that found by previous researchers to find out effective factors of the number of projects backers. We also include the reward characteristics, reward quantity and reward mean price as our independent variables, to conduct the analysis. Because of the different success rate of projects categories in average, we also conducted to analyze the factors effect of projects backers when referring projects in different categories.

The results show that variables including projects goal, comments, updates and FAQs, rewards amount and reward mean price cause effect in varying degrees which would be useful for creators when create projects of different categories. Projects goal is significantly positive related to the backers of all projects as well as projects in different categories. However, according to previous study, it is suggested that an appropriate funding goal of project makes a project more likely to succeed (Mollick 2014). Both updates and comments are significantly positive related to the backers which is the same as the updates effect of projects pledged funds across the funding period (Jin et al. 2018). We picked up two factors as the new indicators, the number of rewards and reward mean price, and find out that the previous one is significantly positive while the latter one is significantly negative associated with the projects backers, which is also proved that projects with more rewards options are attractive to backers (Gerber and Hui 2013) and backers are willing to select reward with lower price (Lin et al. 2016).

All the result can help creators to predict the how many projects backers they can attract based on previous information. Creators can make more valuable strategies with low cost and utilize the resource effectively based on the results that different factors play different roles in various projects categories. Besides, this research result also help projects creators get further acknowledge about backers funding behaviors so that they can take valuable actions before and during funding period. As the important factors affecting backers' funding decision, creators should do more on backers' relationship management such as posting updates and communicated with them in order to enhance a sense of responsibility to improve backers' confidence.

References

Barba-Sánchez, V., Atienza-Sahuquillo, C.: Entrepreneurial motivation and self-employment: evidence from expectancy theory. Int. Entrep. Manag. J. **13**(4), 1097–1115 (2017)

Buda, A.: Life time of correlation between stocks prices on established and emerging markets. arXiv preprint arXiv:1105.6272 (2011)

Cohen, J.: Statistical Power Analysis for the Behavioral Sciences. Routledge, London (2013)

EY: EY FinTech Adoption Index (2017). https://www.ey.com/Publication/vwLUAssets/ey-fintech-adoption-index-2017/$FILE/ey-fintech-adoption-index-2017.pdf

Gerber, E.M., Hui, J.: Crowdfunding: motivations and deterrents for participation. ACM Trans. Comput. Hum. Interact. (TOCHI) **20**(6), 34 (2013)

Insights, C.: Fintech trends to watch in 2018 (2018). https://www.cbinsights.com/research/report/fintech-trends-2018/

Jin, B.-H., Li, Y.-M., Li, Z.-W.: Study on Crowdfunding Patterns and Factors in Different Phases (2018)

Kickstarter: Kickstarter Stats — Kickstarter (2018a). https://www.kickstarter.com/help/stats

Kickstarter: Kickstarter Support (2018b). https://help.kickstarter.com/hc/en-us/

Kuppuswamy, V., Bayus, B.L.: Crowdfunding creative ideas: the dynamics of project backers in Kickstarter. In: Hornuf, L., Cumming, D. (eds.) A Shorter Version of this Paper is in "The Economics of Crowdfunding: Startups, Portals, and Investor Behavior" (2017)

Kuppuswamy, V., Bayus, B.L.: Crowdfunding creative ideas: the dynamics of project backers. In: Cumming, D., Hornuf, L. (eds.) The Economics of Crowdfunding, pp. 151–182. Springer, Cham (2018)

Lin, Y., Lee, W.-C., Chang, C.-C.H.: Analysis of rewards on reward-based crowdfunding platforms. Paper Presented at the 2016 IEEE/ACM International Conference on Advances in Social Networks Analysis and Mining (ASONAM) (2016)

Mollick, E.: The dynamics of crowdfunding: an exploratory study. J. Bus. Ventur. **29**(1), 1–16 (2014)

Neter, J., Wasserman, W., Kutner, M.H.: Applied linear regression models (1989)

Riedl, J.: Crowdfunding technology innovation. IEEE Comput. **46**(3), 100–103 (2013)

Statista: Crowdfunding - Statista Market Forecast (2018). https://www.statista.com/outlook/335/100/crowdfunding/worldwide

Lin, T.: The Determinants of Successful Crowdfunding Projects: An Empirical Study of a Taiwanese Crowdfunding Platform, Flying V. National Taiwan University (2015)

Chien, W.: A Study of Project's Performances on Taiwanese Crowdfunding Platform. National Taiwan University (2016)

Zhang, J., Liu, P.: Rational herding in microloan markets. Manage. Sci. **58**(5), 892–912 (2012)

Retaining Knowledge and Human Resource Management in IT Sector: How We Are SMEs Doing This?

Ana T. Ferreira-Oliveira[1](✉) and Ana F. Bouças[2]

[1] CISAS, Escola Superior de Tecnologia e Gestão do Instituto Politécnico de Viana do Castelo, Rua Escola Industrial e Comercial Nun'Álvares, 4900-347 Viana do Castelo, Portugal
ateresaoliveira@estg.ipvc.pt

[2] Escola Superior de Tecnologia e Gestão do Instituto Politécnico de Viana do Castelo, Rua Escola Industrial e Comercial Nun'Álvares, 4900-347 Viana do Castelo, Portugal

Abstract. Retention of human and intellectual capital is crucial for IT companies. However little is known on how IT that are start-ups and SMEs (Small and Medium enterprises) integrate the human capital expectations, specifically regarding sector's IT reputation on Big Companies. Despite the fact that human resources management (HRM) is very much explored by the scientific community in big companies, with large samples and large human resources department, research in SMEs remains scarce. This paper intends to describe and discuss the strategies that are used on these companies to retain knowledge. We present a qualitative study with semi-structured interviews and focus group with four technological SMEs from the North region of Portugal. Results curiously suggest that some social responsibility practices are more operationalized in SMEs than the practices considered central to HRM. This possibly can be explained by their lower complexity in implementation (e.g. flexibility of timetables vs. developed training programme) assisted by the flexibility and size of SMEs. It was also concluded that the informality of practices is associated with positive socio-relational mechanisms. In these contexts with low technical knowledge of HRM, it may have more positive results than a formalization system that can be a burden that is sometimes unaffordable for SMEs. The companies with the highest prediction of sustainable growth in the medium term present the beginning of a formal HRM system based on organizational trust mechanisms and, additionally, internal social responsibility practices can act as anchors that start the process of building a more technical and conscious human resources management that can work on the development of retaining knowledge.

Keywords: Knowledge retention · Human and intellectual capital · IT · SMEs

1 Introduction

Human resources management has been considered a subtask within the core areas of business management, and over the past few years has gained prominence regarding its

Á. Rocha et al. (Eds.): WorldCIST 2020, AISC 1159, pp. 35–44, 2020.
https://doi.org/10.1007/978-3-030-45688-7_4

power and strategic positioning (Oliveira 2014). With specific regard to SMEs, the difficulties and constraints in management are known, which translate into a multipurpose logic of functions, including the manager himself, who unfolds into diversified functions with different technical requirements. In a specific context of full employment, as we are actually living in Portugal, associated with societal concerns with the increase in job quality, we are facing a moment that seems to be relevant in the change of organisational behaviour.

In addition, it is known the tradition of IT big companies and their understanding of the relevance of strategic human resources management systems. However, little is known about IT SME's and their specific struggle on retaining human capital. Therefore, we intend to describe and discuss the strategies that are used on IT SMEs to retain knowledge. We also intend to raise awareness among members of small and medium-sized enterprises in the technological sector of the need to modify the management of their human resources in order to ensure the creation and maintenance of quality jobs.

Human resources management plays an essential role in the creation of sustainable competitive advantage; however, not all companies, especially small and medium-sized enterprises, understand its importance, or have the technical capacity to implement the most appropriate human resources management practices for the development of their formalised human resources systems (Veloso and Keating 2008). Faria and Machado (2019) suggest that the difference between large companies and SMEs in HRM is related to some of the peculiarities of SMEs such as: a) the role of the human resources manager that is mostly assigned to the owner of the company or to a director who accumulates the position with other functions; b) the absence of structured, standardized and formalized mechanisms such as job analysis, performance management or the recruitment and selection processes; c) training actions are often focused on the short term, and the survey of training needs is very informal or non-existent; d) and, finally, most SMEs are the result of an idea that is put into practice by someone with significant technical competence which, not always, corresponds to good people managers (Cunha et al. 2010). In this sense, several authors state that since small and medium-sized enterprises tend to neglect the existence of a formal and structured HR department (Barrett and Mayson 2007), they end up experiencing difficulties in specific issues associated with technical knowledge in HR. Thus, SMEs tend to adopt informal practices, often recruiting through the personal networks of the owner's contacts (Barrett and Mayson 2007) and are more concerned with the integration of new elements that identify with the organisational culture as opposed to the specific technical skills for the required functions (Heneman et al. 2000). If we think about some specific HRM practices, the capacity of a small and medium company to develop HR strategy is low. In the management of human resources in SMEs, we mostly find practices such as new employees onboarding processes, recruitment and selection, training and performance evaluation.

The OECD (2015) indications and guidelines on the relevance of quality in work divide the concept into quality of earnings, labour market security and quality of the working environment. Human resource management as traditionally described in the literature has not by itself integrated internal social responsibility, nor are numerous research studies known that focus on human resource management as the system that can and should integrate internal social responsibility policies and practices. This work considers that internal social responsibility and human resources management are conceptually distinct concepts, but in practice their integration may act as a lever for development, enabling organizations technically with the understanding that internal social responsibility policies are the pressing issues of human resources management. Corporate social responsibility is divided into two dimensions: internal and external. In its external dimension, CSR includes the network of relations with local communities, customers and suppliers, shareholders and investors, in compliance with universally recognised human rights, as well as in the overall management of the environment. In its internal dimension, socially responsible practices relate to human resources management, health and safety at work, adaptation to change and management of environmental impact and natural resources (Leite 2009). It is important to note that the implementation of socially responsible practices has been studied and often associated with positive results such as increased productivity and competitiveness of the company. With regard to internal CSR practices related to flexible working hours and reduction of working hours, a study carried out in Sweden states that working fewer hours is healthier and increases productivity: the employees involved in the study who worked fewer hours performed their tasks with greater quality and pride than those who worked normal hours (Gomes 2017). The internal development policies and practices, the transparency and openness in the company's communication, the organizations flexibility and the relevance attributed to the psychological and social attitudes developed in a relational context, namely the organizational trust and knowledge sharing are crucial elements in the organizational commitment of SMEs (Curado and Vieira 2019, Ferreira-Oliveira, Keating and Silva 2018). Timely and regular feedback, encouragement and recognition by managers are issues that are part of the HRM literature and are therefore also central to Internal Social Responsibility.

In this article we adopt the practices and attitudes that may constitute the empirical essence of HRM and the strategy for implementing HRM as organizing and driving elements of Internal Social Responsibility.

2 Method

2.1 Participants

The four technological participating firms were contacted on a convenience basis and taking into account the ease of access for the research team. The only inclusion criterion was the need for the company to be incorporated as a small or medium-sized enterprise and to operate in the North of Portugal. Key informants were identified in each company who would contribute to the study with the most relevant information. Following this assessment and considering the structure of the companies in question the interviews and focus group were addressed to managers and employees of each company (Table 1).

Table 1. Characterization of participants

N° of companies	**4**
N° of Employees per company	14–63
Total number of employees covered	109

2.2 Search Strategy, Data Sources and Procedure

The dissemination of the study was conducted using direct contact at the company's premises, telephone contact and email contact at a subsequent stage. Each company received an email with the objectives of the study, its informed consent and ethical considerations to safeguard the methodological issues of the study. A day and a timetable were scheduled with the participants for the interview and all were conducted face-to-face at each company's headquarters. All interviews and focus group were recorded in full and lasted, on average, approximately 60 min. The interviews were followed by verbatim transcriptions. The script was based on Oliveira (2014) regarding the implementation of human resources management practices. A thematic area was added related to the existence of internal social responsibility practices, the motivation for their implementation and their perceived impact. The script was developed with the aim of identifying four main themes: 1. Organization - history, creation and development, organizational structure of the company, production process and company strategy; 2. Human Resources Management - recruitment and selection of employees, reception and socialization, training and performance evaluation; 3. Human Resources Strategy Implementation Process - HR Function - its autonomy and power, the definition of the strategy, the definition of practices and processes of change in HR, critical processes (success versus perceived failure), expected results versus found results, final results of the process; 4. Social Responsibility Practices - implemented practices, motivation for implementation, organizational confidence - Development mechanisms of trust, between leadership and employees and between peers perceived impact, the strategy for the future related to CSR.

The starting points for our initial template were the interview topic guides. The initial template consists of five main themes, subdivided into specific sub-themes. For the analysis of the interview data, it was use the template analysis. This technique is based on the codification of theoretical themes that the researcher considers important for the research and their organization in a hierarchical structure in order to represent the existing relationships between them. Some of these themes are defined before the interviews are applied, but new themes can be changed, eliminated and introduced during the interview analysis (King 2012). As King suggests, the steps considered in performing the template were the following: a) definition of the themes a priori; b) transcription and reading of the interviews; c) data coding; d) construction of the initial template; e) application of the template to the remaining data; f) interpretation of the final template (King 2012).

3 Results

INITIAL TEMPLATE	Final template
ORGANIZATION	1. Organization
HISTORY OF THE ORGANIZATION CREATION AND DEVELOPMENT ORGANIZATIONAL STRUCTURE PRODUCTION PROCESS STRATEGY	1.1. Creation and development 1.1.1 Financing projects for entrepreneurs 1.2. Organisational Structure 1.3. Hierarchical horizontality 1.4. Transparency of management
HUMAN RESOURCES MANAGEMENT	2. Human resources management
ENTRY AND SOCIALIZATION RECRUITMENT AND SELECTION TRAINING PERFORMANCE EVALUATION	2.1 Entry and socialization 2.1.1 Absence of structure 2.1.2 Presentation of the hosting manual 2.1.3. relevance of the team 2.2 Recruitment and Selection 2.2.1 Recruitment by reference 2.2.2 External recruitment 2.2.3. ideal candidate 2.3 Training 2.3.1 Absence of a training plan and budget 2.3.2 Internal training 2.3.3 External training 2.3.4 Defined budget 2.4 Performance evaluation 2.4.1 Absence of performance evaluation 2.4.2. Observation 2.4.4 Evaluation of performance linked to business results
HR STRATEGY IMPLEMENTATION PROCESS	3. HR Strategy Implementation Process
RESPONSIBLE FOR HR, HIERARCHICAL LEVEL AND POWER STRATEGIC DEFINITION AND INTERFERENCE OF WHO HAS RESPONSIBILITY FOR HR HOW DOES THE DEFINITION OF PRACTICES OCCUR? HOW DO CHANGE PROCESSES USUALLY OCCUR? CRITICAL PROCESSES (SUCCESS VS. PERCEIVED FAILURE) EXPECTED RESULTS VS. RESULTS FOUND FINAL RESULTS OF THE PROCESS	3.1 Absence of human resources management department 3.2 Responsible for HR, hierarchical level and power 3.2.1 CEO 3.3 Strategic definition and interference of the HR Manager 3.3.1 Strategic reorganisation by the company in the face of HRM problems 3.4 Definition of practices 3.4.1. Leadership 3.4.2 Employee suggestions

(*continued*)

(*continued*)

INITIAL TEMPLATE	Final template
SOCIAL RESPONSIBILITY PRACTICES	4. Social responsibility practices
IDENTIFY IMPLEMENTED PRACTICES WHAT MOTIVATED YOU TO DEVELOP AND IMPLEMENT CSR PRACTICES? WHAT IS THE PERCEIVED IMPACT OF IMPLEMENTING CSR PRACTICES? WHAT ACTIONS WOULD YOU LIKE TO SEE IMPLEMENTED? WHAT IMPACT DO YOU WANT TO ACHIEVE ON PEOPLE MANAGEMENT?	4.1 Practices implemented 4.1.1 Work-family reconciliation 4.1.2. flexibility of timetable 4.1.3 Salary bonuses and direct benefits 4.1.4 Team Building/team spirit promotion activities 4.1.5 Local community development 4.1.6. Eco-sustainable 4.1.7 Lack of awareness of the implementation of CSR practices 4.2 Motivation 4.2.1 Values of the CEO 4.2.2 Employee performance 4.2.3. suggestions from employees 4.3 Perceived impact 4.3.1 Confidence building mechanisms 4.3.2 Trust between managers and employees 4.3.3 Peer confidence 4.3.4 Organisational culture 4.4 Contribution of CSR practices to the future of the company 4.4.1 Indirect benefits 4.4.2 Ecological footprint 4.4.3 Formalisation of a human resources management department

The answers to the first theme, "Organization" (1), identified two sub-themes: Creation and development (1.1) and Organizational structure (1.2). Within the creation and development of the company, companies in the Information Technology sector referred to the financing of projects for entrepreneurs ("It was born in 2008 following the *Poliempreende*, an IPVC project (…) we won the regional competition and with the support that existed, we advanced with the company". P1). Regarding the "Organisational structure" (1.2), the following subtopics were mentioned: hierarchical horizontality (1.2.1) "… the hierarchy is horizontal (…) each has its functions and people know what they are". P2) and Management Transparency (1.2.2) ("There is also total transparency in the company's spendings, we present what is spent in each department on HR, health insurance, marketing (…) the annual objectives and billing and the benefits to come depending on the achievement of the proposed objectives." P1).

The reactions to the second theme of the interview "Human Resources Management" (2) explored four sub-themes. In the first sub-theme Entry and integration (2.1), it was found that technological small and medium-sized companies do not have structured practices (2.1.1) concerning the entry and integration of new employees ("Entry is not structured, it is all very natural." P4) and, when it involves some structure, it essentially refers to the presentation of the welcoming handbook (2.1.2) ("A welcoming handbook and the passwords for the programs is given." P3); or the relevance of the team in the process of socialization of the new element ("there is effectively a follow-up and we invest some time to be with the person to understand if he or she fits the profile we seek (…) we have a spectacular team, with a treatment that I consider ideal, and this requires some effort on the part of all, so all participate naturally in the reception of the next element" P2).

Within the second sub-theme Recruitment and Selection (2.2), it was found that small and medium-sized companies resort to recruitment by references (2.2.1) ("[…] by references from other people." P4) and external recruitment (2.2.2). Here, companies in the technology and design sectors report having access to a wide range of applications, and have several spontaneous applications in databases ("[…] what we usually do is analyse spontaneous applications. P2; "[…] we have a database of spontaneous applications." P1). In addition, it is also perceived that some companies prioritize soft skills over hard skills in an ideal candidate (2.2.3) when recruiting and selecting a new employee ("Over the years we have made a lot of effort to set up the team. It is more important that the candidate adapts to our way of being than being very good at technical level. That's quick to learn." P2).

With regard to the third sub-theme Training (2.3), it was found that small and medium-sized enterprises that provide training report an absence of a training plan and a stipulated budget (2.3.1) ("The last time we have training was three years ago (…) "If we speak of the mandatory 35 h, in fact, it has not happened.)." P2). Others mention that the only training they offer to their collaborators is internal training (2.3.2) and that this is mostly linked to a first phase of the integration of the collaborator ("Some training is given in the welcome" P4); Also in this sub-theme, it was found that small and medium enterprises also use external training (2.3.3) ["As we are in the digital business, we buy online courses and make them available to all employees. We also have an internal library, with current books, where employees can even order books." P1)]. Finally, one of the companies has a budget for training (2.3.4) ("We have an annual fee of 400€ per employee for the time being." P1).

Sub-theme four focuses on Performance Evaluation (2.4) and is divided into three subtopics: the absence of performance evaluation (2.4.1) ("[…] we have no employee performance evaluation." P4); from the observation by managers and department heads ("I don't have it, but if I had to describe it, I would say it is by observation. P2). Finally, the performance evaluation articulated with the business results (2.4.3) ("The performance evaluation part is very much based on a first phase in which we show the team the objectives achieved in the past year and the benefits we add and the budget and objective for this year and the benefits that will result if we achieve them." P1).

The responses to the third theme "Change Implementation Process" (3) identified four sub-themes: Absence of a Human Resources department (3.1), HR Officer, hierarchical level and power (3.2), Strategic reorganisation in the face of human resources management problems (3.3) and Definition of practices (3.4). For the first sub-theme Absence of a human resources management department, small and medium-sized enterprises do not have a structured human resources department in their structure ("Given the size of the company, we are unable to retain one person in charge of this department alone." P1). The HR Manager, hierarchical level and power (3.2), is the CEO ("I am the HR Manager, the managing partner." P2). Regarding the Strategic Reorganization in face of human resources management problems (3.3), it was verified that there is a critical reflection on the part of the companies in relation to the implemented practices ("We felt that there was a failure because people did not understand what was expected from them and that was how we managed to fill it (…) every two months we have an individual moment with each employee to understand if everything is ok and if there is any personal or professional issue that is affecting their motivation". P1). Finally, with regard to the Definition of Practices (3.4), it is concluded that this task sometimes falls to Management (3.4.1) ("Policies and practices on human resources are defined by both partners of the company." P2) and takes into consideration the opinion of employees (3.4.2) ("When structural changes are of another essence (…) I make suggestions and the employees decide the sharing of tasks between them." P1).

The replies to the fourth and final theme "Internal Social Responsibility Practices" (4) allowed four sub-themes to be identified: Practices implemented (4.1), Motivation (4.2), Perceived impact (4.3), Future actions (4.4). Regarding the first sub-theme Practices implemented (4.1), the implementation of practices in different dimensions was verified: work-family conciliation (4.1.1), with small and medium enterprises reporting taking facilitating measures in this sense ("An employee who has to stay at home because of his child, won't have its salary deducted. A father had a child recently and is going to be working from home for a while". P2); Flexible working hours (4.1.2) ("The employees can work from home one day a week." P1); Salary bonuses and direct benefits (4.1.3) ("We provide health insurance for all employees" P1); Activities to promote team spirit/team-building (4.1.4) ("We also celebrate the company's anniversary on a weekend in which all employees participate and do different types of activities." P2); Contribution to the development of the local community (4.1.5) ("[…] intends to give back to the community what it gives us every day. We are a company, we work in community, our clients buy our systems and as such we have this responsibility (…) the GAF often needs someone to give a lecture to the students (…) we have protocols with Universities and Professional Schools where we participate in projects and receive interns." P1).

The second sub-theme Motivation (4.2), allowed us to list three subtopics: companies refer that what leads them to implement social responsibility practices has to do with the values of the CEO (4.2.1) ("We can only be well with ourselves if the people around us, in this case my employees, are happy and fulfilled. It is important for me to have happy people working with me." P1); with the performance of employees (4.2.2) ("Above all, it allows me to make people happier and, in my view, when people are good and happy, they are more productive, and also well-disposed and motivated." P3); and, finally, based on suggestions from employees (4.2.3) ("Now we have rotational breaks (…) what led to the establishment of this practice was the request of employees." P2).

Regarding the third sub-theme Perceived impact (4.3), the mechanisms of trust development arise (4.3.1.) ("We have a very high level of trust among all (…) we always try to make everyone feel part of the team." P2.), the trust between management and employees ("I trust my team." P3), the trust between peers (4.3.3.) ("I think that's why we have a spectacular team, there's a relation between the people I consider ideal." P2) and the organisational culture (4.3.4) ("[…] we are also creating an entity and culture for the company, we are part of projects that help the community, that really make a difference in people's lives, and employees feel and identify with it". P1).

With regard to the fourth and final sub-theme Contribution of CSR practices to the future (4.4), it was found that companies intend to expand their action mainly in terms of their contribution to indirect benefits (4.4.1) ("[…] health insurance for family members." P1), the ecological footprint (4.4.2) ("[…] in being more energy efficient." P2) and the formalization of a human resources management department (4.4.3) ("I am the one who deals with recruitment and selection and definition of strategies. I think the results are not very positive (…) I feel the employees disconnected and inattentive." P4.

4 Discussion and Conclusion

This article aims to describe and discuss the strategies that are used IT SMEs to retain knowledge. The results point to a poorly structured human resources management system and to the presence of some awareness and implementation of good internal social responsibility practices. With regard to the human resources management system, there was a lack of structure and planning both with regard to entry and integration, as well as recruitment and selection, training and performance assessment. With regard to the process of implementing changes in HR strategy, it was found that none had a human resources department in its structure and that the functions associated with that department are often managed by managers or financial department. However, the novelty of this work is based on the internal social responsibility practices already implemented in SMEs. We are faced with an awareness of companies and some practices already implemented. Although the subject of CSR is unknown to some, the practices are no longer and, in this sense, internal CSR business practices have been included in the following categories: Work-family conciliation; Flexibility of hours; Team-building promotion activities; Local community development and Eco-sustainability. We found in this work that internal social responsibility practices are numerous times more operationalised by SMEs than the core practices of HRM. This aspect can be explained by its lower complexity in the implementation (e.g. flexibility of schedules vs. training) assisted by the flexibility and small size of SMEs that can be a relevant aid in the implementation of these practices. It was also concluded that the implementation of internal CSR practices in SMEs is associated with the personal and professional values of the owner, together with a willingness to reflect the company's culture in its employees and convey to them the message that the company cares about them and the world.

It is relevant to continue specific research in SMEs as these are not only a smaller version of large companies (Storey 1995). They have contextual and dynamic specificities that need to be investigated. In fact, the informality of practices is well known, however we are witnessing an international panorama of small and medium-sized enterprises that are born with ideologies based on a more integrative vision of their human resources and that privilege positive and informal dynamics with good organizational results. It is relevant to continue the research in order to understand to what extent informality can be a problem or counterproductive, if the human resources management processes interspersed with some informality in these contexts may have more positive results than a formalisation load that is sometimes unaffordable for SMEs. However, we must not make the mistake of assuming that HRM systems are neither relevant nor necessary. It is important for SMEs to be able to create formal, contextual, adaptable and flexible practical structures that, together with psychological attitudes such as organisational confidence or transparent management behaviour, enable effective, productive human resource management that is appropriate to existing resources.

References

Oliveira, A.: Perceções do sistema de gestão de recursos humanos e o papel da confiança organizacional. Tese de Douturamento, Universidade do Minho, Portugal (2014)

Veloso, A., Keating, J.: Gestão de Recursos Humanos em PME's de elevada tecnologia. Psicologia **2**, 35–58 (2008)

Faria, A., Machado, C.: Human resources management in a small – and medium-sized enterprise. In: Machado, C., Davim, J. (eds.) Management Science. Management and Industrial Engineering. Springer, Cham (2019)

Cunha, M., Rego, A., Cunha, R., Cabral-Cardoso, C., Marques, C., Gomes, J.: Manual de Gestão de Pessoas e do Capital Humano. Sílabo Press, Lisboa (2010)

Barrett, R., Mayson, S.: Human resource management in growing small firms. J. Small Bus. Enterp. Dev. **14**(2), 307–320 (2007)

Heneman, R.L., Tansky, J.W., Camp, S.M.: Human resource management practices in small and medium-sized enterprises: Unanswered questions and future research perspectives. Entrepreneurship Theory Pract. **25**(1), 11–26 (2000)

OECD: The quality of working lives: Earnings mobility, labour market risk and long-term inequality (2015)

Leite, C.: A Responsabilidade Social das Empresas em Portugal. Master thesis. Faculdade de Psicologia e Ciências da Educação da Universidade de Coimbra, Portugal (2009)

Gomes, T.: A influência da gestão de recursos humanos na motivação dos colaboradores. Dissertação de Mestrado. Instituto Superior de Contabilidade e Administração do Porto, Portugal (2017)

Curado, C., Vieira, S.: Trust, knowledge sharing and organizational commitment in SMEs. Pers. Rev. **48**(6), 1449–1468 (2019)

Ferreira-Oliveira, A.T., Keating, J., Silva, I.: Decision making on human resource management systems. In: Trends and Advances in Information Systems and Technologies, pp. 1040–1045 (2018)

King, N.: Doing template analysis. In: Qualitative Organizational Research: Core Methods and Current Challenges, pp. 426–450 (2012)

Storey, J.: Foreword. Hum. Resour. Manag. J. **5**(4), 3 (1995)

Proposing Ontology-Driven Content Modularization in Documents Based on the Normalized Systems Theory

Vojtěch Knaisl(✉) and Robert Pergl

Faculty of Information Technology, Czech Technical University in Prague, Prague, Czech Republic
knaisvoj@fit.cvut.cz
https://ccmi.fit.cvut.cz

Abstract. A problem of evolvability is widely discussed in the current world, and still, it has not been fully addressed yet. Our approach tries to improve evolvability in a domain of documents. Our approach is based on principles and recommendations from the Normalized Systems Theory. We try to redefine the process of how the document is created and maintained by involving ontologies. We offer a solution which should increase evolvability for a sort of documents which is created by a template and which is often updated. We demonstrate our solution to an example of a Data Management Plan document.

Keywords: Evolvable documents · Normalized Systems Theory · Ontology · Modularity

1 Introduction

The problem of low evolvability of documents is one of the main challenges. The expansion of computers brought a huge digitalization across all aspects of our society. Nevertheless, the old document stays as it was but just in digital form. This is not a problem until we want to reuse the document again; we want to share parts between documents, or we want to update the content of the document in the future.

Someone may assume that the form of traditional document as a package of A4 pages that combine actual content with formatting will be overcome. But it seems that in the near future it will not happen. Meanwhile, the companies created billions of documents for their needs, and now they face a problem of how to manage these amounts of documents. Especially, they face a problem with low evolvability and low reuse. Significant parts of the documents are duplicated. This leads to a state where a mistake in the document is distributed to many copies of the document, and a possible fix is almost unable to apply. The trigger of change can relate to changed law, too, or just to improving the document. Generally, every change is very expensive to apply, and the cost increases with the size of the system.

Á. Rocha et al. (Eds.): WorldCIST 2020, AISC 1159, pp. 45–54, 2020.
https://doi.org/10.1007/978-3-030-45688-7_5

Low evolvability is not just a topic in documents. Thus, many people work are involved in solving this particular issue. One promising approach offers the Normalized System Theory [8]. The theory contains principles ad recommendations to enhance the evolvability of systems. The first application in software engineering brought compelling results [10], and now we can find attempts to apply it in other domains (e.g., requirements engineering [16]). The two fundamental principles related to document domain are fined-grade modular structure and separation of concerns. The paper tries to address these principles in the area of documents.

In Sect. 2, we describe the principles and frameworks on which we built our approach. In Sect. 3, we discuss the problem of document modularization on which we offer a solution in Sect. 4. In Sect. 5, we take the proposed solution and apply it to the real-world use case. Section 6 compares our proposal with already available solution.

2 Methodology

2.1 Normalized Systems Theory

The core of Normalized Systems Theory [8] is to deal with the evolvability of systems. The theory defines that the system is evolvable as far as it is free of combinatorial effects. A combinatorial effect is then defined as a change whose impact is not solely related just to the kind of change but also to the size of the system on which the change is applied.

The cornerstone of the theory is modularity. Authors claim that it is one of the reasons where most of the systems fail. Therefore, the systems have low evolvability. The theory exactly says: "The system should be composed of very fine-graded modules". If we break this rule, more combinatorial effects may appear. NS Theory is based on 4 principles - *Separation of Concerns*, *Data Version Transparency*, *Action Version Transparency*, and *Separation of States*.

For the case of evolvable documents, only the first two principles (*Separation of Concerns*, and *Data Version Transparency*) are applicable. The second two principles (*Action Version Transparency*, and *Separation of States*) are only workflow-related [11,14].

Principle Description:

- *Separation of Concerns* states that all concerns should be separated from each other.
- *Data Version Transparency* states that you should be able to update each module without any impact on other linked modules.

The NS Theory was originally intended to be used for information systems and software development [10]. Due to its generality, other usages appeared. Currently, we can find applications in domains, such as requirements engineering [16], study programs [12], etc.

2.2 Evolvable Documents

Our research extends the previous work, which was done mainly by Gilles Oorts [11]. His primary focus is to improve the evolvability of study programs with a possibility to generalize some principles and recommendations to a broader range of types of documents [12].

The main problem of study programs is their very high number of combinations, which they can be composed together. You have to not only create a study plan (a document) for a concrete student, where you need to describe the subjects, but you need to also create a curriculum (also a document) for the university's accreditation purposes, etc. Thus you have information about a subject which you need to put into different types of documents with a different level of detail.

His approach was to apply principles and recommendations from the NS Theory into an area of documents. Based on that, he split the information of the subject into modules that he could then combines together. The modularization was done based on his knowledge of the domain of study programs. Thus if we took his approach, we wouldn't be able to apply it one-by-one into different documents.

2.3 Ontology-Based Modularization of User Interfaces

Our document modularization approach uses ontologies. We may find this practice also in other domains. Heiko Paulheim from Universität Mannheim uses the power of ontologies to modularize user interface (UI) [13]. By the accurate identification of relationships in ontological analysis, he tries to reduce dependencies between modules to build more independent and loosely coupled modules. He demonstrated his approach to an example of an emergency management application where he split components based on an ontology for emergency management. Our approach inspires in his method by taking the idea of using the ontology for better modularization.

3 Problem of the Modularization

Suppose we assume, modularization is the right way how to make the system evolvable. Then the fundamental question is where to define borders for modules and how to identify them. Wrongly separated modules can cause a problem with evolvability in the future. Imagine, we have a change request on the system. It is a simple change that contains just one concern. Unfortunately, the modularization was done in the wrong manner. The concern is spread over more components and parts of the system. With increasing size and complexity of the system, this can lead to a situation that is intellectually and administratively unmanageable. Currently, the modularization is done by domain experts. We use their knowledge and experience to define boundaries. In sort of systems (like software systems, user interfaces, etc.), the modules and related concerns are easier to identify because these systems are structured naturally. Nevertheless, some systems (e.g., documents) are typically very unstructured. Therefore, to find the right modules is not very straightforward.

The main benefits of the very fined and right modularized system are the ability to reuse modules and the guarantee of good evolvability in the future. Modules in these systems contain just one concern each. The problem of mixing concerns together leads to worse evolvability in the future. Speaking of documents, we have two types of cross-cutting concerns, which we have to be aware of - document cross-cutting concerns and content cross-cutting concerns [14].

The document cross-cutting concerns relate to a layout, a reference mechanism, a relative embedding of text parts, language, etc. These concerns can be addressed by tools that support us in accurately capturing document modules. We do not have to identify these concerns for each document separately. The concerns are the same for all documents, and we have to be just aware of them.

The content cross-cutting concerns are specific to each document. They have to be identified by domain experts. Thus, precise identification depends just on the knowledge and experience of a concrete expert. Very basic modularization is currently handled by chapters and sections which help readers to orientate in the document. If we stick to the principle that every section should talk just about one topic, we increase the reusability of the module in the future. Unfortunately, that does not have to be enough.

Further, the problem can appear in large, complex domains where one domain expert is not enough. To make appropriate modularization, experts have to collaborate. Sometimes, you have to collaborate with the target audience too. All these people have to understand each other, which can be problematic. Otherwise, we are unable to define the modules properly. This situation is not new, and we may know it, for example, from requirements engineering [5]. To improve understanding between people, we used to use ontologies and ontological models.

4 Our Approach

Our approach adds one more step in the beginning. Instead of the creation of a document immediately, firstly, we do an ontological analysis. This helps us better identify entities and relations with which we are working in the document, and further, it helps better identify the modules.

The whole process is described in the picture below Fig. 1. It is split into two sides - the left side relates to gathering the knowledge from users, and the right one is focusing on specifying a concrete document template. Nevertheless, as we already mentioned, initially, the process of creating an evolvable document starts with creating an ontological model. This model should help with identifying content cross-cutting concerns.

For gathering the knowledge from the user, our idea is to have a questionnaire based on the ontological analysis. The questionnaire should retrieve a knowledge which is needed for filling a document template with information specific for the concrete dofcument. The questionnaire contains hierarchical questions that

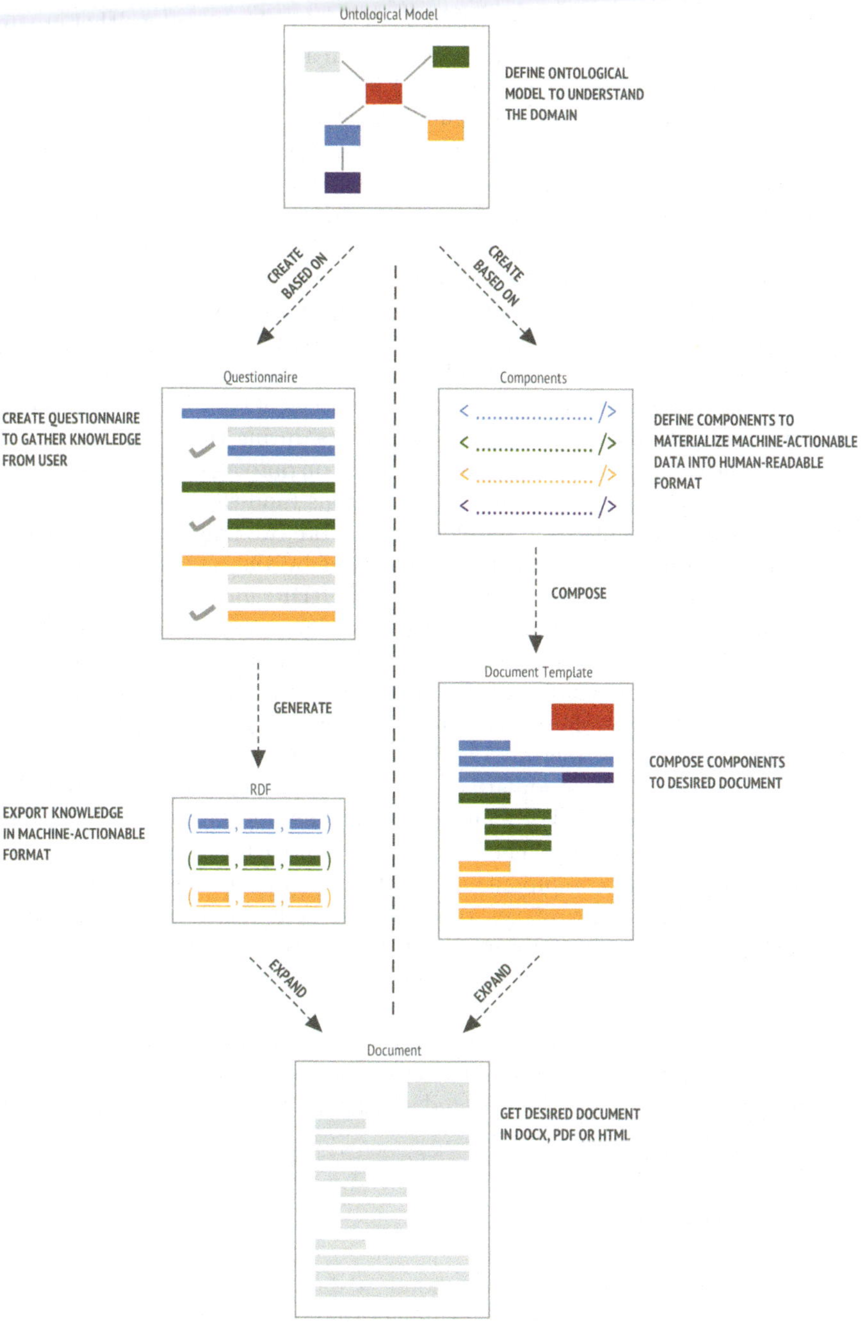

Fig. 1. Process overview

should help users to orientate in the domain. The questions can be either created manually or can be generated from the ontological analysis. The manual curation can later cause a problem because we need to synchronize the changes in ontological analysis with changes in questions. The generated questionnaire may be a better solution. The problem related to question generators is already discussed in conferences [4, 15]. Therefore, there is a possibility to reuse some proposals or tools for that. But the first evaluation counts with the manual question creation due to its simplicity for prototyping.

For modeling ontology, we use experienced Web Ontology Language (OWL). It's widespread ontological language, and we can find many ontologies captured in this language. However, OWL itself doesn't guarantee any quality of a produced ontology. We have to be aware of creating ontologies or reuse some already existing. Consideration for the future could be a usage of OntoUML [6]. It is more descriptive and provides better guidance for users during the analysis.

Thus we have the questionnaire created based on ontological analysis (or even generated from ontological analysis); our questions (and their answers) are connected to specific elements from ontology. Therefore, we may produce relatively simply output from the questionnaire annotated by the ontology. The output data is in the Resource Description Framework (RDF) format. This format will guarantee that data will be readable not just for humans but also for machines. Nowadays, machine-actionability is a more and more important aspect of storing data [18]. An increasing amount of data leads to a state where we are unable to process it manually. Therefore, this can be an advantage for further usage.

The right side of the picture Fig. 1 relates to the document cross-cutting concerns. To support the modularity, we should split the document template into modules (components). For identifying the components, we use the insights from the ontological analysis. The components are then composed into a final document template. The components can be captured in one of the existing templating languages. They offer a possibility to split the document into modules, which is a core feature for our case. But unfortunately, they don't support RDF directly. Thus we have to convert the content from RDF into the so-called template context.

As far as we have content in RDF and prepared document template, we can do the final step, which is a combination of the captured user's knowledge in RDF format together with the selected document template and generate the desired document.

5 Case Study

For validation of our approach, we choose a data management plan (DMP) document [1]. The DMP is a formal document that outlines how data is handled. It is often required to get subsidies by funding agencies. This document lives with a researcher during the duration of the grant. The DMP is a living organism; thus, it is often changed during the phases of the project. Further, each funding agency usually has its own template for the data management plans. Often, they don't differ much in content. They mostly require the same information to be included in. Unfortunately, they expect them in a different form. Each funding agency wants to have there its name, its logo, etc. They want to combine the information into different chapters. Thus the re-usability of the DMP (made for one funding agency) for another funding agency approaches zero.

We may see that if we had to manage the DMP as a standard document (e.g., in Microsoft Word), we would get into a situation that is intellectually and administratively unmanageable. We think that our approach helps to avoid such difficulties. Due to splitting the knowledge gathering and the actual document template, the problem of continually changing information relates just to the left side of our project. Since we have a questionnaire there, the researcher can easily just change his answers and regenerate the target document. The second problem was the problem of different templates for different funding agencies. In a standard document, the researcher has to completely rewrite the document (or copy the text parts) into a new document according to different funding agencies' requirements. This situation can happen, for example, when the researcher failed in getting a subsidy from one funding agency, and he wants to apply for subsiding from different funding agencies. Our approach helps him in a way that he just chooses different document templates and regenerates the target document.

The actual document template is composed of modules. Thus, we can share the common modules between document templates. The well-done modularity, which was accomplished by our ontological analysis, helps us in reusing the parts. Therefore, we avoid duplications in creating the document template specific to a particular funder. Although we are still able to do customization and customize the document template to satisfy the funders' requirements.

We can observe that our proposal includes 2 main phases which are planned to be handled by two different personas. In our case, it would be a data steward and researcher. If we go bottom-up, the researcher wants to submit a data management plan. He fills the questionnaire, selects the document template, and he can generate his document. He doesn't have to worry about the templates and related things. From another perspective, we have a data steward who is an expert in the data stewardship domain, and he is capable of an ontological analysis on the data management plan. Based on the analysis, he designs document modules and composes them into a document template. Further, he should be able to create or generate a questionnaire from the analysis.

6 Related Work

6.1 DITA

Darwin Information Typing Architecture (DITA) is an open standard for writing modular technical documentation [7]. DITA was developed by IBM in 2001, and currently, it is maintained by OASIS Dita Technical Committee. Compares to our approach, DITA also focuses on reusing the content in documents. It tries to achieve it by splitting the content into modules, which are called topics in DITA's terminology. The original use case was maintaining the documentation for company products. Thus, IBM intended to reuse modules for various types of documentation, such as internal manual, user manual, etc. They want to also reuse modules due to the similarity of their products. Therefore, it made sense to share common parts of the documentations across products.

DITA's main focus is on providing the ability to manage the document modules and on delivering the way how to customize these modules. The module can contain conditions which control the shown and hidden parts of the text. The option is heavily used in determining the complexity of the module. The author can drive the level of abstraction, which helps to maintain the text for beginners and for advanced people in one place.

Unfortunately, DITA has no mechanism or guidance on how to modularize the document. The author has to aware of the right modularization of the concrete document. This may end in the presence of undesirable combinatorial effects, which decreases the evolvability. There is also no separation of actual content and the document template, which decreases the evolvability according to the NS Theory, too.

7 Conclusion

Our approach was to improve the evolvability of documents. Our solution has no intention to solve the problem of the evolvability of documents in general. We focused on just a subset of all types of documents. We successfully demonstrated to an example of the Data Management Plan (DMP) document that our approach is valid and can help with the evolvability of DMP or other documents that show similar signs.

We claim that you may save time due to increased evolvability, which was achieved by splitting the cross-cutting concerns and by using the ontological analysis to do a proper modularization. On the other hand, we can observe that our approach includes additional time to set it up. As we discuss in the beginning, a user can benefit from our approach, if he can benefit from increased evolvability. Another use case is when the user has to fill the document in more "instances" (e.g., more DMPs for more projects or funding agencies). Our approach may help him to achieve it without copying documents.

7.1 Future Work

Currently, we are in a state that we formed the process, we showed it where our approach may help and simulated, and demonstrated on one concrete example of a document - the Data Management Plan. To fully verify our approach on real users, we will implement the prototype that will simulate the process. We would like to customize an already implemented solution (Data Stewardship Wizard [2]) and enhance it with our proposed ideas.

As a starting point, we will use an ontology of the RDA working group [9]. This ontology was created bottom-up based on the user's feedback in a data stewardship community [17]. As the desired template, we have chosen a Horizon

2020 template for a data management plan [3]. Our plan is to put the content into RDF, create a document template, and try to make changes and watch how we are able to catch them.

Acknowledgment. This research was supported by the grant of Czech Technical University in Prague No. SGS17/211/OHK3/3T/18. The work on the Data Stewardship Wizard is partially funded by IOCB of the CAS and ELIXIR infrastructure.

References

1. Data management plan. https://library.stanford.edu/research/data-management-services/data-management-plans. Accessed 14 Nov 2019
2. Data stewardship wizard. https://ds-wizard.org. Accessed 14 Nov 2019
3. Horizon 2020 template. https://ec.europa.eu/research/participants/docs/h2020-funding-guide/cross-cutting-issues/open-access-data-management/data-management_en.htm. Accessed 14 Nov 2019
4. Alsubait, T., Parsia, B., Sattler, U.: Generating multiple choice questions from ontologies: lessons learnt. In: CEUR Workshop Proceedings, vol. 1265, pp. 73–84, January 2014
5. Castañeda, V., Ballejos, L., Caliusco, M., Galli, M.: The use of ontologies in requirements engineering. Glob. J. Res. Eng. **10** (2010)
6. Guizzardi, G., Fonseca, C., Benevides, A., Almeida, J., Porello, D., Prince Sales, T.: Endurant types in ontology-driven conceptual modeling: towards Ontouml 2.0. In: Conceptual Modeling, pp. 136–150, October 2018. https://doi.org/10.1007/978-3-030-00847-5_12
7. Harrison, N.: The darwin information typing architecture (DITA): applications for globalization. In: IPCC 2005. Proceedings. International Professional Communication Conference, pp. 115–121, July 2005. https://doi.org/10.1109/IPCC.2005.1494167
8. Mannaert, H., Verelst, J., De Bruyn, P.: Normalized systems theory: from foundations for evolvable software toward a general theory for evolvable design. In: KOPPA BVBA, 01 edn., 17 October 2016 (2016)
9. Miksa, T.: RDA DMP ontology. https://github.com/RDA-DMP-Common/RDA-DMP-Common-Standard. Accessed 14 Nov 2019
10. Oorts, G., Huysmans, P., Bruyn, P.D., Mannaert, H., Verelst, J., Oost, A.: Building evolvable software using normalized systems theory: a case study. In: 2014 47th Hawaii International Conference on System Sciences, pp. 4760–4769, January 2014. https://doi.org/10.1109/HICSS.2014.585
11. Oorts, G., Mannaert, H., Bruyn, P.: Exploring design aspects of modular and evolvable document management. In: Lecture Notes in Business Information Processing, pp. 126–140, April 2017. https://doi.org/10.1007/978-3-319-57955-9_10
12. Oorts, G., Mannaert, H., Bruyn, P., Franquet, I.: On the evolvable and traceable design of (under)graduate education programs. In: Advances in Enterprise Engineering X, pp. 86–100, May 2016. https://doi.org/10.1007/978-3-319-39567-8_6
13. Paulheim, H.: Ontology-based modularization of user interfaces. In: EICS 2009 - Proceedings of the ACM SIGCHI Symposium on Engineering Interactive Computing Systems, pp. 23–28, January 2009. https://doi.org/10.1145/1570433.1570439
14. Suchánek, M., Pergl, R.: Evolvable documents - an initial conceptualization. In: PATTERNS 2018, pp. 39–44. IARIA (2018)

15. Teitsma, M., Sandberg, J., Maris, M., Wielinga, B.: Using an ontology to automatically generate questions for the determination of situations. In: Database and Expert Systems Applications, pp. 456–463, August 2011. https://doi.org/10.1007/978-3-642-23091-2_39
16. Verelst, J., Silva, A., Mannaert, H., Ferreira, D., Huysmans, P.: Identifying combinatorial effects in requirements engineering. In: Lecture Notes in Business Information Processing, May 2013. https://doi.org/10.1007/978-3-642-38117-1_7
17. Walk, P., Neish, P., Miksa, T.: RDA DMP common standard. https://www.rd-alliance.org/groups/dmp-common-standards-wg. Accessed 14 Nov 2019
18. Wilkinson, M., Dumontier, M., Aalbersberg, I.J., Appleton, G., Axton, M., Baak, A., Blomberg, N., Boiten, J.W., Bonino da Silva Santos, L.O., Bourne, P., Bouwman, J., Brookes, A., Clark, T., Crosas, M., Dillo, I., Dumon, O., Edmunds, S., Evelo, C., Finkers, R., Mons, B.: The fair guiding principles for scientific data management and stewardship. Sci. Data **3** (2016). https://doi.org/10.1038/sdata.2016.18

Exercise of the Rights to Communication, in Conventional and Digital Media, in the Republic of Ecuador

Abel Suing(✉), Carlos Ortiz, and Juan Carlos Maldonado

Grupo de Investigación en Comunicación y Cultura Audiovisual Departamento de Ciencias de la Comunicación, Universidad Técnica Particular de Loja, 11-01-608 Loja, Ecuador
arsuing@utpl.edu.ec

Abstract. The Constitution of the Republic of Ecuador in harmony with the Universal Declaration of Human Rights establishes that the social communication system must ensure the exercise of the rights to communication, information and freedom of expression. This protection framework generates positions in companies and social sectors, but rarely have the voices of journalists and employees been heard to identify how communication rights are fulfilled. The purpose of the research is to learn about rights and freedoms related to communication. The methodology is qualitative, through semi-structured interviews 11 communicators working in audiovisual and digital media of Ecuador are approached, between June 29 and July 15, 2019. The hypothesis is: restrictions occur that are overcome in a creative way. Solutions have often been found on platforms and digital media. It remains as a line to explore the informative spaces through the Internet as part of a new communication ecosystem of Ecuador.

Keywords: Right to communication · Regulation · Political communication · Freedom of expression · Freedom of information

1 Introduction

The Republic of Ecuador recognizes the right to free expression since its foundation. The country's first Constitution, in 1830, stipulated in its article 64 that all citizens have the right to "freely express and publish their thoughts through the press, respecting decency and public morals, and always subjecting themselves to the responsibility of the law" [1]. The Andean country goes through constant political instabilities to the point of issuing twenty Constitutions, but in all of them there is freedom of expression.

In the 1970s, in harmony with the postulates of the New Order of Information and Communication, as well as national communication policies, a first effort is made to regulate the communication transmitted by modern media [2]. The first Broadcasting and Television Law was published in the Official Gazette No. 785, dated April 18, 1975, in the dictatorial chaired by General Guillermo Rodríguez Lara [3].

In the 21st century, in a turn towards plural spaces and citizens of communication since until then there were no public media, a debate based on the Universal Declaration

Á. Rocha et al. (Eds.): WorldCIST 2020, AISC 1159, pp. 55–63, 2020.
https://doi.org/10.1007/978-3-030-45688-7_6

of Human Rights begins to incorporate a "new" right: the right to communication. Then it became clear that the model "embodied in the hegemonic interpretation of individualist capitalism about freedom of expression [was] insufficient to process contemporary conflicts, needs and demands related to communication" [4].

In October 2008, Ecuador became the first country in Latin America that recognized the rights to communication in its Constitution, between articles 16 to 20 communication and information are guaranteed as substantial elements of the democratic structure of the State, as well as Articles 66.6 and 18 recognize freedoms of expression and information, essential for the paradigm of good living, one of the axes of the Constituent Assembly of 2007 and 2008. In addition, art. 384 determines that the social communication system will ensure the exercise of the rights of communication, information and freedom of expression.

The right to communication, fruit and derivation of new social emancipations, "encompasses the right to communicate and receive truthful information and freedom of expression (transmission of ideas, thoughts and opinions)" [5], congregates the freedoms of information and expression, two rights closely linked to communicational activity. "Freedom of communication, information and expression are interdependent, inalienable, irreducible and constitutional subjective rights of the same hierarchy" [6], but in practice, freedom of expression has exceptions that always involve protection of vulnerable groups and do not remove right to receive information [7, 8].

On the basis of the 2008 Constitution, an Organic Communication Law (LOC) was issued in 2013 with the purpose of "ensuring the full and effective exercise of freedom of expression and information" [9]. However, there were voices of alertness and concern about contravening freedoms in the field that he tried to regulate. A group of international organizations "fervently opposed the sanction of the LOC (…) this type of legislation would directly threaten the exercise of press freedom by the media" [10].

The LOC was made up of incompatible provisions "with the standards for use in the area of freedom of expression and information (…) omissions that at least doubt that the norm [had] sincere will to guarantee pluralism and (…) communication freedoms" [9]. One of the structures that generated the most controversy was the Superintendence of Communication, which in principle was thought of as a technical surveillance body [5]. During its exercise, the Superintendence issued repeated resolutions inconsistent with the inter-American standards of freedom of expression [11], which meant administrative sanctions to mass media and journalists [7], in conclusion, the norm produced discouragement in society and showed itself incompatible with the democratic principles.

The LOC, which was renovated in 2019, led to some impediments. In the study on the perception of the levels of professional autonomy of journalists from Ecuador, from researchers Oller, Chavero and Ortega published in 2016 [12], it is pointed out that the limits to freedom of expression are in the legislation (punitive LOC that causes self-censorship), in professional ethics (respect for people and their rights) and in the interests or pressures of the media or the company (the editorial line of the media), it should be noted that "the fear of being sanctioned by the regulator bodies (…) cause, on the one hand, the self-censorship of journalists and, on the other, the decrease in emphasis on research and analysis" [12].

The interests of the owners may condition the commitment, of the media, to "remain at the service of citizens and the debate of ideas" [13]. It has not been strange, in the classical liberal conception, that the rights of citizens have been undermined. Before 2008, slightly less than 100% of the media in Ecuador were privately owned, therefore, a market logic prevailed in the formation of public opinion.

As a result of the LOC, the country would have started the "evolution towards a journalistic culture that abandons the neoliberal concept that prevailed in the last decades of the twentieth century (...) and that embraces a socio-democratic model" [12].

The fear of the punishment of journalists is based on the fact that the rule "was only applied in its punitive facet, in relation to complaints to media and journalists, while the most innovative and transformative aspect was left aside, as was the citizen media development" [14].

The authorities of the Superintendence of Communication, perhaps due to the political turmoil and the constant criticism of private media, did not exploit other means such as "less restrictive alternative measures for the fundamental rights that can be secured" [11].

In the context of the regulation of traditional communication, digital media emerge, presenting new dimensions of relationship with audiences. Broadcasts over the Internet do not compete with open-air radio and television because they are aimed at different audiences, however, there is evidence that Ecuadorian digital media have respected communication rights, even several online newspaper managers believe that the law is serving "the social change of communicative practices" [15].

The reasons for this behavior would be that "communication law is in tune with the spirit of the times" [15]. In social networks there is no restriction, the language used may be full of intentions [16], but also of possibilities to express opinions, the Internet is a space that requires more reflection and debate for countries to place homogeneous conditions to regulate the right to communication.

The described framework, for the promotion and protection of rights, has been debated from various sides and generates positions in the communication companies, social and political sectors, but on few occasions the voices of journalists and media managers are heard and inserted to identify in practice how communication rights are fulfilled.

The purpose of the research is to know the experiences of local communicators about the validity of rights and freedoms related to communication. The hypothesis is: restrictions occur for a broad exercise of communication rights, journalists and media employees must propose creative solutions to overcome the limitations they find in the work of informing.

2 Methodology

The methodology is qualitative, through semi-structured interviews, eleven communicators working in audiovisual and digital media of Ecuador are approached, between June 29 and July 15, 2019.

The interviewees answered questions related to freedom of expression and its guarantees, the exercise of the rights related to communication, its impressions on the LOC and the factors that will allow citizens to increase the validity of freedom of information and communication. The design and instruments are close to the definition of research through case studies [17].

The specific questions are: Have you exercised your profession with guarantees for freedom of expression, without restrictions on your communication rights? Can you tell us about an event or time when you saw your freedom of expression and press rights limited, or been censored? Did you overcome the limitations, restrictions, censorship? How do you do it?

The characteristics of the interviewees are: two women and nine men; seven are graduates in social communication, two in social sciences and two do not have a university degree; the average age of the communicators is 39 years; the average time of work in media and communication activities is 19 years; previous work experiences come from print media, radio, local television, institutional communication and correspondents for national and international media; the current fields of work are local television, radio and digital native media.

3 Results

Regarding compliance with the provisions of article 19 of the Universal Declaration of Human Rights, on freedom of opinion and expression, some journalists indicated that they live partially, that "there is freedom of expression, but sometimes this has not been well regulated" [18], for others it is not fulfilled.

> The "worker, employee of a media outlet has some restriction to issue a news as it is presented because there is priority for advertising and preferences for people or institutions. Then you have no freedom of expression to fully issue what you have investigated, because it will always depend on the approval of the person in charge, whether editor or a superior to the journalist" [19].

Journalists established a division over the previous and current government. "Five years ago it was not fulfilled (…) in the government of Rafael Correa the communicators were censored and had no freedom of expression, if they exercised there would be repression against them or the media" [20].

In Ecuador "a gag was imposed on freedom of expression and freedom of the press" [21], even "when the President of the Republic was claimed something, [the citizen who did it] was subject to some repression, including going prisoner, then, the rest was frightened" [22].

With the current government, the communicators are not afraid of being sanctioned, they live "with tranquility a freedom of expression that puts it into practice" [23], "I think that freedom of expression has been resumed a little, citizens can express themselves and nothing happens" [22], but "the speech of open doors and freedoms clashes against the persecution of journalists who have investigated acts of corruption of the previous government" [24].

There are cases of peasants who "claim their right to water through peaceful demonstrations, exercise the right to free expression, however, criminalize them, prosecute them and constitute a transgression and violation of article 19" [25].

In their daily work journalists feel restrictions, which they identify as coming from the government as from individuals and that did not occur only in recent years, they say that "in the time of Rodríguez Lara [1970s] there was no freedom of expression; there was a nefarious government where all the informative entities were controlled and also the most recent during the 10 years of Correa (…), what he did most was to restrict freedom of expression" [23]. "There were restrictions. The notes should be released only for the purpose of reporting. Some journalists no longer gave an opinion" [20].

Recent examples of limitations on freedom of expression caused by the public sector are the following testimonies:

- Radio Centinela del Sur "nine fines were added, totaling almost $ 30 000, for contradicting the national government, when in fact we always told the truth (…) journalists and media were censored, here Ecotel TV and Ecotel Radio were closed because they didn't they agreed with the political line of the government" [23].
- "On one occasion I gave way to actually impose the issue, because it was the medium, it was my responsibility and, well, it was the President of the Republic, but I left the issues, from there they never contacted me again, not even for a press conference, erased (…). I was required, I wanted to impose issues" [23].
- "I remember that I was invited to be part of a panel with President Correa. There we were limited to a series of questions that we should ask and imposed an agenda. I refused, I was censored. Perhaps that is the greatest expression of not having freedom of the press" [21].
- "When the Saturday [weekly management reports of President Correa] were held, we had to comply with the schedules and try to join the ideology of the Government. The person who did not have that similarity was simply terminated his contract" [26].
- "There are authorities that close and do not want to answer all the questions. The former prefect [of Zamora Chinchipe] Salvador Quizhpe, could be an example of press limitation" [27].
- "The Superintendency of Communication recriminated us, a couple of years ago, for having played more foreign music than Ecuadorian on Valentine's Day. They fined me. I turned to a lawyer and explained that the LOC was full of gaps and does not specify whether one by one meant a schedule or all the daily schedule, then the legal side could not respond in this regard and we saved thousands of dollars" [25].

Journalists also express restrictions on freedom of expression in the private media, they point out that "information is often disseminated, ensuring economic interests, kinship or affinity with a person or institution (…), because a means of communication subsists from advertising" [19], on other occasions it is requested "to preserve the image of some politician or entity, to try to see the positive side, the good side" [26].

The limitations would be justified in the "editorial lines that have to be obeyed. There are policies to comply. At some point there was talk of deontological codes that we all have to comply with" [26], two perspectives are expressed from ethics, "[professional] codes to be respected and the media have deontological codes that must be framed in the law. I believe that by joining these two personal and institutional codes we have been able to get ahead perfectly in journalism" [21].

Another reading is self-censorship, which according to one of the interviewees "will never end. When you work for a media outlet there is the owner's editorial line, if he decides that a news affects him is not published and the journalist cannot claim, otherwise he is fired" [24].

Some examples of violations of communication rights, occurred in privately owned media are:

- "I was fired from the media where I worked, in January 2019, because I had been denouncing irregularities within the National Election Council (…) I became a stone in the way of who ran the province and pressed to get me out of the way" [24].
- "A piece of news was made that did not please the authorities of an entity and asked me to withdraw from that activity. I did not receive the support of immediate bosses, they only look after the interests of the media, regardless of whether the information can serve the public" [19].
- "In the Ecotel TV network, a private medium, we had to deal with the Government, we always tried to make up the information a little because of the apathy they felt. We had prior censorship. The workers had to manage against the government line" [26].
- "It happened on UV Televisión, I was making a note denouncing an abuse of a company and the media chose not to publish, for fear that a problem might happen. It's frustrating because of the work you do" [20].
- "There is one person could not interview, specifically the former Mayor Castillo. While working at UV Televisión I was told: Mayor Castillo does not want you to interview him (…). The next day I submitted my resignation" [21].
- "I had the opportunity to interview and witness anomalies in a sports activity, I was able to collect the information citing the sources of complaint, contrasting the information, but the people involved did not find it because they were affected as authorities, then they approached the medium and asked that I not continue to lead the sports segment. Rights were violated, but I had the support of some colleagues and all anomalies became known over time" [19].

The facts presented happened in conventional media, while in local media based on the Internet there is another dimension in the restrictions on freedom of communication.

The difference is that digital media can be removed from the reach of the public faster, with fewer administrative procedures, formal protocols or legal procedures than traditional media. A case is cited below:

> "We have had many people, especially politicians, who have tried to buy us (…) there was a campaign against our brand, so that we will not expose corruption. We can't fight politicians who have a lot of money. We did not yield and we were discharged. They cut our freedom of expression. They have used tools to make us decline (…) They have sent us some "mercenaries", to intimidate us, so that we do not publish certain types of information" [18].

In spite of the reprisals, to the managers and managers of the digital media, their perspective is positive, of greater hope than in the manifested by the journalists who work in traditional media.

"We must all have a regulation, because as human beings we exceed the limits. That towards the LOC, of course, a little manipulated for the benefit of the government. If it had been used well it would have improved communication (…). In the absence of an entity that regulates the partial action of some is noted" [18].

As a result of the issuance of the LOC, and in relation to freedom of information and communication, journalists were consulted if they considered that the norm promoted the right to communication, they agree with respect to:

- "The aspiration of the LOC to guarantee rights based on the Constitution" [28].
- The location of schedules for the programs,
- The care of the rights of children, youth, older adults, people with different abilities, of another ethnicity, etc.
- The promotion of the right to communication.
- The prevalence of deontological codes.
- "The spirit of the law promoted further responsibility" [28].

However, there are negative ratings. "I think the LOC helped something, in the aspect of remuneration, in an acceptable percentage, but we still don't have that full and absolute freedom" [19].

"The right to communication was promoted, but the Correa government did the opposite because it accredited many more means to monopolies" [25].

"I believe that there should be this law, that regulates, that guarantees the right to freedom of expression but that is not so punitive, but that is more humane, attached to the international law of freedom of expression" [22].

The LOC, the right to communication "was a romantic speech, the government took advantage of the public media to turn them into government media" [24], specifically it was intended "to put a gag on the media" [21].

Regarding the conditions or factors that should exist for citizens and journalists to exercise their rights to communication, the respondents stated that journalists need to inform, educate and entertain with the truth, exercise freedom of expression with ethics and responsibility, and stay away from politics.

Institutional conditions are also demanded, such as access to public information, because this way it is guaranteed to inform citizens better, and also citizens' access to the media to give their opinions, to express themselves.

To achieve the stated purposes, it would be important to "start by raising awareness among the owners, who are in front of a media outlet, because they do not have the real knowledge of what journalism is" [19]. "That the owners of the media take away their ambitious personal interests" [24]. Specifically, "our commitment is to focus on the benefit of society [18].

4 Conclusion

Freedom of information and expression constitute purposes, goals for the media and journalists in the Republic of Ecuador. The legal framework is based on guarantees.

The Constitution and the LOC recognize the rights to communication, and create an important reference in the region because they host the reflections derived from the MacBride Report.

In practice, journalists express limitations, point to censures that come from the media themselves and from political sectors, and there is little access to public information. According to researchers Oller, Chavero and Ortega, the journalistic culture of the country is experiencing a transition towards a socio-democratic model, which implies greater demands, autonomy and awareness of communication professionals.

From this logic it is understood that the identification of commercial interests, as well as the intentionality to take a position, from private companies, against the government derives from the discomforts expressed by the communicators.

It would be understood that in the face of biased practices and the violation of the right to information, the regulatory authorities seek balances in the information agendas, both from private and public media, but this aspiration is not fulfilled.

Journalists have seen in the LOC and the Superintendency of Communication the instrument to punish, eliminate voices that disagree with the government and curtail freedoms. There are several episodes that you have in common to hide information for the benefit of few.

It is noteworthy that regulations have been issued for communication during qualified totalitarian regimes, in the dictatorship headed by General Guillermo Rodríguez and in the government of President Correa. It would seem that the free flow of information and thought is contrary to control intentions. In both times the reactions of citizens have been rejection.

The absence of freedom of expression in traditional media occurs due to external pressures and self-censorship imposed by publishers and owners, journalists have been dismissed, and in extreme cases to close the media after legal proceedings.

When it comes to digital media, the pressures come from outside and come, in a short time, to the closing of accounts, addresses, blockages and more, which mean the exit of the cyberspace medium.

Restrictions occur for a broad exercise of communication rights. This includes a bias communication in the freedom broad sense when this is not compatible with the government.

Testimonials confirm the hypothesis. Restrictions occur for a broad exercise of communication rights. The solutions, the ways in which communicators and media officials have overcome the restrictions have often come from platforms and digital media. It remains to explore in the future the information spaces through the Internet as part of a new communication ecosystem of Ecuador.

References

1. Paz y Miño, J.: Derecho de libre expresión y código de ética en Ecuador. Chasqui **18**, 43–47 (1986)
2. García, N., Ávila, C.: Nuevos escenarios para la comunicación comunitaria. Oportunidades y amenazas a medios de comunicación y organizaciones de la sociedad civil a partir de la aplicación del nuevo marco regulatorio ecuatoriano. Palabra Clave **19**(1), 271–303 (2016). https://doi.org/10.5294/pacla.2016.19.1.11

3. Wambra: Ley de Comunicación y la no derogación de la Ley de Radiodifusión y Televisión y su próxima reglamentación en Ecuador (2013). https://wambra.ec/ley-de-comunicacion-y-la-no-derogacion-de-la-ley-de-radiodifusion-y-television-y-su-proxima-reglamentacion-en-ecuador/
4. Jurado, R.: Un nuevo paradigma latinoamericano en la regulación de la comunicación. Chasqui **112**, 86–90 (2010)
5. Pérez, J., Barredo, D.: La Ley Orgánica de Comunicación del Ecuador: la comunicación como servicio público y la comunicación responsable. Derecom **20**, 61–82 (2016)
6. Aguirre, G.: El derecho a la comunicación: entre la libertad negativa y el servicio público. Universidad y Sociedad [seriada en línea] **7**(1), 124–131 (2014)
7. Castro, J.: Libertad de Expresión y Límites. Ius Humani, Revista de Derecho **6**, 11–25 (2017)
8. Charney, J.: Libertad de expresión y pluralismo informativo: compatibilidades y tensiones en el contexto de la televisión. Revista Derecho del Estado **42**, 117–148 (2019). https://doi.org/10.18601/01229893.n42.05
9. Alegría, A.: La Ley Orgánica de Comunicación de Ecuador, ¿un avance en el ejercicio efectivo de las libertades expresión e información y en la participación ciudadana? Revista de Derecho Político **95**, 291–326 (2016)
10. Carrillo, R.: El proceso de debate, elaboración y sanción de la Ley Orgánica de Comunicación del Ecuador (2013): actores implicados en la disputa por la palabra. e-l@tina **68**(17), 1–21 (2019)
11. Alegría, A.: ¿Vulnera la Ley Orgánica de Comunicación de Ecuador los estándares de la corte interamericana de derechos humanos en materia de libertad de expresión? Cuadernos Manuel Giménez Abad **10**, 185–273 (2015)
12. Oller, M., Chavero, P., Ortega, E.: La percepción de los niveles de autonomía profesional de los periodistas de Ecuador. Anuario Electrónico de Estudios en Comunicación Social "Disertaciones" **9**(1), 61–83 (2016). https://doi.org/10.12804/disertaciones.09.01.2016.04
13. Derieux, E.: Liberalismo, económico y libertad de expresión. Chasqui **77**, 44–49 (2002)
14. Tornay, M.: Creando el Derecho a la Comunicación desde abajo: radios comunitarias sostenibles en Venezuela, Ecuador y España. Commons. Revista de Comunicación y Ciudadanía Digital **7**(2), 133–163 (2018). http://dx.doi.org/10.25267/COMMONS.2018.v7.i2.05
15. Mora, J., Ortiz, F., Ávila, B., Romero, A.: La LOC y la educación en principios mediante la mediación social ecuatoriana en las prácticas comunicativas digitales. Mediaciones Sociales **17**, 117–134 (2018). https://doi.org/10.5209/MESO.58406
16. Briones-Hidrovo, G.: Nociones de libertad de expresión en disputa: la opinión pública publicada en la prensa ecuatoriana. Question **1**(59), e080 (2018). https://doi.org/10.24215/16696581e080
17. Stake, R.: Investigación con estudios de casos, 4ta edn. Morata, Madrid, España (2007)
18. Lafebre, J.: Personal communication, 5 July 2019
19. Benítez, A.: Personal communication, 10 July 2019
20. Erraez, E.: Personal communication, 8 July 2019
21. Paladines, L.: Personal communication, 3 July 2019
22. Criollo, W.: Personal communication, 13 July 2019
23. Coronel, F.: Personal communication, 15 July 2019
24. Yauri, G.: Personal communication, 3 July 2019
25. Ojeda, G.: Personal communication, 13 July 2019
26. Eras, I.: Personal communication, 10 July 2019
27. Jiménez, M.: Personal communication, 29 June 2019
28. Pozo, D.: Personal communication, 1 July 2019

Communication in Project Management: An Action Research Approach in an Automotive Manufacturing Company

Ingrid Souza[1,2](✉), Anabela Tereso[1], and Diana Mesquita[1]

[1] Production and Systems Department/Centre, ALGORITMI University of Minho, Campus de Azurém, 4804-533 Guimarães, Portugal
engingridsouza@gmail.com, {anabelat,diana}@dps.uminho.pt

[2] Bosch Car Multimedia Portugal, S.A., Rua Max Grunding, 35 – Lomar, 4705-820 Braga, Portugal

Abstract. Over the years, countless companies have implemented project management to ensure the integration of people with processes and deliveries, with a direct impact on quality and cost. In the automotive industry, this approach helps stakeholder and challenge management. However, to achieve excellence, communication between project managers and their teams, between stakeholders and sponsors, or simply in the workplace must be efficiently managed.

The objective of this study was to help a PMO from an automotive manufacturing company, which has been experiencing difficulties in terms of communication management, for five years. This study highlights the improvements performed in order to change the situation.

An action research approach was applied, taking in consideration the context of the work developed using the principle of "learning by doing". Participant observation was carried out, namely in meetings and the daily practices of the team. Finally, questionnaires were released in order to obtain the KPIs used in the study.

Subsequently, following the best communication management practices according to PMBOK and using the PDCA method, an improvement strategy was defined. Finally, KPIs were developed providing measurements to the communication improvements.

Keywords: Communication management · Stakeholder management · Automotive manufacturing

1 Introduction

Many modern organizations carry out their business tasks and activities using a project-based approach. Developing projects requires a specific approach to develop tasks, perform resources mobilization, integrate different stakeholders, and manage different processes, among others.

Á. Rocha et al. (Eds.): WorldCIST 2020, AISC 1159, pp. 64–73, 2020.
https://doi.org/10.1007/978-3-030-45688-7_7

According to the Project Management Body of Knowledge (PMBOK) guide 6th edition [1], project management can be organized in five processes groups: initiating, planning, executing, monitoring and controlling and closing. In addition, it can be organized in ten project management knowledge areas, one of them being Communications Management.

Muszynska [2] pointed out that communication management is considered to be crucial for the success of the project. Ruão [3] suggests that, on one hand, the communication importance is highlighted by most of the stakeholders but, on the other hand, the communication processes and practices formalized in the company's project management methodology are neither followed nor prioritized by project managers. Almeida, Tereso, Faria, and Ruão [4] recognized that knowledge sharing is also crucial for industrialization projects. And the improvement of project management practices is important in this kind of projects [5].

According to Project Management Institute (PMI) [1], Project Communications Management consists of two parts. The first is focused on developing a strategy to ensure an effective communication process with the stakeholders. The second part is addressed to the activities related to the implementation of this strategy.

These discussions have gained space in scientific communities, mainly because they deal with a sensitive subject. As everything that involves people, it is necessary to keep their engagement. In project management, people are essential for teams, in order to work on project tasks and each member must be aware of their role and responsibilities [3]. Furthermore, it is a challenge to work with others, to ensure the achievement of common goals and it is not possible to perform this without communication.

This study was carried out in a Project Management Office (PMO) of an automotive manufacturing company, established for a department concerned with printed circuit board assembly and interconnection technologies. Due to the increasing number of innovation technology projects the PMO states the process has been very heavy and communication management is not given the proper focus and needs to be improved.

This situation reflected on communication channels, the department documentation and consequently in the team motivation. So this paper described an action research done inside this PMO in order to improve communication among project teams in the section. Several methods were used in order to identify the challenges and difficulties of the communication management process and measures were taken to improve the situation, considering the perspectives of the project teams.

2 Research Methodology

In the context of this study, two Research Questions (RQ) were established, namely:

- RQ1: How is it possible to improve communication management of project teams in an automotive company section?
- RQ2: How much has communication management of project teams improved in an automotive company section?

Thus, to answer these questions, the methodology was carefully chosen considering all aspects of the company, such as the reality of the department and the time available to develop the work.

Considering the practical perspective of the study, the research approach chosen was action research, an approach aimed at action and knowledge creation [6]. The project was developed from the practical perspective of an internship at the company. The chosen methods were applied and provided measurable results.

The techniques and procedures used for data collection and analysis were:

- Observation: in the context of this study, the diagnostic phase is based on the communication structure within the project team that was observed, researched and analyzed. This step was performed with the participation in meetings and the observation of the daily practices, providing a participant observation.
- Semi-structures interviews: they were performed by a pre-questionnaire deployment. It was applied in order to collect information for the final survey, which considered the feedback from the stakeholders.
- Document analysis: it was performed based on the research questions. In order to update and create procedures and work instructions for the department, the existing documents were consulted, characterized and organized.
- Questionnaire deployment: the goal was to get inputs from the entire section through a qualitative survey in order to understand their perceptions and opinions about the current issue, and thus to begin improving procedures using Key Performance Indicators (KPIs) based on the survey.

Using all these methods and considering the PMBOK project communication management approach, which focus on how to develop a strategy to ensure an effective communication process with stakeholders, and then to address activities related to the implementation of this strategy, the Plan, Do, Check, Act (PDCA) method was adopted.

It is an appropriate tool to apply in a communication process that can be used repeatedly. By applying the PDCA cycle, communication maturity and effectiveness will be enhanced over time. This definitely helps in achieving the desired outcome that is effective communication [7].

Data analysis was performed in the PDCA Check step, after each questionnaire deployment. An excel file was elaborated to analyze the collected data. In this way, it was possible to understand the exact type of improvements needed to achieve the main goal of improving communication and showing improvement through KPIs.

3 Findings

Cookson [8] argues that measuring the results of internal communication is essential to ensure that communication is aligned with expectations and strategies. The Continual Self Improvement [7] official blog also explains that using the PDCA cycle to ensure this type of communication and deliver expected results is an advantage when aligned with KPIs. Therefore, in addition to measurement, it is possible to ensure that communication management improvements are properly applied.

The findings of this study were obtained through the PMBOK good practices application using the PDCA method to address issues. The development and its results are detailed below.

3.1 Following PMBOK Good Practices

The first step in ensuring solid work with the PDCA application was analyze the key stakeholders and understand their complaints. Some stakeholders may have limited ability to influence the project work or outcomes. Others may have significant influence on expected outcomes [1].

Depending on their interests, stakeholders may take a strategic and engaged position on projects, or may be at odds with their achievement and, in extreme cases, even try to disrupt it, contributing to failure [9].

In a real project, the project manager will list their stakeholders and rank them based on their level of influence. This way, after classification, the project manager can understand which stakeholder needs the most attention, which will help provide solution to potential issues. Also, and very importantly, the project manager will know which one has the influence of disrupting the project, so he/she needs to be closely managed, performing an analysis as presented in Fig. 1, this management became easier.

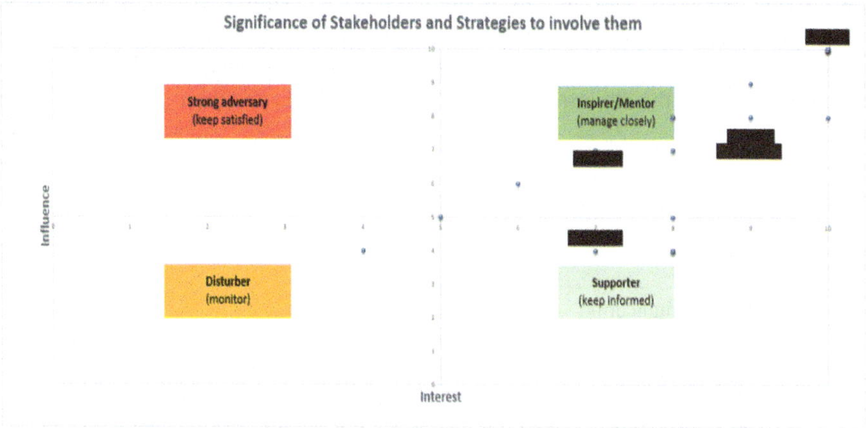

Fig. 1. Stakeholder analysis performed in the case study department

This analysis provided a very clear perception that in the studied department the approach and engagement of the stakeholders was important since most of the stakeholders are in the "Inspire/Mentor" zone or in the "Supporter" zone, so they need to be managed closely or keep informed. Fluid communication is important to improve project success [10], so the research focused on improving several aspects of communication, like communication channels knowledge, information availability, and procedures knowledge.

3.2 1st PDCA Running - Survey Deployment Methodology: Plan

When the strategy is defined with objective of achieving goals, establishing KPIs can provide actions to obtain a plan of measures. According to this strategy, Kerzner [11] argues that KPIs are the base for surveys.

Based on the explanations of Kerzner [11], a survey deployment proved to be a good way to obtain the solution to the communication problem. Therefore, this primary approach consisted on the first step of the PDCA, which is Plan.

To develop a consistent survey, a prior survey was prepared, considering the suggestions of the stakeholders of the department under study. The chosen ones were three, classified as "multipliers". Their suggestions helped to develop a clear final survey for all stakeholders, a total of thirty people.

3.3 1st PDCA Running - Survey Deployment Methodology: Do

After this planning, to develop the first survey, we arrived at the final questionnaire applied. This is the second step of the PDCA, that is, Do.

Thus, the final survey was prepared in order to obtain KPIs from each question. For this, meetings with the PMO team were performed in order to develop questions that would provide the right measure to the indicator.

Therefore, the questionnaire was divided in blocks. The 1st block of questions (question 1 divided in two related sub-questions: 1.1 and 1.2), referred to as indicators, aimed to measure situations related to project management. The 2nd block was dedicated to communication channels (composed by questions 2 and 3). Finally, the 3rd block referred to the procedures in the department (question 4). The KPIs and their goal are presented in Table 1.

Table 1. Key Performance Indicators and their goal.

Question	KPI name	Proposal
1	1.1 Communication Channels Knowledge within MFT3 and 1.2 Relevant Information Availability	- Question 1 is divided in two KPIs. The first one has the goal to measure which are the communication channel most used by people in order to find project documents - The second one aims to measure the number of people who consider project-related information relevant within these channels
2	Information storage	This KPI has the goal to measure which is the channel most used in order to archive project documents
3	Information access	The goal of this KPI is to measure how many people have access to the information needed
4	Procedures knowledge	The goal of this KPI is to measure how many people know the documents that have been developed

The survey was answered by twenty one people of the department of this study.

3.4 1st PDCA Running - Survey Deployment Methodology: Check

This PDCA step, called Check, is reserved to the survey analysis. Due to the situation of the department the operationalization of the survey assures predictive validity since it is applied two times (at the beginning and after the improvements are done).

Question 1 (1.1). Represents the KPI "Communication channels knowledge". It is a simple question, just to know if people use the correct channel to search and provide information. Each participant answered which communication channel is used to find projects information. Most participants answered that they know O: drive in the scope of finding project information. This is an excellent answer because O: is the central place that works as a repository for the department, as C: in a personal computer, to provide all the information, as long as everyone has the correct access.

Question 1 (1.2). The critical point is the open question related to question 1, which represents the KPI "Relevant Information Availability", where 39% of participants do not consider relevant the information found in the communication channel.

Question 2. In question 2, "Information storage KPI", the analysis comes across question 1, where drive O: is mainly used, however, in this case to archive project documents.

Question 3. "Information access KPI", provides the result that most people have correct access to places where information is available. Here, the attention-grabbing topic is a deviation identified by a stakeholder that "being involved in the middle of a project" is the reason for not having correct information access.

Question 4. The second block of questions is composed by just one question, 4, "Procedures Knowledge KPI". It aims to understand if people know the procedures developed and under development within the section. However, 67% of employees do not have this information, which represents a clear communication failure.

3.5 PDCA Running - Survey Deployment Methodology: Act

This last step of PDCA, Act, is dedicated to actions implementation in order to address the issues found. This stage, besides the fact that it is the key to implement continuous improvement, ensures that the former phases were correctly developed.

The output is the action plan focused on improvements implementation. With the diagnostic, an analysis was performed in order to get effective changes. The action plan utilized was based on 5W1H method. That consists of asking a systematic set of questions to collect all the data necessary to draw up a report of the existing situation with the aim of identifying the true nature of the problem and describing the context precisely [12].

Besides that, this involvement made them realize that their feedback is important, and they have space to share opinions testifying that it is part of a good communication. The stakeholder's feedback resulted in the final questionnaire that was the 2nd PDCA running, the continuous improvement.

3.6 PDCA Running – Continuous Improvement

The 2nd PDCA running starts ensuring the best implementation of the actions proposed with 5W1H. The aim was to extend the frontier and engage all section in the communication management improvement cause. The actions were established in order to improve the results stated on the 1st survey. For each question one or more actions were prepared asking: What, Why, When, Where, Who and How will be done?

2nd PDCA Running: Plan – Section Engagement. After the survey results, an agreement with the PMO team leader and the head of section was stablished. It states that, for all activities that were being developed, the whole section should be informed trough an email sent with the information about what was done, updated, changed, etc. Moreover, anything else that was relevant and needs to be communicate face to face, should be done during the monthly department meeting. The official channel became O: to share documentation and Docupedia page to find documentation.

2nd PDCA Running: Do – 2nd Survey Deployment. The timing between the 1st survey deployment and the 2nd was five months and sixteen days. The questions are exactly the same in order to obtain the KPIs results. The goal of this second questionnaire was to measure the improvements regarding the methods implemented and work developed.

2nd PDCA Running: Check – KPIs Evolution. The 2nd analysis regarding the questionnaire applied had the goal to obtain the KPIs evolution.

Communication Channels Knowledge. In terms of measurement, Docupedia use in order to find projects documentation increased 116% compared to April 11, 2019. Also, O: drive increased 7%, while the other channels presented a significantly decrease 53% and 40%, namely SharePoint (that works as a cloud service in the department) and BGN room (internal communication channel used for document approval deployment), respectively. The number of people who does not know how to find the information has reduced to 0%. The evolution can be observed in Fig. 2.

1. Communication Channels Knowledge within MFT3 KPI

	April 11, 2019	September 27, 2019	Evolution
Docupedia	38%	82%	116% ↑
O:	71%	76%	7% ↑
Sharepoint	38%	18%	-53% ↓
BGN room	10%	6%	-40% ↓
Don't know	5%	0%	-100% ↓

Fig. 2. Communication Channels Knowledge KPI evolution

1.1. Relevant information availability KPI

	April 11, 2019	September 27, 2019	Evolution
Yes	61%	88%	44% ↑
No	39%	12%	-69% ↓

Fig. 3. Relevant Information Availability KPI evolution

Relevant Information Availability. In terms of measurement, the number of people who consider all available information to be relevant has increased 44%. The evolution can be observed in Fig. 3.

Information Storage. In terms of measurement, Docupedia use in order to archive projects documentation increased 21% compared to April 11, 2019. Also, O: drive, SharePoint and BGN room increased 52%, 107% and from 0% to 6%, respectively. People who consider not clear where project information needs to be archived has reduced by 75%. The evolution can be observed in Fig. 4.

2. Information Storage KPI

	April 11, 2019	September 27, 2019	Evolution
Docupedia	24%	29%	21% ↑
O:	62%	94%	52% ↑
Sharepoint	14%	29%	107% ↑
BGN room	0%	6%	0%->6% ↑
It isn't clear	24%	6%	-75% ↓

Fig. 4. Information Storage KPI evolution

Information Access. In terms of measurement, the number of people who consider having access to all project information needed for their daily work has increased 34% compared to April 11, 2019. The evolution can be observed in Fig. 5.

3. Information Access KPI

	April 11, 2019	September 27, 2019	Evolution
Yes	70%	94%	34% ↑
No	30%	6%	-80% ↓

Fig. 5. Information Access KPI evolution

Procedures Knowledge. In terms of measurement, the number of people that know which are the procedures (processes and work instructions) that have been defined in department has increased 79%. The evolution can be observed in Fig. 6.

4. Procedures knowledge KPI

	April 11, 2019	September 27, 2019	Evolution
Yes	33%	59%	79% ↑
No	67%	41%	-39% ↓

Fig. 6. Procedures Knowledge KPI evolution

4 Conclusions

It is relevant to mention the relevance of PMBOK guide use in order to understand the importance of following the good practices for project management. These good practices, besides providing a rich field of information and tips to project management daily work, had impact in the work developed by the section and in the employees directly.

At the beginning of this study, the PMO had a lot of running projects and many tasks to develop delayed. They worked just with urgent priorities. The documents to be prepared were stopped, the project team never knew where to find the project documentation. When new documents were developed, no one knew where to store them.

In order to start addressing these problems, a good practice was applied, following PMBOK. It was divided in two parts. The first part focused on developing a strategy to ensure an effective communication process with the stakeholders and the second part was dedicated to address the activities related to implementation of this strategy. To begin this research a stakeholder analysis was implemented in order to methodically evaluate the stakeholders and understand how each one should be dealt with. As part of the strategy developed, the PDCA was adopted. After that, a survey was prepared, as the Plan step of the PDCA cycle. The goal was to engage the stakeholders in this continuous improvement process and obtain KPIs from the survey questions.

The 2nd step of PDCA, Do, was implemented with the survey deployment. The goal was to provide voice to stakeholders and to openly listen to their complaints about documents organization and communication channels.

After the survey deployment, the 3rd step of PDCA, Check, was implemented, analyzing the suggestions. Using the findings from the survey it was possible to set a line about what was more important and urgent to improve. For this, an action plan in 5W1H format was defined, closing this first PDCA running with the Act step.

After almost five and a half months fulfilling the 5W1H, a new PDCA was running again. The Plan step started with planning the implementation, on time, of all actions of 5W1H. The survey was deployed again in order to measure the improvements, as the Do step. And the results, as the Check step, are the KPIs evolution obtained.

The communication channels were reduced. At this point the projects documentation can be found in just one channel. Besides that, the quantity of procedures updated and created increased and the team motivation increased.

Therefore, and answering to the research questions.

- RQ1: How is it possible to improve communication management of project teams in an automotive company section? It was possible to improve communication management of project teams in an automotive company section following the PMBOK good practices. That is, a strategy was developed to ensure an effective communication process with the stakeholders and after that, the activities related to implementation of this strategy were addressed.
- RQ2: How much has communication management of project teams improved in an automotive company section? The communication management of project teams in an automotive company section improved 34% in Information Access, 79% in Procedures

Knowledge, 44% in Relevant Information Availability, 116% in Docupedia use and 52% of O: drive storage use.

From this study it is possible to identify two main questions for future research: Continue the PDCA running is a good option in order to lead to further improvements on the KPIs and increase the documentation update, impacting directly the team motivation? Can employees communication competences be develop by performing coach, using different tools and methods exposed in the literature, with a change management perspective, in order to develop different mindsets to the project communications management process? Finding the answer to these questions will lead to interesting research results in our opinion.

References

1. PMI: A Guide to the Project Management Body of Knowledge (PMBOK® Guide). Project Management Institute, Pennsylvania (2017)
2. Muszynska, K.: Communication management in project teams – practices and patterns. In: Management, Knowledge and Learning - Joint International Conference 2015 - Technology, Innovation and Industrial Management, pp. 1359–1366 (2015)
3. Lopes, A., Ruão, T., Pessôa, C.: Gerir Identidades e Culturas em Organizações Temporárias: O Papel da Comunicação. Figueira & A.T. Peixinho (Eds.), Narrativas Mediáticas e Comunicação: construção da memória como processo de identidade organizacional, pp. 221–254 (2017)
4. Almeida, A., Tereso, A., Faria, J., Ruão, T.: Knowledge sharing in industrialization project management practices. In: Rocha, Á., Adeli, H., Reis, L.P., Costanzo, S. (ed.) Trends and Advances in Information Systems and Technologies. WorldCIST 2018. Advances in Intelligent Systems and Computing. Springer (2018)
5. Fernandes, D., Tereso, A., Fernandes, G.: Improvement of industrialization projects management: an automotive industry case study. In: Rocha Á., Adeli H., Reis L., Costanzo, S. (ed.) New Knowledge in Information Systems and Technologies. WorldCIST 2019. Advances in Intelligent Systems and Computing, pp. 112–121. Springer (2019)
6. Saunders, M., Lewis, P., Thornhill, A.: Research Methods for Business Students, 5th edn. Financial Times Prentice-Hall, Harlow (2009)
7. Tan, H.M.: Continual Self Improvement - 4 Steps to Manage Your Communications. https://continualselfimprovement.wordpress.com/2015/03/06/4-steps-to-manage-your-communications/
8. Cookson, J.: The Ultimate Guide to Employee Communication Goals and KPIs for HR Management. Poppulo (2019)
9. Stakeholder Map: Stakeholder Analysis, Project Management, templates and advice
10. Kerzner, H.: Project Management - A Systems Approach to Planning. Scheduling and Controlling. Wiley, Hoboken (2009)
11. Kerzner, H.R.: Project Management Metrics, KPIs, and Dashboards: A Guide to Measuring and Monitoring Project Performance. Wiley, Hoboken (2011)
12. Humanperf: What is the 5W1H method?, https://www.humanperf.com/en/blog/nowiunderstand-glossary/articles/5W1H-method

A Capacity Management Tool for a Portfolio of Industrialization Projects

Caio Lima(✉), Anabela Tereso, and Madalena Araújo

Production and Systems Department/Centre ALGORITMI, University of Minho, Campus of Azurém, 4804-533 Guimarães, Portugal

caiofurtado17@gmail.com, {anabelat,mmaraujo}@dps.uminho.pt

Abstract. The management of a project portfolio is a complex decision process because it encompasses the achievement of multiple objectives. A critical point that increases the complexity in the decision-making process of a portfolio manager is the allocation of human resources to manage the projects of the portfolio, project managers, which is crucial to the organization's performance. In this case, the project manager can manage more than one project simultaneously and it is necessary to assign project managers to the projects, considering that project activities have an amount of work to be accomplished. The main objective of this work was to provide support for this capacity management problem, which aims to provide an easier decision-making process for the capacity management of an industrialization project portfolio. Therefore, it was developed: a hybrid model that creates a schedule respecting the resource constraints and the established due dates; a recommendation system that considers project managers' allocation and projects requirements; and, an automatic status report that allows identifying the project portfolio capacity usage.

Keywords: Industrialization projects · Project Management · Portfolio Management · RCPSP

1 Introduction

Project Portfolio Management (PPM) is a complex decision process, once it seeks to achieve multiple objectives, as better profitability and better use of organizational resources. Abrantes and Figueiredo [1] describe that the change in a project that belongs to a portfolio implies consequently a cascading effect in the portfolio, which results in a scenario of uncertainty and instability. This unsteady scenario may result in resource conflicts, leading to the challenge of gathering useful information about the capacity status, strategic objectives, and resources for the decision-making process. This is similar to the case under study.

Oh, Yang and Lee [2] argue that strategies could be implemented in project portfolio management through a decision-making process that focuses on expected objectives. The authors further add that building a strategic portfolio that considers the business objectives and constraints is so important as it is challenging. Due to the complexity and

Á. Rocha et al. (Eds.): WorldCIST 2020, AISC 1159, pp. 74–83, 2020.
https://doi.org/10.1007/978-3-030-45688-7_8

the dynamic environment in project portfolio management, the application of traditional tools and techniques in the most complex contexts may be inappropriate, so it emerges the need to develop new tools and techniques to better manage the project portfolio providing clear results and to identify critical information, laying down an easier path in the decision-making process [3].

The present work was carried out at a company that belongs to the automotive industry and is responsible for industrialization projects. The main work purpose was to develop a tool that aims to help in the capacity's management of an industrialization project portfolio. Therefore, to develop the tool it was necessary to develop three main work packages: implement an exact mathematical model and a hybrid model to create the project schedule; develop a recommendation system for assigning projects to a Project Manager (PjM); and, create an automatic status report about the portfolio.

This paper is organized as follows. After the introduction, Sect. 2 presents the literature review relevant to the context of the study. Section 3 brings forward the methodology adopted to develop the tool. Section 4 describes the problem statement and Sect. 5 presents the developed tool. Finally, the main conclusions and suggestions for further work are presented in Sect. 6.

2 Literature Review

Industrialization projects are understood as projects that are related to the design of the manufacturing line to produce a certain product, which aims to reduce costs and increase the production capacity of the manufacturing line [4]. This kind of environment has become more complex and developed challenges in management; thus, many organizations have implemented the quality-gate model as a way to manage the new product development and to ensure that the objectives are met to advance to the next phase [5].

Project Management (PM) is commonly related to processes and tools used to accomplish a temporary and unique work within a specific time, budget and scope, being them the elements of the golden triangle, also known as success criteria for project management. Radujković and Sjekavica [6] claim that the golden triangle overlooks other aspects that the PjM is responsible for managing, which is beyond these elements. Ponsteen and Kusters [7] dissert about a crucial element for PM success, the resource allocation for project execution and management. A class of problems that focus on generating a suitable schedule of a set of activities with scarce resources is well-known as Resource-Constraints Project Schedule Problem (RCPSP). In this class of problems, preemption is not allowed. Tian and Yuan [8] argue that RCPSP arises in the context of approximating the project schedule methods with real problems and describe that RCPSP problems have been classified as NP-Hard. NP-Hard problems have been solved recurring to a considerable number of exact and heuristic solutions, purposely developed in recent years as we can see in Chakrabortty et al. [9] and Kolisch and Hartmann [10]. Thus, RCPSP solutions have been following the trend, and many approaches have been developed for RCPSP as Mixed-Integer Linear Programming (MILP) formulations, heuristics and metaheuristics. Borak and Karl [11] add that the use of heuristic and metaheuristic methods, in recent years, has attracted a lot of attention, which has resulted in the development of hybrid models, that is, the integration of these techniques with exact models for solving problems classified as NP-Hard. This kind of techniques are also called *matheuristics*. Again, RCPSP can profit from the application of hybrid models, therefore considering the many approaches to the RCPSP. For the present work it is important to present a classic Integer Linear Programming (ILP) formulation for the RCPSP and the selected *matheuristic* on which the hybrid model was based.

The ILP formulation used in this work can be described as follows. The RCPSP comprises a number of activities (n) to be scheduled and a number of resources (m) available. Then, the project is defined as a set of $n + 2$ activities, where activity 0 and $n + 1$ are dummy activities, which are the beginning and ending activities of the project, respectively. The set of activities is represented by A = {1,..., n}, which must be scheduled according to the available resources, which belong to R = {1,..., m}. The precedence relations are given by a set P of pairs with index $(m, n) \in P$, that means the activity m is predecessor of n. The processing times or activity durations are represented by the vector d belonging to $\mathbb{N}^{n+2}$, where the jth term, d_j, is the duration of the activity. Each activity (j) needs an amount $r_{j,k}$ of resource k to be processed. Each resource k has a capacity of R_k. The decision variable is $x_{j,t}$, indicating that activity j starts at period t, where $x_{0,0}$ is equal to 1. The variable Z is the project makespan [12]. The set of activities that belong to the critical path of the activity network is represented by CP $= \{cp_1, \ldots, cp_n\}$. Thereby, the RCPSP mathematical formulation adopted in this paper can be represented as follows:

$$Minimize\, Z \tag{1}$$

$$\sum\nolimits_{t=e_j}^{l_j} x_{j,t} = 1, \quad j = 0, 1, \ldots, n+1 \tag{2}$$

$$\sum\nolimits_{t=e_m}^{l_m} t * x_{m,t} + d_m \le \sum\nolimits_{t=e_n}^{l_n} t * x_{n,t}, \ \forall (m, n) \in P \tag{3}$$

$$\sum\nolimits_{j \in J_t} \sum\nolimits_{q \in S_{jt}} r_{j,k} * x_{j,q} \le R_{k,t}, \ k = 1, 2, \ldots, K; t = 1, 2, \ldots, T \tag{4}$$

$$\sum\nolimits_{t=e_j}^{l_j} t * x_{j,t} + d_j \le Z, \ \forall \ j \, without \, successor \tag{5}$$

$$\sum\nolimits_{n \in CP} d_n \le Z \le \sum\nolimits_{j=0}^{n+1} d_j, \ j = 0, 1, \ldots, n+1; n = cp_1, \ldots, cp_n \tag{6}$$

The objective function (1) represents the shortest time the project can be completed. Constraint (2) ensures that each activity can only be started once. Constraint (3) ensures that the precedence relationships in the activity network are respected. Constraint (4) ensures that at each instant of time (t), the resource consumption of activity (j) of each resource k will not exceed the allowed capacity $R_{k,t}$. Constraint (5) ensures that project completion Z will occur after the completion of the project end activities, i.e., the last project activities. Constraint (6) ensures that the shortest project execution time is longer than the critical project path duration and less than the sum of all project activity durations.

The *matheuristic* that the hybrid model used was based on the *Fix-and-Optimize Variable Neighborhood Search* (VNS). The basic idea is to refine a solution by iteration, with each iteration exploring the neighborhood of the current solution to look for a better one. In Maniezzo e Stutzle [13], the *Fix-and-Optimize* VNS is used, initially with a viable solution that considers the strong constraints of the problem, then changes are made to some variables of the mathematical model of the problem and, finally, the solver is called to optimize the problem.

3 Methodology

The present work aimed to develop a tool that gives support to the management of capacities in industrialization projects in an automotive industry, i.e., a tool that creates schedules without over-allocating the PjMs and provides the capacity status of project portfolio capacity according to PjM availability. Therefore it became necessary to understand the context where the research was done. Thus, as the research strategy aims to explore, describe or explain the object of study [14], for this research the strategy adopted was case study, since the present work was developed in a business environment. The methods adopted for data collection were document analysis, observation, interviews and focus group. Using the triangulation of these methods it was possible gathering all organization's information about previously developed studies, management practices and the lifecycle of the industrialization projects in which this work was based. Thereby, all initial information was gathered and the context in which the tool would be applied was studied.

The Design Science Research (DSR) methodology, commonly used in the development of computational artifacts [15], was adopted in this study for the tool development process. The focus group was the research method used to evaluate the tool developed, in several cycles of the DSR methodology, allowing potential tool users to suggest and test the tool, helping to achieve the main expected tool requirements.

4 Problem Statement

The main objective of this work is the development of a tool that aims to support the capacity management of a portfolio of industrialization projects, that is, provide an easier decision-making process to reach a better Capacity Utilization Rate (CUR) [16]. Hence, three main objectives were established for the tool development. The first objective was to develop project schedules with project allocation profiles without over-allocating PjMs. The second objective was to develop a recommendation system for assigning PjMs to projects, thus allowing an allocation leveling of available PjMs to manage projects. Finally, the third objective was to develop an overview of project portfolio CUR.

To understand the developed tool, it is important to perceive the context of the industrialization projects management where this study was done. Projects in this organization have four categories (A, B, C, and D) and each category has a certain level of complexity and requirements for the PjM that can manage each type of project category.

Although projects have a category distinction, all projects are managed with a quality-gate system, where each Quality-Gate (QG) has a set of activities that need to be performed to be able to go to the next QG. It is important to note that each QG has project milestones (due dates), so it may be necessary to compress or decompress the activities that are between the QG's.

A standard project has a set of 45 activities, that are between five QG's and this is the standard activity network to all project categories. This project activity network, as mentioned, is standard and for projects with less complexity it may not be necessary to perform some activities, so these activities have duration and effort equal to zero. On the other hand, projects with greater complexity require higher effort and/or duration of project activities.

The relationship between effort and duration comprised in this paper follows Tereso et al. [17] description. Given a work content of an activity "a" (W_a) and the amount of resources allocated to that same activity, named as effort work (E_a), the duration of an activity (D_a) is given by expression (7).

$$D_a = W_a / E_a \tag{7}$$

Therefore, in scenarios where it is necessary to compress or decompress activities, the calculations accomplished by the tool always preserve the work content of each activity, so the duration and effort values can be changed, however, the amount of work is always preserved.

5 Developed Approach

This section intends to present how the main aspects of the tool, as well as the main objectives of this work, were achieved.

5.1 Project Schedule

To achieve the objective of creating project schedules, each project was identified as consisting of five subprojects, in this case, five QG's. The main constraint for activities scheduling of each QG is that the activities must perform until the due date or in the shortest feasible time and respect the resource availability. In case the activities must be scheduled in the shortest feasible time, the tool creates a schedule according to the ILP formulation for RCPSP previously described. Otherwise, if the QG has a due date, it may be necessary to compress or decompress project activities. The algorithm illustrated in Fig. 1 demonstrates the way that the schedules are developed in case of being necessary to compress or decompress activities. Using the preservation of work content, the activities have the duration, and so the effort, changed to reduce the Critical Path Method duration (CPMd) of the project in order to reach the due date. Due to restrictions of the organization in which this study was developed the activities have an effort with a lower bound of 5% (0.05) and upper bound of 80% (0.80).

In this way, the tool changes to the duration (D1/D2) of activities, and hence the efforts (E1/E2), that belongs to the critical path of the project in order to approximate the Ending Date of the project (EDp) of the schedule baseline to the Due Date of the project (DDp). Once the dates are closer, i.e., there is a scenario where it is possible to schedule activities within the expected due date, an interactive application of the presented ILP formulation, verification and adjustment process is accomplished with the intent to build a feasible scenario and find the best schedule of activities.

```
Compress
Calculate schedule baseline and keep DDp EDc values
Do
  If CPMd > DDp then
    For each i ∈ CPM activities Do
      Reduce CPM Activities Duration
      Calculate new effort (E2)
      If E2 < 0.8 then
        Keep E2 e D2
      Else
        Keep previous duration and effort
  Else
    For each i ∈ CPM activities Do
      Reduce CPM Activities Duration
      Calculate new effort (E2)
      If E2 < 0.8 then
        Keep E2 e D2
      Else
        Keep previous duration and effort
    Calculate new time schedule (DDp)
While DDp > EDc
```

```
Decompress
Calculate schedule baseline and keep DDp EDc values
Do
  If CPMd << DDp then
    For each i ∈ CPM activities Do
      Increase CPM Activities Duration
      Calculate new effort (E2)
      If E2 > 0.05 then
        Keep E2 e D2
      Else
        Keep previous duration and effort
  Else
    For each i ∈ CPM activities Do
      Increase CPM Activities Duration
      Calculate new effort (E2)
      If E2 > 0.05 then
        Keep E2 e D2
      Else
        Keep previous duration and effort
    Calculate new time schedule (DDp)
While DDp < EDc
```

Fig. 1. Algorithm to compress and decompress

Thus, Fig. 2 illustrates the entire project schedule construction process performed by the tool. The algorithm (the exact formulation or the hybrid model) is selected in each QG to create the project schedule.

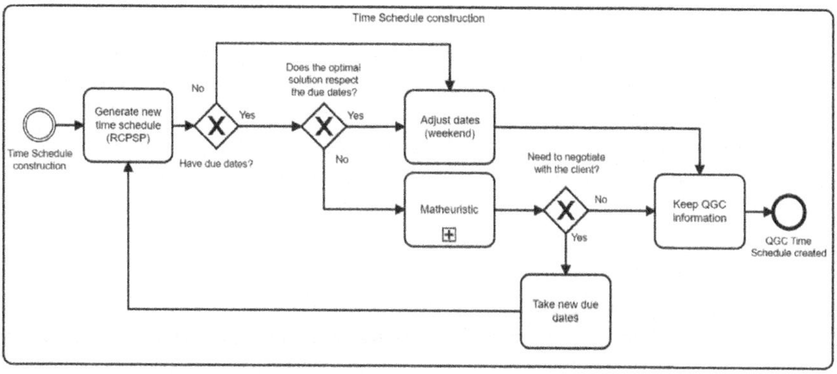

Fig. 2. Construction process of project schedule

The time it takes the tool to find the solution may vary, due to the level of activities' to compress/decompress to respect the due dates. The average run time for the project with 45 activities without due dates is 35.56 s, and with due dates it is 168.70 s. As literature refers, the use of metaheuristics helps to reduce time to find solutions, with less complexity that can be easier to implement in the organizational environment. Therefore, being the algorithm easier to model and program, any further improvements of the algorithm can be easily accommodated in order to make it more competitive concerning computational time.

5.2 Recommendation System

In order to achieve the objective of creating a recommendation system, it was necessary to develop a formula that reflected the real PjM allocation during the considered period. Usually, as dispersion measures are sensitive to outliers [18], the developed formula seeks to mitigate the variation in the amount of PjM's work over the considered period.

$$CV = S/\overline{X} \tag{8}$$

To understand the PjM allocation calculation, it is necessary to explain the concept of Coefficient of Variability (CV). This measure expresses the data variation regarding the average [19], as expressed in (8). After this value is calculated, it is identified whether the PjM allocation is representative or whether adjustments to the average PjM allocation are required. The average adjustment is made according to the coefficient of variability, i.e., the higher the CV, the higher the average adjustment will be required. This adjustment assumes the standard deviation as weight (9).

$$NC = \overline{X} \pm (CV * S) \tag{9}$$

Finally, although the tool considers the requirements necessary to manage the project and, thus, filters which managers may be eligible (have the necessary requirements and are not over-allocated) to manage the project, the decision-making process of assigning a project to a PjM is considered a non-linear process, because there are variables that are beyond those considered in the tool, as organization's objectives, for instance. In this way, it is necessary a holistic view of the organization itself, which increases the complexity of the decision-making process. Therefore, at the end of the recommendation process, a PjM ranking list is displayed, so that the portfolio manager can choose from among the possible managers who is the besto manage the project and takes the decision in line with other objectives that are not considered by the tool.

5.3 Portfolio Report

To achieve the third objective, the tool creates an automatic report and it is always updated when a new project is inserted. This report aims to provide the user with an overview of the portfolio capacity utilization. Figure 3 illustrates some graphs used in this report.

Briefly, this report contains graphs that identify the number of projects by category and how these projects are distributed within the sectors of the organization. It also provides a project portfolio resource allocation status and this information can be segmented by sector. Then, it is also possible to find out which is the most contributing sector to the identified scenario.

It is also important to mention other two aspects of the report. The first aspect is the recommendation of the minimum appropriate number of PjMs required, given the

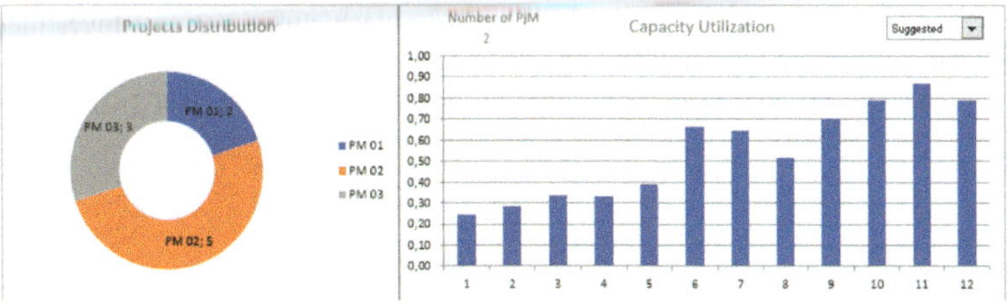

Fig. 3. Some graphs of the portfolio report

number of projects considered in the tool. The second aspect is that the report aims to be iterative, enabling scenario simulations regarding project portfolio capacity to facilitate the project portfolio manager in the decision-making process.

6 Conclusions and Future Work

The present work aimed to develop a tool that would help in the portfolio management of industrialization projects. The work has been developed in a case study context and had three main objectives.

The first objective was to develop schedules in line with the required due dates. This objective was achieved by using an ILP formulation for RCPSP and a developed hybrid model that adjusts the critical path activities to compress or decompress and thus interoperate with the ILP formulation to create a schedule that respects project due dates and the resource (PjMs) availability. The second objective has been achieved with a project recommendation system to PjMs so that it did not result in an over-allocation of PjMs. This objective was achieved using some statistical concepts that can quantitatively identify the PjM's allocation and through a ranking list shows which PjMs can better manage that project. The last objective was to create a report that provided a sector status and give support to the project portfolio managers in the decision-making process. This objective has been achieved in the elaboration of a report that enables the identification of capacity status as well as the identification of which sector has contributed the most to the identified scenario. Also, this report makes possible to accomplish the simulation of different scenarios for capacity utilization and the suggestion of a minimum number of PjMs for the portfolio.

As suggestions for future work, a critical point would be to implement improvements to the algorithm, such as adding other heuristic or metaheuristic techniques that adjust more efficiently and faster to develop the schedule. In addition, for future work, we suggest making improvements to the recommendation system, which considers only two variables at the moment – PjM allocation and project requirements. Other relevant variables can be included, namely taking into account the previous experience of PjMs with different customers or project owners, matching the industrialization project with the most experienced manager with the specific project owner, if available. The extension of the proposed solutions to other kind of specific human resources, as the key stakeholders, i.e., the core team, is also under consideration.

Acknowledgements. This work is supported by: European Structural and Investment Funds in the FEDER component, through the Operational Competitiveness and Internationalization Programme (COMPETE 2020) [Project nº 39479; Funding Reference: POCI-01-0247-FEDER-39479].

References

1. Abrantes, R., Figueiredo, J.: Resource management process framework for dynamic NPD portfolios. Int. J. Proj. Manag. **33**, 1274–1288 (2015). https://doi.org/10.1016/j.ijproman.2015.03.012
2. Oh, J., Yang, J., Lee, S.: Managing uncertainty to improve decision-making in NPD portfolio management with a fuzzy expert system. Expert Syst. Appl. **39**, 9868–9885 (2012). https://doi.org/10.1016/j.eswa.2012.02.164
3. Manole, A.L., Grabara, I.: Methodologies and visualization tools of effective project management. Polish J. Manag. Stud. **14**, 137–149 (2016). https://doi.org/10.17512/pjms.2016.14.2.13
4. Perrotta, D., Araújo, M., Fernandes, G., Tereso, A., Faria, J.: Towards the development of a methodology for managing industrialization projects. Procedia Comput. Sci. **121**, 874–882 (2017). https://doi.org/10.1016/j.procs.2017.11.113
5. Chirumalla, K.: Managing product introduction projects in operations: key challenges in heavy-duty vehicle industry. J. Mod. Proj. Manag. **5**, 108–118 (2018). https://doi.org/10.19255/JMPM01512
6. Radujković, M., Sjekavica, M.: Project management success factors. Procedia Eng. **196**, 607–615 (2017). https://doi.org/10.1016/j.proeng.2017.08.048
7. Ponsteen, A., Kusters, R.J.: Classification of human- and automated resource allocation approaches in multi-project management. Procedia Soc. Behav. Sci. **194**, 165–173 (2015). https://doi.org/10.1016/j.sbspro.2015.06.130
8. Tian, X., Yuan, S.: Genetic algorithm parameters tuning for resource-constrained project scheduling problem. In: AIP Conference Proceedings, p. 040059 (2018)
9. Chakrabortty, R.K., Sarker, R., Essam, D.L.: A comparative study of different integer linear programming approaches for resource-constrained project scheduling problems. Presented at the (2018)
10. Kolisch, R., Hartmann, S.: Experimental investigation of heuristics for resource-constrained project scheduling: an update. Eur. J. Oper. Res. **174**, 23–37 (2006). https://doi.org/10.1016/j.ejor.2005.01.065
11. Borak, S., Karl, W.: Matheuristics. Springer, Boston (2010)
12. Artigues, C., Brucker, P., Knust, S., Koné, O., Lopez, P., Mongeau, M.: A note on "event-based MILP models for resource-constrained project scheduling problems". Comput. Oper. Res. **40**, 1060–1063 (2013). https://doi.org/10.1016/j.cor.2012.10.018
13. Maniezzo, V., Stutzle, T.: Matheuristics 2016 - Proceedings of the Sixth International Workshop on Model-based Metaheuristics. IRIDIA, Bruxelles (2016)
14. Saunders, M., Lewis, P., Thornhill, A.: Research Methods for Business Students (2016)
15. Shrestha, A., Cater-Steel, A., Toleman, M., Rout, T.: Benefits and relevance of International Standards in a design science research project for process assessments. Comput. Stand. Interfaces. **60**, 48–56 (2018). https://doi.org/10.1016/j.csi.2018.04.011
16. Wang, D., Wan, K., Song, X., Liu, Y.: Provincial allocation of coal de-capacity targets in China in terms of cost, efficiency, and fairness. Energy Econ. **78**, 109–128 (2019). https://doi.org/10.1016/j.eneco.2018.11.004

17. Tereso, A., Araújo, M., Elmaghraby, S.: Adaptive resource allocation in multimodal activity networks. Int. J. Prod. Econ. **92**, 1–10 (2004). https://doi.org/10.1016/j.ijpe.2003.09.005
18. Peck, R., Devore, J.L.: Statistics - The exploration & Analysis of data. Brooks/Cole (2010)
19. Keser, İ.K., Kocakoç, İ.D., Şehirlioğlu, A.K.: A new descriptive statistic for functional data: functional coefficient of variation. Alphanumeric J. 4, (2016). https://doi.org/10.17093/aj.2016.4.2.5000185408

Knowledge Management Life Cycle Model Based on PDSA for Agile Companies

Raluca Dovleac[1(✉)], Andreea Ionica[1], Monica Leba[1], and Alvaro Rocha[2]

[1] University of Petrosani, 332006 Petrosani, Romania
raluca.dovleac@yahoo.com
[2] University of Coimbra, Coimbra, Portugal

Abstract. The paper analyzes the ways in which knowledge management contributes to the success of today's companies and leads to an increase in their effectiveness and ability to remain competitive, and furthermore, the tools and practices that they use for their knowledge management process with the role of providing a framework for their knowledge management, which is based on a proposed knowledge management life cycle, inspired by the Agile System Development Life cycle. For this, in the paper, the current life cycle models used in knowledge management have been analyzed, along with the particularities and necessities of agile companies, and based on this, a life cycle model for the knowledge management of agile companies has been proposed, and furthermore, the possibility of integrating quality management tools within each stage of the life cycle has been analyzed, and a set of quality tools has been proposed in the context of the life cycle model.

Keywords: Life cycle · Knowledge management · Agile · Quality

1 Introduction

The current competitive layout has pushed companies of all sizes and from all industries to implement strategic practices in order to ensure their success both in the short and long run.

This paper is focused on addressing two main axes of interest that companies are covering: the implementation of agile practices and methodologies, and thus, becoming agile companies, and the focus on knowledge management and knowledge management systems within them.

Knowledge management refers to the set of information, intellectual capital and communication that a company has to manage in order to benefit from it, and it takes the form of strategies, processes and technology used. This information can be tacit – meaning that it hasn't been captured or transposed in a physical form, or explicit – meaning that the company transposed it in a physical form. Some of the identified benefits of implementing a knowledge management system within a company relate to its contribution in achieving both customer and employee satisfaction, enhancing communication and the results obtained from project management practices [1].

Á. Rocha et al. (Eds.): WorldCIST 2020, AISC 1159, pp. 84–92, 2020.
https://doi.org/10.1007/978-3-030-45688-7_9

Another important aspect that emerged in the context of a company seeking to implement a knowledge management system is the possibility of ensuring the quality of this process' inputs, outputs and practices. This, in turn, led to studies concerned with the role that quality management currently plays in the context of knowledge management and the possibility of extending its influence towards this field in order to obtain better results.

Given this aspect, the paper analyses the set of recommendations in the field of knowledge management regarding quality management practices, in order to provide a model that integrates this practices to ensure a competitive advantage.

Lastly, the paper is concerned with the role that agile companies play in today's society, their necessities and aims at adapting the proposed model to fit their requirements and provide a practical support system for those agile companies looking to implement a knowledge management system.

2 Literature Review

Motivated by the necessity of remaining competitive and ensuring their effectiveness, many organizations are pursuing the area of knowledge management, which, although not new, it now shows an increase in implementation and usage [2–4].

A large number of definitions aiming to cover the topic of knowledge management exist [5], concluding that a general aspect of knowledge management consists mainly of its characteristics, such as: the usage or transfer of knowledge does not lead to its consumption and unavailability, the ability of using knowledge is scarce and the most important source of knowledge for a company is its personnel [6]. The last aspect relates to how knowledge is the result of human activities, typically empirical reflection and experience and can be embodied in multiple forms, such as: rules, concepts, stories, etc. [7].

Knowledge management includes activities which are meant to capture, convert, transfer and measure knowledge in a social context [8]. In the context of the human resources involved in the process of knowledge management, and their contribution with intellectual capital, knowledge assets have been referenced also as being intellectual assets [9].

Knowledge management can therefore be seen as a process [5] that is concerned with acquiring, storing, developing, using, disseminating and transferring information and expertise with the role of improving business performance [10], the decision making process and facilitate innovation and improvement [11].

Given the fact that knowledge management is most often perceived as a process, similarly to other processes, it inhabits a life cycle, known in theory as Knowledge management life cycle. Solid works in the field are covered by Wiig [12], McElroy [13], Bukowitz and Williams [14] and Meyer and Zack [15], but new approaches have emerged throughout time, such as those of: Evans et al. [9], Sağsan [16], Stenholm et al. [17] looking to adapt the knowledge management life cycle to the requirements of today's companies. A comparison in terms of common aspects and differences of the foundational approaches in knowledge management can be observed in Table 1.

Table 1. Comparison of knowledge management life cycle stages

	Create	Acquire	Refine	Store	Use	Transfer
Wiig [12]						
McElroy [13]						
Bukowitz & Williams [14]						
Zack [15]						

As it can be observed from Table 1, all four knowledge management life cycle approaches put an emphasis on the acquisition and usage of information stage, but take different stands when it comes to phases such as: knowledge creation, storing and transfer.

Today's companies are marked by significant changes in the market and consumer preferences and options that push them to adapt in order to remain effective and competitive. In this context, companies are required to show agility and effectiveness in responding to changes in the market and customer preferences, which in turn, leads to redesigning a company's profile to be considered an agile company [18]. Agile companies show an embracing attitude towards changing customer demands and implement adaptive responses to these changes [19].

Agile methodologies have been initially developed and implement in small teams working on single projects, but since, have shown a supportive implementation in companies of all sizes given their noted benefits [20–22].

Agile methodologies are characterized by an iterative approach to product development, short iteration periods, a high degree of flexibility and adaptability to changes in requirements [23, 24]. The development life cycle for the Agile development approach captures the idea of iterative development and reacting to customer requirements. A typical example of the Agile System Development Life cycle (SDLC) can be observed in Fig. 1.

Figure 1 depicts the Agile SDLC, highlighting the particularities of this development approach, such as the iterative development realized through the help of short iteration times, and the possibility of transitioning from the construction phase back to the initiation one, based on customer feedback.

An interesting aspect that emerged in the analysis of the role that knowledge management plays in today's companies, is related to the relationship between quality management and knowledge management, in the form of their impact on the knowledge transfer [25], knowledge creation [26], using a maturity model for quality improvement in knowledge management [27] and the possibility of integrating knowledge management in the Total Quality Management (TQM) practices [28].

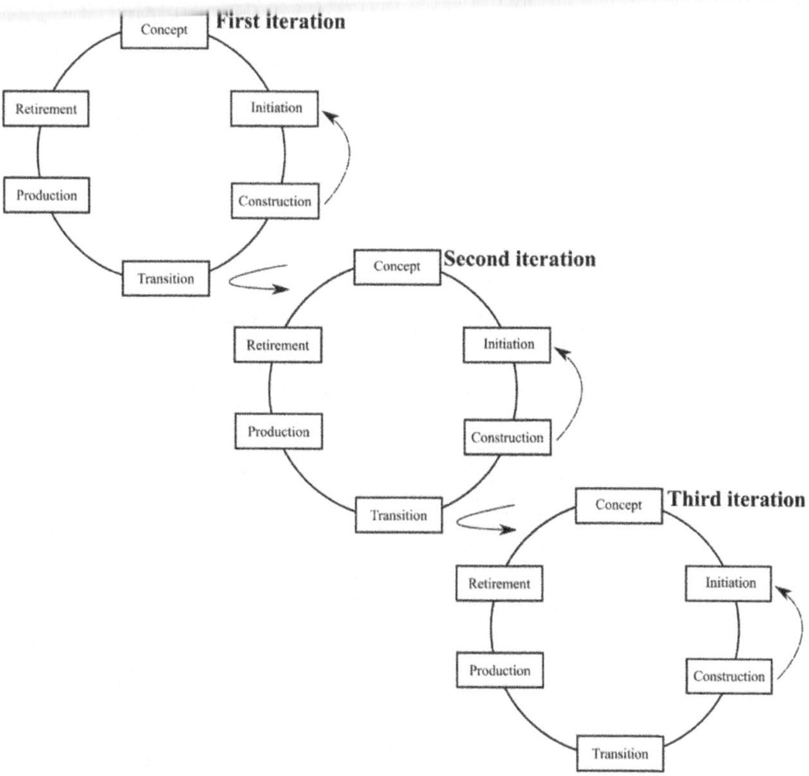

Fig. 1. Agile SDLC

3 Research Context

Given how the number of companies that implement or look to implement knowledge management into their operations has increased, the need for a strategy or framework that would facilitate the implementation of knowledge management is born [9].

The current paper aims to help provide a model for a knowledge management life cycle that takes into account the requirements of today's companies, putting an emphasis on agile companies and the important role of quality management plays in obtaining and ensuring their competitiveness and success. In designing the model for the knowledge management life cycle, the already existing models have been analyzed and adaptations have been made in order to fit the requirements of agile companies.

4 Knowledge Management Life Cycle Based on Agile SDLC

The proposed model integrates quality management practices and tools in the form of shaping and adapting the stages of the knowledge management life cycle in the Plan Do Study Act (PDSA) cycle in order to highlight the relationship between each stage of the knowledge management life cycle and the particularities of the PDSA cycle, and to provide a set of methodological tools that could improve the quality of the knowledge management process.

The proposed model consists of six stages for the knowledge management life cycle as follows: Create, Acquire, Refine, Store, Use and Transfer. The life cycle stages have been selected and adapted based on the already existing models and the necessities of agile companies. Figure 2 shows how these stages of the life cycle can be integrated and represented with the help of the PDSA.

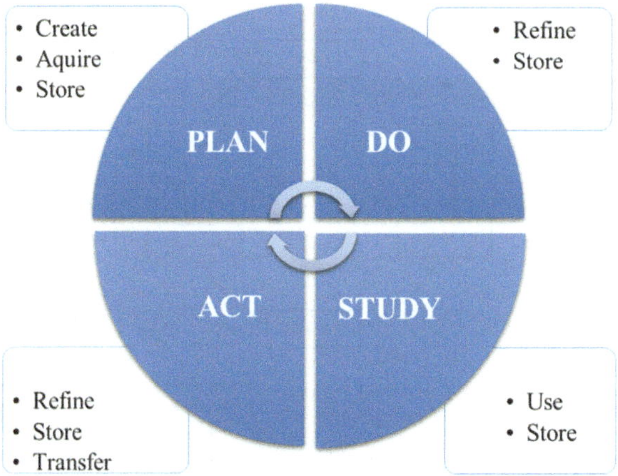

Fig. 2. Stages of the proposed knowledge management life cycle

Given the fact the proposed model is aimed at meeting the requirements of agile companies, it has been adapted to fit the stages of the Agile SDLC, as shown in Fig. 3 below.

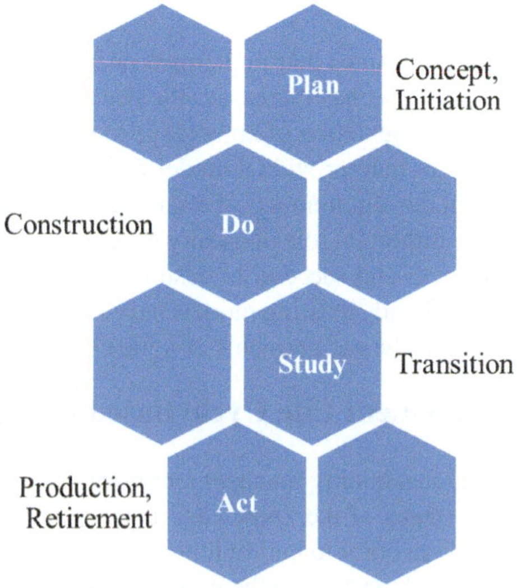

Fig. 3. PDSA cycle model for Agile SDLC stages

The Agile SDLC approach, consisting of six development stages, can be integrated in the PDSA cycle due to the cyclic nature of the life cycle which is similar to the PDSA cycle model. Therefore, in the Plan stage, the corresponding stages of the Agile SDLC are: Concept and Initiation; in the Do stage, the corresponding stage is the Construction one; in the Study stage, the corresponding stage is Transition, and in the Act stage of the PDSA cycle, the corresponding Agile SDLC stages are: Production and Retirement. This model captures the iterative nature of the development approach and the PDSA cycle. Of course, each stage contains one or more aspects concerned with the other stages, such as: the planning phase covers all development stages, from the concept/initiation one, to the retirement and/ or production, where the team decides what to do next, based on the information gathered so far. In the context of the proposed model, the knowledge management life cycle has been adapted to the PDSA cycle, one of the aspects regarding this implying that the development team uses an overview in order to sketch the requirements of each stage of the PDSA cycle, while continuing to work in an Agile SDLC methodology.

Based on this, and taking into account the recommendations for the possibility of implementing quality management tools and practices in the knowledge management process, the paper presents a set of quality management tools that could be implemented in each stage of the knowledge management life cycle, integrated in the PDSA model previously proposed. The list of quality tools, along with the life cycle stages in which they are recommended can be seen in Table 2.

Table 2. Quality tools integrated in the knowledge management PDSA model

	PLAN	DO	STUDY	ACT
Ishikawa				
Checksheet				
Control chart				
Histogram				
Pareto				
Scatter				
Flowchart				

The quality management tools presented have been selected based on their proven benefits in implementation, ease of use, and familiarity for those who have no previous significant experience of working the field of quality management or with quality management tools. The presented tools, are considered to be the basic set of quality control tools which can be used in order to address quality related issues in companies.

As it can be seen from Table 2, the seven basic or classic quality management tools have been used and integrated in the PDSA cycle in order to provide a model that allows the knowledge management life cycle with methodological support regarding the possibility of improvement. Each stage of the PDSA has at least a number of three quality management tools that can be used therefore highlighting the possibility of implementation.

5 Results and Conclusions

The analysis of the requirements of agile companies, and the particularities of the proposed knowledge management life cycle, designed based on the already existing theoretical background, allowed the development of a knowledge management life cycle model adapted for the requirements of today's companies and which takes into account not only their particularities and necessities but also the best practices in the field of knowledge management and the added benefits of quality management through the help of the implementation of quality management tools in the proposed model.

The model, shown in Fig. 4, addresses the requirements of agile companies, by integrating an iterative approach, which allows flexibility and adaptability according to requirements. Furthermore, the model is based on the Agile SDLC approach, in order to facilitate the integration and usability of the model for agile companies.

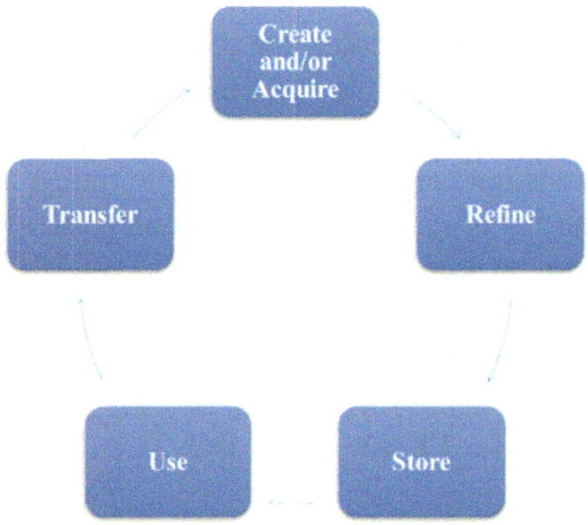

Fig. 4. Knowledge management life cycle model for agile companies

Building on top of the already existing model for knowledge management life cycles, the Agile SDLC and the recommendations regarding the possibility of implementing quality management practices in order to improve the quality of the knowledge creation and transfer processes, a knowledge management life cycle model based on the PDSA cycle has been designed and proposed in the current paper. The possibility of integrating quality management tools in the proposed model has also been explored and detailed, and can therefore be observed in Fig. 5.

The approach used to assign quality tools to each stage of the proposed knowledge management model is similar to the approach used when assigning Agile SDLC stages to the PDSA cycle, meaning that although one or more quality tools can be used across all the stages of the proposed life cycle, the suggested match ensures that the development team works according to Agile principles, and in a way that maximizes their operations' efficiency.

As it can be observed from Fig. 5, the "Histogram" and the "Flowchart" quality tools can be applied throughout the entire life cycle of the proposed mode, the "Ishikawa" and "Check sheet" tools can be used for all phases except the "Use", the "Scatter" diagram

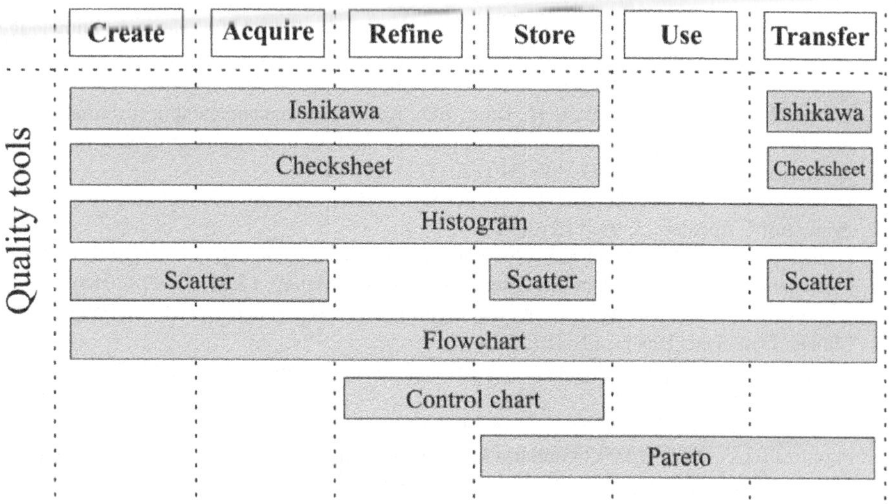

Fig. 5. Quality tools for the knowledge management life cycle model

can be used for the "Create", "Acquire", "Store" and "Transfer" phases, the "Control charts" can be used in the "Refine" and "Store" phase and the "Pareto" diagram can be used in the "Store", "Use" and "Transfer" phases.

Of course various other quality management tools can be further implemented into the proposed more, and more so, quality management practices or principles can also be used and implemented in order to develop the model in accordance with the requirements and necessities of the companies using it.

Further directions include, but are not limited to the implementation, testing and adaptation of the proposed model in the case of agile companies, as well as studies regarding the possibility of implementing additional quality management tools and practices within the model. This could be achieved either by following the path already established within the paper, and focusing on those quality management tools and techniques which are considered to be modern, or by looking at business excellence models and total quality management principles and practices and identifying those factors that could benefit agile companies in the context of the proposed model.

References

1. Lee, K.-W., Corazon, M., Lanting, L., Rojdamrongratana, M.: Managing customer life cycle through knowledge management capability: a contextual role of information technology. Total Qual. Manag. Bus. Excellence **28**(13–14), 1559–583 (2017)
2. Birkinshaw, J., Sheehan, T.: Managing the knowledge life cycle. MIT Sloan Manage. Rev. **44**(1), 75–83 (2002)
3. de Vasconcelos, J.B., Kimble, C., Carreteiro, P., Rocha, Á.: The application of knowledge management to software evolution. Int. J. Inf. Manag. **37**, 1499–1506 (2016)
4. Del Giudice, M., Maggioni, V.: Managerial practices and operative directions of knowledge management within inter-firm networks: a global view. J. Knowl. Manag. **18**(5), 841–846 (2014)

5. Nývlt, V., Prušková, K.: Building information management as a tool for managing knowledge throughout whole building life cycle. IOP Conf. Ser.: Mater. Sci. Eng. **245**, 042070 (2017)
6. Dalkir, K.: Knowledge Management in Theory and Practice. Elsevier, Oxford (2013)
7. Lim, M.K., Tseng, M.-L., Tan, K.H., Buib, T.D.: Knowledge management in sustainable supply chain management: Improving performance through an interpretive structural modelling approach. J. Clean. Prod. **162**, 806–816 (2017)
8. O'Leary, D.E.: Knowledge management across the enterprise resource planning systems life cycle. Int. J. Account. **3**, 99–110 (2002)
9. Evans, M., Dalkir, K., Bidian, C.: A holistic view of the knowledge life cycle: the Knowledge Management Cycle (KMC) model. Electron. J. Knowl. Manag. **12**(2), 85–97 (2014)
10. Gupta, B., Iyer, L., Aronson, J.: Knowledge management: practices and challenges. Ind. Manag. Data Syst. **100**(1), 17–21 (2000)
11. Earl, M.J.: Knowledge management strategies: toward a taxonomy. J. Manag. Inf. Syst. **18**(1), 215–242 (2001)
12. Wiig, K.: Knowledge management foundations: thinking about thinking. How people and organizations create, represent and use knowledge. Arlington (1993)
13. McElroy, M.: The knowledge life cycle. In: ICM Conference on KM, Miami (1999)
14. Bukowitz, W., Williams, R.: The Knowledge Management Fieldbook. Prentice Hall, London (2000)
15. Meyer, M., Zack, M.: The design and implementation of information products. Sloan Manag. Rev. **37**(3), 43–59 (1996)
16. Sağsan, M.: A new life cycle model for processing of knowledge management. In: International Conference on Business, Management and Economics, İzmir (2006)
17. Stenholm, D., Landahl, J., Bergsjö, D.: Knowledge management life cycle: an individual's perspective. In: International Design Conference, Dubrovnik (2014)
18. Bottani, E.: Profile and enablers of agile companies: an empirical investigation. Int. J. Prod. Econ. **125**, 251–261 (2010)
19. Sillitti, A., Ceschi, M., Russo, B., Succi, G.: Managing uncertainty in requirements: a survey in documentation-driven and agile companies. In: 11th IEEE International Symposium on Software Metrics, Como (2005)
20. Boehm, B., Turner, R.: Management challenges to implementing agile processes in traditional development organizations. IEEE Softw. **22**(5), 30–39 (2005)
21. Dybå, T., Dingsøyr, T.: What do we know about agile software development? IEEE Softw. **26**(5), 6–9 (2009)
22. Dikert, K., Paasivaara, M., Lassenius, C.: Challenges and success factors for large-scale agile transformations: a systematic literature review. J. Syst. Softw. **119**, 87–108 (2016)
23. Tarafdar, M., Qrunfleh, S.: Agile supply chain strategy and supply chain performance: complementary roles of supply chain practices and information systems capability for agility. Int. J. Prod. Res. **55**(4), 925–938 (2017)
24. Gunasekaran, A., Yusuf, Y.Y., Adeleye, E.O., Papadopoulos, T., Kovvuri, D., Geyi, D.G.: Agile manufacturing: an evolutionary review of practices. Int. J. Prod. Res. **57**(15–16), 5154–5174 (2019)
25. Molina, L.M., Lloréns-Montes, J., Ruiz-Moreno, A.: Relationship between quality management practices and knowledge transfer. J. Oper. Manag. **25**, 682–701 (2007)
26. Linderman, K., Schroeder, R.G., Zaheer, S., Liedtke, C., Choo, A.S.: Integrating quality management practices with knowledge creation processes. J. Oper. Manag. **22**, 589–607 (2004)
27. Paulzen, O., Doumi, M., Perc, P., Cereijo-Roibas, A.: A maturity model for quality improvement in knowledge management. In: ACIS Proceedings, vol. 5 (2002)
28. Ooi, K.-B.: TQM and knowledge management: literature review. Afr. J. Bus. Manage. **3**(11), 633–643 (2009)

Protocol for Analysis of Root Causes of Problems Affecting the Quality of the Diagnosis Related Group-Based Hospital Data: A Rapid Review and Delphi Process

M. F. Lobo[1,2](✉), M. Oliveira[1], A. R. Oliveira[1], J. V. Santos[1,2,3], V. Alonso[2], F. Lopes[1,2], A. Ramalho[1,2], J. Souza[1,2], J. Viana[1,2], I. Caballero[4], and A. Freitas[1,2]

[1] Department of Community Medicine, Information and Health Decision Sciences (MEDCIDS), Faculty of Medicine, University of Porto, Porto, Portugal
{marianalobo,fernando,alberto}@med.up.pt, mariana_ramos_oliveira@hotmail.com, raquel.oliveira006@gmail.com, jvasco.santos@gmail.com, andrelcramalho@gmail.com, juliobsouza1990@gmail.com, joao.a.viana@gmail.com

[2] Center for Health Technology and Services Research (CINTESIS), Faculty of Medicine, University of Porto, Porto, Portugal
vera.alonsop@gmail.com

[3] Public Health Unit, AceS Grande Porto VIII – Espinho/Gaia, ARS Norte, Porto, Portugal

[4] Institute of Information Systems and Technologies (ITSI), University of Castilla-La Mancha, Ciudad Real, Castilla-La Mancha, Ciudad Real, Spain
ismael.Caballero@uclm.es

Abstract. The Diagnosis Related Group-based administrative hospital database is an important tool for hospital financing in several health systems. It is also an important data source for clinical, epidemiological and health services research. Therefore, the data quality of these databases is of utmost importance for the exploitation of the data. To prevent or to solve these problems, it is paramount to identify the root causes of the lack of data quality. In order to do this process easier, data quality analysts could benefit from having a catalog of potential root causes to explore. Unfortunately, literature has not covered this concern extensively, what motivated us to research. This paper presents a protocol for the analysis of root causes that may affect the quality of these data. The proposed protocol includes two stages: (1) a systematic review to extract root causes from the scientific literature, and (2) a Delphi technique to analyze the relevance of root causes and to map them into data quality dimensions.

Moreover, we provide a pilot study (rapid review) based on a systematic review covering papers published during 2018 and 2019. This aimed to establish proof of concept of the proposed methodology and to examine its feasibility. Eighteen studies were included for data charting, rendering 70 root causes. The Ishikawa framework provided a meaningful way to represent the root causes. Most of these causes were associated with professional knowledge, education and processes performed in the data collection, analysis or treatment.

Á. Rocha et al. (Eds.): WorldCIST 2020, AISC 1159, pp. 93–103, 2020.
https://doi.org/10.1007/978-3-030-45688-7_10

Keywords: DRG-based hospital administrative database · Clinical coding · Data quality · Root causes

1 Introduction

Diagnosis-related group (DRG)-based administrative hospital databases are an essential resource for hospital reimbursement and budgeting in many health management systems. It usually contains administrative, demographic, clinical and reimbursement hospitalization data structured around the concept of DRG and relative weights [1]. Medical records (incl. imagiologic studies, pathology and surgical reports discharge summaries) and administrative records that result from the patient care episode stored at the hospital are the main data sources of these databases [2]. Clinical classification systems, such as the International Classification of Diseases, 9th Clinical Modification (ICD-9-CM), are used to code clinical information, which is then used to compute the relative weight and reimbursement value of the episode through DRGs [3]. Due to several reasons (e.g. the financial interest of hospitals or lack of standards in clinical coding), fallible information is included in this type of databases [2, 4]. This raises several problems in the data quality, potentially compromising its reuse for different purposes (e.g. hospital reimbursement, epidemiological profiles).

In fact, this type of databases also provides an important source of data for some other non-management purposes like clinical, epidemiological and health services research [5–7]. Some of its strengths consist in having a large population and broad (e.g. national) geographic coverage, for instance Healthcare Cost and Utilization Project, which is the largest nationwide collection of longitudinal hospital care data in the United States. When compared with other research data sources (prospective registries or clinical trial data) is considerable less expensive and readily available [8]; and sometimes the only feasible data source available to study specific patient subgroups or to afford international comparisons [9].

In order to obtain the largest benefits from the exploitation of this type of databases, the data should have the highest level of quality possible, so that different type of stakeholders can achieve trustworthy results for their purposes. Thus, we adhere to the definition of "*fitness for use*" given in [10]. This definition states that data quality goes beyond accuracy, and it should be understood as a multidimensional concept, which requires the identification of several data quality dimensions to judge the level of quality of a given dataset. For example, Souza et al. in [11] concluded that the lack of comprehensive reporting of comorbidities might have an important impact in hospital reimbursement, hence highlighting the consequence of data (in)completeness. To the best of our knowledge, there are several studies identifying data quality issues in administrative hospital databases, mainly rooted on data accuracy problems [12–15]; however, even when completeness and accuracy are the two data quality dimensions that usually are the focus on the most data quality problems, there are some other data quality dimensions which can be also addressed when it comes to identify the root causes of the data quality related problem [16]. In fact, a systematized compilation and analysis of their root causes is lacking.

The aim of this work is to provide a protocol to conduct a study that answers the research question:

RQ. What is known about the root causes of problems affecting the quality of DRG-based administrative hospital data, considering both the use of the data for hospital financing and epidemiological and health services research?

The remainder of the paper is structured as follows. Section 2 describes the methodological steps to extract (Subsect. 2.1) and analyze (Subsect. 2.2) root causes of data quality problems. Section 3 provides a pilot study to extract root causes using the most recent published literature. This was implemented to assess the burden of the planned study, its strengths and pitfalls based on a systematic review of studies published between 2018–2019. Finally, in Sect. 4 and 5, we introduce some conclusions that we have raised after implementing this pilot study.

2 Methods

2.1 Systematic Review

This protocol has been designed using the PRISMA-P guidelines for systematic review protocol development [17].

Eligible Criteria. Studies are eligible only if they are published in peer-reviewed journals (including conference papers). The papers must consider data quality of DRG-based administrative hospital database, including inpatient or outpatient hospital episodes, covering aspects related with processes prone to result in data quality issues of these databases, such as processes regarding data generation/acquisition, processing (e.g. clinical coding), storage, and utilization of these databases and their data sources (e.g. medical records), as well as encompassing the different perspective of producers, custodians and users (e.g. clinical coders, physicians, hospital managers, health ministry agents, researchers).

Further eligibility criteria have been defined by language, publication date and accessibility of abstract and full text. We have opted to include peer-reviewed papers, written in English, and published between 1990 and 2019.

Claims or billing data-based studies will not be excluded because data quality problems of these and the DRG-based administrative databases share the same root causes [18–20]. Data quality problems in up-stream data sources (e.g. EHR) are root causes of data quality issues of the database. As such studies covering EHR data quality have also been considered eligible for inclusion. Moreover, data quality issues observed in the DRG-based administrative hospital data may not point to their root causes. Therefore, studies are only eligible when the authors provide factual evidence of root causes of data quality problems of the DRG-based administrative hospital database.

Information Sources and Search. The Medline and SCOPUS have been considered to identify potentially relevant studies. The search strategy has been drafted based on two concept blocks, i.e. (1) administrative hospital databases and (2) data quality. This has then been refined through team discussion and calibrated based on the terms' sensitivity. The final search results have been exported into EndNote X9.2 and duplicates were removed. The online free tool Rayyan [21] has been chosen to assist in the screening phase.

Selection of Sources of Evidence and Data Charting. Two reviewers will evaluate the titles and abstracts, independently. This will be achieved following the above-mentioned eligibility criteria. Disagreements will be solved by consensus. Full text will then be assessed, and the same eligibility criteria will be applied.

A data-charting form has been developed to extract data from the studies, including information regarding the study, and information relevant for the study of root causes of data quality problems. For this task, a root cause has been defined as any identifiable process prone to result in data quality problems of administrative hospital databases and will be abstracted from the studies using copy & paste text as much as possible. Two reviewers will independently chart the data.

Any disagreements on study selection and data extraction will be resolved by consensus and discussion with other team members.

Strategy for Data Synthesis. After sorting duplicates, we will synthesize the list of root causes of data quality problems by mapping them into major and sub-categories. For this, we will borrow the Ishikawa diagram framework to assign each root cause to an Ishikawa diagram branch. Two reviewers will be involved in this task and any conflicts will be discussed. This diagram is a cause analysis tool that identifies many possible causes for an effect or problem graphically sorting ideas into useful categories. Each new root cause that answers a "why?" question to another root cause will generate a hierarchy of root causes within the branch.

2.2 Root Causes Analysis – The Delphi Process

We will analyze the list of root causes of data quality problems using the Delphi technique. This is an iterative process of anonymous self-completed questionnaire with individual feedback [22].

Panel Size and Recruitment. A minimum of 10 participants of the Delphi panel will be identified essentially from our network of contacts, peer recommendations and experts within our research institution and other national and international organizations and then by a snowball process. The panel sample will be slected in order to be multidisciplinary covering clinical coders, physicians, and researchers in data quality, clinical and health services research. Participants will be invited through email after careful revision of their curriculum vitae.

Number of Rounds. In accordance with several Delphi studies, our Delphi process will involve 2-to-3 rounds [23].

In the first round, the Delphi panel will be asked to characterize the root causes according to the dimensions of quality that those root causes affect, using the framework proposed by Wang et al. that encompasses four main data quality dimensions – intrinsic, contextual, representational and accessibility data quality [10]. We will also ask Delphi panelists to rate root causes according to their relevance regarding the impact they have in data quality and consequently their impact in improving data quality.

In the second and third rounds, the Delphi panelists will be asked to review the dimensions and ratings they assigned to root causes. Summary results will be provided to the panelists in between rounds and incorporated in the questionnaire.

Questionnaire, Rating and Consensus. The questionnaire will contain the list of root causes extracted from the systematic review. This will be emailed to each Delphi panelist along with a description of the study objectives, a form for collecting participant consent to complete the entire Delphi process, as well as the willingness to participate in the study and instructions to fill the questionnaire.

For each root cause the Delphi panelists will be asked: 1) to tick the quality dimensions affected by that root cause; 2) to rate the root cause on a scale from 1 to 9 according to their relevance of impact on data quality of DRG-based administrative hospital databases with regards to hospital funding, where 1 is highly irrelevant and 9 is highly relevant; and 3) to rate each root cause according to their relevance to impact data quality of DRG-based administrative hospital data-bases with regards to epidemiologic and health services research, using the same scale. The inclusion of additional root causes will be allowed in the first round.

Consensus will be defined as follows. Quality dimensions affected by the root causes will be determined by the three most common dimensions selected by the panelists. Relevant root causes will be defined as such if in the final round of questioning the root cause achieves a median score of 6–9.

3 Pilot Study – Assessing Feasibility

This section introduces the results from a pilot study implemented based on literature published during 2018–2019. The goal of this effort was to assess the burden and appropriateness of the proposed methodology to extract root causes of data quality problems of the DRG-based administrative hospital database and to map root causes into meaningful categories. The following sections summarize preliminary results of the systematic review and a possible preliminary qualitative analysis using the Ishikawa framework.

3.1 Search and Study Selection

Eighteen studies published between 2018 and 2019 were included. These reflect root causes perceived in data from Algeria, Australia, Canada, England, France, Portugal, and USA. While most studies included are analytic studies mainly focusing on aspects regarding the validity of data to identify a specific group of patients, five qualitative studies were identified rendering approximately half of the root causes extracted (Table 1).

Table 1. Study characteristics.

Ref	Study design	# of root causes	Reported period of data	Country of origin of data	Study aim
[18]	Retrospective observational	1	2013–2016	USA	To investigate if a physician-led documentation initiative within vascular surgery patients results in increased Case Mix Index and contribution margins
[2]	Qualitative	8	N.A.	Portugal	The study investigated the perceptions of medical coders regarding possible problems with health records that may affect the quality of the coded data
[4]	Qualitative	4	N.A.	Portugal	To explore the perceptions of medical doctors regarding problems in quality of coded data related to the transition from ICD-9-CM to ICD-10-CM/PCS
[24]	Retrospective observational	3	2010–2015	USA	To evaluate coding accuracy and its effect on hospital cost for patients undergoing EVAR (endovascular aneurysm repair)
[25]	Retrospective observational	3	2000–2014	France	To assess the accuracy of ICD-10 codes retrieved from administrative claims data in the identification and classification of adult congenital heart disease (ACHD)
[26]	Qualitative	3	N.A.	Canada	The study qualitatively assessed the strengths and barriers regarding clinical coding quality from the perspective of health information managers
[27]	Retrospective observational	3	2012	France	To propose a method to measure the validity of the hospital discharge database to detect complicated ectopic pregnancies with severe bleeding (CEPSB)
[28]	Article commentary	8	N.A.	Australia	To validate the relevancy of clinical documentation improvement (CDI) for the Australian healthcare environment
[29]	Retrospective observational	1	2014–2016	USA	To assess the coding accuracy of medical student documentation
[30]	Retrospective observational	3	2017	USA	To determine the sensitivity, specificity, PPV, and NPV of ICD-10 codes for bleeding events in hospitalized patients previously prescribed with anticoagulants
[31]	Experimental	5	2016	USA	To evaluate if incorporating knowledge of mortality index modeling into the documentation and coding resulted in reductions in the reported mortality index
[32]	Retrospective observational	5	2010	USA	To examine the validity of ICD codes used for identifying and classifying patients with coronary heart disease (CHD) and to examine major sources of error
[33]	Prospective observational	4	2008–2009	France	To develop and assess the validity of novel patient safety indicators in thyroid surgery
[3]	Methodological	3	N.A.	Algeria	To propose a rule-based approach for both describing dirty data occurring in discharge summaries and assisting health practitioners in repairing it
[34]	Experimental & qualitative	10	N.A.	USA	To analyze sources of coding variations to understand how clinical coding professionals arrive at postoperative ileus (POI) coding decisions and to verify existing knowledge that current clinical coding practices lack standardized applications of regulatory guidelines

(*continued*)

Table 1. (*continued*)

Ref	Study design	# of root causes	Reported period of data	Country of origin of data	Study aim
[35]	Retrospective observational	3	N.A.	USA	The authors share four examples of studies that might be included within the expanding role of clinical documentation improvement (CDI) programs
[11]	Retrospective observational	2	2011–2016	Portugal	To characterize the individual impact of comorbidities on APR-DRG classification and hospital funding in the context of respiratory and cardiovascular diseases
[36]	Prospective observational	3	2015	England	To investigate the impact of three alternative coding methods on the inaccuracy of diagnosis codes and hospital reimbursement

3.2 Root Causes Analysis

Seventy root causes of data quality issues were identified. Of these fewer were associated with "Management" and with "Mission" (Fig. 1). The latter reflected the purpose of this type of databases – hospital reimbursement/funding – one refers to the maximization of budget and the other to the lack of incentives to comprehensively report hospital episodes.

Within the Ishikawa branch related with people ("Man"), three subcategories have emerged. These reflect data quality problems rooted in experts' knowledge, preferences, education, or culture among health care professionals, clinical coders and the gaps between the two professional classes. The Ishikawa "Material" branch was subdivided in two categories: one regarding root causes associated with incomplete or missing information and another broad category associated with other information issues (e.g. inconsistent or imprecise information). The "Method" branch was subdivided in five subcategories of processes regarding operations performed in the making or treatment of the data. These covered root causes linked to retrospective collection of data, remote coding (by a clinical coder alone), poor documentation practices (e.g. use of acronyms), among others. Three subcategories were divided within the "Machine" branch. These covered root causes associated with deficiencies in guidelines or consensus (e.g. coding comorbidities), clinical coding classification systems, and tools related with source information (e.g. legibility in paper records). Lastly, the root causes related with "Management" were subdivided into those associated with limitations of resources (e.g. Time constraints and volume of work for coders) and a miscellaneous subcategory of three root causes: 1) Lack of routine and systematic internal clinical coding audits; 2) Unsatisfactory education and engagement of hospital administrator, information technologists and researchers; 3) Condition not formally acknowledged as to complicate the clinical care of the patient.

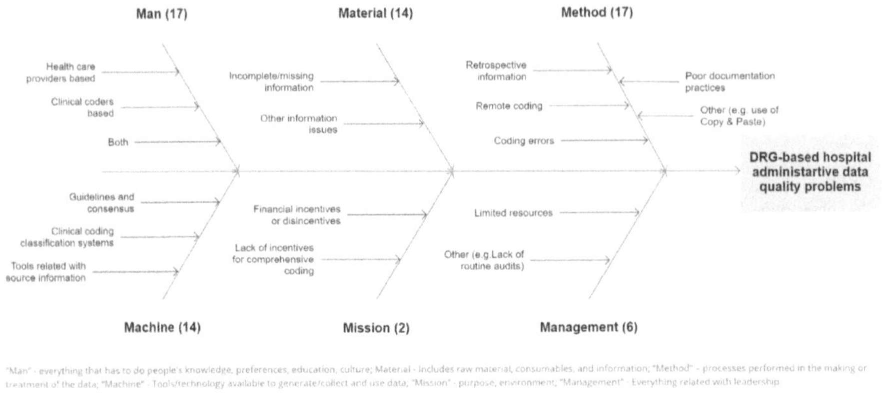

Fig. 1. Ishikawa diagram. (N) indicates the number of root causes within a specific category.

4 Discussion

The presented protocol introduced a method to extract and analyze root causes affecting the data quality of the DRG-based administrative database. As with all methods, there are strengths and limitations to the proposed one.

The pilot study contributed with a comprehensive list of root causes obtained through a systematic process. The few duplicates retrieved and some expected root causes missed (e.g. excessive use of unspecific ICD codes or use of an outdated grouper software for the current clinical codification system) by the pooled of studies included for analysis may indicate that we have not reach that saturation point yet. The Ishikawa can be useful for systematizing the process of root causes analysis – after their classification according to the Ishikawa framework it was more apparent which root causes were duplicates. Moreover, this framework can also be useful to depict the results. Categories may readily elucidate on which actions improve data quality and which root causes may be more easily tackled. The number of root causes extracted per study by each reviewer varied, suggesting that our protocol could be further improved with regards how root causes are to be identified. In fact, discussions raised among reviewers for the pilot study have been critical in providing clarifications in all the steps of the systematic review process and will improve consistency in the remaining studies.

Appraisal of studies quality has not been considered since we privilege comprehensiveness over study quality. The latter will have a small impact in our conclusions since we plan to determine relevance of root causes through a Delphi process.

5 Conclusions

The proposed protocol provides a comprehensive and systematic method to extract and analyze root causes of data quality issues of DRG-based administrative database. The outcome after its full implementation will provide a meaningful and intuitive representation of root causes generalizable to countries collecting this type of data, with a potentially significant impact on improving hospital funding and health research.

Acknowledgements. The authors would like to thank the support given by the Project "POCI-01-0145-FEDER-030766" (1st.IndiQare - Quality indicators in primary health care: validation and implementation of quality indicators as an assessment and comparison tool), funded by Fundação para a Ciência e a Tecnologia (FCT) and co-funded by Fundo de Desenvolvimento Regional (FEDER) through Operacional Competitividade e Internacionalização (COMPETE 2020); and the Project GEMA: Generation and Evaluation of Models for Data Quality (Ref.: SBPLY/17/180501/000293).

References

1. Busse, R., Geissler, A., Aaviksoo, A., Cots, F., Häkkinen, U., Kobel, C., Mateus, C., Or, Z., O'Reilly, J., Serdén, L.: Diagnosis related groups in Europe: moving towards transparency, efficiency, and quality in hospitals? BMJ **346**, f3197 (2013)
2. Alonso, V., Santos, J.V., Pinto, M., Ferreira, J., Lema, I., Lopes, F., Freitas, A.: Health records as the basis of clinical coding: is the quality adequate? A qualitative study of medical coders' perceptions. Health Inf. Manag. J. **49**(1), 28–37 (2019). 1833358319826351
3. Rabia, L., Amarouche, I.A., Bey, K.B.: Rule-based approach for detecting dirty data in discharge summaries. In: 2018 International Symposium on Programming and Systems (ISPS): 2018, pp. 1–6. IEEE (2018)
4. Alonso, V., Santos, J.V., Pinto, M., Ferreira, J., Lema, I., Lopes, F., Freitas, A.: Problems and barriers in the transition to ICD-10-CM/PCS: a qualitative study of medical coders' perceptions. In: World Conference on Information Systems and Technologies, 2019, pp. 72–82. Springer, Cham (2019)
5. Miguel, A., Marques, B., Freitas, A., Lopes, F., Azevedo, L., Pereira, A.C.: Detection of adverse drug reactions using hospital databases—a nationwide study in Portugal. Pharmacoepidemiol. Drug Saf. **22**(8), 907–913 (2013)
6. Santos, J.V., Lisboa, C., Lanna, C., Costa-Pereira, A., Freitas, A.: Hospitalisations with hidradenitis suppurativa: an increasing problem that deserves closer attention. Dermatology **232**(5), 613–618 (2016)
7. Santos, J.V., Pereira, J., Pinto, R., Castro, P.M., Azevedo, E., Freitas, A.: Atrial fibrillation as an ischemic stroke clinical and economic burden modifier: a 15-year nationwide study. Value Health **20**(8), 1083–1091 (2017)
8. Freitas, A., Gaspar, J., Rocha, N., Marreiros, G., Costa-Pereira, A.: Quality in hospital administrative databases. Appl. Math. Inf. Sci. **8**, 1–6 (2014)
9. Lobo, M.F., Azzone, V., Azevedo, L.F., Melica, B., Freitas, A., Bacelar-Nicolau, L., Rocha-Gonçalves, F.N., Nisa, C., Teixeira-Pinto, A., Pereira-Miguel, J.: A comparison of in-hospital acute myocardial infarction management between Portugal and the United States: 2000–2010. Int. J. Qual. Health Care **29**(5), 669–678 (2017)
10. Wang, R.Y., Strong, D.M.: Beyond accuracy: what data quality means to data consumers. J. Manag. Inf. Syst. **12**(4), 5–33 (1996)
11. Souza, J., Santos, J.V., Canedo, V.B., Betanzos, A., Alves, D., Freitas, A.: Importance of coding co-morbidities for APR-DRG assignment: focus on cardiovascular and respiratory diseases. Health Inf. Manag. J. (2019). 1833358319840575
12. Arendt, L.H., Ernst, A., Lindhard, M.S., Jønsson, A.A., Henriksen, T.B., Olsen, J., Thorup, J., Olsen, L.H., Ramlau-Hansen, C.H.: Accuracy of the hypospadias diagnoses and surgical treatment registrations in the Danish National Patient Register. Clin. Epidemiol. **9**, 483 (2017)
13. Assareh, H., Achat, H.M., Levesque, J.-F.: Accuracy of inter-hospital transfer information in Australian hospital administrative databases. Health Informat. J. **25**(3), 960–972 (2017). 1460458217730866

14. Gaspar, J., Lopes, F., Freitas, A.: Detection of Inconsistencies in Hospital Data Coding. In: HEALTHINF: 2012, 189–194 (2012)
15. Marques, B., Sousa-Pinto, B., Silva-Costa, T., Lopes, F., Freitas, A.: Detection of adverse events through hospital administrative data. In: World Conference on Information Systems and Technologies: 2017, pp. 825–834. Springer, Cham (2017)
16. Strong, D.M., Lee, Y.W., Wang, R.Y.: 10 potholes in the road to information quality. Computer **30**(8), 38–46 (1997)
17. Moher, D., Shamseer, L., Clarke, M., Ghersi, D., Liberati, A., Petticrew, M., Shekelle, P., Stewart, L.A.: Preferred reporting items for systematic review and meta-analysis protocols (PRISMA-P) 2015 statement. Syst Rev. **4**, 1 (2015)
18. Aiello, F.A., Judelson, D.R., Durgin, J.M., Doucet, D.R., Simons, J.P., Durocher, D.M., Flahive, J.M., Schanzer, A.: A physician-led initiative to improve clinical documentation results in improved health care documentation, case mix index, and increased contribution margin. J. Vasc. Surg. **68**(5), 1524–1532 (2018)
19. Bastani, H., Goh, J., Bayati, M.: Evidence of upcoding in pay-for-performance programs. Manag. Sci. **65**(3), 1042–1060 (2018)
20. Carter, G.M., Newhouse, J.P., Relles, D.A.: How much change in the case mix index is DRG creep? J. Health Economics **9**(4), 411–428 (1990)
21. Rayan QCRI. https://rayyan.qcri.org/welcome
22. McMillan, S.S., King, M., Tully, M.P.: How to use the nominal group and Delphi techniques. Int. J. Clin. Pharm. **38**(3), 655–662 (2016)
23. Boulkedid, R., Abdoul, H., Loustau, M., Sibony, O., Alberti, C.: Using and reporting the Delphi method for selecting healthcare quality indicators: a systematic review. PLoS ONE **6**(6), e20476 (2011)
24. Ayub, S., Scali, S.T., Richter, J., Huber, T.S., Beck, A.W., Fatima, J., Berceli, S.A., Upchurch, G.R., Arnaoutakis, D., Back, M.R.: Financial implications of coding inaccuracies in patients undergoing elective endovascular abdominal aortic aneurysm repair. J. Vasc. Surg. **69**(1), 210–218 (2019)
25. Cohen, S., Jannot, A.-S., Iserin, L., Bonnet, D., Burgun, A., Escudié, J.-B.: Accuracy of claim data in the identification and classification of adults with congenital heart diseases in electronic medical records. Arch. Cardiovasc. Dis. **112**(1), 31–43 (2019)
26. Doktorchik, C., Lu, M., Quan, H., Ringham, C., Eastwood, C.: A qualitative evaluation of clinically coded data quality from health information manager perspectives. Health Inf. Manag. J. (2019), 1833358319855031
27. Fermaut, M., Fauconnier, A., Brossard, A., Razafimamonjy, J., Fritel, X., Serfaty, A.: Detection of complicated ectopic pregnancies in the hospital discharge database: a validation study. PLoS ONE **14**(6), e0217674 (2019)
28. Hay, P., Wilton, K., Barker, J., Mortley, J., Cumerlato, M.: The importance of clinical documentation improvement for Australian hospitals. Health Inf. Manag. J. **49**(1), 69–73 (2019). 1833358319854185
29. Howard, R., Reddy, R.M.: Coding discrepancies between medical student and physician documentation. J. Surg. Educ. **75**(5), 1230–1235 (2018)
30. Joos, C., Lawrence, K., Jones, A.E., Johnson, S.A., Witt, D.M.: Accuracy of ICD-10 codes for identifying hospitalizations for acute anticoagulation therapy-related bleeding events. Thromb. Res. **181**, 71–76 (2019)
31. Kessler, B.A., Catalino, M.P., Jordan, J.D.: Reducing the reported mortality index within a neurocritical care unit through documentation and coding accuracy. World Neurosurg. **133**, e819–e827 (2019)
32. Khan, A., Ramsey, K., Ballard, C., Armstrong, E., Burchill, L.J., Menashe, V., Pantely, G., Broberg, C.S.: Limited accuracy of administrative data for the identification and classification of adult congenital heart disease. J. Am. Heart Assoc. **7**(2), e007378 (2018)

33. Mercier, F., Laplace, N., Mitmaker, E.J., Colin, C., Kraimps, J.-L., Sebag, F., Bourdy, S., Duclos, A., Lifante, J.-C., CATHY Study Group: Unexpected discrepancies in hospital administrative databases can impact the accuracy of monitoring thyroid surgery outcomes in France. PLoS ONE **13**(12), e0208416 (2018)
34. Resslar, M.A., Ivanitskaya, L.V., Perez III, M.A., Zikos, D.: Sources of variability in hospital administrative data: clinical coding of postoperative ileus. Health Inf. Manag. J. **48**(2), 101–108 (2019)
35. Rodenberg, H., Shay, L., Sheffield, K., Dange, Y.: The Expanding Role of Clinical Documentation Improvement Programs in Research and Analytics. Perspect. Health Inf. Manag. **16**(Winter) (2019)
36. Tsopra, R., Peckham, D., Beirne, P., Rodger, K., Callister, M., White, H., Jais, J.-P., Ghosh, D., Whitaker, P., Clifton, I.J.: The impact of three discharge coding methods on the accuracy of diagnostic coding and hospital reimbursement for inpatient medical care. Int. J. Med. Informat. **115**, 35–42 (2018)

Improving Project Management Practices in a Software Development Team

Sara Pires[1,2(✉)], Anabela Tereso[1], and Gabriela Fernandes[1,3]

[1] ALGORITMI, University of Minho, Campus de Azurém, 4804-533 Guimarães, Portugal
saraspv@hotmail.com, anabelat@dps.uminho.pt
[2] InfoPortugal S.A., Rua Conselheiro Costa Braga 502, 4450-102 Matosinhos, Portugal
[3] CEMMPRE, University of Coimbra, Rua Luis Reis Santos, 3030-788 Coimbra, Portugal
gabriela.fernandes@dem.uc.pt

Abstract. Software development projects continue to deliver results that fall short on organization's expectations. The present research was carried out at InfoPortugal, a technological company specialized in Geographic Information Systems and Tourism and Leisure Solutions, where project management practices were underuse. Therefore, the focus of this paper is to describe the proposals and their implementation to enable the improvement of project management practices in a software development team. Guidelines are provided by a matrix with key areas to be improved and related proposed improvement initiatives. The definition and application of a hybrid project management methodology was the most important improvement initiative to address some of the key problems identified in case under study. Traditional plan-oriented methodologies do not have the flexibility to dynamically adjust to the software development process. While, agile methodologies combine iterative and incremental approaches to adapt to high levels of change, with early and continuous delivery.

Keywords: Software development · Project management · Agile · Planning

1 Introduction

Today, the increased need to reduce costs, the increasing demand for quality, and the need to reduce delivery time can be considered as major challenges of a software development project, where everything goes almost at the speed of light. Many methods, techniques and tools have been developed; however, Project Management (PM) remains a highly problematic endeavor. Projects still do not meet stakeholder expectations and/or do not achieve the expected benefits [1].

PM in software development has shown different results from those known in other areas, having special characteristics such as complexity, compliance, flexibility and invisibility [2].

Planning is essential to the success of a project [3]. However, planning in software development is particularly difficult, and plans are often wrong. Plans should be systematic, flexible enough to handle unique activities [4]. In 2001 came the Agile Manifesto

Á. Rocha et al. (Eds.): WorldCIST 2020, AISC 1159, pp. 104–113, 2020.
https://doi.org/10.1007/978-3-030-45688-7_11

that adapts to high levels of change and adds more value to the project [3]. Overall results show that agile projects have almost four times more success than traditional managed projects [5].

This paper aims to address the following main research question: How to improve PM practices in a software development team? Having a secondary question: What is the impact of lack of project planning on the software development team? In order to answer these research questions, the following specific objectives were defined: (1) Identification of the PM problems/difficulties experienced in the organization; (2) Identification of proposals for improving PM practices that reduce the difficulties experienced in the organization; (3) Implementation of some of the identified proposals to improve PM practices.

This paper is organized as follows. Section 2 presents the literature review on PM in software development projects and traditional and agile approaches. Section 3 presents the research methodology applied. Section 4 describes the case study context and presents and discusses the main research results. Finally, the conclusions and highlights for further research are presented.

2 Literature Review

2.1 Project Management in Software Development

Project, program and portfolio management has been increasingly valued by organizations. Organizations have been increasingly investing in improving PM practices to maintain their competitiveness within a global market in constant change. Software development projects continue to fail despite decades of research. In 2018, 19% of software development projects failed [6]. The causes identified for this failure are the inadequate planning, estimation, metrics and control [7]. To address these problems, there are several studies that present critical success factors, such as decision latency, minimum scope, project sponsors, agile process and talented team [6]. Objectives clearly defined in conjunction with the project mission are identified as the most important key factor to success [8]. A project without a clear structured process is impossible to manage. Therefore, it is important the standardization of a PM methodology (processes, tools and techniques) to reduce risk and uncertainty, and to increase the governance of the software development process [9, 10].

2.2 Traditional (Waterfall) vs Agile Approach

With the growing tendency of using a more agile management in different projects, it is clear that there are two distinct PM approaches - the waterfall and the agile approach [11].

The waterfall model is a sequential software development process, divided into sequential phases that must be completed one after the other. The movement into the next phase is only made when all the previous work phase is completed [12]. This approach can be applied to any project environment, but in situations where projects involve requirements volatility, high degree of uncertainty of change, ambiguity when dealing with complexity in project environment, this waterfall approach presents difficulties in responding quickly. In this scenario the adaptive (agile) approach can and should be considered, since agile development has proved to be adequate to dominate the presented situations and to capitalize the changes as opportunities [11].

VersionOne and CollabNet [13] report on agile methodologies most commonly used in projects in 2017. Scrum is found to be the most popular, 58% of teams claiming to be agile are using Scrum alone or combined with other methods. Extreme Programming is the second most used (10%), and then Kanban (5%).

Many teams are not able to switch to agile methodologies overnight, as agile techniques look very different from those they are used to. For this reason, it makes sense to plan a gradual transition [3] using a hybrid approach. That is, you need to value the specifics of each approach and, if possible, work with both at the same time, as each adds value in its own way.

The use of project management practices and the accomplishment of planning are important factors for the success of a project [14–16]. The plan is based on the assumption that project goals are clearly defined in advance. A project that uses good planning ends 18% to 36% before a poorly planned project [17]. Estimation and planning are critical to the success of any software development project of any size. Good planning reduces risks and uncertainties, increases understanding, improves efficiency, supports better decision making, builds trust, and conveys information [15, 17].

Regarding project management practices, there are four main reasons for applying best practices: improving efficiency; improving effectiveness; standardization; and consistency [16]. Best practices are those actions or activities performed by the company or individuals that lead to a sustained competitive advantage in PM, while adding value to the company, the customer, and stakeholders. Tereso et al. [14] identify the 20 most commonly used tools and techniques. According to this study the kick-off meeting, activity list, progress meetings, Gantt chart and baseline plan are the top five positions in the ranking.

Scrum, Extreme programming and Kanban are methodologies that present a set of agile practices that are fundamental to the adoption of these methodologies (e.g., product backlog, sprint, pair programming, testing and planning game). Their practices can be combined or implemented in isolation, taking into account the needs of each organization.

3 Research Methodology

This research is supported by theories already developed, following therefore a deductive approach. The study followed an action-research strategy that is an interactive process that involves the researcher and practitioners working together in a particular cycle of activities [18], focusing on change. In order to achieve the research objectives, in the first phase, a literature review was carried out regarding PM, focusing on software development projects. Subsequently, systematic data were collected through semi-structured interviews, document analysis, informal conversations and observation of PM practices, with particularly attention to planning. The analysis of the collected data was performed, in order to understand the main problems, present in the development team. Following this detailed analysis, improvements in PM practices and their implementation are suggested. Finally, the contribution of the implementation of new PM practices in the organization is analyzed.

4 Case Study Analysis

4.1 Context

One of InfoPortugal's areas of expertise is software development. The team works closely to the Design knowledge area to create Web Geographic Information Systems (WebGIS) solutions, websites and apps for tourism and for planning and territory management. The organizational structure is based on "One Man Show", that is, all decisions go through the company's executive director assuming a role similar to the project manager. The software development and design team leaders have the role of informing and giving feedback on the status of projects to the executive director. There are no systematic and well-defined processes, so management processes depend largely on the leaders' project decisions. The PM methodology is not clearly defined, making the execution of projects more complex. Project planning takes place weekly at a meeting between the executive director, the development team leader and the design team leader to monitor the status of projects. The task completion dates are set, and the planning is clearly top-down. The company uses some PM tools that help with internal organization and communication such as Redmine, Slack, Gitlab and Openproj.

4.2 Results

Results were obtained through three iterations throughout the investigation. In each iteration four steps were performed: diagnosis, planning, action and evaluation. In the first iteration the primary objective was to identify the problems felt in the software development team, in the second iteration the primary objective was to analyze the number of projects that were on time and on budget, in order to benchmark the best PM practices used, to finalize, the last iteration had the focus on the definition of a hybrid PM methodology in the development team. Data collection was achieved through techniques such as: observing PM practices, both on a day-to-day team basis and weekly meetings with management, and through semi-structured interviews with the software development team. Informal conversations held twice a week with team members in order to gather feedback on the tasks and projects they were carrying out and documentation on the active projects studied, including consideration of proposals and budgets, as well as the number of hours associated to every task.

Six semi-structured interviews were conducted with the four software programmers and the leaders of the development and design team. Interviews were analyzed with NVivo software and the most frequent answers to be addressed in this study were consolidated and synthesized. Regarding the difficulties and problems felt in PM, Table 1 and Table 2 summarizes the interviewee answers:

The team development and design leaders identified 17 PM practices applied in the organization while team members identified only 5. Although there are PM practices in place, these are not identified by the employees because they are not involved in the process. Overall, respondents suggested 65% use of agile practices in contrast to 35% of traditional practices. During the analysis period, 9 projects were active, 7 external and 2 internal. In the active projects it was found that 77% of the projects were developed by only one programmer and 23% by two programmers. It was found that the cause of

Table 1. Difficulties/problems felt in PM

67%	No clear PM methodology
50%	Lack of communication
50%	Shared leadership
17%	Manage projects by recording work hours
17%	Timely report to customer

Table 2. How project planning is done

83%	No initial planning, tasks defined and created in Redmine
67%	Various interpretations of functional analysis. Requirements not detailed enough for team understanding
33%	The focus is lost with the exchange of projects
33%	Schedule should be done by the team leader and the team

this situation is the small number of existing programmers for the high number of active projects. It was only possible to collect information regarding the estimate made for external projects. For these, a comparison was made between estimated and actual cost and time. Of the 7 projects analyzed, 4 have a longer development time than planned. The main causes are lack of resources, project start-up later than planned, difficulty in interpreting functional analysis, layout complexity, poorly budgeted projects, and clients request changes that are not planned. In addition to these causes, another factor focused exclusively on the development phase is related to unrealistic estimates. During the observation period it was found that 63% of the estimates set at the weekly meeting were not met. These are some factors that not only affect the duration, but also the cost of the projects. In relation to cost, the two projects whose difference was relatively larger than the budgeted value were projects B1 and C1, the first being over budget and the second under budget. These two projects represent 34% and 32% of the total costs, so the reason for this disparity was investigated. In the case of Project B1, which was a public tender, it was concluded that it was poorly budgeted, given the high workload associated with the project. In addition, during the development of the project, several customer change requests were made. Project C1 was not a public tender, the price was not limited, so the budget took into account the degree of uncertainty associated with the requested requirements. Taking a final balance, it was concluded that 43% of the projects analyzed were on time and within budget. Through the results were identified Key Areas (KA) that should be improved, which are summarized in the following Table 3.

Table 3. Identified key areas that should be improved

KA1 – Methodology	KA2 – Communication	KA3 – Process
KA4 – Leadership	KA5 – Planning	KA6 – Focus
KA7 – Control	KA8 – Requirements	KA9 – Commitment
KA10 – Transparency	KA11 – Motivation	KA12 – Stakeholders

4.3 Discussion

This section discusses the proposed improvement initiatives for the key areas identified. These proposed improvements were based on the literature review and the organization context. Then the links between the proposed improvements and the identified key areas for improvement are shown in Table 4 and the link between the proposed improvements and effective implementations are presented in Table 5.

P1 – Define a PM methodology: As there is no official methodology in use, the risk of moving to an agile methodology is reduced. In the interviews, the willingness of the interviewees to apply agile methodologies was emphasized. Thus, it was proposed to use Scrum, in a first phase with the team training in the methodology and then with the rigorous application of methodological processes. This proposal should be applied to internal projects, taking into account the type of projects and the lack of objectives. As the customers are organization collaborators, it is easier to communicate directly with them and involve them in the process. However, many teams are not able to switch to agile methodologies overnight [3], as InfoPortugal's external customers require a more traditional approach at an early stage, with a timely proposal and project costs, a functional analysis and design that must be previously approved. Thus, in external projects, it is proposed to implement a hybrid methodology, combining the waterfall model and Scrum methodology. Finally, we proposed to use Kanban for projects that have already been delivered and require maintenance.

P2 – Restructure of the software development team: To implement an agile methodology, the hierarchical structure of the software development team will need to be revised and the choice focused on the matrix type, based on the "Tribes" presented in Spotify Squads model [19].

P3 – Redefine Leaders' Responsibilities: The development team leader becomes Team Facilitator/Scrum Master and is responsible for assisting the team in all projects, sometimes referred to as the "servant leader". And all projects should have a Product Owner, which varies by project type.

P4 – Set clear project objectives: Throughout the research it was found that the lack of well-defined objectives was constant. Thus, the researchers proposed the definition of objectives at three distinct levels: development team, project and collaborator.

P5 – Define good metrics: Define metrics that contribute to the control of the various projects and help the team detect possible deviations from the planned or identify the causes of software project failure. For example, Schedule Performance Index (SPI) and Cost Performance Index (CPI), the number of projects on time and within budget and the number of hours spent fixing bugs after projects are completed.

P6 – Define clearly the requirements: We proposed converting customer-driven documents to something simpler and clearer, targeted at programmers, avoiding loss of information and multiple interpretations. In agile methodologies, the requirements are presented in user stories, which are easily interpreted by the team.

P7 – Engage the team in PM processes: Implementing an agile approach requires the project team to adopt an agile mindset. Agile success teams embrace the growth mindset, where people believe they can learn new skills. A policy of transparency in the organization is essential, sharing the objectives, plans and information needed for project development. The team should be involved in the PM processes.

P8 – Explore Software Tools: The organization presents a set of tools that can be explored and used continuously and systematically. As OpenProj is obsolete it is proposed to change this software to ProjectLibre.

P9 – Develop the Responsibility Matrix: With the high number of projects, it sometimes became complex to realize which human resources were allocated to each project. Thus, it was proposed to create a matrix of responsibilities, adapted to this context.

Table 4 illustrates the link between the identified proposals and key areas for improvement.

Table 4. Link between the identified proposals and key areas for improvement

	P1	P2	P3	P4	P5	P6	P7	P8	P9
KA1: Methodology	x	x	x					x	
KA2: Communication	x			x	x	x		x	x
KA3: Process	x			x	x	x	x		
KA4: Leadership		x	x						
KA5: Planning	x					x	x	x	x
KA6: Focus				x	x		x		
KA7: Control	x				x			x	x
KA8: Requirements	x					x		x	
KA9: Commitment	x			x			x		
KA10: Transparency	x			x	x		x	x	x
KA11: Motivation				x					
KA12: Stakeholders	x				x	x			

Regarding the effective implementations in the organization, these were carried out taking into account the current projects and the prior approval of the leaders and team consent.

I1 - Change of the software used in the development of the project proposals: ProjectLibre, which is a branch of OpenProj, was used, allowing the quick adaptation of this tool to the team. The team mentioned that they use the software and it was found that this choice was characterized by something "very useful and practical", "the adaptation was quick" and "we use it a lot and it makes a lot of sense in the development of the project proposals". It is also "adopted in projects in other organization's areas".

I2 - Introduction of the Responsibility Matrix: The creation of this matrix enabled the visualization of the resources that are allocated to each project and it is also possible to visualize the state of the project. After its inception it was found to be widely used at the time when the team had several external projects, however with the decrease in the number of projects, the responsibility matrix was not used.

I3 - Streamlining PM processes: In order to plan a gradual transition, agile practices have been added to improve learning and alignment between the team and remain project stakeholders. In order to encourage teamwork, increased motivation and increase project delivery speed, project development with just one programmer is currently avoided. Thus, we adopted the strategy of working in pairs whenever possible and use pair programming. Estimates are set by the team together with the team leader. Sprinting began, with well-defined tasks shared within the team. For this purpose, the Redmine is used.

I4 - Development of a Kanban Board: One of the projects had a high number of tasks. Therefore, the Kanban board was implemented. The main objective was to simplify the process and help the team improve their organization and increase visibility, delivering value faster. The opinion of the team members was similar. They said that it allowed for "more organization", "it is great in segregating work", "it is easier to change the state of tasks", which allows a "project overview" and "this is the first thing I do before I start the project".

I5 - Bottom-up planning: A pilot project was developed, with the main objective of doing bottom-up planning. The tasks for the first sprint were selected based on the priority tasks defined by the Product Owner. The control was done through the burndown chart. The metric used to track project progress was the SPI. In the first sprint the team worked with 75% of the planned rate. The main factor was an interruption at the end of the sprint. The programmers were told that they would have to change projects, so they did not have time to finish all the tasks they had planned. In the second sprint it was found that the team worked with 145% of the planned rate. This was because there were no interruptions or change of projects during the sprint and tasks that were pending in the previous sprint were quickly completed after doubts were cleared with the Product Owner. The conclusions from project observation and follow-up are the following: (1) setting objectives at the initial meeting made it possible for stakeholders to clearly understand what was intended, as well as aligning the entire team on the desired goal(s); (2) programmers' definition of planning (time estimates and selection of tasks present in each sprint) allowed for increased commitment and motivation to meet set deadlines; (3) more real time estimates; (4) the creation of a planning enables its control; (5) focus and motivation are lost with the exchange of projects.

Table 5 relates the proposals identified with what was effectively implemented during this research project.

Table 5. Link between the proposed improvements and effective implementations

Proposed solutions	Implementations
P1: Define a PM methodology	I3: Streamlining PM processes I4: Development of a Kanban board
P2: Restructure of the software development team	(Not implemented)
P3: Redefine Leaders' Responsibilities	I5: Bottom-up planning
P4: Set clear project objectives	I5: Bottom-up planning
P5: Define good metrics	I5: Bottom-up planning
P6: Define clearly the requirements	I5: Bottom-up planning
P7: Engage the team in PM processes	I3: Streamlining PM processes I4: Development of a Kanban board I5: Bottom-up planning
P8: Explore Software Tools	I1: Change of the software used in the development of the project proposals I3: Streamlining PM processes
P9: Develop the Responsibility Matrix	I2: Introduction of the Responsibility Matrix

5 Conclusions and Future Work

During this research, the best practices in software development PM were studied from a theoretical perspective as well as in practice through the action-research research strategy. Keeping in mind the research question, there is a strong possibility of improving software development project practices through streamlining processes, making it easier to adapt to the changing market of today.

The main contribution of this research was at the practical level. This paper gives software PM professionals guidelines on how to improve PM practices in organizations with an indefiniteness of PM processes. Identifying the problems felt in the software development team made it possible to select a set of key areas that are common to most small and medium-sized companies in the industry. The choice of proposals focused on solving the identified key areas and through the implementations, it was possible to validate that the proposed solutions work. Lack of planning makes it impossible to monitor and control the status of projects, there is a lack of commitment to meet deadlines, there is a lack of organization, there are various interpretations of requirements and the team is not aligned with the intended project objectives. However, during the implementations of the solutions, it was observed that the involvement of employees in the planning, especially in the definition of time estimates, increased motivation, commitment to meet the estimated deadlines and the achievement of more realistic deadlines. The fact that this study is directed to PM in the software development area limits the study results to this area only.

In the future, the organization should define priority actions to apply to this development team that can also be tested in other sections within the organization. Through the results presented in this research, it is clear the need to implement agile methodologies and have a transparency policy, involving the team in the whole process. Thus, it is suggested the rigorous application of Scrum in a pilot project and analyze the results. As well as the application of the proposed hybrid methodology in a project with low degree of risk and uncertainty.

References

1. Fernandes, G., Ward, S., Araújo, M.: Identifying useful project management practices: a mixed methodology approach. Int. J. Inf. Syst. Proj. Manag. **1**, 5–21 (2013)
2. Brooks, F.: No Silver Bullet: Essence and Accidents of Software Engineering. Computer (Long. Beach. Calif). 20, 10–19 (1987)
3. PMI: Agile Practice Guide. Project Management Institute, Inc., Newtown Square, Pennsylvania USA (2017)
4. Kerzner, H.: Project Management: A Systems Approach to Planning, Scheduling and Controlling. Wiley, New York (2009)
5. The Standish Group: Chaos manifesto 2015. The Standish Group International, Incorporated (2015). https://www.infoq.com/articles/standish-chaos-2015
6. The Standish Group: CHAOS Report: Decision Latency Theory: It Is All About the Interval. The Standish Group International (2018)
7. Jones, C.: Software project management practices: Failure versus success. CrossTalk. 5–9 (2004)
8. Wateridge, J.: IT projects: a basis for success. Int. J. Proj. Manag. **13**, 169–172 (1995). https://doi.org/10.1016/0263-7863(95)00020-Q
9. Fernandes, G., Ward, S., Araújo, M.: Improving and embedding project management practice in organisations—a qualitative study. Int. J. Proj. Manag. **33**, 1052–1067 (2015)
10. Fernandes, G., Ward, S., Araújo, M., Loureiro, I., Braga, A.: Perceptions of different stakeholders on improving and embedding project manage-ment practice in organisations. Procedia Technol. **16**, 957–966 (2014)
11. Fernandes, G., Moreira, S., Pinto, E.., Araújo, M., Machado, R..: A Framework for Managing Collaborative University-Industry R&D Projects within a Program – A Qualitative Study. In: EURAM 2019 Conference 26–28, pp. 1–40 (2019)
12. Rahmanian, M.: A comparative study on hybrid IT project managment using traditional project management and agile approach. Int. J. Comput. Inf. Technol. **03**, 1096–1099 (2014)
13. VersionOne & CollabNet: The 12th annual State of Agile Report. https://explore.versionone.com/state-of-agile/versionone-12th-annual-state-of-agile-report
14. Tereso, A., Ferreira, M., Ribeiro, P., Fernandes, G., Loureiro, I.: Project management practices in private organizations. Proj. Manag. J. **50**, 1–17 (2018)
15. Cohn, M.: Agile estimating and planning. Prentice Hall, Massachusetts (2005)
16. Kerzner, H.: Project management best practices: Achieving global excellence. Wiley, Hoboken (2015)
17. Wysocki, R.K.: Effective Project Management. http://index-of.co.uk/Project%20Management/Effective%20Project%20Management%20Traditional,%20Agile,%20Extreme%20by%20Robert%20K.%20Wysocki%207th%20Edition.pdf
18. College, T., Coughlan, P., Coghlan, D.: Action research for operations management. Int. J. Oper. Prod. Manag. **22**, 220–240 (2001)
19. Kniberg, H., Ivarsson, A.: Spotify - Scaling Agile. https://creativeheldstab.com/wp-content/uploads/2014/09/scaling-agile-spotify-11.pdf

Integrated Model of Knowledge Management and Innovation

Víctor Hugo Medina García(✉), Jennifer Paola Ochoa Pachón(✉), and Kelly Alexandra Ramírez Guzman

Faculty of Engineering, District University "Francisco José de Caldas", Bogotá, Colombia
vmedina@udistrital.edu.co, yeola-18@hotmail.com, kramirez@correos.udistrital.edu.co

Abstract. Knowledge management is a discipline that facilitates the creation, storage, transfer and application of knowledge in organizations.

The purpose of this research it was to study model linkages between knowledge management conditions and innovation capability and their effect on business performance. All this, starting from the combination of both internal conditions (strategic purpose, flexible structure, information and communication technologies, and internal environment) and external conditions (competitive environment), which guide the development and renewal process of new capacities that imply both appropriability and knowledge acquisition effects as well as results generation.

Keywords: Knowledge management · Innovation and communication technologies · Technological innovation · Open innovation

1 Introduction

The last few years have witnessed fundamental changes that have transformed organizational reality, highlighting organizational knowledge as one of the main sources of business achievement. These changes contribute to increasing interest in knowledge management, which has been reflected in the rising number of studies that consider knowledge as an important factor for business success [1]. There have been major changes in business environment among others: the globalization of markets, a great technological development and a rapid exchange of information between people and/or companies. Under this new context, organizations must become each time more competitive, and one of the main sources of current competitive advantage is knowledge. As a result, many companies are now involved in applying what is known as "knowledge management" [2].

One of the key stages of knowledge management is the choice of acquiring from outside or internally generating knowledge assets. The holding of one or other assets, as well as achieving a balance between them, can influence the results obtained. The ability to acquire knowledge and generate innovation can be considered a way of achieving sustainable superior results. Therefore, knowledge management is revealed as a

Á. Rocha et al. (Eds.): WorldCIST 2020, AISC 1159, pp. 114–123, 2020.
https://doi.org/10.1007/978-3-030-45688-7_12

dynamic process, generating business results. Such results derive from acquiring, generating, transferring and obtaining new knowledge within the organization, which must be materialized in a company's capacity to generate adaptation to the environment under uncertainty conditions that allow it to improve its competitive position and, therefore, obtain greater results [1].

To study the relationships between technological innovation and organizational knowledge management (OKM) is considered a new area of research in the scope of Business Management. This has led in recent years to a growing number of works analyzing these relationships, both from a theoretical and empirical point of view. Despite a significant literature increase, there is still a lack of literature in this area that clearly shows the impact of OKM practices on the company's competitive advantage and results. It is also made clear that its effect on technological innovation is not very clear, mainly due to the fact that, beyond technological innovation, those practices affect a wide range of business aspects related to improving quality, efficiency or customer service. In addition, since technological innovation can be understood differently, innovative results, innovative process, technological knowledge and is treated from different theoretical viewpoints, the relationships with knowledge management practices varies in complexity and diversity [3].

2 Background of Innovation and Knowledge Management

To understand this subject, the main related concepts are described below.

2.1 Knowledge Management and Technologies

The role of information technologies as enablers of productive and business improvements, especially technologies related to knowledge management, cannot be overlooked. However, the simple adoption of information technologies does not necessarily accomplish the proposed organizational purposes. Some studies have tried to identify environmental, organizational and individual factors to determine the keys to adoption and implementation of information technologies in big companies. Enterprises vary considerably in their ability to assimilate, integrate and use the full value of technology; therefore, it is important to differentiate between technology adoption and technology implementation or assimilation, that is, the extent and scope of its use within the organization [4].

Turning to a more particular aspect such as information technologies, these in business practice have advanced according to the pace marked by technological progress. According to Nonaka Takeuchi's knowledge generation or creation model [5], ICT properties and functions are applied and integrated, thus obtaining the matrix of knowledge and ICT processes (Table 1) which collects and classifies the different technologies according to the knowledge processes they support [6].

In addition, the support that IT can provide lies in technological and cultural instances to help the dynamics of knowledge management process. These can be [7]:

- Knowledge generation: These are the tools and techniques that focus on the exploration and data analysis to discover interesting patterns within them. Some tools/techniques

Table 1. Knowledge and ICT processes matrix [1].

Process	ICT	Impact of ICTs
Socialization	Yellow pages and knowledge maps, Intranet, Corporate portal, Virtual communities, Video conference, Groupware	It allows tacit knowledge to be obtained from other tacit knowledge through observation, imitation and practice
Exteriorization	Datawarehouse, Simulation software, Multimedia systems, Knowledge portals, Workflow, Intranets, Email, Groupware	It allows the formal description or representation of tacit knowledge and its making available to the entire organization
Combination	Internet, Groupware, Data warehouse, Intranet, Corporate Portal, Forums and Email, Document Management	It allows the access, use, joint and storage of different explicit knowledge. This facilitates the generation of new knowledge
Interiorization	Forums, Virtual Reality, simulation, Data mining, Artificial intelligence. Yellow Pages, Communities of practice	It allows access to explicit knowledge and its assimilation and understanding through reflection, simulation and implementation

are Data Mining, Knowledge Discovery in Databases, Intelligent Decision Support Systems, Expert Systems, Intelligent Agents, among many others.

- Knowledge generation facilitator: These are tools and techniques that facilitate the free flow of knowledge within an organization. Some tools/techniques are Lotus Notes, NetMeeting, Email, Intranets/Extranets & Portals, IdeaFisher, IdeaProcessor, Discussion Groups, Message Service, among others.
- Knowledge Measurements: These are tools and techniques that facilitate the 'visualization' of knowledge. They can be categorized into three categories: knowledge activities, knowledge-based outcomes, and knowledge investments [7].

2.2 Knowledge Management for Entrepreneurial Innovation

Companies attach importance to finding out what they know, as well as to making the best use of the knowledge they possess, in order to define how to acquire it, represent it, retain it and manage it. An organization must know what it must know, and its employees must also know where they can locate the needed knowledge. In this sense, management directs strategic efforts to generate, codify, store and transfer knowledge.

The interpretation of innovation as a process by which new things emerge in science, technology and art. Consequently, the category of "organizational innovation" arises. It can also basically refer to technological innovations at the company level: those of products and processes. New and improved products (goods and services) and processes, which require that "at a minimum, the product or process must be new to the company (or significantly improved), that is, it need not be new to the world". It also identifies the existence of organizational innovation linked to management aspects of management sciences and organizational structures [8].

In this sense, knowledge management for organizational innovation is understood as the set of strategic efforts to generate, codify, store and transfer knowledge so that new things arise in management sciences, and that give added value or competitive advantages to the organization.

2.3 The Development of Innovation Models

In the course of time, many models have been developed with the aim of explaining what the innovation process consists of, and several authors have broadly classified them into stage models, technology push models, integrators, decision models, linear models, first- to fourth-generation innovation models, taking as the most current models of innovation derived from science, the market, links, and technological and social networks [9].

The evolution of innovation models has generated a coexistence of models and in some cases mergers have occurred (see Fig. 1).

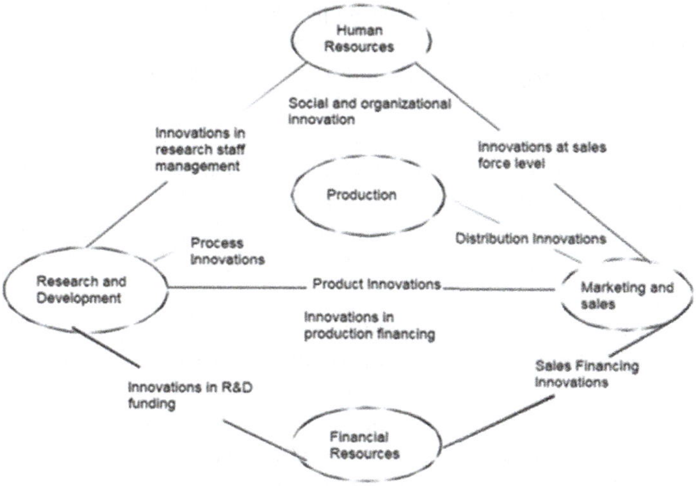

Fig. 1. Innovation processes. Source: [8].

The most recent models in the topic are the network models, "Systems Integration and Networking", which emphasize vertical relationships and collaborations with competitors. Innovation is seen as a networked process of learning and accumulation of know-how.

The progress of the models has led to the use of electronic tools that increase the speed and efficiency of both internal and external processes, and to consider innovation as a knowledge-based process.

In addition, the non-linearity of the innovation process entails an emphasis on cooperation and competitiveness, cooperation allows organizations to access a greater amount of knowledge, resources, and therefore, ideas. The current trend in innovation is towards cooperative innovation given the high need for resources and investment that the process entails [10].

Synergy and percolating vulnerable behavior become key factors for the generation and success of innovation [11].

The most current innovation models recognize the importance of knowledge as the differentiating factor of organizations, specifically, in the management and use of scientific and technical knowledge, thus organizational learning is a variable that promotes innovation.

The success and competitiveness of organizations are based, under the latest models of innovation, on organizational knowledge and intangible assets of the organization. The ability to create, distribute and exploit knowledge and information is rapidly becoming the major source of competitive advantage, causing the emergence of collaborations and networks intensively [12].

3 Integration of Innovation and Knowledge Management

Knowledge management is the creation of value from the intangible assets of organizations, from a technical point of view is supported by the use of information technology, databases, and corporate support systems, and from a social approach knowledge management promotes psychological conditions and behaviors that allow improving organizational processes supported by knowledge.

Improved problem-solving capabilities, scope and maintenance of competitive advantages are the result of the development and application of all types of knowledge in organizations, through knowledge management systems [13].

The conditions that allow the development of the company's knowledge-intensive capabilities are the strategic purpose, the flexible structure, the information technologies and the internal environment, in an internal way, allowing the flow of knowledge, while the external factor is the competitive environment that supports the information requirements through the capture, dissemination, assimilation and application of knowledge.

Integrating knowledge management and innovation (see Fig. 2) raises innovation as a process that occurs as a consequence and driven by the conditions of the organization's knowledge management system and leads to favorable organizational results.

4 Knowledge Management as a Key Element in Open Innovation

A traditional innovation model implies a closed and linear perspective of knowledge generation, development and commercialization, through its own structures. In this way, such knowledge is created and transferred internally and prevented from being transferred to competitors. This model of closed innovation defends, first, that the only strategic knowledge is that developed from internal sources.

Open innovation assumes that companies can and should maintain close relationships with third parties, both in the process of accumulation of knowledge and in the commercialization of knowledge. Through open innovation, on the one hand, the process of innovation is accelerated and reduces the associated costs and risks and, on the other

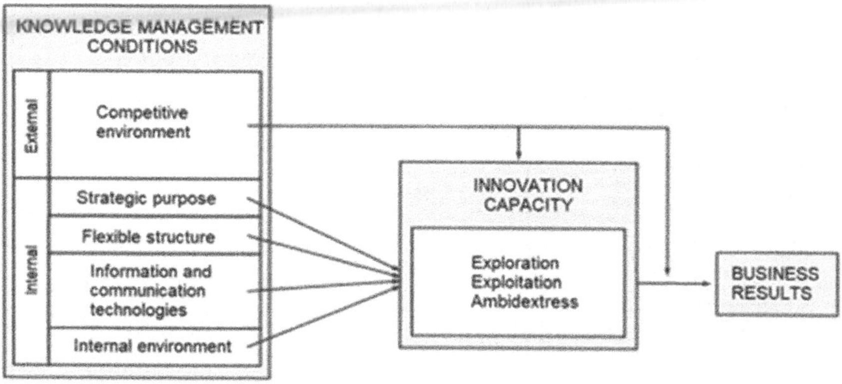

Fig. 2. Integrator model of knowledge management and innovation: Source: [1].

hand, new possibilities are opened for commercial knowledge exploitation. Open innovation will be linked to a set of knowledge management decisions aimed at promoting this new concept of innovative activity [14].

Open innovation implies a philosophy of innovation management based on the articulation of a collaborative process and participation with other agents, which requires the preparation of agreements that consider, in addition to cost sharing, the sharing of ownership of those results. The development of these interactions between different agents generates organizational design problems in this innovation management model: on the one hand, coordination problems of these agents and, on the other, incentive problems in creating and appropriating the innovations' results [15].

The discipline known as knowledge management focuses on the study of the most relevant decisions about this asset and has become one of the most widespread approaches in the strategic management field of a company. Integration of open innovation [14] and knowledge management is based on the Nonaka and Takeuchi Knowledge Creation Spiral model [5].

The model, in which the internal and external success factors of open innovation are considered, is shown below (Fig. 3):

The open model represents a new social context in which innovations can be generated, with implications both for the internal processes carried out in each organization and for the collaborative relationships established. Workgroups that are created in an open innovation project have distinctive characteristics that must be considered, and that require specific management adapted to these new particularities. However, not all open innovation activities have a positive effect on innovation-related outcomes.

Opening innovation in organizations is a process that follows the steps of organizational change:

Thawing: is about creating a sense that a change is needed by communicating it to stakeholders and actors within the company, both internal and external.

Boost: it refers to putting change into practice, by establishing new processes and behaviors consistent with the new organizational vision.

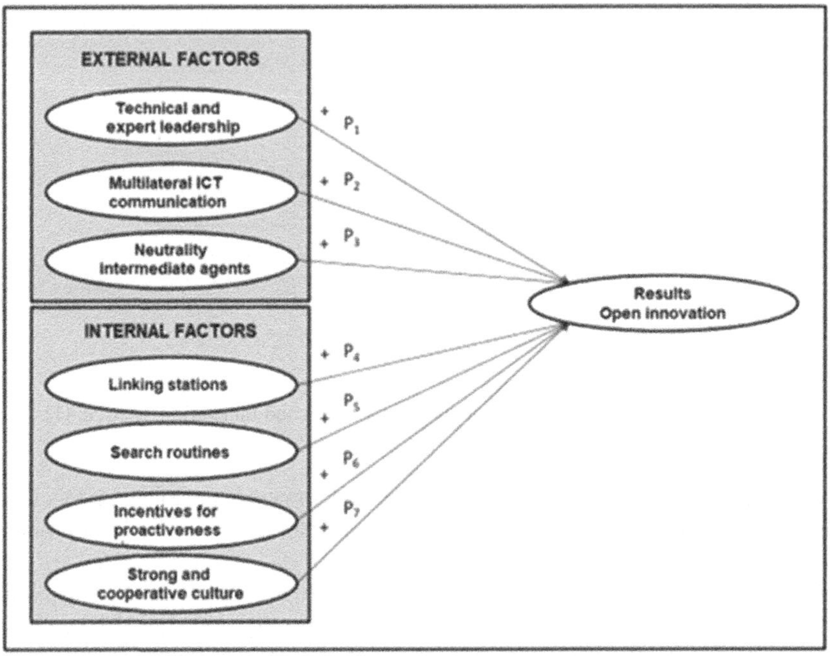

Fig. 3. Integrated Innovation and Knowledge Management Model. Source: The authors.

Institutionalization: consolidation phase of the improvements obtained, avoiding backward steps in the implementation of a new system.

The previous steps to open innovation must be carried out throughout the entire organization, its structure, processes and knowledge management systems. Adequate development of AI will require the use of knowledge management systems capable of integrating external knowledge, as well as disseminating, sharing, and transferring knowledge both within the organization and regarding its environment.

The process of implementing open innovation is identifying and managing existing knowledge both inside and outside the company.

Five phases have been defined within the innovative process (see Fig. 4). These are cyclical, since the generation of new ideas will generate knowledge within the organization [16].

- *Define possible stages in the innovating process*. As the process advances and organizational, financial, cultural, and technical changes progress, these goals will need to be further specified.
- *Identify relevant knowledge*. Once the process framework is established, it is important to identify the knowledge that will be needed to develop open innovation.
- *Selecting a suitable integration mechanism among the different actors involved in the process*. There are several mechanisms for this purpose, among which the techniques for processing ideas and decision groups stand out.

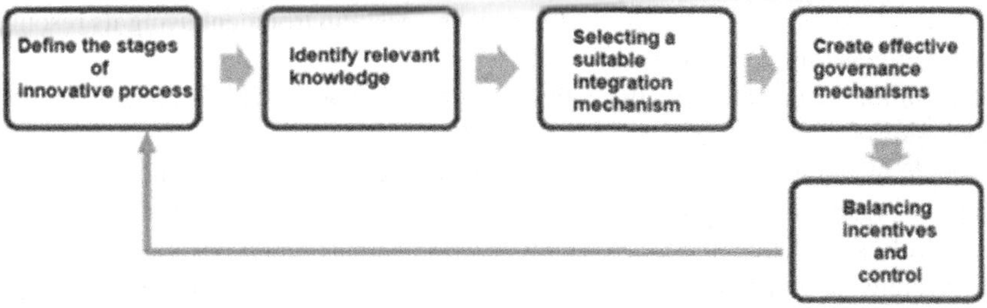

Fig. 4. Phases of open innovation. Source: [16].

- *Create effective governance mechanisms* to address questions about the process, control, or relationship between the organization and external stakeholders.
- *Balancing incentives and control.* Organizational management must establish the right balance between, on the one hand, ensuring the quality and profitability of the work done for the company and, on the other hand, encouraging external participants who have collaborated in the innovative process.

5 Considerations for Model Validation

For the validation of the model, time and evaluation of the results are required in the medium term, so it cannot be presented in this document yet.

In the model presented in this research, the dynamic analysis of the systems with the knowledge management methodology supported by the European Guide [17] is combined and aims to create an evaluation instrument for the knowledge management network in the introduction of incremental innovations in products, services and processes for a company that display a range of possibilities for the development and innovation of social technologies. Therefore, this leads organizations to increase their competitiveness and reduce the socio-economic difficulties that afflict both.

In addition, it is possible to deduce that the competitiveness of an organization increases when the lines of knowledge and methodologies act synergistically together with contextual factors such as the country's commercial policies, fiscal policies, investment security and others, achieving an economic and productive development in the organization.

6 Conclusions

Organizational knowledge is the key to success in strategies, and knowledge management is the methodology that enables collaboration in managing and developing knowledge.

The innovation process must be considered as an integrative element to all business processes and should be guided by strategic business management.

Under present society conditions, characterized by uncertainty, complexity, change and globalization, organizations are required to know themselves and their environment, to be able to take up the challenges of the new business environment using its resources, knowledge generation and innovation.

Innovation is the added value obtained through knowledge management within organizations.

People in charge of knowledge management programs must become entrepreneurs within the organization and be persistent in the processes of capturing, distributing and applying old knowledge in order to create new knowledge and innovate.

Open innovation proposes a new way of managing the innovation process integrated with knowledge management. It means a deep change in business model, systems and mechanisms, practices and, in general, in the mentality of those companies that decide to adopt it.

References

1. Prado, A., Fischer, L.: Conditions of knowledge management, innovation capability and firm performance An explicative model. Journal "Pensamiento y Gestión" **35**(35), 25–63 (2013)
2. García-Pintos, E.A., García Vázquez, J.M., García, P.P.: Incidencia de las políticas de recursos humanos en la transferencia de conocimiento y su efecto sobre la innovación. Investigación Europea Dirección y Economía la Empresa **16**(1), 149–163 (2010)
3. Manzanares, M.J.D., Gómez F.G.: Gestión del conocimiento organizativo, innovación tecnológica y resultados. una investigación empírica. Investig. Eur. Dir. y Econ. la Empresa. **14**(2), 139–167 (2008)
4. Medina, V.H., Gil, J.A., Liberona, D.: Knowledge management model as a factor of educative quality: Towards an excellence model. In: Journal LNBIP 185 - Lecture Notes in Business Information Processing, Ed. Springer, Berlin (2014)
5. Nonaka, L., Takeuchi, H., Umemoto, K.: A theory of organizational knowledge creation. Int. J. Technol. Manage. **11**, 833–884 (1996)
6. Péres, D., Dressler, M.: Tecnologías de la información para la gestión del conocimiento. Acimed **14**(1), 1–19 (2006)
7. Salazar A.A.P.: Modelo de implantación de gestión del conocimiento y tecnologías de información para la generación de ventajas competitivas, Valparaíso Univ. Técnica Federico St. María, p. 91 (2000)
8. Oberto, A.: Gestión de conocimiento para la innovación organizacional: una visión desde Ibero América. Enl@ce Rev. Venez. Inf. Tecnol. y Conoc. **2**(1), 11–29 (2005)
9. Balmaseda, V., Elguezabal, Z., Herriko, P.V.: Evolución de las propuestas sobre el proceso de innovación: ¿qué se puede concluir de su estudio?, **14**, 127–138 (2008)
10. Velasco, E.: La gestión de la innovación. Elementos integrantes y su aplicación en empresas innovadoras en el País Vasco. Serie de Economía y Empresa, Universidad del País Vasco (2010)
11. Boh, L.E.: Understanding of Business Organizations and their Environment as Systems of Increasing Complexity: Features and Implications, pp. 363–377 (2016)
12. González, V., Roberto, L.: Factores que han impulsado la innovación en la instrumentación industrial, un estudio de caso (2004)
13. García, V.M., Estrada, M.L.: Tarazona Bermúdez, G.M.:. Investigación en Ingeniería apoyada por la gestión del conocimiento y la internet social, libro, Doctorado en Ingeniería - Universidad Distrital Francisco José de Caldas. Amadgraf Impresores Ltda. 1ª Ed. Bogotá, Colombia (2019)
14. González-Sánchez, R., García-Muiña, F.E.: Innovación abierta: Un modelo preliminar desde la gestión del conocimiento. Intang. Cap. **7**(1), 82–115 (2011)

15. Rodríguez, L., et al.: "Innovación abierta: desafíos organizacionales de este modelo de gestión de la innovación para las empresas" (2010)
16. San Martón Albizuri, N., Rodríguez Castellanos A.: A conceptual framework for open innovation processes: integration, distribution and cooperation in knowledge. Rev. Estudios Interdisciplinarios en Ciencias Sociales, **14**, 83–101 (2012)
17. The European KM Forum Consortium, IST Project No 2000-26393 (2001)

Trust and Reputation Smart Contracts for Explainable Recommendations

Fátima Leal[1], Bruno Veloso[2], Benedita Malheiro[2,3](✉), and Horacio González-Vélez[1]

[1] CCC/NCI – Cloud Competency Centre, National College of Ireland, Dublin, Ireland
{fatima.leal,horacio.gonzalez-velez}@ncirl.ie
[2] INESC TEC, Porto, Portugal
bruno.m.veloso@inesctec.pt
[3] ISEP/IPP – School of Engineering, Polytechnic Institute of Porto, Porto, Portugal
mbm@isep.ipp.pt

Abstract. Recommendation systems are usually evaluated through accuracy and classification metrics. However, when these systems are supported by crowdsourced data, such metrics are unable to estimate data authenticity, leading to potential unreliability. Consequently, it is essential to ensure data authenticity and processing transparency in large crowdsourced recommendation systems. In this work, processing transparency is achieved by explaining recommendations and data authenticity is ensured via blockchain smart contracts. The proposed method models the pairwise trust and system-wide reputation of crowd contributors; stores the contributor models as smart contracts in a private Ethereum network; and implements a recommendation and explanation engine based on the stored contributor trust and reputation smart contracts. In terms of contributions, this paper explores trust and reputation smart contracts for explainable recommendations. The experiments, which were performed with a crowdsourced data set from Expedia, showed that the proposed method provides cost-free processing transparency and data authenticity at the cost of latency.

Keywords: Smart contracts · Trust and reputation · Explainable recommendations · Transparency · Authenticity · Traceability

1 Introduction

In large e-commerce crowdsourcing platforms, recommendation engines suggest products based on past costumer feedback, current costumer preferences and interactions, and inter-costumer similarities. Nevertheless, the majority of these recommendations are left unexplained. This weakness leaves the user clueless about the reasons behind the recommendation and may lead to distrust in the platform, making the case for the need to explainable recommendations. According to Tintarev and Masthoff [16], an explanation is any content added to the

Á. Rocha et al. (Eds.): WorldCIST 2020, AISC 1159, pp. 124–133, 2020.
https://doi.org/10.1007/978-3-030-45688-7_13

recommendation which aims to increase transparency, trust, and decision-making effectiveness. For crowdsourcing platforms, the explanation of recommendations should be a mandatory system requirement since they are largely based on past costumer feedback. In this context, the paper explores the generation of explanations for collaborative recommendation systems in the tourism domain.

Collaborative recommendation systems create recommendations based on the information of users with similar preferences [5], using memory-based or model-based filters. The most popular memory-based algorithm used to predict user preferences is k-Nearest Neighbours (k-NN). Moreover, with k-NN, not only it is possible to know which users are responsible for the computed recommendations, but they can be further improved by building explicit trust and reputation models of the users [7]. Model-based collaborative filtering, on the other hand, hinders the traceability of the computed recommendations. In this respect, Leal et al. [8] present a trust based post-filtering approach which traces the origin of the generated recommendations. Finally, to strengthen the confidence on the provided recommendations and corresponding explanations, it is necessary to ensure the authenticity and traceability of the data used for modelling. This problem becomes more serious since the majority of these platforms are centralised, allowing potential malicious data tampering. To overcome this obstacle, this work relies on blockchain smart contracts since the adoption of blockchain technology prevents against malicious behaviours and ensures the reliability of transactions [12]. Thus, this proposal adopts blockchain smart contracts to marshal and enable trust and reputation, providing transparent and trustworthy explanations for collaborative recommendation filters. Given their tamper-proof nature, smart contracts aim to ensure data authenticity, *i.e.*, avoiding the manipulation from unethical stakeholders. The experiments, which were performed with data from tourism crowdsourcing platforms, show the advantages of the proposed approach.

The rest of this paper is organised as follows. Section 2 reviews crowdsourcing platforms which incorporate blockchain technology. Section 3 describes the proposed methods for the creation of reliable explainable recommendations. Section 4 reports the results of the experiments and tests performed. Finally, Sect. 5 summarises and discusses the outcomes of this work.

2 Related Work

The development of Artificial Intelligence based systems should be guided by Accountability, Responsibility and Transparency (ART) design principles [3], *i.e.*, systems should explain and justify their decisions (accountability), incorporate human values into technical requirements (responsibility) and describe the decision-making process and how data is used, collected and governed (transparency). Blockchain, with its transparency, traceability, trust, immutability, desintermediation, and security characteristics [11,15], is one of the most promising techniques to achieve ART compliant data processing.

Concerning recommendations, on the one hand, explanations reduce vulnerability and add transparency by giving users detailed information why systems have arrived at a particular decision, and, on the other hand, blockchain technology ensures data authenticity and traceability. As a result, this work proposes a method to explain recommendations using trust and reputation smart contracts, achieving both decision-making transparency and data authenticity and traceability. According to Pilkiton [14], blockchain grants: (*i*) decentralised immutable and traceable reputation; (*ii*) unique users, *i.e.*, only registered users can contribute with ratings or reviews; and (*iii*) portable and transverse reputation.

Crowdsourcing platforms not only advertise offers, but also promote the voluntary feedback sharing, which influences the behaviour of the other costumers. However, this crowdsourced information, which is voluntarily and freely shared, raises reliability questions. Therefore, it is relevant to assess the reliability of crowdsourced information, namely, by using trust and reputation models. Specifically, service reputation can be inferred from the reputation of the service contributors, which, in turn, can be based on the analysis of the individual contributions, *e.g.*, reviews or ratings. Therefore, higher reputation indicates higher service quality, allowing, for example, the provider to increase the price [13].

The application of blockchain together with trust and reputation to crowdsourcing platforms is relatively new. The works found in the literature were published since 2017 and include three works on trust [2,4,10] and two on reputation [1,9]. Regarding the blockchain technology used, four use the Ethereum open source framework [1,4,9,10] and one relies on a proprietary solution [2]. Table 1 compares the identified crowdsourcing platforms in terms of trust and reputation model, blockchain technology and domain of application.

Table 1. Crowdsourcing platforms using blockchain technology [17]

Platform	User model	Blockchain technology	Domain
[2]	Trust	Tweetchain	Twitter
[1]	Reputation	Ethereum	Job finding
[10]	Trust	Ethereum	Payment system
[4]	Trust	Ethereum	Health
[9]	Reputation	Ethereum	Job finding

This literature survey shows that there are few crowdsourcing platforms concerned with the authenticity and processing transparency of crowdsourced data. In particular, in the tourism domain, where crowdsourcing is extremely popular, no such mechanisms are found in the prevailing platforms. Therefore, our work contributes with trust and reputation smart contracts to provide authentic explanations for recommendations, improving the quality of the user experience.

3 Proposed Method

Recommendations can be explained using reliable trust and reputation models of the stakeholders. The proposed method explores blockchain technologies to guarantee the authenticity and traceability of the trust and reputation models created, *i.e.*, that they have not been tampered. In this context, we developed a stream-based blockchain engine illustrated in Fig. 1 which integrates a private Ethereum network. The engine includes four modules: (*i*) the event-driven trust model updater; (*ii*) the smart contract holder (private Ethereum network); (*iii*) the memory- and model-based collaborative predictors; and (*iv*) the evaluator (Root Mean Square Error and Recall@N metrics). Our method applies incremental event-driven updating (data streaming), *i.e.*, the contributor trust model is updated every time a new event occurs.

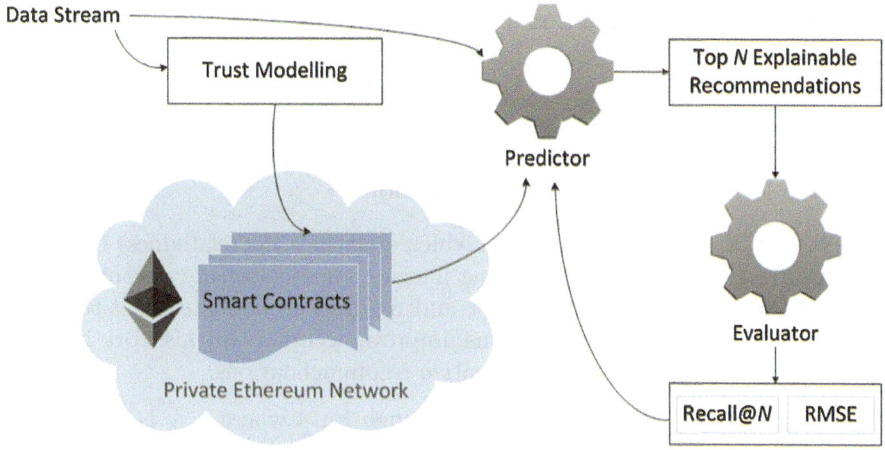

Fig. 1. Explanation and recommendation engine

3.1 Memory-Based Trust and Reputation

Leal et al. [7] proposed a trust and reputation model for a memory-based collaborative filter which quantifies the pairwise trustworthiness between the active user and its neighbour users by analysing their mutual influence. It takes into account the number of times the active user selects, from the top ten recommendations, recommendations based on a given neighbour. The social reputation of an user is based on the relevant user-neighbour trustworthiness. This trust and reputation model is then used to generate higher-quality memory-based recommendations and, in this work, also explanations.

Trustworthiness ($T_{u,k}$) between a user u and k is computed through Eq. 1. The trustworthiness between both users increases when the active user u selects recommendations based on neighbour k. In this context, $n_{u,k}$ represents the number of items actually recommended to u due to k and $N_{u,k}$ the number of times k was chosen as a neighbour of u.

$$T_{u,k} = \frac{n_{u,k}}{N_{u,k}} \tag{1}$$

Neighbour-based Reputation ($R_{u,k}$) quantifies the reputation of u using the pairwise trustworthiness of the neighbours shared between u and k. In Eq. 2, c represents a neighbour common to u and k; $T_{u,c}$ is the pairwise trustworthiness between the user u and c; and N_c is the number of common neighbours between u and k.

$$R_{u,k} = \frac{\sum_{c=1}^{N_c} T_{u,c}}{N_c} \tag{2}$$

System-wide Reputation (R_u) represents overall reputation of user u, considering all users which share at least one neighbour with the u. Equation 3 presents R_u, the system-wide reputation of u, where $R_{u,k}$ is the neighbour-based reputation between users u and k and N is the total number of users which share neighbours with u.

$$R_u = \frac{\sum_{k=1}^{N} R_{u,k}}{N}, R_{u,k} \neq 0 \tag{3}$$

3.2 Model-Based Trust and Reputation

Leal et al. [8] proposed a trust model which quantifies the relatedness between a trustor (u) and the trustee (k), taking into account the set of relevant co-rated items. The authors applied this trust and reputation model as an *a posteriori* filter to sort the predictions and, thus, improve accuracy. In this work it is also used to explain model-based collaborative recommendations.

Trustworthiness ($T_{u,k}$) is calculated through Eq. 4 where $n_{u,k}$ is the number of relevant items co-rated by users u and k and N_u is the total number of items rated by the active user u.

$$T_{u,k} = \frac{n_{u,k}}{N_u} \tag{4}$$

System-wide Reputation (R_u) corresponds to the average user trustworthiness given by Eq. 5, where k is a user which has co-rated at least one relevant item with user u, $T_{u,k}$ is the trustworthiness between u and k and N is the number of users which have co-rated at least one relevant item together with user u.

$$R_u = \frac{\sum_{k=1}^{N} T_{u,k}}{N}, T_{u,k} \neq 0 \tag{5}$$

3.3 Blockchain

Trust and reputation models aim to improve prediction accuracy and reliability. However, when the models are centrally stored, they can be easily manipulated, *e.g.*, to meet hidden interests. Therefore, this paper explores a blockchain-based solution to ensure the authenticity of the trust and reputation models. Since reputation is based on the inter-user trustworthiness, it suffices to guarantee the authenticity of the inter-user trust model. The proposed solution represents the inter-user trust model as a smart contract in the blockchain. This smart contract holds the trust matrix and supports a collection of transactions to alter and access the stored model. Equation 6 displays the smart contract trust matrix $\boldsymbol{T}$, where $T_{u,k}$ represents the trustworthiness between users u and k given by Eq. 1 or Eq. 4, depending on the collaborative predictor employed. The inter-user trustworthiness values are updated with each incoming rating, allowing the tracing of the inter-user trustworthiness over time.

$$\boldsymbol{T} = \begin{array}{c} \begin{array}{ccccc} \boldsymbol{u_1} & \boldsymbol{u_2} & \boldsymbol{u_3} & \cdots & \boldsymbol{u_n} \end{array} \\ \left[\begin{array}{ccccc} \ddots & T_{1,2} & T_{1,3} & \cdots & T_{1,n} \\ T_{2,1} & \ddots & T_{2,3} & \cdots & T_{2,n} \\ T_{3,1} & T_{3,2} & \ddots & \cdots & T_{3,n} \\ \vdots & \vdots & \vdots & \ddots & \vdots \\ T_{n,1} & T_{n,2} & T_{n,3} & \cdots & \ddots \end{array}\right] \end{array} \begin{array}{c} \\ \boldsymbol{u_1} \\ \boldsymbol{u_2} \\ \boldsymbol{u_3} \\ \vdots \\ \boldsymbol{u_n} \end{array} \tag{6}$$

The blockchain technology protects the trust and reputation model from data tampering. Blockchain uses consensus mechanisms to achieve an agreement between the collection of network miners concerning data transactions. Concretely, it consists of a set of rules and procedures to keep data coherency among network nodes. The proposed method uses a private blockchain network which implements the Proof-of-Work (PoW) consensus algorithm. With this algorithm, miners compete against each other to complete the transactions and be rewarded. Once a transaction is approved, it is stored in a new block. With the inter-user trustworthiness values stored in the blockchain, the stream-based recommendation engine generates and explains reliable recommendations.

3.4 Explainable Recommendations

Explainable recommendations address the question of "*why has this item been recommended?*" by sharing with the user the reasons for the recommendation. In a transparent system, it is important to explain the recommendation process, clarifying if it took into account user preferences or rather hidden interests. In this context, the proposed method ensures the authenticity and relies on the inter-user trust model to generate and explain recommendations.

Explanations can rely on different information sources. Specifically, this work uses the preferences of identical users (collaborative filtering), and the user trust and reputation smart contracts. The set of users identical to the active user corresponds, in the case of memory-based filters, to the top nearest neighbours and, in the case of model-based filters, to those users with larger number of relevant co-rated items. Once the engine generates predictions for the active user, it will select and explain the top N recommendations. The trust and reputation smart contracts constitute a trust-based, traceable, and immutable source of information for explainable recommendations.

The recommendations are explained with the help of the trust model stored dynamically in blockchain. Table 2 exemplifies the type of support information available to generate explanations for active user u: the identification and average rating ($\overline{r}$) of the resource and the number (#), average trustworthiness ($\overline{T}$) and system-wide reputation ($\overline{R}$) of the users behind the resource recommendation. Based on these data, the active user obtains the following explanation for the top recommendation: "*Hotel a was recommended to you because seven users with preferences identical to yours, an average system-wide reputation of* 79 % *and in whom you have an average trust of* 80 % *gave it an average rating of* 4.8".

Table 2. Aggregated explanation data

Hotel		Users		
ID	$\overline{r}$	#	$\overline{T}$ (%)	$\overline{R}$ (%)
a	4.8	7	80	79
b	4.8	7	72	80
c	4.7	10	70	72
d	4.4	8	65	60
e	4.2	6	62	62

Table 3. Detailed explanation data

Users Supporting a			
User ID	$r_{s,a}$	$T_{u,s}$ (%)	R_s (%)
11	4.8	80	79
82	5.0	83	78
34	4.7	76	82
14	4.6	82	74
15	4.9	79	81
67	4.7	80	79
20	4.9	80	80

These generic explanations can be further detailed by specifying the individual contributions of each support user. Table 3 displays the detailed information supporting the recommendation of hotel a to the active user u: the support user s identification, the rating given by s to a ($r_{s,a}$), the trust ($T_{u,s}$) between the active user u and the support user s and the system-wide reputation of the support user s (R_s). Based on these data, the active user obtains the following complementary explanation: "*Specifically, user* 11, *with a reputation of* 79 % *and whose tastes you trust* (80 %)*, rated this hotel with a* 4.8*; user* 82, *with a reputation of* 78 % *and whose recommendations you trust* (83 %)*, gave it a* 5.0*; etc.*"

4 Experiments and Results

The experiments involved two main approaches: explainable recommendations with and without smart contracts. The stream-based collaborative recommendation engine adopted is that of [7,8]. The configuration of the implemented private Ethereum network is composed of one regular node and one miner node.

The experiments were conducted on an Openstack cloud instance with 16 GiB RAM, 8 CPU and 160 GiB of hard-disk space. The data were ordered temporally and, then, partitioned. The initial model uses the 20 % of the data set, and the remaining 80 % is inputted as a data stream. The model is incrementally updated with each incoming stream event. The blockchain technologies employed were: (*i*) EthereumJ[1] as Ethereum client; and (*ii*) solidity[2] for the smart contracts. Due to the blockchain transaction time, the experiments were repeated four times.

Data Set. HotelExpedia was proposed by Leal et al. [6]. It contains the following features: (*i*) 6030 hotels; (*ii*) 3098 reviewers, including anonymous reviewers; and (*iii*) 381 941 reviews from 10 different locations. Since these experiments were performed exclusively with data from the 1089 identified reviewers, *i.e.*, we discarded the anonymous users and their inputs. Each user classified at least 20 hotels and each hotel contains at least 10 reviews. Specifically, we use the user and hotel identification and, as multi-criteria ratings, the *overall*, *cleanliness*, *hotel condition*, *service* and *room comfort*.

Blockchain Results. The incremental updating results are depicted in Table 4. It compares the explainable recommendations supported by the trust and reputation model in terms of the average incremental accuracy and processing time with and without blockchain smart contracts. The recommendation accuracy was evaluated using Recall@10 since these are online stream experiments. As expected, the prediction accuracy results remain unchanged regardless of the adoption of blockchain technology. The difference lies in the average event processing time ($\overline{\Delta t}$). Blockchain requires time to mine blocks, which in the case of Ethereum corresponds to an average 10 s to 20 s per block. This time depends on the built in block mining dynamic difficulty algorithm of Ethereum.

Table 4. Comparison of incremental recommendation results

Blockchain	Predictor	Recall@10	RMSE	$\overline{\Delta t}$ (s)
Yes	Memory-based	0.719	0.117	11.000
	Model-based	0.514	0.161	9.000
No	Memory-based	0.719	0.117	0.200
	Model-based	0.514	0.161	0.125

Finally, we determined the latency and throughput of the implemented private Ethereum network with the stream data set (128 605 events). The throughput represents the number of successful transactions per minute and the latency corresponds to the difference in minutes between the submission and completion of a transaction. The average throughput was 7.37 event/min with an average latency of 0.14 min, which indicate a high blockchain efficiency.

[1] https://github.com/ethereum/ethereumj.
[2] https://solidity.readthedocs.io/.

5 Conclusions

Crowdsourcing platforms rely on voluntary contributions, such as ratings, reviews or photos, to generate recommendations. While research has shown that using trust and reputation models of contributors improves the accuracy of the recommendations, these recommendations are mostly opaque to the end-user and are prone to data tampering. To address this problem, this work explores the generation of explanations together with blockchain technology. Our proposal integrates: (*i*) the trust and reputation models presented by Leal et al. [7,8]; (*ii*) a private Ethereum network to store the user trust model; and (*iii*) a collaborative recommendation and explanation engine for stream processing supported by user trust and reputation smart contracts. This solution provides transparency to the end-user and allows tracing the inter-user trustworthiness and the individual system-wide reputation over time.

The proposed method was tested and evaluated with the Expedia data set, using RMSE and Recall@10 as evaluation metrics and employing incremental updating. Additionally, concerning blockchain technology, we analysed the incremental recommendation accuracy, execution time, throughput and latency. While the recommendation accuracy remained unchanged, there was a latency of 8.4 s/event and a throughput of 7.37 event/min. This latency was considerably higher than without blockchain smart contracts, indicating that the price to pay for data authenticity and traceability is latency. Nevertheless, given that the Ethereum blockchain network requires by default between 10 s to 20 s to process transactions, the obtained latency was a good result. To sum up, this paper describes a recommendation and explanation engine for crowdsourcing platforms supported by trust and reputation smart contracts, providing the end-user with transparency and authenticity. As future work, we intend to detect automatically malicious users as well as bots by tracing the evolution, including the change frequency, of individual reputations.

Acknowledgements. This work was partially financed by: (*i*) the ERDF – European Regional Development Fund through the Operational Programme for Competitiveness and Internationalisation - COMPETE 2020 Programme within project ≪POCI-01-0145-FEDER-006961≫, and by National Funds through the FCT – Fundação para a Ciência e a Tecnologia (Portuguese Foundation for Science and Technology) as part of project UID/EEA/50014/2019; and (*ii*) the Irish Research Council in the framework of the EU ERA-NET CHIST-ERA project SPuMoNI: Smart Pharmaceutical Manufacturing (http://www.spumoni.eu).

References

1. Bhatia, G.K., Kumaraguru, P., Dubey, A., Buduru, A.B., Kaulgud, V.: WorkerRep:building trust on crowdsourcing platform using blockchain. Ph.D. thesis, IIIT-Delhi (2018)
2. Buccafurri, F., Lax, G., Nicolazzo, S., Nocera, A.: Tweetchain: an alternative to blockchain for crowd-based applications. In: International Conference on Web Engineering, pp. 386–393, Springer, Cham (2017)

3. Dignum, V.: Responsible artificial intelligence: designing AI for human values. ITU journal: ICT Discoveries **1**(1), 1–8 (2017)
4. Fernández-Caramés, T.M., Fraga-Lamas, P.: Design of a fog computing, blockchain and IoT-based continuous glucose monitoring system for crowdsourcing mhealth. In: Multidisciplinary Digital Publishing Institute Proceedings, p. 37 (2018)
5. Koren, Y., Bell, R.: Advances in collaborative filtering. In: Ricci, F., Rokach, L., Shapira, B. (eds.) Recommender Systems Handbook, chap. 3, pp. 77–118. Springer, Boston (2015), ISBN 978-1-4899-7637-6
6. Leal, F., Malheiro, B., Burguillo, J.C.: Prediction and analysis of hotel ratings from crowd-sourced data. In: World Conference on Information Systems and Technologies, pp. 493–502. Springer, Cham (2017)
7. Leal, F., Malheiro, B., Burguillo, J.C.: Trust and reputation modelling for tourism recommendations supported by crowdsourcing. In: World Conference on Information Systems and Technologies, pp. 829–838. Springer, Cham (2018)
8. Leal, F., Malheiro, B., Burguillo, J.C.: Incremental hotel recommendation with inter-guest trust and similarity post-filtering. In: World Conference on Information Systems and Technologies, pp. 262–272. Springer, Cham (2019)
9. Li, M., Weng, J., Yang, A., Lu, W., Zhang, Y., Hou, L., Liu, J.N., Xiang, Y., Deng, R.H.: Crowdbc: A blockchain-based decentralized framework for crowdsourcing. IEEE Trans. Parallel Distrib. Syst. **30**(6), 1251–1266 (2018)
10. Lu, Y., Tang, Q., Wang, G.: Zebralancer: private and anonymous crowdsourcing system atop open blockchain. In: 2018 IEEE 38th International Conference on Distributed Computing Systems (ICDCS), pp. 853–865. IEEE (2018)
11. Nam, K., Dutt, C.S., Chathoth, P., Khan, M.S.: Blockchain technology for smart city and smart tourism: latest trends and challenges. Asia Pac. J. Tour. Res., 1–15 (2019)
12. Nofer, M., Gomber, P., Hinz, O., Schiereck, D.: Blockchain. Bus. Inf. Syst. Eng. **59**(3), 183–187 (2017)
13. Önder, I., Treiblmaier, H., et al.: Blockchain and tourism: three research propositions. Ann. Tour. Res. **72**(C), 180–182 (2018)
14. Pilkington, M.: 11 blockchain technology: principles and applications. Research Handbook on Digital Transformations, vol. 225 (2016)
15. Rejeb, A., Rejeb, K.: Blockchain technology in tourism: applications and possibilities. World Sci. News **137**, 119–144 (2019)
16. Tintarev, N., Masthoff, J.: Designing and evaluating explanations for recommender systems. In: Recommender systems handbook, pp. 479–510. Springer, Cham (2011)
17. Veloso, B., Leal, F., Malheiro, B., Moreira, F.: Distributed trust & reputation models using blockchain technologies for tourism crowdsourcing platforms. Procedia Comput. Sci. **160**, 457–460 (2019)

Predicting an Election's Outcome Using Sentiment Analysis

Ricardo Martins(✉), José Almeida, Pedro Henriques, and Paulo Novais

Algoritmi Centre/Department of Informatics, University of Minho, Braga, Portugal
ricardo.martins@algoritmi.uminho.pt, {jj,prh,pjon}@di.uminho.pt

Abstract. Political debate - in its essence - carries a robust emotional charge, and social media have become a vast arena for voters to disseminate and discuss the ideas proposed by candidates.

The Brazilian presidential elections of 2018 were marked by a high level of polarization, making the discussion of the candidates' ideas an ideological battlefield, full of accusations and verbal aggression, creating an excellent source for sentiment analysis.

In this paper, we analyze the emotions of the tweets posted about the presidential candidates of Brazil on Twitter, so that it was possible to identify the emotional profile of the adherents of each of the leading candidates, and thus to discern which emotions had the strongest effects upon the election results. Also, we created a model using sentiment analysis and machine learning, which predicted with a correlation of 0.90 the final result of the election.

Keywords: Sentiment analysis · Emotion analysis · Natural processing language · Machine learning

1 Introduction

It is undeniable that social media have changed how people contact to each other, enabling them to maintain relationships that previously would be difficult to maintain for various reasons, such as distance, the passage of time, and misunderstandings.

Barack Obama used social media as the main platform of his presidential campaign in the United States in 2008, and since then, it is well established that social media created a strong influence on voters' decisions in that election and many others. Facebook and Twitter are now considered essential tools for any political campaign, along with other social media platforms that have been created since then.

The influence of social media increases when candidates with low funding levels and low levels of traditional media exposure try to defeat adversaries with more resources. Social media allow the candidates to post their political platforms as well as inflammatory posts against their opponents. In many cases, political candidates use social media mainly to carry provocative attacks against

Á. Rocha et al. (Eds.): WorldCIST 2020, AISC 1159, pp. 134–143, 2020.
https://doi.org/10.1007/978-3-030-45688-7_14

their opponents. Their followers, in turn, use social media to broadcast their reactions of anger and satisfaction in posts that can be shared thousands of times.

In electoral campaigns highly marked by their massive presence on social networks - as was the Brazilian presidential election in 2018 – by using Natural Language Processing (NLP) and Machine Learning (ML) it is possible to identify what voters think and feel about the candidates and their proposals for the various areas of the government. Thus, some interesting research question that arises are: "How do the emotions about each candidate influence the electoral decision?" and "Is it possible to predict the result of an election only by knowing what voters "feel" about a candidate?"

In this paper, we present an approach to predict the results of the election. As a case study, we collected messages from social media about the Brazilian presidential election of 2018, where we used Sentiment Analysis, NLP and ML to identify which emotions dominated the electorate and their correlations with the number of messages about the candidates and finally predict the results.

The remainder of the paper is as follows: Sect. 2 presents some studies in sentiment analysis in elections that inspired this work, while Sect. 3 describes the process of dataset creation for our tests. Section 4 presents our experiments, explains the steps used in the analysis and discusses the results obtained from a set of tests performed. The paper ends with the conclusion and suggestions for future work in Sect. 5.

2 Related Work

The analysis of emotions on Twitter to explain elections is not a new approach. Several researchers have already done work in this area, each with different approaches and results. Wang [11] developed a system to analyze the tweets about presidential candidates in the 2012 U.S. election as expressed on Twitter. His approach analyzes the text's polarities (positive, neutral or negative), the volume of posts and the word most used. This is the same approach used by Heredia et al. [3], that trained a Convolutional Neural Network using an annotated lexicon - Sentiment140 - to detect the text's polarities.

These approaches do not go deeper on the reasons of polarities, and thus, does not identify which sentiments influence the voters' decision, which is the objective of our work.

Tumasjan [10] has developed an analysis of tweets for the German elections which, similar to Wang's work, used the polarities of phrases to analyze the messages. However, unlike the previous work, Tumasjan's study relates the volume of messages to the final result of the election. This approach is highly influenced by financial questions (as richer the candidate is, more publicity about him can be posted in social media), and for this reason, we did not consider trustworthy. Bermingham [1] used this same approach, using the data from the Irish general election of 2011, but, different than Tumasjan, he has expanded the model for training by using polls as parameters for training the predictions, which has inspired our work in the training dataset creation.

While the existing works focused only on the aspect of the text's polarities, the work of Martins [5] inspired our decision to consider the basic emotions contained in the text as a factor of influence in the decision of the voter, and use it to predict results.

3 Dataset Creation

Initially, to draw from the data a general idea of the Brazilian voter, it would be necessary to generalize the emotional profile of these voters, regardless of their region. The idea is that, according to the emotions contained in the texts, it would be possible to determine relevant information about the characteristics of the Brazilian voters. Thus, a pipeline was created for dataset creation, as presented in Fig. 1.

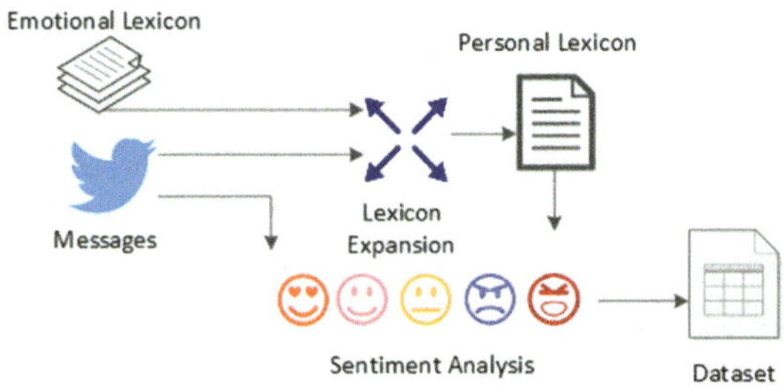

Fig. 1. Dataset's creation pipeline

This pipeline begins with a collection of messages about the candidates. For this purpose, we collected tweets from 145 cities in Brazil, with each state being represented by at least its four largest cities, and delimiting a radius of 30 km for each city. The option for geolocated tweets is to avoid posts from countries different than Brazil, where the author probably would be not able to vote in Brazilian's election. To select what would be considered relevant or not, we defined that only the tweets containing the main candidates' names would be collected. Tweets containing two or more candidates' names were analyzed for all candidates mentioned in the text. Moreover, we considered relevant only tweets - not retweets. This decision was inspired by the necessity to avoid viral posts or the ones from digital influencers. In other words, we wanted to know the opinion from the message's author about a candidate, not the opinion of an author who the author likes.

For gathering the tweets, we developed a script in Python using the official API provided by Twitter, and collected all geolocated messages from the 145 cities mentioned earlier in the period from May 2018 to October 2018, which contained at least one of the following names in their texts: "Bolsonaro", "Ciro Gomes", "Marina Silva", "Alckmin", "Amoêdo", "Álvaro Dias", "Boulos", "Meirelles" and "Haddad", divided in two groups: first round and second round.

3.1 Out of Scope

After an initial analysis of the messages collected, we decided not to handle the ambiguity in the texts during the analysis. The reason for choosing not to address this problem was justified by the tiny amount of messages that could lead to erroneous interpretations. Since it was a mandatory requirement for the messages to contain the name of at least one candidate, the nature of the Twitter messages - which limits each post by 280 characters - already considerably inhibited this type of problem.

Furthermore, this initial analysis showed that the existence of the candidate's name in the text made the context of the message as political and egalitarian in the emotional sense, as derogatory nicknames for the leading candidate candidates bring negative emotions. So, we avoided that exacerbated emotional expressions of some candidates affect others.

3.2 Lexicon Expansion

When working with sentiment analysis, a common approach is to use a dictionary-based algorithm to identify the emotional words in texts. However, according to Feldman [2], "the main disadvantage of any dictionary-based algorithm is that the acquired lexicon is domain-independent and hence does not capture the specific peculiarities of any specific domain." Thus, it is essential to know some particularities about the domain which the texts represent, to avoid misunderstandings and enable analysts to make a better classification of the sentiments contained in the texts.

With this problem in mind, we adapted the solution presented by Martins [4], where the texts were represented by a vector of words and these vectors were used to analyze the similarities of the words contained in an emotional lexicon, to expand it. A major concern when creating these vectors was about the polarization among the candidates. Our idea was that the texts about a candidate do not influence the emotional words of other candidates. For this reason, we adopted the strategy of creating a personal emotional lexicon for each candidate and thus analyzing the candidate's sentiments individually according to their respective emotional lexicon.

An overview of the entire process of lexicon expansion is presented in Fig. 2.

Word Vectors. The process of lexicon expansion begins with grouping all tweets collected by the candidate's name, removing their stopwords and creating the word vectors. For this purpose, we developed a script in Python using the Word2Vec algorithm, presented by Mikolov [7], having as parameters: size of 50; window 5 and trained for 200 epochs. Later, the emotional lexicon is introduced, to feed the word vectors with emotional seed words. For this step, we used the NRC lexicon [8] to provide the emotional words for the word vectors. The reason for this lexicon's choice is that it provides emotional words in Portuguese and also contains indications for polarities (positive and negative) and annotations

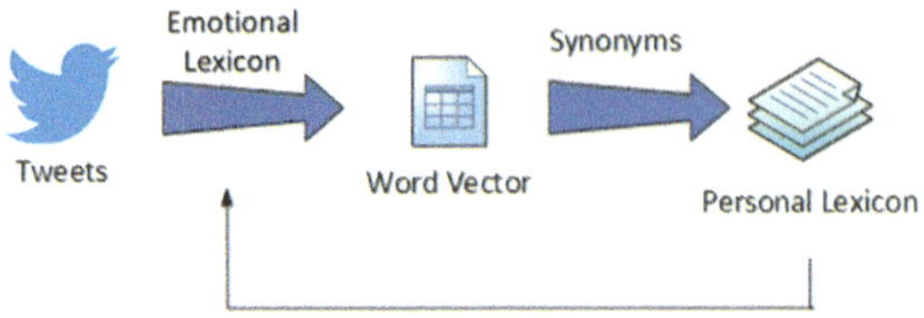

Fig. 2. Lexicon expansion process

for the eight basic emotions according to Plutchik's theory [9], which defines sentiments as *anger, anticipation, disgust, fear, joy, sadness, surprise* and *trust*.

Each word in the emotional lexicon was analyzed in the word vector of each candidate, to identify similarities. For all similar words found with a value higher than 0.7, these words inherited the emotional values from the lexicon's word and - recursively - were analyzed in the word vector to search for new similarities. For each similar word found in the word vectors, we added this word in a "new emotional lexicon" containing the original emotional lexicon and their respective similarities and emotional annotations according to the word vectors.

An important issue to emphasize in this approach is that when finishing the creation of this lexicon, we have a contextual emotional lexicon, because contextual words and its similarities were used in its creation. This context is provided because all messages contain at least one candidate's names. Thus, the context of the lexicon is about politics.

Results. When the lexicon expansion process was finished, the result was a set of personal lexicons about politics, containing the basic lexicon data increased by similarities found in the text and the synonyms of the words. The characteristics of each personal lexicon are presented in Table 1. Due to space limitation, Table 1 only presents the top 5 most known politicians, but in our study, all politicians that were candidate were considered in this analysis.

Table 1. Characteristics of the personal lexicon created

Lexicon	Words	Anger	Anticipation	Disgust	Fear	Joy	Sadness	Surprise	Trust	Positive	Negative
Original Lexicon	13911	8,85%	5,92%	7,48%	10,44%	4,88%	8,46%	3,76%	8,72%	16,36%	23,47%
Geraldo Alckmin	15251	8,89%	5,91%	7,50%	10,39%	4,82%	8,45%	3,79%	9,14%	16,56%	23,45%
Jair Bolsonaro	22082	9,16%	5,73%	7,63%	10,50%	5,01%	9,24%	3,38%	10,10%	17,49%	24,11%
Ciro Gomes	15316	8,95%	6,00%	7,47%	10,31%	4,99%	8,65%	3,77%	9,11%	16,78%	23,50%
Fernando Haddad	18264	9,27%	5,70%	7,47%	10,65%	4,87%	8,67%	3,46%	9,19%	17,26%	23,37%
Marina Silva	14842	8,76%	5,88%	7,40%	10,37%	4,85%	8,44%	3,76%	8,79%	16,38%	23,32%

3.3 Preprocessing

After creating a personal lexicon for each candidate, the next step consisted of creating a text preprocessing pipeline to remove unnecessary information from the texts. This pipeline, as presented in Fig. 3, begins with tokenization, which converts the texts into a list of single words, or *tokens*. Then, the process is divided into two parallel tasks: Part of Speech Tagging (POS-T) and Stopwords Removal. The POS-T process is responsible for identifying grammatical pieces of information for each word in the text, such as adjectives, adverbs and nouns, while the Stopword Removal removes any occurrence in the text of a defined word or list of words.

This strategy of paralleling POS-T and Stopwords removal was used because POS-T needs the text in the original format, to classify the words in their respective grammatical categories correctly.

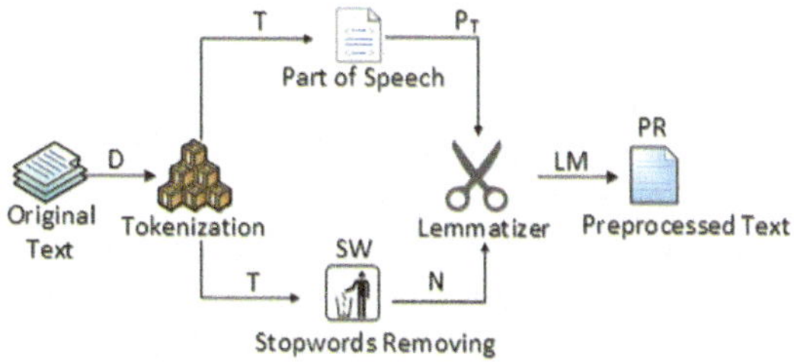

Fig. 3. Preprocessing tasks

Concerning text cleaning, in POS-T, every word in a grammatical category other than noun, verb, adverb or adjective is discarded. This is important because only these grammatical categories carry emotional information that can be used in further steps. So, more formally, the tokenization process converts the original text D in a set of tokens $T = \{t_1, t_2, ..., t_n\}$ where each element contained in T is part of the original document D. These tokens will feed the POS-T, which will label each token with semantic information. Finally all nouns, verbs, adverbs and adjectives will be collected in a set P, where $P_t = \{p_{(t,1)}, p_{(t,2)}, ..., p_{(t,k)}\}$ and $0 \leq$ k $\leq$ n and $P_t \subset T$.

The Stopwords list is a manual and predefined set $SW = \{sw_1, sw_2, ...sw_y\}$ of words, intended to avoid the analysis of common and irrelevant words. There are many examples of Stopwords lists on the internet and in libraries for Natural Language Processing (NLP). In our approach, after the Stopwords Removing process, the result list is a set $N = T - SW$.

After the parallel preprocessing tasks finish, the result document ST must contain a set of words where $ST = P \cap N$.

Later, in LM a lemmatizer process reduces the words to their lemma. This step is important because allows considering all inflected words as only one, producing the set of preprocessed texts $PR = \{LM(ST_1), LM(ST_2), ..., LM(ST_z)\}$.

The final result of this pipeline is a new emotional lexicon that considers the similarities of words used in expressions that cite the candidates, and their synonyms, and that is a personal representation of the sentiments about each candidate.

For this preprocessing step, we developed a Python module using Spacy[1] for automatizing the Tokenization, POS-T, Stopwords Removal and Lemmatization processes. We chose to use this toolkit in the development because it provides support for Brazilian Portuguese in all steps described earlier.

3.4 Sentiment Analysis

Once we created new personal lexicons for all candidates, the next step in the dataset creation was to analyze the emotions contained in the texts about each candidate.

For this purpose, we developed a tool that counts the frequency of each emotional word in a text. The result of this analysis is the final dataset, containing the emotional analysis of each Twitter message for each candidate, on a scale from 0 to 100 for each Plutchik's primary emotion.

This approach - a bag-of-word approach - was adopted because we intended to identify which emotions were more relevant to the voters when deciding their candidate, besides to generate a "candidate fingerprint" through the words used to describe them. Moreover, the absence of emotional corpora about politics in Portuguese restricted the possibility of using other techniques to identify the emotions in our texts.

4 Data Analysis

Once the dataset was created, we attempted to use data analysis to identify some particularities about the data, and how these particularities could explain the results of the elections. We used several techniques to identify correlations between the results of the elections and the data analyzed.

Once we identified the emotions that influenced the first-round results and how they did so, the next objective was to predict the results for the second round. To reach this objective, we decided to use an approach based on machine learning. The goal of this approach was to train a model that could accurately predict the percentage of votes for each candidate based on the emotions previously identified, the percentage of votes cast for each candidate in the first round, and the emotions contained in tweets on the day of the second-round vote.

4.1 Training Dataset

During the creation of a dataset for training the model, an important issue was identified: how to "translate" the emotions into a percentage of votes. Once the first round results were known, the relationship between candidates' emotional profiles and the percentage of votes cast on that day could also be determined. However, it was necessary to obtain many more examples to train a model.

[1] https://spacy.io/.

To bypass this obstacle, we chose to use the public information on electoral polls available from public institutes. The chosen institutes were: Instituto Brasileiro de Opinião e Pesquisa (Ibope)[2], Instituto Datafolha[3], Vox Populi[4] and Paraná Pesquisas[5] which are the most important polling institutes in Brazil.

To create the training dataset, we collected the voting intention results of 42 polls for candidates that we had in the dataset, which resulted in 324 examples for training, with 107,039.52 tweets analyzed. Later, knowing the period of each poll, we analyzed the average of each basic emotion for each candidate in the same period from the database. We then transferred these emotions to a new file, indicating the candidate's name, the number of candidates in the poll, period, emotions, and institute.

Despite the number of Tweets messages of each candidate are different, this new dataset contains each candidate's grouped emotions during the period of each poll. Thus, all candidates had the same number of registers in the dataset. This approach ensured that the most cited candidates in messages did not bias the dataset.

4.2 Predicting Results

After creating the training dataset, the next step was to train a model for predicting the results for the second round. For this purpose, we analyzed five different machine learning algorithms, to identify the best correlation between data and results. In all cases, the dataset was separated 70% for training and 30% for testing, using the Mean Absolute Error (MAE) as errors standard measure to have a comparison basis between traditional pools and twitter Sentiment Analysis.

The algorithms (using implementations for R and all tuned for the best fit) chosen for the analysis and their results after training the models are presented in Table 2.

Table 2. Algorithms evaluation

Algorithm	Correlation	MAE
Simple Linear Regression	0,2639	10,6302
SVM	0,3677	9,3608
Decision Table	0,5385	9,226
Extreme Gradient Boost	**0,9096**	**0,9787**
Random Forest	0,8088	6,322

[2] http://www.ibope.com.br.
[3] http://datafolha.folha.uol.com.br/.
[4] http://www.voxpopuli.com.br.
[5] http://www.paranapesquisas.com.br/.

The best result was obtained by an Extreme Gradient Boost algorithm, which had a correlation of 0,9096 (very strong correlation) between the results in the polls and the emotions in the same period, and a mean average error of 0,9787.

Once the model was created, the goal was to predict the percentage of votes for each candidate in the second round. In our experiment, we decided to predict the values based only on the emotions expressed on the day of second-round voting until 17:00. This time limitation is because voters could vote only until 17:00. Voters made their decisions based on their emotions during the voting process. Therefore, emotions that were expressed before the second-round election day were not crucial in this analysis.

After collecting the tweets for each second-round candidate - Jair Bolsonaro and Fernando Haddad - on October 20 and analyzing tweets about them using the same process presented in Sect. 3.3, we got the values presented in Table 3.

Table 3. Emotions in second round's day

Candidate	Anger	Anticipation	Disgust	Fear	Joy	Sadness	Surprise	Trust
Jair Bolsonaro	13,14%	11,14%	8,18%	12,71%	12,55%	14,80%	6,97%	20,50%
Fernando Haddad	12,39%	13,50%	6,41%	10,67%	14,95%	14,22%	7,57%	20,29%

When applying these values to the model created previously, we got a prediction of **54,58% for Jair Bolsonaro and 43,98% for Fernando Haddad and 0,9787% of MAE** that can be considered as a **correct prevision** because the official results for the second round were 55,15% for Jair Bolsonaro and 44,87% for Fernando Haddad, whose values are in the accepted error margin.

5 Conclusion

Social media have changed the way people interact and express their thoughts about everything. Because of this changing reality and the vast quantity of data available, sentiment analysis is becoming a powerful, fast, and relatively inexpensive tool that is extremely useful for analyzing many different types of scenarios and predicting future results.

Furthermore, although there have been many studies about the influence of social media on elections, there are not approaches using sentiment analysis to identify voters' emotions and predict future election results, while taking into account the results of previous studies.

The correlation between voters' emotions and the percentage of votes shows how vital is to know the audience's sentiments to plan effective strategies for interacting with them. Moreover, the very strong correlations that we found between the basic emotions and poll results', as well our model's successful prediction of the second-round results of Brazilians elections, strongly suggest that sentiment analysis can become a viable and reliable alternative to traditional opinion polls, with the advantages of being much faster and less expensive.

However, it is important to emphasize that there are no studies yet published about what the acceptable threshold is for replacing traditional polls with sentiment analysis. Also, it has not yet been established how many tweets must be analyzed to replace a traditional opinion poll with a sufficient degree of certainty. In future, we plan to identify the relationship between the number of textual opinions and the reliability of our model, to define a threshold for the safe use of sentiment analysis instead of an opinion poll.

Acknowledgements. This work has been supported by FCT - Fundação para a Ciência e Tecnologia within the Project Scope: UID/CEC/00319/2019.

References

1. Bermingham, A., Smeaton, A.: On using Twitter to monitor political sentiment and predict election results. In: Proceedings of the Workshop on Sentiment Analysis Where AI Meets Psychology (SAAIP 2011), pp. 2–10 (2011)
2. Feldman, R.: Techniques and applications for sentiment analysis. Commun. ACM **56**(4), 82–89 (2013)
3. Heredia, B., Prusa, J., Khoshgoftaar, T.: Location-based Twitter sentiment analysis for predicting the U.S. 2016 presidential election (2018)
4. Martins, R., Almeida, J., Novais, P., Henriques, P.: Creating a social media-based personal emotional lexicon. In: Proceedings of the 24th Brazilian Symposium on Multimedia and the Web, WebMedia 2018, New York, NY, USA, pp. 261–264. ACM (2018)
5. Martins, R., Almeida, J.J., Henriques, P.R., Novais, P.: Predicting performance problems through emotional analysis (short paper). In: OASIcs-OpenAccess Series in Informatics, vol. 62. Schloss Dagstuhl-Leibniz-Zentrum fuer Informatik (2018)
6. Maziero, E.G., Pardo, T.A.S., Di Felippo, A., Dias-da Silva, B.C.: A base de dados lexical e a interface web do tep 2.0-thesaurus eletrônico para o português do brasil. In: VI Workshop em Tecnologia da informação e da linguagem humana (TIL), pp. 390–392 (2008)
7. Mikolov, T., Sutskever, I., Chen, K., Corrado, G.S., Dean, J.: Distributed representations of words and phrases and their compositionality. In: Advances in Neural Information Processing Systems, pp. 3111–3119 (2013)
8. Mohammad, S., Turney, P.D.: Crowdsourcing a word-emotion association Lexicon. Comput. Intell. **29**(3), 436–465 (2013)
9. Plutchik, R.: Emotions: a general psychoevolutionary theory. In: Approaches to Emotion 1984, pp. 197–219 (1984)
10. Tumasjan, A., Sprenger, T.O., Sandner, P.G., Welpe, I.M.: Predicting elections with Twitter: what 140 characters reveal about political sentiment. In: Fourth International AAAI Conference on Weblogs and Social Media (2010)
11. Wang, H., Can, D., Kazemzadeh, A., Bar, F., Narayanan, S.: A system for real-time Twitter sentiment analysis of 2012 US presidential election cycle. In: Proceedings of the ACL 2012 System Demonstrations, pp. 115–120. Association for Computational Linguistics (2012)

Data Science in Pharmaceutical Industry

António Pesqueira[1], Maria José Sousa[2(✉)], Álvaro Rocha[3], and Miguel Sousa[4]

[1] Alexion Pharmaceuticals, Zurich, Switzerland
antonio.pesqueira@live.com
[2] Business Research Unit, Instituto Universitário de Lisboa, Lisbon, Portugal
mjdcsousa@gmail.com
[3] Universidade de Coimbra, Coimbra, Portugal
amrocha@gmail.com
[4] Essex University, Colchester, UK
miguel.ac.sousa@gmail.com

Abstract. Data Science demand from Medical Affairs (MA) functions in the pharmaceutical industry are exponentially increasing, where business cases around more modern execution of activities and strategic planning are becoming a reality. MA is still lagging in terms of implementing data science and big data technology in the current times, which means a reflecting immaturity of capabilities and processes to implement these technologies better. This paper aims to identify possible gaps in the literature and define a starting point to better understand the application of Data Science for pharmaceutical MA departments through the identification and synthesis of data science criteria used in MA case studies as presented in the scientific literature. We applied a Systematic Literature Review of studies published up to (and including) 2017 through a database search and backward and forward snowballing. In total, we evaluated 2247 papers, of which 11 included specific data science methodologies criteria used in medical affairs departments. It was also made a quantitative analysis based on data from a questionnaire applied to Takeda, a Pharma organization. The findings indicate that there is good evidence in the empirical relation between Data Technostructure and Data Management dimensions of the Data Science strategy of the organization.

Keywords: Data science · Pharmaceutical industry · Literature review · Big data technologies

1 Introduction

Over the last ten years, the Pharmaceutical Industry has been under greater scrutiny from regulators, healthcare professionals (HCPs), and patients. Resulted from those factors, pharmaceutical companies are now relying tremendously on strategic functions like Medical Affairs (MA).

For quite some time, MA' primary role was defined mainly as a scientific exchange, information support, managing daily regulatory reporting requirements or driving medical evidence generation (e.g., Phase IV studies, Real World Evidences or collaborative research) with a strong focus on priority diseases and developed products (Dyer 2011).

Á. Rocha et al. (Eds.): WorldCIST 2020, AISC 1159, pp. 144–154, 2020.
https://doi.org/10.1007/978-3-030-45688-7_15

In nowadays, MA is becoming a central function and core element of all pharma operations.

As MA was set up to operate independently from sales pressures, their role has grown from the last decade to creating strategic relationships with healthcare professionals (HCPs), Key Opinion Leaders (KOLs), and other stakeholders (e.g., payers, regulators, investigators, and others).

Also, the MA function is growing the importance and capacity in the technology processes improvements, technology adoption, and primarily being the focus with improving HCPs, KOLs, and stakeholder engagement activities.

Even though Data Science is a crucial concept within the pharmaceuticals industry, to the best of our knowledge, no study explored which criteria are being used in the context of Data Science applicable to MA. There is a need for research to assess which Data Science practices proposed by practitioners are beneficial. In this context, the main goal and contribution of this paper are to report the state of the are in the field of Data Science in MA employing a systematic literature review. We followed the guidelines proposed by Kitchenham and Charters (2007). In this paper, we detail our study and also point the gaps and future directions for research in Data Science for MA functions.

2 Theoretical Framework

2.1 Introduction of Data Science and Big Data Technologies

Data Science has attracted intense and growing attention from significant healthcare and life sciences organizations, including the big pharmaceutical companies that maintain a traditional data-oriented scientific and clinical development fields, as very far parts of the business and management structures., where data is not shared across different departments like market access or marketing.

The progressing digital transformation stimulates a considerable growth of digital data. Consequently, the data volume is forecasted 44 trillion gigabytes until 2020 (EMC Digital Universe 2014).

Data is an asset for any business organization, and having the capacity to understand all the connected trends, patterns, and extract meaningful information and knowledge from the data is referred to as data science.

The topics of data science technologies encompass two different aspects. Data science refers to traditional statistics that are produced on argumentation analysis or specific, methodical problems, with additional capacity for exploratory analysis and integration of data crunching and data mining. On another hand, data science technologies also are resulted from traditional software development that has a strong basis on traditional platforms like data warehouses, having the main capacity to aggregate several quantities of data managed and stored on distributed development platforms that later integrate into distributed computation or integrated software.

It is fundamental for the strategic decision-making process of a pharmaceutical organization to identify challenges, capitalize on opportunities, and to predict future trends and behaviours of HCPs, KOLs, and other stakeholders (Grom 2013).

The critical challenges for medical affairs are the management of the exponentially growing data, its meaningful analysis, deploying low-cost processing tools and practices while minimizing the potential risks relating to safety, inconsistency, redundancy, and privacy.

Data science is gaining middle ground in all MA establishments for the efficient utilization of resources: storage and time and efficient decision making to exploit new methods and procedures.

2.2 Medical Affairs in the Pharmaceutical Industry

A pharmaceutical industry model includes two main pillars: an R&D function being responsible for developing new medicines/molecules and a commercial team in charge of marketing and selling those products during a post-clinical phase and after all clinical development and trials are completed.

Medical Affairs (MA) serves as a connecting bridge between R&D and Sales/Marketing, facilitating the transition of products and knowledge from R&D to the market access and commercialization stages. Despite many changes over the last years, all the stakeholders (payers, regulators, HCPs) continued to demand high levels of scientific knowledge and to have better interactions in terms of transparency and information sharing with the industry in its interactions. Also, here, the role and importance of MA in a more complex healthcare marketplace environment are increasing exponentially (Jain 2017).

In the past, pharmaceutical companies considered MA just a support function and one that could even slow down marketing and commercial activities.

Furthermore, in this fasting moving market dynamics, the way MA engage with all the involved stakeholders is becoming more fundamental than ever, where MA is considered as the best function to provide scientific and clinical expertise to support approved medicines, work closely with R&D in the developing new drugs throughout post-approval activities and to be better prepared to respond to customer demands and to develop and maintain stronger long-term relationships with key opinion leaders, scientific societies, payers and patient groups (Plantevin et al. 2017).

To increase the transparency and efficiency of all the activities developed and its relationship with physicians, it has undergone significant changes in the way MA can understand all the surrounding data and the way that can quickly manage all the hidden connections and patterns to understand the engagement outcomes and stakeholders needs.

Many of these changes have led to an increase in the responsibility of MA but still not the full capacity to make a practical usage of the data.

3 Review Method

According to Kitchenham and Charters (2007), a systematic review is an evidence-based technique that uses a well-structured and repeatable methodology to identify, analyze and interpret all the relevant academic papers related to a specific research question or phenomenon of interest.

A fundamental assumption of this technique is the involved protocol, which is the plan that will describe the conduct of the systematic review. It includes the research questions, search process, selection process, and data analysis procedures.

A summary of the protocol used in this review is given in the following sub-sections.

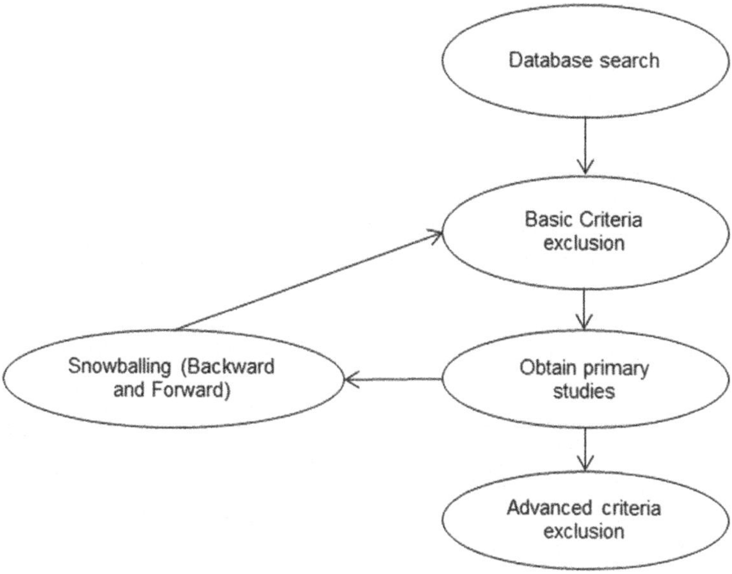

Fig. 1. Overview of the search and selection process

3.1 Research Questions

This research aims to identify possible gaps in the literature and define a starting point to define Data Science for Medical Affairs practitioners, employees or representatives, through the identification and synthesis of the Data Science criteria used in Medical Affairs projects as presented in the scientific literature. Given this, we formulated the following research question (RQ): What are the most used statistical techniques in Medical Affairs case studies, research papers, or academic investigation articles, where data scientists were used, and conventional data science tools were selected?

3.2 Search Strategy

To minimize the probability of missing relevant articles, publications, we used a combined search strategy, which is based on database search and snowballing (backward and forward). First, we defined a search string and used it to search databases containing scientific papers in the context of data science. After it was applied the basic exclusion criteria, and the resulting papers were defined as the starting set for the snowballing process. After executing the snowballing iterations, we applied the advanced criteria exclusion, which is related to the actual data extraction and quality assessment. We show an overview of the search and selection process in Fig. 1.

We decided to use this strategy to avoid missing publications, papers, or articles due to limitations and inconsistencies of digital libraries. They have different formats to handle the Boolean expressions, as discussed in Brereton et al. (2007), and we were not sure how reliable is their ability to handle searches with long strings. Finally, there is evidence in the literature on the risks of missing papers using only one approach (Badampudi et al. 2015).

3.3 Search Terms

Starting with the research questions, suitable keywords were identified using synonyms and related terms. The following keywords were used to formulate the search string:

- Population: Data Science and Medical Affairs. Alternative keywords: Medical Data Science, Data Science in Medical Affairs, Medical Affairs Data, and Data Science in Pharmaceutical.
- Intervention: Data Science. Alternative keywords: data science in medical affairs and medical data.
- Context: Industry or academia. Our target population was papers performed in the industry or academy, and we intended to capture papers in that context regardless of the type of research performed.

To define a first version of the search string, the keywords within a category were joined by using the Boolean operator 'OR,' and the two categories were joined using the Boolean operator 'AND.' This was done to target only papers in the context of data science related to medical affairs. To simplify the strings and include additional synonyms, we defined the following search string:

("data science" OR "medical affairs" OR "medical" OR "data" OR "medical data science" AND (medical AND (data OR science) AND (data science OR science OR (medical AND (affairs OR data clinical OR medical science OR data affairs)) OR "data science in medical affairs".

3.4 Data Sources

Since our goal is to cover the literature published in Data Science, we chose the following digital databases for data retrieval: ACM Digital Library; Science Direct; Springer; Web of Science; Wiley Online Library; Google Scholar. We did not include IEEExplore because it could not handle our search string due to its size. On the other hand, Web of Science and Google Scholar also indexes IEEE papers.

3.5 Selection Criteria

Before applying the selection criteria given the topic of the review, we defined generic exclusion criteria: Published in non-peer reviewed publication channels such as books, thesis or dissertations, tutorials, keynotes, and others. OR Not available in English OR A duplicate.

We implemented the first two criteria in the search strings that were executed in the digital libraries, wherever possible.

Afterward, the remaining papers were evaluated through two sets of selection criteria: basic and advanced.

3.5.1 Basic Criteria

We applied the necessary criteria to evaluate if papers are relevant to the aims of our paper by reading the titles and abstracts. These criteria were applied to papers that passed the generic exclusion criteria and were identified through database search or snowballing. In this context, we included papers that: Are related to data science AND Are related to medical affairs.

Following the procedure presented in Ali et al. (2014), we decided that papers are classified as: Relevant, Irrelevant or Uncertain (in the case, the available information on the title and abstract is inconclusive).

Only the papers evaluated as relevant were select for inclusion in the next section of this paper.

3.5.2 Advanced Criteria

The advanced criteria are related to the actual data extraction, in which the full-text of the papers were thoroughly read.

The studies published in multiple papers and only including the extended version of the study. Additionally, all the papers that were not relevant to assess the request questions were excluded as they did not contain any relevant information. In other words, a paper was only included if it contained examples of data science applied and used in a medical affairs context.

3.5.3 Snowballing

The snowballing approach was, first, performed on the set of papers identified through the database search and included using the necessary criteria. For each paper in the set, we applied the backward and forward snowballing.

To execute the forward snowballing, we used Springer and Google Scholar to identify the title and abstract of the papers, citing our set of selected papers. We applied the basic criteria, to include these papers. To execute the backward snowballing, first, we distributed the papers to be evaluated, and the reviewer was responsible for applying the advanced exclusion criteria.

This was done by evaluating the title in the reference list and, if necessary, the place of reference in the text. Afterward, the included studies were evaluated using the essential criteria, in which the reviewer assessed each paper.

3.6 Data Extraction

We used a spreadsheet editor to record relevant information. With the spreadsheet, we were able to map each datum extracted with its source. From each paper, we extracted general information such as the year and name of the publication channel and data related to the RQs. The following data were extracted from the papers: i. type of article (journal, conference, magazine and scientific journal, webpage, and others), ii. name of the publication channel, iii. year of publication, iv. data science used the method, v. statistical analyze applied, vi. number of cases, vii. research type, viii. research question type, ix. empirical research type, x. research validation.

For question (vii), we used the classification presented by Wieringa et al. (2006): validation research, evaluation research, solution proposal, philosophical papers, opinion papers, or experience papers. For (viii), we used the classification presented by Shaw (2003): method or means of development; a method for analysis or evaluation; design, evaluation, or analysis of a particular instance; generalization or characterization; or feasibility study or exploration. For question (ix), we used the classification presented by Tonella et al. (2007): experiment, observational study, experience report, case study, or systematic review. For (x), we used the classification scheme presented by Shaw (2003): analysis, evaluation, experience, example, persuasion, or blatant assertion. Also, as we can see from Fig. 2, we initially identified 2247 papers through the different data sources, and then we start applying the criteria.

The final seed set result was 32 remaining papers, where 864 duplicates were removed from the essential criteria definitions and then 1351 duplicates removed for the seed set creation.

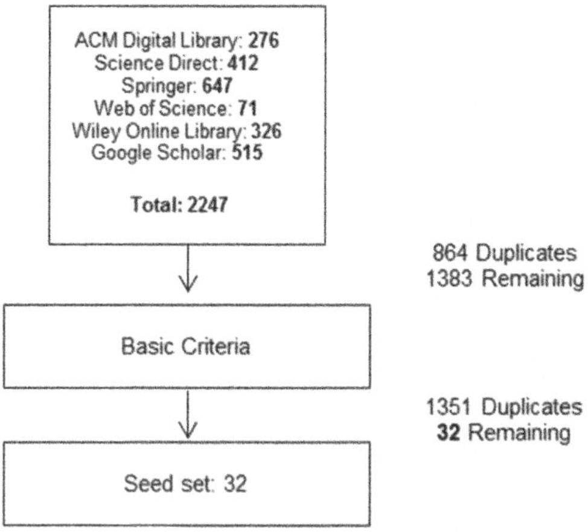

Fig. 2. Overview of the database search

4 Results

In this section, we present the results for the systematic review process and the research questions as well. In Fig. 3, we present an overview of the number of studies passing through the different stages of the study. We show details of the results of the database search in Fig. 2 and of the results of the snowballing process, in which we iterated twice, in Fig. 3.

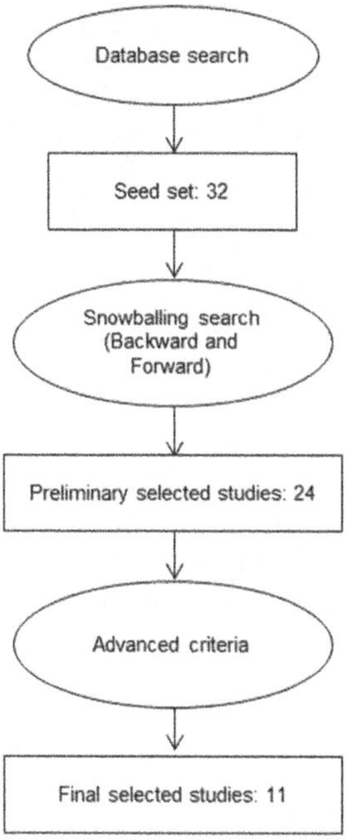

Fig. 3. Number of papers in study selection.

In Fig. 4, we show the number of papers per year. In Fig. 5, we show the distribution of papers per type of publication channel.

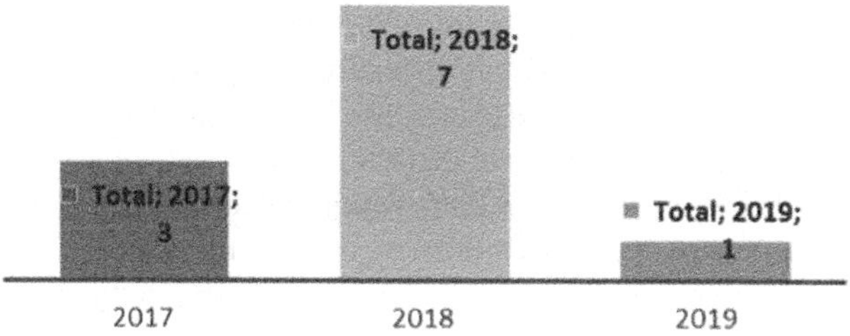

Fig. 4. Number of papers per year.

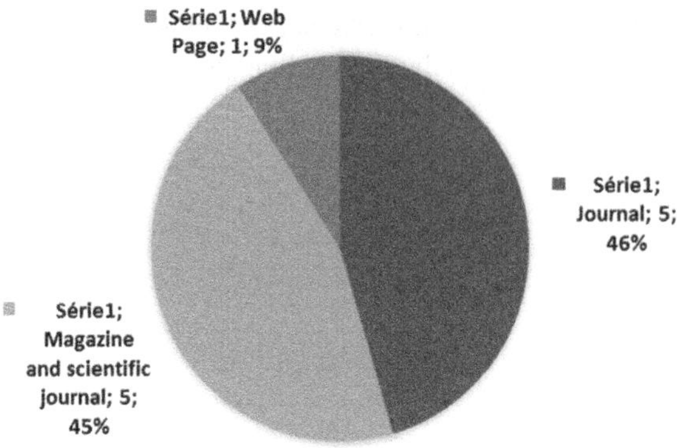

Fig. 5. Distribution of papers per type of publication channel.

5 Conclusions

This study presents a systematic literature review on the application of data science to medical affairs. We used a hybrid approach composed of database search and snowballing. The primary search fetched 1383 unique results, from which 32 papers were selected as a seed set for snowballing. After the data extraction, only 11 papers were included in the study. Data from these papers were analyzed to answer each research question. There is a variety of data science techniques used in medical affairs. Furthermore, some papers used a multilevel approach to perform more advanced statistical analysis and using high-performance computing capacity from open software tools.

Data scientists or medical affairs practitioners can use the results of this study as a guide for them to apply data science techniques on their projects or compare them with their methodologies. Moreover, based on the results of this study, we recommend that there is a strong need to publish more papers presenting how data science is applied in medical affairs studies and research projects and to conduct empirical studies to assess the results of applying this practice.

For future work, we intend to execute a survey with medical affairs practitioners and compare the collected data with our results in this study.

Data science is a new interdisciplinary specialty, which requires strong practical ability and adaptive organizational culture to effectively implement the described techniques and models to support medical affairs in daily activities.

In conclusion, the role of Medical Affairs within a pharmaceutical company serves to spearhead the dissemination (and in some cases, the generation) of unbiased clinical and scientific information about medicine to the healthcare community and to offer medical and scientific expertise.

The purpose of this article was to demonstrate, and communicate the value of data science in a medical affairs function in enhancing the knowledge of medicines and the associated therapeutic areas in which a company focus its research efforts, in providing thorough understanding of its medicines: interpret emerging scientific trends, clinical data and the competitive landscape and align internal stakeholders on a balanced benefit/risk proposition.

All of the described research questions were indeed validated in this article where data science can play a decisive role in giving the necessary tools and processes to a medical affairs department in communicating to the medical and scientific communities in an accurate, fair and balanced manner about the benefits and the risks of the medicines, enabling prescribers and other healthcare decision-makers to make informed decisions with patients and use medicines safely and effectively.

Data Science also gives concrete support to medical affairs in working cross-functionally with colleagues from Marketing, Sales, Regulatory and Access to guide the acquisition and integration of clinical data so that existing clinical evidence is communicated accurately, reflecting the value of the medicines, to help to inform the right capital allocation decisions in the advancement of the lifecycle of the brands and the company's pipeline and to ensure launch readiness, organizing and training medical affairs colleagues and providing them with the tools to excel within the pre-, and post-launch period.

Acknowledgements. We would like to thank to AISTI (Associação Ibérica de Sistemas e Tecnologias de Informação/Iberian Association for Information Systems and Technologies) for the financial support granted in this study.

References

Dyer, S.: Medical Science Liaison – Aligning the activities and goals of Medical Science Liaison teams for strengthened corporate sustainability. The Medical Science Liaison Corporation (2011)

Badampudi, D., Wohlin, C., Petersen, K.: Experiences from using snowballing and database searches in systematic literature studies. In: Proceedings of the 19th International Conference on Evaluation and Assessment in Software Engineering (EASE 2015). ACM, New York, Article 17 (2015). 10 p.

Grom, T.: Medical affairs: beyond the science. Showcase feature (2013). http://www.pharmavoice.com/article/medical-affairsbeyond-the-science/. Accessed 30 Jan 2019

Jain, S.: Bridging the gap between R&D and commercialization in the pharmaceutical industry: role of medical affairs and medical communications. Int. J. Biomed. Sci. **3**(3), 44–49 (2017)

Kitchenham, B., Charters, S.: Guidelines for performing systematic literature reviews in software engineering. Technical report EBSE-2007-01. School of Computer Science and Mathematics, Keele University (2007)

Ali, N.B., Petersen, K., Wohlin, C.: A systematic literature review on the industrial use of software process simulation. J. Syst. Softw. **97**(C), 65–85 (2014)

Tonella, P., Torchiano, M., Bois, B.D., Systä, T.: Empirical studies in reverse engineering: state of the art and future trends. Empir. Softw. Eng. **12**(5), 551–571 (2007)

PharmaForum. Medical Affairs The heart of a data-driven, patient-centric pharma (2017). https://pharmaphorum.com/views-and-analysis/medical-affairs-heart-data-driven-patient-centricpharma/. Accessed 30 Jan 2019

Plantevin, L., Schlegel, C., Gordian, M.: Reinventing the Role of Medical Affairs (2017). http://www.bain.com/publications/articles/reinventing-the-role-of-medical-affairs.aspx

Brereton, P., Kitchenham, B.A., Budgen, D., Turner, M., Khalil, M.: Lessons from applying the systematic literature review process within the software engineering domain. J. Syst. Softw. **80**(4), 571–583 (2007)

Wieringa, R., Maiden, N., Mead, N., Rolland, C.: Requirements engineering paper classification and evaluation criteria: a proposal and a discussion. Requirements Eng. **11**(1), 102–107 (2006)
Shaw, M.: Writing good software engineering research papers. In: Proceedings 25th - International Conference on Software Engineering, pp. 726–736 (2003)
Strategic Benchmarking Research (2014). http://pt.slideshare.net/bestpracticesllc/pop-253-a-report-summary-strategic-kol-management. Accessed 30 Jan 2019

DICOM Metadata Quality Analysis for Mammography Radiation Exposure Characterization

Milton Santos[1,4], Augusto Silva[2,4], and Nelson Pacheco Rocha[3,4](✉)

[1] Health Sciences School, University of Aveiro, Aveiro, Portugal
mrs@ua.pt
[2] Department of Electronics, Telecommunications and Informatics, University of Aveiro, Aveiro, Portugal
augusto.silva@ua.pt
[3] Medical Sciences Department, University of Aveiro, Aveiro, Portugal
npr@ua.pt
[4] Institute of Electronics and Informatics Engineering of Aveiro, University of Aveiro, Aveiro, Portugal

Abstract. The usage of ionizing radiation on human tissues for medical purposes has been object of regular analyses using Digital Imaging and Communication in Medicine (DICOM) metadata. Particularly, the DICOM metadata related to mammographic studies has been used to support the monitoring of individual and population exposure. The objective of this work was to analyze the quality of DICOM metadata to characterize radiation exposure in mammographic studies performed during the first year of activity of a mammography equipment. Although DICOM metadata allow to characterize the radiation dose in mammographic studies, the results show that it is pertinent to effectively improve the quality of the stored metadata.

Keywords: DICOM metadata · Radiology · Medical imaging · Data quality · Mammography · Radiation dose exposure

1 Introduction

The technological evolution related to the provision of medical imaging healthcare led to the existence of large amounts of digital information, which promotes a significant change in the expectations of both patients and healthcare providers.

A considerable part of the information used and produced during medical imaging procedures is stored in Radiology Information Systems (RIS) and Picture and Archiving Communication Systems (PACS). These imaging data embedded in the Digital Imaging and Communication in Medicine (DICOM) format [1] can represent a very relevant source of information for secondary studies. However, the usefulness of the available information may be impaired if it is not complete, sufficiently detailed, consistent, relevant, timely and accurate in a given context [2, 3].

Á. Rocha et al. (Eds.): WorldCIST 2020, AISC 1159, pp. 155–164, 2020.
https://doi.org/10.1007/978-3-030-45688-7_16

Concerning the analysis of the quality of the DICOM metadata included in mammographic studies to retrieve the radiation exposure of the patients, the scientific literature does not report significant evidence. Therefore, the objective of the research work reported by this paper was to identify the extent to which the quality of the available DICOM metadata may contribute to the characterization of the radiation exposure of mammography studies.

In section two, the use of DICOM metadata to characterize medical imaging studies and radiology departments performance, but also some scenarios resulting from the lack of clinical data quality are addressed. In section three we present the materials and methods used for DICOM metadata quality analysis for mammography radiation exposure characterization. Section four is devoted to results presentation and discussion. Sections five gathers the conclusions and some prospects for future work.

2 Background

The DICOM metadata can be used in biomedical research context [4] but also to retrieve efficiency metrics for medical imaging productivity such as, for example, performance indicators [5–7].

DICOM metadata have been used with different objectives, namely to support protocols usage and optimization [8–10], equipment performance monitoring [11], or measurements of the exposure to which patients are subject during mammographic studies [12–15].

The RIS and PACS information transaction processes are critical for carrying out medical imaging studies in an appropriate way [16], namely those studies that require data to be shared with PACS databases. In this respect, some authors have identified lack of data quality [17, 18] that can hinder health care delivery [19].

Quality and completeness of metadata that are part of the DICOM images may be compromised, namely by the inappropriate completion of DICOM attributes [20–22]. On the other hand, the fact that only a small number of attributes are mandatory has impact on the value of the information produced.

The analysis of data quality has been subject of considerable research, namely related to Electronic Health Records (EHR) secondary use [23], the definition of frameworks for single-site and multisite data quality assessment to support EHR-based clinical research [24], and broader assessments of distributed data networks [25].

The quality of the available data related to medical imaging is often difficult to characterize. However, one can expect that poorly documented imaging processes are very likely to induce losses in efficiency and effectiveness of medical imaging. Moreover, the performance of the medical imaging services impacts the respective stakeholders [26].

3 Methods and Materials

We herein describe an exploratory and retrospective study based on DICOM metadata related to mammography studies and stored in the PACS of a medium size healthcare facility.

Authorization to access the data was granted by the administration and ethics commission, upon established guarantees of the information confidentiality, as well as the patients' privacy.

The DICOM metadata from the mammography studies performed on the healthcare facility during the first year of equipment activity were analysed. Inclusion criteria were defined as all mammography studies that were stored in the local PACS, publicly disclosing the relevant DICOM attributes and that could be indexed with Dicoogle [27].

In terms of the experimental plan, the authors start with the indexing and extraction of DICOM metadata from the PACS archive. Access to the DICOM metadata was achieved using the Dicoogle tool installed on a local virtual machine.

In a second step the authors characterize the mammography Information Object Definition (IOD). Relevant metadata was extracted and exported to Microsoft Excel format, making possible a subsequent statistical analysis using the Statistical Package for Social Sciences (SPSS).

Posteriorly, the data quality analysis (e.g., missing data or improperly filled fields identification) and the data normalization were performed (e.g., among the DICOM Patient Age or Patient Birth Date attributes in files where the Patient Age field was empty or with "0" as a value). Moreover, the authors identified unexpected values related to Patient identifications and radiation dose (e.g., incorrect values or empty fields).

Finally, the exposure characterisation was performed based on the DICOM attribute Entrance Dose in mGy (0040,8302) (which stores the values of the Entrance Surface Air Kerma - ESAK) and the DICOM attribute Organ Dose (0040,0316) (which stores the values of the Average Glandular Dose - AGD) [28], as well as DICOM attributes Exposure (0018,1152), KVP (0018,0060), Body Part Thickness (0018,11A0), Compression Force (0018,11A2) and View Position (0018,5101).

4 Results

4.1 From DICOM Metadata Indexing and Extraction to Mammography Image IOD Characterization

The indexation of the DICOM metadata stored in the PACS of the selected healthcare facility took 648 h. The indexing process covered more than 4152 GB of information resulting from the aggregation of the DICOM metadata belonging to 7525275 images, corresponding to studies performed on 63789 patients (Table 1).

From the initial sample, the authors analyzed the DICOM metadata belonging to 379 mammography studies performed on 351 patients that correspond to the studies carried out during the first year of equipment activity. For that, a specific query was defined.

In addition to the public 124 DICOM attributes belonging to different modules that constitute the IOD of the mammography images, the following sequences were also identified: Anatomic Region Sequence, Icon Image Sequence, Referenced Study Sequence, Request Attributes Sequence, Source Image Sequence, VOI LUT Sequence, and View Code Sequence (Fig. 1).

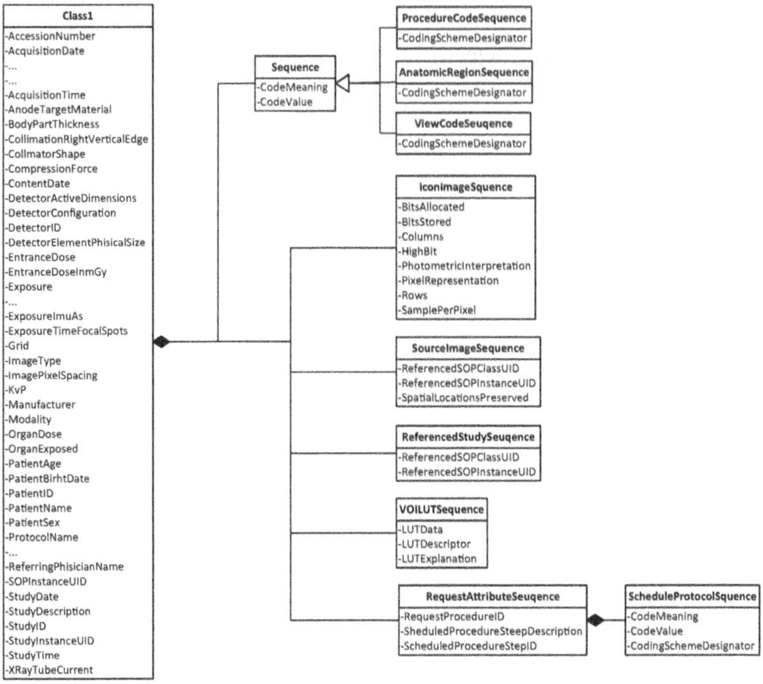

Fig. 1. Partial mammography image IOD based on public DICOM attributes (UML class diagram).

Table 1. Results from de DICOM metadata indexing process over the whole PACS.

Indexed Directories	Dimension (Gb)	Index Size (Gb)	Indexing Time (h)	Number of images	Number of Patients
A	745.81	6.54	117		
S	672.80	5.39	105		
C	577.00	5.06	90		
D	581.00	5.10	91	7525275	63789
E	234.00	2.06	36		
F	246.00	2.15	38		
G	703.00	6.17	110		
H	393.00	3.45	61		
Σ 8	4152.61	35.92	648	7525275	63789

4.2 Data Quality Analysis and Sample Normalization

The process of DICOM metadata quality analysis and data normalization was done on the specific and relevant DICOM metadata for the purpose of the study and identified in Table 2.

Table 2. DICOM attributes under analysis for DICOM metadata quality characterization.

X Ray Tube Current	Exposure Time	Compression Force
KVP	Filter Material	Body Part Thickness
View Position	Field Of View Shape	Body Part Examined
Patient Sex	Exposure In uAs	Anode Target Material
Patient Orientation	Exposure Control Mode Description	Study Date
Patient Name	Organ Dose	Patient Age
Patient ID	Exposure	SOP Instance UID
Patient Birth Date	Entrance Dose In mGy	Study Description

From the initial sample, 47 images were identified with the Patient ID attribute incorrectly filled in, 29 images with the Patient Age attribute empty, 128 images with illegibly Entrance Dose in the mGy DICOM attribute, and 40 empty fields. On the other hand, 11 files related to studies performed on operative parts (e.g., tissues resulting from the surgical removal of breast tumors) were removed.

The data normalization process resulted in a sample of 1302 images related to 342 studies performed on 321 patients (Table 3), which results from the exclusion of 16.5% images related to 9.8% studies performed on more than 8.0% patients.

Table 3. Initial sample normalization process results. Files excluded and final sample.

DICOM attribute/tag	Unexpected value	n
Patient Age (0010,0020)	"0" as a value	29
Patient ID (0010,0020)	Incorrect value	47
Entrance Dose In mGy (0040,8302)	Empty field	40
Entrance Dose In mGy (0040,8302)	Illegible value	128
Studies on operative parts		11
Final number of validated files	1302	
Total of validated studies	342	
Total validated patients	321	

At a strictly DICOM metadata level, the poor data quality may jeopardize the patient-centered imaging care delivery and the definition of medical imaging strategies to promote a better patient exposure protection, especially taking into account factors such as the age or exposure levels.

Moreover, DICOM metadata quality also impacts the quality of information stored in DICOM Structured Report (DICOM SR), or in other information systems.

4.3 Population Characterization

The percentage distribution of patients by age group and gender is shown in Fig. 2. In this figure, we found that the percentage of male patients is residual. We also found that the patients who contribute the most to the final sample belong to age groups between 40 and 59 years.

4.4 Exposure Characterization

The analysis of mammographic studies exposure levels was done using a sample consisting of 321 patients, 342 studies and 1302 images.

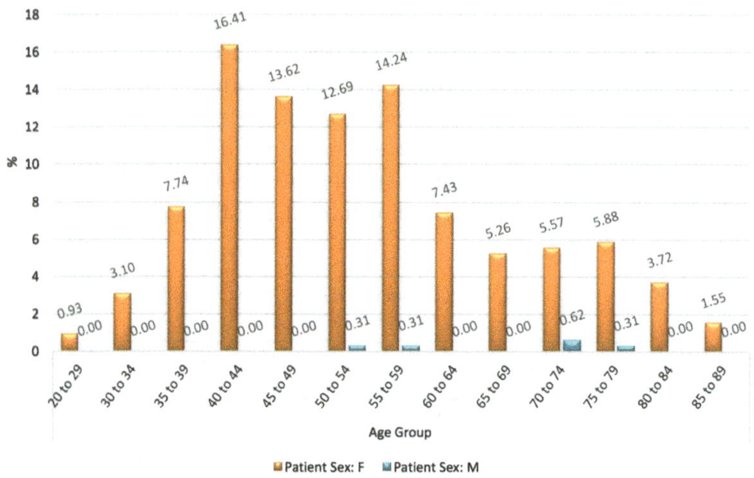

Fig. 2. Distribution, by age group and gender, of patients with mammography studies performed.

From the analysis of the DICOM attribute View Position, the DICOM attribute that allows the registration of the mammographic projection designation (e.g., Craniocaudal - CC or Mediolateral oblique - MLO), the authors identified 633 images related to CC, representing 48.6% of the mammographic images, and 666 images (51.2%) related to MLO projection. It was also identified an image related to the projection Medio-Lateral and two Superoinferior oblique projection images - SIO (Table 4).

Table 4 shows the average values and standard deviation of the values related to breast thickness (Body Part Thickness), compression force, ESAK (Entrance Dose In mGy DICOM attribute), AGD (Organ Dose DICOM attribute), mAs (Exposures DICOM attribute) and Kv. When we analyze Table 4, we find that the average exposure dose in the CC projections (ESAK = 7.85 mGy and AGD = 1.67 mGy) is lower than those obtained in the MLO projections. However, in these projections, the mean compression as well as the average thickness of the breast after compression, are higher than those obtained in CC projections and have a higher standard deviation.

When analyzing the standard deviation values identified for ESAK values (4.48 mGy in CC projections and 4.96 mGy in MLO projections) and mAs (standard deviation values

Table 4. Average values and standard deviation of mammography studies dose descriptors and technical factors.

Projection	n -%-	Body Part Thickness (0018.11A0) [mm] - Standard Deviation-	Compression Force (0018.11A2) [N] - Standard deviation-	Entrance Dose In mGy (0040.8302) [mGy] - Standard deviation-	Organ Dose (0040.0316) [mGy] - Standard deviation-	Exposure (0018.1152) [mAs] - Standard deviation-	KVP (0018.0060) [kv] - Standard deviation-
CC	633 - 48.62 -	50.65 - 12.19 -	98.89 - 43.16 -	7.85 - 4.48 -	1.67 - 0.73 -	85.80 - 56.04 -	28.42 - 1.54 -
ML	1 - 0.08 -	28.00	160	3.70	1.09	48.10	27.00
MLO	666 - 51.15 -	57.58 - 16.03 -	108.05 - 44.71 -	8.86 -4.96-	1.70 -0.68-	91.86 -55.92-	28.72 - 1.57 -
SIO	2 - 0.15 -	68.50 - 17.68 -	100 - 28.28 -	17.63 - 1.01 -	2.37 - 0.14 -	200.00	26.00

of 56.0 mAs in CC projections and 55.92 more MLO projections) we can infer that these results may deserve special attention. In fact, these values may be related to some variability of the sample population or variability in the selection of exposure factors but may also result from the use of different automatic exposure protocols selected in accordance with the study objectives and radiographer performance.

On the other hand, the analysis performed by age group allowed a better characterization of the variability of the values obtained, both concerning dose descriptors and other parameters associated to the study execution, such as the variability of breast compression and thickness. These data are shown in Figs. 3 and 4.

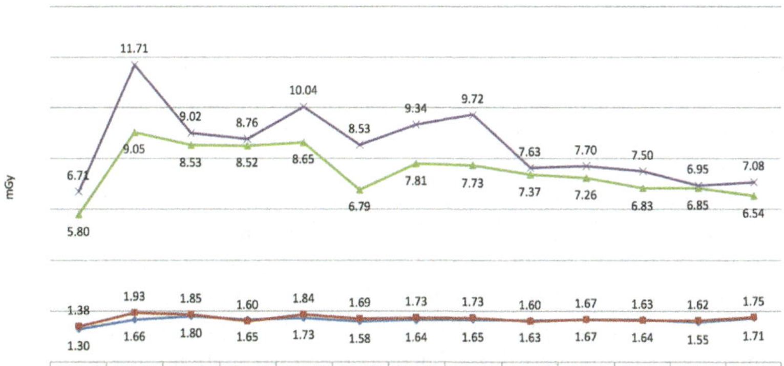

	25 to 29	30 to 34	35 to 39	40 to 44	45 to 49	50 to 54	55 to 59	60 to 64	65 to 69	70 to 74	75 to 79	80 to 84	85 to 89
CC MGD Average (mGy)	1.30	1.66	1.80	1.65	1.73	1.58	1.64	1.65	1.63	1.67	1.64	1.55	1.71
MLO MGD Average (mGy)	1.38	1.93	1.85	1.60	1.84	1.69	1.73	1.73	1.60	1.67	1.63	1.62	1.75
CC ESAK Average (mGy)	5.80	9.05	8.53	8.52	8.65	6.79	7.81	7.73	7.37	7.26	6.83	6.85	6.54
MLO ESAK Average (mGy)	6.71	11.71	9.02	8.76	10.04	8.53	9.34	9.72	7.63	7.70	7.50	6.95	7.08

MGD & ESAK Average values / Age group

Fig. 3. ESAK and AGD analysis per age group.

The radiation dose to which the patients were submitted during mammography might be inferred from the ESAK and the AGD together with other exposure parameters, such as compression and breast thickness. This analysis may contribute to the correct interpretation of the exposure to which the patients were exposed.

Within the scope of the values related to the compression used, it was also possible to identify situations that may require better attention on the part of health professionals,

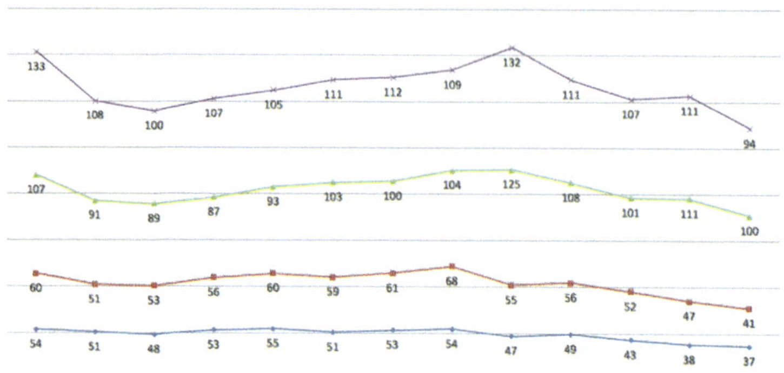

	25 to 29	30 to 34	35 to 39	40 to 44	45 to 49	50 to 54	55 to 59	60 to 64	65 to 69	70 to 74	75 to 79	80 to 84	85 to 89
MLO Average Compression (N)	133	108	100	107	105	111	112	109	132	111	107	111	94
CC Average Compression (N)	107	91	89	87	93	103	100	104	125	108	101	111	100
MLO Average Thickness (mm)	60	51	53	56	60	59	61	68	55	56	52	47	41
CC Average Thickness (mm)	54	51	48	53	55	51	53	54	47	49	43	38	37

Compression & Thickness Average values / Age group

Fig. 4. Average breast Compression and breast Thickness per age group analysis.

namely concerning the standard deviation of 43.16 N in the CC incidence and 44.71 N in the MLO incidences. However, it is pertinent to consider that the compression performed during the study may not depend only on the professional performing the study [29, 30].

5 Conclusion

As a conclusion, the study reported in this paper shows that it is possible to characterize the exposure to which the population is subject in mammography studies based on the DICOM metadata stored in a PACS archive. However, it is pertinent to effectively poll the quality of the stored metadata.

PACS administrators and or Radiology Department Managers should enforce DICOM conformance statements where clear fulfilment of dose related metadata should be in line with more recent DICOM recommendations. Moreover, manufacturers must fulfil these recommendations through public DICOM attributes.

As future work, we anticipate the relevance of verifying if the quality of the metadata changes over time, namely those that characterize the patients and the exposure to which they are submitted during the imaging studies and evaluating the repercussion of using dose monitoring information systems.

Acknowledgments. This work was financially supported by National Funds through FCT – Fundação para a Ciência e a Tecnologia, I.P., under the project UI IEETA: UID/CEC/00127/2019.

References

1. Association NEM, ed. Digital Imaging and Communication in Medicine (DICOM) Part 3: Information Object Definition (2019). http://dicom.nema.org/dicom/2003/03_03PU.PDF. Accessed 31 Dec 2019

2. Wyatt, J.C., Liu, J.L.Y.: Basic concepts in medical informatics. J. Epidemiol. Community Health. **56**(11), 808 LP–812 (2002). https://doi.org/10.1136/jech.56.11.808
3. Tayi, G.K., Ballou, D.P.: Examining Data Quality, vol. 41. ACM Press, New York (1998). https://doi.org/10.1145/269012.269021
4. Freymann, J., Kirby, J., Perry, J., Clunie, D., Jaffe, C.: Image data sharing for biomedical research—meeting HIPAA requirements for de-identification. J. Digit. Imaging **25**(1), 14–24 (2012). https://doi.org/10.1007/s10278-011-9422-x
5. Hu, M., Pavlicek, W., Liu, P.T., et al.: Informatics in radiology: efficiency metrics for imaging device productivity. Radiographics **31**(2), 603–616 (2011). https://doi.org/10.1148/rg.312105714
6. Jahnen, A., Kohler, S., Hermen, J., Tack, D., Back, C.: Automatic computed tomography patient dose calculation using DICOM header metadata. Radiat. Prot. Dosimetry **147**(1–2), 317–320 (2011). https://doi.org/10.1093/rpd/ncr338
7. Prieto, C., Vano, E., Ten, J.I., et al.: Image retake analysis in digital radiography using DICOM header information. J. Digit. Imaging **22**(4), 393–399 (2009). https://doi.org/10.1007/s10278-008-9135-y
8. Klausz, R., Shramchenko, N.: Dose to population as a metric in the design of optimised exposure control in digital mammography. Radiat. Prot. Dosimetry **114**(1–3), 369–374 (2005). https://doi.org/10.1093/rpd/nch579
9. Chen, B., Wang, Y., Sun, X., et al.: Analysis of patient dose in full field digital mammography. Eur. J. Radiol. **81**(5), 868–872 (2012). https://doi.org/10.1016/j.ejrad.2011.02.027
10. Ko, M.-S., Kim, H.H., Cha, J.H., Shin, H.J., Kim, J.H., Kim, M.J.: Dose reduction in automatic optimization parameter of full-field digital mammography: breast phantom study. J. Breast Cancer **16**(1), 90–96 (2013)
11. McCullagh, J.B., Baldelli, P., Phelan, N.: Clinical dose performance of full field digital mammography in a breast screening programme. Br. J. Radiol. **84**(1007), 1027–1033 (2011). https://doi.org/10.1259/bjr/83821596
12. Morán, P., Chevalier, M., Ten, J.I., Fernández Soto, J.M., Vañó, E.: A survey of patient dose and clinical factors in a full-field digital mammography system. Radiat. Prot. Dosimetry **114**(1–3), 375–379 (2005). https://doi.org/10.1093/rpd/nch514
13. Chevalier, M., Morán, P., Ten, J.I., Fernández Soto, J.M., Cepeda, T., Vañó, E.: Patient dose in digital mammography. Med. Phys. **31**(9), 2471–2479 (2004)
14. Alizadeh Riabi, H., Mehnati, P., Mesbahi, A.: Evaluation of mean glandular dose in a full-field digital mammography unit in Tabriz, Iran. Radiat. Prot. Dosimetry **142**(2–4), 222–227 (2010). https://doi.org/10.1093/rpd/ncq218
15. Bor, D., Olgar, T., Aydin, E., Tükel, S.: Variations in breast doses for an automatic mammography unit. Diagn. Interv. Radiol. **14**, 122–126 (2008)
16. Khorasani, R.: Computerized physician order entry and decision support: improving the quality of care. RadioGraphics **21**(4), 1015–1018 (2001)
17. Troude, P., Dozol, A., Soyer, P., et al.: Improvement of radiology requisition. Diagn. Interv. Imaging **95**(1), 69–75 (2014). https://doi.org/10.1016/j.diii.2013.07.002
18. Cohen, M.D., Curtin, S., Lee, R.: Evaluation of the quality of radiology requisitions for intensive care unit patients. Acad. Radiol. **13**(2), 236–240 (2006). https://doi.org/10.1016/j.acra.2005.10.017
19. Clément, O.: Iatrogénie des produits de contraste. J. Radiol. **86**(5, Part 2), 567–572 (2005). https://doi.org/10.1016/S0221-0363(05)81409-X
20. Güld, M.O., Keysers, D., Schubert, H., Wein, B.B., Bredno, J., Lehmann, T.M., Kohnen, M.: Quality of DICOM header information for image categorization. In: Siegel, E.L., Huang, H.K. (ed.) Medical Imaging 2002: PACS and Integrated Medical Information Systems: Design and Evaluation. SPIE, vol. 8 (2002). https://doi.org/10.1117/12.467017

21. Santos, M., Bastião, L., Costa, C., Silva, A., Rocha, N.: Clinical data mining in small hospital PACS: contributions for radiology department improvement. In: Information Systems and Technologies for Enhancing Health and Social Care, pp. 236–251. IGI Global (2013). https://doi.org/10.4018/978-1-4666-3667-5.ch016
22. Santos, M., Silva, A., Rocha, N.: Characterization of the stakeholders of medical imaging based on an image repository BT. In: Rocha, Á., Correia, A.M., Adeli, H., Reis, L.P., Costanzo, S., (eds.) Recent Advances in Information Systems and Technologies: Volume 2, pp. 805–814. Springer, Cham (2017)
23. Botsis, T., Hartvigsen, G., Chen, F., Weng, C.: Secondary use of EHR: data quality issues and informatics opportunities. Summit Trans. Bioinforma. 2010, 1–5 (2010). http://www.ncbi.nlm.nih.gov/pmc/articles/PMC3041534/
24. Kahn, M.G., Raebel, M.A., Glanz, J.M., Riedlinger, K., Steiner, J.F.: A pragmatic framework for single-site and multisite data quality assessment in electronic health record-based clinical research. Med. Care **50** Suppl(0), S21–S29. https://doi.org/10.1097/mlr.0b013e318257dd67
25. Kahn, M.G., Brown, J.S., Chun, A.T., et al.: Transparent reporting of data quality in distributed data networks. EGEMS (Washington, DC) **3**(1), 1052 (2015). https://doi.org/10.13063/2327-9214.1052
26. HealthImaging: Medicare Paid $38 Million in Erroneously Documented Imaging Claims (2011)
27. Costa, C., Ferreira, C., Bastião, L., Ribeiro, L., Silva, A., Oliveira, J.: Dicoogle - an open source peer-to-peer PACS. J. Digit. Imaging **24**(5), 848–856 (2011)
28. Tsalafoutas, I.A.A., Panaretos, C., Thalassinou, S., et al.: A survey on average glandular doses in digital mammography units of four different manufacturers. Physica Med. **30**, e112 (2017). https://doi.org/10.1016/j.ejmp.2014.07.317
29. Miller, D., Livingstone, V., Herbison, G.P.: Interventions for relieving the pain and discomfort of screening mammography. Cochrane Database Syst. Rev. (1) (2008). https://doi.org/10.1002/14651858.cd002942.pub2
30. Perry, N., Broeders, M., de Wolf, C., et al.: European guidelines for quality assurance in breast cancer screening and diagnosis. Fourth edition—summary document. Ann Oncol. **19**(4), 614–622 (2008). https://doi.org/10.1093/annonc/mdm481

The Use of Social Media in the Recruitment Process

Viktoriya Sharaburyak[1], Gonçalo Moreira[1], Mafalda Reis[2], Pedro Silva[2], and Manuel Au-Yong-Oliveira[1,3](✉)

[1] Department of Economics, Management, Industrial Engineering and Tourism, University of Aveiro, 3810-193 Aveiro, Portugal
{sharaburyak,goncalo67,mao}@ua.pt

[2] Department of Languages and Cultures, University of Aveiro, 3810-193 Aveiro, Portugal
{mafalda.oliv,pfmsilva}@ua.pt

[3] GOVCOPP, Aveiro, Portugal

Abstract. Social media platforms have been increasingly used by companies, over the past few years, in order to make their recruitment and selection process more efficient. The aim is to gather more data about the applicants' competences and personality traits, assuring that they are hiring the right candidates. Nevertheless, this method may be controversial since it can lead to different legal and ethical issues such as discrimination and invasion of privacy. Additionally, some problems may arise regarding the validity and fairness of the information found in such platforms. As Facebook is the most frequently used social platform among students and recently employed workers, we chose it as the main focus of our survey (which had 212 answers), in order to understand the applicants' point of view about this subject. The main conclusions were that the majority of the respondents affirmed being aware of this practice, despite not agreeing with it; and, also, they do not believe that this is a good tool to evaluate their potential. Finally, our participants' answers also led us to conclude that there could be contradictions regarding their judgement about the accessing of personal information by firms on Facebook. 76,9% of our respondents affirmed that they usually assess the profile of people who send them friend requests. Our respondents consider the use of Facebook during the selection process as being unethical, even though they assess others' profiles and search for information themselves.

Keywords: Social media · Selection · Process · Hiring · Legal · Ethical · Facebook · Candidates

1 Introduction

The evolution of technology diminished the boundaries between work and personal life, leading to a higher level of intimate information being available about individuals [1]. Social media platforms revolutionized the way people interact and their growing popularity led companies to take advantage by including them in their recruitment and selection process [6].

Á. Rocha et al. (Eds.): WorldCIST 2020, AISC 1159, pp. 165–174, 2020.
https://doi.org/10.1007/978-3-030-45688-7_17

This procedure became frequent as the costs of hiring the wrong employees are high and firms need to make sure they select the most fitting employees. Therefore, they are using social media as a strategic tool to achieve competitive advantage [1].

Many authors see the use of social media in the recruitment process as a good practice. In contrast, its usage in the selection process to evaluate and screen applicants is seen by others with more scepticism. Since there is little research on the impact of social media in the selection process, the authors believe that this practice has yet "too many risks and few demonstrated gains" [14].

Facebook is the most popular and the most used social platform by employers in their selection process, as they believe that this social website provides relevant information regarding the applicants' soft skills and personality traits [1].

There is a lot of discussion related to how this method can benefit or damage companies, since some of the problems related to legal and ethical questions may arise. It is important to understand to what extent the analysis of a social media profile offers enough information in terms of quantity and quality, and the fairness of this procedure [6].

Throughout this article, we analyse the current use of social media in selection, some legal and ethical considerations related to this practice, as well as some best practices for the applicants and firms alike.

In an effort to understand the candidates' perspective on this subject, the main questions that guided our paper were the following:

- Are candidates aware of the use of this method in the selection process?
- Do candidates believe that their social media profiles reflect their personality and professional potential?
- Is this method considered legal and ethical?

2 The Use of Social Media in Recruitment and Selection: A Review of the Literature

Nowadays, companies are dealing with an employee turnover problem, hence they are trying to reduce costs by investing in better selection methods [2]. Selection is used by companies as a strategy to avoid hiring the wrong employee and the costs associated to low levels of productivity, inadequate behaviours, and a conflicting work environment.

Firms need to ensure that they have the relevant information regarding a candidate's previous experience, skills, and personality traits, in order to predict future outcomes [1]. Recruiters use social media profiles to collect personal information such as birth date, contacts, address [2], professional background, and interests [1]. In addition, social media may reveal data regarding creative abilities, teamwork [3], communication skills, as well as confirming interview data [1].

While LinkedIn is more used to evaluate professional skills, Facebook analyses are more focused on the candidates' soft skills and how they would fit into the organization, trying to understand if their values match [1].

Among social networks, Facebook is the most popular platform [6], having "almost 2.38 billion monthly active users" [8], and more organizations are using Facebook as a tool to evaluate potential employees [1, 9].

The Facebook profile may be the first interview between the organization and the candidate, and there is a chance that, "if you don't like a person there, you probably won't like working with them" [4]. On the other hand, the information found on the profile can also be used by the recruiter "to guide the interview and lead conversations" [1], as well as, "to build connection" [3]. In some cases, social media can facilitate the evaluation of candidates that are geographically distant [5].

Although more firms are including social media information in their selection process, and sometimes, even to eliminate candidates, "only a handful of empirical studies have examined the validity of such practices", as well as its fairness [10].

2.1 Legal and Ethical Considerations

The use of social media in the selection process brings many legal and ethical questions [10] and "although information posted on the Internet is generally considered "public", many legal systems bar the consultation of particular types of information by those making hiring decisions" [12].

Firstly, the "potential for discrimination" has been a concern, since "in the western world, discrimination on the grounds of such things as gender, age, marital status, race, disability and so on are illegal in employment" [1]. This personal data seen by the recruiter on the candidate's profile may, even if unconsciously, influence their decision [11].

Legally it is difficult to define the "boundaries of privacy". Meanwhile, some courts defend the candidates' right of privacy, and others defend that "individuals accept the risk that personal information will go public when they communicate with other people" [3].

The General Data Protection Regulation already predicts heavy fines for the inappropriate use of personal data; "fines could reach €10 million or 2% of the company's annual turnover globally" [13].

Another issue is whether employers should ask the candidates' permission, or at least inform them when accessing their social media profile with the intent to evaluate them [1].

On the other hand, it is difficult to assess the validity of the data collected since it may be inaccurate or distorted "in an effort to be humorous or more accepted at a particular point in time" [3], or used by the applicant to manipulate others' perceptions [1].

In this context, other questions emerge, such as, to what extent should the candidates' movies, hobbies, or book interests be included in their job evaluation [3]?

2.2 Things to Avoid on Social Media

Many recruiters claim that the type of content found on social media influences their decisions and leads to some applicants' rejections [5, 10, 14]. Therefore, applicants need to be aware that they may lose a job due to inappropriate information published in their profiles [6], such as:

- "Provocative or inappropriate photographs" [2, 7].
- "Information about drinking or using drugs" [2].
- "Discriminatory remarks related to race, gender, religion" [2].
- "Criminal behaviour" [7].
- "Lying about qualifications" [2].
- "Badmouthing a previous company or fellow employees" [2].
- "Poor communication skills" [2, 7].

2.3 Selected Best Practices for Candidates According to the Literature

- Currently, many companies feel the necessity for job applicants to own a social media account. Therefore, creating a social media profile is the candidates' first step to appearing active on the Internet [9].
- Displaying some interests, such as music, books, among other activities, is a good strategy for the candidate to "show openness to experience" [9].
- In order to seem knowledgeable and relevant to the employer, the applicant might "share professional related content", as remarkable facts or "inspirational quotes from business leaders" in the profile [9].
- Being cautious of what to share in the profile. The applicant should avoid publishing inappropriate content [9].
- Candidates should check the privacy settings on their social media profile, in order to limit who can view what they post, their shares, and other definitions [9].

2.4 Selected Best Practices for Companies According to the Literature

- Employers should be cautious about how they perceive the information found on social networks, ensuring its validity, since it does "not give a full picture of the individual" [6].
- Before using social media screening, companies should balance if the potential benefits and risks involved are worth the legal problems they might face [6].
- The human resources department should be aware of the federal and international legislation regarding the use of social media in the hiring process, as laws are being adapted to new technologies and their impact in several domains [15].
- The individual responsible for collecting social media data should not be the same person that makes the hiring decision, to ensure that information such as age, sex, religion, among others, that might lead to discrimination, cannot influence the final decision [6].

- Companies should be transparent during the process by asking for permission or warning job applicants regarding the use of social media as a screening tool [14].
- Companies should create a formal policy, in which all procedures are "carefully documented, to create a paper trail that evidences the organization's decision-making process in relation to social media" [14].

3 Methodology

After the literature review, we can affirm that the usage of social media has become a frequent method used by companies to evaluate and screen applicants, even though there is little research regarding this subject and its benefits.

Moreover, there are still many doubts concerning the validity and fairness of this usage, which leads us to seek out more about these concerns in an attempt to understand the applicants' point of view.

Considering that Facebook is one of the most used platforms in the recruitment process, we decided to analyse this particular social website in our research.

Since our goal was to acknowledge the impact of this phenomenon, we decided to use Google Forms to obtain primary data. We developed an online survey, of our own authorship, entitled "Facebook as a tool for the candidates' selection?".

The target audience chosen for this survey was mainly university students, as we are aware that this public is more likely to intend to enter the job market in the next few years and as recent workers. Besides being the ones who use social networks more frequently, they are also the ones who share a bigger amount of personal data online.

The survey was anonymous and was available in Portuguese, as we wanted to know more about the Portuguese students' point of view. The survey was online for two weeks, starting on the 22nd October and ending on the 3rd November.

In order to reach a wider and richer network, the group members agreed to share the survey in various Facebook groups and to send it via private message to a large amount of people considered suitable for the research.

This survey consisted of 13 questions, 12 of which were closed-ended questions, having one open-ended question connected to a previous answer, so that we could have a clearer perspective. A significant amount of the questions had the options: "Yes" and "No". We tried to develop precise questions, in order to obtain concrete answers from the respondents. The most relevant questions of our survey are described in the "Results" section of this article.

4 Results

With this survey we obtained 212 answers. We were able to achieve 58% of the answers from university students, since that was the main target audience we defined. The majority of answers came from students from the University of Aveiro, due to our proximity with this public. Furthermore, 19,8% were workers, and 17,5% were working students (both of these segments are also relevant as they are in the job market too).

We were able to reach 78,3% of the respondents with the ages between 18–24 years, and the second biggest age group was between 25–30 years, with a percentage of 16,5% of the total.

As we realized that Facebook is one of the most used tools by companies, we tried to understand the repercussion that it has in the audiences' life on a daily basis. To the question "What impact has Facebook on your daily life?", 32,5% responded that they use and update their Facebook page frequently, 57,5% use and update it occasionally, whereas only 9,9% of the respondents do not use and update their profile frequently.

In order to acknowledge the alertness to this practice among our target audience, we decided to ask people if they think companies are using Facebook in their recruitment and selection process, to which 67,5% responded "Yes".

We also attempted to figure out if individuals are aware if companies have ever used Facebook as a method to evaluate their profile. The majority of respondents (90,6%) stated that they do not know if that has ever happened to them.

Moreover, the literature review showed us that companies have already rejected applicants due to content they found on their social media profile. Taking this into account, we asked respondents "Are you cautious about the content you share on your Facebook profile and how it may impact your future career?". To this question, 79,7% responded affirmatively.

Linked to the previous question, we decided to investigate if Facebook users consider having or to be present in inappropriate photos/videos that they would not want their future employer to see. The responses revealed that 83% consider not having any of these types of content.

We decided to create an open-ended answer for those who responded affirmatively to the previous question to get a bigger picture of what they consider to be inappropriate content. A very significant number of individuals declared having photos involving alcohol and parties. On a smaller scale, we obtained answers regarding "embarrassing" photos and inappropriate humour posts.

Furthermore, we intended to understand our sample's point of view regarding the effectiveness of Facebook to evaluate the candidates' potential. Figure 1 shows that 73.6% of the respondents answered that they do not consider it to be a good tool to evaluate competencies and social traits.

Our sample was inquired about the validity/fairness of this practice, too. As can be seen from Fig. 2, a substantial number of people (37.3%) considered this method "unethical and an invasion of privacy", and 31.1% believe their Facebook profile does not give evidence to who they really are (giving a total of 68,4% disagreeing with the use of Facebook, as an evaluation tool, and concerning ethics and validity). Only a relatively small percentage (20.3% in total, or 13,7% plus 6,6% - see Fig. 2) agrees with the use of Facebook for such purposes.

It was of interest to us to know if our audience has ever revised their privacy profile settings, since this would show us if they are comfortable with having their content public. The responses revealed that 82.5% of those inquired have made adjustments to their privacy settings.

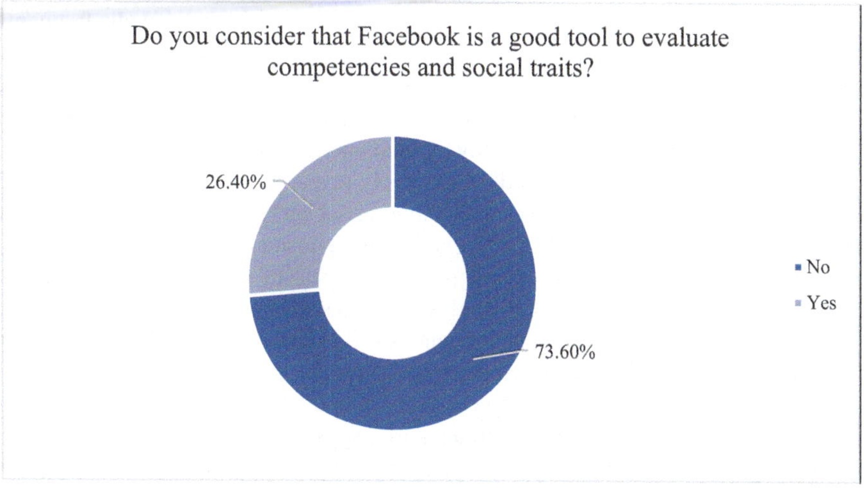

Fig. 1. "Do you consider that Facebook is a good tool to evaluate competencies and social traits?"

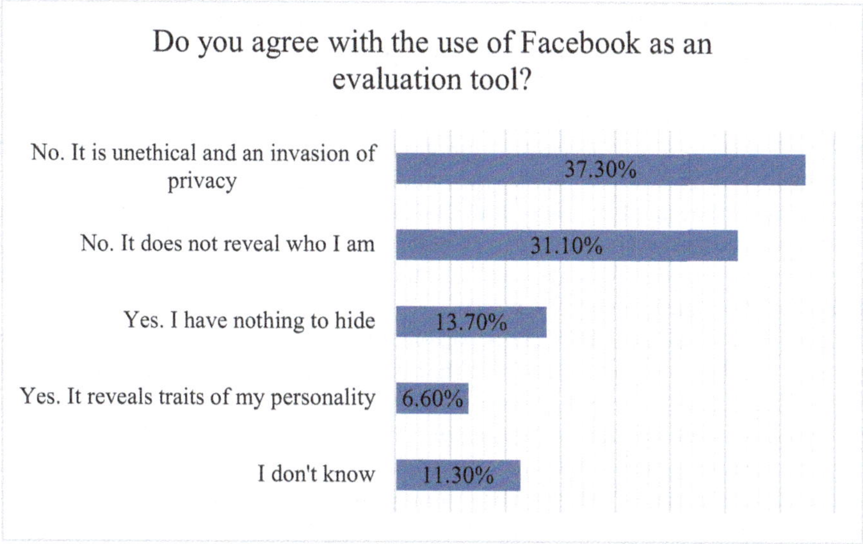

Fig. 2. "Do you agree with the use of Facebook as an evaluation tool?"

5 Discussion

The survey we performed gave us a better insight regarding the university students' perspective on this phenomenon. "Students need to be more aware of their online presence and how their posts, shares, "likes," tweets, and other modes of communication can affect the outcome of their future with an employer" [6].

We know that companies are increasingly relying on the content existing on the applicants' social media profile to assess their potential. However, our survey showed that most of the respondents (57,5%) do not use or update their social media frequently. This made us realize that companies are possibly dealing with outdated information that may not be accurate and does not evidence who applicants really are.

On the other hand, as this practice has been progressively used by firms, we wanted to understand to which extent university students believe this is a common approach. Surprisingly, the majority responded affirmatively revealing that this selection process method is already acknowledged as a reality among our survey audience. However, there is still a substantial percentage (32,5%) that has an opposite opinion.

Although students are aware of this practice the survey revealed that only a small percentage (9,4%) has experienced this. This reality is still uncommon in Portugal, where a great number of small and medium companies still have not adopted new technologies in their hiring process and do not prioritize social media in their strategies.

With the survey, we realized that people are already being careful with the content that they share online. Yet, there is still an alarming percentage (20,3%) affirming that they do not pay attention to the content shared in their profile, which can have a negative impact on their professional futures, leading some recruiters to reject them after a first sighting of their profiles.

The most common inappropriate content present in social media profiles is related to drinking behaviours and excessive parties. Many companies consider these behaviours the most undesirable ones, however these are precisely the most frequent to appear in our survey audience.

On the one hand, there is a need to change Facebook users' mentality and on the other hand there is also a necessity, from the recruiters' side, to be more flexible and careful with how they deal with this type of content.

The use of Facebook to assess a candidate's skills and personality traits was not something our survey respondents agreed with, concluding that this is not an appropriate method. However, multiple companies use this social platform to obtain this type of information through a candidate's profile.

This practice can even be more debatable after exploring our survey participants' opinions on if they agree or disagree with the assessment of their Facebook profile during the selection process. The results showed that the majority of the respondents disagree and consider it unethical and an invasion of their privacy.

Since the use of Facebook to screen candidates can be considered unethical or even unacceptable, companies risk not having future employees' permission to access their online information. Firms could even be involved in legal issues as candidates could make a complaint against them. Another proof that our respondents do not feel comfortable with recruiters searching their Facebook profile is that 82,5% stated to have revised the privacy settings thus restricting who is able to see their posts, shares or friends.

Lastly, our participants' answers also led us to conclude that there could be contradictions regarding their judgement about the accessing of personal information by firms on Facebook. 76,9% of our sample affirmed that they usually assess the profile of people who send them friend requests. Our respondents consider the use of Facebook during the selection process as being unethical even though they assess others' profiles and search for information themselves.

6 Conclusion and Future Developments

The use of social media in the hiring process is still a method that lacks research. It is viewed with scepticism and many issues related to legality and ethicality continue to emerge.

Paying attention to what is published online has never been so important to the students' future professional success as it is nowadays, as many individuals can lose a job opportunity because of something they have posted or shared on their Facebook page, or in another social media platform.

Our paper had the goal to create more awareness regarding this practice, because "one of the easiest ways to help spread knowledge about social networking sites and potential screening processes by hiring departments is to educate upcoming university students about their postings and the lasting image they can have" [6].

People need to be extra cautious about their online activity, by certifying that their online persona matches the person who they really are, avoiding inappropriate information that can be seen by companies as "personality red flags" [4].

Furthermore, companies need to be careful if they decide to make social networks part of their hiring process. It is necessary to be aware of the laws around this matter, as well as to understand how this will be perceived by candidates. To achieve a fairer process, firms need to create formal and clear procedures.

In conclusion, this phenomenon is still relatively recent in Portugal and there is still little literature associated to it. Therefore, we believe it would be interesting to search for Portuguese companies that are currently using social media as a selection tool and to try to interview executives there in order to better understand what benefits they have with this method, possible drawbacks they have had, and what type of procedures they are using to identify the most accurate and trustworthy information.

Acknowledgements. We would like to thank the participants of our online survey, for their time and effort, since we could not have completed this research study without their testimonies.

References

1. Hoek, J., O'Kane, P., McCracken, M.: Publishing personal information online: how employers' access, observe and utilise social networking sites within selection procedures. Pers. Rev. **45**(1), 67–83 (2016)
2. Balint, B., Rau-Foster, M.: Cybersnooping: I see what you did there. J. Organ. Cult. Commun. Confl. **19**(1), 72–81 (2015)
3. Smith, W.P., Kidder, D.L.: You've been tagged! (Then again, maybe not): employers and Facebook. Bus. Horiz. **53**(5), 491–499 (2010)
4. Hill, K.: Facebook Can Tell You if a Person is Worth Hiring. Forbes (2015). https://www.forbes.com/sites/kashmirhill/2012/03/05/facebook-can-tell-you-if-a-person-is-worth-hiring/#4afc46b164ce. Accessed 16 Oct 2019
5. Slavić, A., Bjekić, R., Berber, N.: The role of the internet and social networks in recruitment and selection process. Strateg. Manage. **22**(3), 36–43 (2017)
6. Hazelton, A.S., Terhorst, A.: Legal and ethical considerations for social media hiring practices in the workplace. Hilltop Rev. **7**(2) (2015). Article No. 7

7. Driver, S.: Keep it Clean: Social Media Screenings Gain in Popularity. Business News Daily, 7 October (2018). https://www.businessnewsdaily.com/2377-social-media-hiring.html. Accessed 17 Oct 2019
8. Clement, J.: Global social media ranking. Statista (2019). https://www.statista.com/statistics/272014/global-social-networks-ranked-by-number-of-users/. Accessed 25 Oct 2019
9. Schawbel, D.: How to Shape Your Facebook Profile to Help You Land a Job (2012). http://business.time.com/2012/03/01/how-to-shape-your-facebook-profile-to-help-you-land-a-job/. Accessed 17 Oct 2019
10. Becton, J.B., Walker, H.J., Schwager, P., Gilstrap, J.B.: Is what you see what you get? Investigating the relationship between social media content and counterproductive work behaviors, alcohol consumption, and episodic heavy drinking. Int. J. Hum. Resour. Manage. **30**(15), 2251–2272 (2019)
11. Kluemper, D.H., Rosen, P.A.: Future employment selection methods: evaluating social networking web sites. J. Manag. Psychol. **24**(6), 567–580 (2009)
12. Landers, R.N., Schmidt, G.B.: Social media in employee selection and recruitment: an overview. In: Landers, R., Schmidt, G. (eds.) Social Media in Employee Selection and Recruitment. Springer, Cham (2016)
13. Mateus, C.: Empresas que vigiem nas redes sociais candidatos a emprego violam a lei. Expresso, 09 Oct 2018 (2018). https://expresso.pt/economia/2018-10-09-Empresas-que-vigiem-nas-redes-sociais-candidatos-a-emprego-violam-a-lei. Accessed 11 Nov 2019
14. Landers, R.N., Schmidt, G.B.: Social media in employee selection and recruitment: current knowledge, unanswered questions, and future directions. In: Landers, R., Schmidt, G. (eds.) Social Media in Employee Selection and Recruitment. Springer, Cham (2016)
15. Black, S.L., Washington, M.L., Schmidt, G.B.: How to stay current in social media to be competitive in recruitment and selection. In: Landers, R., Schmidt, G. (eds.) Social Media in Employee Selection and Recruitment. Springer, Cham (2016)

Complex Human Emotions in Alzheimer's Interviews: First Steps

Héctor F. Gomez A[1(✉)], Elena Malo M[2], Richard E. Ruiz O[3], and Carlos Martinez[4]

[1] Universidad Técnica de Ambato, Ambato, Ecuador
hfgomez@uta.edu.ec
[2] Universidad Técnica Particular de Loja, Loja, Ecuador
semalo@utpl.edu.ec
[3] Instituto Superior Tecnológico Loja, Loja, Ecuador
rruiz@instututos.gob.ec
[4] Universidad Regional Autonoma de los Andes, Ambato, Ecuador
ce.martinez@uniandes.edu.ec

Abstract. In this paper, we intended to test the hypothesis that there is a common denominator in complex emotions which human experts can identify, specifically in determining target scenarios. The underlining concept behind our work is that the system aims to compose complex emotions from simplex emotions. Our propose does not completely ignore the need for human operators since they (the human operators) make the final decision to confirm that the input video sequence corresponds to the simplex and complex emotion. However, the proposed system is designed in such a way that, theoretically once it is fully installed and operational, it would work on its own by automatically generating message alerts for the human operator. This entire process is based on the identification and labelling of what we call elementary or basic activities, namely events that are recognizable by artificial vision algorithms in this case *HER*. To test our hypothesis, we carried out an experiment on human emotions in Alzheimer's interviews video. The results of experimentation showing that is possible identify complex emotions in video, include in important context how Alzheimer.

Keywords: Human activities · Behavior · Ontology

1 Introduction

Alzheimer's disease, which is the most common form of dementia, is estimated to affect up to 50% of people aged over 85. When we first posed the question, "Can a computer be a caregiver?" just a few years ago, the question seemed rather rhetorical. Nevertheless, the problem is still not easy to answer as technical innovation, social expectations, and ethical considerations, become of increasing importance. Recent research suggests that a lot of inter-personal communication is non-verbal. Facial expressions, physical gestures, and eye contact are important factors when trying to interpret non-verbal communication. Non-verbal communication is thus rooted in idiosyncratic and, sometimes, random

Á. Rocha et al. (Eds.): WorldCIST 2020, AISC 1159, pp. 175–179, 2020.
https://doi.org/10.1007/978-3-030-45688-7_18

symbols that can vary from region to region. However, the generic characteristics of such communication are often considered to be global. Non-verbal communication is of particular importance in high-velocity and time-constrained situations involving threat. Nowadays, for instance, police officers must increasingly develop perceptual shorthand to distinguish emerging threatening situations from non-threatening ones. When two people establish communication, facial expressions can be simulated, ritualized or be completely voluntary [1]. Facial expressions are also closely related with emotional states and are spontaneously expressed, or used to mask an instinctive emotional state [2]. The resulting leaked expression lasts a faction of a second, but is almost always indicative of specific emotional or psychological states. Facial expressions represent an emotional state that could be perceived by analyzing real-time images captured from cameras that are deployed in various locations for security purposes. The content could then be sent to a control center for human operators to take further measures. Such types of intelligent surveillance system are critical for providing a safer environment and ensuring that maximum security measures are used. In addition, these systems can be used for a wide variety of applications. The major difference between traditional and the proposed surveillance system is that the latter focuses on averting crime by introducing a time element [3]. The problem of automatic recognition in these types of environments is, however, challenging for current face recognition technology. Furthermore, there are still limitations for technology to be able to accurately predict a person's 'state of mind' amongst a multitude using facial expression analysis [4]. Only when these general, group-oriented systems are available, can researchers concentrate on developing systems that use machine learning algorithms to analyze and learn about the distinctive patterns of affective reactions of individual users [5]. Affective Computing is a field of research that is related to the analysis of human emotions by means of technology. Despite issues related to confidentiality and the explosive growth of surveillance networks and effective computing over the past decade, there exist a number of challenges for future security technologies, namely advancing and building upon current automatic and biometric identity-based technologies. The next generation technology of such devices may consist of systems that are capable of successfully matching a person's face to images in a database, and the interpretation or prediction of an individual's original motives and/or emotional state and their subsequent behavior in a potentially threatening scenario such as terrorism. The ultimate objective of such a system would therefore have to take into consideration real-time multi-modal input from congested urban locations and combine the input to produce precise predictions of the fundamental motives or states of individual agents. Human-computer interface researchers denote this category of effective system as a computer environment that possesses a plethora of information that is both accessible to a human actor and accessible by a computer system through an advanced interface [6]. Our proposal is to create a laboratory training environment through the use of high-fidelity macro, micro and subtle human expressions (MSE), as seen through the work of Ekman. Although, programs have been developed to identify micro-expressions in the human face, our proposal contributes to the analysis of emotions- using the hypothesis that is based on combining the Macro (M), Micro (M) and Subtle (S) expressions, i.e. with the aim of achieving the closest approximation to that emotion or feeling which a person wishes to show through their facial expressions.

We relate MMS (Fig. 1) with the environment in which they are created. Indeed, we base our view on the general hypothesis that people not only hide their true emotions, but they also wish to hide their feelings of being influenced by others in any given situation. We developed the online software HER to identify facial expressions. HER uses facial expressions to classify simple (anger, fear) and complex human emotions (nervousness, depression) in video or stilled images. HER is based on the work of [7, 8] on emotions and underlying emotions; it is using Binary and Canny filters for face recognition. After defining the concepts, second step that will be reused in the ontology [9]. The ontology was instancing with human emotions in YouTube videos, the action of appreciating a human emotion in a video is a primitive event that associates a physical object with a primitive state, in this case a person showing an emotion.

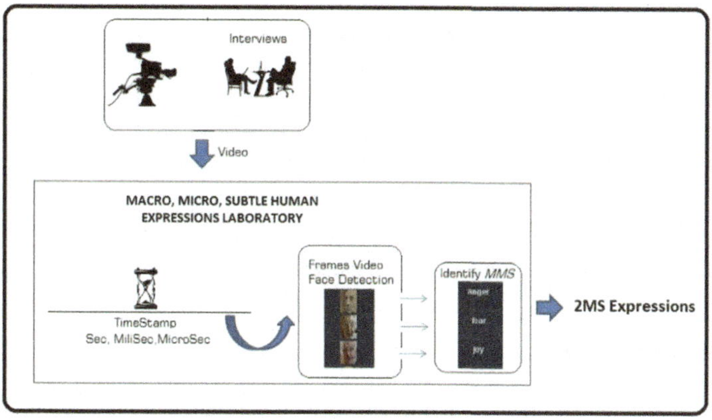

Fig. 1. MMS Laboratory. The input videos or images and the output 2MS Expressions

2 Methodology

Firsts step (Fig. 2) is to find simplex human emotion using HER. As mentioned in the introduction, micro-patterns are small patterns comprised of several sufficiently repeated simplex human emotions. In order to obtain these, we must have a series of positive case sequences (i.e. happy, anger, etc.). Each sequence represents the emotion behavior of a person and is obtained by labeling the individual activities of the monitored person for each second of video surveillance, e.g. "happy", "normal", "happy", "happy", "happy", "normal", "normal", "normal". It may be assumed that HER can recognize these human emotions. Second step is instance the ontology:

Taking into consideration that a primitive event (Fig. 2) is nothing more than the relationship between a physical object and a primitive state, it is necessary to define the temporal context of the evaluation of the event, to define the complex temporal relationships immersed in the process of appreciation of emotions. To fit the model and relate the video concept to the OWL-Time instant definition (Fig. 3), the frame concept is introduced, in which the primitive event is appreciated.

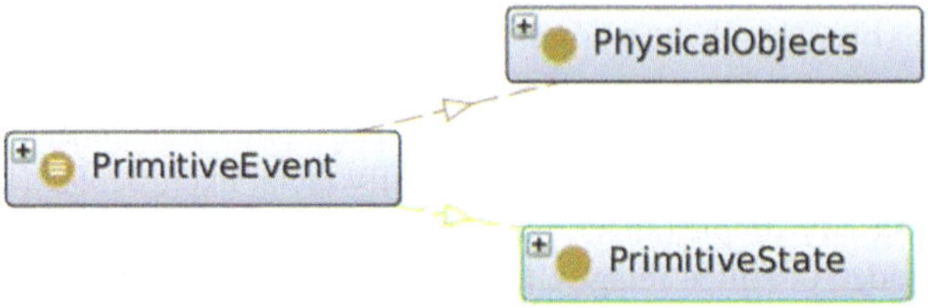

Fig. 2. Appreciation of events (simplex emotion) in videos

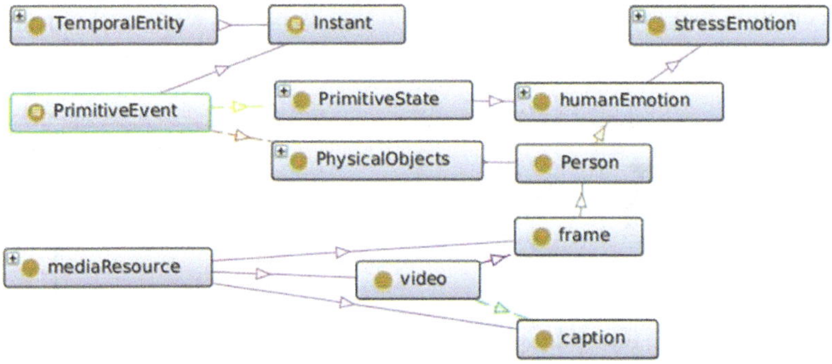

Fig. 3. Human emotion model in YouTube videos.

3 Experimentation

Video recordings of Youtube Alzheimer's interviews. In all cases, the human operator examines video streams from different cameras. Finally, in our experimentation we determined the sensitivity analysis using F1Score algorithm[1]. The human expert analysed the results all time, and he selected the more representative frames and videos to the experiment.

HER obtained better results to Alzheimer's interviews videos (Table 1). The simplex emotions are very important to infer new complex emotions in the patients. These emotions were instancing in the ontology.

Table 1. Sensitivity analysis Alzheimer's interviews

Emotion	TP	TN	FP	FN	Precision	Recall	F1Score
Sadness	45	41	9	5	0,83	0,9	0,86
Fear	45	44	6	5	0,88	0,9	0,89
Happy	47	43	7	3	0,87	0,94	0,90
Surprise	40	46	4	10	0,90	0,8	0,85

[1] More details available in: http://www.monperrus.net/martin/understanding+the+f1-score.

4 Conclusions

In this paper, we intended to test the hypothesis that there is a common denominator in complex emotions which human experts can identify, specifically in determining target scenarios. The underlining concept behind our work is that the system aims to compose complex emotions from simplex emotions. Our propose does not completely ignore the need for human operators since they (the human operators) make the final decision to confirm that the input video sequence corresponds to the simplex and complex emotion. However, the proposed system is designed in such a way that, theoretically once it is fully installed and operational, it would work on its own by automatically generating message alerts for the human operator. This entire process is based on the identification and labelling of what we call elementary or basic activities, namely events that are recognizable by artificial vision algorithms in this case HER. By using these labelled sequences, i.e. where Alzheimer's patients usually occur to instance the ontology. The results of experimentation showing that is possible identify complex emotions in video, include in important context how Alzheimer.

References

1. Russell, J., Ferdández-Dols, J.M.: The Psychology of Facial Expression. Cambridge University Press, New York (1997)
2. Porter, S., ten Brinke, L.: Reading between the lies: identifying concealed and falsified emotions in universal facial expressions. Psychol. Sci. **19**(5), 508–514 (2008)
3. Wang, R., Fang, B.: Affective computing and biometrics based HCI surveillance system, Shangai (2008)
4. Pentland, A.: Looking at people: sensing for ubiquitous and wearable computing. IEEE Trans. Pattern Anal. Mach. Intell. **22**, 107–119 (2000)
5. Pantic, M., Rothkrantz, L.: Affect-sensitive multimodal monitoring in ubiquitous computing: advances and challenges. ICEIS **1**, 466–474 (2001)
6. Lisetti, C., Schiano, D.: Automatic facial expression interpretation: where human-computer interaction, artificial intelligence and cognitive science intersect. Pragmatics Cogn. **8**, 185–235 (2000)
7. Morel, S., Beaucoisin, V., Perrin, M., George, N.: Very early modulation of brain responses to neutral faces by a single prior association with an emotional context: evidence from MEG. Neuroimage **61**, 1461–1470 (2012)
8. Righart, R., Gelder, B.: Rapid influence of emotional scenes on encoding of facial expressions: an ERP study. Soc. Cogn. Affect. Neurosci. **3**, 270–278 (2008)
9. Francois, A.R.J., Nevatia, R., Hobbs, J., Bolles, R., Smith, J.: VERL: an ontology framework for representing and annotating video events. IEEE MultiMedia **12**, 76–86 (2005)

Contribution of Social Tagging to Clustering Effectiveness Using as Interpretant the User's Community

Elisabete Cunha[1] and Álvaro Figueira[2(✉)]

[1] Instituto Politécnico de Viana do Castelo, Av. Gaspar de Castro, Viana do Castelo, Portugal
elisabetecunha@ese.ipvc.pt
[2] CRACS/INESCTEC, University of Porto, Rua do Campo Alegre, 1021/55, Porto, Portugal
arf@dcc.fc.up.pt

Abstract. In this article we discuss how social tagging can be used to improve the methodology used for clustering evaluation. We analyze the impact of the integration of tags in the clustering process and its effectiveness. Following the semiotic theory, the own nature of tags allows the reflection of which ones should be considered depending on the interpretant (community of users, or tag writer). Using a case with the community of users as the interpretant, our novel clustering algorithm (k-C), which is based on community detection on a network of tags, was compared with the standard k-means algorithm. The results indicate that the k-C algorithm created more effective clusters.

Keywords: Cluster analysis · Social tagging · Information retrieval

1 Introduction

In recent years there has been a change in the way information is displayed online. The generalized access to the World Wide Web allowed an easy production, editing, distribution and sharing of the information, resulting in a massive increase of data. Systems were created with a purpose to collect and share that information, as well as allowing users to tag, or comment the collected data. Meanwhile, the automatic organization of that information is one of the biggest challenges in the current Web context. Despite the existence of several clustering algorithms, the commitment between effectiveness (forming groups that make sense) and efficiency (doing so in an acceptable running time) is still difficult to achieve.

This work intends to assess if a document clustering system improves its effectiveness when integrating a social classification system by the use of tags. We selected the k-means algorithm as our starting point because it fits text clustering; it is seen as simple, efficient and it has great effectiveness improvement potential since it relies on the initial seed selection and on the kind of data. To validate our work on the way social tagging can be integrated on the clustering algorithm, we began by identifying how the nature of tags, under the Semiotic Theory of Huang and Chuang [1], may contribute to, on one

Á. Rocha et al. (Eds.): WorldCIST 2020, AISC 1159, pp. 180–190, 2020.
https://doi.org/10.1007/978-3-030-45688-7_19

hand, create the conditions to understand which tags have true impact on the grouping of the documents depending on the interpretant: i.e., if it is the community of users or if it is the author of the tags, and on the other hand, to rebuild this investigation design since it will allow a proper selection of the repositories concerning the two above mentioned interpretants.

One of the approaches to tag integration consists in analyzing a network of tags to specifically determine which seeds will be chosen. This approach originated a new clustering algorithm based on the k-means algorithm, which we named k-Communities (k-C) [2], and that is based on community detection on a network of tags. Preliminary results from small scale databases indicated the k-C algorithm had the best results [2, 3]. In this paper we will present a new test: the interpretant being the user community. The results are critically analyzed according to the external criteria using a hypothesis test.

2 Social Tagging

To analyze the nature of tags we use the framework proposed by Huang and Chuang [1] which makes a connection between tagging and the semiotics theory, considering tagging as a system of signs. Therefore, each resource, together with its tag and the person who attributed it are seen as belonging to a "sign system". The 10 classes of signs were then designated by Huang and Chuang [1] as shown on Fig. 1.

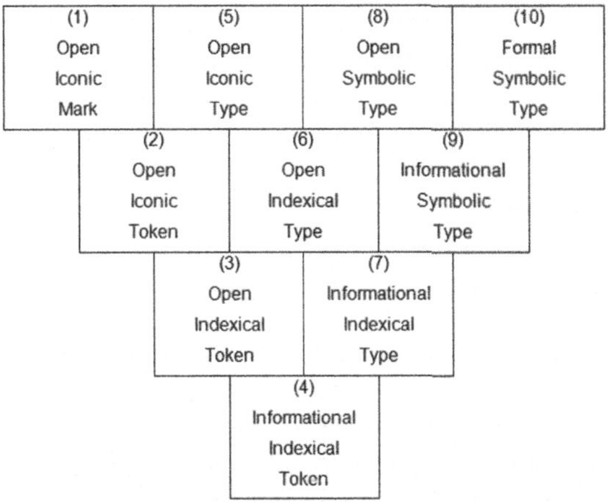

Fig. 1. Original pierce classes, adapted by Huang and Chuang [1].

2.1 How Can Tagging Contribute to Improve Document Clustering?

The search for a social consensus on how the resources of a certain system should be organized is at least controversial since it can be done in many valid ways. In a semiotics theory perspective, we may depend on the interpretant to understand how the resources should be organized. As a user within a community, the interpretation of tags may occur according to the communities' interests but on the tag writer's point of view the interpretation may be based on its personal preferences. There is no classification system suited for all, in every culture and at the same time. Similarly, when regarding online communications, the way the resources are organized will be different to the three kinds of actors involved in tagging and even to the same kind of tagging actors the tags may be interpreted differently.

Within the context of this work we aim to understand how social tagging can contribute to improve the quality of a documents clustering process and therefore it is necessary to understand how the signs of tagging appear in the tagging activity.

2.2 Interpretation According to the User Community

Starting from a list of tags we will analyze how the signs (3), (6) and (8) relate, since signs (6) and (8) have sign (3) as a replica. On sign (3), starting from the tag list, are used techniques which provide clustering methods. Sign (6) presents tags as shapes not listed on the vocabulary (for instance, an acronym), being more specific tags than those dealt on sign (8), the natural language terms (therefore more general and so more popular).

We analyzed a list of tags on an attempt to understand how we can interpret the grouping of documents under the perspective of the user community. For that we considered a sample of 5000 Wikipedia documents taken from the online repository provided by the NLP Group. We noticed that, in average, 57% of tags were given by a single tagger. Furthermore, we found that the mean and median of the tags attributed by 2 taggers is approximately 10%. As the number of taggers which attributes the same tag increases, the percentage of the mean and median tends to a value close to 0%. Still, some of the tags which are only attributed once are related to tags already attributed by more than one tagger. For example: musika (1 vote) and music (15 votes); type_font_designers (1 vote), type (2 votes), font (14 votes) designer (4 votes). To a sample of 25 resources we verified that, in average, approximately 15% of tags (given by only one user) are already represented on tags given by more than one user.

2.3 Interpretation According to the Tag Writer

Signs (4), (7) and (9) are related, since signs (7) and (9) have sign (4) as a replica. Sign (4) is usually embedded in systems which provide in a single site the individual tags given on several systems. On sign (7), the given tags are used to categorize the preferred resources on a given system (it evolves sign (6) since it needs to point out the subject and it is presented through sign (5)). Conversely, sign (9) acts as sign (8) with the difference that it needs connection to the resource.

Therefore, considering the existence of this signs, clustering may be differently interpreted. Hence, resources get grouped differently, according to different tag writers.

3 Community Detection

The area of *Community Detection* is relatively new and of great importance in areas such as Computer Science, Biology and Sociology, where systems are frequently represented by graphs [4]. Complex networks, in this case tag networks, are represented through graphs where vertices are organized in communities (clusters) with many edges linking the vertices within the same community, and a few linking vertices of different communities.

3.1 The Girvan and Newman Algorithm

The algorithm proposed by Girvan and Newman [5] starts by removing the edges with a larger betweenness centrality. To compute the *betweenness centrality* we calculate the number of shortest paths which run through a certain edge. When we can find more than one shortest path, all of these are equally weighted so that the total is one. Lastly, the new graph centralities are recalculated and the edge with the greatest centrality is removed.

Modularity is a quality function, the most popular being also the proposed by Girvan and Newman [7], which was introduced to set a stop criteria on their community detection algorithm, but since then has become an essential element on many clustering methods [4]. This function is based on the idea that it is not expected to find a group structure on a random graph, so the possible existence of groups is revealed through comparing the density of the edges on the subgraph, and the density we expect to find on the subgraph if the vertices are connected, regardless of the community's structure. The expected density of the edges depends on the selected null model (for instance, a copy of the original graph maintaining some of its properties, but without its community frame) [4].

The Girvan and Newman algorithm is a hierarchical algorithm. Hence, the modularity is calculated on each dendrogram evolution and in the end the best Q is chosen. In this algorithm, at every iteration it is necessary to calculate the variation ΔQ of the modularity obtained from merging any two communities of the partition being analyzed in order to allow choosing the best merge. After deciding which communities to merge, it's necessary to update the matrix e_{ij} stating which fraction of the edges between clusters i e j of the partition is being executed.

Clauset *et al.* [8] proposed an optimization of Newman's algorithm. The authors showed that the algorithm can be executed in a more efficient way using information structures to disperse matrixes. This method allows to analyze the structures of large dimension graph communities containing up to 10^6 vertices.

Wakita e Tsurumi [6] noticed that due to the tendency to form very large communities, the algorithm proposed by Clauset *et al.* [8] was inefficient since it generated low balanced dendrograms, leading to an execution time close to its worst case scenario. To improve the situation, they proposed a modification that, on every step, intends to merge communities returning the highest value of ΔQ's product by a factor called "consolidation ratio", resulting in similar size communities [4]. This change allows the analysis of systems which contain up to 10^7 vertices.

4 The Similarity Measure of the k-C Algorithm

Clustering algorithms which are simultaneously effective and efficient are still a challenge. The k-means algorithm is known for its efficiency, but the quality of the generated clusters depends on the selection of the initial seeds since it's random choice may result in a bad cluster optimization. The algorithm starts by selecting k random seeds and then calculates the distance from each document to every seed, grouping each document to its nearest seed. The new centroid will be the means of the document vectors on each cluster. The process is repeated until the "convergence" is achieved [9]. Intending to improve performance, several methods have been proposed. Among them is the k-means++ proposed by Arthur and Vasilvitskii [10], where the seeds are selected using specific probabilities before executing the k-means algorithm.

However, the number of partitions remains a problem. Considering that the formed clusters should satisfy people's interests, we propose analyzing the tags given to the documents. Hence, a community detection process will be used to see how the information may be related and, consequently, the number of partitions should mirror the collective intuition on how the documents should be organized. In this sense, we propose that, once the community detection algorithm is executed, the number of communities (with more than one document) is k and that the seeds will be the documents with a greater degree within its own community. Moreover, one of the most used the similarity measure to implement the k-means algorithm is the Euclidian distance but for text clustering is cosine similarity [11]. However, when using cosine similarity, the new centroid will still be calculated using Euclidian distance. Consequently, the outliers will influence the position of the new centroid. In this sense, when using cosine similarity, the vectors are commonly normalized which, according to Zhong [12] gives more importance to the direction than to the magnitude (leading to the name Spherical k-means).

As expressed by the authors of the k-C algorithm, and seen from the corresponding article describing it [2], the vectors are not normalized and the new centroid will be the document's vector which is more similar to the documents within its cluster, than using cosine similarity.

5 Experimental Process

The methodological procedures to assess the algorithms consisted on a wiki's repository where the interpretant is the community of users. To assess the obtained results on the wikis repository we will use the external criteria using a hypothesis test. Lastly, the k-C algorithm will be compared to the Spherical k-means algorithm.

In the case studied, the D_{wiki} dataset is comprised of 1170 documents that we divided into 10 stratified repositories: one to be used for the test and the remaining to be merged and used for training. The procedure is then to be repeated using a 10-fold cross-validation. Throughout the study, our goal is to compare the results of the two algorithms for the test data and for the training data and, at the end, determine which algorithm is more stable. Therefore, we formulate the following hypothesis test. First, for the test and training data:

- **H1:** There are differences between the results of the external evaluation measure when using the Spherical k-means algorithm, or the k-C algorithm.

And then, for the stability of the algorithms:

- **H2:** There are differences between the Spherical k-means algorithm and the k-C algorithm when comparing the absolute differences of the results of the test data, and the training data from external evaluation measure.

We use the Wilcoxon Signed Ranks as a statistical test. This choice relied on the studies of Demsar on the statistical tests used to compare classifiers [13], since, this being a non-parametric test, it allows a paired comparison of the results without demanding that they follow a normal distribution and is suited to the number of results available (for instance, the more potent pared t-test can only be used with more than 30 data and the assurance of a normal population).

5.1 Case Study: The Interpretant Is the User's Community

The Wikipedia repository was obtained online (http://nlp.uned.es/social-tagging/wiki10+) consisting of 20764 documents to which were given tags by an average of 34.5 users. From this repository we collected 12000 documents and generated the tag cloud presented on Fig. 2. The tag "Wikipedia" is the one most frequently given, even if it is the least important (in this context) to detect the communities of documents which share the same tags.

Fig. 2. Tag cloud from a 1200 Wikipedia wikis.

Then, we choose to reduce the repository by collecting only the documents which contained at least one among the five most given tags: "art", "biology", "health", "physics", "programming" or "typography". The current repository consists of 1170 documents and it's tag cloud obtained with gc(2;0,25) is presented on Fig. 3, where we can see once again that the "Wikipedia" tag is the most emphatic.

Fig. 3. Tag cloud produced using six specific tags.

Since the “Wikipedia” tag is not relevant to organize our documents (in this case, we do not intend to organize the documents according to its provenance) we choose to eliminate the tags “wiki” and “Wikipedia” from our repository obtaining the tag cloud presented on Fig. 4. It is easily perceived that a large percentage of documents were tagged with the “programming” tag. On the other hand, from the tags we selected, the “typography” tag is the least emphatic, demonstrating the reduce number of documents with this tag.

Fig. 4. Tag cloud from the reduced repository without tags *wiki* and *Wikipedia*.

Having set the repository and its respective tags, we organized manually the documents through the tags given by the users on 6 classes: “art”, “biology”, “health”, “physics”, “programming” and “typography”. Whenever a document had been given more than one of these tags, it was placed on the class which was derived by more users. On Table 1, we can see the number of documents which remained on each class and its respective tag cloud.

Table 1. Created classes for the D_{wiki} repository.

Classes	# Doc	Tag Cloud
Art	280	
Biology	90	
Health	200	
Physics	150	
Programming	400	
Typography	50	

Analysis of the Test Data Results. The Spherical k-means algorithm was executed with k = 6, since the manual classes were 6. When executing the k-C algorithm the criteria used to add new centroids was $cos(x_i, C_i) = 0, \forall C_i$.

Concerning the k-C algorithm, the chosen criteria to add new centroids did not change the number of seeds detected by the Wakita-Tsurumi community detection algorithm which varied between 4 and 5, opposing the 6 manually obtained. We can advance two possibilities to justify the fewer number of classes: first, this may happen because the documents tagged with "*biology*" and "*physics*" were placed on an unique community (looking at the Table 1 it shows that the tag "*science*" appears with great highlight on the documents tagged with "*biology*" and "*physics*", leading to the grouping in a single class); secondly, in 50% of the cases, the documents tagged with "*typography*" appear in a single community, wherein the remaining cases they are grouped with the documents tagged with "*art*". As we can see in Fig. 5, the average of performance measures using 10 data sets with 117 documents shows that the k-C is the one which obtained the best results for all measures.

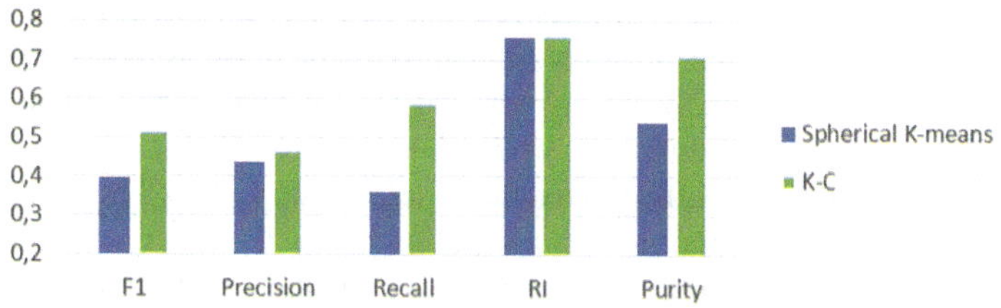

Fig. 5. Average measures.

The results of the test data showed that the k-C algorithm gives the best results, statistically significant, for the F1, Recall and purity (Table 2). We use the following note: (a) k-C < Sk-means (b) k-C > Sk-means on Tables 2, 3 and 4.

Table 2. The p value – test

External measures	F1	Precision	Recall	Rand Index	Purity
p	**0.017 (b)**	0.445 (b)	**0.07 (b)**	0.799 (b)	**0.007 (b)**

Analysis of the Training Data Results. The 10 training repositories have 1053 documents each. As proceeded on the test data, the Spherical k-means was executed with k = 6. To the k-C algorithm, the chosen criteria to add new centroids did not change the number of seeds detected by the Wakita-Tsurumi community detection algorithm. The number of seeds was 4 in all 10 data sets (opposing the 6 which were manually obtained). Analyzing the results shown on Fig. 6, we can see that, in average, the k-C algorithm presents better results to the *Recall* measurement with a difference of approximately 17%, meaning that the k-C algorithm has less FN pairs (False Negatives) then the Spherical k-means algorithm. On the contrary, the Precision measurement presents better results on the Spherical k-means algorithm when compared to the k-C algorithm with a 13% difference. We recall that these results are coherent with the k-C algorithm having less 2 clusters than what was expected by the manual organization, so the Precision result in the k-C algorithm is considering that the pairs of the merged groups are FP pairs (False Positives). The results of the F1 and Rand Index measurements don't suggest significant differences between the two algorithms. On the other hand, the Purity indicates that the k-C algorithm has more documents organized like those manually organized, with a difference of approximately 7%.

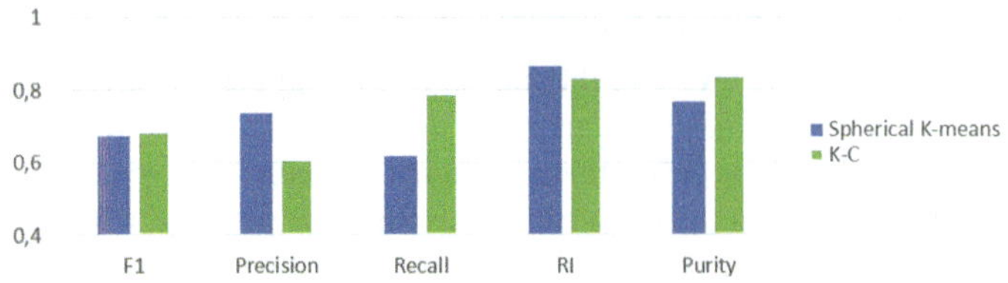

Fig. 6. Training – Average measures.

Intending to ascertain if the differences detected are statistically significant, we ran the Wilcoxon Test to each of the presented assessment measures, using the same hypothesis test used when analysing the test data. As we can see in Table 3, the k-C algorithm obtained significantly better results than the Spherical k-means algorithm to the Recall and Purity measures. The spherical k-means results are only statistically better than the k-C algorithm for Precision on the training data. However, it should be noted that the k-C algorithm determines only 4 clusters, while the manual classes are 6, so the Precision measure will be affected in the k-C algorithm.

Table 3. The p value – training.

External measures	F1	Precision	Recall	Rand Index	Purity
p	0.508 (b)	**0.005** (a)	**0.005** (b)	0.093 (a)	**0.009** (b)

Analysis of Algorithm Stability. To assess the stability of both the Spherical k-means and the k-C algorithms, we calculated the absolute differences between the results of the test data and the training data to each one of the used external evaluation measures.

As seen on Fig. 7, the k-C algorithm presents in average the smallest differences between training and test results to each of the used evaluation measures. To conclude if the observed differences are statistically different, we applied the hypothesis test formulated at the beginning of this section.

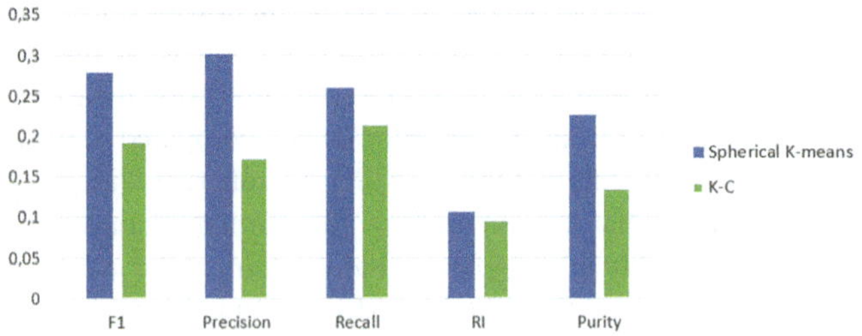

Fig. 7. Comparison between training and testing.

The hypothesis test indicates that, as seen on Table 4, the k-C algorithm is only more stable than the Spherical k-means algorithm with statically significant differences on the Precision and Purity measurements (differences on both measures at $p < 0{,}05$).

Table 4. The p value – stability.

External measures	F1	Precision	Recall	Rand Index	Purity
p	0.093 (b)	**0.013** (b)	0.575 (b)	0.508 (b)	**0.047** (b)

6 Conclusions

This paper presents new tests to the k-C algorithm, which were based on a case study. In this case study we selected a repository whose interpretant was a community of users.

The results on the test data showed that the k-C algorithm presents the best results, statistically significant to the F1, *Recall* and *Purity* measurements. On the training data set, this significant difference was only seen on the *Recall* and *Purity* measurements. However, it is important to state that the k-C algorithm was using k = 4 while the manual classes were 6. This happened because the community detection algorithm only detected 4 communities and the criteria to adding new centroids was very restrictive, only allowing new additions to the set of seeds when the cosine similarity to the rest of the seeds was zero. Therefore, we consider that further research is required on the need to adjust the community detection algorithms to the methodology we propose to form clusters. After comparing the algorithms regarding their stability, we noticed that the average of the absolute differences between test and training data indicate that the k-C algorithm is more stable, whatever the used measurement. However, after running a hypothesis test, we concluded that it is only possible to guarantee that the difference between the algorithms is statistically significant on the *Precision* and *Purity* measures, where in the other cases we found no reason to reject the null hypothesis.

Acknowledgements. This work is financed by National Funds through the Portuguese funding agency, FCT - Fundação para a Ciência e a Tecnologia within project: UID/EEA/50014.

References

1. Huang, A.W., Chuang, T.: Social tagging, online communication, and Peircean semiotics: a conceptual framework. J. Inf. Sci. **35**, 340–357 (2009)
2. Cunha, E., Figueira, Á., Mealha, Ó.: Clustering documents using tagging communities and semantic proximity. In: 8th Iberian Conference on Information Systems and Technologies (CISTI), Lisboa, Portugal, vol. I, pp. 591–596 (2013)
3. Cunha, E., Figueira, Á., Mealha, Ó.: Clustering and classifying text documents - a revisit to tagging integration methods. In: 5th International Conference on Knowledge Discovery and information Retrieval (KDIR 2013), Vila Moura, Portugal, pp. 160–168 (2013)

4. Fortunato, S., Castellano, C.: Community structure in graphs. In: Encyclopedia of Complexity and Systems Science, pp. 1141–1163 (2009)
5. Girvan, M., Newman, M.E.J.: Community structure in social and biological networks. In: Proceedings of the National Academy of Science, no. 12, pp. 7821–7826 (2002)
6. Wakita, K., Tsurumi, T.: Finding community structure in mega-scale social networks: [extended abstract]. In: Proceedings of the 16th International Conference on World Wide Web, Banff, Alberta, Canada, pp. 1275–1276. ACM (2007)
7. Newman, M.E.J., Girvan, M.: Finding and evaluating community structure in networks. Phys. Rev. E **69**(2), 026113 (2004)
8. Clauset, A., Newman, M.E.J., Moore, C.: Finding community structure in very large networks. Phys. Rev. E **70**, 066111 (2004)
9. MacQueen, J.B.: Some methods for classification and analysis of multivariate. In: Proceedings of the Fifth Berkeley Symposium on Mathematical Statistics and Probability, pp. 281–297. University of California Press (1967)
10. Arthur, D., Vassilvitskii, S.: k-means++: the advantages of careful seeding. In: Proceedings of the Eighteenth Annual ACM-SIAM Symposium on Discrete Algorithms, Society for Industrial and Applied Mathematics, New Orleans, Louisiana, pp. 1027–1035 (2007)
11. Feldman, R., Sanger, J.: The Text Mining Handbook Advanced Approaches in Analyzing Unstructured Data, 1st edn., p. 410. Cambridge University Press, Cambridge (2007)
12. Zhong, S.: Efficient online spherical k-means clustering. In: Prokhorov, D. (eds.) Proceeding of the IEEE International Joint Conference on Neural Networks (IJCNN 2005) (2005)
13. Demsar, J.: Statistical comparisons of classifiers over multiple data sets. J. Mach. Learn. Res. **7**, 1–30 (2006)

NutriSem: A Semantics-Driven Approach to Calculating Nutritional Value of Recipes

Rabia Azzi[1](✉), Sylvie Despres[2], and Gayo Diallo[1]

[1] BPH Center - INSERM U1219, Team ERIAS, Univ. Bordeaux, 33000 Bordeaux, France
{rabia.azzi,gayo.diallo}@u-bordeaux.fr
[2] LIMICS - INSERM UMRS 1142, Univ. Paris 13, Sorbonne Université, 93017 Bobigny Cedex, France

Abstract. The proliferation of recipes and other food information on the Web makes it difficult to identify recipes and foods for people who want to eat healthily. There are various tools related to calculating recipes nutritional values (NVs) but often the results obtained for the same recipe are varied. In this article we present *NutriSem*, a framework which allows automating the nutritional qualification of cooking recipes. It consists of four steps: (i) lexical enrichment of terms denoting ingredients; (ii) generating of nutritional calculus from lexical pattern and composition table requests; (iii) calculation and allocation of the final score; (iv) translation of the calculated score into a graphical scale. The core of the approach is based on mappings established between text corpora (cooking recipes) and structured data (food composition tables). A Knowledge Graph resource is used to enhance the quality of the mappings and therefore allow a better nutritional qualification of a given recipe.

Keywords: Food recipes · Lexical pattern · Knowledge graph · Nutritional score computation

1 Introduction

Over the last 30 years, various scientific, clinical and epidemiological studies have shown that dietary and eating behavior are major determinants of many chronic noncommunicable diseases and constitute a major issue [1]. In response, the European Union authorities and several other countries are nowadays working on elaborating the notion of "nutritional profile". Thus, various nutritional profiling systems have been developed on the basis of established scientific grounds. The most successful one are those based on the definition of a nutritional score, which is a result obtained by combining all the nutrients involved in the composition of a given food [2]. The calculation of such a score is complex because of

Á. Rocha et al. (Eds.): WorldCIST 2020, AISC 1159, pp. 191–201, 2020.
https://doi.org/10.1007/978-3-030-45688-7_20

the lack of immediate access to nutritional information contained in food composition tables, which include the food category, the actual edible parts and the status of the consumed food. These characteristics can vary greatly therefore impacting significantly the resulting calculated score. The automation of the score calculation requires a precise match between the foods mentioned in the recipes and those listed in the composition tables. However, such a matching is often inaccurate because the vocabularies used to describe these two resources differ. To solve this issue, we propose to rely on a Knowledge Graph (KG) allowing, besides a better pairing, extracting additional knowledge useful to score calculation. These additional knowledge include the status of the food and its category. Indeed, the calculation of the nutritional values (NVs) often takes into account for each ingredient the way it is prepared (e.g., cooked, raw) and the form in which it is used (e.g., whole, without skin). The category (e.g., fruit, vegetable or meat) to which the food belongs to is also involved in the score calculation.

The main contribution of the work introduced in this paper is a generic framework of nutritional qualification of cooking recipes, the *NutriSem* approach. *NutriSem* includes a textual component, the corpus of recipes, a structured component constituted of knowledge about food composition (for example, the Nutrinet[1] resources in the France context), a nutrition related KG which formalizes the main entities about the domain (for example the [3] resource) and a set of rules for extracting relevant lexical patterns. The core of the approach is based on mappings established between text corpora (cooking recipes) and structured data (food composition tables). The detail of the overall approach is structured as follows: (i) lexical enrichment of terms denoting the ingredients, (ii) generation of data for nutritional calculus from lexical pattern and composition table requests, (iii) calculation and allocation of the score, (iv) translation of the calculated score into a graphical scale. A KG is used in order to enhance the quality of the computed mappings between cooking recipes and food composition tables. This allows therefore a better nutritional qualification of the given recipe.

In the following section, we first introduce the pre-treatment process applied on the dataset used in *NutriSem*, then we describe the followed workflow.

2 Workflow

2.1 Dataset Pre-treatment

Recipe Dataset. The recipe data contains varying metadata fields as shown in the Fig. 1: the quantity and label of each ingredient of the recipe and the detail of the preparation instruction. A pre-processing task is performed to make it usable. From Fig. 1, it could be observed that ingredients are not always labelled the same in the ingredients list (left pane) and on the preparation list (right pane) (for instance "bouquet garni" against "bouquet garni de thym

[1] http://www.etude-nutrinet-sante.fr.

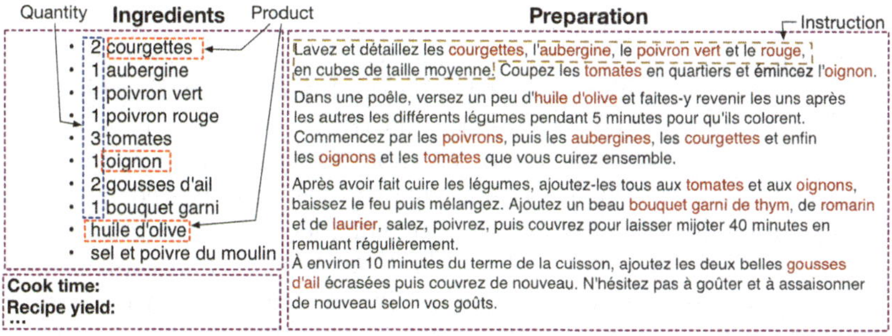

Fig. 1. Recipe metadata information. The recipe is semi-structured in XML format. The elements of the recipe taken into account are the list of ingredients and the list of instructions for the preparation of the recipe.

(bunch of thyme)"). This makes their recognition difficult. In some cases the ingredients appear as: (1) a verb in the preparation (salt/season); (2) a shortcut (fish fillet/fillet); (3) implicit (the instruction "fry the onions in the preparation", deduce that it is in the olive oil ingredient); (4) a mixture (mix the ingredients of the dough, etc.). In addition, some ingredients which appear in the preparation could be missing in the ingredients list (for instance accompanying suggestions for the dish). To tackle the above issues, an analysis is conducted to identify the different kinds of structure patterns to consider and we have defined the *Percentage of Appearance of an Ingredient in a Preparation score* ($PAIP$). To do so, we randomly selected 150 kitchen recipes among the collection of 30,000 gathered recipes. From the selected recipes, we isolated the ingredients from the detail of the preparations, producing 150 ingredient lists and 150 preparation instruction lists. A lemmatization process using the TreeTagger Part-Of-Speech tool[2] is then applied for each term in the resulting lists. Finally, the $PAIP$ score is calculated following the formula:

$$PAIP = \frac{\sum_{i=1}^{n} (\frac{Ning}{Ting})}{Trcp}$$

where n is the number of recipe, $Ting$ is total number of ingredients, $Ning$ is the number of recognized ingredients and $Trcp$ is the total number of recipes. The resulting $PAIP$ was 45%, denoting that large part of the ingredients are not recognized directly in the preparation (55%). This low percentage can be explained by the provenance of the recipes, which are elaborated on the Web by Internet users without any formalized instructions.

[2] https://www.cis.uni-muenchen.de/~schmid/tools/TreeTagger/.

Food Composition Table. The food composition table (*FCT*) consists in a list of foods characterized by the nutrients that they contain. Several food composition tables are accessible and usable on the Internet (for example, Ciqual,[3] USDA,[4] FCEN,[5] etc.). Most of essential nutrients are common to all composition tables. The differences concern the non-exhaustive food coverage of the tables. For the same food-constituent pair, the values indicated in two tables can be very different. For instance, the value of energy for kiwifruit in Nutrinet and USDA is respectively 49.7 and 60 kcal/100g. This is often explained by the provenance of the products which could be different from country to another. So, the kiwifruit consumed in the United States is very different (in terms of species, place of production, etc.) from that consumed in France. Moreover, when the data is difficult and/or expensive to obtain with chemical analysis the loans become frequent from other tables. In this case, the table quality is less adapted to the context. Nutrinet (68 nutrients for each food) is one of the *FCT* which meet the criteria defined by [4].

The Knowledge Graph Resource. Although our approach can relies on any KG which described relevant nutritional entities, in this study, the *KiMO* modular digital kitchen KG, expressed in OWL 2 is used [3]. It is composed of seven different modules: `CUISINE`, `PREPARATION`, `ALIMENT`, `UNITE`, `MATERIEL`, `NUTRITION` and `PERSON`. However, in the context of our work, only `ALIMENT` and `PREPARATION` are relevant. The `ALIMENT` (Food) module is LANGUAL compatible (a multilingual faceted thesaurus, dedicated to describing food products in a systematic way) [16] and contains the relevant concepts relating to foods (about 11035), their mode of conditioning and their characteristics. As for the `PREPARATION` module (about 6,164 concepts), it formalizes the basic preparations (for example, cooking juice, connecting element, etc.) used in a recipe. More specifically, it describes food transformations *via* culinary actions and types of cooking.

2.2 Nutritional Score Calculation

A challenge for us is to be able to handle properly contents of preparation instructions in order to extract relevant information. In the food context, [12] propose a software toolkit with SPARQL-based service that can be used to create a unified food KG that links the various silos related to food while preserving the provenance information. The most difficult part of writing a SPARQL query is to compose a `Where` clause that requires a complete understanding of the concepts included and their relationships by various properties which has led to the establishment of other approaches, to make it easier for users to write these query statements. Our adopted approach follows a two-layer processing method: a lexical layer and a rules layer. They are detailed in the following section.

[3] https://ciqual.anses.fr/.

[4] https://fdc.nal.usda.gov/.

[5] https://aliments-nutrition.canada.ca/cnf-fce/index-fra.jsp.

In the lexical processing layer, we have built a lexicon about all the preferred and alternative (synonyms) labels that occur in the KG. Thus, it became possible to perform queries on this lexicon by using a dedicated designed algorithm.

Ingredients List Lexical Enrichment. It consists in enriching the list of available ingredients in order to improve the result that could be obtained during the named entities recognition (NER) phase. The chosen approach consists in exploiting two axes for the analysis of the entities:

- **An Ontological axis**: generalizing any encountered specific entity following the subsumption relationships ($is - a$). For instance, $Tomate$ ($Tomato$) $\rightarrow$ $VegetableFruit$ ($Vegetable\ and\ Fruit$) $\rightarrow$ $Vegetable \rightarrow Food$).
- **A Structural axis**: exploiting the topological neighbouring of entities. For instance, $Tomato$(`NOUN`), $Peel$(`VER`), *etc.*).

For each recipe, a tokenization process is performed for annotating (NER) all the ingredients and preparation instruction as detailed in Fig. 2. The lexical search of the lemmatized motif of ingredients in the KG and preparation instructions gives access to the hierarchy of concepts associated with this motif, as well as to other information, for example, the module to which it belongs ($Food$); the concept identifier (class) in the ontology; the "basic product" by exploring all the "ObjectProperty"; etc. To identify the product status in the recipe (cooked, raw, etc.) and the category to which it belongs to ($Fruit$, $Vegetable$, $Meat$, etc.), we rely on querying the KG using the SPARQL language.

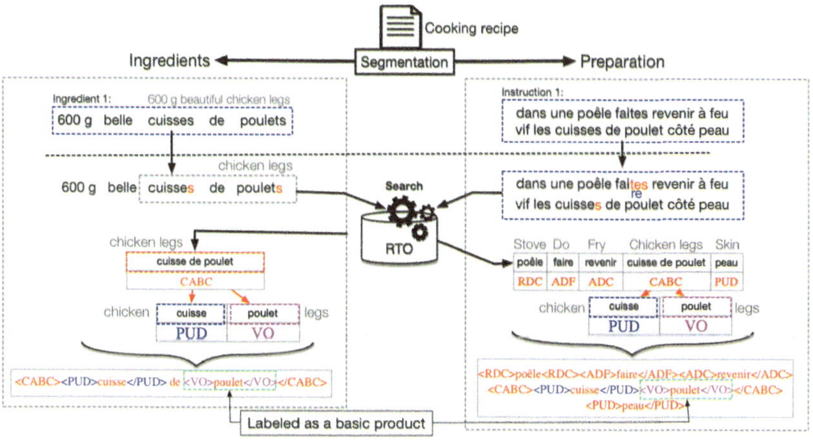

Fig. 2. Annotation of recipes using the KG. First, each lemmatized motif of ingredient and preparation instruction is tagged by the concept identifier in the KG. Then, the "base product" (e.g., chicken legs/chicken) is identified. Finally, in the case where the base product belongs to the first list, all the concepts of the second list corresponding to the same instruction are associated with this "base product"

Generating the Nutritional Calculus. The second step of the algorithm is to produce a list of figure that is defined as a `"product_id/category/weight"` template. This `"product_id"` corresponds to the one in the database Nutrinet. To identify this `"product_id"` we starting from the list of enriched products in the first step, and create a the pattern as `"basic product/part/cooking action/other modifications"`. This pattern is used to filter the results returned from our query in the database.

The identification of each pattern term in the list of "enriched product" terms involves the lexicon querying. In the case of no any pattern term is identified, a default value is assigned. For instance, "part" takes the default value "integer".

Then the database of Nutrinet is looked up to identify the unit of each ingredient. These units are eventually converted into grams, following a procedure provided by our dietitian experts.

Calculation and Allocation of the Final Score. The NVs calculation automatize a combination of the method proposed by [6] and labeling with the *Nutri-score* adopted in France as part of the National Health Nutrition Program. The yield factor calculation is performed from the recipe weight before and after being cooked; This yield factor is applied at the recipe level and the nutrient retention factors at the ingredient level. For example, the edible percentage of a chicken legs is only 60%. The proposed algorithm, which has been validated by the dietitians experts illustrated as follows (Algorithm 1):

Algorithm 1: Calculating NVs of recipe

```
Input : IngList, NutList, YieldCoef, Portion
Output: NVs of the recipe.
foreach Ing ∈ Ilist do
    foreach Ing ∈ IngList do
        NewIng_value ← Pre-pro1 (Ing, YieldCoef)
        foreach Nut ∈ NutList do
            IngNut_value ← Pre-pro2 (Ing, NewIng_value)
        end
        NutRecipe_value ← Pre-pro3 (IngNut_value)
    end
    FV_value ← Pre-pro4 (IngList)
    DishNetWeight ← Pre-pro5 (IngList)
    PortionWeight ← Pre-pro6 (DishNetWeight, Portion)
end
return NutRecipe_value, FV_value, DishNetWeight, PortionWeight
```

Where FV is the fruits and vegetables, WFV is the weight of fruits and vegetables, WOI is the weight of other ingredients and *Pre-pro* is the pre-processing operations.

$$Pre\text{-}pro2: \frac{NewIng_{value} * Ing_{value}}{100} * Ing_{YieldCoef} \;\; ; \;\; Pre\text{-}pro3: \sum_{i=0}^{n} IngNut_{value}$$

$$Pre\text{-}pro4: \frac{WFV + (2 * WFV\ dried)}{WFV + (2 * WFV\ dried) + WOI} \;\; ; \;\; Pre\text{-}pro5: \sum_{i=0}^{n} NewIng_{value}$$

$$Pre\text{-}proc6: \frac{DishNetWeight}{Portion} \;\; ; \;\; Pre\text{-}pro1: Ing_{value} * Ing_{YieldCoef}$$

Eventually, the final score is computed following the profiling system used by the [7]. It is outlined by Algorithm 2:

Algorithm 2: The recipe Score attribution

```
Input  : FV_value, ScoreTab, NutRecipe_value
Output: Score of the recipe
foreach NutRecipe ∈ NutRecipe_value do
 | NutRecipe_point ← PRE-PROCESSING(NutRecipe, FV_value, ScoreTab)
end
NegElements_points ← Σ_{i=0}^{n} NegElement_point)
PosElements_points ← Σ_{i=0}^{n} PosElement_point)
if NegElement_value < 11 then
 | Score_value ← (NegElements_points − PosElements_points)
else
 | if FV_point = 5 then
 |  | Score_value ← (NegElements_points − PosElements_points)
 | end
 | if FV_point < 5 then
 |  | Score_value ← (NegElements_points − (Fiber_point + FV_point))
 | end
end
```

The theoretical score range from −15 (most favorable) to +40 (worst). Colors and letters are assigned according to the final score: A/green (from −15 to −1); B/green mint (from 0 to 2); C/yellow (from 3 to 10); E/orange (from 11 to 18); D/red (from 19 to 40).

Translating the Score into a Graphical Scale. For each recipe, the score is visualized using the official French nutritional guidelines (PNNS) scale suggested by Serge Hercberg, a nutritional expert, as part of its report to the French Ministry of Health on recommendations related to nutritional prevention.[6] The scale is in the form of 5 discs of different colors on a range from green to red, coupled with a corresponding letter on the scale of school "grades" from A to E. The size of the disks depends on the score reached by the recipe. We used this representational paradigm as the ability of people to classify products according to their nutritional quality has been evaluated in a previous study by [8]. This study shows that nutritional information systems significantly increase the ability of individuals to classify products according to their nutritional quality. Thus, the results obtained are in favor of the PNNS, which gives better results regardless of age, sex and socio-economic category.

[6] https://tinyurl.com/yx2f7bwz.

3 Evaluation and Discussion

The *NutriSem* approach has been evaluated a randomly selected sample of 10 recipes from the corpus of 45 recipes previously manually assessed by three dietitian specialists from the Nutrition Lab of the Paris 13 University (EREN[7]). In addition, three publicly available systems (*Nutryaccess*,[8] *Monmenu*[9] and *Myfitnesspal*[10]) have been compared to *NutriSem* on five criteria. The Table 1 presents the results obtained for the first six recipes of the sample. The first column of the table corresponds to the recipes NVs calculated by the dietitian experts, the second corresponds to the results obtained by *NutriSem*, while the third column is those obtained by the *Monmenu*.

Table 1. Results for the first six recipes

Recipe	Ene			Pro			Fat			Sug			Sod			Fib		
	ER	To	Mon	ER	To	Mon	ER	To	Mon	ER	To	Mon	ER	To	Mon	ER	To	Mon
R1	149	171	127	11	13	9	5	6	8	1	1	2	231	118	–	1	1	1
R2	225	237	435	6	7	9	1	1	19	4	4	57	153	61	–	1	1	–
R3	237	254	300	3	3	6	1	2	4	26	24	35	77	29	–	1	1	–
R4	458	413	485	5	5	4	6	17	24	40	30	63	116	71	–	1	2	–
R5	83	100	33	1	1	1	1	1	2	1	1	2	202	211	–	2	2	–
R6	108	119	71	8	12	8	1	2	3	1	1	2	517	444	–	2	1	–

Legend: Ene: energy Kcal/100g, Pro: protein g/100g, Fat: saturated fats g/100g, Sug: simple sugars g/100g, Sod: sodium Mg/100g, Fib: fibers g/100g, ER: dietitian experts, To: *NutriSem*, Mon: *Monmenu*, R1: blanquette de veau, R2: crêpes, R3: gâteau aux pommes, R4: moelleux au chocolat, R5: ratatouille, R6: salade niçcoise

From Table 1, it could be observed that for the recipe *ratatouille*, unlike the other systems, the results obtained by *NutriSem* for **R1** are close to those obtained by the dietitian experts. The significant difference between the results can be explained by several factors. The first one is related to the food composition tables used to calculate the NVs of the recipe. Indeed, the three other systems compared to *NutriSem* provide no information on the nature and source of the data used when we use the Nutrinet table developed by dietitian experts from the Nutritional Epidemiology Research group in Paris (France). Nutrinet meets higher quality criteria. The second deals with the treatment of the recipe. On the one hand, while in *NutriSem* each recipe is handled in a automated way, *Nutryaccess* and *Myfitnesspal* follows a semi-automatic approach. Indeed, each ingredient in the recipe is manually selected by the user from a list of ingredients. Therefore, the choice depends on the culinary skills of the user.

[7] https://eren.univ-paris13.fr/index.php/en/.
[8] http://www.nutryaccess.com/.
[9] http://www.monmenu.fr/.
[10] https://www.myfitnesspal.com/.

NutriSem and *Monmenu* perform an automated mapping between the list of ingredients of any recipe and those of composition table. Another limitation of the three other systems is the approach of the NVs computing for each recipe which consists in simply adding together the NVs of each ingredient. Therefore, it does not take into account the status of the product in the final dish (whether it is cooked or raw) and the variation due to the preparation process of the dish. The third factor is the conversion method applied to switch from one unit to another (for example, the average weight of an apple in grams). No information is available on the conversion method that the three systems apply. Household units differ according to the country because the dimensions of food and household measures are different. Therefore, units must be handled with caution as it could introduce significant errors in the nutrient content of a recipe in particular. Finally, another observed limitation corresponds to the nutritional information provided to the user concerning the prepared dish. For example, in most cases no information is provided by the three systems for the value of sodium or the percentage of fruits and vegetables in the dish, which is essential. In contrast, in addition to such information, *NutriSem* provides a score reflecting the overall nutritional quality of the consumed dish.

According to the performed evaluation, *NutriSem* could be improved in several ways. First, it could be interesting to evaluate the system in a wider context, with a focus on each step. For example, quantifying the impact of the identification of the products composing the ingredients, or the cooking method on final score would be interesting to investigate. Although the mapping process between the recipe and the entries in the composition table yielded promising results and proven to be conclusive in various cases, standardizing of the labels of the Nutrinet composition table could lead to better results. In this context, LANGUAL [16] could be a useful resource to rely on. Moreover, taking into account the food sensory qualifications by using the sensory module of the *KiMO* KG [3] would enable personalized suggestions for people suffering from chronic diseases by taking into account their tastes. This will induce a better acceptability and observance of their diets.

4 Conclusion and Future Work

Diet remains a daily activity, eating habits have changed a lot in recent years. These profound changes have both positive and negative health and nutritional benefits and limitations. That is why it is essential to rethink and reinvent the concepts about the simplification and health education on a daily basis. This paper described *NutriSem*, a nutritional qualification methodology and system for cooking recipes. It relies on a KG for ingredients identification and nutritional score computing. The conducted evaluation shows promising results compared to three other publicly available systems that it outperforms. *NutriSem* will be integrated into a cardiovascular risk prevention ecosystem to provide content and tools for better informed healthy lifestyle to users.

References

1. Afshin, A., John, P., Fay, K.A., et al.: Health effects of dietary risks in 195 countries, 1990–2017: a systematic analysis for the global burden of disease study 2017. Lancet **393**, 1958–1972 (2019). https://doi.org/10.1016/s0140-6736(19)30041-8. Elsevier BV
2. Azaïs-Braesco, C., Goffi, C., Labouze, E.: Nutrient profiling: comparison and critical analysis of existing systems. Public Health Nutr. **9**, 613–622 (2006). https://doi.org/10.1079/PHN2006966. Cambridge University Press
3. Despres, S.: Construction d'une ontologie modulaire pour l'univers de la cuisine numérique. In: 25éme Journées francophones d'ingénierie des connaissances, Clermont-Ferrand, France, pp. 01–28 (2014)
4. Greenfield, H., Southgate, D.A.T.: Food composition data: production, management, and use. In: Food and Agriculture Organization of the United Nations, pp. 243. B.A. Burlingame et U.R. Charrondière, Rome (2007)
5. Nouvel, D., Antoine, J.E., Friburger, N.: Food composition data: production, management, and use. In: 5th Language and Technology Conference, LTC 2011, Poznan, Poland, pp. 226–237 (2014). https://doi.org/10.1007/978-3-319-08958-419
6. Charrondiere, U.R., Burlingame, B., Berman, S., Elmadfa, I.: Food composition study guide: questions, exercises and answers. In: Food and Agriculture Organization of the United Nations, Rome, Italy, pp. 282 (2011)
7. Hercberg, C.: Propositions pour un nouvel élan de la politique nutritionnelle française de santé publique. In: Rapport 2013 dans le cadre de la Stratégie Nationale de Santé, France, pp. 128. France (2013)
8. Ducrot, P., Méjean, C., Julia, C., et al.: P224: Compréhension objective vis-à-vis de différents systèmes d'information nutritionnelle simplifiés sur la face avant des emballages des aliments: étude NutriNet-Santé. Nutrition Clinique et Métabolisme **28**, 186–187 (2014). https://doi.org/10.1016/s0985-0562(14)70866-0. Elsevier BV
9. Magesh, P., Thangaraj, C.: Comparing the performance of semantic image retrieval using SPARQL query, decision tree algorithm and lire. J. Comput. Sci. **9**, 1041–1050 (2013). https://doi.org/10.3844/jcssp.2013.1041.1050. Science Publications
10. Harish, D.V.N., Srinivas, Y., Rajesh, K., Anuradha, P.: Image annotations using machine learning and features of ID3 algorithm. Int. J. Comput. Appl. **25**, 45–49 (2011). https://doi.org/10.5120/3024-4090. Foundation of Computer Science
11. Fadzli, S.A., Setchi, R.: Semantic approach to image retrieval using statistical models based on a lexical ontology. In: Knowledge-Based and Intelligent Information and Engineering Systems, pp. 240–250. Springer, Heidelberg (2010). https://doi.org/10.1007/978-3-642-15384-6_26
12. Haussmann, S., Seneviratne, O., Chen, Y., Neeman, Y., et al.: FoodKG: a semantics-driven knowledge graph for food recommendation. In: Lecture Notes in Computer Science, pp. 146–162. Springer (2019). https://doi.org/10.1007/978-3-030-30796-7_10
13. He, J., Liu, L., Yu, F., Han, Y.: A method of RDF fuzzy query based on no query language service with permutated breadth first search algorithm. Procedia Comput. Sci. **100**, 321–328 (2016). https://doi.org/10.1016/j.procs.2016.09.163. Elsevier BV
14. Hamada, S.E.: Enrichment lexical knowledge with interword based features. In: IEEE International Conference on Systems, Man and Cybernetics. IEEE (2003). https://doi.org/10.1109/icsmc.2002.1176105

15. Amato, F., Mazzeo, A., Elia, A., Maisto, A., Pelosi, S.: Morphosemantic strategies for the automatic enrichment of Italian lexical databases in the medical domain. Int. J. Grid Util. Comput. **8**, 312 (2017). https://doi.org/10.1504/ijguc.2017.088262
16. Møller, A., Ireland, J.: LanguaL 2017 - Multilingual Thesaurus. Technical report. Danish Food Informatics (2018). https://doi.org/10.13140/rg.2.2.13274.64964

Using Machine Learning Models to Predict the Length of Stay in a Hospital Setting

Rachda Naila Mekhaldi[1(✉)], Patrice Caulier[1], Sondes Chaabane[1], Abdelahad Chraibi[2], and Sylvain Piechowiak[1]

[1] Laboratory of Industrial and Human Automation Control, Mechanical Engineering and Computer Science, Polytechnic University of Hauts-de-France, Campus Mont Houy, 59300 Valenciennes, France
{rachdanaila.mekhaldi,patrice.cualier,sondes.chaabane, sylvain.piechowiak}@uphf.fr

[2] Alicante, 50 rue Philippe de Girard Seclin, 59113 Seclin, France
abdelahad.chraibi@alicante.fr

Abstract. Proper prediction of Length Of Stay (LOS) has become increasingly important these years. The LOS prediction provides better services, managing hospital resources and controls their costs. In this paper, we implemented and compared two Machine Learning (ML) methods, the Random Forest (RF) and the Gradient Boosting model (GB), using an open source available dataset. This data are been firstly preprocessed by combining data transformation, data standardization and data codification. Then, the RF and the GB were carried out, with a phase of hyper parameters tuning until setting optimal coefficients. Finally, the Mean Square Error (MAE), the R-squared (R^2) and the Adjusted R-squared (Adjusted R^2) metrics are selected to evaluate model with parameters.

Keywords: Length of stay prediction · Machine learning · Healthcare system

1 Introduction

Healthcare institutions, academic researchers and industry organizations in various areas are working in coordination to improve the quality of care and the management of healthcare systems. The Length Of Stay (LOS) is considered as one of the basic indicators to evaluate the performance of care services and care quality which explain the growing interest of predicting LOS in hospitals these past years. The LOS represents the interval time between the admission of the patient and his discharge [1]. Estimating the LOS at the admission time provides an approximation of the patient's discharge date involving an appropriate planification of care activities. As a result, expecting the acute value of the LOS at the time of patients' admission is useful to highlight a planning strategy for the hospital's logistics.

Á. Rocha et al. (Eds.): WorldCIST 2020, AISC 1159, pp. 202–211, 2020.
https://doi.org/10.1007/978-3-030-45688-7_21

According to [2], LOS is also used to comply with hospital budget constraints and for reimbursement purposes. In France, data gathered from the program of the medicalization of the information systems (PMSI in French which stands for *Programme de Médicalisation des Systèmes d'Informations*) is used for LOS prediction. This program aims to define the activity of care institutions, either in the public or private sector. Its objective is to calculate the budget to be allocated to the units concerned in order to reduce the gap of resources between health institutions [3]. Thus, hospitals are forced to adapt their strategy for planning admissions and reduce the patient's LOS. Further, this program encourages health institutions to review their policy in terms of cost efficiency and resource optimization. This leads to improve the admissions' scheduling and to manage the beds' occupancy in the hospital [4]. Estimating the LOS at the admission time also helps to reduce the waiting time for a patient, to get a better plan of care activities, to optimize resources in hospitals, etc.

The LOS is a complex variable that depends on various factors that could be related to the clinical and social context of the patient and his care [5]. The prediction of LOS highly depends on the type of service in which the patient is admitted. Indeed, LOS models in an emergency or ambulatory department differs from that of a scheduled department [6].

Understanding the factors impacting the LOS is crucial to predict its value. Artificial intelligence methods involving ML and data mining are used in this area of research. To first analyze factors impacting the LOS and, second, to make a LOS prediction model based on these factors.

The objective of this paper is to implement ML process for LOS prediction starting with data preprocessing to model validation. The Random Forest (RF) and the Gradient Boosting Model (GBM) are implemented and compared. This allows to highlight the difficulties faced to deal with medical data and to analyse regression methods in the prediction of LOS. In the first part, we review factors impacting LOS prediction and applicable methods in the area of data mining and ML. Second, we present our contributions. At the end, model evaluation, discussion and perspectives are presented.

2 Related Work

Studies on using data mining and ML methods in LOS prediction in hospitals have been increasingly addressed in the literature. In this section, we will present a range of research related to the problem described above. Firstly, factors impacting the LOS are presented. Next, ML Methods used for the prediction problem are highlighted.

2.1 Factors Impacting the Length of Stay

Identify factors that constitute LOS models is the first task to carry out in the field of LOS prediction. Awareness of factors and elements that determine LOS is being the purpose of many surveys as it promotes the development of efficient clinical pathways and optimizes resource utilization and management [7]. In [8], the patient's age, his gender, hematocrit, serum and creatinine measurements are used to predict LOS in the cardiac surgical intensive care unit. Researchers in [9] indicated that the influential factors might differ between urgent operation and non-urgent operation cases in the prediction of prolonged LOS before surgery. The factors used in this study are demographic information, medical history, vital signs, laboratory data, operation physician data, operation, and nursing data. D. Loshin in [10] explored factors affecting the LOS in teaching hospitals of a middle-income country. The study includes factors related to patient's information such as age, gender, marital status, occupation and place of residence. In addition to these factors, the patient's admissions' history, type of admission, type of treatment, patient condition at discharge and type of payment for hospital fees are considered. In the study conducted by [11], information about admissions, discharges, transfers, caregivers are prescriptions collected. These variables are combined with medical information, laboratory analysis, diagnosis, demographic information, and patient notes. LOS is determined in [12] by the patient's age, gender, and patient admission condition. This study focuses on the LOS prediction in a surgery service. The authors add operation information and hospitalization information to create the model.

From these works concerning factors impacting LOS, we have noticed that in general, patient's information (demographic and medical) are widely used in the conception of the LOS model. Other factors such as medical patient's information differ from a service to another. They are determined by the type of the medical service.

2.2 Machine Learning Methods in LOS Prediction

Our aim is to apply ML to predict acute patient's LOS in a hospital setting. Therefore, this section highlights the ML algorithms applied for this problem.

To analyze the huge amount of medical data, several methods of data mining were used. Preceding studies have used multiple supervised learning techniques to create a prediction model of LOS based on factors impacting this variable. In [8], the ALM model (Asset Liability Management) was applied to select variables to be used in the ML process. From 36 variables, only 8 were selected. Then, the odds ratio was used to study the dependence between the qualitative variables. An Artificial Neural Networks model was implemented for the prediction. [9] in their study, they built three different models based on a total of 68 variables. The models were: Decision Tree, Support Vector Machine and the Random Forest. They marked the best result with the Random Forest. In another study, supervised ML algorithms were carried out to study factors affecting LOS and make a prediction model. Among the Classification and Regression Tree (CART), Automatic Interaction Detection (CHAID) and the Support Vector Regression, the best result was obtained from the CART model [13]. The purpose of the study in [14] was to determine factors influencing LOS and predict it in the general surgery department. The first step was to delete the repeated records and deal with outliers and missing data. This problem was solved by involving medical experts in the survey and the result of the literature. In the second step, the regression linear algorithm, the decision tree, the naïve Bayes and the K-nearest neighbor were used. The best performance was achieved with the decision tree. To predict if the patient will have a long stay or a short one in the Intensive Care Unit (ICU) department. Authors in [11] used the MIMIC3 (Medical Information Mart for Intensive Care) [15] dataset to implement neural network and the random forest algorithms. The accuracy attended was equal to 80% with both algorithms. In another work [10] proposed a model based on supervised ML to predict the LOS for patients suffering from coronary disease in the cardiac department. A decision tree was employed to extract rules from the data. Support Vector Machines (SVM) and Artificial Neural Network were developed for the prediction problem.

From this past researches, ML algorithms, in particular, supervised learning algorithms are widely used in the field of LOS prediction. Neural networks, decision trees, SVM and random forest are the most cited algorithms. In our study, we implemented the random forest as the best algorithm in term of performance in LOS prediction in the literature. We chose the gradient boosting model as one of ensemble algorithms for regression problem as a second tested ML technique.

3 Methodology

Our work aims to use ML methods to predict the LOS in hospitals. In this section, we provide a brief description of the dataset used and present our methodological process.

3.1 Dataset Description

In our study, real case dataset still being under preparation due to administrative procedures. For this, we choose to employ *Microsoft dataset for the length of stay prediction* [16] because of the data availability. Further, the variables describing the LOS in this dataset are similar to those found in the literature.

Microsoft dataset for LOS prediction is an open-source dataset that provides categorical data (e.g. gender) and numerical data (e.g. glucose). It contains a total of 100 000 observations representing the patient's information specified by 28 variables including the LOS as the number of days. Given the state of the art, all variables according to the laboratory analysis and medical history are kept. The date of admission and the date of discharge are replaced by a binary flag indicating if it's weekend or not. The sequential variable "*eid*" is removed because it doesn't report any extra information. This variable is a just sequential number to count the observations. The dataset contains unbalanced information and outliers. It doesn't include any missing data.

3.2 Working Methods

Our ML process involved three main steps: data preprocessing, ML models conception and model evaluation.

Data Preprocessing: Depending on the type of the variable, a preprocessing method is applied. Data transformation and data standardization are applied to numerical data. The one-hot-encoding is used to represent categorical data.

Data Transformation: Data transformation is required when the distribution of data is "not normal". Histograms and the distribution of each variable are drawn. If the variable's distribution is not normal, a transformation using the logarithmic function is used. The following figures show the histograms and the distribution of the variable "*neutrophils*" before (Fig. 1) and after (Fig. 2) transformation.

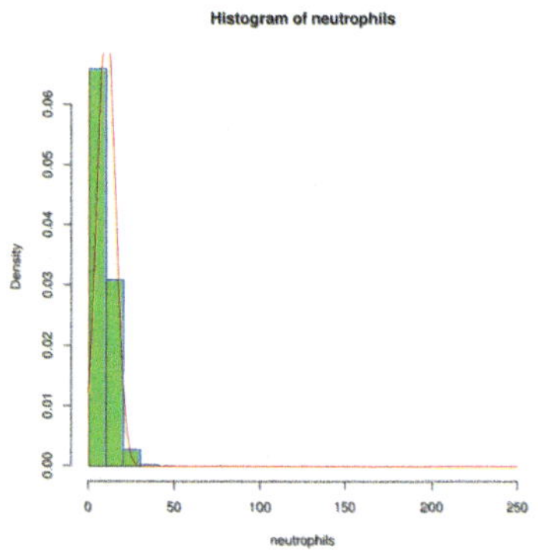

Fig. 1. Histogram before transformation

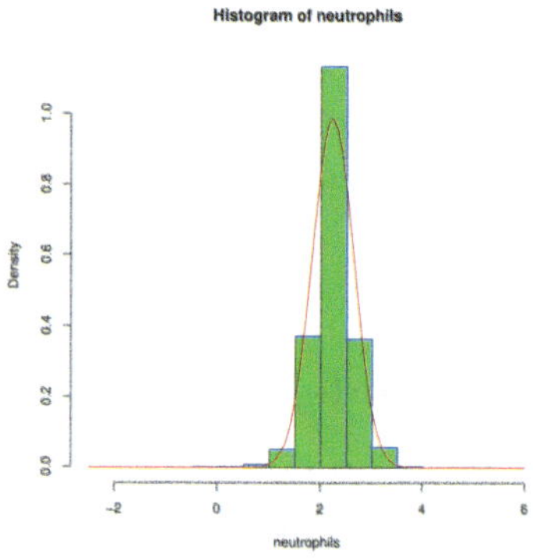

Fig. 2. Histogram after transformation

Data Standardization: Z-score standardization is applied to numerical data. The aim is to map the source data into a target structural representation [10]. The function used is as follows where X_i represents an observation, μ is the average of the variable X and σ is its standard deviation:

$$Z_i = \frac{X_i - \mu}{\sigma} \quad (1)$$

Data Codification: To handle the categorical variables, the one-hot-encoding approach is applied. The first step is to transform the categorical variables to integer variables. The second step is to transform the integer representation to a binary representation. This method is widely used to ignore the natural ordered relationship between integer values [17]. To illustrate this method, the variable "*rcount*" is used. This variable represents the number of readmissions during the previous 180 days. The table below illustrates this method for the varibale "*rcount*".

The Table 1 is an exemple of applying the one-hot-encoding approach.

Table 1. Application of the one-hot-encoding approach on varibale "*rcount*".

rount	0	1	2	3	4	5+
rcount_0	1	0	0	0	0	0
rcount_1	0	1	0	0	0	0
rcount_2	0	0	1	0	0	0
rcount_3	0	0	0	1	0	0
rcount_4	0	0	0	0	1	0
rcount_5+	0	0	0	0	0	1

3.3 Learning Methods

The second step in an ML process is to develop learning algorithms. In this study, a particular interest to apply the random forest and the gradient boosting model to our regression problem is given. The random forest is a supervised ML algorithm used for classification and regression. We choose the random forest as our learning algorithm because its performance is one of the best in the literature in LOS prediction. Also, this algorithm was used by Microsoft to develop ML services for LOS prediction. The second used algorithm (i.e. Gradient Boosting Model) is based on training many decisions trees models in a gradual, additive and sequential manner. As for the Random Forest, this algorithm is tested in Microsoft Project.

For each algorithm, to further improve our training model, tuning parameters is conducted. For the RF, the numbers of trees (NT) that defines the number of trees used in each step of the RF is varied. The second parameter varied for the RF is the maximum depth (MD) which represents the maximum depth of each tree chosen. For the GBM, the number of trees that will be fit in series in each step to reduce prediction errors represented by NT and the learning rate (LR) that corresponds to how quickly the error is corrected from each tree to the next and is a simple multiplier are varied [18]. For the chosen parameters, there is a compensation between them.

4 Experimental Results and Evaluation

The last step in this study is model evaluation. This step allows to test a model with the different varied parameters and save the best model. Python and R programming language were used in the implementation phase and a processor Intel(R) Core(TM) i7-8650U with a 8 GB of RAM. Regression metrics such as *MAE*, R^2 and *Adjusted*R^2 are used to evaluate the system's performance. MAE represents the absolute difference between the predicted values and the observed ones and it's robustness to existing outliers in the dataset. The R^2 and *Adjusted* R^2 highlight how well the target variable explain the variability in the attributes.

In what follows, the curves of the variation of metric's values depending on the tuning parameters are given.

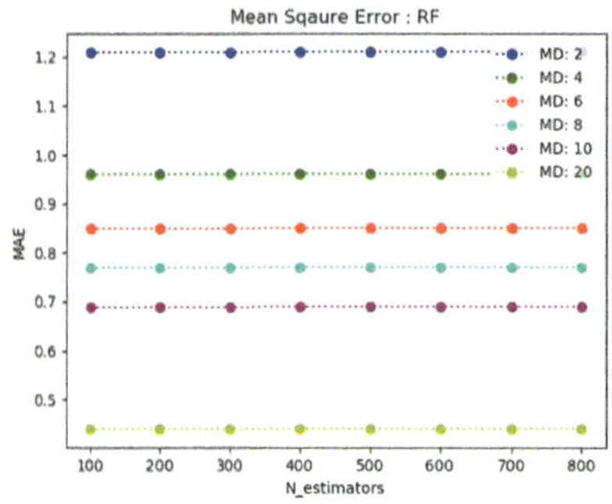

Fig. 3. MAE: RF

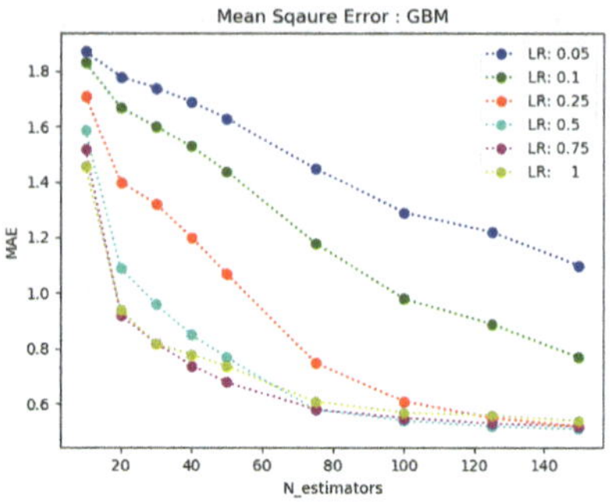

Fig. 4. MAE: GBM

The results of MAE for the two parameters variations are presented in the figure above (Fig. 3 and Fig. 4). For the random forest, the lowest value is 0.44 and for the GBM the lowest value is 0.55. These values correspond to NT = 800 and MD = 20 for the RF and to NT = 100 and LR = 0.75 for the GBM. In this case, the MAE obtained by the RF is better than the one obtained by the GBM. For the RF, the parameter MD is more significant to reduce the MAE than the NT. In opposition to the GBM, both of the parameters (e.i. MD and LR) impact the MAE. More the NT and LR are bigger, less the MAE is.

In the following, figures of R^2 (Fig. 5 and Fig. 6) and adjusted R^2 (Fig. 7 and Fig. 8) are presented for both algorithms. First, we compared the values of R^2 for the RF and the GBM. Then, to ignore the importance of variables in the construction of the models, we have used the metric adjusted R^2.

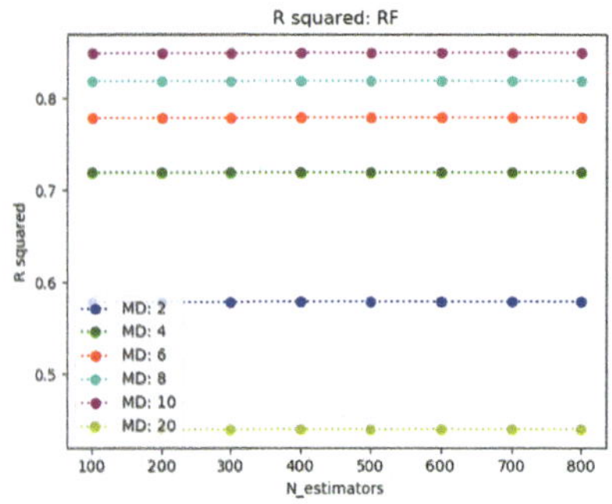

Fig. 5. R^2: RF

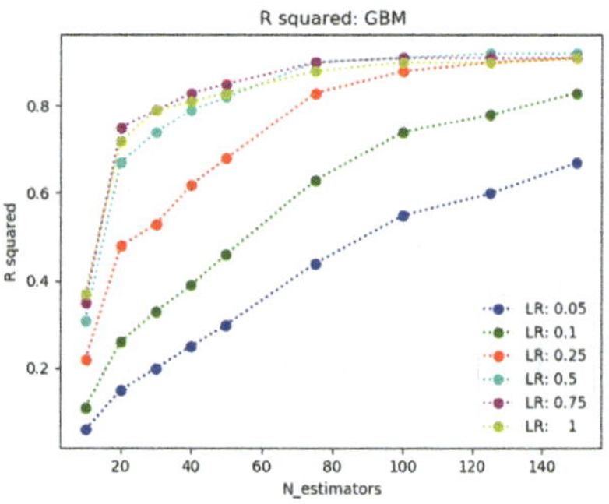

Fig. 6. R^2: GBM

Concerning the R^2, the importance of parameters does not change comparing to the one concluded from the MAE. For the RF, the highest value corresponds to NT = 800 and MD = 10 and it's equal to 0.92. For the GBM, we have noticed that the best result is equal to 0.91. According to R2, the difference between the two models is not significant.

The results achieved for the adjusted R^2 by the RF and the GBM are shown in Fig. 7 and Fig. 8.

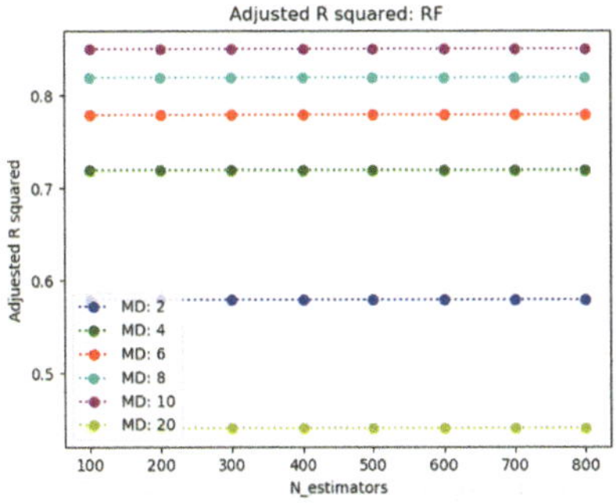

Fig. 7. Adjusted R^2: RF

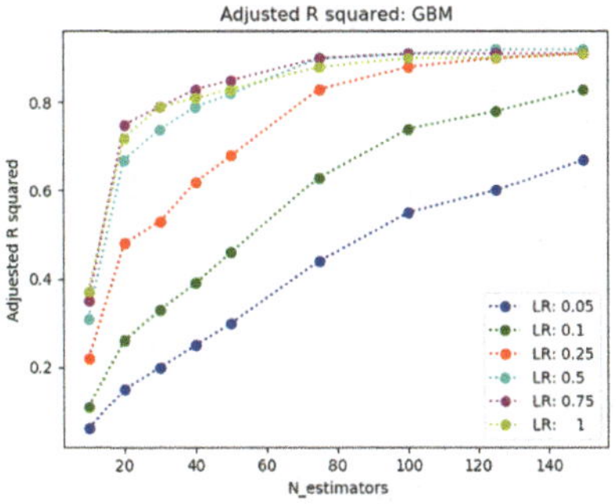

Fig. 8. Adjusted R^2: GBM

Comparing to the metric of R^2, we did not notice any difference with the results of adjusted R^2.

time, the GBM is more convenient than the RF. The best models are saved for both algorithms to be trained on a real dataset later. Regarding theses results and to improve the performance of the models, the role of medical experts is crucial in the preprocessing step. Indeed, they could decide if a value represent an outlier or not. In fact, medical data specially results of lab analysis contains huge number of outliers. Using the statistical methods to deal with that is not always enough. Also, to improve the performance of models, unbalanced classes must be handled by applying statistical techniques or omitting some observations. For the ML phase, the combination of the right values of parameters used to train the model is extremely important. Also, the objective and the requirement of the study must be defined first for choosing the ML method (e.g. real time prediction or not).

5 Conclusion

In this paper, we investigated the factors impacting the LOS and the most known ML methods in the literature. Data transformation, data standardization and data codification were involved in the preprocessing step. Then, we explored two ML algorithms to implement a LOS prediction model. The system was trained on a dataset containing similar information found in the literature. ML and ensemble techniques used in this research had a similar performance. For both algorithms, the predictive models perform with an MAE lower than 0.45 and a R^2 and *Adjusted* R^2 greater than 0.92. These are a satisfying result regarding the nature of the dataset that we used.

One potential limitation of our study is the unavailability of a real dataset. In fact, we had to use Microsoft dataset as an example due to the difficulties that we met in access to the real data. In our case, this dataset allows to highlight the complexity of medical data. For better results, medical experts must be involved to determine factors impacting the LOS and in the preprocessing step. The values of the LOS in the used dataset were limited to a small set. An alternative to enhance the performance of the system is to transform the LOS from a numerical data to a categorical one based on the survey, the existing work and medical experts.

Acknowledgments. The authors would like to thank the region of Hauts-de-France (FEDER funds) and the company Alicante (https://www.alicante.fr/) for the funding and their support and contributions.

References

1. Khosravizadeh, O., Vatankhah, S., Bastani, P., Kalhor, R., Alirezaei, S., Doosty, F.: Factors affecting length of stay in teaching hospitals of a middle-income country. Electron. Phys. **8**(10), 3042–3047 (2016)
2. Carter, E.M., Potts, H.W.: Predicting length of stay from an electronic patient record system: a primary total knee replacement example. BMC Med. Inform. Decis. Mak. **14**(1), 1–13 (2014)
3. Fédération Hospitalière de France: Le PMSI: des objectifs et une standardisation des données pour le service public hospitalier et les établissements privés de soin (2019)
4. Turgeman, L., May, J.H., Sciulli, R.: Insights from a machine learning model for predicting the hospital Length of Stay (LOS) at the time of admission. Expert Syst. Appl. **78**, 376–385 (2017)
5. Shea, S., Sideli, R.V., Dumouchel, W., Pulver, G., Arons, R.R., Clayton, P.D.: Computer-generated informational messages directed to physicians: effect on length of hospital stay. J. Am. Med. Inform. Assoc. **2**(1), 58–64 (1995)
6. Rigal, M.: Management des lits et durée moyenne de séjour: Exemple de recherche d'optimisation au Centre Hospitalier d'Avignon. Ph.D. thesis (2009)
7. Rowan, M., Ryan, T., Hegarty, F., Hare, N.O.: The use of artificial neural networks to stratify the length of stay of cardiac patients based on preoperative and initial postoperative factors. Artif. Intell. Med. **40**, 211–221 (2007)
8. Lafaro, R.J., Pothula, S., Kubal, K.P., Inchiosa, M.A., Pothula, V.M., Yuan, S.C., Maerz, D.A., Montes, L., Oleszkiewicz, S.M., Yusupov, A., Perline, R., Inchiosa, M.A.: Neural network prediction of ICU length of stay following cardiac surgery based on pre- incision variables. Plos One **10**(12), 1–19 (2015)
9. Chuang, M.T., Hu, Y.H., Tsai, C.F., Lo, C.L., Lin, W.C.: The identification of prolonged length of stay for surgery patients. In: Proceedings - 2015 IEEE International Conference on Systems, Man, and Cybernetics, SMC 2015, pp. 3000–3003 (2016)
10. Loshin, D.: Data consolidation and integration. In: Master Data Management, pp. 177–199. Elsevier (2009)
11. Gentimis, T., Alnaser, A.J., Durante, A., Cook, K., Steele, R.: Predicting hospital length of stay using neural networks on MIMIC III data. In: IEEE 15th International Conference on Dependable, Autonomic and Secure Computing, 15th International Conference on Pervasive Intelligence and Computing, 3rd International Conferece on Big Data Intelligence and Computing and Cyber Science and Technology Congress (DASC/PiCom/DataCom/CyberSciTech), pp. 1194–1201 (2017)

12. Li, J.S., Tian, Y., Liu, Y.F., Shu, T., Liang, M.H.: Applying a BP neural network model to predict the length of hospital stay. In: LNCS, vol. 7798, pp. 18–29. Springer, Heidelberg (2013)
13. Pendharkar, P.C., Khurana, H.: Machine learning techniques for predicting hospital length of stay in Pennsylvania federal and specialty hospitals. Int. J. Comput. Sci. Appl. **11**, 45–56 (2014)
14. Aghajani, S., Kargari, M.: Determining factors influencing length of stay and predicting length of stay using data mining in the general surgery department. Hospital Pract. Res. (HPR) **1**(2), 53–58 (2016)
15. Johnson, A.E., Pollard, T.J., Shen, L., Lehman, L.W.H., Feng, M., Ghassemi, M., Moody, B., Szolovits, P., Anthony Celi, L., Mark, R.G.: MIMIC-III, a freely accessible critical care database. Sci. Data **3**, 1–9 (2016)
16. Microsoft: Predicting Hospital Length of Stay (2017)
17. Brownlee, J.: Why One-Hot encode data in machine learning? (2017)
18. Carnevale, R.: Understanding Gradient Boosting, Part 1 - Data Stuff (2015)

Prediction of Length of Stay for Stroke Patients Using Artificial Neural Networks

Cristiana Neto[1], Maria Brito[1], Hugo Peixoto[1], Vítor Lopes[2], António Abelha[1], and José Machado[1(✉)]

[1] ALGORITMI Research Center, Department of Informatics, University of Minho, Braga, Portugal
cristiana.neto@algoritmi.uminho.pt, a73580@alunos.uminho.pt, {hpeixoto,abelha,jmac}@di.uminho.pt

[2] Centro Hospitalar Universitário de São João, Porto, Portugal
vitor.nuno.lopes@chsj.min-saude.pt

Abstract. Strokes are neurological events that affect a certain area of the brain. Since brain controls fundamental body activities, brain cell deterioration and dead can lead to serious disabilities and poor life quality. This makes strokes the leading cause of disabilities and mortality worldwide. Patients that suffer strokes are hospitalized in order to be submitted to surgery and receive recovery therapies. Thus, it's important to predict the length of stay for these patients, since it can be costly to them and their family, as well as to the medical institutions. The aim of this study is to make a prediction on the number of days of patients' hospital stays based on information available about the neurological event that happened, the patient's health status and surgery details. A neural network was put to test with three attribute subsets with different sizes. The best result was obtained with the subset with fewer features obtaining a RMSE and a MAE of 5.9451 and 4.6354, respectively.

Keywords: Healthcare · Data Mining · Machine Learning · Artificial neural network · Stroke · Patient · Length of stay · Feature selection

1 Introduction

It is known that the healthcare industry produces huge amounts of data every day that incorporates various sectors and areas of expertise. The information can go from hospital resources to the patients' health status and diagnosis of diseases [1]. Thus, the resultant data is represented in different types and formats, making it very heterogeneous [2]. The lack of structure and poor standard practices can often lead to a lack of quality in the produced healthcare data [3].

In a healthcare facility, data is collected and stored at a rapid pace, which promote the arising of Knowledge Discovery in Databases (KDD). KDD provide more in-depth knowledge obtaining hidden patterns that may exist in the collected data. Data Mining (DM), the most important step of KDD, focus on

Á. Rocha et al. (Eds.): WorldCIST 2020, AISC 1159, pp. 212–221, 2020.
https://doi.org/10.1007/978-3-030-45688-7_22

the extraction of knowledge from large quantities of data, aiming at discovering important information to the industry [4]. Whereas a traditional data analysis performed by a human being is not possible, the application of Machine Learning (ML) algorithms can easily interpret the data and its details. Since machines can be taught how to properly look at data, their predictions often lead to low error values [5].

The application of DM techniques on healthcare data brings a many advantage to the industry. Medical institutions can discover new and useful knowledge that otherwise would remain unknown [6]. By using these methods, the quality of the healthcare services can be improved, and new management rules can be implemented in order to increase the productivity. This also brings new ways of preventing fraud and abuses that could highlight inappropriate patterns on insurance claims or medical prescriptions [2].

Strokes are a leading cause of long-term disabilities, poor quality of life and mortality worldwide. A patient that suffers a stroke can face permanent complications and psychological issues throughout their remaining life [7]. Depending on its intensity, a stroke frequently leads to the patient's death, making it one of the most recurrent epidemiology of this century. The aging of the population and unhealthy lifestyles increase its risks and unfavorable outcomes [8].

When the blood supply to the brain gets interrupted, the brain cells do not receive the needed blood flow for their normal function and start to die. This is called a brain attack, or a stroke [9]. The patients are hospitalized and submitted to surgeries depending on the type of the stroke. Therapeutic interventions aim to minimize the length of their hospitalization, since it can be costly both to the patient and their family, as well as to the hospital [10]. Thus, it becomes necessary to predict the length of stay (LOS) for stroke patients. It can depend of many factors, such as the stroke's intensity and the patient's health and recovery. Therefore, this is the goal of study: given some input data, with information related to this neurological event, patient and surgery details, create a model that predicts the LOS with the lowest possible error.

Artificial neural network (ANN) is the ML algorithm used on this study to predict the number of days of a stroke patient's hospital stay. These neural networks aim to mimic the function of the human brain, hence the use of biological designations such as neurons and synapses [11]. They differ from conventional algorithmic approaches in a way that they do not follow a set of instructions to solve a certain problem. They are often called "black boxes", since it is not possible to understand how they really work [12]. The truth is that neurons are very powerful units that can assume many roles in the storage of information, images recognition and classification problems [13].

Regarding the structure of this paper, it is divided in six major sections. After the current introduction, the second section presents related work. The methodology used is then described in next section. Section four describes the steps of the Cross Industry Standard Process for Data Mining (CRISP-DM) methodology that was followed during the DM process. In section five the obtained results are presented and discussed. Finally, section six includes conclusions and future work.

2 Related Work

Researches associated to the prediction of LOS related to various specialty areas have become a relevant topic of study, since there are many advantages that the application of ML algorithms can bring to this area. The prior studies range from general medical divisions to specific medical diseases/treatments.

Hasanat et al. [14] aim to predict the LOS for patients in order to control the hospital costs and improve its efficiency. The researchers selected a subset of features using the information gain metric and tested it with various ML models. The Bayesian network model obtained the best result for accuracy with a value of 81.28%, while the algorithm C4.5 resulted in a 77.1% of accuracy.

Lella et al. [15] created a novel prediction model for the LOS of patients admitted to hospitals. The Growing Neural Gas model obtained an accuracy value of 96.36% which was a best result than the ones produced by the likes of ZeroR, OneR, J48 and Self Organizing Map (SOM).

Combes et al. [16] presented an approach to estimate the LOS in an emergency department using models based on linear regression. The results were satisfactory, with an error of approximately 2 h in 75% of cases.

Khajehali and Alizadeh [22] developed a study with the aim of explore the important factors affecting the LOS of patients with pneumonia in hospitals. This study concluded that Bayesian boosting method led to better results in identifying the factors affecting LOS (accuracy 95.17%).

Rezaianzadeh et al. [23] carried out a study with the aim of determine the predictors of LOS in cardiologic care wards developed and carried out based on DM approaches. The median and mean LOS was 4 and 4.15 days, respectively. The factors associated with the increase in the LOS (more than 4 days) were: the ST segment elevation myocardial infarction (STEMI) diagnosis at the time of referral, being in the 50–70 years old group, history of smoking, high blood lipids, history of hypertension, hypertension at the time of admission, and high serum troponin levels.

Lee et al. [24] attempted to identify potential predictors of intensive care unit (ICU) LOS (LOS) for single lung transplant patients. Several conclusions were obtained through this study, including: the median ICU LOS was 5 days, and this was highly correlated with the duration of mechanical ventilation; patients with pulmonary hypertension had the longest ICU LOS.

Machado et al. [25] used real data to identify patterns in patients' profiles and surgical events, in order to predict if patients will need hospital care for a shorter or longer period of time after surgery for perforated peptic ulcer. The best accuracy obtained was 87.30% using JRip.

3 Methodology

For the development of this study, the methodology selected was the Cross-Industry Standard Process for Data Mining, commonly known as CRISP-DM.

This process acknowledges six fundamental steps, namely: Business Understanding, Data Understanding, Data Preparation, Modeling, Evaluation and Deployment [17]. It is considered the most popular methodology for DM projects, since it is very flexible and promotes transitions between the different phases.

The ML software Weka was used for the general analysis of the distribution and balance of the data. The data preparation tasks defined also resorted to this software. During this phase, many techniques were used, such as:

- One Hot Encoding, that consists in transforming all the categorical variables in binary attributes. Let's say that an attribute has 4 different classes. This technique turns the classes into different variables with values of 0 or 1. This way, the prediction models will be able to make better predictions [24];
- Attribute selection with WrapperSubsetEval (the algorithm chosen was Logistic Regression) in conjunction with the search method BestFirst (with the direction as forward);
- Attribute selection with CfsSubsetEval in conjunction with BestFirst (with the direction as backward).

On the other hand, the selection of attributes, as well as the implementation, training and testing of the ANN were made on the platform RStudio using the programming language R in order to induce the Data Mining Models (DMM).

4 Data Mining Process

4.1 Business Understanding

When stroke patients are admitted to the hospital after having suffered from a neurological event, information about their health status and lifestyle habits are recorded, as well as details about the stroke. After being submitted to surgery, the medical professionals register the needed information about the surgery that was performed and complications that the patient is feeling. Long stays at the hospital can be costly to the patients as well to the medical institutions. Therefore, healthcare institutions efficiency and productivity can be put at risk if the lengths of stay are unexpected.

Thus, the goal of this study is to help the medical professionals and management teams by aiming at predicting the LOS for stroke patients that are being admitted to hospitals.

4.2 Data Understanding

The data that was provided to this study was collected from a hospital located in Portugal. It is about patients that suffered from a stroke and had to be admitted to that hospital in order to be submitted to surgery. It includes various details about their health, such as the assessment of diabetes and smoking habits. It also reports on the characteristics of the neurological event that happened.

Moreover, data about the surgery performed and any lingering complications is also contained on the dataset.

There were 36 attributes with 203 samples. From a first analysis, it was evident that some attributes were not relevant to the study at hand. Their reasons differ, but they offered no important information for the prediction process. The majority of the features presented percentages of missing values below 26%. The only exceptions were two attributes that present the ankle-brachial index for both the right and left ankle. These two variables have 98% of values that are missing. The tasks that were done in order to properly prepare the data for the submission to the ML algorithms are described on the next section.

4.3 Data Preparation

The data preparation phase acts as a way of manipulating the data in order to obtain better results from the models. Thus, the data is submitted to various methods where the attributes' relevance for the study, the percentages of missing values, the derivation of new variables, the data transformation and the selection of attributes are all considered. This way, the data is properly prepared to be submitted to the ML algorithms. This step is especially important for the knowledge extraction process, because it can be the determining factor for a satisfactory prediction.

After a close inspection of the dataset, two attributes were identified as having no relevance for the end goal. The patient identification and the medical observations offer no meaningful information to the prediction process, since the former only serves as a way of identifying the patient and the latter is represented as a text box, where the medical professionals wrote down what they thought was needed. It has no specified structure.

The next task was the substitution of the class designations for incremental integers. It's easier for the model to process integers instead of actual words or entire sentences. This helps by reducing the computing time that a model takes to make a prediction on the given data.

Additionally, two new attributes were derived from existent ones in the dataset. The first variable is related to the number of days between the neurological event that the patient suffered and the hospital admission. The other one represents the length of the patients' hospital stay.

There are features that have high percentages of values that are missing. For this reason, they cannot be worked with, since it's impractical to use them for prediction proposes. Those variables are the index for the right and left brachial-ankle pressure. On the other hand, any attribute that represented a date was also removed, since the new ones that were created offer more meaning to the DM process.

The missing values that were present in the dataset were replaced with the mean and the mode of the classes for the numerical and categorical attributes, respectively. Moreover, the numerical variables were normalized as a way of having the results between 0 and 1. This normalization was performed using the

Eq. (1). Also, the technique One Hot Encoding, introduced before, was applied to the dataset.

$$NormalizedValue = \frac{(X - X_{min})}{(X_{max} - X_{min})} \quad (1)$$

Since the dataset contains many attributes, it's possible that there are variables that contribute more to the prediction process than others. For this reason, two attribute selection methods were applied. As a result of these techniques, a subset of attributes is selected, in which every attribute is individually evaluated by addressing its relevance to the prediction of the target variable chosen.

By the application of the attribute evaluator WrapperSubsetEval (the algorithm chosen was Logistic Regression) in conjunction with the search method BestFirst (with the direction as forward), 14 variables were selected. Whereas, the evaluator CfsSubsetEval with BestFirst (with the direction as backward) selected just 7 variables. According to the Weka documentation [18], this evaluator assesses the worth of the features by considering their individual ability of prediction with the degree of redundancy between them. Both these evaluators and search methods are available in this software.

As follows, three use cases were defined, as presented on Table 1. The first corresponds to the original dataset with all the features, after the application of the data preparation techniques. The second use case is represented by the subset of attributes that were selected by the WrapperSubsetEval, while the third contains the variables produced by the evaluator CfsSubsetEval. These will be put to test in order to determine which set of attributes is more meaningful to the prediction process. The set that obtains the lowest error is then considered to be the most appropriate to predict correctly the target variable.

Table 1. Summary of the used datasets

Use case	Number of attributes	Selection technique
1	33	–
2	14	WrapperSubsetEval
3	7	CfsSubsetEval

4.4 Modeling

The ML algorithm chosen to process the data and predict the desired results is ANN. This method allows scientists to make analysis on complex data [11]. Knowledge that was hidden can be extracted offering meaningful information to a business.

Many different configurations were tried out during the development of this study. However, the one that consistently obtained the best results was a neural network using the backpropagation algorithm. This is a very popular method where the output produced by the NN is evaluated in comparison to the correct output. When the results do not meet the expectations, the weights between the defined layers are modified. The process is then repeated until the error value is satisfactory [19].

The neural network created contained three hidden layers with different sizes. The first layer had 40 nodes, whereas the second included 20 and the third 10. The choice for the number nodes on each layer depends on the problem at hand. While different combinations were put to test, this composition registered the lowest errors and the least computing time. The learning rate was defined as 0.01, while the error threshold was 0.001. These tunings of the NN allowed for a more customized modelling that had in mind the data provided and the problem that is to be solved.

4.5 Evaluation

The measures chosen to evaluate the performance of the prediction models created were Root Mean Squared Error (also known as RMSE) and Mean Absolute Error (abbreviated as MAE). Their definition is presented on Eq. (2) and Eq. (3), respectively. The first (RMSE) is a rule that measures the average magnitude of the error. It is the square root of the average of squared differences between the prediction made and the desired result. It has the advantage of giving more importance to large errors [20]. On the other hand, MAE, scores the average magnitude of the errors with no consideration for their actual direction. By removing the square vales of the error and considering its absolute value, the bias towards outlying points is removed [21]. The results obtained for these metrics are presented on the next section, as well as the analysis and discussion of the results.

$$RMSE = \sqrt{\sum_{i=1}^{n} \frac{(\widehat{y_i} - y_i)}{n}} \tag{2}$$

$$MAE = \frac{1}{n}\sum_{i=1}^{n} |x_i - x| \tag{3}$$

5 Results and Discussion

Three different use cases were defined with the propose of identifying which attributes were more relevant to the prediction process. The results obtained for the performance metrics RMSE and MAE are shown in Table 2 for the defined use cases.

In the first case, all the features after the application of the data preparation methods are submitted to the NN. This use case will serve as a comparison term between the whole set of features and different subsets with fewer attributes. This will result in an analysis that will determine if it is better to use smaller set of attributes or all of them.

Table 2. RMSE and MAE values for the different use cases

Use case	RMSE	MAES
1	6.2951	4.6350
2	7.6601	4.5478
3	5.9451	4.6354

Thus, the second study case contains only 14 variables. These were defined using an appropriate evaluator that selected the most relevant attributes for the prediction of the length of the hospital stay for stroke patients. Aiming at reducing even further the size of the subset of features, a third use case was created where only 7 attributes of the original dataset are included. These three distinct situations will enable a study on the different prediction capabilities of the models created using various sets of attributes as input data for the neural net.

Table 2 shows that the third use case (the one with fewer attributes) obtained the best value for the first performance metric, RMSE. With a value of around 5.9451, it is the lowest RMSE recorded. Both the first use case as well as the second use case registered higher values for that measure. The first use case produced a result of approximately 6.2951, whilst the second use case resulted in the highest value (7.6601). This indicates that the smallest subset of attributes is a better input for the net, since it improves its predictions on the desired output. The set of all attributes also produce a better result than the subset of features that were selected by the evaluator WrapperSubsetEval.

Nonetheless, the results obtained for the MAE metric rank the models in a different way. The second use case produced a better result for this measure with a value of 4.5478. The other use cases did not obtain much worse results. In fact, the difference between them is not significant at all. The first use case had a MAE value of 4.6350 and the third a result of 4.6354. As can be seen, the different between the best and worst values is not bigger than 0.09.

As follows, since the divergence between the MAE values is not considered to be meaningful, the best set of attributes were selected by the CfsSubsetEval evaluator (third use case), which obtained the best RMSE result. In this case, the results were improved using feature selection, since it enabled reducing the noise from the data and selecting the most useful features to use in the training.

6 Conclusion and Future Work

Considered to be a dangerous epidemiology that is to be persistent on generations to come, strokes can lead to the death of patients. If not, the probability of suffering from physical complications and psychological issues is huge. Due to the interruption of the blood flow, important brain zones are put at risk of never recovering.

The application of ML algorithms can greatly improve the healthcare services performed, as well as increase the possibility of the patient's survival. ANNs aim to mimic the functioning of the human brain, making them an interesting model to prediction problems. In this research, given information about the patient's health, the stroke and the performed surgery, the goal was to produce a model capable of predicting the LOS for stroke patients.

By testing three use cases with different sizes of feature sets, it was possible to define an optimal neural network configuration where the lowest error values were registered. It was concluded that the third use case, that is the one with fewer variables, obtained better results than the others attribute sets.

The future work includes getting more data detailing different aspects of the patient's health, stroke and surgery and test them with other neural network tunings.

Acknowledgments. This article is a result of the project Deus Ex Machina: NORTE-01-0145-FEDER-000026, supported by Norte Portugal Regional Operational Program (NORTE 2020), under the PORTUGAL 2020 Partnership Agreement, through the European Regional Development Fund (ERDF). The work has also been supported by FCT – Fundação para a Ciência e Tecnologia within the Project Scope: UID/CEC/00319/2019.

References

1. Kautish, S., Abbas Ahmed, R.K.: A comprehensive review of current and future applications of data mining in medicine & healthcare. Int. J. Eng. Trends Technol. **38**(2), 60–63 (2016)
2. Durairaj, M., Ranjani, V.: Data mining applications in healthcare sector: a study. Int. J. Sci. Technol. Res. **2**(10), 29–35 (2013)
3. Sukumar, S.R., Natarajan, R., Ferrell, R.K.: Quality of Big Data in health care. Int. J. Health Care Qual. Assur. **28**(6), 621–34 (2015)
4. Zhang, S., Zhang, C., Yang, Q.: Data preparation for data mining. Appl. Artif. Intell. **17**(5), 375–381 (2010)
5. Simeone, O.: A very brief introduction to machine learning with applications to communication systems. IEEE Trans. Cogn. Commun. Netw. **4**(4), 648–664 (2018)
6. Dennison, T., Qazi, F.: Data mining in health care. In: Conference: Proceedings of the 2005 International Conference on Data Mining. CSREA Press, Las Vegas (2005). 89-9
7. Clarke, D., Forster, A.: Improving post-stroke recovery: the role of the multidisciplinary health care team. J. Multidiscip. Healthc. **8**, 433–442 (2015)
8. Dreyer, R., Murugiah, K., Nuti, S.V., Dharmarajan, K., Chen, S.I., Chen, R., Wayda, B., Ranasinghe, I.: Most important outcomes research papers on stroke and transient ischemic attack. Circ. Cardiovasc. Qual. Outcomes **7**(1), 191–204 (2014)
9. Gund, M.B., Jagtap, P.N., Ingale, V.B., Patil, R.Y.: Stroke: a brain attack. IOSR J. Pharm. **3**(8), 1–23 (2013)
10. Evrim, G., Turhan, K., Arzu, G., Vesile, Ö., Kursad, K.: The factors affecting length of stay in hospital among acute stroke patients. J. Neurol. Sci. **34**(2), 143–152 (2017)

11. Maind, S.B., Wankar, P.: Research paper on basic of artificial neural network. Int. J. Recent. Innov. Trends Comput. Commun. **2**, 96–100 (2014)
12. Benitez, J.M., Castro, J.L., Requena, I.: Are artificial neural networks black boxes? IEEE Trans. Neural Netw. **8**(5), 1156–1164 (1997)
13. Romaniuk, R.: Analysis of electrical patterns activity in artificial multi-stable neural networks. In: Proceedings of SPIE, vol. 11176 (2019)
14. Al Taleb, A.R., Hoque, M., Hasanat, A., Khan, M.B.: Application of data mining techniques to predict length of stay of stroke patients. In: 2017 International Conference on Informatics, Health & Technology (ICIHT), pp. 1–5. IEEE (2017)
15. Lella, L., Di Giorgio, A., Dragoni, A.F.: Length of stay prediction and analysis through a growing neural gas model. In: 2015 Artificial Intelligence and Assistive Medicine, pp. 11–21 (2015)
16. Combes, C., Kadri, F., Chaabane, S.: Predicting hospital length of stay using regression models: application to emergency department. In: 10éme Conference Francophone de Modélisation, Optimisation et Simulation - MOSIM 2014 (2014)
17. Silva, E., Cardoso, L., Portela, F., Abelha, A., Santos, M. F., Machado, J.: Predicting nosocomial infection by using data mining technologies. In: New Contributions in Information Systems and Technologies, pp. 189–198. Springer, Cham (2015)
18. Neto, C., Brito, M., Lopes, V., Peixoto, H., Abelha, A., Machado, J.: Application of data mining for the prediction of mortality and occurrence of complications for gastric cancer patients. Entropy **21**(12), 1163 (2019)
19. Skorpil, V., Stastny, J.: Neural networks and back propagation algorithm. In: ELECTRONICS 2016, Sozopol, Bulgaria, pp. 173–178 (2016)
20. Twomey, J.M., Smith, A.E.: Performance measures, consistency, and power for artificial neural network models. Math. Comput. Model. **21**(1–2), 243–258 (1995)
21. Kneale, P., See, L., Smith, A.: Towards defining evaluation measures for neural network forecasting models. In: Proceedings of the Sixth International Conference on GeoComputation, University of Queensland, Australia (2001)
22. Khajehali, N., Alizadeh, S.: Extract critical factors affecting the length of hospital stay of pneumonia patient by data mining (case study: an Iranian hospital). Artif. Intell. Med. **83**, 2–13 (2017)
23. Lee, K.H., Martich, G.D., Boujoukos, A.J., Keenan, R.J., Griffith, B.P.: Predicting ICU length of stay following single lung transplantation. Chest **110**(4), 1014–1017 (1996)
24. Rodríguez, P., Bautista, M.A., Gonzalez, J., Escalera, S.: Beyond one-hot encoding: lower dimensional target embedding. Image Vis. Comput. **75**, 21–31 (2018)
25. Machado, J., Cardoso, A.C., Gomes, I., Silva, I., Lopes, V., Peixoto, H., Abelha, A.: Predicting the length of hospital stay after surgery for perforated peptic ulcer. In: International Conference on Information Technology & Systems, pp. 569–579. Springer, Cham (2019)

Artificial Intelligence in Service Delivery Systems: A Systematic Literature Review

João Reis[1(✉)], Marlene Amorim[1], Yuval Cohen[2], and Mário Rodrigues[3]

[1] Department of Economics, Management, Industrial Engineering and Tourism, GOVCOPP, Aveiro University, Aveiro, Portugal
{reis.joao,mamorim}@ua.pt

[2] Department of Industrial Engineering, Afeka College of Engineering, Tel Aviv, Israel
yuvalc@afeka.ac.il

[3] IEETA & ESTGA, Universidade de Aveiro, Aveiro, Portugal
mjfr@ua.pt

Abstract. Artificial intelligence (AI) is transforming the 21st century service industries. With increased availability of virtual channels, new approaches to resource management are required for effective service delivery. A notable example is Amazon, which is reshaping itself with AI-based technologies, relying on robot service delivery systems, either through faster inventory checks or product delivery that reached unprecedented speed. This study provides an overview of the existing theory concerning the next generation of AI technologies that are revolutionizing the service delivery systems (SDS). To this end, we have systematically reviewed the literature to identify and synthesize the existing body of knowledge and update academics and practitioners regarding the latest AI developments on the SDS's. This article argues that AI technologies are driving the service industry and have had promising results in reducing the service lead time while is being more cost-effective and error-free. Future studies should contribute to strengthen the theoretical production, while AI is being continuously reinforced with new empirical evidence.

Keywords: Artificial intelligence · Service delivery systems · Systematic literature review

1 Introduction

The field of AI has shown an upward trend of growth in the 21st century [1], and have been increasingly reshaping the service industry by performing several tasks and constituting a major source of innovation [2, 3].

The AI-based technologies can be put in use in human services to help companies alleviating considerable administrative burden and free up time for more critical responsibilities by improving decision-making, and creating cheaper and faster delivery services [4, 5]. In that extent, Amazon has developed a retail store in Seattle that enables shoppers to take products from shelved and walk directly without checking out to pay,

Á. Rocha et al. (Eds.): WorldCIST 2020, AISC 1159, pp. 222–233, 2020.
https://doi.org/10.1007/978-3-030-45688-7_23

the store is called *Amazon Go*, and relies on computer vision to track shoppers during the buying process [6].

In the service domain, robots are also encompassing a wider range of advanced technologies that has the potential to overcome the traditional capabilities of industrial robots [7]. The aforementioned argument is reinforced by Wirtz *et al.* [8, p. 907], which states that "modern robotics, in combination with rapidly improving technologies, like AI, are bringing opportunities to a wide range of innovations, that have the potential to dramatically change service industries". A few examples are presented by McKinsey Global Institute, that is now arguing that autonomous drones that uses AI technologies as *deep learning* are completing last-mile deliveries and/or AI-enhanced robots are tracking inventory in warehouses and recognizing empty shelves in zero-error perspective [9].

This article is organized as follows. Section 2 presents the preliminary concepts. Section 3, it is explained the methodological process. Section 4, we discuss the most relevant results from the systematic review. We conclude this article by presenting the contributions to theory and practice, and guidelines for future research.

2 Terminology

AI has been labelled in a context of digital transformation that enabled major business improvements to augment customer experience, streamline operations or create new business models [10, 11]. AI technologies are able to develop cognitive abilities, or enhance human capabilities [12]. Consequently, AI developments in service delivery may potentially increase the added value to customers. In line with the above, we review in this section the AI technologies that are revolutionizing the service delivery systems. In short, artificial intelligence is being coined in the literature as human behaviours, which can be performed by machines, systems or networks [13]. According to Diebolt *et al.* [14, p. 1] AI combines two properties: "self-learning by the successive and repetitive processing of data, as well as the capacity to adapt, that is to say the possibility for a scripted program to deal with multiple situations likely to vary over time". Following, we provide an overview of four AI-related technologies used in SDS (Fig. 1).

The first AI technology we have identified is the *Machine Learning* (ML) that is a subset of artificial intelligence, which often uses statistical techniques to allow computers to learn with data, even if they have not been programmed to do so [15]. Probably, one of the best-known examples of ML is in healthcare – IBM's Watson Health [16]. Bini [16] argues that Watson Health has been fed everything that has ever been written in any language and at any time related to the diagnosis and treatment of cancer. The program has the ability to match all the information that has been incorporated to it and cross-checks with all the patient-specific information. As a result, the program recommends a treatment, which basically supports the doctor's decision-making process. Under the umbrella of ML, we have identified the *artificial neural networks* (NNs) that is the process through which machines learn from observational data, i.e. the activation of sensors that perceive the environment, and figuring out its own solutions [17, 18]. Several NNs applications are being used in the service industry, such as healthcare e.g., by predicting sociodemographic variables of depression among geriatric population [19], or in business e.g., to evaluate customer loyalty and improve satisfaction [20]. The advent

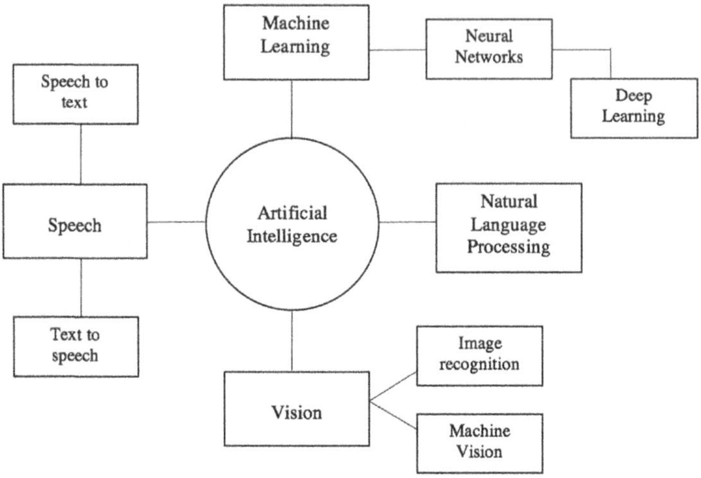

Fig. 1. Artificial intelligence in service delivery systems

of *deep learning* (DL) represents a major development in the field of neural networks and achieved tremendous success in real-world application [21], including the SDS. DL is considered as part of the broader family of machine learning [22] and is considered as the ability to exploit hierarchical feature representation learned solely from the available data, instead of features designed by hand according to domain-specific knowledge [23]. According to Sui *et al.* [23] deep models have made significant advances, outperforming regular classification models in multiple domains such as medical diagnoses. The aforementioned argument is corroborated by Cao *et al.* [24] which argues that "developed from artificial neural networks, deep-learning-based algorithms show great promise in extracting features and learning patterns from complex data", examples includes medical image classification, genomic sequence analysis, as well as protein structure classification and prediction. A number of deep learning architectures have been applied to other fields, such as computer vision, automatic speech recognition, and audio and music signal recognition and these have been shown to produce cutting-edge results in various tasks [25].

Speech recognition enables devices to recognize, adapt and translate voice information in understandable forms, including e.g. voice user interfaces such as voice dialling [26]. Speech recognition software was initially designed for individuals with physical disabilities, which had been adopting assisting technologies with writing difficulties [27]. Those technologies are also used in the healthcare services that include voice recognition systems used by radiologist to record and convert the voice into text. This process is called speech recognition–voice-to-text (VTT) even though there are no consensus about its advantages, while there are many benefits and pitfalls [28, 29]. Speech recognition–text-to-speech (TTS) technologies automatically speaks textual information although it produces mostly reasonable-sounding speech, however, it does not yet sound quite human [30].

Computer vision is a technology that aims to reproduce the effect of human vision by electronically perceiving and understanding an image [31]. Computer vision has several different applications, for instance, it provides an effective alternative for automated, non-destructive and cost-effective technique to inspection and grading fruits and vegetables to accomplish quality requirements [32]. Therefore, great advances were made in computer vision in improving visual perception, increasing the capabilities of intelligent agents and robots in performing complex tasks, combined with visual patter recognition, which also paved the way to self-driving cars [25]. Although computer vision is often identified as *machine vision* [33], our understanding is that *machine vision* is being increasingly addressed to industrial processes where the outcome is based on the image analysis done by a vision system [34].

Natural language processing (NLP) is the study of mathematical and computational modelling of various aspects of language, which includes spoken language systems that integrate speech and natural language [35]. In practical terms, NLP facilitates the rise of virtual assistants by making dialogue more intuitive [36], example of this use are the translating languages of Google translate [16] or virtual assistant technologies like Apple's Siri, Amazon's Alexa, and so on. Improvements in NLP, when combined with other AI technologies i.e. machine learning – based voice recognition systems, achieved a few years ago 95% accuracy [37]. In turn, speech recognition systems have the ability to recognize human speech, related applications include voice search, call routing, voice dialling and speech-to-text [26], as was mentioned before.

3 Methodology

We have conducted a systematic literature review, which is considered by Fink [38, p. 3] a "systematic, explicit, and reproducible method for identifying, evaluating, and synthesizing the existing body of completed and recorded work produced by researchers, scholars and practitioners". Dilevko [39, p. 451] also argues that conducting research literature reviews promotes the "scientification of the literature review" while, according to Petticrew and Roberts [40, p. 9], it "adheres closely to a set of scientific methods that explicitly aim to limit systematic error (bias), mainly by attempting to identify, appraise and synthesize all relevant studies". We therefore argue in favour of the suitability of the method for this specific research, since the objective is to narrow the field of study, and synthetizing the existing literature in order to evaluate the artificial intelligence phenomenon in service delivery systems.

To keep transparency and easy reproduction of results [41] we have used a single databased – Scopus, which is one of the largest abstract and citation databased of peer-reviewed literature. The search query was conducted with the following words: "Artificial Intelligence" AND "Service Delivery" limited to document title, abstract and keywords. The database search was conducted on January 31st, 2019, and yielded 131 manuscripts. The selected documents spanned from 1992–2018 due the raising expectations regarding the impact of AI technologies on the everyday practice of medicine [42]. More recent publication peaks are related to new AI technologies, such as speech recognition, natural language generation, machine learning and so on (Fig. 2).

We have also included peer-reviewed articles and conference papers available in full-text format. To avoid wrong interpretations, we have selected literature in English and to narrow the field of study we have selected manuscripts from the following subject areas:

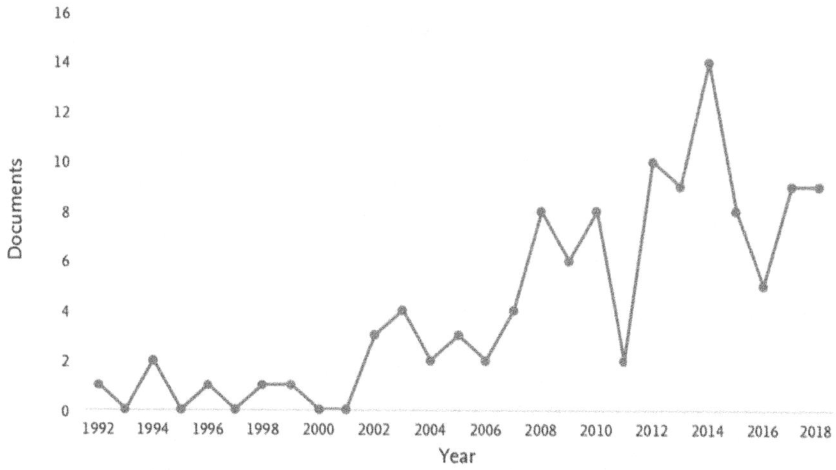

Fig. 2. Documents by year – range 1992 to 2018

industrial engineering and management, decision sciences, healthcare and, computer and social sciences. The introduction AI technologies in service encounter was ultimately enhanced by the digitalization of health records in the United States [43]. More recently, new developments in the healthcare industry continued and are mainly due to artificial intelligence robots [44]. Followed by the US we have found India, which focused AI studies on computer science i.e. cloud computing [45, 46] and also on healthcare e.g. e-Medicare services [47]. Figure 3 shows the reviewed documents by country.

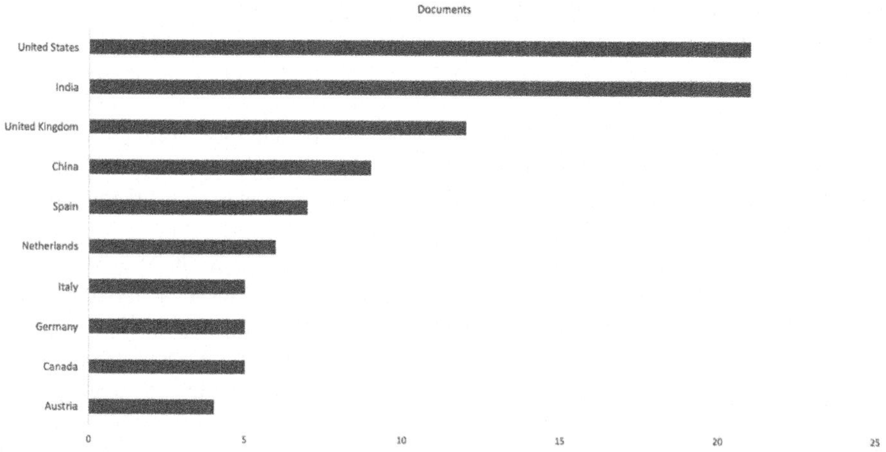

Fig. 3. Documents by country or territory – document counts for up to 10 countries/territories

The remaining 112 references included 28 journal articles and 84 conference papers. Although we did not want to rely firmly on material published by practitioners in a systematic review we added 5 articles for corroboration and triangulation purposes, as noted in the findings section.

Data as then analysed according the technique of content analysis [48], with the help of a dedicated software for qualitative analysis (NVivo), which allowed to build and hierarchize categories and subcategories so as to identify emerging patterns and ideas [49]. The methodological section also presents some limitations, namely: the literature review is restricted to the selected words, therefore, we acknowledge the possibility of having some articles missing. Due to space limitations it was also not possible to list all the selected references, which may be provided on request by contacting the first author.

4 Findings

4.1 Next Generation of AI in SDS – Technological Synergies

As previously explained, AI is composed of several associated technologies (Fig. 1). In a broader context, the literature labels several autonomous agents that are independent in their actions to adapt their service based activities as situation changes [50]. In a nutshell, we found some additional technological advancements that are of particular interest, as some articles raises the indication that leading companies are actually combining AI-based technologies to complementary technologies, such as robotics, with the intent of potentiating or even modernising their SDS's. Therefore, in the private sector, technological advances such as AI, biometrics and robotics are set to become the norm and challenge conventional thinking, which is revolutionising service industries [51]. At a boarder domain, this view is shared by Mikhaylov *et al.* [52], arguing that AI also has significant potential in the public sector, that is undergoing in a transformation with robotics and automation, changing the way how services are provided. Therefore, building on AI-based technology, several companies (e.g. Amazon) are testing autonomous vehicles and taking advantages of current advances in technology for robotics and drones [53]. In light with the above, well-defined strategies are currently advanced by practitioners [54–56] that will certainly attract further academic attention. Some examples are followed illustrated to enlighten readers:

Autonomous ground vehicles (AGVs), these are driverless vehicles that need little or no human intervention to deliver services or products, these SDS requires customers to open parcel lockers using a personal code, but are unlikely to be widely available until the 2020s [55]. Autonomous ground vehicles solutions are being coined to several markets. For instance, fully autonomous ground vehicles (FAGV) that operates in a real worlds environment were firstly developed for use in factories and other industrial/business sites based on behaviour-based artificial intelligence control [57]. Examples of AI usage are: artificial neural networks, capable of performing the reactive aspects of autonomous driving, such as staying on the road and avoiding obstacles [58]. In recent years, the rapid development of several technologies such as AI, cognitive science, automatic control, ground mapping, sensor technology and other fields are continually revolutionizing the field of unmanned driving [59].

Service robots, which are machines capable of conducting autonomous decision making based on data they gather from sensors in order to adapt to specific situations and learn from previous episodes [8]. They are used in logistics (e.g. to move goods around warehouses), medicine (e.g. exoskeletons or surgical robots) or sales (e.g. to give customers information) [60]. Service robots soon encompassed a wide spectrum of advanced technologies and hold the potential for surpassing industrial robots in both scope and diversity [7]. A notable contemporaneous example is the use of service robots in smart homes, as an efficient solution for domestic healthcare. As population is growing older, seniors are subjected to isolation and health issues are risen, this created an opportunity to domestic healthcare [61]. Ramoly *et al.* [61] argues that in response, scientists currently observed the emergence of both service robotics and smart environment to "monitor the activity of users, understand situations from different sources of data, intervene through actuators or robots, interact with a person, maintain a company to the user, and/or alert the medical staff if necessary" (p. 414). An example to outdoor conditions and rough-terrain is provided by Raibert *et al.* [62], which argues that Boston Dynamics is developing new breed of rough-terrain robots that capture mobility, autonomy and speed of living creatures. These robots use AI-enable technologies, such as computer vision, to autonomously conduct operations without human intervention.

Service drones, are autonomous aircrafts that carry packages to the destination along the most direct route and with relatively high average speed [54]. Joerss *et al.* [54] also states that, like droids and AGVs, they also need to be supervised due to AI absence. In that regard, Levy [63] argues Amazons' Prime Air drone-delivery service, which is still in the prototype phase, has to build its AI separately, since its autonomous drones cannot count on cloud connectivity. Apart from the term drone, *Unmanned Aerial System* (UAS) is a term that refers to flying platform and the ground station that controls platforms [64]. While most people think of UAS as sophisticated military technology, business across industries realize that drones have multiple commercial applications (i.e. drone-delivery services) for retail stores [65]. The future of service drones is promising, not just in the service industries, but also in research. Recently, a team Japanese researcher's has given the next step towards robotic pollination. They have equipped a radio wave-controlled drone to pollinize flowers due to bee declines. Gross [66] argue that there is little doubt that with current technologies it would be relatively straightforward to make pollinating drones independent of the radio controller, by using AI technologies to navigate between plants (e.g. computer vision assisted obstacle avoidance) and decide where to pick up pollen and where to deposit it.

Although the mentioned articles from practitioners (i.e. McKinsey) have not been identified in the systematic review. In the absence of existing AI-related cases, we were required to illustrate reality with practitioners' knowledge. In our opinion, scholars need to strengthen the existing theory with new evidence, while practitioners develop new technologies associated with AI.

4.2 Cost-Effective and Error-Free? Reality or Myth?

It is more or less acceptable to say that digital ecosystem business models possibly enable better sales and earnings growth driven to cost-effective environments [51]. Governments around the world are attempting to grow benefits of digital ecosystems to transform traditional public administration into a modern and citizen-centred to ensure cost-efficient delivery of public services [67].

Despite the investment made with digital technologies, in the specific sphere of AI we found contradictory results. On Pezzullo's *et al.* [28] article, the authors admit that cost savings in voice recognition dictation may not be observed. The authors have surveyed radiologists that expressed dissatisfaction with voice recognition with feelings of frustration and increased fatigue, result from more error ridden radiology reports and increased costs when compared with traditional transcription services. Chang *et al.* [29] somewhat found the same results, as they refer that voice recognition presents many benefits, but there are also many pitfalls. The pitfalls they refer are due to error rates and strike a balance between quality and speed of reports generated by the machine. On the other hand, Gartner *et al.* [68] discusses the introduction of diagnosis-related groups (DRGs), which in prospective payment systems has put some pressure on hospitals to use efficiently their resources. Hospitals where DRG systems are being used, patients are classified into groups with related characteristics, as they are expected to use the same amount of resources. These classifications versus used resources, in turn, will influence the hospital revenue. It was in that extent that Gartner *et al.* [68] gave their contribution, as the authors shown that accurate DRG classification using AI methods and programming-based resources increased the hospital's contribution margin, along with the number of patients and the utilization of resources, such as operating rooms and beds. Huang *et al.* [69] studied inpatient length of stay (LOS) perdition using AI-based technologies to improve the health services delivery, as well as to assess reasonable representation of resource consumption.

Overall, it is our understanding that AI may be *de facto* cost-efficient to most of the reviewed cases; however, in some presented cases, the respective authors expressed dissatisfaction with AI-enables technologies due to errors rates, which, in turn, increased costs when compared to the traditional services. Which makes us believe that there may be some AI-enabled technologies that need to be suitable for each type of SDS. Despite the advances in AI, particularly in the field of computer science, it is expected that advances in the coming decades will indeed make the AI error-free and cost-effective in SDS.

5 Concluding Remarks

Despite the latest AI progresses, few organizations have incorporated AI-related technologies into their service delivery systems. In fact, McKinsey Global Institute refers a survey to 3,073 respondents, where only 20% said they had adopted one or more AI-related technology at scale or in a core part of their business [36]. The aforementioned arguments can be revealing of the importance of theoretical research in this area of knowledge, thereby shedding light on possible best practices and successful implementation of AI in the SDS of the world's leading companies.

This article also provides some theoretical contributions, as we have proposed an exploratory theoretical model for the use AI technologies in SDS. We also discussed the possibilities of using distinct existing technologies that, if associated with AI, may eventually strengthen the SDS. Some clarification is also needed regarding the efficiency of AI technologies, in particular with regard to costs and reduction of failures in certain SDS. To practitioners, this article provides an overview of current developments from worldwide leading companies, such as Amazon, which is testing artificial intelligence technologies in SDS (e.g. Amazon Go), but also because it continues to develop new prototypes to improve each time plus the service delivery (e.g. Prime Air drone-delivery service).

Future research may provide insightful contributions to strengthen the theoretical production, while AI is being continuously reinforced with new empirical evidence. The theoretical production should therefore deepen the empirical knowledge established by practitioners in the area of AI synergies with other technologies (e.g. robotics) to improve the SDS. Finally, further research may focus on perceiving the areas where AI can be most successful in reducing costs in SDS. This line of research aims to provide avenues for scholars and academics and stimulate scientific research in less exploited areas, where AI has lower financial revenues and has greater error prospect in SDS.

References

1. Liu, J., Kong, X., Xia, F., Bai, X., Wang, L., Qing, Q., Lee, I.: Artificial intelligence in the 21st century. IEEE Access **6**, 34403–34421 (2018)
2. Pan, Y.: Heading toward artificial intelligence 2.0. Engineering **2**(4), 409–413 (2016)
3. Huang, M., Rust, R.: Artificial intelligence in service. J. Serv. Res. **21**(2), 155–172 (2018)
4. Farha, R., Leon-Garcia, A.: Market-based hierarchical resources management using machine learning. In: International Workshop on Distributed Systems: Operations and Management. Springer, Heidelberg (2007)
5. Eggers, W., Fisherman, T., Kishnani, P.: AI-augmented human services: using cognitive technologies to transform program delivery. Deloitte Insights (2017)
6. Grewal, D., Roggeveen, A., Nordfält, J.: The future of retailing. J. Retail. **93**(1), 1–6 (2017)
7. Engelhardt, K.: Service robotics and artificial intelligence: current research and future directions. ISA Trans. **29**(1), 31–40 (1990)
8. Wirtz, J., Patterson, P., Kunz, W., Gruber, T., Lu, V., Paluch, S., Martings, A.: Brave new world: service robots in the frontline. J. Serv. Manag. **29**(5), 907–931 (2018)
9. Chui, M., Francisco, S.: Artificial intelligence the next digital frontier? vol. 47. McKinsey and Company Global Institute (2017)
10. Reis, J., Amorim, M., Melão, N., Matos, P.: Digital transformation: a literature review and guidelines for future research. In: World Conference on Information Systems and Technologies, pp. 411–421. Springer, Cham (2018)
11. Warner, K., Wäger, M.: Building dynamic capabilities for digital transformation: an ongoing process of strategic renewal. Long Range Plan. **52**(3), 326–349 (2019)
12. Wirtz, B., Weyerer, J., Geyer, C.: Artificial intelligence and the public sector—applications and challenges. Int. J. Public Adm. **42**(7), 1–20 (2018)
13. Li, D., Du, Y.: Artificial Intelligence with Uncertainty. CRC Press, Boca Raton (2017)
14. Diebolt, V., Azancot, I., Boissel, F., Adenot, I., Balague, C., Barthélémy, P., Boubenna, N., Coulonjou, H., Fernandez, X., Habran, E., Lethiec, F.: "Artificial intelligence": which services, which applications, which results and which development today in clinical research? Which impact on the quality of care? Which recommendations? Therapie **74**(1), 155–164 (2019)
15. Heinonen, K., Kietzmann, J., Pitt, L.: AI and machine learning in service management. Special Issue Call for Papers from J. Serv. Manag. www.emeraldgrouppublishing.com/authors/writing/calls.htm?id=8053. Accessed 16 Jan 2019

16. Bini, S.: Artificial intelligence, machine learning, deep learning, and cognitive computing: what do these terms mean and how will they impact health care? J. Arthroplast. **33**(8), 2358–2361 (2018)
17. Nielsen, M.: Neural Networks and Deep Learning. Determination Press, USA (2015)
18. Schmidhuber, J.: Deep learning in neural networks: an overview. Neural Netw. **61**, 85–117 (2015)
19. Sau, A., Bhakta, I.: Artificial neural network (ANN) model to predict depression among geriatric population at a slum in Kolkata, India. J. Clin. Diagn. Res. **11**(5), VC01 (2017)
20. Ansari, A., Riasi, A.: Modelling and evaluating customer loyalty using neural networks: evidence from startup insurance companies. Futur. Bus. J. **2**(1), 15–30 (2016)
21. Doya, K., Wang, D.: Fostering deep learning and beyond. Neural Netw. **97**, iii–iv (2018)
22. SFR-IA Group, French Radiology Community: Artificial intelligence and medical imaging 2018: French Radiology Community white paper. Diagnostic and Interventional Imaging **99**(11), 727–742 (2018)
23. Sui, J., Liu, M., Lee, J., Calhoun, V., Zhang, J.: Deep learning methods and applications in neuroimaging. https://www.journals.elsevier.com/journal-of-neuroscience-methods/call-for-papers/deep-learning-methods-and-applications-in-neuroimaging. Accessed 19 Feb 2019
24. Cao, C., Liu, F., Tan, H., Song, D., Shu, W., Li, W., Zhou, Y., Bo, X., Xie, Z.: Deep learning and its applications in biomedicine. Genomics Proteomics Bioinform. **16**(1), 17–32 (2018)
25. Perez, J., Deligianni, F., Ravi, D., Yang, G.: Artificial Intelligence and Robotics. arXiv preprint arXiv:1803.10813 (2018)
26. Husnjak, S., Perakovic, D., Jovovic, I.: Possibilities of using speech recognition systems of smart terminal devices in traffic environment. Procedia Eng. **69**, 778–787 (2014)
27. Lewandowski, L., Wood, W., Miller, L.: Technological applications for individuals with learning disabilities and ADHD. In: Computer-Assisted and Web-Based Innovations in Psychology, Special Education, and Health, pp. 61–93. Academic Press (2016)
28. Pezzullo, J., Tung, G., Rogg, J., Davis, L., Brody, J., Mayo-Smith, W.: Voice recognition dictation: radiologist as transcriptionist. J. Digit. Imaging **21**(4), 384–389 (2008)
29. Chang, C., Strahan, R., Jolley, D.: Non-clinical errors using voice recognition dictation software for radiology reports: a retrospective audit. J. Digit. Imaging **24**(4), 724–728 (2011)
30. Campbell, J., Campbell, W., Lewandowski, S., Weinstein, C.: Cognitive services for the user. In: Cognitive Radio Technology, pp. 313–335. Newnes (2006)
31. Sonka, M., Hlavac, V., Boyle, R.: Image Processing, Analysis, and Machine Vision. PWS Publishing, California (1999)
32. Brosnan, T., Sun, D.: Improving quality inspection of food products by computer vision—a review. J. Food Eng. **61**(1), 3–16 (2004)
33. Jayas, D., Ghosh, P., Paliwal, J., Karunakaran, C.: Quality evaluation of wheat. In: Computer Vision Technology for Food Quality Evaluation, pp. 351–376. Academic Press (2008)
34. Rao, A.: Future directions in industrial machine vision: a case study of semiconductor manufacturing applications. Image Vis. Comput. **14**(1), 3–19 (1996)
35. Reshamwala, A., Mishra, D., Pawar, P.: Review on natural language processing. IRACST Eng. Sci. Technol. Int. J. (ESTIJ) **3**(1), 113–116 (2013)
36. Bughin, J., Hazan, E., Ramaswamy, S., Chui, M., Allas, T., Dahlstrom, P., Henke, N., Trench, M.: Artificial intelligence – The next digital frontier. McKinsey Glob Institute (2017)
37. Agarwal, P.: Public administration challenges in the world of AI and Bots. Public Adm. Rev. **78**(6), 917–921 (2018)
38. Fink, A.: Conducting Research Literature Reviews. From the Internet to Paper, 3rd edn. Sage, London (2010)
39. Dilevko, J.: Guest editorial: reading literature and literature reviews. Libr. Inf. Sci. Res. **4**(29), 451–454 (2008)

40. Petticrew, M., Roberts, H.: Systematic Reviews in the Social Sciences: A Practical Guide. Blackwell Publishing, Malden (2006)
41. Buchanan, D., Bryman, A.: The Sage Handbook of Organizational Research Methods. SAGE Publications Ltd., Thousand Oaks (2009)
42. McCauley, N., Ala, M.: The use of expert systems in the healthcare industry. Inf. Manag. **22**(4), 227–235 (1992)
43. Huang, S., Parker, M., Hsu, S., Pandya, A., Tripathi, S.: Knowledge-based evaluation of nursing care practice. In: IEEE International Conference on Information Reuse & Integration, pp. 171–174. IEEE (2009)
44. Yoon, S., Lee, D.: Artificial intelligence and robots in healthcare: what are the success factors for technology-based service encounters? Int. J. Healthc. Manag. **12**(3), 218–225 (2018)
45. Chandra, D.m., Bhadoria, R.: Cloud computing model for national e-governance plan (NeGP). In: 2012 Fourth International Conference on Computational Intelligence and Communication Networks, pp. 520–524. IEEE (2012)
46. Krohn-Grimberghe, A., Gupta, A., Chadha, A., Vyas, R.: Cloud computing in optimizing supply chain management of automotive component industry. In: Proceedings of the 1st International Conference on Internet of Things and Machine Learning, p. 71. ACM (2017)
47. Gupta, A., Verma, A., Kalra, P.: Tele health therapy: an ambient technology. In: 2015 Global Conference on Communication Technologies, pp. 749–754. IEEE (2015)
48. Blaxter, L.: How to Research. McGraw-Hill Education, New York (2010)
49. Skålén, P.: Service marketing control as practice: a case study. Qual. Mark. Res. Int. J. **14**(4), 374–390 (2011)
50. Kerr, D., O'Sullivan, D., Evans, R., Richardson, R., Somers, F.: Experiences using intelligent agent technologies as a unifying approach to network management, service management and service delivery. In: International Conference on Intelligence in Services and Networks, pp. 115–126. Springer, Heidelberg (1998)
51. Coetzee, J.: Strategic implications of Fintech on South African retail banks. S. Afr. J. Econ. Manag. Sci. **21**(1), 1–11 (2018)
52. Mikhaylov, S.J., Esteve, M., Campion, A.: Artificial intelligence for the public sector: opportunities and challenges of cross-sector collaboration. Philos. Trans. R. Soc. A: Math. Phys. Eng. Sci. **376**(2128), 20170357 (2018)
53. Van Doorn, J., Mende, M., Noble, S., Hulland, J., Ostrom, A., Grewal, D., Petersen, J.: Domo arigato Mr. Roboto: emergence of automated social presence in organizational frontlines and customers' service experiences. J. Serv. Res.J. Serv. Res. **20**(1), 43–58 (2017)
54. Joerss, M., Schröder, J., Neuhaus, F., Klink, C., Mann, F.: Parcel delivery: The future of last mile. McKinsey & Company (2016)
55. Bouton, S., Hannon, E., Haydamous, L., Heid, B., Knupfer, S., Naucler, T., Neuhaus, F., Nijssen, J., Ramanathan, S.: An integrated perspective on the future of mobility, part 2: transforming urban delivery (2017)
56. Manyika, J.: A Future that Works: AI, Automation, Employment, and Productivity. McKinsey Global Institute Research, Technical report (2017)
57. Gomi, T., Ide, K., Matsuo, H.: The development of a fully autonomous ground vehicle (FAGV). In: Proceedings of the Intelligent Vehicles' 94 Symposium, pp. 62–67. IEEE (1994)
58. Pomerleau, D., Gowdy, J., Thorpe, C.: Combining artificial neural networks and symbolic processing for autonomous robot guidance. Eng. Appl. Artif. Intell. **4**(4), 279–285 (1991)
59. Zhang, X., Gao, H., Guo, M., Li, G., Liu, Y., Li, D.: A study on key technologies of unmanned driving. CAAI Trans. Intell. Technol. **1**(1), 4–13 (2016)
60. McKinsey Report: Shoulder to shoulder with robots. Tapping the potential of automation in Poland. McKinsey Report (2018)
61. Ramoly, N., Bouzeghoub, A., Finance, B.: A framework for service robot in smart home: An efficient solution for domestic healthcare. IRBM **39**(6), 413–420 (2018)

62. Raibert, M., Blankespoor, K., Nelson, G., Playter, R.: Bigdog, the rough-terrain quadruped robot. IFAC Proc. Vol. **41**(2), 10822–10825 (2008)
63. Levy, S.: Inside Amazon's artificial intelligence flywheel (2018) https://www.wired.com/story/amazon-artificial-intelligence-flywheel/. Accessed 3 Mar 2019
64. Custers, B.: Drones here, there and everywhere introduction and overview. In: The Future of Drone Use, pp. 3–20. TMC Asser Press, The Hague (2016)
65. Cohn, P., Green, A., Langstaff, M., Roller, M.: Commercial drones are here: the future of unmanned aerial systems (2017). https://www.mckinsey.com/industries/capital-projects-and-infrastructure/our-insights/commercial-drones-are-here-the-future-of-unmanned-aerial-systems. Accessed 3 Mar 2019
66. Gross, M.: How Will Robots Integrate into our World?. Elsevier, Amsterdam (2017)
67. Sarker, N., Wu, M., Hossin, A.: Smart governance through Bigdata: digital transformation of public agencies. In: International Conference on Artificial Intelligence and Big Data, pp. 62–70. IEEE (2018)
68. Gartner, D., Kolisch, R., Neill, D.B., Padman, R.: Machine learning approaches for early DRG classification and resource allocation. INFORMS J. Comput. **27**(4), 718–734 (2015)
69. Huang, Z., Juarez, J., Duan, H., Li, H.: Reprint of "Length of stay prediction for clinical treatment process using temporal similarity". Expert Syst. Appl. **41**(2), 274–283 (2014)

Benefits of Implementing Marketing Automation in Recruitment

John Wernbom, Kristoffer Tidemand, and Eriks Sneiders(✉)

Department of Computer and Systems Sciences, Stockholm University, Postbox 7003, 164 07 Kista, Sweden
johnwernbom@gmail.com, kristoffer.tidemand@gmail.com, eriks@dsv.su.se

Abstract. Marketing and recruitment operate in similar settings: both strive to address people who most often are not looking for being addressed. While IT-support in marketing has been given its share of attention, the benefits of tools and automation in recruitment process are poorly researched. This paper explores these benefits. Eleven benefits have been identified from the literature and our interviews with recruiters. Five of these benefits are shared between recruitment and marketing, whereas six address relationship building and ethics more specific to recruitment.

Keywords: IT in organizations · IT-support in business · IT in recruitment · IT in decision making

1 Introduction

Recruitment refers to the process of employing new people to work for a company or organization. One of the biggest difficulties in the recruitment business is passive candidates: about 70% of recruitment candidates are not actively looking for a job [1]. Hence, the recruitment problem is pretty much that of marketing: how do recruiters reach people who are not looking for to be recruited?

The benefits that automation of marketing offers are increased productivity, better decision support, higher return on market-related investments, increased customer satisfaction and customer loyalty [2]. We assume that IT-support in recruitment is similar to that in marketing, but we do not really know it. There exists a study on adopting technology in order to address challenges in various hiring functions involving sales engine, sourcing, interviewing, providing offers, post-offer follow-ups, and finally joining and induction [3]. There does not exist, to the best of our knowledge, a broader study on benefits of tools and automation in the recruitment process. Our paper contributes to filling this knowledge gap. We started with a literature study, then did interviews with recruiters, and finally compiled a list of eleven benefits that automation brings to the recruitment process. Five of these benefits are shared between recruitment and marketing, while six address relationship building and ethics more specific to recruitment.

Á. Rocha et al. (Eds.): WorldCIST 2020, AISC 1159, pp. 234–241, 2020.
https://doi.org/10.1007/978-3-030-45688-7_24

2 Previous Studies

We searched for publications on Google Scholar and Springer with queries such as "use of automation within recruiting", "automated marketing within recruitment", "automated communication in recruiting" that covered automation in the recruitment process. We soon discovered that academic research was scarce although the subject was somewhat thoroughly discussed in commercial blogs. We broadened the scope and explored research on automation and technology in general in relation to recruitment and human-resource management.

The benefits and qualities of automation in recruitment, identified in the literature, are following:

A1: Reaching a large number of people. According to a survey by Brandteg & Følstad [4], 68% of the respondents believe chatbots improve productivity; in this context productivity means reaching a large amount of people. Only 10% believe chatbots are suitable for building relationships with respondents. The conclusion of this survey could be that chatbots have potential in the initial moment of communication, e.g. when a candidate or client is visiting a website for the first time. Also, chatbots are a fast and easy tool for gathering intelligence.
A2: Elimination of prejudice. Algorithms could do the screening of candidates in an automated manner without human interference and thus eliminate prejudice, for example maintain gender equality [5]. But it could also maintain prejudice, depending on the parameters of the algorithm. The system needs to be calibrated.
A3: Assisting the recruiter. For automation to be successful, it should work as an aid to the recruiter rather than a substitute [6]. AI and tools for automation have limited ability to build relationships and evaluate "soft values" such as leadership, teamwork and social competence. Automation therefore should function as an aid to the recruiter; automation should ease the workload but not take full control over it.
A4: Better candidate experience. There is a general problem that many recruiters fail in their communication towards the candidate. Answers and feedback drag on, and sometimes recruiters fail to respond at all [7]. This leads to frustration among candidates. According to Wright [8], automated response with the help of AI, can offer a more satisfying candidate experience, due to continuous feedback to the candidate about their status in the recruitment process.
A5: Human presence is important. Even though AI and automated responses are said to be suitable, Wright [8] also acknowledges that lack of human presence is regarded as a destructive factor by the candidate. The conclusion is that the content of automated messages must be of such a high standard that it is perceived as relevant and human made.
A6 and A7: Time and cost efficiency. Faliagka et al. [9] tested possible advantages of automated recruitment process through implementation of some algorithms on LinkedIn. The result showed that algorithms performed at the level of professional recruiters in matching an employment with a suitable candidate. This could potentially lead to cost- and time savings, due to the redundancy of human interference. Ajaz [10] mentions how digitization of the recruitment process can lead to cost- and time savings due to self-service by the candidate. The candidate could for instance update or add information on

career sites, though one could argue this example is somewhat more due to digitization than automation.

A8: Better, faster and cheaper. Gupta et al. [3] made a survey asking 52 recruiters about benefits of technology (primarily artificial intelligence and robotics) in recruitment. 77% of the recruiters agreed that the technology in hiring process reduces hiring costs; 79% agreed that the technology improves the "quality of hire" metrics; 75% agreed that the technology improves the "competency/skills gap analysis" metrics; 73% agreed that the technology improves the "professional development" metrics; 96% agreed that the technology makes recruitment process "better, faster, cheaper".

A9: Improved transparency. In the same survey, 65% of the recruiters believed that technology improves the recruitment process transparency for both candidates and recruiters.

A10: Emancipation of human resources. Automated filtering of CV simplifies big data management; recruiters do not have to do the work. The automated filtering includes feedback to candidates, background checks, competence comparison, as well as offer some "customer service" that would require a high degree of human involvement [10].

A11: Faster feedback to candidates. Besides what is mentioned in A10, automated filtering also provides faster candidate feedback due to the automation itself; a human recruiter does not have to send feedback to each and every candidate [10].

3 Interviews

We contacted ten companies in Sweden for interviews; four of them eventually agreed to be interviewed. Two of the four were generic recruitment companies; one of the four did recruitment in IT-sector; another one of the four was a financial-IT company. In total, five people were interviewed. They had different titles, tasks and obligations due to the diversity of company cultures and routines. The only suitability requirement for the respondents was hands-on experience in recruitment and automated marketing techniques. We asked the companies about (i) their recruitment process, (ii) the tools they use in this process, (iii) and how the tools and automated marketing techniques are relevant for their recruitment process.

The benefits and qualities of automation in recruitment, identified from the interviews, are following:

B1: Time savings. Several respondents use automated e-mail delivery. One respondent describes it as the most automated tool they have at their disposal: "There is functionality, that you could get an e-mail from us, from our system once a month with all our new job openings. I guess that is the most automated way that we use the system from a marketing point of view. We also do e-mail campaigns with the system." The e-mail tools are able to manage what is being sent to whom. This is described as particularly useful when it comes to feedback and nurturing mails, since the recruiters do not have to do anything. They do not have to write or administrate any mails each and every time, which in turn is regarded as a time saving benefit.

B2: Satisfying candidate experience. E-mail is a recurrent subject in this study. In particular e-mail automation that could be scheduled and sent in various steps with some degree of personalization. The benefit of automated e-mail is not only limited to the benefits of B1. It also creates a better recruitment experience for the candidate.
B3: Personalized content. It is possible to personalize messages – but only to some extent. "If you could find a way to identify someone's competences as good as you would like it to be ... and then push out a message, then it would start getting really interesting. Are we there yet? No."
B4: Reaching relevant target groups. Algorithms on social media create accurate advertising, which generates visibility to relevant target groups. Different target groups are made based on assumptions about each group. The target groups can be multiple, based on the data that the potential candidates share on social media. "If we are going to find an IT person, we make four different target groups where target group A is everyone who has studied IT, target group B is everyone who has put a 'like' on NASA, target group C is everyone who follows IBM on Facebook, and target group D is everyone who has put a 'like' on the game Fortnite. [...] If it shows that people who like Fortnite do not like our ad and do not interact with it, we decrease the ad budget for that target group. But if it shows that people who have been studying IT interact with the content, we increase the budget. This way we increase the chance to reach people who are interested in the job position. Then, what happens is that people hopefully receive the information, relevant people, relevant target groups as you call it."
B5: Relevant propositions. A tracking system facilitates building relationships with the candidates, through improved data management. This leads to more relevant propositions. Earlier one would use an Excel-sheet or even papers to keep track of this data. "Our current system with its modules allows us managing our candidates and tracking the candidates' history, our communication with the candidates, so that we can flag the candidates depending on what they might be interested in, and then establish communication through the system based on that. As the time goes, our propositions are likely to become more and more relevant to individual candidates."
B6: Reaching a large number of candidates. Chatbots enable communication with a large number of candidates with relatively little effort. The communication is customizable. "We could choose how the bot would communicate, it could mimic my writing style, I could customize how it would maintain a dialog with the candidates, [...] we could customize which profiles it would target."

4 Final Set of Benefits

After we had acquired the two sets of benefits and qualities of automation in the recruitment process, set "A" from the literature and set "B" from the interviews, we aggregated them into the final set "C" shown in Table 1. Figure 1 illustrates the correlations between the two sets: two linked "A"-"B" benefits or qualities contribute to the same "C" benefit or quality.

Table 1. Final benefits and qualities of automation in the recruitment process.

Benefit or quality	Literature, interviews	Comments
C1. Self-service information acquisition by a large number of people	A1, B6	Chatbots have been used for answering questions automatically
C2. Gathering intelligence about people's interests and concerns	A1, B6	The source of such intelligence is chatbots' log files
C3. Transparency of the recruitment process	A2, A9, B4	Data management and processing tools are not biased unless requested to be so
C4. Elimination of prejudice against candidates	A2, A9, B4	Comes with increased transparency
C5. Timely response to candidates	A1, A4, A11, B6	Self-service information acquisition provides immediate response, scheduled messages provide timely response. "I forgot", "I didn't have time" do not occur
C6. Time- and cost-efficient recruitment process	A1, A6, A7, A8, A10, B1, B6	Automation of recruitment routines means time saving for recruiters. In high-wage countries it means also cost saving
C7. Automation of routine tasks re-allocates recruiters' time and effort for solving complex tasks	A1, A6, A7, A10, B1, B6	More human resources for non-automated tasks. If the task requires competence, this means more value for money
C8. Reaching relevant target groups in the beginning of the recruitment process	B4	Advertising in social media is customized to address a particular group. Still, there are limitations, see "Human presence is important" below
C9. More relevant proposition	B5	A good data-management and communication-tracking system allows building a relationship with a candidate, and eventually creating a more relevant job offer

(*continued*)

Table 1. (*continued*)

Benefit or quality	Literature, interviews	Comments
Qualities		
C10. Personalized automated communication towards candidates	A5, B2, B3	It is possible to customize the communication towards a candidate, and address many candidates personally. Personalization of automated communication has its limits
C11. Human presence is important	A3, A5, B3	Humans are better than computers in judging "soft values" such as leadership, teamwork and social competence. Lack of human presence during recruitment process is considered destructive by candidates. Personalization of automated communication has its limits

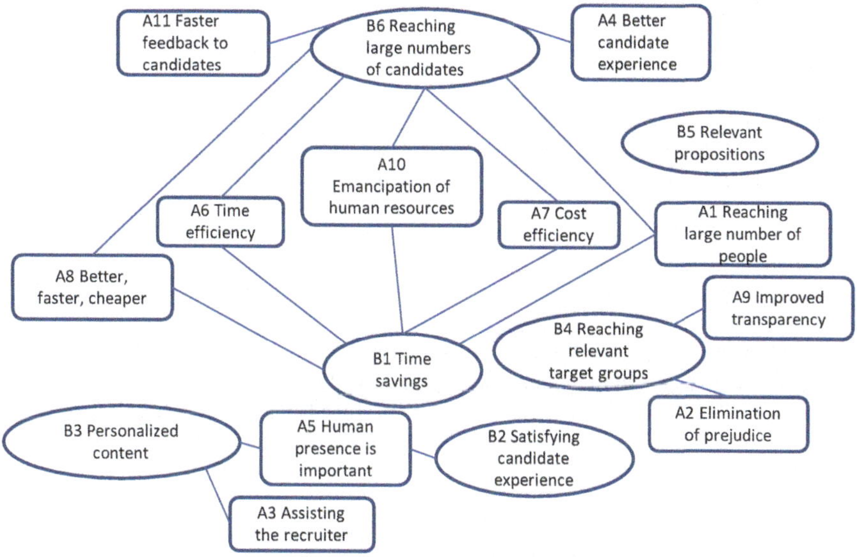

Fig. 1. The benefits and qualities that recruiters may gain from automation in the recruitment process, obtained from the literature (set "A") and the interviews (set "B").

5 Conclusions

We obtained 17 benefits and qualities of automation in the recruitment process – 11 from the literature and 6 from the interviews – and aggregated them into 11 final benefits and qualities. Our initial assumption was that automation in marketing and automation in recruitment are similar because both operate in similar settings – they bring a message to people who are not looking for that message. Similarities do exist: the benefits C1, C2 and C8 address automated communication with a large audience of potential customers, which is also a source of business intelligence; the benefits C6 and C7 address time-, cost-, and manpower-efficient process.

There are differences, though. Good recruitment creates a relationship between a recruiter and a candidate, which facilitates a more relevant job proposition (C9). A pleasant and productive communication experience (C5, C10, C11) is a part of the relationship building. Such relationships are not built by marketing.

Furthermore, good recruitment means fair and ethical candidate selection process (C3, C4). Although ethics in marketing do exit [11], marketing does not work with fair prioritizing of customers.

Evaluating "soft values" (e.g., leadership, teamwork and social competence) is important in recruitment (see C11), and it is mostly a manual process; automatic evaluation of some other "soft values" with respect to an individual's receptivity to the message gets increasingly significant in political marketing on social-media [12].

References

1. LinkedIn: Why & How People Change Jobs. https://business.linkedin.com/content/dam/business/talent-solutions/global/en_us/job-switchers/PDF/job-switchers-global-report-english.pdf. Accessed 22 Jan 2020
2. Heimbach, I., Kostyra, D.S., Hinz, O.: Marketing automation. Bus. Inf. Syst. Eng. **57**, 129–133 (2015). https://doi.org/10.1007/s12599-015-0370-8
3. Gupta, P., Fernandes, S.F., Jain, M.: Automation in recruitment: a new frontier. J. Inf. Technol. Teach. Cases **8**(2), 118–125 (2018). https://doi.org/10.1057/s41266-018-0042-x
4. Brandtzaeg, P.B., Følstad, A.: Why people use chatbots. In: Kompatsiaris, I., et al. (eds.) Internet Science. INSCI 2017. LNCS, vol. 10673, pp. 377–392. Springer, Cham (2017). https://doi.org/10.1007/978-3-319-70284-1_30
5. Bogen, M., Rieke, A.: Help Wanted: An Examination of Hiring Algorithms, Equity, and Bias. http://www.upturn.org/hiring-algorithms. Accessed 22 Jan 2020
6. Grana, J.: Why AI and Humans Need to Work Together to Recruit Better. https://www.interseller.io/blog/2018/07/30/ai-in-recruiting/. Accessed 22 Jan 2020
7. 4 Causes of a Bad Candidate Experience. https://www.recruitment-software.co.uk/4-causes-of-a-bad-candidate-experience/. Accessed 22 Jan 2020
8. Wright, A.D.: Job Seekers Are Frustrated With Automated Recruiting. https://www.shrm.org/resourcesandtools/hr-topics/technology/pages/candidates-soured-too-much-technology.aspx. Accessed 22 Jan 2020
9. Faliagka, E., Tsakalidis, A., Tzimas, G.: An integrated e-recruitment system for automated personality mining and applicant ranking. Internet Res. **22**(5), 551–568 (2012). https://doi.org/10.1108/10662241211271545

10. Ajaz, K.F.: The changing dynamics of talent acquisition, an analysis of recruitment trends, marketing strategies and emerging software and services. Acad. Arena **6**(4), 67–73 (2014). https://doi.org/10.7537/marsaaj060414.01
11. McKinley, M.M. (ed.): Ethics in Marketing and Communications. Palgrave Macmillan (2011). https://doi.org/10.1057/9780230367142
12. Badawy, A., Addawood, A., Lerman, K., Ferrara, E.: Characterizing the 2016 Russian IRA influence campaign. Soc. Netw. Anal. Min. **9**(1) (2019). Article no. 31. https://doi.org/10.1007/s13278-019-0578-6

Data Fusion Model from Coupling Ontologies and Clinical Reports to Guide Medical Diagnosis Process

Adama Sow(✉), Abdoulaye Guissé, and Oumar Niang

Information Processing and Intelligent Systems Lab (LTISI),
Computer Science and Telecommunications Engineering Department,
Polytechnic School of Thiès, Thies, Senegal
{asow,aguisse,oniang}@ept.sn

Abstract. In this article, we focus on access to data that can help clinicians in the medical diagnostic process before proposing appropriate treatment. With the explosion of medical knowledge, we are interested to structure them into the informations collection step. We propose an ontology resulting from a fusion of several existing and open medical ontologies and terminologies. On the other hand, we exploit real cases of patients to improve the list of signs of each disease. This work leads to a knowledge base (KB) associating all human diseases with their relevant signs. Cases are also stored in the KB. Each disease is described by all the signs observed and verified in all the patients carrying this same disease. The association of sickness and its signs is thus continuously nourished as there are new cases of diagnosis.

Keywords: Medical ontologies · Linked data · Knowledge engineering · Medical decision · E-health system

1 Introduction

Medical diagnosis is a process that consists of a continuous collection of the medical informations that the clinician makes before integrating and interpreting it for the management of his patient's health problems. Collection step is as important as it's complex for the clinician, especially when it necessitates quickly recourse to masses of medical knowledge that are constantly exploding on an international scale. It's into the perspective of assisting clinicians in the exploitation of this knowledge, that our research is located. Our goal is that guides access to relevant medical informations at each of diagnostic process step.

This article focuses on modeling of knowledge base (KB). It's about producting a data model targeting knowledge available in both formal and non-formal resources. The goal is to merge the strengths of all these resources to provide access to a variety of shared knowledge facilitating identification and association of human diseases to of their available relevant characteristic signs such as symptoms and symptoms and clinical signs.

Á. Rocha et al. (Eds.): WorldCIST 2020, AISC 1159, pp. 242–251, 2020.
https://doi.org/10.1007/978-3-030-45688-7_25

The core of this KB is an ontology produced from a coupling of several existing and open medical ontologies and terminologies. The obtained ontology covers a multitude of descriptive informations of human diseases but also describes the typology and the semantic of signs collected from a patient. Indeed, in existing ontologies we find on the one hand ontologies of diseases associated for each to a list of symptoms whose exhaustiveness is to be clarified, and on the other hand ontologies which conceptualize all the signs that can be appeared in a patient but without no link with diseases. These last ontologies include clinical signs whose values are obtained from in-depth examinations. Ontologies of diseases aiming at a generic conception do not take into account clinical signs which are nevertheless known into sign ontologies.

We propose to enrich our ontology associating each disease with their clinical signs. This is made possible by exploiting clinical reports for real cases of patients. Each clinician systematically archive all data of any patient in his medical folder (MF). Although the MFs are confidential, we have been able to obtain, in collaboration with local hospitals, anonymous descriptive case reports. Analysis of their content makes possible identification of all symptoms observed on a patient, as well as the clinical signs that made it possible to confirm an accurate diagnosis. However, these last signs being specific to a given patient, they are associated with a disease by the case that carries them. Cases are stored in the knowledge base. Each disease of our ontology is described by all the signs observed and verified in all the patients carrying this same disease. The association of sickness and its signs is thus continuously nourished as there are new cases of diagnosis.

Thus, in the Sect. 2, we perform a state study on medical ontologies and their use in diagnostic systems. Then, in Sect. 3, we present the constitution of our ontology starting from the selection of the relevant information for the medical diagnosis to come from the federation of target ontologies and the clinical cases. In the Sect. 4 and 5, before to conclude, we describe our modeling approach to build and enrich our ontology from these ontologies but also from textual clinical cases reports.

2 Related Work

In order to establish the diagnosis [2], it is important for the clinician to cross-check all informations on the patient's state of health. It's precisely the patient's opinion about his condition to identify his pains, physical examinations made by the clinician during consultations, and in-depth examinations (clinics and paraclinical) whose allow the identification of the most complex and implicit signs.

In medical diagnostic support systems [3], this collection phase is a cognitive activity where the semantics of information is controlled through knowledge known in medical jargon. It's for this very purpose medical ontologies have been conceived [1,5]. The bioportal site alone totals more than 816 medical ontologies indexed over 39 million of resources. These are common medical vocabularies based on shared concepts facilitate the interoperability of documents between stakeholders in the field and especially the development of knowledge. Medical ontologies, that we are interested, consist modeling medical entities such as human diseases, their characteristic signs, their known treatments, or the hospital processes of patient care.

The diagnostic process is based on the reasoning around diseases and their characteristic signs, the current difficulty, with regard to existing disease ontologies, lies in the fact that these signs are listed in a non-exhaustive manner [4] and not very formal [6]. Only the most common symptoms are stated in these ontologies and their presence varies from one patient to another. Moreover, the clinical signs take values at a patient are not even taken into account. There are, however, ontologies specific to the conceptualization of the signs [1,4] but they are not associated with diseases.

We are not aiming here to build an ontology from non-formal resources but our goal is to merge the strengths of several existing ontologies in order to have an ontology sufficiently provided in terms of diseases and to associate with each of them all of its relevant signs and appearing in most of the patients who have been affected by these same diseases. This association has already been the subject of research. Indeed, [6], propose in their ontology project Disy, to give to the clinicians latitude to cite for each disease all its signs.

The work of [4] offers a coupling of ontologies in order to group together for each disease all of its signs present in these target ontologies. For our part, our proposal is similar to that of the latter authors in that we are also looking for a federation of ontologies of human diseases and signs in order to constitute a news that is adapted for medical diagnosis. However, despite this desire for coupling, current ontologies are still not large enough to describe in detail the diseases with all of their characteristic signs. To overcome this, we try to focus on the analysis of real cases of patients who have already been diagnosed and whose clinicians have transcribed the entire process in textual reports. This analysis then makes it possible to list new signs, until now not yet taken into account in existing medical ontologies.

We do not address here the complex research questions on the integration of ontologies as demonstrated in the works [7] and [8], our work is based on the fusion of ontologies of different specialties that we let's gather to have all the information useful for medical diagnosis. These ontologies are then enriched with cases already diagnosed by clinicians.

3 Targeted Informations

3.1 What Informations into Medical Ontologies?

We propose constitution of an ontology from of a coupling of a set of ontologies around a structure unifying all human diseases as well as their characteristic signs. The diseases correspond to the possible diagnosis. The signs are those can be identified on a patient in order to conclude on a specific diagnosis that can refer to one or more diseases. Diseases are organized in a hierarchical way. They and their derived forms are grouped into categories, which may themselves be subcategories of diseases. The diseases are lexicalized in order to have for each disease the set of the most known nominative terms and their synonyms. For each disease, it will be important to keep all definitions in order to have the most shared semantics. Most of the known signs of each disease are formally listed from those available in the target medical ontologies.

We analyze and exploite here medical ontologies made available to the public via the BioPortal platform. We chose DOID[1] and MESH[2] as disease ontologies, as well as SYMP,[3] and CSSO[4] as ontologies of signs.

DOID ontology (Disease Ontology) serves us as a reference ontology. It proposes a hierarchy of *10389* human diseases and disease categories. We can see each disease has a unique identifier, and is classified in one or more categories. The disease of *Hepatitis A* belongs to the category of (*"skin diseases"*) and to the category of (*"viral infectious diseases"*). However, from one identifier to another, there is no description to say that a given identifier refers to a disease or a category of diseases. But, considering the hierarchical graph, all the leaf concepts correspond to the diseases and those who have threads constitute categories.

Each disease in DOID refers to the same disease in other ontological bases such as that of the Medical Subject Headings (MESH) terminological resource. It's one of the reference thesauri in the biomedical field. It's known for the multitudes of synonymous terms proposed as denominations of a disease. Each of the diseases has a preferential term (*prefLabel:hepatitis A*) which is the most used denomination, but also of several synonymous terms (*altLabel:Viral hepatitis A, Viral hepatitis type A, Hepatitis Infectious, Hepatitides Infectious, Infectious Hepatitis, Infectious Hepatitides*). These terms correspond to different hepatitis A nominations around the world. Definitions disease available in MESH will be conserved in our ontology result. Otherwise, the DOID proposes also one tag as definition (*obo:IAO_*) in a semi-formalized language goes a little further in the description of the disease. It's easy to decompose this description from groups of verbal words such as *results_in, located_in, caused_by (or has_material_basis_in), transmitted_by or has_symptom* which refer to the characteristic signs of a disease, corresponding respectively to the manifestation of the disease, to its location in the human anatomy, to the agent at the origin of the disease, to its modes of transmission, and to his symptoms.

It is with this in mind that we have to consider the SYPM and CSSO ontologies. The first onc is developed in the same project as the DOID, and in the same way as this one for the diseases, SYMP proposes a hierarchical structure complete of all the clinical signs and symptoms, which are also classified in categories of signs. SYMP affixes to each sign a definition referring to how it manifests itself in the patient. The second also brandishes the same goal as the SYMP but it is

[1] http://purl.bioontology.org/ontology/DOID.
[2] https://www.nlm.nih.gov/mesh/.
[3] http://purl.bioontology.org/ontology/SYMP.
[4] http://purl.bioontology.org/ontology/CSSO.

a little less accomplished. Only the third of SYMP signs are taken into account in CSSO. However, the latter brings a plus, a terminology for each sign.

3.2 What Informations into Clinical Cases Reports?

After having established our goal of federation of ontologies, we are discovering to what extent we can enrich this ontology from the analysis of medical reports of diagnostic cases that have already been validated by clinicians. We consider here a case of a patient diagnosed with Hepatitis A[5]. This disease is of viral origin designating inflammation of the liver. It's listed among sexually transmitted infections is in the top ten (10) of most dangerous diseases in Senegal[6].

The case of Hepatitis A is transcribed in a textual report which includes the symptomatic description of the state of health of a patient whose diagnosis is then confirmed after a set of in-depth examinations. These types of reports explode in the registers (digital or not) of clinicians and gives a visibility on the signs necessary for diagnosis confirmation. Indeed, this report crosses all the characteristic signs allowing to conclude on Hepatitis A disease, and we will discover that its analysis extends our ontology because it makes it possible to associate a given disease with the set of relevant signs.

First, several types of signs present in this report. For Hepatitis A, only 9 general symptoms appear in the ontology. These are fever, fatigue, loss of appetite, nausea, vomiting, abdominal pain, clay colored bowel movements, joint pain, and jaundice. And only 7/9 are therefore identifiable for this case and correspond to the first observable signs in the patient. Other signs, although listed in the ontology (from SYMP and CSSO) and not associated with Hepatitis A, correspond to the general signs indicating its sex, age, and excesses, but also to antecedents and clinicals signs. These result from in-depth examinations and refer to nominative terms and values. In the end, more than 16 signs are added to those who describe Hepatitis A in the ontology.

4 Knowledge Base Model

The data structure (Figs. 1 and 2) shows that our ontology, while listing all the signs that may be present in a patient from the ontologies of target signs, only associates the most common signs to a given disease. Therefore, the specific signs described in the contents of a case are also stored in the ontology but their values can be recorded only in the case of type *"MedicalDiagnosisCase"*, which is associated with it with all the signs present in its content as well as their values, and the disease (or diseases) to which it corresponds. Consequently, a disease will always be related to all these common symptoms via the ontology of diseases, and to a set of specific signs according to the number of real cases already diagnosed.

[5] Example from http://www.immunologyclinic.com/.
[6] http://www.who.int/countries/sen/en/.

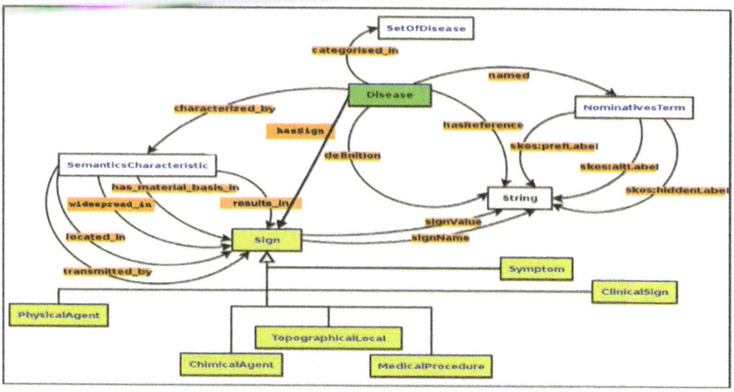

Fig. 1. View data structure of our medical ontology liking diseases and their signs

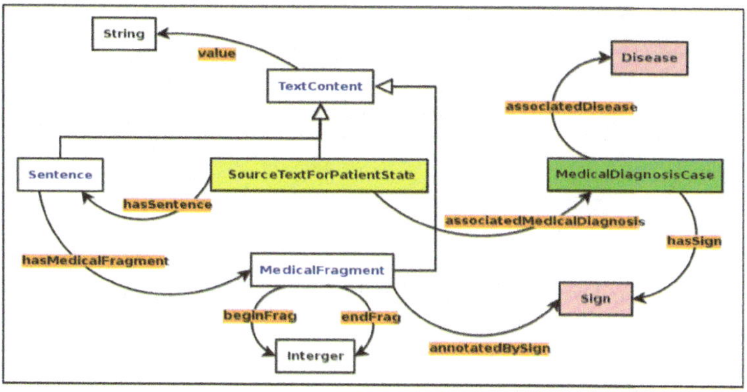

Fig. 2. Overview data structure federing diseases, signs, and formal cases

The structure is disease-centric (*Disease Class*) with all informations classes necessary for understanding the disease as well as the recommendation of potential diagnosis. Each disease is identified (*categorized_in*) in one or more categories (*SetOfDiseases Class*). Each disease is associated (*named*) with a set of nominative terms (*NominativeTerms*) synonyms, from the preferred term (skos:prefLabel), to alternative terms (skos:altLabel, skos:hiddenLabel). Each disease is associated (characterized_by) with a set of semantic characteristics (SemanticCharacteristics Class) and through the relations *transmitted_by, located_in, caused_by, results_in* refer respectively to different types of signs such as PhysicalAgent, TopographicalLocate, PhysicalAgent, ChimicalAgent (morphological elements) or MedicalProcedure (Medical Procedure). Signs type Symptom or ClinicalSign are associated to a given disease with *hasSign* property.

Each sign has a name and possibly a value, especially in the case of measurable clinical signs. Each of the signs classes, list all the possible signs, but a given disease is associated only with the most common signs, the other signs are attached on a specific patient case for same diagnosis, and varie from one case to another. As a result the data structure stores both an ontology of diseases and signs, but also the case base of validated diagnoses.

owl:Class	Disease, SetOfDisease, NominativeTerms, Sign, Symptom, ClinicalSign, MedicalDiagnosisCase	
rdfs:SubClassOf	ClinicalSign (Sign), ClinicalSign (Sign)	
owl:ObjectProperty	**rdfs:domain**	**rdfs:range**
categorzed_in	Disease	SetOfDisease
hasSign	Disease or MedicalDiagnosisCase	Sign (Symptom or ClinicalSign or ...)
named	Disease	NominativeTerms
associatedDisease	MedicalDiagnosisCase	Disease
...	...	...
owl:DataProperty	**rdfs:domain**	**rdfs:range**
signValue	Sign	xsd:string
definition	Disease	xsd:string
skos:prefLabel	NominativeTerms	xsd:string
...	...	...
swrl : formalism of cases		
hasSign(?c1,?x1) ∧ hasSign(?c1,?x2)∧ ... ∧ hasSign(?c1,?xn) → associatedDisease(?c1,?d)		

Fig. 3. Overview data structure implementation

The cases base formalizes selected data from the clinical reports from an extraction process of the signs present on each report. Thus, each case (Fig. 2) returns to a textual transcription of real state of health of a patient (*SourceTextForPatientState*), and is materialized by (*MedicalDiagnosisCase*) concept. The latter makes the link between the disease of the diagnosis (*associatedDisease*) and their descriptive signs (*hasSign*).

Different data formats of the ontologies we have selected are implemented with the W3C standards of the Semantic Web around the RDF, SKOS, RDFS and OWL languages. So to facilitate the recovery of targeted data on each of these resources, we propose a structure implementation (Fig. 3) using the same technologies and which inherits from them the same conceptual formalisms. These languages are designed to integrate around a basic RDF graph.

The cases are formalized as SWRL[7] that are compatibles with RDF technologies. Each case is described in two sides: the premise that refers to identified signs for the case and the conclusion corresponds to the diagnosed disease. Thus, for the extraction of signs, textual clinical reports are annotated with NLP tools such as NooJ[8] and Clamp[9] with regard to the ontology of diseases and signs.

[7] https://www.w3.org/TR/rif-overview/.
[8] http://www.nooj-association.org/.
[9] https://clamp.uth.edu/.

Table 1. Description of resulting Diseases and Signs Ontology

Element types	Ontological object	Target ontologies	Number of elements
Diagnosis			
Maladies	Diseases Class	DOID	6442
Categories	SetOfDiseases Class	DOID	3947
Synonyms diagnosis terms	prefLabel, altLabel, hiddenLabel	DOID, MESH	27586
Signs			
Symptoms and Clinical Signs	Symptom Class and CinicalSign Class	SYMP	942
Other Signs	PhysicalAgent, ChemicalAgent, TopographicalLocate, MedicalProcedure	DOID	6020
Synonyms signs terms	prefLabel, altLabel	CSSO	4710

4.1 Selected Data from Ontologies

Data structure (Fig. 2) is loaded by querying the different target ontological resources with the SPARQL query language. These are directly executed on SPARQL EndPoint, open query interfaces for browsing RDF graphs. Here we use BioPortal's. In total we have five (5) SPARQL query patterns that recovery (in the Table 1):

- all the diseases which constitute the leaves of the classes starting from the DOID, as well as their definitions starting from MESH;
- all disease categories from the DOID where we select their name, description, and parent categories;
- all nominative terms synonyms of diseases from the DOID, but especially from MESH, are the preferred label, as well as alternative labels for each disease;
- all the basic characteristic signs for each disease from semi-formalized descriptions of the DOID;
- all the nominal terms synonymous of signs: the preferential labels are extracted from SYMP, the alternative labels are extracted from CSSO ontology.

4.2 Selected Data from Medical Textual Reports

In the experimental setting of this work, we use on the one hand, sample of five (5) real cases of patients who have already been diagnosed. The examples chosen are different cases on tropical diseases[10] and allow us to visualize the

[10] Examples in http://medecinetropicale.free.fr/.

Table 2. Symptoms and Clinical Signs into sample clinical reports

Disease	Symptoms linking with diseasee	Symptoms present in example case	New added symptoms	Clinical signs
Hepatitis A	9	7	7	9
Dengue	10	5	11	20
Malaria	6	4	8	24
Syphilis	5	2	12	14
Chikungunya	9	5	7	29

contribution of cases. Indeed, on the Table 2 we can notice that for each disease, there is a precise number of general symptoms indicated by the ontology but the totality of them are not present in this patient. In addition, new symptoms identified in the ontology and not associated with the disease are emerging, as well as clinical signs specific to each patient. For example, for the case that refers to Hepatitis A, of the 9 symptoms that appear in the ontology and associated with this disease, only 7 are identified, and in addition 7 other new symptoms are detected as well as the clinical signs.

On the other hand, we used a larger sample of 156 cases of the same Hepatitis A disease. The Table 3 shows symptom appetition rates and test intervals measuring clinical signs. This shows the importance of clinical reports, especially in the context of medical diagnostic assistance. It's possible to classify the characteristic features of each disease by order of appearance in most patients who have already been diagnosed. For clinical signs, the intervals indicate, as reports are added in the case base, what are the most frequent minimum and maximum values. The cases base formalizing clinical reports therefore offers additional visibility to the basic description of the diseases carried by medical ontologies. The case base produced, formalizing clinical reports, therefore offers additional visibility to the basic description of the diseases carried by medical ontologies.

Table 3. Rate appearance of some symptoms and clinical signs on Hepatitis A clinical reports dataset

Symptoms	fever	fatigue	loss-of-appetite	vomiting	abdo-pain	c-c-b-movs	joint-pain
% appearance	39.1	64.1	20.5	50.0	12.8	11.5	19.2
Clinical signs	liver-big	liver-firm	bilirubin	aspartate	alk-phos	albumin	protime
% or interval appearance	76.9	38,5	0.3 to 8	14 to 648	26 to 296	2.1 to 6.4	0 to 100

5 Conclusion

In this article, the problem is focused on the establishment of a medical diagnostic support system based on open and shared ontology resources. It's a question here of the constitution of a central ontology federating a set of ontologies and medical terminologies targets, which answer the need for information in order to facilitate the task of the clinician in the identification of the potential diagnosis, among which he will have the latitude of choose or validate the most reliable knowingly. We have therefore proposed a data structure model and a federation methodology facilitating the recovery of human diseases and their most relevant characteristic signs, from targeted ontologies but also from an analysis of real cases of confirmed diagnosis. In the end, we have an ontology of diseases, signs, and all these cases which should serve as a knowledge base (KB) into a search engine to facilate access informations to clinicians. In a future work, we will show the interactive functionalities to navigate into this KB to collecte relevant informations step by step during the medical diagnostic process.

References

1. Anbarasi, M.S., Naveen, P., Selvaganapathi, S., Mohamed Nowsath Ali, I.: Ontology based medical diagnosis decision support system. Int. J. Eng. Res. Technol. (IJERT) (2013)
2. Balogh, E.P., Miller, B.T., Ball, J.R.: Improving Diagnosis in Health Care, National Academies of Sciences, Engineering, and Medicine. The National Academies Press, Washington, DC (2015)
3. Reyes-Ortiz, J.A., Jimenez, A.L., Cater, J., Malendés, C.A.: Ontology-based Knowledge Representation for Supporting Medical Decisions, Recherche in Computer Science (2013)
4. Mohammed, O., Benlamri, R., Fong, S.: Building a diseases symptoms ontology for medical diagnosis: an integrative approach. In: IEEE International Conference on Future Generation Commnication Technology (FGCT 2012), Décembre 2012
5. Hoehndorf, R., Schofield, P.N., Gkoutos, G.V.: The role of ontologies in biological and biomedical research: a functional perspective. Brief. Bioinform. J. **16**, 1069–1080 (2015)
6. Oberkampf, H., Zillner, S., Bauer, B.: Interpreting patient data using medical background knowledge. In: Proceedings of the 3rd International Conference on Biomedical Ontology (ICB0), Austria, 21–25 July 2012
7. Xie, J., Liu, F., Guan, S.-U.: Tree-structure based ontology integration. J. Inf. Sci. **37**(6), 594–613 (2011)
8. Alizadeh, M., Shahrezaei, M.H., Tahernezhad-Javazm, F.: Ontology based information integration: a survey. In: Proceedings of arXiv e-prints (2019)

Computational Analysis of a Literary Work in the Context of Its Spatiality

Ivan Dunđer(✉), Marko Pavlovski, and Sanja Seljan

Faculty of Humanities and Social Sciences, University of Zagreb, Ivana Lučića 3, 10000 Zagreb, Croatia

ivandunder@gmail.com, {mpavlovs,sanja.seljan}@ffzg.hr

Abstract. Spatiality is a term used to describe the attributes of a given space, its various cultural identities in an established time, differentiated from the notion of territoriality. While territoriality is naturally bound by the established limits of the national territory of a state, spatiality overcomes geographical distinctions and focuses on the identity or identities of a space defined solely on the basis of its "cultural territory", unlimited by international territorial boundaries. On the example of a literary work, the spatial elements used to achieve such a definition of space can be investigated through the adoperation of words that form the motives of such a work. The frequent use of topoi of the space defined in a literary work establishes the rhythm of its narrative. The authors of this paper shall make a computational qualitative and quantitative corpus analysis of a literary work that thematizes an urban identity in the context of its space and time. The literary method of repeating the topoi of a city, or simply its name and all other types of words derived from its lexical root will be analyzed with natural language processing techniques that will expose the aforementioned method with mathematical precision. The results of this analysis will afterwards be interpreted in a way that it is possible to exemplify how those words, used for the literary method of establishing spatial elements in the analyzed literary work and the rhythm of its narrative, help achieve a sense of spatiality and a singular united cultural identification of an extraterritorial space by its literary inhabitants.

Keywords: Computational corpus analysis · Natural language processing · Corpus linguistics · Qualitative and quantitative content analysis · Spatiality · Time · Identity · Literary work · Information and communication science

1 Introduction

Spatiality, a term that refers to all the elements that constitute a given space, its attributes, topoi and identity, is mostly used in sociological studies. Recently, the term has been adapted to literature, as a way to define the space a literary work is set in, but also all the elements that constitute it, its cultural identity. The need to define spatiality, as a term different from the terms of space and time, came with the advent of postmodernism and specifically the postmodern literary period.

Á. Rocha et al. (Eds.): WorldCIST 2020, AISC 1159, pp. 252–261, 2020.
https://doi.org/10.1007/978-3-030-45688-7_26

While various authors and literary critics disagree on the duration of the period, some critics believe spatiality to be one of the key terms of understanding the postmodern literary praxis. In this paper the authors will interpret space, time and spatiality through a computational quantitative and qualitative corpus analysis of Fyodor Dostoyevsky's *Crime and Punishment*, one of the prime examples of the realism literary period, with special accent on all of the elements that define and compose the cultural identity of space, so that in the end one could find new spaces of research, if not in the real world, then in the literary dimension with its reflection of reality. Defining spatiality in a novel that belongs to a literary era preceding the very same term, *spatiality* could give insights into the author's awareness of a sense of place without it being an element of the literary technique.

While the term *space* is often used in connection with the term *time* to define the context of a given setting in literary fiction (its "space and time"), spatiality in literature could be defined as a term used to describe the attributes of a given space, its various cultural identities; in plural because, according to [1], "identity is not a singularity" and "identities, not identity", in an established time and place.

Spatiality must be differentiated from the notion of territoriality. In a number of disciplines, but most surely in literary theory, territoriality is bound by the established limits of the national territory of a state, while spatiality overcomes these geographical distinctions, focusing on the identities of a space defined solely based on its "cultural territory", which is of course unlimited by international territorial boundaries.

Those spatial elements can be investigated through the author's use of words that form the motives of such of a literary work, and on a particular example the authors of this paper could analyze the words that constitute such an achievement of the given definition of space. The rhythm of a literary work's narrative, in this case Fyodor Mikhailovich Dostoyevsky's *Crime and Punishment*, could be established by a frequent use of topoi that define the set place of the analyzed novel.

Furthermore, the authors shall make a computational qualitative and quantitative corpus analysis of the chosen literary work that thematizes an urban identity or identities in the context of its space and time, analyzing it with natural language processing (NLP) techniques. The chosen computational method which also makes use of descriptive statistics will expose the literary method of repeating the topoi of a city, or simply its name and all other types of words derived from its lexical root with mathematical precision.

The results of this analysis will afterwards be interpreted in such a way that it is possible to exemplify how those words, used for the literary method of establishing spatial elements in the analyzed literary work and the rhythm of its narrative, help achieve a sense of spatiality and a singular united cultural identification of an extraterritorial space by its literary inhabitants. Further explanation and elaboration of the motivation behind such a research will be given in detail in the following chapter "Motivation and related work". This chapter will also offer an overview of related literature to the topic of space, in relation to a given space and, specifically, the term of spatiality (in some authors also: *spaciality*) differentiated, but not completely, from the sociology-based definition.

According to Matvejević, the concept of identity should be used in its plural form. Identity interpreted in plural is derived from the Latin proverb "Idem, nec unum", meaning "Identical, but not one" [1]. In the aforementioned context it belongs to the domain of culture, specifically, it refers to various cultural identities of a given space. Identities – the term, used in plural, was propagated by both [1] and [2, 3] in his literary work. The two authors wrote novels that could be defined as spatial novels on the basis of the given theme.

2 Motivation and Related Work

This section explains the motivation of a computational qualitative and quantitative corpus analysis of a literary work in the context of its spatiality, exemplified by Fyodor Mikhailovich Dostoyevsky's novel *Crime and Punishment*. The term spatiality, when applied to a literary work, helps to understand the various cultural identities of space described in the mentioned work, and as such improves the perception of the time setting. Spatiality, seen as a combination of varied spatial elements that define a given space, can be analyzed in a mathematical way, improving the understanding of an author's ability to achieve a sense of presence with the literary tools at his disposal.

This does not interfere with the artistic value and does not diminish the aesthetic element of a literary work. Furthermore, this does not "tear apart the soul of the literary work" but improves the comprehension of the literary method used by the author to define spatiality. In other words, an extensive qualitative and quantitative corpus analysis of the analyzed work can give valuable insights into how the author achieves the *genius loci*, i.e. the sense of space. The sociology of space, or spatial sociology, defined earliest by [4], is a sub-discipline of sociology. While spatiality in the context of spatial sociology refers to the study of natural and social space with its "real" individuals and organizations, the term can be applied to literary works as a means to establish the far-sight of an author regarding the future of the social and nature space defined in his literary work and his ability to reflect, with literary methods, the described space in the context of its time. It could also be interesting to establish the importance of space between spatial sociology, its concrete examples and the space as defined in a literary work.

Reference [5] asserts: "Spatial determinists may believe that space always has social effects, but such effects are not automatic and are indirect. The presence of gold "under" natural space has considerable economic effects, but only in societies in which gold is valued. Similarly, when homeless people must raise families in a dilapidated motel room, the social space they live in has behavioral effects, but only because welfare agency benefits are too meagre to enable recipients to afford livable dwellings. Thus, space almost never has total and direct causal power, and then functions as an intervening causal variable."

The power of space or as [6] famously wrote, life in a set of relations, could be total and direct in a literary work but, if this may not prove true, opposed to the possibilities of spatial sociology, one can measure this hypothesis in a literary work with the uttermost accuracy.

Performing a mathematically precise qualitative and quantitative corpus analysis of a literary work enables measuring of the frequency of all used words and noting the words that are most important to the author in constructing his literary work. Such analyses heavily rely on computational natural language processing methods, which can be used for various purposes, and measuring the frequency of spatial elements and their impact on the given corpus is just one of the many possibilities. As this corpus analysis can be undergone in a large number of languages, future research is going to measure the closeness of translations of a literary work in comparison to the original. The results of such a measurement could be a relevant source of establishing the quality of a translated text and could be applied in higher education, in literary studies but also in other fields of research that require text analysis.

Still, the question is why to transport a sociological term to a literary work and why do a qualitative and quantitative analysis of its corpus in the context of its spatiality? What is the difference between spatial sociology and literary spatiality evaluation, other than one being based on real spaces and the other on literary constructions? One of the possible answers could be: the imagination of an author, especially of the classical greats of literary history. Although it is highly debatable if imagination can be measured in any way with existing tools, the ramifications of an author's imaginative possibilities could be evaluated by applying computational natural language processing techniques on his literary work.

Spatiality in literature was covered in its widest range by theatre studies, as space, its definition and possible ways to interpret it in the literary genre are one of the key elements of such studies. "To understand space means to define the existential place of Man's energy in a given time" [7]. As a sociological term, spatiality was examined in literary theory, mostly associated with works of contemporary literature written in the English language or by English-speaking theorists in a wide range of social studies and philosophy [8], in the USA [9], but also in Croatia, notably in the work by [10, 11]. Furthermore, as mentioned before, literary works where spatiality is accentuated can be found in the opus of Matvejević, notably in [12], and Magris, notably in [2] and [3]. Paraphrasing [13], today we start to read spaces as our cultural writing and to identify and estimate spatially for ourselves and for the others.

3 Research

The following subsections will describe the chosen dataset, the research methodology and applied natural language processing techniques.

3.1 Dataset and Preprocessing

The authors of this paper chose to conduct the experiment on Fyodor Dostoyevsky's novel *Crime and Punishment*; an English translation by Constance Garnett, which was used as the dataset. It was made available through the Project Gutenberg, acquired in October 2019 and retrieved from http://www.gutenberg.org/files/2554/2554-h/2554-h.htm. After the acquisition of the dataset the necessary preprocessing was carried out. The process included a few steps. First, the dataset was converted to a plain text file with UTF-8 file encoding which ensured that all the characters were saved in an appropriate way. Then, the dataset was cleaned, i.e. redundant text chunks (e.g. Translator's preface,

License agreement etc.) and unnecessary whitespace characters (e.g. running blank space characters, ending blank space characters, consecutive blank spaces, tabs) had to be removed. Afterwards, sentence/segment splitting was performed. Preprocessing was mainly done by applying sed, awk and Perl scripts, created by the authors of this paper.

3.2 Research Methodology and Applied Techniques

Once the preprocessing steps were done, descriptive statistics were calculated, such as total number of words and characters, maximum and minimum length of sentences/segments, total number of sentences, arithmetic mean, standard deviation etc.

Then, natural language processing techniques and analyses focusing mainly on word statistics [14] and concordances were applied: e.g. calculating the distribution of words over frequencies [15, 16] and the total number of words-unique words (vocabulary size) ratio [17, 18], analyzing the most frequent words [19], plotting the "physical" position of certain words in the given dataset [20, 21], examining log-frequencies etc. Also, concordancing (the process of listing a word with its contexts) was done [22]. This enabled the authors, not only to understand the elements of spatiality, but also, to interpret to some extent the motivation and significance of the literary method of repeating the topoi of a particular city – in this case, Saint Petersburg, which in Constance Garnett's translation appears in the form "Petersburg".

The authors would like to point that the conducted analyses were overall based on the frequency of words in the dataset, which cannot solely explain the qualitative characteristics of the novel, and, therefore, more extensive analyses should be carried out in the future – such as sentiment analysis [23], vector representation of words [24] or collocation extraction [25]. These extensive analyses could allow scholars to achieve an even deeper understanding of the text in question, its genre, setting, style, theme, tone etc. This could be achieved by using the aforementioned objective, time- and cost-efficient computational methods, which could be applied on a variety of texts, even outside the domain of fiction.

4 Results and Discussion

The descriptive corpus statistics and initial research results after preprocessing of the dataset are given in Table 1.

Table 1 shows that the dataset after performing the described preprocessing steps consisted of roughly 200 thousand words, separated by 14447 sentences/segments, comprised of more than 1.1 million characters, including spaces. Average sentence size was ca. 77 characters, i.e. 14.11 words. Almost 21000 words were unique (word types, i.e. distinct words), indicating relatively rich vocabulary, i.e. large number of specific content words. This is not completely reflected in the total number of words-unique words (vocabulary size) ratio, which is still less than 10. In other words, on every 10 words comes one distinct word. The largest sentence consisted of 108 words, whereas the smallest sentence (in this case segment) was only 1 word long, which was expected, as

Table 1. Descriptive corpus statistics and initial research results after preprocessing.

Category	Value	Category	Value
Total number of characters (with spaces)	1107908	Total number of words	203849
Vocabulary (types, i.e. distinct unique words)	20845	Total number of words-unique words ratio	9.78
Number of sentences/segments	14447	Maximum sentence/segment size	108 (words)
Minimum sentence/segment size	1 (words)	Average sentence/segment size	76.69 (characters), 14.11 (words)
Standard deviation	11.86 (words)		

such segments were mostly identifications of book parts and book chapters. This resulted in an overall standard deviation equaling approximately 12 words.

The reported statistics will serve as a reference point for future exploratory analyses of other translations of the same literary work. Table 2 presents the distribution of words over frequencies.

Table 2. Distribution of words over frequencies.

Freq.	No. of words	% of vocabulary	% of words
1	11555	55.43	5.67
2	3035	14.56	2.98
3	1503	7.21	2.21
4	891	4.27	1.75
5	601	2.88	1.47
6	448	2.15	1.32
7	345	1.66	1.18
8	267	1.28	1.05
9	186	0.89	0.82
10	164	0.79	0.80
>10	1850	8.88	80.74

More than 11500 words appear only once in the entire dataset and make up ca. 55% of all unique words and almost 6% of the total number of words in the preprocessed dataset. Words that appear only once are also called *hapax legomena*. *Dis legomena* (appear only twice) are represented with ca. 15% in the total number of unique words.

Words that appear only three times, i.e. *tris legomena*, make up almost 3% of all words in the dataset. The authors of this paper would also like to emphasize that almost 81% of all words appear more than 10 times. This is largely due to the very frequent use of function words such as articles, auxiliary verbs, conjunctions, prepositions, pronouns etc., which are used for constructing complex sentences in the analyzed novel. Very few content words (e.g. nouns, verbs, adjectives, and adverbs) are among the top 100 most frequent words, as shown in Table 3.

Table 3. Analysis of the 100 most frequent words.

Rank	Frequency	Word	Rank	Frequency	Word
1	7785	the	15	1828	her
2	6932	and	16	1793	but
3	5386	to	17	1754	not
4	4868	he	18	1747	s
5	4585	a	19	1701	with
6	4399	i	20	1694	she
7	4047	you	...		
8	3776	of	38	785	Raskolnikov
9	3444	it	81	402	Sonia
10	3284	that	90	347	Razumihin
11	3175	in	95	325	Dounia
12	2805	was	98	304	Ivanovna
13	2094	his	...		
14	2064	at	450	52	**Petersburg**

In the analyzed translation of the chosen novel, the name of the city Saint Petersburg appears only in the form of "Petersburg", which is found in total 52 times in the dataset and is overall ranked 450th. Figure 1 shows the concordance plot of the word "Petersburg". The barcode-like plot is a visual representation of concordances and marks with a vertical line the position in the dataset where a word of interest occurred. In this case, the word "Petersburg" represents the KWIC, i.e. the key word in context [26].

Fig. 1. Concordance plot of the word "Petersburg".

The presented plot indicates relatively uneven distribution of occurrences of the word "Petersburg" throughout the dataset. Also, the analyzed word appears more frequently in the second half of the novel.

Figure 2 implies that the chosen literary work contains rich vocabulary, but also that very specific terminology is used, since most of the distinct words (ca. 55%) appear only once. This phenomenon is depicted by a histogram with a nonlinear scale which is used to graphically present a large range of quantities.

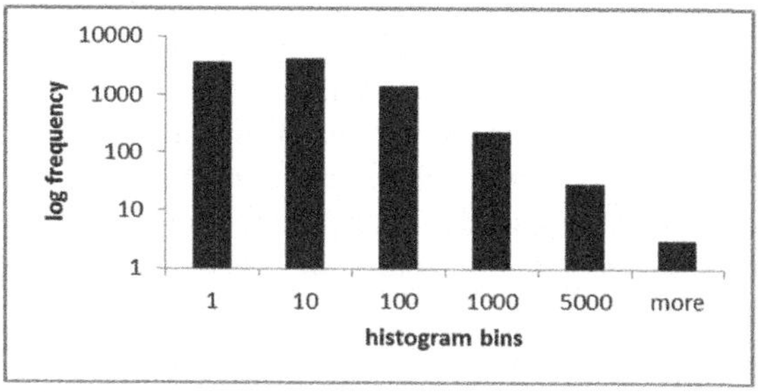

Fig. 2. Histogram of log-frequencies of words with respect to specific bins.

In this case, it is a histogram showing log-frequencies of words with respect to specific bins [27]. It uses a logarithmic scale to plot one variable that covers a large range of values (in this case – the frequency), while at the same time another variable has only a restricted range (i.e. bins). This visualization does not alter the data itself, responds very efficiently to skewness towards large values [15] and therefore points out data features that would not be easily observable if both variables had been plotted in a linear way. In general, visualization of data is very helpful and often recommended as a crucial stage of data exploration before any statistical tests are applied, as it can show main tendencies in the data, helps to discover interesting patterns, reveals potential outliers and other problematic aspects of a data set [28].

5 Conclusion

Although still a work in progress, NLP methods and techniques have proven to be very useful in literature analysis. From the preliminary results of the qualitative and quantitative corpus analysis of the chosen literary work in the context of its spatiality, the authors of this paper can establish that *Crime and Punishment* is not a spatial novel in the strict sense of the term. It is actually defined more by its characters and their relations than the space the novel is set in. The most frequent spatial element (the city of Saint Petersburg, also Sankt-Peterburg, or simply Petersburg) is always contextualized, and this possibly adds to the semantic value of the word. The distribution of words over frequencies indicates rich vocabulary in the novel, but also that a very specific terminology is used, since most of the distinct words appear only once. Further research is planned across different translations of the same literary work.

Based on the results of this research, the authors can conclude that *Crime and Punishment* is not a spatial novel: names of characters are more frequently used than the topoi of Saint Petersburg. Therefore, it could be concluded that the computational analysis confirms that the novel is centered on the relations between its characters. Saint

Petersburg is mentioned 52 times (it is the most frequent toponym in the novel), and comparing it to the times the names of the characters are mentioned, it is possible to assert that, differentiated from Magris' or Matvejević's work, space is not the main theme throughout the novel, like the Danube was in the work of Magris. From the example of the 52 concordances of the word "Petersburg", it is interesting to note that it is always contextualized in a phrase, but it is rarely personificated and mostly serves as the topos of the city itself, with no other meaning. Although, from a perspective of spatial identity, it is defining to note that most of the concordances concern arriving or leaving Saint Petersburg, and not defining the city in its spatiality but using it as a symbol of the need to migrate elsewhere. Moreover, the detailed concordance analysis shows that the analyzed topos occurs most frequently in a pejorative context. This demonstrates that a computational method applied on a literary text can be used as an unbiased and fast approach to text analysis.

References

1. Matvejević, P.: Sulle identita dell'Europa. Fondazione Laboratorio Mediterraneo, Naples (1994)
2. Magris, C.: Danubio. Garzanti Libri, Milan (2007)
3. Magris, C.: Microcosmi. Garzanti Libri, Milan (2006)
4. Simmel, G.: Soziologie: Untersuchungen über die Formen der Vergesellschaftung. Duncker & Humblot, Leipzig (1908)
5. Gans, H.J.: The sociology of space: a use-centered view. City Community **1**(4), 329–339 (2002)
6. Foucault, M.: Des Espaces Autres. Architecture Mouvement Continuité **5**, 46–49 (1984)
7. Pavlovski, B.: Prostori kazališnih svečanosti. Naklada MD, Zagreb (2000)
8. Casey, E.S.: The Fate of Place. A Philosophical History. University of California Press, Berkeley (1997)
9. Hillis Miller, J.: Topographies. Stanford University Press, Stanford (1995)
10. Grgas, S.: Američki studiji danas: Identitet, kapital, spacijalnost. Naklada MD, Zagreb (2015)
11. Grgas, S.: Ispisivanje prostora: čitanje suvremenoga američkog romana. Naklada MD, Zagreb (2000)
12. Matvejević, P.: Mediteranski brevijar. Grafički zavod Hrvatske, Zagreb (1987)
13. Kort, W.A.: Place and Space in Modern Fiction. University Press of Florida, Gainesville (2004)
14. Baroni, M.: Distributions in text. In: Lüdeling, A., Kytö, M. (eds.) Corpus Linguistics: An International Handbook 2, pp. 803–821. Walter de Gruyter GmbH, Berlin (2008)
15. Gries, S.Th.: Useful statistics for corpus linguistics. In: Sánchez, A., Almela, M. (eds.) A Mosaic of Corpus Linguistics: Selected Approaches, pp. 269–291. Peter Lang, Frankfurt am Main (2010)
16. Jaworski, R., Dunđer, I., Seljan, S.: Usability analysis of the concordia tool applying novel concordance searching. In: 10th International Conference on Natural Language Processing (HrTAL 2016). Lecture Notes in Computer Science (LNCS/LNAI), p. 11. Springer. Berlin (2016, in print)
17. Kumar, G.B., Murthy, K.N., Chaudhuri, B.B.: Statistical analysis of Telugu text corpora. Int. J. Dravidian Linguist. **36**(2), 71–99 (2007)

18. Bharati, A., Rao, P., Sangal, R., Bendre, S. M.: Basic statistical analysis of corpus and cross comparison among corpora. In: Proceedings of the International Conference on Natural Language Processing (ICON 2002), p. 10. Centre for Language Technologies Research Centre – International Institute of Information Technology, Hyderabad, India (2002)
19. Dunđer, I., Horvat, M., Lugović, S.: Word occurrences and emotions in social media: case study on a Twitter corpus. In: Proceedings of the 39th International Convention on Information and Communication Technology, Electronics and Microelectronics (MIPRO 2016), pp. 1557–1560. Croatian Society for Information and Communication Technology, Electronics and Microelectronics – MIPRO, Rijeka, Croatia (2016)
20. Dunđer, I., Horvat, M., Lugović. S.: Exploratory study of words and emotions in Tweets of UK start-up founders. In: Proceedings of The Second International Scientific Conference "Communication Management Forum" (CMF 2017), p. 18. The Edward Bernays College of Communication Management, Zagreb, Croatia (2017)
21. Dunđer, I., Pavlovski, M.: Computational concordance analysis of fictional literary work. In: Proceedings of the 41st International Convention on Information and Communication Technology, Electronics and Microelectronics (MIPRO 2018), pp. 0644–0648. Croatian Society for Information and Communication Technology, Electronics and Microelectronics – MIPRO, Rijeka, Croatia (2018)
22. Kilgarriff, A., Kosem, I.: Corpus Tools for Lexicographers. In: Granger, S., Paquot, M. (eds.) Electronic Lexicography, pp. 31–55. Oxford University Press, Oxford (2012)
23. Dunđer, I., Pavlovski, M.: Behind the dystopian sentiment: a sentiment analysis of George Orwell's 1984. In: Proceedings of the 42nd International Convention on Information and Communication Technology, Electronics and Microelectronics (MIPRO 2019), pp. 0685–0690. Croatian Society for Information and Communication Technology, Electronics and Microelectronics – MIPRO, Rijeka, Croatia (2019)
24. Dunđer, I., Pavlovski, M.: Through the limits of newspeak: an analysis of the vector representation of words in George Orwell's 1984. In: Proceedings of the 42nd International Convention on Information and Communication Technology, Electronics and Microelectronics (MIPRO 2019), pp. 0691–0696. Croatian Society for Information and Communication Technology, Electronics and Microelectronics – MIPRO, Rijeka, Croatia (2019)
25. Seljan, S., Dunđer, I., Gašpar, A.: From digitisation process to terminological digital resources. In: Proceedings of the 36th International Convention on Information and Communication Technology, Electronics and Microelectronics (MIPRO 2013), pp. 1329–1334. Croatian Society for Information and Communication Technology, Electronics and Microelectronics – MIPRO, Rijeka, Croatia (2013)
26. Church, K., Gale, W., Hanks, P., Hindle, D.: Using statistics in lexical analysis. In: Zernik, U. (ed.) Lexical Acquisition: Exploiting On-Line Resources to Build a Lexicon, pp. 115–164. Lawrence Erlbaum Associates, Hillsdale (1991)
27. Pavlovski, M., Dunđer, I.: Is big brother watching you? A computational analysis of frequencies of dystopian terminology in George Orwell's 1984. In: Proceedings of the 41st International Convention on Information and Communication Technology, Electronics and Microelectronics (MIPRO 2018), pp. 0638–0643. Croatian Society for Information and Communication Technology, Electronics and Microelectronics – MIPRO, Rijeka, Croatia (2018)
28. Evert, S., Schneider, G., Brezina, V., Gries, S.Th., Lijffijt, J., Rayson, P., Wallis, S., Hardie, A.: Corpus statistics: key issues and controversies. Panel discussion at Corpus Linguistics 2015, Lancaster, United Kingdom (2015)

Data Extraction and Preprocessing for Automated Question Answering Based on Knowledge Graphs

Aleksei Romanov(✉), Dmitry Volchek, and Dmitry Mouromtsev

ITMO University, St. Petersburg, Russia
romanov@itmo.ru, dvolchek@yandex.ru, mouromtsev@mail.ifmo.ru

Abstract. This article discusses the algorithms and methods of data representation for knowledge graphs. The proposed algorithms make it possible to automate the process of extracting and processing data from users' requests within the mixed learning process and reduce the role of an expert in the preparation of question-answering data sets necessary for training models of dialogue systems. The results show that the method of enrichment of the knowledge graph leads to an increase in the number of links and the accuracy of vector representation models.

Keywords: Semantic technology · Ontology · RDF · Embeddings · QA system

1 Introduction

Dialog systems that are a technological trend have recently received growing attention. Nowadays, the most popular language technology is an intelligent virtual assistant. Depending on the application context, such systems are called a dialog interface, question-answering system, or chatbot. However, they all share the same concept because they are made to perform specific tasks by way of interactive communication with machines and queries in a natural language [1]. Among the most famous dialog systems are digital voice assistants, Apple Siri, Amazon Alexa, Google Assistant, Yandex Alice. Their emergence stimulated the growth of many text-based dialog systems designed to solve concrete problems. It can be argued that dialog systems have become the de facto standard for many industries, from retail to booking, and, in particular, customer support.

One of the areas of dialog system application is education, especially the development of e-learning and distance learning technologies in the implementation of educational programs. The Federal Law on Education in the Russian Federation describes possible ways of integration and organization educational process with the use of distance learning technologies and e-learning based on Unified Educational Information Environments (UEIE).

Development of such projects and their direct integration in educational processes of the leading Russian universities leads to a significant increase in both

Á. Rocha et al. (Eds.): WorldCIST 2020, AISC 1159, pp. 262–270, 2020.
https://doi.org/10.1007/978-3-030-45688-7_27

the number of educational materials and the number of students, which in turn generates large amounts of data. It also increases the number of support staff for such systems, including curators of online courses and support specialists competent in a specific field or subject. An increase in the number of employees involved in this process entails economic costs and efficiency evaluations.

Generalization of UEIE information, including study materials, tasks, tests, exercises, user messages on forums, support requests in the form of one knowledge base that relies on a semantic network, can solve the efficiency problem. Unlike databases available in any UEIE, knowledge bases not only store factual information, but also make conclusions on existing or newly introduced facts and, thereby, perform semantic or meaningful information processing.

1.1 Motivation

The knowledgebase can be a base for the development of an intelligent system, one example of which is the creation of a question answering system. Their development is a goal of many corporations, in particular, in banking. Potentially, dialog systems will save money by automating repetitive tasks while supporting various processes of user interaction with the system.

In education, it is the communication between a student and a teacher on the educational forum or by e-mail. However, these questions are often repeated. Thus, they can be formalized to automate the answering process. The second important advantage is the ability to provide feedback 24 hours a day, unlike support staff. Conversational systems also allow gaining experience by analyzing conversations.

The current problem of interactive systems is the closed nature of many developments and commercialization. In particular, the important problem is the preprocessing of data for interactive system models to learn from.

Therefore, the problem is to generalize and develop methods and algorithms that will ensure a continuous process of developing a working model of the dialog system using raw, semi-structured data of support resources.

1.2 Dialog Systems

Recent advances in dialogue systems are overwhelmingly driven by deep learning techniques that have been used to improve a wide range of big data applications, such as computer vision, natural language processing, and recommendation systems. For conversational systems, deep learning can use a huge amount of data to learn meaningful function representations and response generation strategies, while requiring a minimum amount of manual processing.

Goal-oriented dialog systems based on the **pipeline method** typically consist of the following four components [2]:

1. Natural Language Understanding, NLU. It splits user text into predefined semantic slots and converts it to a machine-readable form.

2. Dialog State Tracker, DST. It defines the current state, which is used to select the next action.
3. Dialog Policy Learning, DPL. It selects the next action based on the current state of the dialog.
4. Natural Language Generation, NLG. It displays the selected action due to the generated natural language response.

As a rule, non-target dialogue systems do not try to solve a specific problem and give an answer to a question, but their initial goal is to maintain a dialogue with a person on an arbitrary topic. The approaches to constructing **non-target** dialog systems, including the following:

1. **Generative approach** that is based on neural networks. This approach requires large amounts of data for training, while the question answering pairs must be fully correct. The advantage of the approach is the ability to generate answers close to the natural language in cases when the answer is not contained in the data set.
2. The **search approach** is based on the principle that the answer is inside the dataset, and the model goal is to find the most suitable one.
3. The **ensemble approach (hybrid)** involves combining methods to improve the responses generated by the neural network through providing an additional context by using the options offered by the search approach.

1.3 Knowledge Graphs

The approach based on knowledge graphs is actively developing and becoming extremely popular among modern dialog systems [4]. A knowledge graph is a set of structured information ordered in a graph format, where nodes are entities and edges are relationships. The knowledge graph $\mathcal{G}$ consists of triples of the form 'subject–predicate–object' – tuples (h, r, t), where $h(head)$ – subject, $r(relation)$ – predicate, $t(tail)$ – object. Such triples are called true triples and they form a set $\mathcal{S} = \{(h, r, t)\}$. Herewith $h, t \in \mathcal{E}$ and $r \in \mathcal{R}$, where $\mathcal{E}$ – set of entities, and $\mathcal{R}$ – set of relations.

To create a system based on knowledge graphs, it is necessary to present data in the appropriate triples format, as well as to process user requests from proposals, in the triples format, for example, from a question: 'In what year was St. Petersburg founded?' triple can be obtained 'St. Petersburg – founded – year'. Such a triple is then compared with the data available in the knowledge graph and place of the year property, values will be substituted.

Thus, knowledge graphs allow us to model both abstract logical statements and schemes, and to fill these schemes with particular objects of the real world [13]. Moreover, knowledge graphs allow machines to reason and derive new knowledge not previously described in the graph. Formal semi-structuring and powerful logical apparatus distinguish knowledge graphs from traditional relational databases, which are structured, that is, have established relationships and relationships.

2 Method

The primary use case of developing and maintaining methods and algorithms for the automated processing of semi-structured support resources is to automate the preparation of a training set for the dialog system. The described processing is based on natural language processing and machine learning methods, which allows selecting the text from e-mails with the subsequent extraction of question answering pairs.

2.1 Data Preprocessing

The basic algorithm is collecting and preprocessing data. Automatically collected emails require format conversion. Initially, these messages are transmitted in the Multipurpose Internet Mail Extension (MIME) format. The basic format for email messages is defined by the RFC822 standard. According to it, each message consists of head and content. The head stores service information, namely the address necessary to send the message, while the messages can be stored in any format, and the standard only determines the format for transmission.

Messages are collected in the base64 byte-string format. It is a sequence of printed ASCII characters (Latin characters, numbers, and special characters). Thus, each email must be decoded in the UTF-8 format because it is the most common and convenient data encoding format. Next, the message text is read line by line and cleared of extraneous data. It includes HTML and CSS markup based on tags.

The main task of the subsequent processing is to separate the user's signature from the message because its availability and format are not regulated by representation standards in any way. For this, a series of heuristics have been developed to classify message lines:

1. **Regular Expressions** is a simple but effective way [12] to find standardized information: web addresses, telephones, markers of the message signature department, email address, full name.
2. **String selection** is a criterion for the length of message signature lines, determined on the basis of marked up messages.
3. **Keywords** are a criterion based on the assumption that a certain set of words is used in the message signature.

Thus, to identify keywords, the received messages are divided into tokens, each of which is compared with a list of stop words, and then reduced to normal form. As a result, all documents (email) are presented as a bag of words that are further analyzed.

Based on the obtained heuristics the problem of binary classification is solved. The problem can be solved by common methods of machine learning with a different number of processed last k lines of the message. The best result is obtained with $k = 10$.

2.2 Question Extraction

The algorithm for extracting question answering pairs based on the following criteria:

- the sentence ends with a question mark;
- the sentence contains a word or phrase with a semantic load defined as a pair of a noun and a verb by way of extracting universal dependencies [5] (an example of the corresponding representation is shown in the Fig. 1);
- the sentence contains pronominal adverbs.

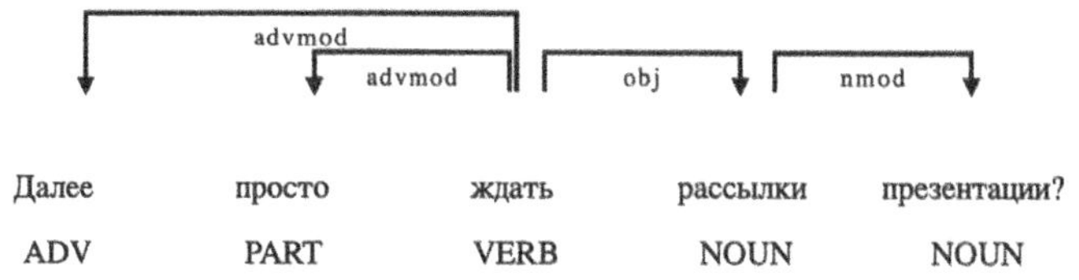

Fig. 1. Representation of a sentence using universal dependencies for Russian. The translation into English: Should I wait for the email with a presentation?

These criteria are combined:

- last symbols e '?' and e contains a noun and a verb;
- first three or more characters e contained in $\mathcal{ADV}_q$ and e contains a noun and a verb.

The algorithm produces identifiers of messages and questions. Message identifiers are used for searching the respective answers. Such information is contained in the RFC822 transmission format of messages.

2.3 Knowledge Graph Enrichment

Based on the obtained data we introduce the process of creating a basic ontological model for a knowledge graph of support resources, taking into account its metadata enrichment, topic modeling, named entities recognition (NER), and semantically related questions.

The sequence of actions for applying topic modeling methods includes tokenization and stemming or lemmatization of requests. As a result, each request is presented as a bag of words $\mathcal{D}$. To ensure the accuracy of the model, words with a frequency of more than one are selected to cut off common words, and focus the topic model within a specific context.

In the next step, we define the determination of the sentence semantic proximity based on word proximity. To define it, the Russian language thesaurus RuWordNet [7] based on the RuThes thesaurus [6] that was converted to WordNet [8] was used. The thesaurus contains 111500 words and expressions in Russian in total. To mark parts of speech, an approach based on Universal Dependencies [5] was used.

The semantic vector can be obtained by way of the pairwise comparison of the words included in the sentence with each other, where each coordinate of the vector corresponds to the maximum value of the proximity of pairwise words [9]. The obtained value is normalized based on the criterion of the synonymy threshold that is proposed to be equal to 0.8025 in the paper [10].

For all question pairs in the knowledge graph $\mathcal{G}$, the algorithm produces the added triplets of the form $(\mathcal{S}_1, \texttt{SimilarToTheQuestion}, \mathcal{S}_2)$ with a set threshold for proximity of sentences γ. This parameter can be established empirically only and must take high values to ensure that only true identical questions are close.

NER extraction allows us to enrich the knowledge graph with metadata specific to certain types of questions. It is proposed to use the BERT model [11] available for Russian and having the best results at the time of the experiment. Identified entities are added to the knowledge graph $\mathcal{G}$, and are associated with an email and/or question depending on the location of the named entity. One entity can be associated with any questions or email.

Finally, we are using the TransE [3] vector representation algorithm to prove the relevance and necessity of the knowledge graph and its enrichment with the specified metadata. The main idea of the algorithm is to represent the triples (h, r, t) of the set $\mathcal{S}$ in a certain vector space of dimension R^k, where k is the hyperparameter of the model. All entities h, t form the set $\mathcal{E}$, all relationships r form the set $\mathcal{R}$. Embeddings are usually designated by the corresponding bold characters $\mathbf{h}, \mathbf{r}, \mathbf{t}$. Then, for the true triples, the condition $\mathbf{h} + \mathbf{r} \approx \mathbf{t}$ must be satisfied while being in contradiction with the false triples. The set of false triples $\mathcal{S}'$ is formed by substitution of h or t (but not both) of the true triplet for a random entity, that is:

$$\mathcal{S}' = \{(h', r, t) \, |h' \in \mathcal{E}\} \cup \{(h, r, t') \, |t' \in \mathcal{E}\}.$$

The search for such vector coordinates is performed using gradient descent while minimizing the loss function

$$\mathcal{L} = \sum_{(h,r,t) \in \mathcal{S}} \sum_{(h',r,t') \in \mathcal{S'}'_{(h,r,t)}} [\gamma + d(\mathbf{h} + \mathbf{r}, \mathbf{t}) - d\left(\mathbf{h}' + \mathbf{r}, \mathbf{t}'\right)]_+ ,$$

where d is a measure of similarity, $[x]_+$ means that only positive terms are considered, $\gamma > 0$ is an indent hyperparameter.

3 Results

The modeling was performed using four machine learning methods, and the results were evaluated using a confusion matrix presented on the Fig. 2.

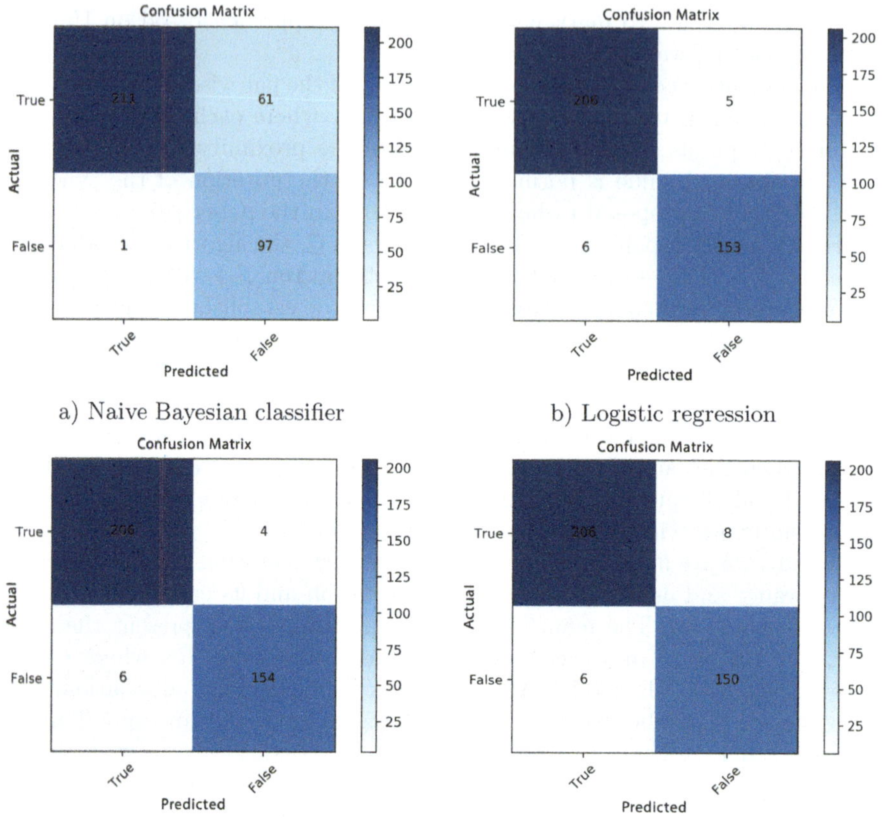

a) Naive Bayesian classifier

b) Logistic regression

c) SVM

d) AdaBoost

Fig. 2. Confusion matrix for the signature string classifier, $k = 10$.

The best result is achieved when $k = 10$ using the support vector machine (SVM) $\texttt{F1} = 0.9763$. The Fig. 2 shows the confusion matrices for the selected classification methods based on model verification on a test data set from 370 marked up messages (212 messages contain a signature, 158 do not contain a signature). At the same time, the results of the basic algorithm adapted for Russian (translated keywords and regular expressions adapted for the Cyrillic alphabet) with the same parameters shows the worst result $\texttt{F1} = 0.7332$.

Thus, the processing sequence of the developed algorithm allows receiving the processed support requests. Compared to the basic algorithm, the obtained metrics allow estimating the proposed approach effectiveness. The accuracy increased from $\texttt{F1} = 0.7332$ to $\texttt{F1} = 0.9763$. The increase was about 33.16%.

Table 1. Metrics for the developed algorithm, $k = 10$.

Algorithm	Metrics		
	`F1`	`Precision`	`Recall`
Naive Bayesian Classifier	0.8719	0.7757	0.9953
Logistic Regression	0.9740	0.9763	0.9717
SVM	**0.9763**	0.9810	0.9717
AdaBoost	0.9671	0.9626	0.9717

Based on described methods we perform a series of experiments, based on the different states of the knowledge graph (Table 1):

- The basic structure of the knowledge graph – $\mathcal{G}_1$;
- The knowledge graph $\mathcal{G}_1$ enriched with named entities – $\mathcal{G}_2$;
- The knowledge graph $\mathcal{G}_2$ enriched with semantic proximity of questions – $\mathcal{G}_3$;
- The knowledge graph $\mathcal{G}_3$ enriched with the topic of requests as a result of topic modeling – $\mathcal{G}_4$;
- The knowledge graph $\mathcal{G}_4$ enriched with metadata of requests – $\mathcal{G}$.

The Table 2 shows the results of the model accuracy evaluation.

Table 2. Model accuracy for different knowledge graphs.

Method	Metrics	
	`F1`	`EM`
DeepPavlov [14]	0.2341	0.1172
$\mathcal{G}_1$	0.2763(+18.03%)	0.1493(+27.42%)
$\mathcal{G}_2$	0.2764(+0.04%)	0.1493(+0.00%)
$\mathcal{G}_3$, $\gamma = 0.73$	0.2902(+4.00%)	0.1531(+2.53%)
$\mathcal{G}_4$	0.4179(+44.00%)	0.2136(+39.51%)
$\mathcal{G}$	0.5057(+21.01%)	0.2250(+5.31%)

4 Conclusions

Thus, the problems stated in the article were solved making it possible to prove the effectiveness of the developed algorithms and method in the experimental study based on the accuracy assessment criteria.

The proposed methods were compared based on various embedding approaches and concerning basic experiment without knowledge graphs. The proposed methods showed higher accuracy results. The lack of direct analogs or methods that can be used to develop dialog systems in Russian for the automated processing of resource data emphasizes the relevance of the paper.

Therefore, the goal of improving the accuracy of answers in dialog systems during automated processing of resource data by the means of the semantic annotation method and metadata enrichment as well as embedding algorithms of knowledge graphs was achieved.

Acknowledgments. This work was supported by the Government of Russian Federation (Grant 08-08).

References

1. Dale, R.: The return of the chatbots. Nat. Lang. Eng. **22**(5), 811–817 (2016)
2. Bordes, A., Boureau, Y.L., Weston, J.: Learning end-to-end goal-oriented dialog. arXiv preprint arXiv:1605.07683 (2016)
3. Bordes, A., Usunier, N., Garcia-Duran, A., Weston, J., Yakhnenko, O.: Translating embeddings for modeling multi-relational data. In: Advances in Neural Information Processing Systems, pp. 2787–2795 (2013)
4. Zhang, Y., Dai, H., Kozareva, Z., Smola, A.J., Song, L.: Variational reasoning for question answering with knowledge graph. In: Thirty-Second AAAI Conference on Artificial Intelligence, April 2018
5. Nivre, J., De Marneffe, M. C., Ginter, F., Goldberg, Y., Hajic, J., Manning, C. D., McDonald, R., Petrov, S., Pyysalo, S., Silveira, N., Tsarfaty, R.: Universal dependencies v1: a multilingual treebank collection. In: Proceedings of the Tenth International Conference on Language Resources and Evaluation (LREC 2016), pp. 1659–1666, May 2016
6. Dobrov, B.V., Lukashevich, N.V.: Thesaurus RuTez as a resource for solving problems of information retrieval. In: Proceedings of the All-Russian Conference of Knowledge-Ontology-Theory (UMBRELLA 2009), Novosibirsk, vol. 10 (2009)
7. Lukashevich, N.V., Lashevich, G.E.: RuWordNet thesaurus: structure and current state. In: Knowledge-Ontology-Theory (UMBRELLA 2017), pp. 48–57 (2017)
8. Miller, G.A.: WordNet: a lexical database for English. Commun. ACM **38**(11), 39–41 (1995)
9. Pawar, A., Mago, V.: Calculating the similarity between words and sentences using a lexical database and corpus statistics. arXiv preprint arXiv:1802.05667 (2018)
10. Rubenstein, H., Goodenough, J.B.: Contextual correlates of synonymy. Commun. ACM **8**(10), 627–633 (1965)
11. Devlin, J., Chang, M.W., Lee, K., Toutanova, K.: BERT: pre-training of deep bidirectional transformers for language understanding. arXiv preprint arXiv:1810.04805 (2018)
12. Carvalho, V.R., Cohen, W.W.: Learning to extract signature and reply lines from email. In: Proceedings of the Conference on Email and Anti-Spam, vol. 2004, July 2004
13. Romanov, A., Volchek, D., Chirkin, A., Mouromtsev, D., Sender, A., Dergachev, A.: Implementing a natural language processing approach for an online exercise in urban design. In: Piotrowski's Readings in Language Engineering and Applied Linguistics, pp. 139–154 (2018)
14. Burtsev, M., Seliverstov, A., Airapetyan, R., Arkhipov, M., Baymurzina, D., Bushkov, N., Gureenkova, O., Khakhulin, T., Kuratov, Y., Kuznetsov, D., Litinsky, A.: DeepPavlov: open-source library for dialogue systems. In: Proceedings of ACL 2018, System Demonstrations, pp. 122–127, July 2018

Ontology Learning Approach Based on Analysis of the Context and Metadata of a Weakly Structured Content

Dmitry Volchek(✉), Aleksei Romanov, and Dmitry Mouromtsev

ITMO University, St. Petersburg, Russia
dvolchek@yandex.ru, romanov@itmo.ru, d.muromtsev@gmail.com

Abstract. This article describes ontology learning approach based on the analysis of metadata and the context of weakly structured content. Today, there is a paradigm shift in ontological engineering. It consists of the transition from manual to automatic or semi-automatic design. This approach is called ontology learning. When an author creates a document, one holds in one's head a model of a certain subject area. Then, analyzing the document, it is possible to restore the model of this subject area. This process is called reverse engineering. Current articles describe ontology learning approaches based on content analysis. We propose to use not only the content, but, if it is possible, its metadata and the context for ontology learning purposes. As the main results of the work, we can introduce the model for the joint presentation of content and its metadata in a content management system. To extract the terms, the ensemble method was used, combining the algorithms for extracting terms both with and without contrast corpus. Metadata was used to expand candidates attribute space. In addition, methods for constructing taxonomic relations based on the vector representation of words and non-taxonomic relations by analyzing universal dependencies are described.

Keywords: Ontology learning · MOOC · Online courses · Semantic technologies · Ontology · Embeddings

1 Introduction

Over a fairly long period and to this day, there has been a rapid increase in information volumes. This is good on the one hand, as it allows humanity to expand its horizons and learn more about the world around us. To use knowledge for our own benefit. On the other hand, a lot of information leads to a number of significant problems. The greater the amount of accumulated information and knowledge, the more difficult it is to structure, integrate, update this data. It is worth mentioning the problem of information search in a huge number of different sources. That's why technological companies such as Google, Facebook and others pay so much attention to their search services, creating new algorithms to increase the relevance of search results. One of the latest innovations is the idea

Á. Rocha et al. (Eds.): WorldCIST 2020, AISC 1159, pp. 271–280, 2020.
https://doi.org/10.1007/978-3-030-45688-7_28

of personalised search query results. For example two different users get different information as a result of the same query, based on already known information and preferences of the exact user.

Another direction is the precise answers to user's questions, when the search engine outputs as a result not only a list of documents or web pages, but a direct answer to the request. Such an opportunity is provided by the technology of related data, when, in addition to the information itself, its semantic meaning is known. In this case, not only people can work with the information presented in this form, but also machines to provide information automatic processing. To implement the idea of related data, structures are used called knowledge graphs – these are large information knowledge bases that are directed graphs, where the nodes are some entities and the edges are the relationships between these entities. A significant difference between knowledge graphs and classical graphs in mathematics is that the edges of the knowledge graph can characterise various types of relations between entities. To build and populate the knowledge graph, you must first create its structure, which is called the ontological model.

Nowadays, there are a fairly large number of such models, and the scope of their application varies from education and logistics, to the design of smart manufacturing, business processes, search services and so on. Thus, the task of creating or expanding existing ontological models for various subject areas is of current interest.

1.1 Motivation

Ontological modelling is a complex process that requires a serious range of competences, and technological resources. Various methods, related problems and ways to solve them in the context of ontology development are the objects of many studies.

In particular, the article [1] describes various theoretical aspects of ontological modelling, including the classification of ontologies, principles of construction, and the possibility of automatic creation. The authors of research [2] put forward the assumption that the specificity of each individual subject area is prevailing in the ontology design process, which entails the development of separate design approaches in each individual case. In [3], the authors also note the need to use various approaches in the process of ontological modelling of a particular domain in order to take into account all its features and to create an ontology that will directly take into account all the features of the domain. In the last decade, there has been a significant shift in the paradigm of ontology design. In the works of 15–20 years ago [4] and [5] the key research subject and protagonist was an ontologist. Two different ontologists can create different ontological models of the same subject area because of the individuality, experience, psychology and other aspects. Thus earlier works consider the methodology of ontology design to solve such problems.

Later works [6] prefer to focus on the data of the subject area (Fig. 1). This approach allows to automate the process of ontology design. A person acts as an expert of the subject area, validates and adjusts the process, if necessary.

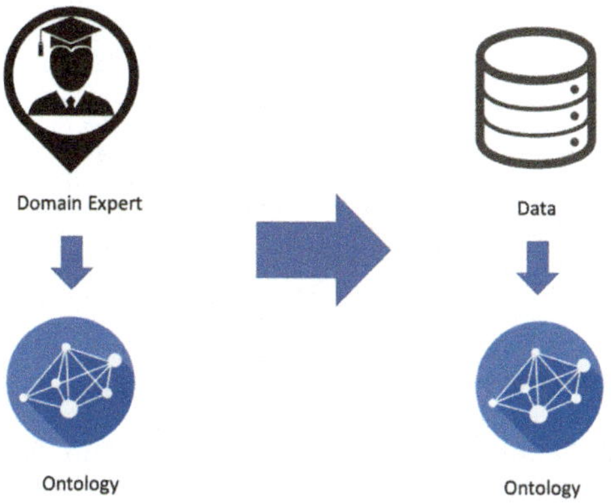

Fig. 1. Ontology engineering process transformation

1.2 Ontology Learning

The approach when the ontological model is built not by an expert in the subject area, but directly on the basis of the available data, is called (on analogy with machine learning) ontology learning. The main idea is to build an ontological model in a semi-automatic mode, and the role of an expert is to validate the results obtained, design learning algorithms, select hyper parameters, markup data sets, and so on.

Moreover, the data on the simulated domain that are presented in a structured form are simply mapped on the ontological model. The situation is quite different with weakly structured content, but such data on the simulated subject area is usually the predominant amount. In practice, these are usually text documents storing information about the subject area itself. These can be regulations, protocols, technical documentation, project documentation, descriptions of entities or processes, and so on. Therefore, to extract the ontological structure from such text documents, it is necessary to use natural language processing (NLP) methods. From the point of view of the general ontology learning algorithm, the following sequence of steps, which is called the "layer cake" [6], is considered to be a standard, it is worth noting that, with small variations, such a scheme is also suitable for creating the ontological model directly by the subject domain expert (Fig. 2).

Since learning is based on text documents, it is first necessary to extract keywords or terms that would describe the essence of the simulated subject area. The next step is the search for synonyms in order to eliminate the duality of interpretation of a particular object. After the establishment of all semantic similarities, the terms of the subject area become the concepts of this area.

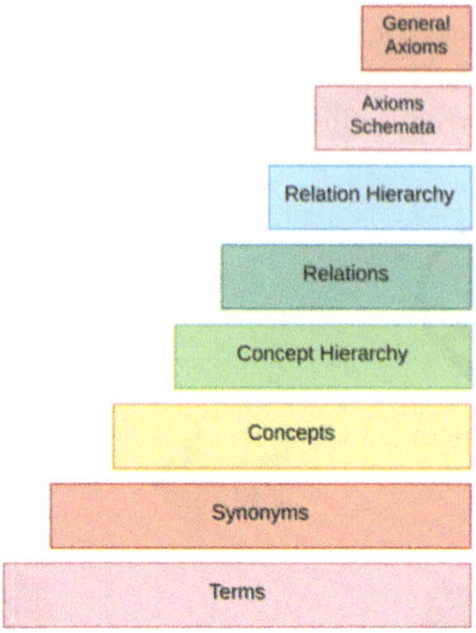

Fig. 2. Ontology learning layer cake (picture from [6])

Next, the taxonomic relationships between the extracted concepts are constructed on the basis of the Class - Subclass principle. In addition to the hierarchy of concepts, it is necessary to distinguish non-taxonomic relations, which introduce the main semantic completeness of the created ontological model. After validation of the obtained relations, the domain expert forms axioms that complete the ontological model. It is worth mentioning that at the final stage it is possible to add rules, restrictions, inference and so on.

2 Method

Ontology learning is based on data about a simulated subject area, and often this data is stored in the so-called content management systems (CMS). They can be Internet portals, business process support systems, online education platforms, and so on. At the same time, such systems store a huge amount of metadata (the number of views of the document, the user's time with the document, the number of users repeated views of a particular document, and many others). This metadata can indirectly characterise the importance of the terms contained in this document from the point of view of modelled subject area.

2.1 Concept Extraction

Existing term extraction algorithms do not take into account metadata about the document from which the extraction of terms takes place. On the other hand, context is also important. If a term occurs not only in the document itself, used in ontology learning, but also in some external sources, then it is more likely to be a concept of the subject area under consideration.

Existing methods for extracting terms from a text in a natural language operate on the so-called concept of "Termhood" of a particular candidate for terms.

By termhood is meant how much this or that candidate for terms describes the subject area under consideration. Globally, algorithms for extracting candidates for terms can be divided into two classes:

- performing extraction directly from the document;
- using a contrasting corpus, that is, a set of documents that are not specific to the modelled subject area. Then, if the candidate is found both in the document under consideration and in the contrast corpus, its termhood is reduced (it is assumed that this is a general term that is not specific to the subject area under consideration).

In this paper, we propose an approach that allows one to take into account metadata for constructing a classification model (whether the candidate in terms is a term of a simulated domain). The formation of the candidate attribute vector is shown in Fig. 3.

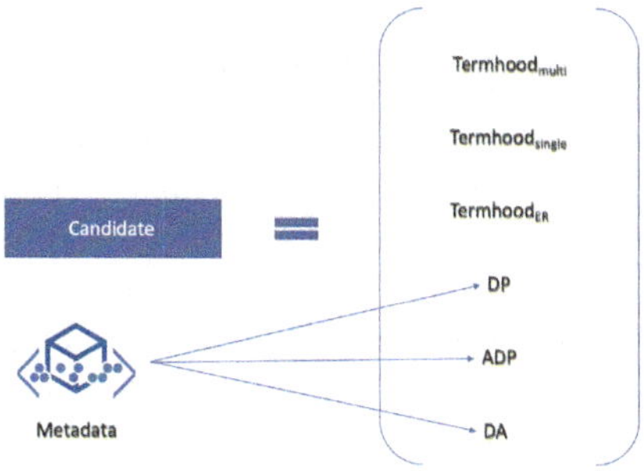

Fig. 3. Candidate for terms vector

The first 3 attributes characterise the termhood of the candidate for terms according to extraction algorithms. Another attributes represent metadata of the document, used to extract the candidate.

- $Termhood_{multi}$ – the termhood of the candidate as a result of applying algorithms based on the use of a contrasting corpus;
- $Termhood_{single}$ – terminology of the candidate as a result of applying algorithms that do not use a contrasting corpus;
- $Termhood_{ER}$ – the number of external sources that contain the candidate;
- DP – the number of views of the document from which the candidate was extracted;
- ADP – the relative number of views of the document;
- DA – time since the last document update.

For the classification algorithm to work correctly, it is necessary to normalize the attribute values:

$$x' = \frac{x - x_{\min}}{x_{\max} - x_{\min}}, \quad (1)$$

where x' is the normalized attribute x, $x_{\min}$ and $x_{\max}$ – min and max values of x attributes of all objects respectively.

2.2 Taxonomic Relations Extraction

The most popular are the following algorithms for constructing taxonomic relations: Term subsumption and analysis of formal concepts (FCA) [7]. Both of these approaches are based on statistical techniques, which gives a significant advantage in terms of domain and language independence. However, it is not always possible to achieve an acceptable level of accuracy in building relationships. In this paper, it is proposed to use a vector representation of words and subsequent hierarchical clustering in order to extract taxonomic relations.

It is proposed to use Word2Vec technology [8] for the vector representation of words. This technology compares favorably with similar ones, in the sense that when performing operations on vectors, meaningful constructions can be obtained. For example, the authors of [8] themselves describe a situation where, when the vectors "Russia" and "River" are added together, a vector extremely close to the "Volga" vector will be obtained.

2.3 Non-taxonomic Relations Extraction

To extract non-taxonomic relations, it is proposed to use a universal dependencies analysis mechanism. To do so, it's necessary to analyze each sentence and build a so-called dependency tree. An example of such a tree is presented in Fig. 4.

As a result of construction, we can consider the resulting dependency tree as a graph linking individual words in a sentence. Further, for analysis, the concepts extracted at the first stage are used, namely, various pairs of concepts are considered and their joint use in the sentence is revealed. Under consideration are sentences containing only pairs of concepts. There are several situations for each such sentence:

- the absence of any relationship between the concepts;
- only one concept has a connection;
- the connection is determined between the two concepts.

If it is possible to determine a connection between the two concepts, then there is a triple "Object - Predicate - Subject", which is added to the list of axioms.

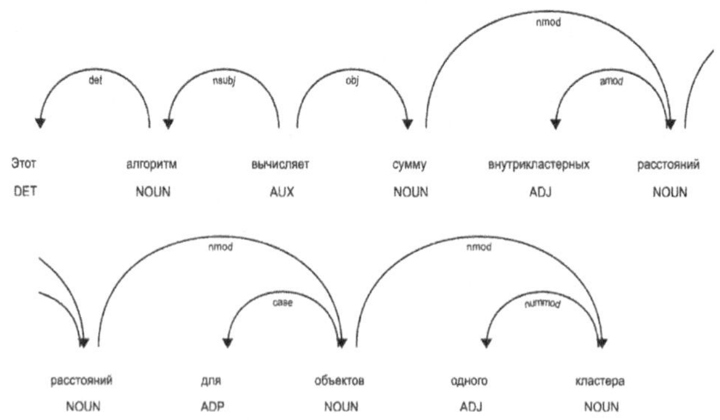

Fig. 4. UD Tree (English translation: This algorithm computes the sum of intracluster distances for objects of a specific cluster.)

3 Results

As the data for the experiment, massive open online courses of the Digital Culture block of disciplines at ITMO University were used. Courses are hosted on the Open EdX platform, which allows you to extract not only content, but also metadata for later use in ontology learning.

For the joint presentation of content and metadata, a basic ontological model was developed, as shown in Fig. 5.

As the next step, CMS data was mapped to the developed basic ontology. To do this, we extracted and annotated data from Open EdX in accordance with the basic ontology. As a result, a knowledge graph was obtained that stores information not only about the structure of the platform, the courses located on it, but also the metadata for each individual element of the course.

The results of mapping are presented in the Table 1.

Table 1. Data mapping.

Course	Sections	Subsections	Units	Documents	Users
Introduction to DC	18	53	424	23	1407
Data storage and processing	6	14	86	13	2775
Introduction to ML	5	13	91	13	2358
Advanced ML	5	12	84	12	2358
Statistics	12	37	222	20	1119
Applied AI	6	16	70	9	2358
Advanced data storage	7	25	145	12	1213

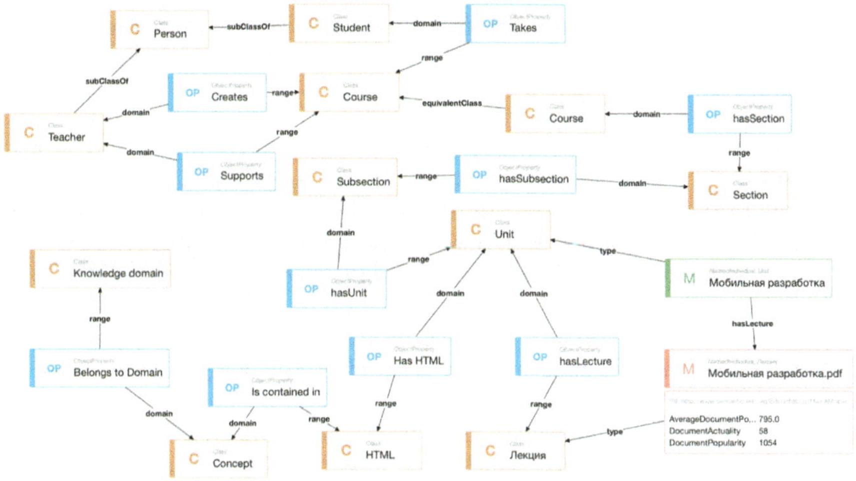

Fig. 5. Basic ontological model

To train the classification model, we selected 20 documents containing 437 terms. Each document is a lecture (subsection) of a specific course and consist of about 30k characters (4k tokens) on average. As a test sample, 5 documents containing 143 candidates for terms were used. To solve the classification problem, several algorithms were considered. The results are shown in Table 2.

Table 2. Candidates classification

Method	Precision	Recall	f_1
LogReg	0.56	0.52	0.54
Naive Bayes	0.52	0.45	0.48
Decision Tree	0.52	0.54	0.53
KNN	0.51	0.49	0.49
SVM	**0.63**	**0.55**	**0.59**

Candidates for terms extracting example is presented in the Fig. 6.

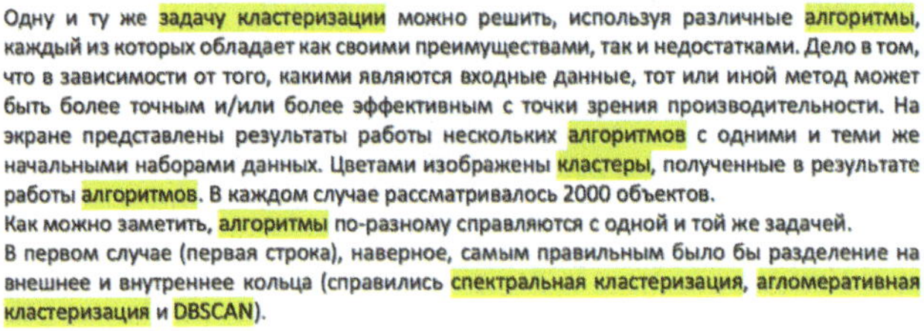
Одну и ту же задачу кластеризации можно решить, используя различные алгоритмы, каждый из которых обладает как своими преимуществами, так и недостатками. Дело в том, что в зависимости от того, какими являются входные данные, тот или иной метод может быть более точным и/или более эффективным с точки зрения производительности. На экране представлены результаты работы нескольких алгоритмов с одними и теми же начальными наборами данных. Цветами изображены кластеры, полученные в результате работы алгоритмов. В каждом случае рассматривалось 2000 объектов.
Как можно заметить, алгоритмы по-разному справляются с одной и той же задачей.
В первом случае (первая строка), наверное, самым правильным было бы разделение на внешнее и внутреннее кольца (справились спектральная кластеризация, агломеративная кластеризация и DBSCAN).

Fig. 6. Candidate extraction. English translation of key words: Clustering task, algorithms, clusters, spectral clustering, agglomerative clustering, DBSCAN.

It is possible to see that the algorithm works mostly correct and provides valid results.

To build taxonomic relations, we trained the model for the vector representation of the words Word2Vec skipgram (space dimension: 300, window width 10), after which hierarchical clustering was performed. The result (english translation) is shown in Fig. 7.

Number of clusters should be chosen by the ontologist, but it is possible to see that concepts on the left are related to the clustering algorithms and concepts on the right are related to distance.

To build non-taxonomic relations, a model was trained based on the analysis of universal dependencies, which allows determining relations between two concepts within the sentence. An example is shown in Fig. 8.

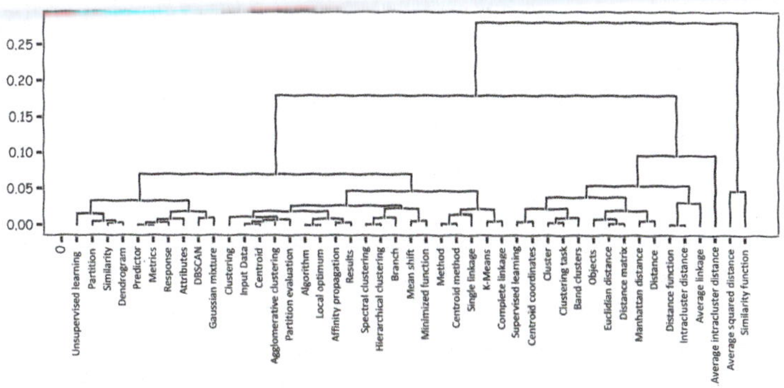

Fig. 7. Taxonimic relations

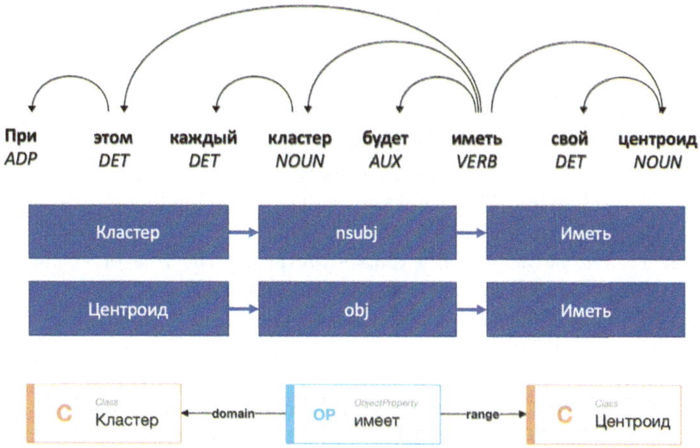

Fig. 8. Non taxonomic relations (English translation: Each cluster will have its centroid. **Cluster – has – Centroid**)

To evaluate the obtained ontological model, an application-based approach was used. It can be concluded that, with the relatively low involvement of the domain expert, the developed ontology performs its functions. As an application, we used an extension for the Open EdX platform, called XBlock, to provide the opportunity of individualization of the education.

4 Conclusions

Thus, the problems stated in the article were solved making it possible to prove the effectiveness of the developed algorithms and method in the experimental study.

The tasks associated with developing ontologies, extracting data from Open EdX, and automatically extracting concepts using NLP algorithms and classification were completed. In addition, a method was developed for constructing taxonomic relations based on a vector representation of words followed by hierarchical clustering, as well as a method for extracting non-taxonomic relations by analyzing universal dependencies.

As a possible continuation of the work, we can consider the task of feature selection (from both termhood indicators and metadata involved), a more detailed study of the vector representation of words and the use of vector operations to represent terms consisting of several words. Moreover, additional training is possible for a model for constructing non-taxonomic relations, which allows us to extract relations not only between two concepts in one sentence, but also if there is a more complex relation connecting 3 or more concepts.

References

1. Konstantinova, N., Mitrofanova, O.: Ontology as a knowledge storage system. Portal, Information and Communication Technologies in Education. http://www.ict.edu.ru/ft/005706/68352e2-st08.pdf. Accessed 21 Mar 2018
2. Gavrilova, T., Gulyakina, N.: Visual knowledge processing techniques: a brief review. Sci. Tech. Inf. Process. **38**(6), 403–408 (2011)
3. Gavrilova, T., Gladkova, M.: Big data structuring: the role of visual models and ontologies. Procedia Comput. Sci. **31**, 336–343 (2014)
4. Gavrilova, T., Khoroshevsky, V.: Knowledge Bases of Intelligent Systems. Piter, Sankt-Petersburg (2000)
5. Chastikov, A., Gavrilova, T., Belov, D.: Development of expert systems. CLIPS Environment, vol. 608. BHV-Petersburg, Saint-Petersburg (2003)
6. Asim, M.N., Wasim, M., Khan, M.U.G., Mahmood, W., Abbasi, H.M.: A survey of ontology learning techniques and applications. Database **2018**, 1–24 (2018)
7. Ganter, B., Wille, R.: Formal Concept Analysis: Mathematical Foundations. Springer, Berlin (2012)
8. Mikolov, T., Sutskever, I., Chen, K., Corrado, G.S., Dean, J.: Distributed representations of words and phrases and their compositionality. In: Advances in neural information processing systems, pp. 3111–3119 (2013)

Paving the Way for IT Governance in the Public Sector

Veronica Telino[1], Ricardo Massa[1], Ioná Mota[1], Alex Sandro[1], and Fernando Moreira[2](✉)

[1] Federal University of Pernambuco, Recife, PE 50.740-560, Brazil
[2] REMIT, IJP, Universidade Portucalense, Porto & IEETA, Universidade de Aveiro, Aveiro, Portugal
fmoreira@upt.pt

Abstract. Public sector organizations often have information technology governance proposals to implement or solutions that are not enough to solve their information technology problems. This instability wastes technology investments, adds no value, and affects the performance of industry organizations. In this study, our main goal is to identify aspects of the environments that facilitate implementation and ensure the continuity of the governance process in organizations, based on people's behavior, as the human factor is critical to ensuring that the organization's goals are met. We conducted a case study through the application of interviews and questionnaires to eight managers of a Brazilian public university. We collect data from the perspective of the Information Technology and Innovation Diffusion Theory and the Technology Acceptance Model. The results point that aspects related to manager behavior are involvement, communication, reality, training, and diversity. These aspects correspond to the basis of our proposal. The setting of the environment is the product of the arrangement of these aspects, the recommendations of the literature and the interpretation of the dimensions of the questionnaire. To make the organizational environment conducive to governance, we must involve managers and their teams in change. Together they can pave the way, break paradigms, and turn governance into a habit.

Keywords: IT governance · Human capital · Organizational change

1 Introduction

Information Technology (IT) can be a success factor for government agencies [1]. Such technologies may assure that financial and operational organizations' information is precise, reliable, available and up to date. However, in the public sector, the inefficient usage of IT is persistent and causes waste of money invested in such technologies. Moreover, their aggregated value is not assured and the performance of organizations of this segment is often compromised, thus weakening public governance. Hence, decisions pertaining to IT need to be well defined, managed and supervised by the high-management and not only by the technology area. IT governance may help in this sense since it is a management instrument.

Á. Rocha et al. (Eds.): WorldCIST 2020, AISC 1159, pp. 281–290, 2020.
https://doi.org/10.1007/978-3-030-45688-7_29

Sometimes, organizations count with proposals that have not yet been implemented or with solutions that are not enough to solve the IT governance problem. This paper intends to fill this gap. Bearing in mind that human factors needed to be taken into account when solving the problem of the inefficient usage of IT, we have collected information about the use of IT from public managers through a case study.

People seams to usually resist change in their work environments. This resistance to change within organizations is considered natural [2]. Resistance to change is a phenomenon which can be considered as organizational risk [3]. However, to evolve in this sense, it is necessary to engage people in such changes and to break paradigms.

Our goal is to present a set of environmental criteria that is conducive to governance improvements in the public sector, based on the opinion and behavior of managers that decide about these technologies in their daily activities. The proposal may accelerate governance acceptance and stability. In this light, the research question we are trying to answer is "*How do we configure an environment conducive to governance with the help of people's opinions?*"

This paper consists of the introduction, background and related work, method, analysis and synthesis of the findings, results, conclusions and further work.

2 Background

2.1 IT Governance as an Organizational Innovation

Governance and organizational innovation are two themes that can exist simultaneously and cooperatively in organizations. In order to see governance as an innovative organizational process we introduce a few definitions and establish correspondence relationships between them.

Governance is indispensable for IT to work transparently as far as executives, administration boards, shareholders and other stakeholders in its environment. Its goal is to improve the plans to govern IT in public administration, as argued by [4]. It can be thought of an action requested by higher governing bodies and auditing units and internal and external public bodies control that allows them to define an integrated IT policy, a decision-making system about these resources, the usage of better management practices as well as fulfilling the demands from regulating bodies.

Innovation is something new or that the user has not yet seen. It is the successful introduction of products, services, processes, methods and systems that did not exist in the market or that contain some new feature (or at least different) from what was being used [5].

There is organizational innovation. It is the implementation of new management techniques, business practices or significant changes in work as well as in external relations of the organization, with the aim of improving knowledge use, workflow efficiency or the quality of assets and services [6, 7]. The differential of organizational innovation lies in the fact that it is at the forefront of the organization and that it results from management strategic decisions [7]. These works together with Schumpeter[1] about the

[1] Joseph Alois Schumpeter, Austrian economist and political scientist, is considered to be one of the most important economists of the first half of the 20th century. He was one of the first to consider technological innovation as a catalyst of capitalist development.

novel ways of organizing companies allows us to identify some points of intersection with governance: (i) they may be considered new management practices or techniques; (ii) the possibility of implementing such practices is a direct result from decisions made by the organizations strategic sectors; (iii) they may cause significant changes in the organization of IT related work; (iv) they may improve financial gains; (v) they promote quality gains in products and services offered by the organization.

In addition to the process of change through innovation, researchers refer to digital transformation, as well as a process that involves change and consideration at various levels of the organization. They warn that organizations need to change their internal structure to remain competitive in the marketplace [8]. Researchers believe that managers must change their business strategy to meet the new digital reality, which implies adaptations of process and operations management [9]. Both show the innovation and the need for ownership of digital transformation as conditions to overcoming challenges and not missing opportunities in organizations.

In this way, the usage of IT and its governance may be considered as organizational innovation and digital transformation. The purpose of this observation is to study behavior and to assess peoples' point of views regarding the inclination to practice IT governance.

Among the studies on digital innovation and its implications for the organization, we can mention [10], which refers to digital innovation in terms of new products and services offered by organizations and their value creation and [11] that investigating technological innovation, its processes and results.

2.2 IT Governance Based on the IT and Innovation Diffusion Theory

To consider governance as innovation in IT usage allows us to use the IT and Innovation Diffusion Theory (IT and IDT) as well as Technology Acceptance Model (TAM) adaptations in both questionnaire elaboration and for data analysis to estimate innovation capability and the adoption of IT governance by organizations. The theory provides concepts and empirical results that can be applied to the technology assessment study, its adoption and implementation. The theory is widely applied in IT. The works of classical diffusion [5] and of IT [12] are fundamental for our assumptions, although there are other studies devoted to the diffusion of innovation [13, 14].

Innovation or new technologies diffusion is a social phenomenon through which innovation is disseminated by the individuals that can potentially adopt them – innovation users. Its adoption depends on the innovation itself, on its adopters, on the communication channels through which it will be disseminated, on the time and environment that may or may not stimulate its adoption. Without dissemination, innovation has no impact. The adopters are managers that make IT-related decisions in the organization. The organizational innovation proposed in our research follows the scope of the dominant paradigm proposed by [15], in terms of investigating how potential adopters we have at the University, the process of assimilation by the likely adopting managers and promoting the diffusion of governance within them. Researchers should value the human factor in the process.

According to the IT and IDT, for someone to make the decision to adopt governance, they need to be persuaded to get to know it, what may take some time. In order to speed things along and ensure the implementation progress, the initial adopters must be subsidized. The study on innovation behavior may inform the organization's high management and avoid future governance discontinuity problems.

Using application of the IT and IDT, we have researched IT governance in the organizational context, and we observe it to be a complex innovation due to the difficulties that public organization face to evolve in this process. Difficulties might be related to peoples' knowledge and beliefs barriers. Understand such barriers is important to adopt governance, since the innovation capacity of an organization is determined by its ability to recognize information value, assimilate information about it and to apply them for productive goals. Knowledge barriers must be reduced [12]. This approach defined a set of variables that characterize such technologies. Innovation diffusion theory is the most popular theory used by researchers [16]. Researchers claim that the extraction and knowledge of the variables defined by used theories should be considered, since they represent important features of the innovation, adopters' profile, implementation and strategies used by the organizations. Most of them correspond to items found in the questionnaire, since they are relevant to or adaptable to the research. Such items are organized according to the dimension they investigate, in order to make results interpretation easier.

2.3 Technology Acceptance Model

The TAM is largely used in IT acceptance research [17]. There are other versions of the TAM studied in the literature. We adapted for our study the final version of TAM [18]. The model speculates that people's attitudes with respect to adopting new technologies is a function of their beliefs about perceived utility and the perception of the ease of use of these new technologies. Perceived utility means how much an individual believes that, by using some technology, their productivity will increase. Perceived ease of use is how much an individual believes that using some technology does not require effort on their part.

Since these two variables are key to the acceptance process of new technologies, it is our belief that an adaptation of TAM allows us to treat as well as measure the acceptance of the use of IT in the organization, taking into consideration that new technologies and technological innovations have the same meaning and IT governance is assumed to be innovation. Technological Innovation is a term referring to processes and products innovations.

Besides these two variables, TAM suggests that when the potential user is presented to new technology, various factors interfere in their decision about when and how to use it. We have adapted the variables defined in TAM for the questionnaire, with the aim at knowing the manager's behavior with respect to accepting and adopting governance.

We have not applied the model since it is out of the scope of the research and its specifications prescribe causal relations between its variables. Our study was developed according to a constructivist perspective.

2.4 Related Work

Some studies related directly to this research are [4, 19–24]. The studies investigate, discussing the application, define plans, and point out problems or develop models related to public sector IT governance problems.

3 Method

We have carried out a case study in a Federal University of Higher Education in Brazil founded in the 1940s. This is a reference university in the country.

In the search for more complete results, we adopted the mixed method in our approach. We used a qualitative source of data collection by performing semi-structured interviews and a quantitative source by applying questionnaires. This methodological alternative leads to the combination of aspects of both qualitative and quantitative research. Thus, the compensatory objectives regarding the limitations and weaknesses of each approach complement each other, producing better outcomes [25]. We used the semi-structured interview protocol. Its scripts include a combination of more or less structured, flexible questions required from all participants and without a predefined order [26]. The eight managers with different profiles invited to participate in this study were chosen due to their influence in IT-related decisions. The interviews were conducted face-to-face, in the period from February 22 to June 13, 2017. Before data collection, we validated the collection instrument through a pilot study. During the interview meetings, managers discoursed freely about each question, could discuss other topics, and authorized the researchers to tape their audios. We omit the names of the interviewees. Then, two researchers reviewed in parity the five and a half hours of the speech audios to transcribe it into a structured document. In the next step, the document was analyzed, codified, and categorized into aspects that represent strengths and weaknesses associated with the IT evolution.

In the closed questions of the questionnaire, the participants answered by using a four-point Likert scale. We have chosen to use a four-point scale, to ensure easiness, speed and precision in answers. We have not used the neutral point to force respondents to take one side or another.

3.1 Analysis and Synthesis of the Findings

We could identify thirty-eight aspects related to IT usage during the interviews. Some of those help foster IT and its governance at the University, whereas others compromise this advance. We extracted five aspects that are influenced by the managers' behavior according to the existing literature as well as to the interpretation of the results of each dimension we have investigated. Interventions in such aspects may cause the organizational environment more receptive to governance. The aspects will be presented below.

Dimensions Investigated. Managers' behavior is investigated according to the following six dimensions: (i) managers' *perceived usefulness* of IT governance: translates the manager's perception of how useful IT governance practices are for their work; (ii) managers' *perceived ease of use* or *ease-use* of IT governance practices: translates how easy managers find to use IT governance practices; (iii) managers' awareness of the level of adhesion to governance *practices* currently adopted in the *organization*: translates how much the manager believes that the organization is on the right path with respect to currently existing IT governance practices; (iv) managers' awareness about the correctness of their actions with regards to IT governance *practices* adopted in the *unit* where they currently work: translates how aware managers are about the correctness of their actions with regards to IT governance practices adopted in the unit where they currently work; (v) the degree of *satisfaction* with IT in the manager's *unit*: translates the degree of satisfaction with IT in the manager's unit; (vi) how quickly the *manager* is *innovator* and is able to adopt new technologies: translates how quickly the manager is able to adopt new technologies, that is, investigates the personal characteristics of those involved in order to find out how long they take to adhere to an innovation. In summary, data show whether the manager finds IT governance practices useful and easy to use; whether they believe that the organization is on the right path; whether they use practices correctly in their unit; whether their unit is currently satisfied with IT and whether they are innovators. The following Table 1 presents the descriptive statistics and Cronbach's alpha (α) for each dimension. Part of the information was taken from IBM SPSS statistical package output, version 22.0.

Table 1. Summary of statistics by dimension.

Dimension	Median	Mode	Minimum extreme	Maximum extreme	α	α based on items	N.SPSS output
Perceived usefulness	1	1	1	3	0.971	0.968	15
Perceived ease of use or ease-use	2	2	1	4	0.684	0.766	14
Practices-organization	2	2	1	4	0.706	0.761	12
Practices-unit	2	2	1	4	0.887	0.882	17
Satisfaction-unit	2	2	1	3	0.823	0.856	5
Manager-innovator	2	2	1	4	0.791	0.797	10

4 Results

4.1 Environment Configuration for Governance

The elements present in the proposed configuration include the investigated dimensions, the literature's recommendations, and five aspects of interest surveyed as well as how they connect, according to the respondent's opinions. The following Fig. 1 presents the configuration named *drICT*, an acronym of the selected aspects diversity, reality, involvement, communication, and training that should be done prior to the governance implementation in the organizational environment.

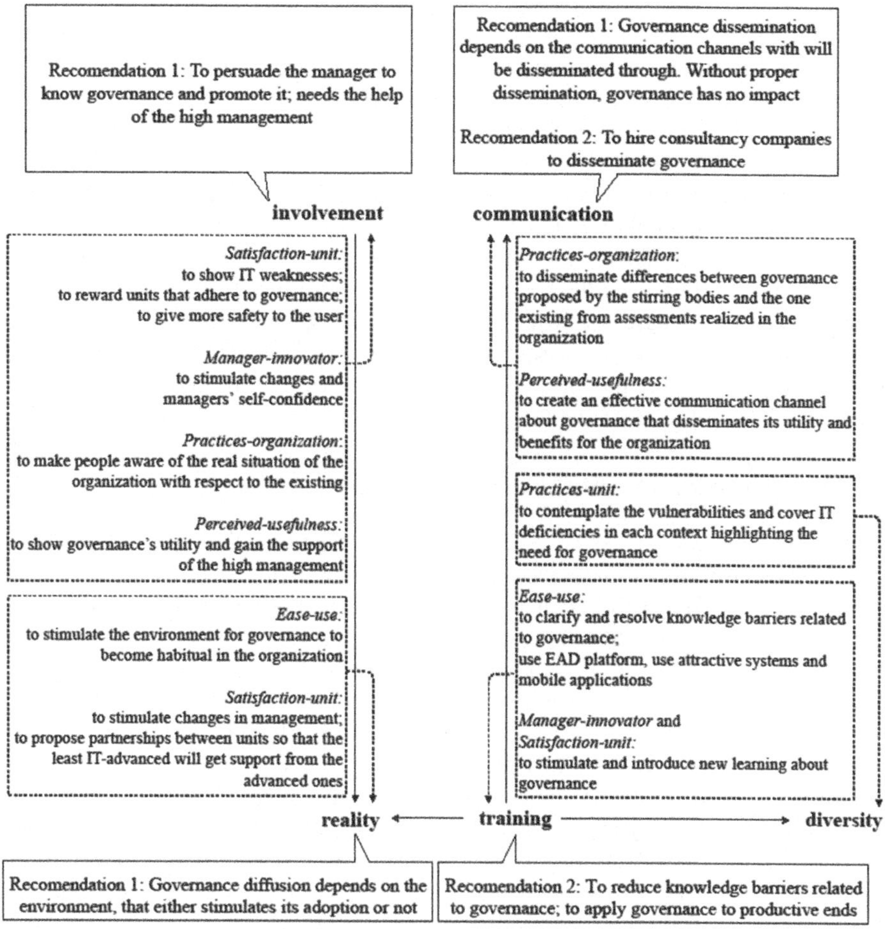

Fig. 1. *drICT* configuration

The configuration does not replace planning or organizational governance projects. Its creation aims at accelerating this process' implementation and ensuring continuity.

The *drICT* configuration shows the interconnectedness of the aspects considered according to interviewees. Connection may help the organization management and define starting points for the environment preparation work. Aspects may change according to the organization and the researchers' perception if different strategies are adopted. Interconnectedness allows us to observe that training improves reality, communication and diversity, and involvement improves reality. This means that one-off advances resonate with the whole.

The introduction of the recommendations 1 and 2, adapted from [5, 12], in the configuration scheme is intended to show that our findings are contextualized in relation to the theories used.

If the organization decides to start change with aspect reality, it will act in the Ease-use and Satisfaction-unit dimensions. Ease-use is established if governance becomes customary in the organization's environment. Satisfaction-unit can be guaranteed with changes in management and partnerships between units. The suggestion is for less advanced IT units to be supported by more advanced units. Support makes it easy to use and easy use can become a habit.

If the organization starts with training, the action will be in the Ease-use, Manager-innovator and Satisfaction-unit dimensions. As training interferes with reality, communication and diversity, the organizational environment will benefit to a greater extent. However, even in this case, the evolution towards stable governance can happen at a slow pace. This is because we are dealing with change of habit, a non-trivial process for most people. We see that each aspect has its dimensions and recommendations associated, except for the diversity that was recommended to be considered in *drICT* only the qualitative findings of our research.

5 Final Conclusions and Further Work

In this paper, we have presented a step-by-step approach to making the organizational environment favorable to the adhesion and continuity of IT governance through the survey of the opinions of public managers involved with our research problem. Results allow the necessary measures concerning the efficient use of IT to be implemented and the time for this adhesion to be shortened.

We believe that these pre-requirement measures may: (i) make governance a habit; (ii) fulfill the needs of units that are most vulnerable to IT problems; (iii) to improve communication about IT related issues in the organization, as well as, with the units directly responsible by technologies; (iv) to promote a feeling of inclusion in units that feel less benefited by IT management; (v) to promote open IT practices that count with, open, transparent decisions that aim to encompass the whole organization; (vi) to manage individual needs to make it easier to use IT in units where the costs inherent to technologies are different and information maintenance are specific; (vii) to increase investments in IT training plans; (viii) to form multiplier governance units starting from pilot units; (ix) to strengthen and make IT more professional to act as management units; (x) to train more people and more IT managers; (xi) to make people rethink their attitudes with respect to their organizations and leave their comfort zone; (xii) to make users adopt a more daring attitude with respect to IT and strive to achieve success for themselves and for their organization; (xiii) to shorten IT units response time to their users; (xiv) to pay attention to services and more urgent needs; (xv) to involve users in the decision-making process, making them feel like they own the organization IT; (xvi) to make the IT unit management believe in innovation; (xvii) to become more familiar with IT related procedures; (xviii) to show the importance of IT and the value it can add to the organization.

The choice of the aspect to be tackled can be made flexible in order to cater for different management expectations. Configuration is easily applicable and may be replicated in other instances, since besides tackling elements common with other organizations, the step-by-step to its elaboration may be share and serve as reference to organization with similar interests and environments.

As public servants, we want to contribute to the advancement of public governance, desired by courts and auditing bodies in emerging or developing countries, such as Brazil. Suggestions for further work include: (i) investigating how associated pairs of aspects benefit from this association; (ii) proposing new dimensions or expanding the existing ones based on surveys carried out in other organizations; (iii) generating other configurations that promote governance adhesion and continuity; (iv) testing and sharing of the used techniques in new research so that structures about other themes can be created and developed.

References

1. Thomas, M.A., Cipolla, J., Lambert, B., Carter, L.: Data management maturity assessment of public sector agencies. Govern. Inf. Q. **36**(4), 101401 (2019)
2. Vrhovec, S.L.R.: Responding to stakeholders' resistance to change in software projects – a literature review. In: 39th MIPRO 2016, pp. 1157–1161. IEEE, Opatija (2016)
3. Vrhovec, S.L.R., Hovelja, T., Vavpotič, D., Krisper, M.: Diagnosing organizational risks in software projects: stakeholder resistance. Int. J. Project Manage. **33**(6), 1262–1273 (2015)
4. Nunes, D., Rosa, I., Da Silva, M.M.: Improving IT strategic plan for the Portuguese public administration. In: International Conference on Business Informatics 2013, vol. 1, pp. 125–130. IEEE, Vienna (2013)
5. Rogers, E.M.: Diffusion of Innovation, 5th edn. Free Press, New York (2003)
6. Banco Nacional do Desenvolvimento. http://www.bndes.gov.br. Accessed 31 May 2016
7. De Carvalho, M.M.: Inovação: estratégias e comunidades de conhecimento. Atlas, São Paulo (2009)
8. Verhoefa, P.C., Broekhuizena, T., Bartb, Y., Bhattacharyaa, A., Donga, J.Q., Fabiana, N., Haenlein, M.: Digital transformation: a multidisciplinary reflection and research agenda. J. Bus. Res. (2019)
9. Reis, J., Amorim, M., Melão, N., Matos, P.: Digital transformation: a literature review and guidelines for future research. In: Rocha, Á., Adeli, H., Reis, L.P., Costanzo, S. (eds.) Trends and Advances in Information Systems and Technologies. WorldCIST 2018. Advances in Intelligent Systems and Computing, vol 745. Springer, Cham (2018)
10. Suseno, Y., Laurell, C., Sick, N.: Assessing value creation in digital innovation ecosystems: a social media analytics approach. J. Strateg. Inf. Syst. **27**, 335–349 (2018)
11. Nambisan, S., Lyytinen, K., Majchrzak, A., Song, M.: Digital innovation management: reinventing innovation management research in a digital world. Manage. Inf. Syst. Q. **41**(1), 223–238 (2017)
12. Fichman, R.G.: Information technology diffusion: a review of empirical research. In: 13th ICIS 1992, vol. 39, pp. 195–206. AIS eLibrary, Dallas (1992)
13. Vega, A., Chiasson, M.: A comprehensive framework to research digital innovation: the joint use of the systems of innovation and critical realism. J. Strateg. Inf. Syst. **28**, 242–256 (2019)
14. Shin, D.: A living lab as socio-technical ecosystem: evaluating the Korean living lab of internet of things. Govern. Inf. Q. **36**(2), 264–275 (2019)
15. Fichman, R.G.: Going beyond the dominant paradigm for information technology innovation research: emerging concepts and methods. J. Assoc. Inf. Syst. **5**(8), 314–355 (2004)
16. Jha, S., Bose, I.: Innovation research in information systems: a commentary on contemporary trends and issues. Inf. Manage. **53**(3), 297–306 (2016)

17. Davis, F.D.: Perceived usefulness, perceived ease of use, and user acceptance of information technology. MIS Q. **13**(3), 319–340 (1989)
18. Lai, P.C.: The literature review of technology adoption models and theories for the novelty technology. J. Inf. Syst. Technol. Manag. **14**(1), 21–38 (2017)
19. Ajayi, B.A., Hussin, H.: IT governance from practitioners' perspective: sharing the experience of a Malaysian University. J. Theor. Appl. Inf. Technol. **88**(2), 219–230 (2016)
20. Mamboko, P., Zhou, M., Tsokota, T., Mhaka, C.: IT governance: status and level of implementation in Zimbabwean urban local authorities. Eur. J. Bus. Manag. **7**(1), 173–179 (2015)
21. Qassimi, N.A., Rusu, L.: IT governance in a public organization in a developing country: a case study of a governmental organization. Procedia Comput. Sci. **64**, 450–456 (2015)
22. Levstek, A., Hovelja, T., Pucihar, A.: IT governance mechanisms and contingency factors: towards an adaptive IT governance model. J. Manag. Inform. Hum. Resour. **51**, 286–310 (2018)
23. Vugec, D.S., Spremić, M., Bach, M.P.: IT governance adoption in banking and insurance sector: longitudinal case study of COBIT use. Int. J. Qual. Res. **11**, 691–716 (2017)
24. Turel, O., Liu, P., Bart, C.: Board-level information technology governance effects on organizational performance: the roles of strategic alignment and authoritarian governance style. Inf. Syst. Manag. **34**, 117–136 (2017)
25. Flick, U.: Introdução à Metodologia de Pesquisa: um guia para iniciantes. Penso, Porto Alegre (2013)
26. Merriam, S.B.: Qualitative Research: A Guide to Design and Implementation. Revised and expanded from Qualitative research and case study applications in education. Jossey-Bass, San Francisco (2009)

ICT and Big Data Adoption in SMEs from Rural Areas: Comparison Between Portugal, Spain and Russia

João Paulo Pereira[1,2(✉)] and Valeriia Ostritsova[3]

[1] Instituto Politécnico de Bragança, Campus de Santa Apolónia, 5300-302 Bragança, Portugal
jprp@ipb.pt

[2] UNIAG (Applied Management Research Unit), Bragança, Portugal

[3] Kuban State Agrarian University, Krasnodar, Russian Federation
valeriyaostricova@mail.ru

Abstract. One of the components of the success of modern economic systems of any level and scale is the use of the latest information and communication technologies (ICT). The use of technology allows the increase in the efficiency of the production process and conduct the business to a completely new level. The main objective of this research is to assess the state of ICT in small and medium-sized enterprises, and the possibility of using Big data technologies for analyzing and making forecasts for improving economic performance.

The research of the use of ICT were carried out on the example of small and medium enterprises (SME) in rural areas of three countries: Alto Trás-os-Montes (Portugal), Castela e Leão (Spain) and Krasnodar (Russia). The goals of the work are to conduct a comparative description of the selected areas, to determine the level of the current state of ICT, to bring proposals for improving this level. The main conclusion is that although the objects of research are at a different level of economic development, they have similar experimental data, and the proposed improvement proposals may be useful for each of them.

Keywords: ICT · Big data · SME · Data mining · Rural areas

1 Introduction

Medium and small businesses, as defined by the European Commission [1], include enterprises with up to 250 employees and a maximum annual turnover of 50 million euros. The importance of small and medium-sized enterprises (SMEs) today is indisputable for both developed and developing countries [2, 3].

Providing millions of jobs, the institute of small and medium enterprises is the primary means of sustainable industrial and social diversification of society, thus acting as one of the main drivers of economic development in most countries [4]. However, phenomena such as globalization, the internationalization of national markets, the global economic crisis, financial market volatility, reduced investment, rapidly changing consumer demand are putting increasing pressure on SMEs, encouraging them to find ways

Á. Rocha et al. (Eds.): WorldCIST 2020, AISC 1159, pp. 291–301, 2020.
https://doi.org/10.1007/978-3-030-45688-7_30

to survive and develop in today's business environment. And one of the ways to improve the competitiveness of enterprises is the use of information and communication technologies. Access to information and communication technologies (ICT) improves business efficiency and the global economy in general. However, today the use of ICT in managing business processes of small and medium-sized enterprises in developing countries is rather moderate [3, 5, 6].

The main purpose of creating and introducing new information technologies is to improve the quality and efficiency of organization management, increase productivity, reduce costs for production and management activities, minimize risks, and so on. Most in-demand are new technologies, that allow to solve the greatest number of similar tasks in the complex. They bring tangible financial returns to the creators. However, should be kept in mind that the creation and introduction of new technologies, products and innovations in the market require investment.

So in other words, the economic essence of innovations lies in the fact that they can bring a large income, but their costs and expenses are required for their creation and distribution. Needless to say, servers, cabling, training - all this is expensive. However, there are some newest information technologies, which are characterized by relatively small amounts of required investments and very high returns. A typical example in modern small and medium businesses shows that the majority of successful enterprises are ready on average to invest in their information technology infrastructure about 1/20 of their working capital, that is, about 5% of revenue per year [4].

This work analyzes the position of the business environment in SMEs in order to determine their economic status and the degree of adoption of ICT and digital skills, as well as their future prospects. The following regions of the studied countries are considered: Alto Trás-os-Montes (Portugal), Castela e Leão (Spain) and Krasnodar region (Russia).

From the demographic point of view, the areas analyzed in this study are characterized by low density, population dispersion, aging and migration of young people. The labor market is characterized by a low level of labor force, a shortage of workers with appropriate qualifications, a low level of entrepreneurship and, as a result, a lack of young people. Thus, ICT is considered necessary to improve the functioning of the company, customer service, as well as to increase sales and enter new markets.

The economic structure of the analyzed areas is characterized by a high share of the agricultural sector, and also by some services (in particular, personal and social). Industry and construction have an average weight, and both areas are very concentrated in some areas. Most entrepreneurs (65.4%) started their activities at the first discovery of the opportunity for this. 74.2% admitted that there were some difficulties to start their activities. the greatest difficulties were the lack of the necessary rules and procedures (43.1%) and the search for the necessary funds (21.7%). The surveyed companies also noted that in order to promote the development of SMEs, it would be necessary to establish subsidies, implement measures to increase the population, create budget assistance and reduce the volume of bureaucratic documentation.

2 ICT for SMEs in Emerging Economies

The adoption of ICT by SMEs provides the ability to quickly access, evaluate, process and distribute large volumes of data and information about business infrastructure [7]. Therefore, only SMEs that use state-of-the-art technology have the opportunity to enter the international market and remain competitive despite the challenges of globalization, liberalization, etc. Other obvious strengths of ICT are [8]: New business model; Access to new markets; New marketing tools, products and communications services; New ways of working relationships; Improved network performance; Teamwork; Automation of the production process; Cost reduction; Improving communication both within the organization and between organizations; and More access to information and data processing.

It is safe to believe that small and medium-sized enterprises are the "engine" of the economy today. For example, in 2017, in the EU, SMEs accounted for 99.8% of all European companies and provided over 90 million people with jobs (over 66% of all jobs) [6, 9].

It is obvious that the life of the SMEs is, on the one hand, a rich field of opportunities, and on the other hand a huge amount of difficulties and "traps". This situation is typical for countries with developed market economies, and for countries with economies in transition. However, in rural areas the difficulties are bigger.

Here are some weaknesses and risks. For example, these are safety risks if proper measures are not followed. Also, do not forget about the reputational risks associated with closer interaction of customers in social networks. Improper use of communication technologies and information can lead to poor performance.

The key reason for the limited use of most new technologies by small and medium-sized businesses is the uncertainty of entrepreneurs in obtaining benefits, their commitment to outdated work organization principles, which, combined with limited resources and high risks in implementing software, impedes business development.

3 Methodology

To collect the data for the Portuguese and Spanish regions we used the results of the COMPETIC project[1] (competic-poctep.com), that made 263 surveys in the Spanish region of Castela e Leão (Leão, Zamora, Salamanca, Valladolid, Ávila) and 170 in the Portuguese regions of Alto Tâmega and Terra de Trás-os-Montes.

Field work was conducted from 24 September to 16 November 2018. Sample design: random sample by area, type of company and sector. Sampling error in Portugal: ±7.52% for global data, 95.5% for a confidence level and estimates of equally likely categories (p = q = 50%). In Spain, the sample size was 263 surveys. Sampling error in Spain: ±6.04% for global data, 95.5% for a confidence level and estimates of equally likely categories (p = q = 50%).

[1] COMPETIC Project - Support to entrepreneurs, self-employed and micro-enterprises in rural areas to create and develop their businesses taking advantage of ICT opportunities (operation 0381_COMPETIC_2_E).

The data from Krasnodar region were collected from the site of the Russian Federal State Statistics Service (Rosstat) and from the statistical compendium of the study «Indicators of the digital economy: 2018». The indicators characterizing research and development in the field of ICT, personnel of the digital economy, the activities of the ICT, content and media sectors, the development of telecommunications were reviewed. Statistical data were used reflecting the demand for digital technologies in the business sector and the social sphere. The data from the scientific and practical manual "Information and communication technologies of the South of Russia" were also reviewed. Materials were taken from the Ministry of Communications and Mass Media, Rosstat, the Ministry of Education and Science of Russia, the Ministry of Culture of Russia, the Bank of Russia, OECD, Eurostat, ITU, the United Nations Department for Economic and Social Development, the World Economic Forum.

4 Results

In Spain, the main goals of using ICT in each area are improving the functioning and improving the quality of customer service. Entering new markets, as well as reducing costs, are of course most important in trade and tourism (62.2% and 43.7%, respectively), the study of new contacts and opportunities is in the service sector (64.7%).

In Portugal, much attention is paid to gaining competitive advantages (46.9%). Also, the introduction of ICT is expected to increase market share and sales (36.3% and 44.2%, respectively).

In Russia, the introduction of ICT is expected to progress in related industries, such as big data, quantum technologies, robotics, artificial intelligence and neurotechnology, as well as the industrial Internet and wireless technologies [10].

In the distribution of areas by activity (Table 1) it is clear that the largest process is occupied by the industry, which includes: the extractive industries, industrial processing, electricity, gas, steam and air conditioning, water supply, sanitation, waste management and decontamination [10, 11].

Table 1. Distribution of areas by type of activity.

	Castela e Leão (Spain)	Alto Trás-os-Montes (Portugal)	Krasnodar region (Russia)
Industry	23.7%	26,5%	23,3%
Construction	7.3%	3,5%	9,6%
Trade	9.1%	9,7%	5,6%
Tourism	18,2%	8,0%	6,7%
Social services and education	3,6%	12,4%	8,6
Agricultural sector	10,9%	9,7%	21%
Business services	12,7%	14,3%	12,1%
Other services	14.5%	15,9%	13,1%

The agricultural sector includes activities such as: agriculture, livestock, forestry and fisheries. Trade includes wholesale and retail trade, car and motorcycle repair, transportation and storage, hospitality, information and communications. Business services are also of great value and include: financial and insurance activities, real estate operations, professional, scientific and technical activities, administrative activities and support services. Other services here include: public administration and defense, compulsory social security, sanitation and social services, arts, entertainment and leisure activities, repair of household products, etc.

4.1 General Use of ICT

All three countries studied have already appreciated the benefits of using ICT and are therefore ready to invest more and more in this area. More than a third of companies determine the level of ICT implementation as low or very low. However, the share of small companies willing to invest in ICT in the next couple of years was 33% in Spain, 53.5% in Portugal and more than 60% in Russia, and this figure grew by 12% over the last year. Of course, these indicators can and should be higher.

Table 2 shows the total national use of ICT in companies with less than 10 employees. Consideration of sectors separately gave the following data: in the agricultural sector, the highest percentage of investment probability (41.3%), it is also high for industry (35.5%) and services (32.2%). However, for the trade sector, in which ICT is used only by 50%, the probability of investment is no more than 28.5%. Nevertheless, based on statistics, it should be emphasized that those companies, that indicated that they have a higher level of ICT, use give an even greater prospect of investing over the next two years.

Speaking about the speed of Internet access, the surveyed companies answered the following. In Spain, a larger percentage has a speed from 2 to 10 Mb/s, less than 2 Mb/s have only 5.6% of companies, and 100 Mb/s - 3.6%. In Portugal, 44.4% of companies have a speed of 100 Mb/s, less than 2 Mb/s only 0.6%, 2.5% up to 10 Mb/s and 4.9% - up to 30 Mb/s. In Russia, 57.5% of companies use speeds from 10 to 100 Mb/s, 16.1% from 100 Mb–1 Gb/s. and 19.5% of companies less than 10 Mb/s. 46.2% of respondents do not know what the speed should be (in accordance with the contract).

With regard to the types of Internet use, the most common are information retrieval and email correspondence (70%), followed by banking, ordering to suppliers, product requests, and only 15.3% for video conferencing. Going deeper into the sector breakdown, video conferencing is more used in agriculture (21.9%) and in services (19.3%). Orders and payments to suppliers are used in the services sector (82.2% and 63.7%), sales prevail in trade (80.4%). The transfer of information about products reaches the highest value in industry (7.0%) and the service sector (82.9%). As regards the use of sending and receiving transport documents, it is more popular in industry (54.5%) and in agriculture (53.2%). Finally, banking services have the highest use among service companies (89.4%), and the lowest in tourism (66.2%). Many are using the Internet to contact the authorities, and most of them fill out forms (57.1%) and get information from web pages (49.9%).

Table 2. Total national use of ICT in companies with less than 10 employees, %.

	Castela e Leão (ES)	Alto Trás-os-Montes (PT)	Krasnodar region (RU)
Companies with computers	72,42%	88,4%	60%
Companies with documents scanning system	43,4%	55,6%	63,8%
Companies with data storage solutions	49,4%	43,9%	54,2%
Companies with automatic presence control system	17,7%	5,9%	4,6%
Companies with Internet connection	66,98%	86,4%	86,8%
Companies with video projectors	11,9%	8,2%	7,3%
Companies with backup system	56,8%	42,1%	53,5%
Companies with Intranet	10,5%	13,5%	12%
Companies with fixed broadband connection	85,7%	86,4%	74.5%
Companies with a mobile broadband connection	74,04%	87,7%	99,8%
Companies with Internet connection and website	55,0%	46,2%	43.4%
Companies using social media	53,2%	63,2%	44%
Companies that purchase a cloud computing service used over the Internet	9,3%	32,2%	19.0%
Companies with Open Source SW	11,6%	59,7%	49%
Companies with trusted anti-virus and anti-malware	80,1%	50,6%	87,6%
Use of CRM-, ERP-system	41,5%	32%	14,5%

The use of ICT business management tools and applications is becoming increasingly popular. The first conclusions regarding such applications – 60% do not use any of them. The most implemented applications in the enterprise are: management software (ERP) (40% in Portugal, 46% in Spain and 17% in Russia), CRM systems (37% in Spain, 24% in Portugal and 12% in Russia).

4.2 Security

Speaking about computer security, which is an important aspect in business, only 4.6% of companies did not take care of protecting both personal and business data. 9 out of 10 companies have implemented at least one security measure. All surveyed companies are interested in deploying security measures. Table 3 presents the most popular ways of ensuring security in companies in the regions studied.

In Spain, 89.2% have already implemented them to some extent. 80.1% have antivirus, 79.5% protection of personal data, 78.9% have an updated operating system, and 67.7% make backup copies regularly. The least implemented measure is security plans and updates. With regard to various sectors of the economy, in the service sector, 95.0% of companies have implemented some security measures. On the other hand, there is a lower percentage in trade and agriculture (82.1% and 88.0%, respectively). The most common measures in various sectors are antivirus and operating system update.

Table 3. Security practices.

	Castela e Leão (ES)	Alto Trás-os-Montes (PT)	Krasnodar region (RU)
Technical means of user authentication	72,3%	93,5%	64.4%
Software preventing unauthorized access to malware	15,36%	26,32%	18,5%
Regularly updated antivirus programs	80,1%	50,6%	87.6%
Firewall and spam filter	64,7%	37,1%	57.5%
Computer or network intrusion detection systems	79,5%	40,6%	33.8%
Backing up data to media	67,7%	55,3%	31.2%

Interviewed companies in Portugal also attach great importance to security. 93.5% indicates that information is available only to those who are duly authorized, 62.4% - the operating system is up to date, and 61.2%, who always use the original software.

Also, about half of the companies indicate that they back up regularly, their antiviruses are reliable and constantly updated.

Compared with Portugal and Spain, Russia faces security problems three times less. However, the security surveyed companies do not neglect. About 4% of companies even use biometric user authentication tools [12]. In the transport sector, the highest percentage of antivirus updates (95.9%), about 90% of the use of electronic signatures is allocated to the service and communication sectors. Strong authentication tools are popular in the mining industry (59.1%). However, the largest percentage of backup in the field of communications (44%).

4.3 Challenges for ICT Implementation

The main "brakes" of ICT use in small and medium-sized companies are the following provisions (Table 4).

Table 4. Challenges for ICT implementation.

	Castela e Leão	Alto Trás-os-Montes	Krasnodar region
Lack of technical connectivity	17%	11,3%	7,8%
Lack of skills for work	15,36%	26,32%	18,5%
Lack of government support	10,08%	7,24%	9,7%
High connection costs	25,6%	30,78%	30,4%
These solutions do not meet the needs of the company	11,6%	7,8%	11,3%
For security and privacy reasons	10,68%	8,96%	12,1%
Lack of awareness	8,96%	7,6%	10,2%

Consider, for example, Internet technology. Almost all surveyed companies have access to the Internet. The reasons why some companies do not use this technology are associated with the cost of maintenance and connection (26.6% and 18.6%, respectively). The study also identified the following reasons: suppliers and customers do not use the Internet, employees spend a lot of time, and there is uncertainty about security on the Internet.

Some of the difficulties are internal and related to corporate culture. Any changes in this case require a budget, and many managers simply are not afraid of return on investment. A huge problem is also the lack of sufficiently qualified staff.

On the other hand, there are external barriers that do not allow achieving the desired level of scanning. This, for example, lack of infrastructure or difficulty (impossibility) of access to the Internet, it is worth noting separately the difficulties with the speed of access to the network. Plus, companies point out insufficient supply of adaptation solutions for each company.

5 Conclusion

The following are the main results for Spain. Tourism stands out among other activities by positioning in social networks and web technologies that are practically not used, for example, in the agricultural sector. In commerce, cloud computing is becoming more common, while the largest percentage of open source software comes from services. More established technologies (equipment and the Internet) are widely used in the agricultural sector, but for other indicators only less. Cloud computing and Intranet are almost never used in Spain. DSL is the main type of Internet access (55.9%). Below ten percent frequent access: Wi-Fi, satellite access, mobile broadband and cable TV. Spain has improved in terms of human capital, but still below average. Despite growing demand in the labor market, the supply of ICT specialists continues to be below the EU average. Most Spaniards make good use of various online services. Spain has made the most progress in using the company's digital technology. Most Spanish companies turn to social networks, e-billing, cloud services and e-commerce. Spain is on the 45th place in the global ranking of the use of technology in enterprises.

The following are some results for Portugal. In the field of communications, Portugal continues to improve results (eighth place in the EU in 2018). The employment rate also increased. However, 18% of the Portuguese handicrafts do not have the proper qualifications compared to the EU average (10%). Despite progress in almost all indicators considered in this area, Portugal fell to position 21 in the DESI 2018 rating on the use of Internet services (19th place in the previous edition). Only the rate of use of online banking and purchases has increased. About 25% of Portuguese companies have a high or very high digital usage, compared with the EU average (21.5%), but over the past year in this area, the surveyed area fell from 9th to 11th place in standings. And on the other hand, the percentage of e-commerce business volume (16%) is almost two percentage points below the EU average. By region, the use of different technologies is uneven: desktops, mobile devices, cloud computing, social networks, a website and open source software are more present in Alto Tâmega. The remaining technologies have more presence in the region Trás-os-Montes.

The following are the main results for Russia. According to the ICT development index over the past ten years, Russia has risen from 49 to 45th place in the world ranking. By the level of personnel specialization (younger than 35 years), Russia is in third place (56%), the same values for Spain and Portugal - 36 and 37%, respectively. Cloud services have become one of the most dynamic areas of the IT market in Russia. In 2016, the total market volume reached $ 422.11 million and continued growth in 2017–1818, increasing by 11.8%. Despite the fact that over the past few years there has been an increase in the Russian ICT market, there are a number of factors hindering the development of this market. First, it is the monopolization of the ICT market in Russia. Secondly, the low investment attractiveness of the ICT services market. Thirdly, the ousting from the domestic market of Russian producers of information technology, telecommunications and communications, and an increase in the outflow of specialists and copyright holders abroad. Nevertheless, the ICT-costs of the regions in 2018 increased by 9%.

ICT growth is accelerating due to the active modernization of its infrastructure. Throughout the study, the following proposals were formulated that can produce an effective result:

- Promoting the utility of various ICT-related technologies so that each company can select the best and most useful technologies in its operations;
- Stimulation and support of service centers to promote entrepreneurship (services such as: installation and maintenance of ICT, graphic design, financial and legal advice for companies, etc.);
- Reducing government taxes on computer equipment, providing interest-free loans to businesses in the field of ICT (software) and launching ICT incentive policies, subsidies and motivational programs, government investment in general;
- Modernization of the federal postal infrastructure;
- Development of a unified telecommunication network, including third-generation networks, which allow using video telephony, high-speed Internet access, watching movies and TV programs on a mobile phone, which will attract additional customers connected to cellular networks;
- Development of digital broadcasting, which improves the number and quality of television programs received, to organize the receipt of interactive services;
- Distribution of new forms of services in television as the transition to a digital broadcasting standard (interactive television);
- Distribution of opportunities for remote work;
- Economic support and specialized training based on technology startups.

This study not only provides an estimate of the scale of ICT distribution, but also identifies the main problem areas. The results can be useful in the development of measures aimed at the further development of infrastructure and the involvement of the general population in the use of new technologies. It is also a prerequisite for further research aimed at developing an expanded model of business process management in small and medium-sized enterprises, taking into account the impact of ICT in an emerging market and exploring the possibilities and benefits of its application for small business and society as a whole.

Acknowledgments. Projeto COMPETIC (0381_COMPETIC_2_E) - Co-financiado pelo Fundo Europeu de Desenvolvimento Regional (FEDER) através do Programa Interreg V-A Espanha-Portugal (POCTEP) 2014-2020.

UNIAG, R&D unit funded by the FCT – Portuguese Foundation for the Development of Science and Technology, Ministry of Science, Technology and Higher Education. Project n. UID/GES/4752/2019.

References

1. Commision, E.: Retrieved Recommendation 2003/361/EC: SME definition (2003)
2. Ongori, H., Migiro, S.O.: Information and communication technologies adoption in SMEs: literature review. J. Chin. Entrep. **2**(1), 93–104 (2010)
3. Pereira, J.P., Exposto, J., Ostritsova, V.: Analysis of the ICT use in companies of Castela and Leão and Northern of Portugal. In: 2019 14th Iberian Conference on Information Systems and Technologies (CISTI), pp. 1–6 (2019)
4. Balocco, R., Mogre, R., Toletti, G.: Mobile internet and SMEs: a focus on the adoption. Emerald Gorup Publishing Limited (2009)

5. Chong, S.: Business process management for SMEs: an exploratory study of implementation factors for the Australian wine industry. J. Inf. Syst. Small Bus. **1**, 41–58 (2009)
6. Pereira, J.P., Ostritsova, V.: ICTs adoption in SMEs located in less favoured regions: case study of Northern of Portugal and Castela and Leão (Spain). In: Advances in Intelligent Systems and Computing (2019)
7. Azam, M.S.: Diffusion of ICT and SME performance. Adv. Bus. Mark. Purch. **23A**, 7–290 (2015)
8. Kotelnikov, V.: Information and communication technologies for small and medium-sized enterprises. UN-APCICT (2018)
9. Katua, N.T.: The role of SMEs in employment creation and economic growth in selected countries. Int. J. Educ. Res. **2**, 461–472 (2014)
10. Abdrakhmanova, G., Vishnevskiy, K., Volkova, G.: Digital economy indicators in the Russian Federation. In: Data Book, National Research University Higher School of Economics, Moscow, HSE (2018)
11. Program Interreg V-A Espanha-Portugal (POCTEP). The COMPETIC Project (2014)
12. Abdrakhmanova, G.I., Gokhberg, L.M., Kovaleva, G.G.: Information and Communication Technology by the Population in Russia. National Research University Higher School of Economics (2015)

Towards an APIs Adoption Agile Model in Large Banks

Marta S. Tabares(✉) and Elizabeth Suescun

Universidad EAFIT, Avenida las Vegas, Medellin, Colombia
{mtabares,esuescu1}@eafit.edu.co
http://www.eafit.edu.co/investigacion/grupos/i-d-i-tic/Paginas/inicio.aspx

Abstract. Nowadays, large banks are facing challenges related to the generation of differential value, which they can accomplish by starting up a group of APIs which enable new criteria to define financial services for ever more expert clients in the management of opportunities regarding their finances and in the use of diverse information for decision making. This article presents an API adoption agile model to foster new digital capacities which facilitate the transformation from traditional businesses strongly supported on information systems to APIs. To accomplish this a review has been made of the literature which guides the classification of the most relevant aspects of businesses based on APIs in banking, then, the components of the proposed model are defined, and finally, the proposal is validated with a use case from the area of Collections in one of the most important banks in South America. Finally, the research suggests a series of results through a gap analysis which prove that technological transformations based on APIs can be absorbed in a fast way both by the business process as well as by the clients.

Keywords: Agile model · API · Application Programming Interfaces · Digital transformation · Banking digital transformation · Fintech · Large bank · Banking Collection

1 Introduction

Banking, on a world level, is one of the benchmarks for corporate digital transformation, with great impact on the world economy. This stems from a very assertive vision of utilizing APIs (Application Programming Interfaces) as software elements to enable new businesses. Originally, the APIs have been used within the organizations to integrate diverse systems and to allow data exchange between business areas. This facilitates the interconnection of business services and processes in all the organization. Starting from the basic e-commerce concept, APIs allow for online and real time interactions with suppliers and clients from different devices like PCs, smartphones, tablets, etc. Besides, they expose corporate assets to developers so they can create applications which facilitate the commercialization of information or new digital businesses [1].

Á. Rocha et al. (Eds.): WorldCIST 2020, AISC 1159, pp. 302–311, 2020.
https://doi.org/10.1007/978-3-030-45688-7_31

Currently, banking on an international level has advanced a lot in providing digital banking services for the clients. However, within traditional big banks there are financial services that do not move at the same speed in their transformation process. Normally, these are services which involve more than two players or they are services with financial regulations that require human or a system's regulatory intermediation to be carried out. This motivated the research and identification of breaches which can occur to transform a traditional financial service into a digital service in all its life cycle. The proposal is particularly focused on the process of Banking Collections which can empower ever more the relationship between players of the process and consequently take advantage from a series of digital businesses to revitalize governmental regulations by using APIs.

1.1 Research Objective

To propose an agile reference model so that large banks adopt APIs, specifically in a business context where more than two players intervene in the transactional environment.

1.2 Contribution

The suggested model introduces the concept of Agile Experience defining APIs in a digital context from three perspectives: Knowledge Transfer, Value Proposition and Gap Analysis, and Evolutionary Prototyping. Besides, the model gather experts from business areas, digital experts of the economic sector, and experts in emerging technologies which participate in different model interactions. Thus, this contribution provides answers to the following research questions:

- What new digital elements condition of banking legacy systems which interact with APIs that provide new business opportunities?
- What agile innovative elements can be adopted by interdisciplinary groups when APIs must support digital needs?

2 Research Methodology

For the development of the proposal, the research team used an empirical research methodology [2] in combination with elements of Design Thinking [3] in order to achieve the proposal. In the first phase the problem is identified beginning with a series of meetings with the different areas of the bank involved with the service of collections. This allows to identify the problem and the different points of view of the actors regarding the possible management of the APIs for Collections from the different types presented in numeral 3. In the second phase, a literature review was done in two steps: first, the APIs fundamentals, and second it is recognized the evolution of financial services towards digital services. In the third phase, the proposal is described and the elements of the agile model which can be developed as experimental base. Finally, the results of the experience are evaluated and concluded to project the future work.

3 Literature Review

3.1 API Definition and Basic Uses

An API is a software to software interface defined by the contract (it exposes interfaces to be consumed or used by another software) which determines the communication between applications through a network without user interaction. It is defined by community established standards, it is independent from the other programming languages and allows portability [1]. Its main purpose is to be used by application or services developers. Normally the APIs tend to do tasks such as [4]: Data transmission, which establishes the way in which data is safely transmitted. In general, APIs use HTTP/HTTPS as transport layer. Data exchange, which establishes the format of the exchanged data. The most commonly used are XML and JSON. XML has a little more functionality than JSON but JSON is more popular in the developer's community. Data access, which is related to access management: who has access, to what data, and how to accomplish this. The most used standards are SAML (http://saml.xml.org/saml-specifications) and OAuth 2.0 (https://oauth.net/2/). API design, which refers the way in which APIs are designed. The most common standardized design principles are REST (Representational State Transfer) and SOAP (Simple Object Access Protocol). Specifically, RESTful services have been used to compose high quality Web-scale applications based on the mashups. These require the use of APIs and data sources (i.e. any piece of data in the Web such as: a document, an image, or a tweet) to produce enriched results. Besides, they use content from more than one source in order to create a service displayed in a single graphical interface [5].

3.2 API Classification Proposals

Different authors have classified or typified APIs according to different characteristics such as security and purpose for the business. Table 1 presents some of those classifications.

3.3 Evolution of the Financial in Large Banks

Traditional financial services have been completely or partially automated by means of big software components or information systems which were developed back when there were no advanced software development techniques, that is why they are denominated as legacy systems. Consequently, these systems are not easy to change because of their old-fashioned structure, whereby they are complex and not flexible. Thus, changes frequently required by the business are hard to implement in said systems. Nonetheless, they have significantly served the organizations and have delivered the expected results over time. Information systems such as ERP might be losing their value because they do not have a strategy or a capacity to rewrite the solution in a modern technology, it becomes hard to preserve, it loses commercial value, it does not accept changes, maintenance costs increase, functionality and performance decrease, there is no access

Table 1. API classification proposals

API type	Description
By privacy level [1,6]	
Private/Internal	Closed API, which means, accessible only within the limits of the organization. It is used for internal integration of application, and B2E applications (Business2Employee)
Private/Associated	Used for B2B partner integration. The APIs for partners grant exclusive access left to the discretion of the API supplier. Bilateral agreements over specific data exchange between, for example, a bank and a software supplier [4]
Public	Open to everybody for use. In this group, the following subgroups can be differentiated: APIs of members, APIs of knowledge, and public APIs [4]
By the usage in the context of the business [7,8]	
System/Enrichment	It improves the comprehension of the situation with historical data of the customer relationship management systems (CRM), account records, demographic analysis, health records and similar ones
Interaction API	It helps to identify opportunities for customers, employees and devices involvement. Beside, it includes mechanisms such as mobile location detection, sensor monitoring, predictive analysis and human observation
Action API	It allows near real time measurements. Action type interface examples include push notifications, instrumented devices and human tasks management systems
By the different target audiences [9]	
Integration API	Created or used by corporate architects and developers. The API can be a Web service, a micro service and a REST APIs which must be treated as integration technologies
Innovation API	Fintechs is a newfangled concept and business model which provides alternative financial services for digital customers such as; changing client requirements, governmental regulations, and incoming, among others. "This type of innovation changes the way in which people purchase, pay or self-manage [10]"
Connectivity API	Bank's Consumer. This type of API emerges as a new connectivity and transactional channel between the bank and the final consumer, which allow to rationalize and ensure access on demand to financial services and consequently to the data provided by the information systems located at the backend of the organization
APIs for Banking as a Platform	Digital neobanks. APIs enable platforms that create new capacities for bank and customer [11]

to commercial content and it does not have the capacity to reconfigure itself, which is why the system ultimately becomes obsolete, effectively at the end of its useful life [12]. However, current banking demands digital services which require APIs and other ways of exposing digital services that interact with flexible architectures that support robust solutions of the organization.

Sajic et al. [13] propose alternatives to modernize the transformation of classic banks into modern digital banks. In their investigation they identify how the classic banks (retail services) are still waiting in their agencies for a large number of clients who cannot get used to making all their requests through the virtual services currently offered by said banks. Chen et al. [14] show a comparative case study to contrast and analyze the Industrial and Commercial Bank of China (ICBC) and Citibank in the way they are addressing the digital transformation. Analyze the strategies, organizations, human resources systems and product innovations adopted by these two banks in response to the impact of FinTech. The great evolution of the big banks is the challenge of living with the FinTech [15]. These companies are financial services companies of digital origin. The ecosystem that defines them is made up of 5 elements: technology developers, financial services startups developed under the capabilities of emerging technologies, new consumers of financial services, government, and their link with the large traditional financial institutions. The integration between these elements provides the generation of agile innovation in new financial businesses leveraged by new technological developments in big data, data analysis, mobile applications, blockchain, artificial intelligence, among others [16,17]. This allows them to create new opportunities and services in markets until recently neglected by large banks, moving from traditional financial services to almost personalized services. The fact of the evolution and constitution of markets with cryptocurrencies and digital markets without any financial intermediaries, low costs, and in real time.

3.4 Banking Collections

Collections is a banking service has three actors: the "Bank" who acts as a payment collector for companies who provides different kind of services (e.g. public services). The "Biller", who is a client of the bank, needs to collect for example: rents, installments of memberships, donations, tithes, or bill payments done by individuals or corporations. The "Payer" is a direct consumer of biller' services and can do payments through the "Bank". The biller sends information to the bank about collection novelties (accounts receivable) through different type of electronic channels. The bank randomly receives payments in different time periods defined by the collector. The relationship between them will be kept valid if the contract between the parties is active. Thus, the bank acts as a payment gateway. IT companies such as IBM and ORACLE have been offering payment platforms or "Payment Hubs" with completely flexible and dynamic architectures where APIs, micro services and web services interact with centralized systems in order to provide online alternatives for the bank's clients, in real time, and from any place so they can do their payments [18,19].

The Use Case used to describe the application of proposed model is located in one of the most important banks in South America. Specifically, it deal with Collection System as part of the process of renewing its banking core. To achieve that goal, it analyzes the current state of its business model and makes the adjustments it deems necessary so that the computer system can support them in said achievement. These changes activate needs in technological aspects such as: the unification of databases for customers with B2B contracts, management of standardized files for customers who cannot establish a B2B relationship, and have a more direct relationship with bank customers who are also customers of the Collector. This motivates an investigation around APIs that could be agilely aggravated towards the generation of digital negotiations of greater impact that would complement the renewals of the collection system. To address this particular case, the current status of the use of APIs in digital financial markets and their relationship with Fintechs was identified. This allowed to lead a transfer of knowledge activity to the commercial team of the collection area.

4 Agile Model for APIs Adoption in Large Banks

The API research group defined an agile model for the adoption of APIs. This model is used to make a fast cycle identification of digital products, from the identification of the requirements of the customers or users regarding new API-based businesses to the basic prototyping of new digital businesses. It integrates the following phases: knowledge transfer, value proposition, gap analysis, creating a basic prototype, and the pilot testing (the last two phases are not described in the paper due to lack of space). Each one of phases these identifies, analyzes, selects, and defines knowledge objects that are associated with digital culture, emerging technologies, business processes, digital clients, and API applications. Figure 1 shows proposed model, and how through different iterations the team of experts achieve in brief periods and from different viewpoints, the digital products/services. To execute this model, a continuous transformation interdisciplinary team is required. It should include experts in: business, the digital transformation in the specif economic sector, and the emerging technologies implementation.

4.1 Knowledge Transfer

Knowledge transfer is the key element that the digital transformation team needs to create and promote a permanent collaborative work environment, where teams can interact nimbly to generate new products or services. This activity must be supported on technologies that facilitate the gain and use of knowledge related to the different and particular fields associated with the business processes that will be intervened, as well as the emerging technologies that will be involved. This will allow an agile understanding among them in a simple and productive way through the practice and demonstration of experts [20, 21]. Knowledge transfer among teams of experts is iterative and incremental. During each iteration, teams of experts share their knowledge considering the achievement of the following objectives: Identifying the user requirements (emphasizing), describing user requirements and problems (defining), and challenging assumptions to have an initial brainstorm with all of them. Figure 2 shows topics of interaction that have been discussed with the teams during moments of knowledge transfer. During the development of the use case, four meetings were scheduled so that collections experts

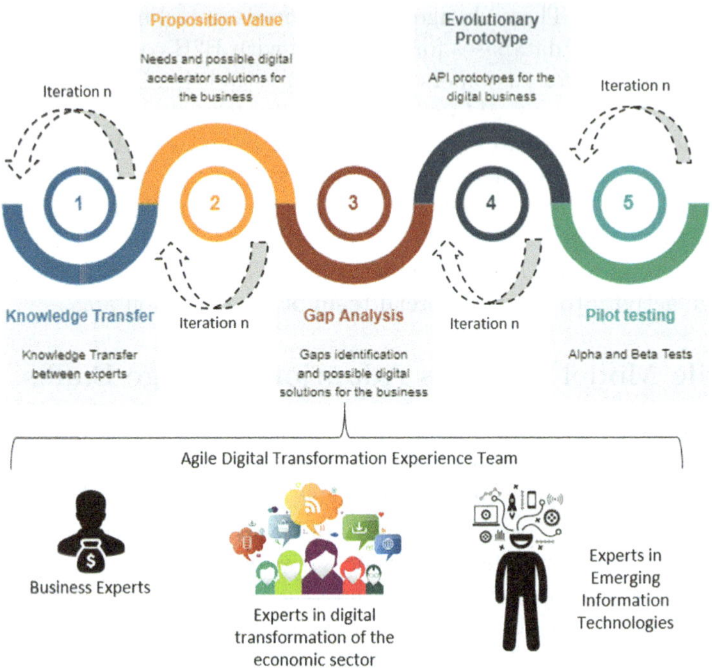

Fig. 1. APIs adoption agile model

explained the current definition of the portfolio of services set out for their clients. Moreover, they described the requirements associated with the improvement and growth of the business (the specifications of those requirements were supported employing documentation produced by the company). During each iteration, theories on digital culture, and emerging technologies that are supporting the banking market with the use of API, on top of the support Fintechs provide worldwide, were used to motivate the experts. This helped the entire group design possible API-supported scenarios in a way that gives feasibility to the future vision of the digital business from different perspectives.

Transfer Iteration	Business Experts	Experts in Digital Transformation	Experts in Emerging Information Technologies
Business Experts		Processes of the Business	Processes of the Business
Experts in Digital transformation	Digital Culture (Technological and business View)		Digital Culture (Business View)
Experts in Emerging Information Technologies	New Technologies experiences	New Technologies experiences	

Fig. 2. Transfer knowledge among experts

4.2 Value Proposition

Once a high-level requirements specification and some possible digital business scenarios are defined, the value proposition can be stated. The methodology is based on the classification and prioritization of the highest feasibility scenarios. For each scenario, a devising process takes place (which might require various iterations) where the teams participate, and a value proposition is finally consolidated (using Canvas methodology [22]) for each scenario. During the development of the use case, four sessions were conducted by the teams of experts. In these sessions, 34 pains were identified to be subsequently consolidated into 16, and a minimum of one solution was generated respectively (digital product or service). Some of them are shown in Fig. 3, which illustrates one of the canvas models of the value proposition that was met. In this, the profile of the client was made to coincide with the value map through the declaration of the products or services that were offered, or that are planned to be offered, which are described as pain relievers and bring joy to clients.

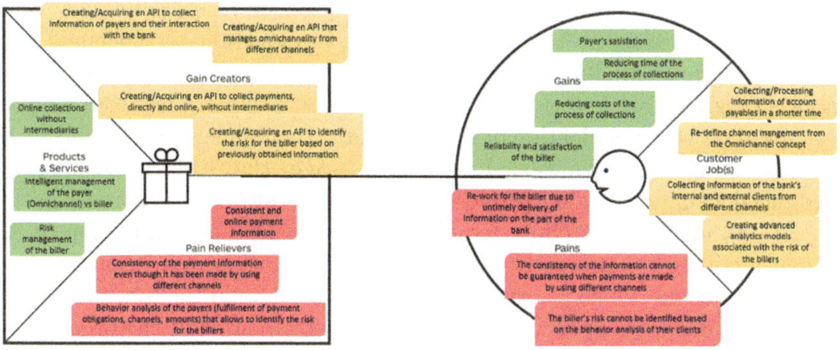

Fig. 3. Canvas Proposition Value applied to a particular scenario at the Collection Service

4.3 Gap Analysis

When the profile of the client or the specific segment of the client, as well as the value map, were identified, a gap analysis was done to meet the desired objective with APIs that could complement the core of Collections at any time (for this particular case). The gap methodology was applied to the collections team of the bank, who were selected to obtain information from the system and the collection business thus getting to know the situation of API pains and opportunities. Apart from gap identification, doing a trend analysis allowed for tying findings of the bank together with references and opportunities that were identified in the literature. A gap analysis that complements the value proposition was done. This analysis focuses the compilation of more detailed information between the current condition (Pain) and the future condition (Relief), by identifying with a further level of detail the gap (what causes the pain), and how it must be closed. This helps understand the effort that the bank will have to undertake in a new digital business supported or enabled by API.

5 Conclusions and Future Work

Based on the experience with expert groups, and specifically with the bank's collection process teams, the agile experience of adopting digital products or services in large banks, answered the research questions as follows:

- The new elements that condition the banks' existing legacy systems and their interaction with API are primarily related to flexible and reactive software architectural designs. Thus, one or more APIs can quickly relate to affected business processes through the agile experience of adopting digital products and services. As the teams complete their experience, the bank assesses the results identified in the gap analysis and motivates the API implementation plan to take the most relevant aspects of that model.
- The elements of innovation that can be adopted by interdisciplinary groups when it is necessary to make API-supported innovation are specified in the agile model presented in this article. It highlights the articulation of different stages ranging from knowledge transfer among the experts involved who have as a product the specification of requirements at a high level and identification of digital business scenarios. In the next stage, the devising scenarios are produced using the value proposition canvas model tool. To reach greater detail and identify the effort required in the creation or acquisition of API, a gap analysis is performed to achieve a classification and prioritization of implementations. Once prioritized, it is possible to build the first prototype of the digital products or services that will be implemented in the organization.

References

1. De, B.: API Management: An Architect's Guide to Developing and Managing APIs for Your Organization, 1st edn. Apress, Berkeley (2017)
2. Kothari, C.R.: Research Methodology: Methods and Techniques. New Age International, New Delhi (2004)
3. Link, P., Lewrick, M., Leifer, L.: The Design Thinking Playbook: Mindful Digital Transformation of Teams, Products, Services, Businesses and Ecosystems. Wiley, New Jersey (2018)
4. EBA Working Group. Euro banking association: Understanding the business relevance of open APIs and open banking for banks, May 2016
5. Garriga, M., Mateos, C., Flores, A., Cechich, A., Zunino, A.: Restful service composition at a glance: a survey. J. Netw. Comput. Appl. **60**, 32–53 (2016)
6. Seelemann, I., Yildirim, K., Gupta, R., Hamdy, S., Gucer, V., Wisnewski, B., Perepa, B.: Getting started with IBM API. Connect concepts and architecture guide, September 2016
7. Jensen, C.T.: APIs for dummies®, IBM limited edition (2015)

8. Jensen, C.T., Ashby, D.: APIs for dummies®, 3rd IBM limited edition (2018)
9. Hines, P.: APIs in banking: unlocking business value with banking as a platform (BaaP), March 2018
10. Premchand, A., Choudhry, A.: Open banking APIs for transformation in banking. In: 2018 International Conference on Communication, Computing and Internet of Things (IC3IoT), pp. 25–29, February 2018
11. Ozcan, P., Zachariadis, M.: The API economy and digital transformation in financial services: The case of open banking. SWIFT Institute Research Paper Series, July 2017
12. Hussain, S.M., Bhatti, S.N., Rasool, M.F.U.: Legacy system and ways of its evolution. In: 2017 International Conference on Communication Technologies (ComTech), pp. 56–59, April 2017
13. Sajic, M., Bundalo, D., Bundalo, Z., Pašalić, D.: Digital technologies in transformation of classical retail bank into digital bank. In: 2017 25th Telecommunication Forum (TELFOR), pp. 1–4, November 2017
14. Chen, Z., Li, Y., Yawen, W., Luo, J.: The transition from traditional banking to mobile internet finance: an organizational innovation perspective - a comparative study of citibank and ICBC. Financ. Innov. **3**(1), 12 (2017)
15. Lee, I., Shin, Y.J.: Fintech: ecosystem, business models, investment decisions, and challenges. Bus. Horiz. **61**(1), 35–46 (2018)
16. Milian, E.Z., Spinola, M.d.M., de Carvalho, M.M.: Fintechs: a literature review and research agenda. Electron. Commer. Res. Appl. 34, 100833 (2019)
17. Thakor, A.V.: Fintech and banking: What do we know? J. Financ. Intermed. **41**, 100833 (2019)
18. Lodge, G.: A real-time hub for a real-time everything future, March 2018
19. Oracle Financial Services. Payments 2.0 unleashing payments innovation as the edge in a digital landscape (2018)
20. Nonaka, I., Toyama, R., Konno, N.: SECI, Ba and leadership: a unified model of dynamic knowledge creation. Long Range Plan. **33**(1), 5–34 (2000)
21. Leonard, D., Barton, D., Swap, W.C.: Critical Knowledge Transfer: Tools for Managing Your Company's Deep Smarts. Harvard Business Review Press, Boston (2014)
22. Bernarda, G., Smith, A., Osterwalder, A., Pigneur, Y.: Value Proposition Design: How to Create Products and Services Customers Want (Strategyzer), 1st edn. Wiley, Somerset (2014)

A Business Performance Management Framework

Ana Carina Brissos Pereira(✉) and Miguel de Castro Neto

NOVA Information Management School (NOVA IMS), Universidade Nova de Lisboa, Lisbon, Portugal
{m2016151,mneto}@novaims.unl.pt

Abstract. In an increasingly competitive market, companies need to look not only at results, but also at how they can improve their performance to achieve them. Knowing the factors that influence business performance allows to identify initiatives that lead to their improvement or mitigate potential risks, ensuring strategic alignment across the organization. This article presents a Business Performance Management (BPM) framework, where the key components that impact business performance are described, which includes Business Intelligence (BI) as an integral part of the technological framework and a Performance Management (PM) cycle as a methodological approach to its implementation for business performance improvement.

Keywords: Business Performance Management · Business Intelligence · Framework

1 Introduction

The successful execution of a business strategy is a well-recognized requirement for an organization's survival in the hypercompetitive marketplace [1]. BPM becomes essential in allowing companies to align operational strategy and objectives with business activities in order to manage overall performance through better supported actions and decision making [2].

BPM is a set of integrated closed-loop management and analytic processes supported by technology that helps businesses define strategic goals, measure and monitor its Key Performance Indicators (KPI) and act proactively to achieve goals [1]. BPM can also be referred to as Enterprise Performance Management (EPM), Corporate Performance Management (CPM), Strategic Enterprise Management (SEM) and, in a simplified version, Performance Management (PM). Note that some organizations use the acronym BPM for Business Process Management which, although related to PM, is a distinct concept which is not within the scope of this article.

This article focuses on a BPM framework presentation where the key components that impact business performance are described, which includes BI as an integral part of the technological framework and a PM cycle as a methodological approach to its implementation for business performance improvement.

Á. Rocha et al. (Eds.): WorldCIST 2020, AISC 1159, pp. 312–323, 2020.
https://doi.org/10.1007/978-3-030-45688-7_32

2 Business Performance Management

BPM involves a variety of integrated operational and analytical processes that complete two sequential activities: First, it facilitates the creation of strategic objectives by defining concrete objectives and meaningful KPIs for the organization; the second, supports the next phase of performance management, where objectives are related to operational metrics and linked to performance actions or initiatives that lead to effective strategy execution [3]. BPM is about improving performance in the right direction [4], allowing the organization to focus on what really creates business value while ensuring its long-term continuity. BPM encourages process efficiency as well as the efficient use of both financial, human and material resources [5].

2.1 Business Intelligence

The concept of BPM becomes inseparable from another important concept that is Business Intelligence (BI) as an integral part of a BPM system.

Business Intelligence (BI) is the process of collecting enough of the right information and in the right manner at the right time and delivering the right results to the right people for decision-making purposes [6]. In other words, BI encompasses a set of procedures, techniques, methodologies and technological tools that allow data collection from various systems (internal and external), processing them for analysis and availability of information in reports and dashboards, which support decision-making and business strategy definition. BI can be defined as the process of transforming data into information and information into knowledge [5]. The main components of a BI architecture are: data sources, ETL processes, data warehouse (DW), On-line Analytical Processing (OLAP) and metadata.

Data sources can be operating and transactional systems such as Enterprise Resource Planning (ERP) – Online Transaction Processing (OLTP) – external sources or data in other BI/DW architectures. These can be structured data in relational databases (DB), or other format such as excel or flat/text files (e.g. csv or txt files), semi-structured or unstructured data such as text documents, PDFs, images, videos, audio or other information that is not clearly organized (or in table format) and cannot be automatically related to other information.

ETL Processes includes three processes: 1) Extract data is the process of identifying relevant data to be collected, which may be either inside or outside the organization and is usually not integrated, is incomplete or duplicate [7]. This data is sent to a temporary workspace called Staging Area (SA), which is never accessible to users.; 2) Transform and cleaning is performed in SA and aims to make the data related to each other, standardizing them through the application of business rules, correcting errors to ensure consistency across the organization for reporting purposes and analysis; 3) Load data into the final DW, the presentation layer, is the last phase of the process where data is already uniformed to be consistently presented and accessible for reporting and analysis.

The final DW or presentation area is one of the key components of a BI architecture and is where data is organized, stored and accessible to users for querying, reporting and analysis. This presentation area includes data marts (DM), which are subsets of the main DW that aim to meet the specific needs of a specific area (or users) of the organization.

The final DW supports next component through data storage and maintenance in multi-dimensional structures for query, reporting and analysis [7].

On-line Analytical Processing (OLAP) is the data access component where data is shared by "OLAP cubes", which allows data exploration, summarized and/or aggregate views from many perspectives in a user-friendly way with fast response time. Data cubes are dimensional models stored in multi-dimensional OLAP structures [7].

Metadata is a very important component which refers to data about data [7]. This area is like a DW encyclopedia [8]. The metadata repository is used to store business and technical information about data, including business rules and data definitions. Ensuring maintenance and regularly updating metadata is essential [7].

3 The BPM Framework

The proposed BPM framework rests on five key-components: Environment, Organizational Culture, Systems and Information Technology, Processes and, in the center of everything, People - the *cuore* (heart) of an organization (see Fig. 1).

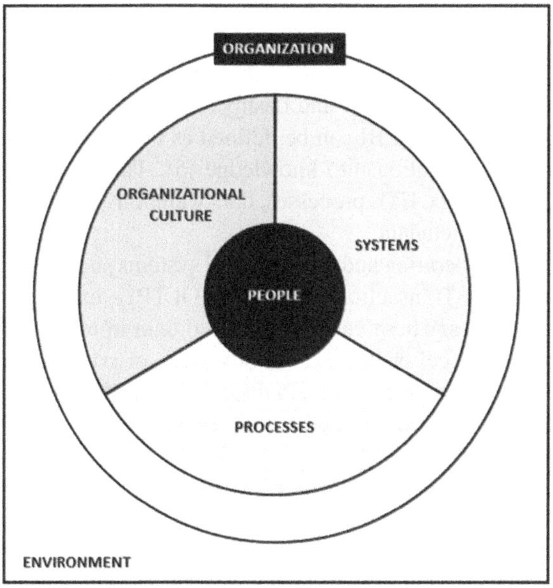

Fig. 1. The BPM framework

3.1 Environment

The environment, or context in which the organization operates, includes all the forces or institutions surrounding the company that may affect performance, operation and resources and directly shape the organizational structure [9].

External factors that influence performance may be political (P), economic (E), socio-cultural (S), and technological (T). PEST analysis can be used to analyze the current and future environment as a process of strategic business management [10].

Political factors include governmental, fiscal, labor, commercial, environmental, market regulatory legislation in which the organization operates and other standards, as well as political activities designed to influence its performance [9] (e.g., European General Data Protection Regulation (GDPR), new accounting standards, fiscal and regulatory policies).

Economic factors represent the overall economic health of the country or region where the organization operates [9] (e.g., inflation rate, unemployment rate, interest rate, exchange rates).

Socio-cultural factors represent the demographic characteristics, norms, customs and values of the population of the country, region or location where the organization operates (e.g., population density, population age, education, geographical distribution).

Technological factors include scientific and technological advances in a specific industry, as well as in society in a broader way, and which generate competitive advantage or disadvantage (e.g., digital cameras, smartphones with integrated camera, Wi-Fi technology, Internet, Internet of Things (IoT), process automation and digitalization).

These external factors create tensions and can constitute threats but also opportunities [11]. In the operational environment, there is also the influence of customers, suppliers, shareholders and the competitive market on the organization's performance.

3.2 Organizational Culture

Organizational Culture is the organization's DNA that inspires the people who are part of it, both in the way they think, behave and act, determining the motivation behind their actions [12], impacting individual performance, then the performance of an organization [13]. The organization's DNA includes Vision, Mission and Values.

Vision is what the organization wants and aims to achieve. It is the "future state", the projection of itself in a short, medium and long-term perspective. It's the answer to the question, "Where does the organization want to go?".

Mission is the reason for the organization's existence, its purpose. It is the answer to "Why does the organization exist?".

Values are the philosophy behind people's behaviors and attitudes that lead the organization to fulfill its vision and mission. It includes the beliefs, assumptions, ethical principles and a whole set of understandings that lead people to action, impacting the performance of the organization. It answers the question "How will we achieve the vision and mission?".

Therefore, Vision, Mission and Values are the basis of everything. It is the foundations, structures and pillars that underpin an entire organization, as in a house. If they are not solid, consistent and coherent, they can cause a collapse of an entire organization. Probably the organization will not resist both internal and external pressures and its continuity and economic viability is questioned.

Also, within this component an increasingly agile culture must be created [14] and foster a performance-oriented culture, continuous improvement with knowledge sharing and transparent communication. An "agile organization" is an organization that has the ability to reinvent, adapt and quickly implement changes and succeed in a turbulent, ambiguous and fast changing context. The type of organizational structure has an impact on the "agility" of the organization, being more agile the lower the hierarchy, the greater the effectiveness of strategic alignment [15]. An Agile organization stands out for both stability and dynamism, and according to a McKinsey & Company study, less than 25% of organizations achieve both.

Communicating results, vision, mission and values are critical to strategic alignment. People will only know where to go if they know what the organization's goals and plans are and what is expected of them and how they will contribute to achieving those results.

3.3 Systems and Information Technology

The Systems and Information Technology component includes business support infrastructures, architectures, applications and software (such as ERPs, CRMs, etc.), but also BI systems and architectures that feed the BPM system (seen here as a DW architecture exclusively dedicated to BPM). A Technological Framework of a BPM System is presented below (see Fig. 2).

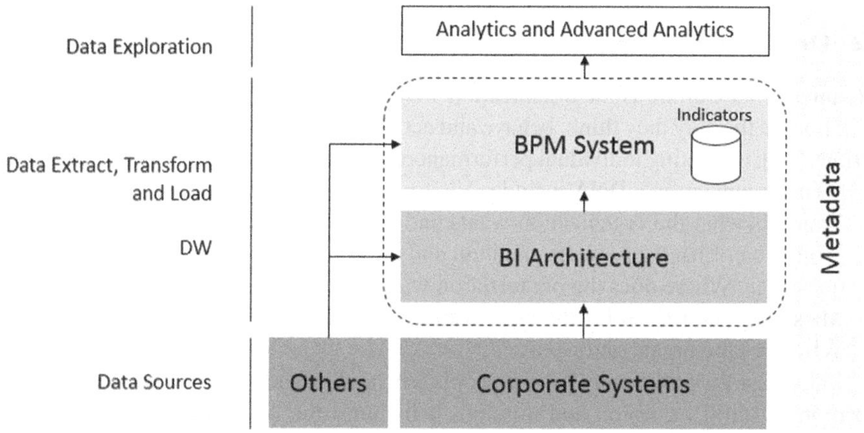

Fig. 2. Technological framework of a BPM system

The Technological Framework of a BPM System consists of a traditional BI architecture that includes the following subcomponents (see Table 1):

Table 1. The subcomponents of the technological framework of a BPM system

Subcomponent	Area/layer	Examples
Corporate systems	Data sources	Sistemas OLTP, e.g.: SAP ERP, CRMs, Ticketing Platforms, etc.
Other sources	Data sources	Excel or flat/text files with manual inputs or indicators/data, survey results, web pages, conversion rates from external sources, etc.
BI architecture (DW or Data Lake)	BI solution	ETL Processes, Staging Area (SA), Data Warehouses, Data Marts or Data Lakes
BPM system (DW)	BI solution	ETL Processes, exclusive DW for BPM and Performance Indicators DB; Indicators Master Data Management
Analytics and advanced analytics	BI solution, analytics	Dashboards, queries and ad-hoc reports, Balanced Scorecards (BSC) and maps, predictive models and others BI Tools: MS Power BI, SAS Visual Analytics; SAS Miner, SAS Guide and others
Metadata	Metadata	Knowledge Management layer with data about data, that includes tables and fields description, the source and metrics formulas, etc.

To highlight in this technological framework:

- The importance of metadata as the "knowledge management layer" of all information (organizational assets), from data source systems to their availability in reports and dashboards. This area ensures asset continuity (information/data and architecture) and the maximization of Return on Investment (ROI), hence the importance of being always up to date;
- The data architecture exclusively dedicated to BPM supports activities of the areas responsible for preparing information to support decision-making and performance management, as well as the entire PM cycle;
- The main objective of the performance indicators DB is to store the current and historical values of all metrics considered relevant to the business and to monitor their performance, with the possibility of simulating various scenarios, as well as ensuring the consistency of reported information. Another benefit is the possibility that, in the medium term, it will allow for deeper analysis of pattern discovery and correlations between variables (the metrics) and the development of predictive models, which may bring new knowledge and/or new business insights. Parmenter (2010), refers to the

need to create DB to record/store in order to guarantee the reliability and consistency of the reported information [16]. Depending on the organization's technology maturity level and even the size itself, this record could be carried out using common tools such as Ms Excel, Ms Access DB or SharePoint. The important thing is to measure and store the indicators values allowing the comparison over time and evaluate the company's performance and strategy execution success.

3.4 Processes

The Processes component includes all the business processes of an organization in various areas, such as accounting, logistics, sales, etc., including IT processes, information management and data governance model, as well as PM and decision-making processes. This component has a large scope that is not the focus of this article, however it is important to highlight the relevance of processes simplification, standardization, automation and digitalization combined with an agile, performance-driven and continuous improvement culture.

3.5 People

People are the key resource of an organization and without them an organization does not exist or work [10]. Being the most valuable asset of an organization, the people component is the heart of an entire BPM System and it is extremely important to define a Human Resources Management (HRM) strategy integrated into the business strategy that enables the organization to achieve their goals [17].

An organization's DNA is made by the people who are part of it, so People and Organizational Culture are two inseparable components. BPM involves, beyond the communication of the vision, mission and values, and business strategy and objectives, a HRM strategy that promotes the improvement of individual performance across the organization through:

1) Greater involvement and people commitment to the organization through the DNA sharing, which gives them clear direction, alignment and understanding and a sense of purpose, giving meaning to performing their tasks [12];
2) Consistent leadership practices throughout the organization and behaviors that reflect their DNA, ensuring honest and sincere communication, discipline and commitment from leaders, creating trust environment [12]; The leader has an increasingly active role in employee development by becoming a leader-coach;
3) Developing leaders by empowering them with the skills needed for their role consisting of physical, intellectual, emotional and spiritual strength that drives personal fulfillment and in turn inspires others to follow [18];
4) Encourage positive behaviors through leaders at various levels of the organization in order to generate greater enthusiasm and motivation and to gain greater involvement and commitment from employees, encouraging them to consistent positive behaviors and ways of working;

5) Attracting talent to the market and/or university community through good projection of the organization's brand as an employer [19] and presentation of "magnetic" offers for an Employee Value Proposition (EVP) stronger than the competitors [20];
6) Talent retention through fair performance (meritocracy) and training plans that enable skills development and encourage the talent. Promote development opportunities and in a consistent way recognize and reward the performance of both individuals and teams, promoting alignment of expectations between employees and the organization [12];
7) Measures allowing family-work balance, such as time flexibility, teleworking (distance work), reimbursement of well-being expenses, employee assistance program (among others), which promote employee well-being and happiness, contributing to higher motivational levels. The role of leaders is critical to the success of policies and measures by encouraging their adoption and/or sharing work tasks to support those who use them [21]. Employee happiness and health become a competitive advantage;
8) Involvement of employees in the community, extending the scope of the organization actions through volunteer initiatives [13].

To highlight, creating a performance culture is finding a balance between performance and people, without sacrificing either [12].

4 The BPM Spiral

All components previously described actively participate in the four steps PM cycle, which begins with (1) Strategize – Strategy definition and alignment; (2) Plan – Strategy execution planning; (3) Monitor/Analyze – KPIs and metrics monitoring and analysis;

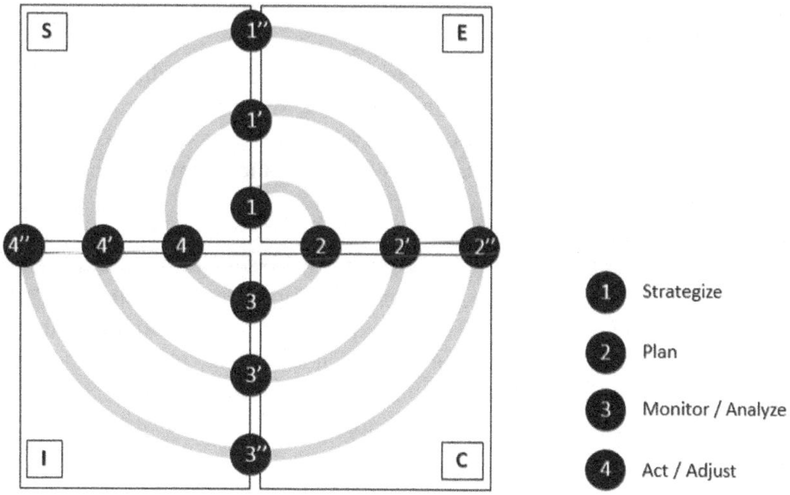

Fig. 3. The BPM Spiral

and (4) Act/Adjust – actions and initiatives to correct and/or adjust for goals' achievement. This cycle is executed not in a continuous "closed-loop" (as proposed by BPM standards group [22]), but in a continuous "spiral", recalling Nonaka's (1991) "Spiral of Knowledge". This means that at the end of the first full cycle, the organization and its components have a higher level of knowledge about their performance, allowing them to know the "state of things" more broadly compared to the beginning of the cycle. The BPM Spiral becomes a methodological approach for a BPM System implementation and a process for business performance improvement. Therefore, the concept of "BPM Spiral" and its analogy with Nonaka's (1991) SECI Model [23] is born (see Fig. 3).

Briefly, tacit knowledge is what is in people's minds and is created by personal and professional experiences, personal beliefs and perspectives; Explicit knowledge is knowledge described in procedures, manuals and other forms that allow summarizing and mapping information, making it easier to communicate and share [23].

The strategy definition (1') arises from socialization [S], where tacit knowledge of the interaction of previous knowledge level of the spiral (4) is converted into new tacit knowledge, through observation, social interaction with shared experiences, that generate strategic thinking, which results in its definition or redefinition/adjustment. 'Strategize' step wants to answer the question "Where do we want to go?".

Execution planning (2') consists in the conversion of tacit to explicit knowledge through the externalization [E] of the strategy (or its adjustment), which is translated into plans, actions and initiatives to be carried out, budgets definition, objectives and clear goals that are communicated and shared (articulation) through BSC, reports and dashboards and other maps. Plan wants to answer the question "How did we get there?".

In monitoring and analysis (3'), there is the conversion of explicit into explicit knowledge through the combination [C], which involves the collection of internal and external information to the organization in different types/formats, which are converted into new knowledge by interpretation and news insights generation. As an example, we have the KPIs monitoring and the exploration of the "root causes" that lead to the information cross-checking to interpret and to reach conclusions. Monitor/Analyze wants to answer the question "What are we doing?".

Acting and adjusting (4') transforms explicit into tacit knowledge through internalization [I], "learning by doing", which creates new knowledge through the formulation of new know-how and new mental maps on a given subject which leads to goals achievement and strategy success. Act/Adjust wants to answer the question "What do we need to do differently?".

This process from 1' to 4' is repeated in new interactions (1" to 4" and so on) without having a specific starting point through organizational interaction by the four patterns. The Tacit-Explicit (externalization/articulation) and Explicit-Tacit (internalization) patterns are important because they require a personal commitment from those who participate, since everything that involves tacit knowledge includes mental models and perceptions that are not easy to articulate and are highly subjective and dependent on individuals [23]. BPM Spiral and its analogy with Nonaka's SECI Model [23] is presented below (see Table 2).

Table 2. BPM Spiral and Nonaka's SECI model [23]

Spiral step	Knowledge type	Conversion by	Example
Strategize	Tacit - Tacit	Socialization	Boards members meeting for strategy discussion
Plan	Tacit - Explicit	Externalization	Business plan and budget definition, objectives, indicators, metrics and goals definitions, actions and initiatives and disclosure through internal communication channels
Monitor/analyze	Explicit - Explicit	Combination	Monthly/quarterly indicators monitoring and analysis through dashboards or maps, data exploration and pattern analysis
Act/adjust	Explicit - Tacit	Internalization	KPIs interpretation and insight generations, actions in order to improve performance and achieve the goals

5 Conclusions

The BPM framework proposed in this paper draws on the existent literature, but also on personal observations and professional experience about business performance. A BPM System includes five key-components: Environment, or context in which the organization operates, Organizational Culture, Systems and Information Technology, Processes and People.

A BPM System is much broader transcending organization's boundaries and includes not only internal but also external factors that directly and indirectly impact business performance. PEST analysis – Political, Economic, Socio-cultural and Technological factors - helps organizations to identify threats but also opportunities in the environment that surround the organization which it is part of.

Organizational culture plays a crucial role in individual performance, as well as organizational performance, inspiring people in the way they behave and act in their daily lives. Organization's DNA – Vision, Mission and Values – must be consistent with what is observed in practice and the leaders should be a mirror of this DNA, ensuring honest and sincere communication. Inconsistencies have an impact on individual performance, business and how the organization is perceived by its employees, customers and other stakeholders. An agile organization involves, besides the organizational structure simplification, the ability to adapt systems and processes in order to act quickly and effectively, addressing the new market and customer's needs. Therefore, it's important to foster a performance-oriented culture, continuous improvement and knowledge sharing.

Systems and Information Technology becomes crucial in monitoring and analyzing business performance, enabling problems anticipation, patterns discovery and variables correlation, among others. An exclusive DW area for BPM that feeds reports and dashboards, allows the recording and storage of indicator values, scenarios simulation and ensures the consistency of reported information. BI is the technological component that integrates a BPM system and supports the "BPM Spiral" processes, allowing the consolidation and treatment of high data volume and its availability for decision-making. On the other hand, BPM uses and leverages BI by adding context and direction, enabling the maximization of ROI for such projects.

The "BPM Spiral" is the process of PM that begins with strategy definition and alignment, its translation into plans, monitoring and analysis and the action and/or adjustment towards the goals' achievement. At the end of each complete cycle, the organization has a higher level of knowledge about its performance.

Finally, people are the *cuore* (heart) of an organization, the main asset and muscle that drives business performance. People's happiness and well-being become a competitive advantage, generating more motivation and enthusiasm, which in turn increase productivity, improving individual and business performance.

In order to continue the development of the presented BPM framework, in-depth studies should be carried out on each of the components, namely the processes component that was out of the scope of this paper. As part of the methodological approach for the BPM system implementation, a checklist to gather information about the "state of the art" should be developed to identify the components to be created or enhanced for business performance improvement.

References

1. Frolick, M.N., Ariyachandra, T.R.: Business performance management: one truth. Inf. Syst. Manag. **23**(1), 41–48 (2006)
2. Ballard, C., Mcdowell, S.: Business Performance Management Meets Business Intelligence, p. 224. IBM Redbooks, San Jose (2005)
3. Iervolino, C.: Business Performance Management, pp. 2–3 (2005)
4. Eckerson, W.: Best Practices in Management: Business Performance Management and Technical Strategies. TDWI Rep. Ser., no. March, p. 32 (2004)
5. Golfarelli, M., Rizzi, S., Cella, I.: Beyond data warehousing. In: Proceedings of the 7th ACM International Workshop on Data Warehousing and OLAP - DOLAP 2004, p. 1 (2005)
6. Zeng, L., Li, L., Duan, L.: Business intelligence in enterprise computing environment. Inf. Technol. Manag. **13**(4), 297–310 (2012)
7. Ong, I., Siew, P., Wong, S.: A five-layered business intelligence architecture. Commun. IBIMA 1–11 (2011)
8. Kimball, R., Ross, M.: The data warehouse toolkit: the complete guide to dimensional modelling. Career Data Anal. (978), 1–447 (2002)
9. Daft, R.: Management (2012)
10. Worthington, I., Britton, C.: The Business Environment (2006)
11. Smith, P.: Business Performance Management – approaches and tensions. ICAEW (2015)
12. Deger, T.: Global Changemaker - Shaping Effective Organizations. Global Changemaker (2010). https://www.global-changemaker.com/
13. Cokins, G.: Performance Management: Integrating Strategy Execution, Methodologies, Risk, and Analytics (2009)
14. McKinsey & Company. How to create an agile organization. McKinsey Co., no. October, pp. 1–16 (2017)
15. Wouter, A.K., De Smet, A., Weerda, K.: Agility: It rhymes with stability | McKinsey & Company. McKinsey Quarterly (2015)

16. Parmenter, D.: Key Performance Indicators (KPI) (2010)
17. Armstrong, M.: Strategic Human Resource Management: A Guide to Action, vol. 16, no. 3. Kogan Page Ltd., London (2006)
18. Barsh, J., Cranston, S., Craske, A.R.: Centered leadership: how talented women thrive in career. McKinsey Q. **4**, 35–48 (2008)
19. Dewhurst, M., Pettigrew, M., Srinivasan, R., Choudhary, V.: How multinationals can attract the talent they need. McKinsey Q. (3), 92–99 (2012)
20. Keller, S., Meaney, M.: Attracting and retaining the right talent, no. November, pp. 1–14. McKinsey Co. (2017)
21. Pina e Cunha, M., et al.: Desafios à conciliação família-trabalho. Nov. SBE, Confed. Empres. Port, p. 72 (2018)
22. Group, B.S.: Industry Framework Document. BPM Stand. Gr. (2005)
23. Nonaka, I., Takeuchi, H.: Knowledge-Creating Company, p. 43. Oxford University, New York (2007). Bloom. Bus. Libr. - Manag. Libr

Supply-Demand Matrix: A Process-Oriented Approach for Data Warehouses with Constellation Schemas

Luís Cavique[1(✉)], Mariana Cavique[2], and Jorge M. A. Santos[3]

[1] Universidade Aberta, BioISI-MAS, Lisbon, Portugal
luis.cavique@uab.pt
[2] Universidade Europeia, Lisbon, Portugal
mariana.cavique@universidadeeuropeia.pt
[3] Universidade Évora, CIMA, Évora, Portugal
jmas@uevora.pt

Abstract. Star schema in data warehouses is a very well established model. However, the increasing number of star schemas creating large constellations schemas add new challenges in the organizations. In this document, we intend to make a contribution in the technical architecture of data warehouses with constellation schemas using an extension of the bus matrix. The proposed supply-demand matrix details the raw data from the original databases, describes the constellation schemas with different dimensions and establishes the information demand requirements.

Keywords: Denormalization forms · Data warehouses · Constellation schemas · Process oriented

1 Introduction

To build a data warehouse, two types of architecture can be found: the Inmon architecture (Inmon 2005) and the Kimball architecture (Kimball and Ross 2013). In a historic perspective, Inmon coined the term 'data warehouse' in 1990 and in 1996 Kimball published the first edition of the Data Warehouse Toolkit (Breslin 2004).

On one hand, Inmon strategy advocates a top-down approach which begins with the corporate data model. On the other hand, Kimball's architecture uses a bottom-up approach based on the dimensional modeling, where the fundamental concept is the star schema. Most of the companies adopt Kimball's strategies, given the reduce costs of creating a star schema, but aspire a corporate model with Inmon's design. A detailed document reporting similarities and differences of the two methods can be found in Breslin (2004).

Since Kimball's strategy is supported by the development of different data marts, by distinct teams, it risks losing the integrated vision of the organization. In this work we choose the bottom-up data warehouse approach and extend the study to the constellation schema, since most of the bibliography focuses only on the star schema (Shin and Sanders 2006; Caldeira 2012; Santos and Ramos 2017).

Á. Rocha et al. (Eds.): WorldCIST 2020, AISC 1159, pp. 324–332, 2020.
https://doi.org/10.1007/978-3-030-45688-7_33

The goal of this paper is to develop a systematic procedure to transform more than one database into a constellation schema, given a set of requirements. This work defines data suppliers and information consumers and balances the supply and the demand of information flow.

The paper is organized in four sections. In Sect. 2, related work is presented. Section 3 presents the supply-demand matrix with a running example. Finally, in Sect. 4, some conclusions are drawn.

2 Related Work

In this document, we develop a procedure to support database denormalization and integration in a fact constellation schema of a data warehousing. In this section, first, we present a way to differentiate types of tables in a database. Then, we introduce a database denormalization process (Cavique et al. 2019). The bus matrix (Kimball and Ross 2013) and its extensions are reviewed. Finally, some aspects of technical architecture are reported.

2.1 Types of Tables

In the database denormalization process it is important to differentiate the types of tables in a database based on their relationships.

We reuse the work of Cavique et al. (2019) which identify three types of tables, using the following nomenclature, as shown in Fig. 1:

- lookup tables for tables only with cardinality equal to 1,
- intermediate tables for tables with cardinality 1 and N, and
- fact tables for tables only with cardinality equal to N.

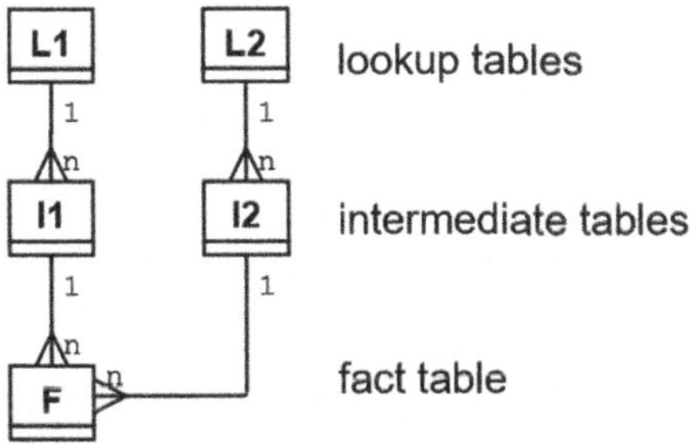

Fig. 1. Lookup, intermediate and fact tables

In this work we also draw all database tables with relation 1:N with the following rule - the table with a single line is drawn on top, while the table with multiple lines is drawn underneath.

In similar approaches, like modeling agile data warehouse with Data Vault (Linstedt and Graziano 2011) the authors also found three different types of structures: hubs, links and satellites.

A fact table in a database corresponds also to a fact table in a data warehouse, for that reason we use the same name.

2.2 Top-Down Database Denormalization Process

Cavique et al. (2019) present a top-down database denormalization process with two denormalization forms.

In the First Denormalization Form (1DF) given a database schema, in order to avoid multiple paths for the same query, a split strategy is applied aiming to find a poly-tree structure. In Fig. 2.a, in order to avoid multiple paths (L1-I1-F1-L2 and L1-I2-F2-L2) table L1 is duplicated and a poly-tree is found.

In the Second Denormalization Form (2DF) given a poly-tree the goal is to find for each fact table its own tree. In Fig. 2.b the poly-tree is divided into two trees which roots are F1 and F2.

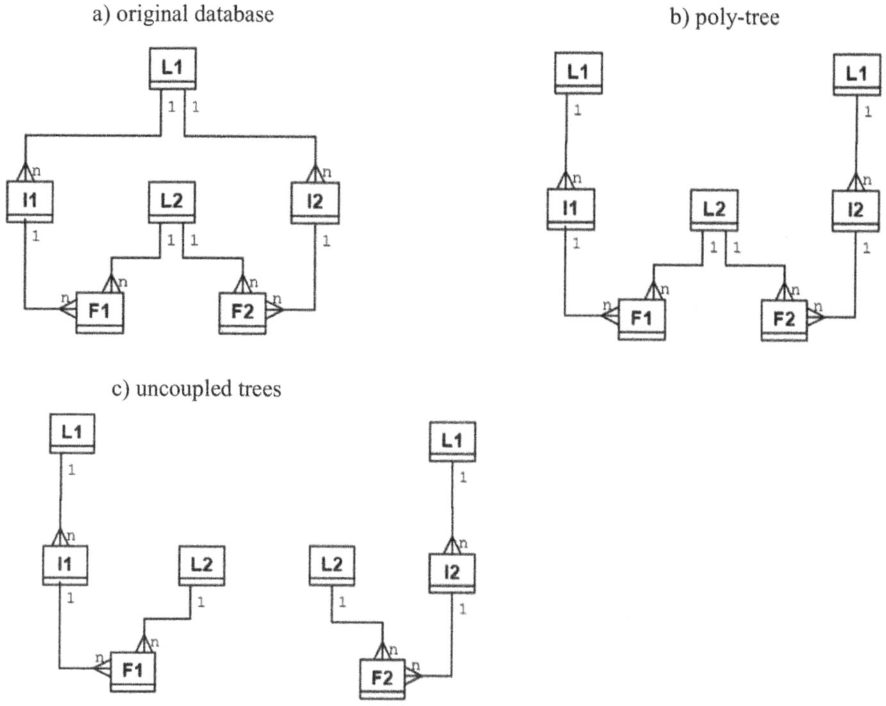

Fig. 2. Denormalization process from the original database to uncoupled trees

2.3 Bus Matrix Evolution

In process-based management, where processes are transversal to departments, the organizations can be represented by a matrix with processes versus departments.

The business process matrix, also called bus matrix (Kimball and Ross 2002), combines the business process with dimensions of the dimensional data model. The bus matrix is usually represented as processes versus dimensions.

The bus matrix is oriented to a single star schema. The complexity of the organizations leads them to store many star schemas or constellations. In order to show a constellation, the bus matrix can evolve in the dimensional data model and includes the fact tables instead of process (Shahzad and Sohail 2009). The constellation matrix is represented as facts versus dimensions, as represented in Fig. 3.

business process	fact table	dimension 1	dimension 2	dimension 3	dimension 4
Process X	A	X			X
	B		X	X	
	C				X

Fig. 3. Constellation matrix

2.4 Technical Architecture

The technical architecture of data warehouse is proposed by Kimball and Caserta (2004) where the concepts of back-room and front-room are proposed.

The back-room corresponds to the data management, in particular the sources subsystems and the staging area. The staging area is divided in two groups: (i) the ETL process of extracting, cleaning, conforming, and delivering data, and (ii) the storage of the dimensional tables ready to delivery atomic or aggregate data.

The front-room corresponds to the presentation area, where the user's community is able to browse and analyze data, using standard reports or ad-hoc queries.

The back-room and front-room work out like two separated data silos. In our work we propose a process from data source to data presentation in order to avoid redundancy or lack of information.

3 Proposed Model

In this section we develop a procedure to find the supply-demand information matrix. First, based on the database denormalization process of Cavique et al. (2019) a new denormalization process is presented. Then, we show how to extract all fact tables from a database. Finally, we present the constellation matrix, followed by the supply-demand matrix with a running example.

3.1 Bottom-Up Denormalization Process

As already mentioned, Cavique et al. (2019) present a top-down database denormalization process with two denormalization forms.

A similar decomposition process, with two phases, can be described using the inverse strategy, i.e. the bottom-up method.

In the First Denormalization Form using bottom-up method (1DF_bu) given a database schema, all the fact tables are identified, i.e., tables only with cardinality equal to N. In Fig. 2.a tables F1 and F2 should be recognized.

To obtain the Second Denormalization Form (2DF), for each fact table, add the tables from the upper level and repeat the process until no more tables can be added. Figure 4 exemplifies for table F1, in the first iteration I1 an L2 are added, and in the second iteration L1 is also added, obtaining Fig. 2.c on then left. The procedure is repeated for table F2 obtaining the two uncoupled trees of Fig. 2.c.

Summarizing the two denormalization strategies, for the First Denormalization Form we have two ways: 1DF via top-down, or 1DF_td, and 1DF via bottom up, or 1DF_bu. The Second Denormalization Form, 2DF, is equal for both pathways.

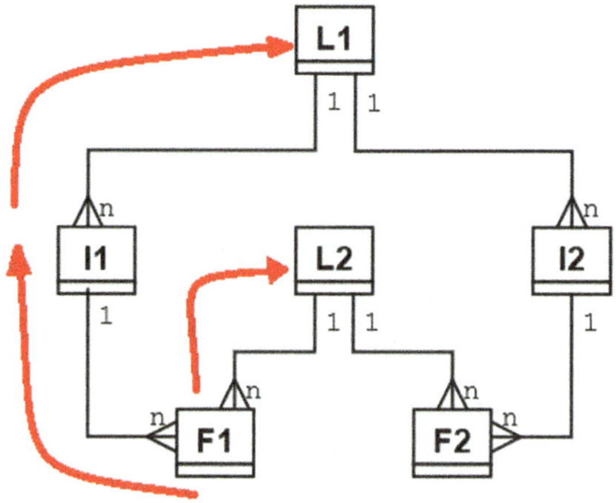

Fig. 4. Bottom-up denormalization process

3.2 Database Reduced Representation

In order to show how to extract all fact tables and dimensions from a database, we are going to exemplify the database reduction with the well-known Sakila database (2019), from the MySQL examples, which supports a DVD rental business.

First all the tables are categorized in lookup, intermediate or fact tables, using the definitions mentioned in Subsect. 2.1. Table 1 shows columns with the name of the table, the type of table and the type of facts. The information about the fact tables is extracted without going into the details of the attributes of each table.

The numeric measures in a fact table fall into three categories (Kimball and Ross 2013): additive facts, semi-addictive facts and non-additive facts. An extra category should also be mentioned, the fact tables without fact.

The reduced information from Sakila database retrieves a single fact table with additive facts, the Payment table.

Table 1 Sakila reduced databaseú

table name	type of table	type of facts
country	lookup	
city	intermediate	
address	intermediate	
customer	intermediate	
store	intermediate	
staff	intermediate	
rental	intermediate	
payment	fact	additive
actor	lookup	
language	lookup	
category	lookup	
film	intermediate	
inventory	intermediate	
film_category	fact	without facts
film_actor	fact	without facts
film_text	fact	without facts

3.3 Constellation Matrix

Given the fact tables of Table 1 and applying the denormalization process described in Subsect. 3.1 it is possible to associate fact tables and dimensions.

In the running example we add a Human Resource database with the previous one. Figure 5 shows the constellation matrix (Shahzad and Sohail 2009) for Sakila and Human Resource databases, where facts and dimensions come together. The type of facts also reports: 'a' means additive, 'na' non-additive and 'wf' without facts.

Business database	Fact table	type	customer	staff	rental	actor	category	film	inventory
Sakila	payment	a	X	X	X				
	film_category	wf					X	X	
	film_actor	wf				X		X	
	film_text	wf							X
Human Resource	payroll	a		X					
	job_history	na		X					

Fig. 5. Constellation matrix for Sakila and HR databases

3.4 Supply-Demand Matrix

The constant arrival of new legislation and new business opportunities generates new requirements to the system that should adapt to change. By data warehouse requirement we mean a report or a data view to analyze or mine. Each requirement should be supported by source data, i.e. one or more fact tables of the constellation schema as shown in Fig. 6. Since a requirement can use more than one fact table, a correlation sub-matrix is shown on the right. This type of correlation is inspired by the House of Quality (Tapke et al. 2003).

Given the fact table Payment with additive facts, it is possible to answer to the requirements of a rental weekly report. The other requirement is a monthly payroll report which is possible to obtain given the fact table Payroll. Annually it is required a job analysis with needs additive facts and non-additive facts from table Payroll and Job_history.

#	fact table	type	customer	staff	rental	actor	category	film	inventory	requirements	1	2	3	4	5	6
1	payment	a	X	X	X					rental weekly report	X					
2	film_category	wf					X	X								
3	film_actor	wf				X		X								
4	film_text	wf							X							
5	payroll	a		X						monthly payroll report					X	
6	job_history	na		X						annual job analysis					X	X

(supply) (demand)

Fig. 6. Supply-demand matrix

Requirement oriented data warehouse is a challenge for the Kimball architecture which uses a bottom-up approach. In Jovanovic et al. (2014) the authors present a method to iteratively design the multi-dimensional schema of a data warehouse from requirements. Our systematic Procedure 1 follows a similar approach, by iterating the finding of new fact tables, followed by the matching with new dimensions and integrating with requirements, until the balance between supply and demand is established.

Procedure 1. Generation of the Supply-demand Information Matrix:
Input: files, databases
Output: supply-demand matrix
1. Iterate
1.1. Find new Fact Tables
1.2. Match with Dimensions
1.3. Integrate with Requirements
2. Until balance between supply and demand is established

To find all the fact tables in a database the procedure Subsect. 3.2 is applied, which classifies each table into lookup, intermediate or fact. To match fact table with dimensions, creating a constellation matrix, the procedure in Subsect. 3.3 can be used. Finally, in order to integrate requirements, the information is mapped in the supply-demand matrix.

The emerging discipline of Organizational Engineering (Magalhães et al. 2007; Aveiro et al. 2011) advocates news principles. Organizational Engineering argues that each organization has its own identity and concerns for its interrelated subsystems. By developing meaningful meetings and business process KPI, organizations tend to be process-dependent rather than people-dependent. As a result, they can easily adapt personnel and they achieve high teams' performance.

Our process-oriented method of finding fact tables, matching dimensions and answer to requirement is iterative and incremental. On each iteration, new fact tables and/or dimensions should be added, to support new requirements. This approach goes beyond Technical Architecture design, with a back-room and a front-room working separately. The supply-demand matrix allows bird's-eye view of the data warehouse by representing the process from data source to data presentation, in order to avoid redundancy or lack of information.

The proposed systematic procedure follows also the Organizational Engineering by avoiding the human dependency, by establishing a set of rules to follow, strengthening aspects of systems engineering rather than constantly recreating new ways to solve the same problems for the purpose of personal appreciation.

In our data warehouse design a process is created from the data supply to the information demand. The process should iterate while the organization is learning and evolving.

4 Conclusions

Although star schema is a very well-established model, the increasing number of star schemas in large constellations adds new challenges in the organizations. Also, the constant arrival of new legislation and new business generates new requirements. Incremental demands, internal and external, cause the loss of the overall vision of the organization.

The goal of our paper is to develop a process-oriented procedure in the technical architecture of a data warehouses with constellation schemas using an extension of the bus matrix, in order to obtain a bird's-eye view of the system by representing the process from data source to data presentation. The integrated vision of supply and demand goes beyond technical architecture using a back-room and a front-room. This process view extracts information about the fact tables, without going into the details of the attributes of each table.

This work is also an attempt to bring together the visions of Kimball and Inmon, using a bottom-up approach to find fact tables and a top-down view to meet the requirements. The effort to match supply and demand of information avoids commitment on reports that does not correspond to actual data, causing the disappointment of the end users, and allows the deletion fact tables that are not used in the requirements.

An additional contribution regarding denormalization forms is reported. Given the top-down denormalization process by Cavique et al. (2019) we propose a bottom-up denormalization strategy, also with two denormalization forms, 1DF_bu and 2DF.

In future work, following the advices of Organizational Engineering we plan to establish KPI in our process-oriented data warehouse matrix.

References

Aveiro, D., Silva, A.R., Tribolet, J.: Control organization: a DEMO based specification and extension. In: First Enterprise Engineering Working Conference, EEWC 2011, Antwerp, Belgium (2011)

Breslin, M.: Data warehousing battle of the giants: comparing the basics of the Kimball and Inmon models. Bus. Intell. J. **7**, 6–20 (2004)

Caldeira, C.P.: Data Warehousing: conceitos e modelos com exemplos práticos, 2ª edição, Edições Sílabo (2012)

Cavique, L., Cavique, M., Gonçalves, A.: Extraction of fact tables from a relational database: an effort to establish rules in denormalization. In: Rocha, Á., Adeli, H., Reis, L., Costanzo, S. (eds.) New Knowledge in Information Systems and Technologies, WorldCIST 2019. Advances in Intelligent Systems and Computing, vol. 930, pp. 936–945. Springer, Cham (2019). https://doi.org/10.1007/978-3-030-16181-1_88

Inmon, W.H.: Building the Data Warehouse, 4th edn. Wiley, New York (2005)

Jovanovic, P., Romero, O., Simitsis, A., Abelló, A., Mayorova, D.: A requirement driven approach to the design and evolution of data warehouses. Inf. Syst. **44**, 94–119 (2014). https://doi.org/10.1016/j.is.2014.01.004

Kimball, R., Caserta, J.: The ETL Data warehouse Toolkit: Practical Techniques for Extracting, Cleaning, Conforming and Delivering Data. Wiley, New York (2004)

Kimball, R., Ross, M.: The Data Warehouse Toolkit: The Complete Guide to Dimensional Modeling, 2nd edn. John Wiley & Sons, New York (2002). ISBN 0471200247

Kimball, R., Ross, M.: The Data Warehouse Toolkit: the Definitive Guide to Dimensional Modeling, 3rd edn. Wiley, New York (2013). ISBN 9781118530801

Linstedt, D., Graziano, K.: Super Charge Your Data Warehouse: Invaluable Data Modeling Rules to Implement Your Data Vault. Create Space Publishing Platform, Scotts Valley (2011)

Magalhães, R., Zacarias, M., Tribolet, J.: Making sense of enterprise architectures as tools of Organizational Self-Awareness (OSA). J. Enterp. Archit. **3**(4), 64–72 (2007)

Sakila. https://database.guide/what-is-a-database-schema/sakila_full_database_schema_diagram/. Accessed Nov (2019)

Santos, M.Y., Ramos, I.: Business Intelligence: da informação ao conhecimento, 3ª edição, FCA - Editora de Informática (2017)

Shahzad, K., Sohail, A.: A systematic approach for transformation of ER schema to dimensional schema. In: Proceedings of the 6th International Conference on Frontiers of Information Technology, FIT 2009 (2009)

Shin, S.K., Sanders, G.L.: Denormalization strategies for data retrieval from data warehouses. Decis. Support Syst. **42**, 267–282 (2006)

Tapke, J., Muller, A., Johnson, G., Siec, J.: House of Quality: Steps in Understanding the House of Quality, IE 361. Iowa State University (2003)

Time-Series Directional Efficiency for Knowledge Benchmarking in Service Organizations

Thyago Celso Cavalvante Nepomuceno[1,2](✉), Victor Diogho Heuer de Carvalho[3], and Ana Paula Cabral Seixas Costa[1]

[1] Universidade Federal de Pernambuco, Recife, PE, Brazil
[2] Sapienza Università degli Studi di Roma, Rome, RM, Italy
nepomuceno@diag.uniroma1.it
[3] Universidade Federal de Alagoas, Delmiro, AL, Brazil

Abstract. Data Envelopment Analysis (DEA) is a linear programming tool that indicates benchmarking peers for inefficient service units to become efficient. Nevertheless, for strategic reasons the benchmarking of best practices and knowledge aggregation from efficient competitors is not usual. A time-series adaptation for directional model is proposed in this work as an alternative. The analysis applied to one branch unit of Brazilian Federal Saving Bank allowed an internal benchmarking of efficient periods of which innovative processes, competitive strategies, human resource changes, and specific incentive structures were adopted. This added knowledge provided an advantage to improve the performance of the service unit. In addition, managers to draw the best strategy in each period can use the model on pre-determined goals.

Keywords: Data Envelopment Analysis · Knowledge Management · Directional Distance Functions · Efficiency · Benchmarking

1 Introduction

Benchmarking techniques emerged in the late 80s as a quality management tool, defined as a manner of improving the performance of a Decision-Making Unit (DMU) by comparing its practices to a benchmark standard unit which are more effective in the results [1]. The information retrieved in this process can be leveraged into knowledge for the better understanding on the managerial aspects for the firm to remain dynamic and competitive. Knowledge Management (KM) is often perceived by the organizations as the product of this benchmarking process essential to the growth, viability and profitability of the enterprise [2]. The transference of best practices and the reuse of knowledge provides an environment propitious to innovation, delivering a superior value to the costumers for considering critical elements of the production process.

Since the 80's, benchmarking has been essential for the absorptive capacity leading to new knowledge generation, knowledge transfer to allow organizations to exchange what

Á. Rocha et al. (Eds.): WorldCIST 2020, AISC 1159, pp. 333–339, 2020.
https://doi.org/10.1007/978-3-030-45688-7_34

they do best, and empirically changing the companies' added knowledge in competitive market advantages [3]. The association of these elements results in an organizational environment conducive to the sharing of ideas, promoting fluidity in organizational knowledge consequently enabling companies to visualize themselves as driving entities of the knowledge spiral [4]. The organizational culture and climate are essential elements for organizations to effectively manage knowledge as people's attitudes directly affect the outcomes associated with knowledge exchange and combination [5]. Results and job-oriented organizational cultures, for example, have the ability to positively affect people's performance in the knowledge management process while tightly controlled cultures have the opposite effect to this process [6]. This highlights the importance of the human component in the fundamental triad of knowledge management (people, processes and technologies) and defines the important role that organizational management plays in creating conditions such as technical and organizational infrastructures, ensuring the maintenance of an environment conducive to knowledge sharing [7].

Where to obtain the critical information to be leveraged in knowledge still represents a special challenge. Today, Data Envelopment Analysis (DEA) models and ramifications are the most used mathematical programming tool to discover and assess efficient benchmarks, and they have been growing exponentially in number of applications and software developments [8, 9]. Nevertheless, for strategic reasons, firms, industries and service organizations often refuse to share critical information about their business success in order to maintain a competitive advantage, which jeopardizes much of the traditional DEA prospects. It is notable that companies that perform high quality management seek to motivate knowledge sharing among their people, ensuring organizational learning as a continuous improvement factor to their processes, and developing benchmarking as an indispensable activity for quality assurance [10]. The issue of refusing to benchmark, however, is common even considering only one big organization with several small branches. Not rarely, this business chain is characterized by harsh competition among the branches for better rewards and recognition from the top administration.

In order to cope with this issue, we propose a time-series based formulation for the Directional Distance Function (DDF) model of efficiency analysis. The linear programming provides time-sorted efficient peers for internal assessments instead of relative efficient competitors. As result, we have a set of periods where the analyzed organization has reached the efficiency in reducing their resources and/or expanding their products. Each of these efficient periods works as the standard benchmark for one or more inefficient periods according to the production configuration and seasonal aspects. With this information, the manager is able to investigate the best practices and strategies which led to the efficiency in the specific period, and the periods where those actions and strategies can be coherently applied. In this way, the quality of knowledge as a final product from the knowledge management processes can be ensured, resulting in an improvement for the efficient decision making processes in the service organizations.

The main contribution of this assessment is the identification and aggregation of strategic knowledge that can be crucial for the business success. Besides the added knowledge from this contribution, the model can provide an absolute number for the resource slack to be spared or product to be expanded in each specific inefficient moment. In addition, the model can also be adapted to include specific goals from the decision maker and then verify, according to similar past production configurations, the optimal number of personal or technical resource required to reach the pre-defined goal efficiently. The application in one branch of the Brazilian Federal Savings Bank highlights the prospects of this contribution.

2 Time-Series Directional Efficiency

Directional Distance Functions provide non-radial measures for the technical efficiency of service units by imposing a feasible direction g_{x_i} $\forall\, i = 1, 2, \ldots, n$ for the input contraction and/or g_{y_r} $\forall\, r = 1, 2, \ldots, s$ for the output expansion of a given decision unit j from m decision units operating under a known production technology set $\Psi_{(x,y)} \in \mathbb{R}^{n+s}$ so that x can produce y (see the general and robust formulations in [11–13]). Traditionally, this is a powerful benchmarking tool to gauge an efficient frontier Ψ^* in the production set by maximizing the feasible contraction and/or expansion β so that $max\, \beta > 0 \,|\, (x - \beta g_{(x)}, y + \beta g_{(y)}) \in \Psi_{(x,\,y)}$. In the absence of data from competitors, however, the benchmarking procedure can be performed through time-series data from the unit under evaluation.

A close formulation for time-series data by [14] adapted the common input-oriented slack-based model with the inclusion of manager goals for business transactions in the same context of this application. The focus, however, set on resource allocation would resort to both time-series and cross-sectional data for m service units operating under the same production technology. The following linear programming algorithm adapts a directional distance model to focus on the internal knowledge benchmarking by the m service units which does not require to assume the same production technology for it compares the performance of the service unit in the moment $t' = 1, 2, \ldots, p$ to the remaining moments $t \in p - 1\, \forall\, t \neq t'$. With sufficient data, this produce robust time-sorted peers for the benchmarking of best practices.

Consider a set of p periods (or moments) where in each period $t = 1, 2, \ldots, p$, the service unit uses $= 1, 2, \ldots, n$ inputs to produce $r = 1, 2, \ldots, s$ in the most favorable direction for the input reduction g_{x_i}, or output improvement g_{y_r}, defined by the data. Different market scenarios for the business organization may justify choosing different directions by the manager under specific incentive structures. In addition, the decision maker can have different preferences over different periods with regard the usage of their internal resources. Nevertheless, the choice for some specific subjective direction is not crucial in this analysis. An input-oriented directional efficiency, as defined by [12] and [15] can be adapted to:

$$\mathrm{D}_t\left(x, y, g_{x_i}, g_{y_r}\right) = max\, \beta$$

$$s.t. \quad \sum_{t=1}^{p} z_t\, y_{tr} \geq y_{t'r};\ \sum_{t=1}^{p} z_t\, x_{ti} \leq x_{t'i} - \beta g_{x_i};\ \sum_{t=1}^{p} z_t = 1,\, z_t \geq 0; \qquad (1)$$

$$\forall\, r = 1, 2, \ldots, s;\ i = 1, 2, \ldots, n;\ t = 1, 2, \ldots, p$$

As result, we have relative measure for the technical efficiency β and an absolute measure for the technical inefficiency $(x_{t'i} - \beta g_{x_i})$ which defines the set of time-sorted benchmarks for the knowledge aggregation. The periods of time presenting $\beta = 1$

(and $(x_{t'i} - \beta g_{x_i}) = 0$) are the efficient periods to be internally investigated. For all the remaining inefficient periods, the programming provides linear combinations toward the closest points in the efficient frontier for the benchmarking of the best practices (see Fig. 1). Including a goal by the manager after the construction of the time-sorted efficient frontier is made by changing the second and third constraint in the linear formulation to include an expected input i' and the desirable output r' by the manager. We have a feasible contraction of a predicted input and/or an expansion of the output compared to this inclusion:

$$D_t\left(x, y, g_{x_i}, g_{y_r}\right) = max\ \beta$$
$$s.t. \quad \sum_{t=1}^{p} z_t\, y_{tr} \geq y_{r'};\ \sum_{t=1}^{p} z_t\, x_{ti} \leq x_{i'} - \beta g_{x_i};\ \sum_{t=1}^{p} z_t = 1,\ z_t \geq 0, \quad (2)$$
$$\forall\, r = 1, 2, \ldots, s;\ i = 1, 2, \ldots, n;\ t = 1, 2, \ldots, p$$

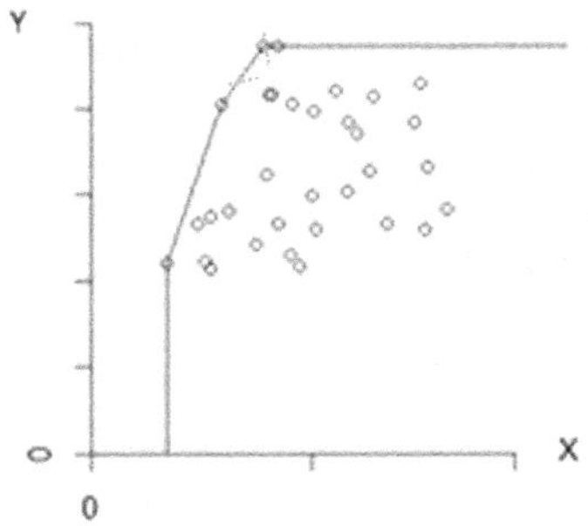

Fig. 1. Time potential improvements for the efficiency benchmarking

3 Assessment

Caixa Econômica Federal (CEF) is the biggest state-owned bank in Latin America. It is the Brazilian government right-hand executor of social policies, having a crucial importance in the country's development. Until recently, the bank considered three main

Table 1. Descriptive statistics

Variables	Average	Standard deviation	Minimum	Maximum	Improvement potential (Max.)
Employees (Input)	25.53	1.74	22	28	–
SPTS results (Output)	158397.75	21594.62	99079	196349	84489.2
Business (Output)	1493.47	234.90	1105	2047	945

blocks of activities: Business Transactions, Risk Assessments and Social Programs and Teller Services (SPTS). For the purpose and limitation of this work, only the results associated to the Business Block and Social Programs and Teller Services Block are considered as the output. As input, the number of employees are considered. The data is regarded to one specific branch located in the state of Pernambuco for the past 3 years (36 months as the units for the assessment, from July 2016 to June 2019). The number of employees diverge from one month to the other due, mainly, to vacations, maternity leave and other medical issues.

Table 2. Summary of the benchmarking results

Closest benchmarks	Inefficient periods	Maximum potential for improvement		Some reviewed strategies (added knowledge)
		Service	Business	
July/16 August/2016 December/2016 March/2017	Oct/17; Nov/16; May/17; Jun/17; Aug/18	30078.0 (Oct/2017)	721.0 (Oct/2017, Nov/2016)	1. Schedule Opening Accounts; 2. Redirecting SP clients; 3. Motorcycle Insurance (in the high incidence of felonies); 4. Increased Payroll Loans; 5. Internet Banking and ATM campaigns and propagandas
January/2017 October/2018	Dec/18; Jul/18; Nov/17; Set/18; Jun/18; Mar/18; May/18; Feb/17; Sep/17; Apr/17; Aug/17; Oct/16;	42148.0 (Dec/2018)	910.5 (Feb/2017)	1. Day-offs Bonus (on specific individual goals and campaigns); 2. Reallocation of 2 employees; 3. Allocating one exclusive employee for screening particular clients before opening; 4. Increased number of meetings; 5. Individual meetings (short talks with the manager)
April/2019 May/2019	Jun/19; Mar/19; Feb/19; Jan/19; Dec/17; – Jan/18; Feb/18; Set/16; Nov/18; Apr/18; Jul/17	84489.2 (Jun/2019)	945.0 (Feb/2018)	1. Adjustment in the unit's layout 2. Changing product natures; 3. Day-offs and Gifts (on specific individual goals and campaigns); 4. Increased number of phone calls for potential clients (to update customer information)

Number of Efficient Benchmark Periods: 8; Number of Inefficient Periods: 28; Median: 0.09164; Eff range: D = 0: 22.2%; 0 < D =< 0.05: 11.1%; 0.05 < D =< 0.1: 22.2%; 0.1 < D =< 0.15: 19.4%; 0.15 < D =< 0.2: 16.7%; 0.2 < D =< 0.25: 0.0%; 0.25 < D =< 0.3: 2.8%; 0.3 < D =< 0.35: 2.8%; 0.35 < D =< 0.4: 0.0%; 0.4 < D =< 0.45: 2.8%;

An output-oriented model under the assumption of variable return to scale was selected for the assessment. Table 1 presents some descriptive statistics about the main variable used in the assessment. The Improvement Potential in the last column represents the most inefficient month considering the results on SPTS output (Jun/2019) and Business Transactions output (Feb/2018). The Fig. 2 illustrates the Isotrans frontier (the efficiency frontier considering the two outputs in this assessment) under the Directional Distance Function technology in the Eq. (2). The points on the frontier represent the efficient months that are the benchmarks for the remaining inefficient months. This information is provided by the Table 2 as a summary of the results.

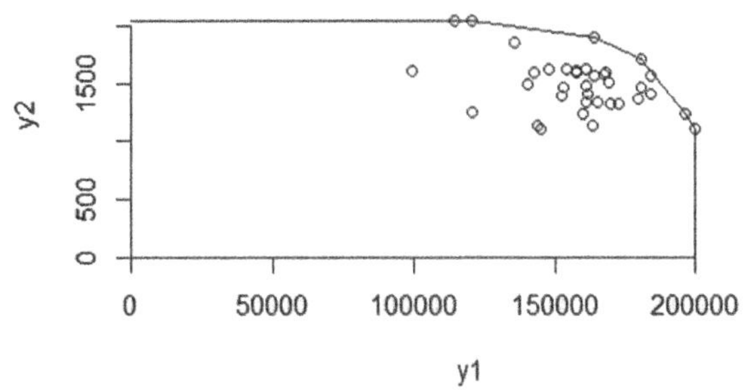

Fig. 2. Transformation Curve (Isotrans) for the Time-series DDF technology

4 Discussion and Final Remarks

Table 2 has the information on the inefficient months and their closest benchmark months, eight in total, the maximum potential improvement (slacks) in the results for the Business Block and Social Programs and Teller Service Block. As an instance, during the first season class, which has the months Oct/17, Nov/16, May/17, Jun/17, and Aug/18, the unit can improve their results up to 721 business transactions and serve up to 30078 more clients. Some of the strategies introduced by the unit under evaluation during or close to the efficient benchmark months are described in the last column of Table 2. This added knowledge was possible through interviews with employees based on the specific efficient months resulted from this application.

These strategies, when adopted by the unit under evaluation, may be significant to gauge an efficient production process resorting to less resource to produce more business transactions and/or services. This short contribution did not apply the definition of managerial goals. This validation by consultations with the decision maker of the service unit presents an interesting extension for the proposed model. Designing new knowledge management strategies, process innovations, planning and control the performance of human resources according to the seasonality and specific environmental aspects of the business are some of the possibilities.

References

1. Carpenter, S., Rudge, S.: A self-help approach to knowledge management benchmarking. J. Knowl. Manag. **7**(5), 82–95 (2003)
2. O'Dell, C., Wiig, K., Odem, P.: Benchmarking unveils emerging knowledge management strategies. Benchmark. Int. J. **6**(3), 202–211 (1999)
3. Cepeda-Carrion, I., Martelo-Landroguez, S., Leal-Rodríguez, A.L., Leal-Millán, A.: Critical processes of knowledge management: an approach toward the creation of customer value. Eur. Res. Manag. Bus. Econ. **23**, 1–7 (2016). https://doi.org/10.1016/j.iedeen.2016.03.001
4. Nonaka, I., Konno, N.: The concept of "Ba": building a foundation for knowledge creation. Calif. Manage. Rev. **40**, 40–54 (1998). https://doi.org/10.2307/41165942
5. Wang, S., Noe, R.A.: Knowledge sharing: a review and directions for future research. Hum. Resour. Manag. Rev. **20**, 115–131 (2010). https://doi.org/10.1016/j.hrmr.2009.10.001
6. Chang, C.L., Lin, T.-C.: The role of organizational culture in the knowledge management process. J. Knowl. Manag. **19**, 433–455 (2015). https://doi.org/10.1108/JKM-08-2014-0353
7. Hooff, B., Huysman, M.: Managing knowledge sharing: emergent and engineering approaches. Inf. Manag. **46**, 1–8 (2009). https://doi.org/10.1016/j.im.2008.09.002
8. Daraio, C., Kerstens, K., Nepomuceno, T., Sickles, R.C.: Empirical surveys of frontier applications: a meta-review. Int. Trans. Oper. Res. **27**, 709–738 (2020). https://doi.org/10.1111/itor.12649
9. Daraio, C., Kerstens, K.H., Nepomuceno, T.C.C., Sickles, R.: Productivity and efficiency analysis software: an exploratory bibliographical survey of the options. J. Econ. Surv. **33**(1), 85–100 (2019)
10. Calvo-Mora, A., Navarro-García, A., Rey-Moreno, M., Periañez-Cristobal, R.: Excellence management practices, knowledge management and key business results in large organisations and SMEs: a multi-group analysis. Eur. Manag. J. **34**, 661–673 (2016). https://doi.org/10.1016/j.emj.2016.06.005
11. Chambers, R.G., Chung, Y., Färe, R.: Profit, directional distance functions, and Nerlovian efficiency. J. Optim. Theory Appl. **98**(2), 351–364 (1998)
12. Färe, R., Grosskopf, S.: New directions: efficiency and productivity, vol. 3 (2006)
13. Daraio, C., Simar, L.: Directional distances and their robust versions: computational and testing issues. Eur. J. Oper. Res. **237**(1), 358–369 (2014)
14. Nepomuceno, T.C., Costa, A.P.C.: Resource allocation with time series DEA applied to Brazilian Federal Saving banks. Econ. Bull. **39**(2), 1384–1392 (2019)
15. Nepomuceno, T.C.C., Daraio, C., Costa, A.P.C.S.: Combining multi-criteria and directional distances to decompose non-compensatory measures of sustainable banking efficiency. Appl. Econ. Lett. **27**(4), 329–334 (2020)

Learning Patterns Identification as a Strategy for Digital Appropriation Skills in Fresher University Students

David Alberto García Arango[1(✉)], Gloria Cecilia Agudelo Alzate[2], Oscar Andrés Cuéllar Rojas[1], Jorge Eliécer Villarreal Fernández[1], and Cesar Felipe Henao Villa[1]

[1] Corporación Universitaria Americana, 050012 Medellín, Colombia
dagarcia@coruniamericana.edu.co
[2] Universidad de Antioquia, 050010 Medellín, Colombia

Abstract. In the current era of the digital and knowledge society with an imminent fourth industrial revolution, many subjects related to the appropriation of Digital Technologies are understood as basic professional skills of the 21st century. In this way of thinking, this article shows the results obtained from a study of 205 last year's high school students about their learning patterns and how this could be related to the acquisition of digital technology appropriation skills in their first university semester. The mixed research approach uses the data from a survey of learning patterns named Inventory of Learning Styles (ILS) to perform a factor analysis and identify clusters that allowed to classify the population into groups according to their affinity level with learning patterns, closely related to skills of appropriation of digital technologies. The research makes possible to establish curricular management strategies to strengthen these skills. With the data analysis, it was possible to identify three groups where one of them has greater affinity with learning patterns oriented to meaningful learning, this group also has a better performance in digital technology appropriation skills.

Keywords: Digital technologies skills · Inventory of Learning Styles · Learning patterns · Factor analysis · Fresher university students

1 Introduction

The incursion of Digital Technologies in productive processes led to ways of thinking reconfiguration about learning and teaching. This can be possible by reconstructing the concept of skill in education and reviewing the challenges that we must face in a context determined by the so-called knowledge society.

Content overcrowding, development of new ways of knowing and exponential growth of data and information, are changing perspectives regarding the desirable skills for citizens, those essential for the modern professional who intends to be successful in the future. Such skills can be seen in more detail in [1].

Á. Rocha et al. (Eds.): WorldCIST 2020, AISC 1159, pp. 340–350, 2020.
https://doi.org/10.1007/978-3-030-45688-7_35

Some projects oriented towards the identification of these skills can be studied in [1] that categorize them as: ways of thinking, ways of working, tools for working, living together in the world. Besides, [2] relates them to UNESCO's purposes towards learning to know, learn to do, learn to be and learn to live together; additionally the OECD [3] associates these skills with the purposes of interacting in heterogeneous groups, using tools interactively, acting autonomously; [4], relates the 21st century skills with learning and innovation, media and technology, life and career. Finally, [5] considers the aspect of learning to learn as fundamental.

Reviewing texts describing this type of skills and comparing them with the new taxonomy of educational objectives as a framework for objectives, evaluation and standards proposed by Marzano and Kendall [6], it is possible to identify that 21st century skills development depends strongly of another skills in the wat to cognitive, metacognitive and internal levels, which suppose the domain of problem solving and research for productivity and in the line of basic principles for citizenship, construction and communication in global environments. In this sense, it is essential to inquire about learning, since it is from there that key aspects involved in the development of competencies. In this paper, learning patterns for a specific context are identified in order to propose ways to support curriculum management policies in universities that allow a better appropriation of digital skills.

In line with the above, this article exposes the development of the identification of learning patterns through the application of the questionnaire proposed by Vermunt [7], as a first diagnostic step in the development of skills in appropriation of digital technologies in fresher university students.

2 Research Context

The 21st century skills proposed by the ATC21s project are divided into four categories (ways of thinking, ways of living in the world, ways of working, tools for working) with subcategories promoting the integral development for students (see Table 1).

Table 1. Categories and subcategories of 21st century skills (see [1])

Category	Associated subcategory
Ways of thinking	Creativity and innovation Critical thinking Problem resolution Learn to learn
Ways of living in the world	Personal and social responsibility Local and global citizenship
Work tools	Technologies appropriation Information management
Ways of working	Communication Collaboration

For this research, there is special interest in the "Work Tools" category and the consequent subcategory of "appropriation of digital technologies" defined as: "ability to explore, create, communicate and detect technologies as tools" [1]. From this definition it is interpreted that the basis of this skill is the ability to explore, create, communicate and produce in other environments mediated by digital technologies.

This basis needed for appropriation of digital technologies can be observed too in a review study carried out by Rodríguez García, Romero Rodríguez and Campos Soto [8] who said that, "... despite of abilities that digital natives present with respect to technologies, their digital competence can be inferior to that of digital migrants". The author manifests the necessity to attain conditions like producing digital content based in learning and creativity skills.

Going deeper into the concept of digital skill, the one proposed by the National Institute of Educational Technologies and Teacher Training (INTEF acronym in Spanish) and according to the Common Framework of Digital Teaching Skills based on European guidelines, defines digital skill as the "creative, critical and safe use of information and communication technologies to achieve objectives related to work, employability, learning, leisure time, inclusion and participation in society" [9, p. 9].

The Common Framework of Digital Teaching Skills establishes five dimensions in which this competence is subdivided: information and informational literacy, communication and collaboration from digital networks, digital content creation, responsible and secure use of the network and technology in general, and technology-mediated problem solving [8].

In this matter, it was proposed to identify learning patterns as a way of approaching teaching of digital skills appropriation, this was possible by using Vermunt's Learning Styles Inventory (ILS) which consists of 120 questions distributed in four domains or dimensions related to learning.

The combination of these patterns in four different complexity levels make possible to characterise the so-called learning patterns, which are selected as "a coherent set of learning activities that students usually use, their beliefs about learning and their motivation for learning, a set which is characteristic of them in a certain period of time" [10]. The questionnaire in mention inquiries about learning conceptions, motivational orientations about learning, learning regulation strategies and learning processing strategies in students [11]. The combination of these levels of appropriation in each orientation generates a different learning pattern as shown in Table 2.

Table 2. Domain and learning patterns (appropriation levels as can be seen in [11])

Domain	Meaning Directed (MD)	Reproduction Directed (RD)	Application Directed (AD)	Undirected (UD)
Mental models of learning	Construction of knowledge	Intake of knowledge	Use of knowledge	Co-operation and stimulating education

(*continued*)

Table 2. (*continued*)

Domain	Meaning Directed (MD)	Reproduction Directed (RD)	Application Directed (AD)	Undirected (UD)
Learning orientations	Personally interested	Certificate and self-test directed	Vocation directed	Ambivalent
Regulation strategies	Self-regulation (Learning processes, results and learning content)	External regulation (Learning processes and results)	External and self-regulated	Lack of regulation
Processing strategies	Deep processing (relating, structuring and critical)	Stepwise processing (memorizing, rehearsing and analyzing)	Concrete processing	Scarce processing

3 Methodology

The study was mixed and cross-sectional, demarcated by a pragmatic-paradigmatic level. In this sense, results and theories that are adjusted to each other give as result the case being analyzed [12].

The data collection instrument has been applied to 206 high school students. The instrument was Inventory Learning Styles (ILS) developed by Vermunt J. D. [12] and translated into Spanish by Martínez-Fernández et al. [13]. The population selected is in the José María Bernal Educational Institution in the municipality of Caldas of the department of Antioquia - Colombia. The institution has more than 3200 students distributed in three locations: elementary, secondary, and vocational media levels. The context of the research takes place in an urban area where the families of the students are of medium-low income. The institution, being of public type, is regulated by the guidelines defined by the Educational Secretary regulated by the Ministry of National Education.

The data was treated with the SPSS software [14] in this way: in first place, a factor analysis was applied with the method of extraction of principal components with their respective KMO and Bartlett's sphericity test. In second place, the regression values generated from the extraction by components, were stored in order to make a hierarchical cluster analysis with linking groups method and a measure of squared Euclidean distance.

In a third stage, scatterplots were generated in order to compare pairs of regressions obtained in the first stage and to identify how the clusters obtained in the second stage are distributed (the validity of the complementary use of factor analysis and cluster analysis hierarchical can be verified in Gorman and Primavera [15]). As a fourth an last stage, the results obtained were analyzed in order to generate research conclusions identifying relationship patterns that affect the teaching of Digital Technologies appropriation skills.

4 Results and Analysis

In this section the results for each stage showed in the methodological aspects are presented.

4.1 Stage I

In factor analysis, the characteristics related to the meaning directed learning pattern (MD) were chosen, whose subcategories in each domain are construction of knowledge, personally interested, self-regulation (learning processes, results and learning content) and the deep processing (relating, structuring and critical). The mentioned characteristics are complementary and define a group of behaviors oriented towards significant learning. The values obtained in the matrix of principal components are presented below (Table 5).

Table 3. Component matrix in mental model of learning – construction of knowledge

Question number	Component 1
85	,576
88	,604
92	,579
96	,597
98	,665
104	,573
116	,569
117	,749
119	,742

Extraction method: principal components analysis. One extracted component

Table 4. Component matrix in learning orientations – personally interested

Question number	Component 1	Component 2
57	,245	,608
65	,697	,300
69	–,527	,653
74	,232	,576
78	,753	,117

Extraction method: principal components analysis. Two extracted components

Table 5. Component matrix in regulation strategies – self-regulation

Question number	Component 1
21	,663
24	,653
31	,597
36	,649
46	,665
50	,633
51	,680
16	,643
119	,742
28	,672
42	,662
54	,590

Extraction method: principal components analysis. One extracted component

Table 6. Component matrix in processing strategies – deep processing

Question number	Component 1
6	,617
10	,695
13	,692
19	,702
25	,738
34	,715
35	,698
29	,752
39	,532
43	,720
49	,633

Extraction method: principal components analysis. One extracted component

Note how motivation characteristic for Table 4 is subdivided into two components; each component is identified with a group of questions that can be interpreted in order to build categories. Regressions called Component3A and Component3B can be interpreted as the components obtained from the characteristics related to motivation for personal interest that does not depend of knowledge type and motivation for personal interest that does depend on the type of knowledge respectively.

Each element of Tables 3 to 6 corresponds to a question from the ILS questionnaire. In total, five components were extracted with their respective regression indexes, with this result, a hierarchical cluster analysis was performed as can be seen in the second stage.

4.2 Stage II

Clustering based on regressions of main components results in three groups that according to the ANOVA test, are independent (see Table 7).

Table 7. Component matrix in processing strategies – deep processing

Component		Square sum	DF	Quadratic Mean	F	Sig.
Component 1	Between groups	75,888	2	37,944	59,65	,000
	Inside groups	129,11	203	,636		
	Total	205,00	205			
Component 2	Between groups	68,452	2	34,226	50,88	,000
	Inside groups	136,54	203	,673		
	Total	205,00	205			
Component 4	Between groups	85,792	2	42,896	73,04	,000
	Inside groups	119,20	203	,587		
	Total	205,00	205			
Component 3A	Between groups Inside groups	62,867 142,13	2 203	31,434 ,700	44,89	,000
	Total	205,00	205			
Component 3B	Between groups Inside groups	8,484 196,51	2 203	4,242 ,968	4,382	,014
	Total	205,00	205			

Table 7 shows the results of cluster independence obtained from Component1 (learning through deep processing), Component2 (self-regulation of learning), Component3A (motivation for personal interest independent of the type of knowledge), Component3B (motivation for personal interest depending of type of knowledge) and Component4 (knowledge as construction of learning).

4.3 Stage 3

Graphs are generated in order to link the results of Stages I and II. For each graph, the clusters are represented by circles with different colors (blue, green and brown) and their respective numbers (1, 2 and 3), the horizontal and vertical lines represent the average value on each axis, forming quadrants that allow to classify the sample according to its grouping tendency. The higher the points are, the higher values in the vertical axis component; in other hand, the more to the right are the points, it is assumed that they have higher values in the horizontal axis component (Figs. 2, 3, 4, 5, 7 and 8).

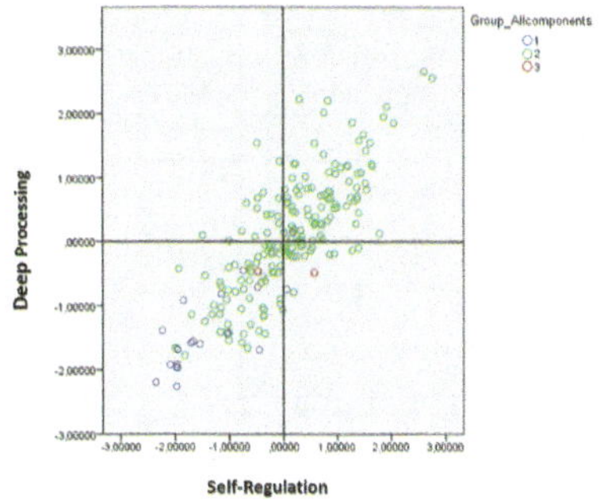

Fig. 1. Comparative graph for deep processing and self-regulation (left).

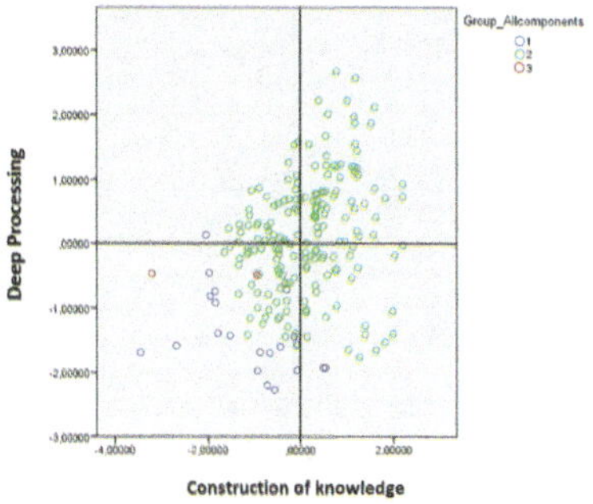

Fig. 2. Comparative graph for deep processing and construction of knowledge (right).

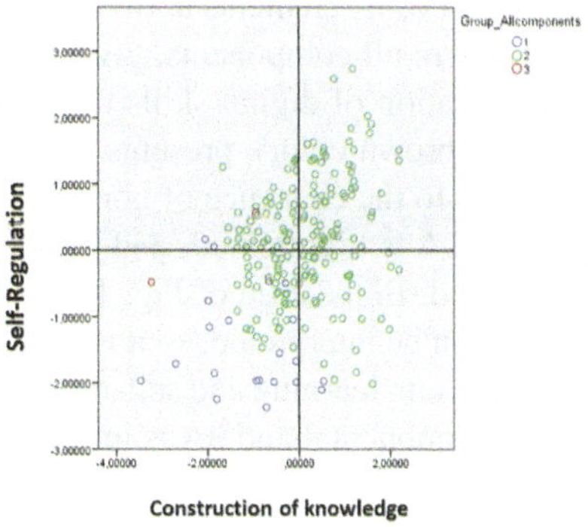

Fig. 3. Comparative graphic for self-regulation and construction of knowledge (left).

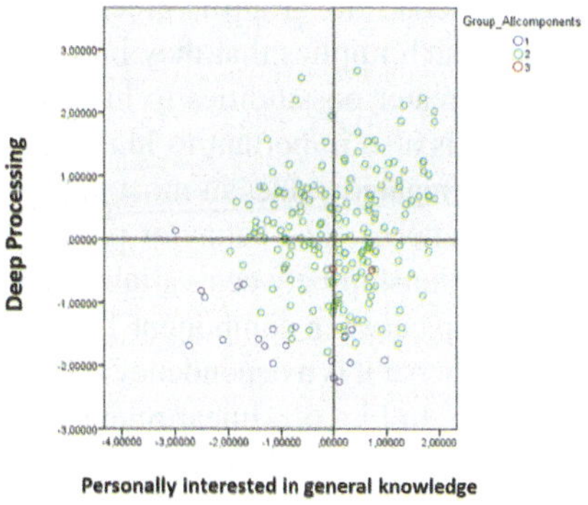

Fig. 4. Comparative graphic for deep processing and personally interested in general knowledge (right)

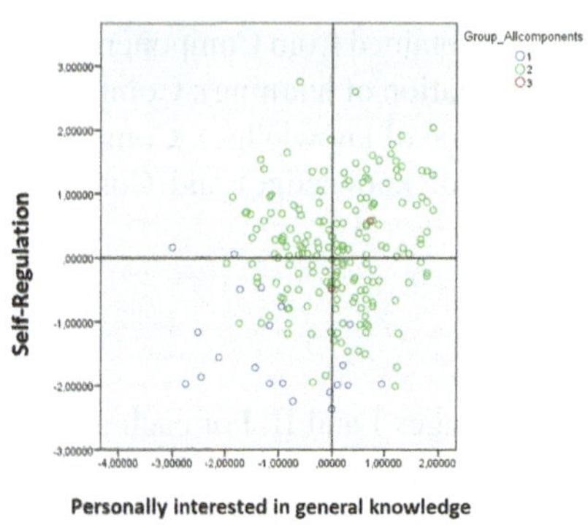

Fig. 5. Comparative graphic for self-regulation and personally interested in general knowledge (left).

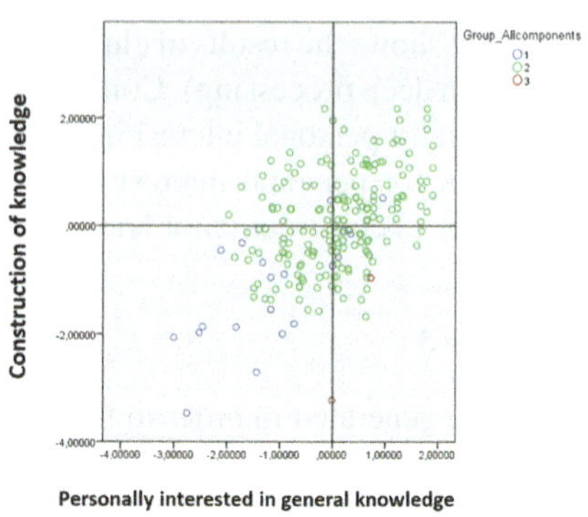

Fig. 6. Comparative graphic for construction of knowledge and personally interested in general knowledge (right).

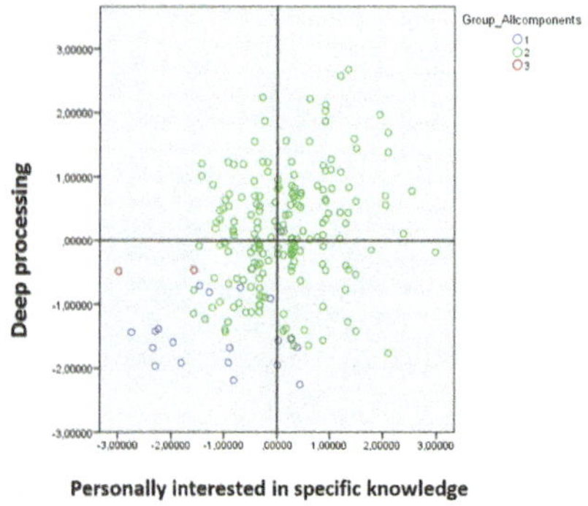

Fig. 7. Comparative graphic for deep processing and Personally interested in specific knowledge (left).

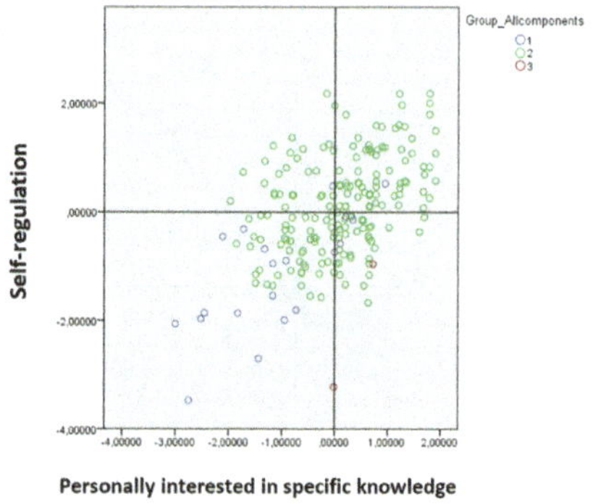

Fig. 8. Comparative graphic for self-regulation and personally interested in specific knowledge (right).

As general rule, group 1 (green), was found in most of its elements above the other groups, which implies that they have with high values in all components, giving these students greater possibilities to have a better appropriation of digital skills than other groups. It is also important to highlight that group 3 (brown color), presents disparate and disaggregated values in most graphs, which is due to the existence of non-directed learning patterns. The behavior pattern of Figs. 1 and 6 is highlighted, and besides a linear relationship between variables has been observed. In the case of Fig. 1, a linear relationship between component 1 and component 2 can be interpreted, so it is possible to establish that it is a dependency between deep processing learning and self-regulation of learning. In Fig. 6, a linear relationship between component 4 and 3A is interpreted, being possible to establish a dependence between learning as a knowledge construction and motivation for personal interest independent of the type of knowledge.

5 Conclusion

The results allow to identify three groups where one of them has greater affinity in relation to learning patterns directed at meaning. This group also has a better performance at digital technology appropriation skills. There is a dependency in the sample studied, between a deep learning processing (elaborated, structured and critical) and self-regulation of learning. Therefore, a self-regulated student in their learning will be more likely to have elaborate, structured and critical processing capabilities.

There is a dependence between the interest in knowledge and the perception that the student has about knowledge. The greater interest in knowledge, the more critical and reflective is the way in which the student conceives knowledge. A future research study is related to conducting this research longitudinally, identifying changes in the combinatorial learning patterns every time that significant experiences with digital technologies are generated. Using this it's possible to introduce curricular adaptations that allow an improvement in the learning pattern that could show if the green group becomes more numerous. This line of research is also proposed in Vermunt and Donche [11].

The sample used, are senior high school students who mostly have a learning pattern aimed at meaning, which is related to the competence of appropriation of digital technologies. In this sense, a learning pattern aimed at meaning generates the conditions to the structured and critical elaboration of knowledge, as well as for the self-regulation of processes and contents in education. The university curricular management should consider the identification of the prevalence of this learning pattern and based on it, implement curricular intervention policies in fresher university students.

References

1. Binkley, M., Erstad, O., Herman, J., Raizen, S., Ripleym, M., Rumble, M.: Defining 21st Century Skills. University of Melbourne, Melbourne (2010)
2. Delors, J.: Learning the Treasure Within; Education: the Necessary Utopia. UNESCO Publications, París (1996)
3. OECD: Doctorate for Education and Skills. Skills beyond school. Testing student and universiy performance globally. OECD, Bruselas (2012)
4. Brown, S.: Best Practices in 21st Century Learning Environments: A Study of Two P21 Exemplar Schools. Brandman University, Irvine (2018)
5. Gordon, J.: Key competences in Europe: Opening doors for lifelong learners across the school curriculum and teacher education. Case Networks Reports, p. 87 (2009)
6. Marzano, R.J., Kendall, J.S.: The New Taxonomy of Educational Objectives. Sage Publications, London (2007)
7. Vermunt, J.D.: The regulation of constructive learning processes. Br. J. Educ. Psychol. **68**, 149–171 (1998)
8. Rodríguez García, A.M., Romero Rodríguez, J.M., Campos Soto, M.N.: De nativos digitales a aprendices digitales: la realidad que se esconde en las universidades españolas, de Innovaciones e investigaciones universitarias hispano-italianas, Sevilla, Geforán S.L., pp. 116-132 (2018)
9. INTEF: Marco común de competencia digital docente. Ministerio de Educación, Cultura y Deporte, Madrid (2017)

10. Vermunt, J.D., Donche, V.: A learning patterns perspective on student learning in higher education: state of the art and moving forward. Educ. Psychol. Rev. **29**, 131 (2017)
11. García Ravidá, L.B.: Patrones de aprendizaje en universitarios Latinoamericanos: Dimensión cultural e implicaciones educativas. Universidad Autónoma de Barcelona, Barcelona (2017)
12. Samaja, J.: Epistemología y metodología. Elementos para una teoría de la investigación científica. Eudeba, Buenos Aires (2012)
13. Martínez-Fernández, J.R., García-Ravidá, L., González-Velásquez, L., Gutiérrez-Braojos, C., Poggioli, L., Ramírez-Otálvaro y, P., Tellería, M.B.: Inventario de Estilos de Aprendizaje en Español. Documento interno del Grup de Recerca PAFIU. Universitat Autònoma de Barcelona, Barcelona (2009)
14. IBM Corp.: IBM SPSS Statistics for Windows, Version 23.0, Armonk (2015)
15. Gorman, B.S., Primavera, L.H.: The complementary use of cluster and factor analysis methods. J. Exp. Educ. **51**(4), 165–168 (1983)

Research of the Competency Model's Influence on Staff Management in Food Industry

João Paulo Pereira[1,2](✉), Efanova Natalya[3], and Ivan Slesarenko[3]

[1] Instituto Politécnico de Bragança, Campus de Santa Apolónia, 5300-302 Bragança, Portugal
jprp@ipb.pt
[2] UNIAG (Applied Management Research Unit), Bragança, Portugal
[3] Kuban State Agrarian University, Krasnodar, Russian Federation
efanova.nv@gmail.com, one.concealed.light@gmail.com

Abstract. This paper presents methods of staff management modification based on competency model and assignment one of three ranks (high, middle, low) for each employee. The research showed that such staff management processes as waiters' group making and waiters' financial motivation can be effectively modified using competency model but should to be complemented to provide more information that could be useful for other processes and prevent the risk of negative impaction on business. Results of experiments showed that restaurants where experienced and inexperienced waiters are balanced have better earnings and visiting dynamics than unbalanced. But proposed method cannot guarantee good restaurant performance and is ineffective in restaurant with high staff turnover. Waiters' financial motivation based on knowledge level and implementation of average check's plan can increase restaurant's average check. But this motivation can be used only if restaurant strategy is based on the high value of average check. Also indicators used in salary calculation method should be protected from artificial increasing.

Keywords: Competency model · Food industry · Staff management · Waiter's motivation · Waiters group · Group making · Average check

1 Introduction

In today's world due to evolving of food service industry staff management process becomes an important aspect, especially for restaurant chain management.

One of this process' components is staff appraisal. Despite of some disadvantages [1, 10], such methods as KPI and 360° can be effectively used [8, 13]. And these methods can be used to appraise linear staff [14], but due to difficulty of its development and maintenance it will take more resources.

An alternative staff appraisal method, adapted for automatic processing data from different sources, is the using an employee's profile, based on competency model. Competency model represents levels of employee's knowledge, skills and abilities. Recent

Á. Rocha et al. (Eds.): WorldCIST 2020, AISC 1159, pp. 351–360, 2020.
https://doi.org/10.1007/978-3-030-45688-7_36

research showed that competency model correlates with some staff efficiency indicators, but it is vulnerable from reliability of data sources [12].

When a company resolves a question of staff appraisal, the next question appears: what components of staff management this appraisal can affect. In another words, which decision-making processes can be based on appraisal results.

First possible process is drafting of waiters' work schedule. Since schedule's flexibility can be motivating factor [9] and overloaded schedule can be a psychological risk [6], this process is important.

Restaurant business is subjected by staff turnover, especially waiters' turnover. So from time to time inexperienced waiters appear in restaurant and experienced waiters leave it. In these cases it is important to avoid days without experienced waiters in schedule because of next statements.

Firstly, it affects customer's waiting time. When restaurant's load increases, inexperienced waiter can fail in making operative decisions, so average time of waiting will be increased. Despite in some cases long waiting time can be used for increasing customer satisfaction [4], it still affects revenue [3].

Secondly, it affects service quality. Not every inexperienced waiter can provide high quality of service, especially when restaurant's load is increased. Service quality affects customer loyalty [5, 7] and service failure can be crucial, especially for fine dining restaurants [11].

Second possible process is the motivation of waiters, viz. financial motivation. The research of Pakistan's banking sector showed that salary is much important factor for employee motivation as compared to other variable factors [16]. The research of employee motivation in Piran also showed that in times of economic crisis money is the most important motivational factor to work in foodservice [9].

Waiter's salary can be divided to two components: formal part, paid by restaurant's owner, and tip part, paid by customer. Tips can be distributed or stayed individual and it may affect motivation [15]. In case of stable tips it is possible to calculate how hard waiter has to work to earn minimum wage [2] to plan formal part.

In cases of lack of stable tips formal part has more priority and it becomes important to develop a way of salary calculation which will satisfy waiters and let to get labor market competitiveness.

These processes modification methods, based on competency model, and results of experiments are described in this paper. The main goal of this research is determination of efficiency of using competency model in described staff management processes.

2 Materials and Methods

The competency model represents a weighted graph where set of vertices is consisted of positions, competences and competence components and set of edges describes the connections between the vertices.

The main goal of using this model is collecting data from different sources in one place to get an instrument of employee appraisal. So an integral method of calculation the final rank based on dissimilar indicators (competence components layer) is required.

Proposed method involves assignment one of three ranks (high, middle, low) for each employee based on next algorithm:

1. Choosing and grouping important indicators for a process. For example, to make waiters group a manager has to know financial efficiency and knowledge level of each waiter. So the first group will be consisted of knowledge levels' which are necessary to provide high level of service. Second group will contain only one indicator – employee's average check (only if organization's strategy is based on average check increasing).
2. Choosing ranks' limitation indicators. These indicators will define a possibility of an employee to get some rank. The employee's length of work is an example of limitation indicator. Limitation indicators also determine an integral appraisal model's flexibility.
3. Constructing a coordinate axis for each group and definition ranks' breakpoints. An expert who can determine which indicator's level is necessary for each rank. Limitation indicators' breakpoints are defined on that stage too.
4. Constructing a coordinate space based on axes and definition of ranks' areas. Employee's indicators will be represented as point in that space, so his rank will be defined by the area in which his point is located.

Hybrid ranks, like high-middle or middle-low, can appear during areas' construction. So it is important to determine the method of construction especially if process doesn't allow hybrid ranks. Two methods of 2-axes space construction are proposed:

1. Rectangle as an area: all breakpoints are creating a line parallel another axis, forming a grid (Fig. 1). This simple method of construction assumes hybrid rank's availability.

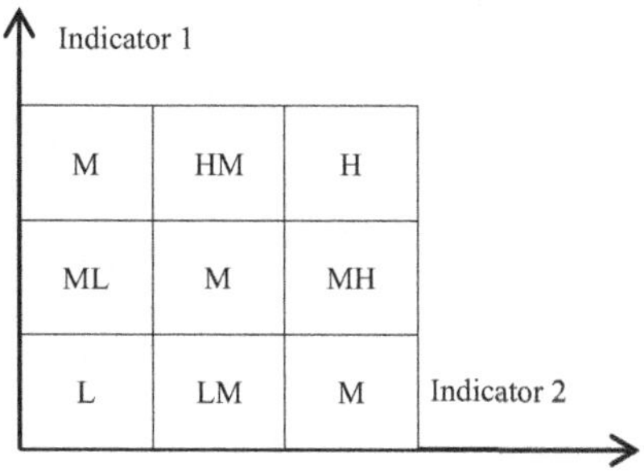

Fig. 1. Example of rectangle as an area

2. Circle sector as an area (Fig. 2). Circle center as indicators' standards is determined. For indicators' discrete levels standard is maximum level. For indicators' infinite levels standard is the most efficient level. For each rank the maximum allowed distance between employee's point and center will define radius of rank's circle.

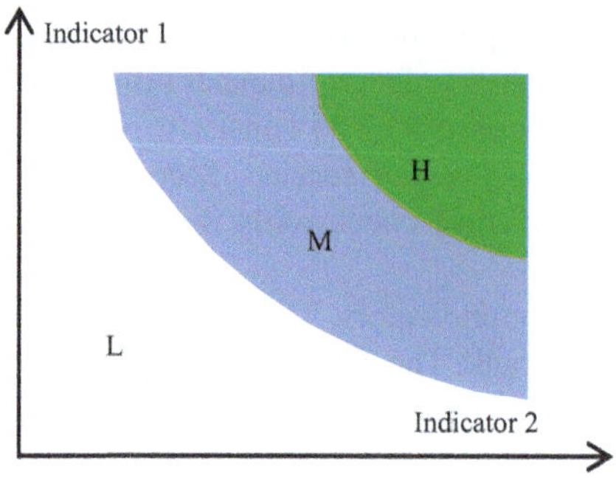

Fig. 2. Example of circle sector as an area

To modify process of group making next method is proposed.
Waiters are represented as vertices (Fig. 3):

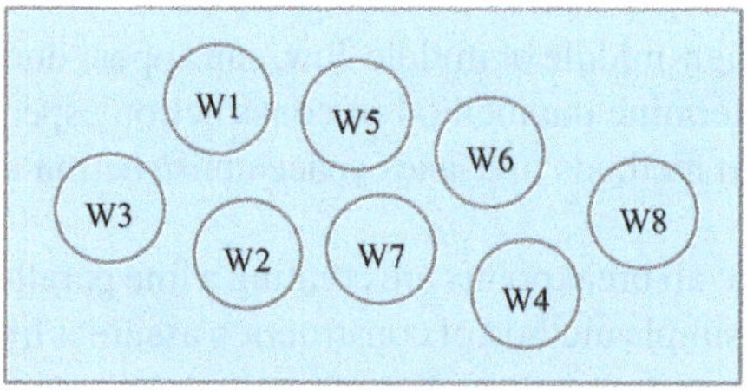

Fig. 3. Waiters as vertices

After ranks determination these vertices can be clustered into groups (Fig. 4).

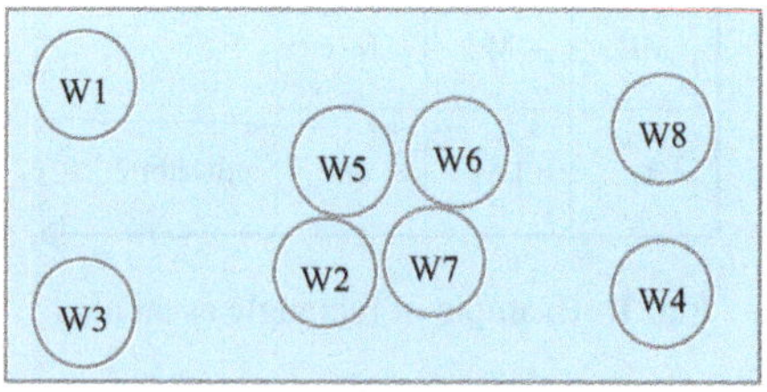

Fig. 4. Vertices clustering

Weighted graph G can be built with these vertices.

$$G = (V, R), \tag{1}$$

where $V = \langle H, M, L \rangle$, a set of vertices of graph (high, middle and low rank employees).

R is a set of edges describing the connections between the vertices. Weight of edge defines a usefulness of two employees' presence in a group.

An edge between high and low rank vertices will have the highest weight, and an edge between two low rank vertices will have the lowest weight. Weight of an edge between two middle rank vertices can be determined with weaknesses (low level indicators) of employees, so the more similar weaknesses they have the less weight of an edge will be.

The main goal is maximization of group's vertices' edges' weights' sum.

If there is no possibility to make about two or three stable waiters group, priority of middle vertices edges decreases, because it is more important to get balance between high rank and low rank employees in a group. This case appears when organization allows flexible schedule of work and the most part of employees has individual schedule.

For this task HML coordinate space will have two axes with construction a rank's area as a circle sector. The first axis is the average knowledge level (results of last three tests are taken into account). The second axis is comparative average check calculated by formula (2) where *ACw* is waiter's average check and *ACr* is restaurant's average check. This indicator means performance of waiter in comparing with performance of all his restaurant's waiters. Limitation indicator is the length of work.

$$CAC = \frac{AC_w}{AC_r}, \tag{2}$$

If there is lack of high rank waiters, algorithm of rank determination will be modified with next steps until reaching of balance between high and low ranks:

1. Removing middle rank limitation breakpoint.
2. Removing high rank limitation breakpoint.
3. Increasing high rank circle radius if it less than middle rank circle radius (the best of middle rank waiters become high rank).
4. Increasing middle rank circle radius (the best of low rank waiters become middle rank).

The main goal of this group making method is providing high quality of service and saving current indicators of restaurant earnings and visiting. The lack of situations when nobody in group has good skills and knowledge level and can help other waiters should lead to it.

Another HML coordinate space is proposed to modify a process of waiters' motivation. The first axis is the average knowledge level (only last test is taken into account). The second axis is the implementation of individual average check's plan. This plan depends on time, place and day of each order. Method of area construction is "rectangle as an area". Waiters' rank defines his salary. Hybrid rank will define the average value between two ranks' salaries.

The main goal of this process' modification is earnings' increasing and saving the labor market competitiveness.

Next experiments will be conducted to test efficiency of these processes' modifications.

For 4 month 15 restaurants with at least 8 waiters will get list with waiters' ranks every month. Also they will get recommendations how to make waiters' groups; they won't be forced to follow it.

Balance of groups will be determinate with next rules:

Good balance – count of high rank waiters is not less than count of low rank waiters.
Normal balance – count of high rank waiters is less by one than low rank waiters. In case of troubles with high rank waiters' pool normal balance is counted as good.
Bad balance – count of high rank waiters is less than count of high rank waiters.

For each day and each month quality of group's balance will be determined. The changing of year-to-year dynamics of restaurant earnings and visiting will be efficiency indicators for month balance quality. Implementations of average check's and earnings' plans will be efficiency indicators for day balance quality.

To test the modification of waiters' motivation process we will apply the proposed method in restaurants, where it's important to have high value of average check, for three month. The average check's year-to-year dynamics will be the efficiency indicator.

3 Results

The following charts show results obtained from the experiments.

3.1 Groups' Making

A changing of dynamics is described with histogram chart. The X-axis is the quality of balance. The Y-axis is the average changing of dynamics in percent. Vertical lines over columns mean the minimum and maximum changing.

Figure 5 shows the changing of dynamics of restaurant visiting.

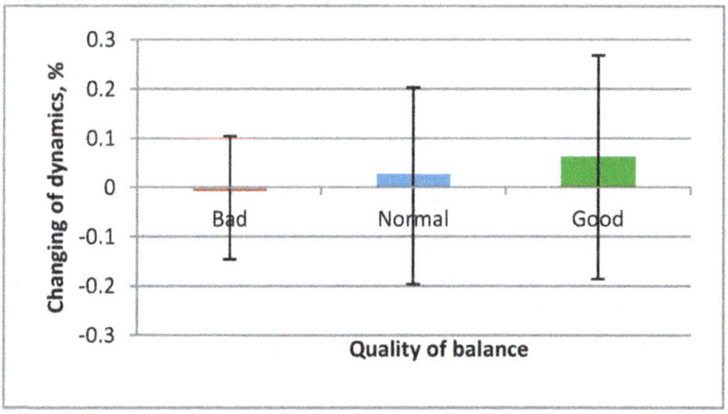

Fig. 5. Visiting's dynamics

There is the linear growth of dynamics' changing from bad balance to good balance. But the minimum values of good and normal balance are less than minimum value of bad balance. It means that even if restaurant has good or normal waiters' groups' balance, it still will be able show bad results.

Figure 6 shows the changing of dynamics of earnings.

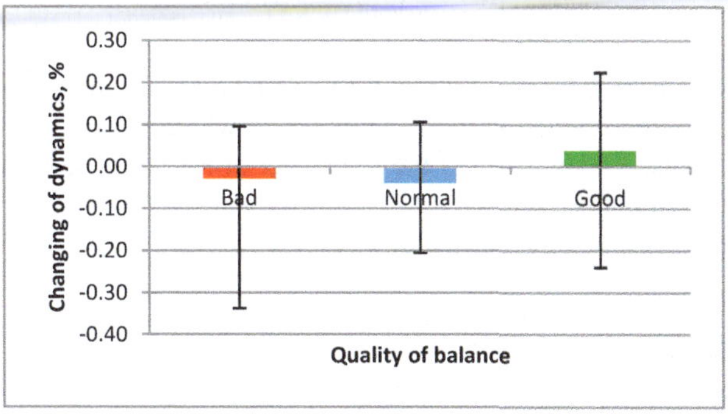

Fig. 6. Earnings' dynamics

This dynamics is different than dynamics of restaurant visiting. The main reason is restaurants strategies. One strategy can be based on increasing of average check that should lead to earnings' growth. Another strategy can be based on providing high quality of service without using aggressive marketing methods. So the possible reason of negative earnings' dynamics of normal balanced restaurants is lack of average check's based strategy.

Implementations of plans are described with line chart presented at Fig. 7. The X-axis is the quality of balance of group. The Y-axis is the percent of implementation of plan by this group. Points mean average percent of implementation of plan, and line allows a pair comparison.

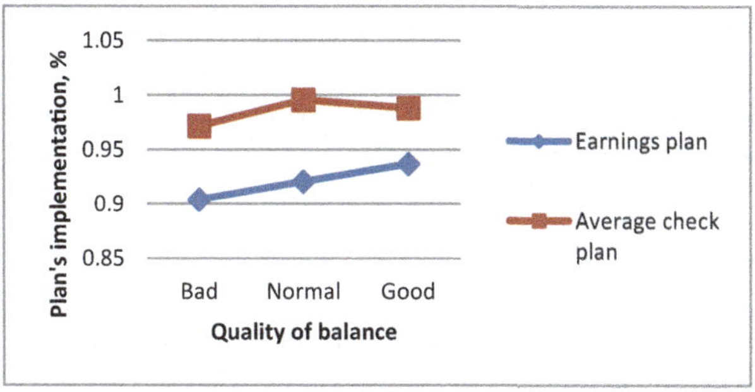

Fig. 7. Implementations of plans

The implementation of the earnings' plan is increases with balance of group, which means that groups with good balance had better performance than groups with bad balance. But groups with normal balance had the best implementation of average check's plan. The most possible reason is the forethought of restaurant managers, who could made normal balance groups for low plan days.

3.2 Waiters' Motivation System

Modified motivation system was applied in five restaurants for three month. Results of this experiment are represented with histogram chart where restaurants were placed on the X-axis, while the Y-axis featured the dynamics of average check in percent. Each column means one month.

Figure 8 shows results of experiment.

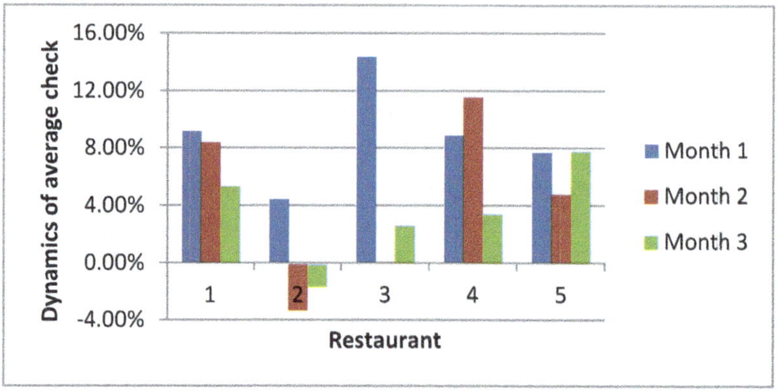

Fig. 8. Average check's dynamics

The first month was the most successful for four restaurants. But the second and the third months were not so good. The second restaurant's waiters had bad performance so they didn't get motivation to increase the average check. The third restaurant refused to use new motivation system, so its waiters were not motivated to increase the average check too. However, there is still only two month with negative dynamics and one month with zero dynamics.

4 Discussion

Experiments demonstrated the competency model's positive influence on the staff management process.

Restaurants with good waiters groups' balance had positive dynamics of earnings and visiting. Restaurants with normal waiters groups' balance had positive visiting's dynamics and negative earnings' dynamics that can be explained with their strategies' features.

But there were bad performance of restaurants with good or normal balance, so groups' making based on the competency model can't guarantee good results: it can only increase the possibility of success. Good performance of some restaurants with bad balance also demonstrates that ignoring of groups' balancing will not guarantee bad performance. So groups making should be perceived as important process, but not as high priority process. Next problems can prevent modification of this process:

1. Staff turnover. In cases when the most waiters don't want to work in a restaurant more than three months lack of high rank waiters will have a place. So adaptation of method for these cases is needed.
2. Purity of source data. If waiter cheats on testing he will be able to get high rank. So group can get low rank waiter instead of high rank waiter. This can lead to bad quality of balance instead of normal of good.
3. Ranks' binding to location. Relative indicator is used for one of axes, so waiter gets his rank relative to his restaurant. If he has high rank in one restaurant, it doesn't mean high rank in another.

In its turn, total restaurant balance was discovered. It is determined by quantitative structure of rank groups. It can determine next indicators:

1. Restaurant's stability. Lack of middle rank waiters or big amount of low rank waiters (small amount of high rank) demonstrates the possibility of falling of restaurant performance in case of destabilization of high rank group. It can be the warning about possible HR accident.
2. Restaurant's stagnation. It is defined by lack of high and low rank waiters. In another words, it is restaurant where all waiters have middle rank. So it is necessary to determine the rank of the restaurant to understand what rank all its waiters have.

Also waiter's rank's dynamics can demonstrate his ability to study and readiness to advancing.

Modified waiters' motivation system demonstrated good results. But the most important problem was discovered: all indicators that are used for axes have a risk of being compromised. Some waiters with bad performance and a wish to have the highest salary can try to artificially increase their indicators. And it will affect other processes where these indicators are used too. This statement was approved during experiment.

So during developing the motivation system based on HML it is important to follow next strategy:

1. Choosing the most protected from compromising indicators.
2. Providing purity of data sources (for example, preventing cheating during testing).

Finally, modification, based on competency model, of researched staff management processes can have a positive impact on business. But methods of modification should be complemented to provide more information that could be useful for other processes and prevent the risk of negative impaction on business.

References

1. Brett, J.F., Atwater, L.E.: 360 degree feedback: accuracy, reactions, and perceptions of usefulness. J. Appl. Psychol. **86**(5), 930–942 (2001)
2. Casteel, K., Smart, C.: How Hard Is Your Server Working To Earn Minimum Wage? (2017). https://projects.fivethirtyeight.com/tipping-workers-minimum-wage/. Accessed 18 Oct 2009

3. De Vries, J., Roy, D., De Koster, R.: Worth the wait? How restaurant waiting time influences customer behavior and revenue. J. Oper. Manag. **63**, 59–78 (2018)
4. Giebelhausen, M.D., Robinson, S.G., Cronin, J.J.: Worth waiting for: increasing satisfaction by making consumers wait. J. Acad. Mark. Sci. **39**(6), 889–905 (2011)
5. Ha, J., Jang, S.S.: Effects of service quality and food quality: the moderating role of atmospherics in an ethnic restaurant segment. Int. J. Hosp. Manag. **29**(3), 520–529 (2010)
6. Hassard, J., Teoh, K., Cox, T.: Managing psychosocial risks in HORECA. EU-OSHA (European Agency for Safety & Health at Work), Bilbao, Spain (2013)
7. Hyun, S.S.: Predictors of relationship quality and loyalty in the chain restaurant industry. Cornell Hosp. Q. **51**(2), 251–267 (2010)
8. Karkoulian, S., Assaker, G., Hallak, R.: An empirical study of 360-degree feedback, organizational justice, and firm sustainability. J. Bus. Res. **69**(5), 1862–1867 (2016)
9. Kukanja, M.: Influence of demographic characteristics on employee motivation in catering companies. Tour. Hosp. Manag. **19**(1), 97–107 (2013)
10. Malysheva, M.A.: KPI: advantages and disadvantages of implementing. Informaciya kak dvigatel nauchnogo progressa: sbornik statey Mezhdunarodnoy nauchno - prakticheskoy konferencii, pp. 134–137 (2017). (in Russian)
11. Namkung, Y., Jang, S.: Service failures in restaurants: which stage of service failure is the most critical? Cornell Hosp. Q. **51**(3), 323–343 (2010)
12. Pereira, J.P., Efanova, N., Slesarenko, I.: A new model for evaluation of human resources: case study of catering industry. In: Rocha, Á., Adeli, H., Reis, L., Costanzo, S. (eds.) New Knowledge in Information Systems and Technologies. WorldCIST 2019. Advances in Intelligent Systems and Computing, vol. 930. Springer, Cham (2019)
13. Pérez-Álvarez, J.M., Maté, A., Gómez-López, M.T., Trujillo, J.: Tactical business-process-decision support based on KPIs monitoring and validation. Comput. Ind. **102**, 23–39 (2018)
14. Popova, M.V.: KPI in the sphere of public catering. Vestnik nauki i obrazovaniya **5**(41), 109–111 (2018). (in Russian)
15. Raspor, A., Rozman, T.: Impact of tipping on workers' motivation: comparison between the hospitality and gaming industries in slovenia. Sociol. Discourse **6**(11), 67–92 (2017)
16. Shafiq, M.M., Naseem, M.A.: Association between Reward and Employee motivation: a case study Banking Sector of Pakistan (2011). Available at SSRN 1857663

Towards Message-Driven Ontology Population - Facing Challenges in Real-World IoT

David Graf[1(✉)], Wieland Schwinger[1], Elisabeth Kapsammer[1], Werner Retschitzegger[1], Birgit Pröll[1], and Norbert Baumgartner[2]

[1] Johannes Kepler University, Linz, Austria
{david.graf,wieland.schwinger,elisabeth.kapsammer,werner.retschitzegger, birgit.proll}@cis.jku.at
[2] team GmbH, Vienna, Austria
norbert.baumgartner@te-am.net

Abstract. Large-scale Internet-of-Things (IoT) environments as being found in critical infrastructures such as *Intelligent Transportation Systems (ITS)* are characterized by (i) massive *heterogeneity* of data, (ii) prevalent *legacy systems*, and (iii) continuous *evolution* of operational technology. In such environments, the realization of crosscutting services demands a *conceptual IoT representation*, most promising, in terms of a domain ontology. Populating the ontology's *A-Box*, however, faces some challenges, which are not sufficiently addressed by now. In this respect, the contribution of this short paper is three-fold: Firstly, in order to point out the complexity of addressed real-world IoT environments, we identify prevalent *challenges for (semi-)automatic ontology population by means of a real world example*. Secondly, in order to address these challenges, we elaborate on related work by *identifying promising lines of research* relevant for ontology population. Thirdly, based thereupon, we *sketch out a solution approach* towards message-driven ontology population.

Keywords: Internet-of-Things · (Semi-)automatic ontology population · Intelligent Transportation Systems

1 Introduction

Conceptual IoT Representation. Large-scale Internet-of-Things (IoT) environments as being found in critical infrastructures such as *Intelligent Transportation Systems (ITS)* are characterized by massive *heterogeneity* of data provided by the variety of individual systems and data sources involved. Prevalent *legacy systems* often lacking structured and semantic information. Moreover, systems

This work is supported by the Austrian Research Promotion Agency (FFG) under grant FFG Forschungspartnerschaften 874490.

Á. Rocha et al. (Eds.): WorldCIST 2020, AISC 1159, pp. 361–368, 2020.
https://doi.org/10.1007/978-3-030-45688-7_37

themselves are subject to continuous *evolution*, meaning that the underlying Operational Technology (OT), employed for monitoring and controlling the ITS' operation, is continuously added, removed or replaced. In such environments, to realize crosscutting services like service quality monitoring of OT [8] or failure reasoning, a *conceptual IoT representation*, being independent from the actual technology used, is an indispensable prerequisite.

(Semi-)Automatic Ontology Population. A promising paradigm to address *heterogeneity*, *evolution* and *legacy systems* are *semantic technologies* in terms of *ontologies* [7]. While the *T-Box*, or the OT types respectively, are manually specified by domain experts in terms of an *object type catalog* providing simple taxonomic relationships, it is simply not feasible from a practical point of view to manually populate a domain ontology's *A-Box* with thousands of objects, particularly in the light of continuous evolution of OT, thus (semi-)automatic ontology population is a must. Moreover, ontology population is aggravated by the fact that the majority of historically grown systems are lacking homogeneous object information in a machine interpretable format. The only machine-readable data source available about underlying OT, which can be used as a basis for population, is often their communication data recorded within various message logs of systems consisting of human interpretable service messages as well as failure messages.

Paper Contribution. Aiming a *conceptual representation of the underlying OT* in terms of a domain ontology, i.e., OT objects and their relationships in between, the contribution of this short paper is three-fold: Firstly, in order to point out the complexity of addressed real-world IoT environments, we identify prevalent *challenges for (semi-)automatic ontology population based on message logs by means of a real world example* in the ITS domain in Sect. 2. Secondly, in order to address these challenges, we elaborate on related work by *identifying promising lines of research* relevant for ontology population through message logs in Sect. 3. Thirdly, based thereupon, we *sketch out a solution approach* towards message-driven ontology population in Sect. 4.

2 IoT Ontology Population Challenges by Example

In order to provide an illustrative picture of prevalent challenges when using messages to populate a domain ontology, we discuss these challenges based on a concrete example in the ITS domain. The IoT environment considered by our work comprises more than 100.000 OT devices of more than 200 different types[1] ranging from simple sensing and actuating devices (e.g., a lightning device) to complex systems consisting of several devices of various types (e.g., a radar systems), which are geographically distributed over 2.220 km highway and 165 tunnels. During operation, these OT devices provide status information for the operators in terms of *a stream of messages* consisting of (i) human interpretable *message text*, (ii) a *unique identification* of the affected device, and (iii) *time* information, recorded within the logs of various systems.

[1] Based on the domain specific *object type catalog* defined by the operating company.

Using these messages as a basis to populate a domain ontology, however, faces some challenges discussed in the following and exemplified by reflecting on the OT of an *emergency call system* being a crucial part for safety in tunnels on a highway. From an OT perspective, such an *emergency call system* consists of (i) two *door-contacts* (trigger a message when a door was opened), (ii) a *SOS button* to push in case of an emergency (triggers a message when pushed), and (iii) a *SOS phone* to make an emergency call (triggers a message when the phone is off-hook as well as when it is on-hook again), thus representing a composition in the T-Box of an ontology. These OT devices, as well as the *emergency call system* itself, are able to individually and independently send messages via various gateways to an operational monitoring and control system logging these service messages (e.g., the *door-contact* notifies "door opened") or failures messages (e.g., the *emergency call system* notifies a "power supply error"). Regarding the A-Box of the ontology, naturally, there exist hundreds of *emergency call systems* (i.e., several of them in each tunnel) as naturally *door-contacts*, *SOS buttons* and *SOS phones* thereof. Whereby all of them finally need to be represented as the corresponding interrelated OT objects of their respective OT types belonging to the corresponding OT object of the composed OT type. This composition information is, however, not available and needs to be learned from messages. In the following, we discuss major challenges surrounding (1) *type-instantiation* and (2) *object-linking*, visualized in Fig. 1.

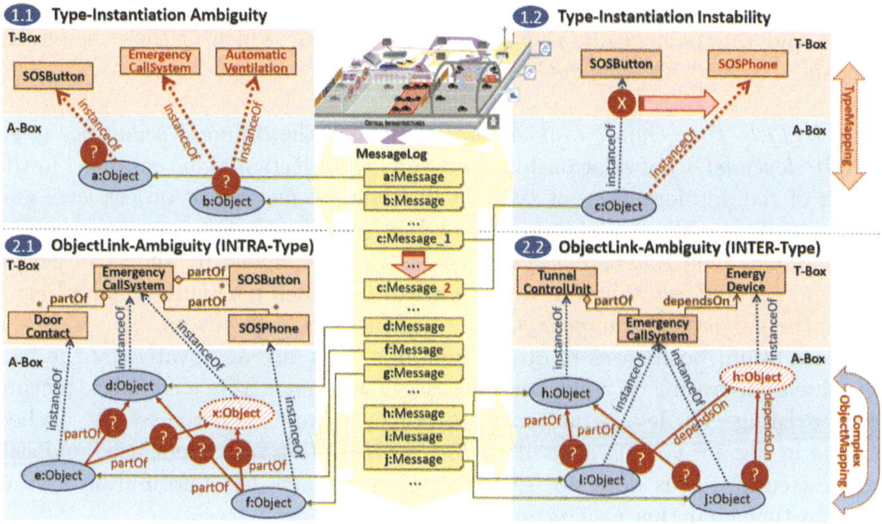

Fig. 1. Challenges of message-driven ontology population

(1.1) Type-Instantiation Ambiguity. Whereas the unique identification of the OT object is given through the affected device, their respective OT type information is not explicitly part of the message but rather just implicitly indicated

through the message text, and to make matters worse, identical message text might be used by different OT types, which makes the type mapping of an OT object to the correct OT type ambiguous (cf. Fig. 1). For instance, the message "SOS button pushed" (cf. "a:Message") indicates the OT type *SOSButton*, whereas, an example of a more generic message found like "maintenance active" (cf. "b:Message") might originate from an *EmergencyCallSystem* or from other OT like an *AutomaticVentilation* unit.

(1.2) Type-Instantiation Instability. As new messages related to the same OT object come in eventually, thus more information about an OT object is collected over time, their respective mapping to OT types might need to be adjusted (cf. Fig. 1). For instance, a message "SOS - main error" (cf. "c:Message_1") originating from an OT object being an initial indication of an OT type *SOSButton*. However, a later message "call active" (cf. "c:Message_2") from the same OT object makes it necessary to revise the OT type mapping since this message indicates that the very same OT object is rather an instance of a *SOSPhone*. Hence, the type mapping might change incrementally after each message requiring to consider all previous messages not the most recent one, only.

(2.1) INTRA-Type Object Link-Ambiguity. As there is a vast number of OT objects and due to messages lacking information of how these OT objects interrelate to each other, complex object mapping is challenging (cf. Fig. 1). For instance, since multiple objects of the OT types *DoorContact*, *SOSButton*, *SOSPhone* and *EmergencyCallSystem* exist, it is unclear which of those belong to the same physical *EmergencyCallSystem* unit.

(2.2) INTER-Type Object Link-Ambiguity. While the different semantics (e.g., *partOf*, *dependsOn*) of relationships between OT objects may be captured in the T-Box of the ontology (i.e., at type level), they are missing at object level and thus are not readily available for the A-Box (cf. Fig. 1). For instance, the OT object of OT type *EmergencyCallSystem* is *energySupplied* by an OT object of OT type *EnergyDevice*, however, it is unclear on which particular *EnergyDevice* object the particular *EmergencyCallSystem* instance depends on.

In addition, both cases of object link-ambiguity are aggravated by the fact that due to operating on a stream of messages we do not have a complete picture of all existing OT devices at a certain point in time (depicted by the dashed objects in Fig. 1). For instance, if an *EmergencyCallSystem* is not used and still works accurate, thus does not send messages, we lack the information at this point in time that this *EmergencyCallSystem* even exist.

3 Identifying Promising Lines of Research

Addressing the discussed challenges, our systematic literature review follows a goal-oriented strategy, meaning that we first aim identifying promising lines of research regarding our goal (i.e., populating a domain ontology) such as being found in the areas of *ontology population* and *semantic annotation* from text or

Table 1. Related approaches

			Ganino et al. [6]	Lin et al. [12]	Liu et al. [13]	Jayawardana et al. [10]	Reyes-Ortiz et al. [17]	Belkaroui et al. [3]	Matzner and Scholta [15][a]	Ni et al. [16]	Jin et al. [11]	Endler et al. [5]	Jafari et al. [9]	Detro et al. [4][b]
Source	Structure	Semi-structured		✓				✓	✓	✓	✓		✓	✓
		Unstructured	✓		✓	✓	✓							
	Dynamicity	Static	✓		✓	✓	✓	✓	✓	✓	✓		✓	✓
		Dynamic		✓								✓		
Techniques	Information retrieval	Text-based			✓	✓	✓	✓						
	Machine learning	Supervised			✓		✓							
		Semi-supervised				✓								
		Unsupervised		✓					✓	✓	✓			
	Data analysis	Distance-based				✓			✓					
		Similarity-based								✓	✓			
	Other methods	Tool-usage (e.g, GATE)	✓											✓
		Complex-event-processing										✓		
		Specific data-mining											✓	
Target	Content-Ontology	Wine events						✓						
		Health events												✓
		Law events				✓								
		IoT data (e.g., sensor values)		✓	✓							✓		
	Resource-Ontology	OT												
		Roles												✓
		Web services					✓							
	Other formalisms	Organizational model							✓	✓	✓			
		Document	✓											
		RBAC model											✓	
Challenges	Type-Instantiation	Ambiguity and instability	✓	✓	✓	✓	✓	✓	✓					
	Obj. Link-Ambiguity	INTRA- and INTER-type		✓	✓			✓	✓	✓	✓	✓	✓	✓

[a] survey,
[b] preliminary work and not yet validated

semi-structured data, surveyed most recently by [14], before considering research areas beyond. Related work is compared primarily based on (i) the *source* data structure, (ii) the *techniques* applied, (iii) the *target* data structure, and (iv) the ability to address our *challenges* (cf. Section 2). A summary of related work along these comparison dimensions is given in Table 1.

Ontology Population and Semantic Annotation. Promising data-driven techniques such as clustering and semi-supervised classification are used by [10] and [17] to populate a domain ontology, the latter being closely related regarding the target data structure by populating an ontology with resources in terms of web services. Both, however, use primarily text documents as data source and do not use semi-structured stream data originating from logs. Closely related is the approach of [3] populating an event ontology for monitoring vineyards grounded on an heterogeneous IoT sensor network aiming to mine causality relationships between events occurred during the life cycle of a wine production. Although the data originates also from IoT sensors, the target ontology primarily focus on events and not on a representation of the underlying IoT environment. Related with respect to techniques used are approaches in the area of semantic annotation, i.e., the process to annotate entities in a given text with semantics [6] (e.g., using ontology classes), such as the work of [12,13], which, however, address primarily the input data source themselves (often in terms of text documents) as target data structure.

Compared to our work, approaches discussed so far mainly differ regarding the source data structure, i.e., they do not consider streaming log data. Hence, in order to identify promising lines of research also with respect to the source data structure, in the following, we focus on related work having similar requirements to ours in that respect namely (i) using semi-structured log data as data source, (ii) extracting knowledge about hidden resources (e.g., people, systems, roles) and their relationships in between, and (iii) being confronted with stream data. All these requirements are also tackled by work in the area of process mining [18], more precisely in one of its sub-field *organizational process mining* [1].

Organizational Process Mining. While the work of [2] provides a systematic review of automated process discovery methods, most promising mining approaches are reviewed by [15] aiming to "derive the underlying organizational structure of a CPS" from event logs. Discussed techniques such as "metrics based on (possible) causality" focusing on temporal succession of activities, or "metrics based on joint cases" focusing on frequency and correlation of resources, seems to be eminently suitable to derive relationships between objects also in the IoT domain. Furthermore, variations of distance measures, e.g., those used by [16], as well as traditional clustering techniques applied to event logs, e.g., those used by [11], are promising approaches to transfer to the IoT domain. In addition, since "time is a key relation between pieces of information" [5], time-based approaches are highly relevant for our work such as the organizational mining approach of [9]. With respect to the source data structure and the techniques used, closely related is the approach of [4] in terms of semantically annotating event log information in the health-care domain.

As Table 1 shows, although approaches discussed so far are related to our work in some of the comparison dimensions, none of them are directly applicable to our requirements since none aim a conceptual representation of OT as target data structure, especially not based on message streams as data source structure.

4 Envisioned Approach

Based on previously discussed challenges and identified promising lines of research, we finally sketch out a solution approach towards message-driven ontology population aiming a conceptual representation of OT, in the following.

Addressing the *type-instantiation challenges*, we envision to employ text-similarity-based techniques like used by [10,17] in terms of mapping message texts originating from a particular OT object to the most similar OT type, textually described by domain specific documents and technical specifications. Thereby a crucial aspect is to incrementally verify already existing type mappings since we have to be aware of new information provided by the message stream.

Addressing the *object link-ambiguity challenges*, we envision the usage of temporal patterns like [16] by applying distance measures or clustering techniques [11] with respect to certain timestamps of messages to identify relationships between OT objects. The rational behind is based on two hypotheses, namely (1) logical relationships between OT objects result in nearly simultaneous messages in case of a cross-device function failure, e.g., a shared energy-supply, and (2) physical relationships between OT objects result in typical functional temporal patterns of messages, e.g., in case of an *emergency call system* within 10 min typically the messages "door opened", "SOS button pushed", and "emergency call active" before again "door opened" are triggered.

First experiments towards this envisioned approach using samples of data have already shown promising results. For this in a first step we are now investigating in more detail on available real-world data expecting to elaborating on a prototype as the following step.

References

1. Appice, A.: Towards mining the organizational structure of a dynamic event scenario. J. Intell. Inf. Syst. **50**(1), 165–193 (2018)
2. Augusto, A., et al.: Automated discovery of process models from event logs: review and benchmark. IEEE Trans. Knowl. Data Eng. **31**(4), 686–705 (2018)
3. Belkaroui, R. et al.: Towards events ontology based on data sensors network for viticulture domain. In: Proceedings of the International Conference on the Internet of Things, pp. 1–7. ACM (2018)
4. Detro, S. et al.: Enhancing semantic interoperability in healthcare using semantic process mining. In: Proceedings of the International Conference on Information Society and Technology, pp. 80–85 (2016)
5. Endler, M. et al.: Towards stream-based reasoning and machine learning for IoT applications. In: Intelligent System Conference, pp. 202–209. IEEE (2017)
6. Ganino, G., et al.: Ontology population for open-source intelligence: a GATE-based solution. Softw. Pract. Exp. **48**(12), 2302–2330 (2018)
7. Graf, D., Kapsammer E., Schwinger W., Retschitzegger W., Baumgartner N.: Cutting a path through the IoT ontology jungle - a meta survey. In: International Conference on Internet of Things and Intelligence Systems. IEEE (2019)
8. Graf, D., Retschitzegger W., Schwinger W., Kapsammer E., Baumgartner N., Pröll B.: Towards operational technology monitoring in intelligent transportation systems. In: International Conference on Management of Digital Eco-Systems. ACM (2019)
9. Jafari, M., et al.: Role mining in access history logs. J. Comput. Inf. Syst. Ind. Manag. Appl. **1**, 258–265 (2009)
10. Jayawardana, V. et al.: Semi-Supervised instance population of an ontology using word vector embeddings. In: Proceedings of the International Conference on Advances in ICT for Emerging Regions, pp. 217–223. IEEE (2017)

11. Jin, T., et al.: Organizational modeling from event logs. In: Proceedings of the International Conference on Grid and Cooperative Computing, pp. 670–675. IEEE (2007)
12. Lin, S., et al.: Dynamic data driven-based automatic clustering and semantic annotation for internet of things sensor data. Sens. Mater. **31**(6), 1789–1801 (2019)
13. Liu, F., et al.: Device-oriented automatic semantic annotation in IoT. J. Sens. **2017**, 9589064:1–9589064:14 (2017)
14. Lubani, M., et al.: Ontology population: approaches and design aspects. J. Inf. Sci. **45**(4), 502–515 (2019)
15. Matzner, M., Scholta, H.: Process mining approaches to detect organizational properties in CPS. In: European Conference on Information Systems (2014)
16. Ni, Z., et al.: Mining organizational structure from workflow logs. In: Proceedings of the International Conference on e-Education, Entertainment and e-Management, pp. 222–225. IEEE (2011)
17. Reyes-Ortiz, J., et al.: Web services ontology population through text classification. In: Proceedings of the Federated Conference on Computer Science and Information Systems, pp. 491–495. IEEE (2016)
18. Van Der Aalst, W., et al.: Process mining manifesto. In: Proceedings of the International Conference on Business Process Management, pp. 169–194. Springer (2011)

A First Step to Specify Arcade Games as Multi-agent Systems

Carlos Marín-Lora[1,2(✉)], Alejandro Cercós[2], Miguel Chover[1,2], and Jose M. Sotoca[1,2]

[1] Institute of New Imaging Technologies, UJI, Castellón, Spain
[2] Universitat Jaume I, Castellón, Spain
CMARIN@UJI.ES

Abstract. The lack of formalities in the development of video games is one of the main obstacles for the incorporation of new professionals to the field. Although there are general proposals to describe and specify video games with techniques such as Game Design Document or Game Description Language, these are usually aimed at implementations in predetermined media, which determines the game specification from the outset to its implementation in the selected platform. This paper proposes a method for the definition, specification and implementation of a video game based on multi-agent systems, where its elements, functionalities and interactions are established independently of the platform used for its development. To prove its validity and capabilities, the classic arcade game Frogger has been used as a demonstrator. This game has been defined in its general form and subsequently implemented on three different platforms following the same specification. Each of these implementations has been made using different engines, languages and programming techniques, but in any case, meeting the requirements of the game and multi-agent systems.

Keywords: Computer games · Game specification · Multi-agent systems

1 Introduction

Despite being an industry with more than forty years of existence, the design and development of video games lack formalities and standards that unify the definition and specification process of its products. In this sense, the design and development of video games still has several open fronts demanding to keep researching in order to achieve optimum performance in the definition and specification, development and implementation processes. Actually, some authors indicate that it would be necessary to define a formal language that standardises concepts and processes present in any development [1,2]. In fact, this absence of formalities makes the initiation in the design and development of games for non-expert people a truly complicated task. In order to find new models to

Á. Rocha et al. (Eds.): WorldCIST 2020, AISC 1159, pp. 369–379, 2020.
https://doi.org/10.1007/978-3-030-45688-7_38

specify video games formally, it has been recognised that multi-agent systems (MAS) have several similarities with video games. They are systems composed of entities, called agents, living and interacting with each other in a shared environment, where each agent has a set of properties and a set of action rules to determine their tasks and goals [32]. These systems are widely used in fields such as control systems or artificial intelligence to solve complex problems, by dividing complex tasks into simpler ones and assigning them to agents. The analogy between the definition of a MAS and the specification of a video game seems apparent: the agents play the role of game elements, usually known as game objects; interacting with each other and sending information about the game logic, the collisions between them or the user interaction, and where each one has a task to fulfil in the game based on a set of behaviour rules.

This paper presents a specification model based on MAS to prototype, define and specify games regardless of the platform on which it is intended to be implemented. In order to demonstrate the validity of the model, the definition of a classic arcade game with frequent features and mechanics is proposed along with its implementation in three different platforms. The reason for choosing a multi-agent model is because the behaviours and interactions in these systems have correspondences with the behaviours and interactions between individuals and their environment in the real world [12]. For this reason, it has been considered that the first steps for non-expert people are more affordable by taking the game as an analogy with the real world [20,24]. In this sense, this proposal is based on the hypothesis that MAS are suitable for the specification and implementation of video games. In order to test this hypothesis, the study is focused on arcade games because their mechanics have been replicated and extended consistently since the first video games. Specifically, the analysis draws from the description of the Frogger game [13] and its implementation in NetLogo [29], a programming environment based on turtle language oriented to the MAS prototyping. Based on that model, the game has been implemented in Gamesonomy a visual programming 2D game engine [14] and in the Unity [30] game engine. Subsequently, an analysis of the differences and particularities of each case has been composed.

Finally, the document presented is organised as follows: Sect. 2 presents the state of the art related to software and video game specification systems using MAS. Next, Sect. 3 details the model that has been followed to formally define the analogy between video games and multi-agent systems. After that, in Sect. 4, the use case is presented with which the validity of the model in different game engines will be analysed, together with the results obtained after that experience in Sect. 5. Finally, in Sect. 6 several conclusions from the realisation of this study are presented, and its possible future derivations.

2 State of the Art

The biggest obstacles faced by upstarts in the development of video games appear at such early stages as the design and specification of a project. In other words, in the process of translating the creative process into requirements lists and tasks specification, as it would be made in software engineering. Over the years, this problem has been tried to be addressed with tools such as the "Game Design Document" (GDD) models, where the creative process must be described and mapped into a breakdown of requirements, tasks and goals [8]. However, there is no standard GDD format to describe the game mechanics and no technical reference language that standardises technical concepts [2], which usually causes gaps between the established creative concept and the software developed [1]. In fact, one of the first problems to address when starting a project arises with the platform selection, taking into account its capabilities and limitations to correctly define the game. Some studies present models to determine which is the appropriate platform for a project [3], while others look for methods to reuse video game specifications between different platforms extracting common features from a several game development platforms [5]. From a research standpoint, the quest for a generic framework has been present in the video game literature for a while. For instance, and despite doing so indirectly, the formal definition of games has received contributions from the AI research for games and from competitions in General Game Playing (GGP) [17,23,27,28] where the same system must learn to play any game based on a Game Description Language (GDL) [10,18]. In the grey area between traditional methods as GDD and others as advanced as GDL, some papers have already studied the possible synergies between the concepts of GDL and MAS [26]. The MAS are composed of sets of agents that interact and make decisions with each other within a shared environment [12,16,32]. Where behaviours have traditionally been defined by decision theory and game theory metrics [22]. Among the uses of MAS, on robotics and autonomous systems they have satisfied critical real-time restrictions [7,21], also the implementation of virtual commerce and trade fairs [15,25], obstacles avoidance in navigation [31] and mailing delivery using mobile robots [9]. In the video games field, they have been used to simulate large number of people in restricted areas [4] or to prototype the development of game engines [19].

3 Video Games as Multi-agent Systems

The goal of this work is to define a methodology for the specification and implementation of video games. The proposal focuses on the analogy between MAS and video games. The MAS have formal specification aspects to define a video game in a generic way, integrating specific aspects of the game such as the game logic with its entities and its behaviour rules, or the game physics with the detection and response of collisions between game elements. The method for the definition of a game must consider the features of the game, the elements that compose it, the behaviours definition and the user interaction. From this point, it is necessary to make a formal analogy between these game concepts and their correspondent concepts in a MAS.

Inside the shared environment, each agent has sets of properties and behaviour rules determining their interaction with each other. In order to prove that a game can be expressed as a MAS, a formal theoretical framework is presented. For this purpose, the definition of agent proposed by M. Wooldridge [32] has been used, where an agent is a computer system located in an environment and is capable of performing tasks autonomously to meet its design objectives. In addition, this method is also based on the work done by Marín-Lora et al. [20], where a game engine is defined as MAS. As a summary, some of the characteristics of the theoretical framework used for this proposal are detailed below:

- The environment to which the agents belong can be in any of the discrete states of a finite set of states $E = [e_0, e_1, e_2, ...]$.
- The environment shared by all agents has a set of properties that determine their status and that can be accessed by any agent in the environment.
- The system agents have a series of generic properties common to all of them: geometric, visual, physical, etc. In addition, they have the ability to assimilate new properties to perform specific tasks.
- Agents have behaviour rules to alter the environment state or an agent state in order to fulfill their plans and objectives.
- The agents have a set of possible actions available with the ability to transform their environment $Ac = [\alpha_0\ , \alpha_1\ , \alpha_2\ , ...]$.
- The run r of an agent on its environment is the interlayered sequence of actions and environment states $r{:}e_0 \rightarrow^{\alpha_0} e_1 \rightarrow^{\alpha_1} e_2 \rightarrow^{\alpha_2} ...\ e_{u-1} \rightarrow^{\alpha_{u-1}} e_u$.
- The set of all possible runs is R, where R^{AC} represents the subset of R that ends with an action, and R^E represents the subset of R that ends with a state of the environment. The members of R are represented as $R{=}[r_0, r_1,...]$.
- The state transformation function introduces the effect of the actions of an agent on an environment τ: $R^{AC} \rightarrow \wp(E)$ [11].

In order to transfer these concepts to a video game, it is necessary to implement the general characteristics of the game and the objects of the game such as those of an environment and its agents, respectively. Taking into account an essential requirement for the validity of this work: the same functions and attributes must exist for each element, regardless of the limitations or characteristics of the game engine or software environment selected for its implementation. This proposal follows the methodology presented in Marín-Lora et al. [20], where the rule specification is structured using a first-order logic semantics [6] based on two predicates: a condition *If()* and an action *Do()*. Where each predicate can run calls to other actions α of the system, or evaluate arithmetic and boolean expressions. An example of these rules and the technical specification of the game could be presented in the following fashion:

Environment
Assets: { Sprites, Sounds }
Properties: { General, Physics, Input, Logic, Audio, Render }
Agents
- Agent1:
  Properties: { *propertyA: true, propertyB: 7.52, propertyC: "Hello"* }
  Scripts: { *script1: { If(A) → Do(B)* }, *script2: { Do(C)* } }
- Agent2:
  Properties: { *propertyA: false, propertyB: -5, propertyC: "Hi again!"* }
  Scripts: { *script1: { If(A>B)→Do(C)* }, *script2: { If(D)→Do(E=E+15)* }}

4 Use Case: Frogger

In order to determine if a game can be implemented in the same way on different game engines based on this proposal, the Frogger is going to be described following a MAS specification. First, the concept and objectives of the game are described. Subsequently, the game specification will be presented following the methodology outlined in the previous section. This game was selected because it implements several arcade games mechanics and there is an open-source implementation in NetLogo, which puts in context and highlights further issues.

The Frogger game presents a scene where the player drives a frog passing through a road crossed by cars and trucks, and a river with trunks and turtles until reaching a safe area of water lilies. Initially, the frog is safely located at the bottom of the screen and can move up, down, right or left. At the beginning of the game, the frog has 5 lives, which can lose if collides with a car, a truck, or if it contacts with the water. Game elements representing cars and trucks come up from the right side of the screen and move with constant speed to the left side. To overcome the road, the frog must not contact with a car or a truck. For the water, it needs to remain on a trunk or turtle noticing that turtles can sink during arbitrary intervals. The player will win the level when it reaches the water lilies at the top screen. The next level would start-up to a maximum of 5.

Environment
Assets: { *frogSprite, truckSprite, carSprite, trunkSprite, turtleSprite, waterlilySprite, jumpSound, collisionSound, drownSound, winSound* }
Properties: { *width: 20, height: 20, centerX: 0, centerY: 0, lives: 5, levelComplete: false, currentLevel: 0, stopwatch: 0, timeLimit: 60 diveProbability: 0.1* }
Agents
- Game Manager:
  Properties: { }
  Scripts: {
  ∘ *LevelComplete: { If(Env.levelComplete) → (If(Env.currentLevel < 4) → Do(Env.currentLevel=Env.currentLevel+1) ∧ Do(resetLevel)) }*
  ∘ *ResetLevel: {If(Env.lives≠Frog.lives)→Do(Env.lives=Frog.lives)∧Do(resetLevel)}*
  ∘ *Stopwatch: { Do(Env.stopwatch=Env.stopwatch+1) ∧ If(Env.stopwatch ≥ Env.timeLimit) → Do(Frog.lives=Frog.lives-1) ∧Do(Env.stopwatch=Env.timeLimit)}*
  ∘ *EndGame: { If(Env.lives == 0) → Do(endGame) }*
- Frog:
  Properties:{*x:0,y:Env.centerY-Env.height/2,lives:5,jumpStep:1,jumpCount:0*}
  Scripts: {
  ∘ *LeftJump: { If(Env.leftKey) → Do(x = x - jumpStep) ∧ Do(rotation = 0) ∧ Do(jumpCount = jumpCount + 1) }*
  ∘ *RightJump: { If(Env.rightKey) → Do(x = x + jumpStep) ∧ Do(rotation = 90) ∧ Do(jumpCount = jumpCount + 1) }*
  ∘ *UpJump: { If(Env.upKey) → Do(y = y - jumpStep) ∧ Do(rotation = 180) ∧ Do(jumpCount = jumpCount + 1) }*
  ∘ *DownJump: { If(Env.downKey) → Do(y = y + jumpStep) ∧ Do(rotation = 270) ∧ Do(jumpCount = jumpCount + 1) }*
  ∘ *VehicleCollision: { If(vehicleCollision) → Do(lives = lives - 1) }*
  ∘ *Boating: { If(boatCollision) → Do(x = x + boat.velocityX * boat.directionX)}*
  ∘ *WaterCollision: { If(waterCollision) → Do(lives = lives - 1)}*
  ∘ *WaterLily: { If(waterlilyCollision) → Do(Env.levelComplete = true)}*
- Car, Truck, Trunk:
  Properties: { *originX: Env.centerX + Env.width / 2, x: originX, y: 0, velocityX: 1, velocityY: 0, directionX: -1, directionY: 0* }
  Scripts: {
  ∘ *Movement: {Do(x = x + velocityX * directionX)}*
  ∘ *Limits: {If(x < -originX) → Do(x = originX)}*
- Turtle:
  Properties: { *originX: Env.centerX + Env.width / 2, x: originX, y: 0, velocityX: 1, velocityY: 0, directionX: -1, directionY: 0, dive: false* }
  Scripts: {
  ∘ *Movement: {Do(x = x + velocityX * directionX)}*
  ∘ *Limits: {Limits: {If(x < -originX) → Do(x = originX)}*
  ∘ *Dive: {If (random(0 , 1) ¡ Env.diveProbabilty) → Do(dive = true)}*

5 Results

From the previous section, a version of the Frogger game has been implemented in three different systems: the MAS prototyping environment NetLogo and in the game engines Gamesonomy and Unity. All following the same guidelines: the game description and the game definition as MAS, but taking into account the differences and limitations of each system to achieve an identical result.

5.1 NetLogo

First of all, it is necessary to point out that NetLogo is not a graphical programming environment. Graphics and user interaction are not ideal to generate a good user experience. Examples of these limitations are the grid-based layout and the looping *ticks*, from which the drawing and logic must be managed explicitly. However, it allows to edit the interface, the scene and the frame-rate without using code: creating buttons that run functions or edit properties using sliders. The scene configuration is limited to the origin of coordinates and the scene dimensions. Also, the vertical and/or horizontal limits have to be determined, for an agent to exit at one end and appear at the other while moving.

For the game interface, four buttons have been arranged and linked to keyboard events and the movement functions, a native function *setup* callback to reload the level, a play function with the native *go* function and three sliders for the frog's lives in a range of 1 to 5, the initial level selection and the highest time available to complete the level from 60 to 10 seconds. Also, four monitors have been arranged that show the remaining lives, the current level, the time left and a jump counter. At the beginning of each level, each agent is created in its initial position by executing the *setup* function. Agents are generated by creating NetLogo agent objects called *turtle*, but for instantiation, this process must be performed with the *breed* function. Also, the game global variables must be defined. In the agent's case, each one except the frog has two new properties: speed and time. The speed determines the time interval that elapses between each jump. This action is controlled by a time variable, which counts the *ticks* remaining until the next iteration. Besides that, it is also arbitrarily determined that turtles can dive and which not by initialising their dive variable. Once in execution, the activation of the start button starts the game loop as long as lives is greater than zero. First, the game cycle records user input events to determine the movement the frog must perform. It should be noted that the configuration of the movement of the frog, unlike the other agents, prevents it from crossing the horizontal limits of the scene. The implementation of the behaviour rules of the agents, it is possible to execute the actions of each agent when a predetermined condition is fulfilled using the native *ask* function. The agent's movement relies on a time property that decreases every *tick* until it reaches 0. When this condition is met, agents can move. The movement is normally performed by the forward function. However, trunk and turtle agents have special functions, since they must interact with the frog agent and in the case of turtles, they can be submerged in the river arbitrarily. The last step of the loop is to check if the frog has reached a waterlily and therefore has exceeded the level, or if it has collided with a vehicle or with the water and therefore has not exceeded the level. For the first case, the water lily texture would be changed to the frog one and the current level would be increased if possible. In the second case, one life would be subtracted and, if there are lives left, the level would be restarted.

5.2 Gamesonomy

Gamesonomy is a web-based 2D game engine to create games without using code. In this platform, the game objects are known as Actors, and their behaviours are defined by decision trees composed by a reduced set of conditions and actions. The game loop evaluates the rules in a continuous cycle. At the beginning of the loop, it is checked if the conditions are met and if they are, the evaluation flow continues its way across the proper tree branch until it reaches an action leaf.

The screen size and resolution are adjusted from the game properties, where global game properties can also be created such as the current level, lives, and stopwatch. It should be noted that Gamesonomy does not has a default option for board limits, so it must be determined for each agent if necessary. However, the game global functions need to be implemented in an actor. In this case, a game manager actor was created to run game behaviour rules. It is located in the middle of the environment, from where the level elements are instantiated. Among them, the setup rule is only executed at the game start. Also, a rule to remove all objects from the scene has been included. This function is activated for an iteration so that each agent in the scene could be self-destructed. In order to arrange the actors, first, it is necessary to create the instantiable ones (vehicles and boats) to be spawned in their proper line. Conversely, the frog actor, which is neither instantiated nor eliminated. Furthermore, each actor has labels that identify them in their interactions. The frog has three rules: one checks the user's inputs to control its movement and the other two check if it collides with a vehicle or water and, if it does, it asks for the restart of the game. In the water case, if it is not in contact with a turtle or trunk actor. The other actors have rules that allow them to move, exit the screen and self-destruct if the game is restarted. For the movement, the timers have been used to control their speed so that from time to time it advances in the direction in which it is looking. Trunks and turtles can move the frog if they are in contact with it, and the latter can also dive in randomly if its dive property is true. The amount of turtles diving is controlled by a diving probability property, which can be set up in the game.

5.3 Unity

Unity is a general-purpose game engine that allows developing 2D and 3D games. In this case, the agent element is called game object. It is the basic component from which the scenes are composed, and it can store components to perform specific tasks. The behaviours are defined on the game objects with C# scripts. In this development, there is also a game manager storing the global properties and handling the state of the game and the game functions. A start function is included in its initialisation script, known as setup in the previous implementations, which is executed on the game load to initialise the local properties. The different types of game objects have been stored as "prefabs" so that the manager can instantiate them at the beginning of the game by accessing the reference of each prefab. The agents' movement function has been generalised so that it is compatible with all of them, also allowing them to access the customised velocity and direction properties. As mentioned previously, the movement function advances a unit and, in the case of turtles and trunks, drags the frog with the same amount of movement. The velocity value depends on the time differential elapsed since the previous loop, the interval between two consecutive executions. In this way, the speed can be measured in exact seconds. In the case of the truck and turtle agents, it has been necessary to create two scripts for specific behaviours: in the truck, the joint script coordinates the movements of the front and rear when leaving the screen, when the front leaves, it appears on the other side but the back no. In the turtle, the diving script is used to determine when a turtle that can dive is done and when it comes out, the turtles that can dive are determined by creating the level in the manager. In the water lilies case, their only function is to check if the frog collides and notify the manager as soon as it does. Finally, the frog agent collects the user's inputs to move on the board and checks the collisions with other agents, calling the game manager if necessary.

5.4 Conclusions and Future Work

This paper proposes a method for the definition, specification and implementation of video games based on MAS, where its elements, functionalities and interactions are established regardless of the platform used for its development. To prove its validity and capabilities, the classic arcade game Frogger has been used as a demonstrator. This game has been defined in a general form and then implemented on three platforms following the same specification. Each of these implementations has been made using different engines, languages and programming techniques, but meeting the requirements of the game and the MAS.

The purpose of this work was to demonstrate that video games and MAS share several features and it is possible to improve video game development processes in this way. From the previous sections, it is extracted that the MAS features fit as a starting point for the video games definition and therefore it is necessary to keep working in methods that potentiate a symbiosis between these concepts. Additionally, the incorporation of agent specification systems into the video game development can ease the understanding of games before they are even implemented, which could ease the access into the sector of professionals who do not have technical experience creating video games.

As future work, it is intended to explore the potential of this method by designing and implementing a MAS-based game engine and generate games that meet the characteristics of the MAS. Trying to explore the capabilities of these interactive systems in games and in virtual and augmented reality experiences, and to provide game development tools to non-expert profiles such as children.

Acknowledgments. This work has been supported by the Ministry of Science and Technology (TIN2016- 75866-C3-1-R, TI2018-098651-B-C54) and the Universitat Jaume I research projects (UJI-B2018-56, UJI-B2018-44).

References

1. Ampatzoglou, A., Stamelos, I.: Software engineering research for computer games: a systematic review. Inf. Softw. Technol. **52**(9), 888–901 (2010)
2. Anderson, E.F., Engel, S., McLoughlin, L., Comninos, P.: The case for research in game engine architecture (2008)
3. Anderson, E.F., McLoughlin, L., Watson, J., Holmes, S., Jones, P., Pallett, H., Smith, B.: Choosing the infrastructure for entertainment and serious computer games-a whiteroom benchmark for game engine selection. In: 5th International Conference on Games and Virtual Worlds for Serious Applications, pp. 1-8. IEEE (2013)
4. Aranda, G., Trescak, T., Esteva, M., Rodriguez, I., Carrascosa, C.: Massively multiplayer online games developed with agents. In: Pan, Z., Cheok, A.D., Müller, W., Chang, M., Zhang, M. (eds.) Transactions on Edutainment VII, pp. 129–138. Springer, Heidelberg (2012)
5. BinSubaih, A., Maddock, S., Romano, D.: A survey of 'game' portability. University of Sheffield, Technical report CS-07-05 (2007)
6. Brachman, R.J., Levesque, H.J., Reiter, R.: Knowledge Representation. MIT press, Cambridge (1992)
7. Botti, V., Carrascosa, C., Julián, V., Soler, J.: Modelling agents in hard real-time environments. In: European Workshop on Modelling Autonomous Agents in a Multi-Agent World, pp. 63–76. Springer, Heidelberg (1999)
8. Callele, D., Neufeld, E., Schneider, K.: Requirements engineering and the creative process in the video game industry. In: 13th IEEE International Conference on Requirements Engineering, RE 2005, pp. 240-250. IEEE (2005)
9. Carrascosa, C., Bajo, J., Julián, V., Corchado, J.M., Botti, V.: Hybrid multi-agent architecture as a real-time problem-solving model. Exp. Sys. Appl. **34**(1), 2–17 (2008)
10. Ebner, M., Levine, J., Lucas, S.M., Schaul, T., Thompson, T., Togelius, J.: Towards a video game description language (2013)
11. Fagin, R., Fagin, R., Fagin, R., Halpern, J.Y.: Reasoning about knowledge and probability. J. ACM (JACM) **41**(2), 340–367 (1994)
12. Ferber, J., Weiss, G.: Multi-agent Systems: An Introduction to Distributed Artificial Intelligence, vol. 1. Reading: Addison-Wesley, Boston (1999)
13. Frogger Game, 15 November 2019. https://en.wikipedia.org/wiki/Frogger
14. Gamesonomy, 15 November 2019. http://www.gamesonomy.com
15. Garcés, A., Quirós, R., Chover, M., Huerta, J.: E-Commerce transaction modeling using moderately open multi-agent systems. In: ICEIS, vol. 4, pp. 167-172 (2007)
16. Garcés, A., Quirós, R., Chover, M., Camahort, E.: Implementing virtual agents: a HABA-based approach. Int. J. Multimedia. Appl. **2**, 1–15 (2010)
17. Genesereth, M., Love, N., Pell, B.: General game playing: overview of the AAAI competition. AI Mag. **26**(2), 62–62 (2005)
18. Love, N., Hinrichs, T., Haley, D., Schkufza, E., Genesereth, M.: General game playing: game description language specification (2008)

19. Marín-Lora, C., Chover, M., Sotoca, J.M.: Prototyping a game engine architecture as a multi-agent system. In: 27th International Conference in Central Europe on Computer Graphics, Visualization and Computer Vision, WSCG 2019 (2019)
20. Marín-Lora, C., Chover, M., Sotoca, J.M., García, L.A.: A game engine to make games as multi-agent systems. Adv. Eng. Softw. **140**, 102732 (2020)
21. Occello, M., Demazeau, Y.: Modelling decision making systems using agents for cooperation in a real time constraints. In: 3rd IFAC Symposium on Intelligent Autonomous Vehicles, vol. 1, pp. 51-56 (1998)
22. Parsons, S., Wooldridge, M.: Game theory and decision theory in multi-agent systems. Auton. Agent. Multi-Agent Syst. **5**(3), 243–254 (2002)
23. Perez-Liebana, D., Samothrakis, S., Togelius, J., Schaul, T., Lucas, S.M.: General video game ai: competition, challenges and opportunities. In: 30th AAAI (2016)
24. Rebollo, C., Marín-Lora, C. Remolar, I. Chover, M.: Gamesonomy vs Scratch: two different ways to introduce programming. In: 15th International Conference on Cognition and Exploratory Learning in the Digital Age, CELDA 2018. IADIS Press (2018). ISBN 9789898533814
25. Remolar, I., Garcés, A., Rebollo, C., Chover, M., Quirós, R., Gumbau, J.: Developing a virtual trade fair using an agent-oriented approach. Multimedia Tools Appl. **74**(13), 4561–4582 (2015)
26. Schiffel, S., Thielscher, M.: A multiagent semantics for the game description language. In: International Conference on Agents and Artificial Intelligence, pp. 44–55. Springer, Heidelberg (2009)
27. Thielscher, M.: A general game description language for incomplete information games. In: Twenty-Fourth AAAI Conference on Artificial Intelligence (2010)
28. Thielscher, M.: The general game playing description language is universal. In: Twenty-Second International Joint Conference on Artificial Intelligence (2011)
29. Tisue, S., Wilensky, U.: Netlogo: A simple environment for modeling complexity. In: International conference on complex systems, vol. 21, pp. 16-21 (2004)
30. Unity 3D Game Engine, 15 November 2019. https://unity.com/
31. Van den Berg, J., Lin, M., Manocha, D.: Reciprocal velocity obstacles for real-time multi-agent navigation. In: 2008 IEEE International Conference on Robotics and Automation, pp. 1928-1935. IEEE (2008)
32. Wooldridge, M.: An Introduction to Multiagent Systems. Wiley, Chichester (2009)

Overview of an Approach on Utilizing Trust and Provenance Metrics Combined with Social Network Metrics on Recommender Systems

Korab Rrmoku(✉) and Besnik Selimi

South East Europe University, 1200 Tetovo, North Macedonia
{kr26918,b.selimi}@seeu.edu.mk

Abstract. In this paper the focus will be to estimate and evaluate the measures and metrics of Social Network Analysis (SNA) and network trust and provenance in order to achieve better recommendations. Since, each of these fields are usually treated separately, the aim is to merge and correlate Social Network Analysis metrics and network provenance in order to develop and enhance the baselines of certain Recommender System (RS). A thorough background research has been conducted in order to distinguish the main components of each field separately, followed by the methodology that is in intention to be our main approach. At the end, we will have an evaluation overview of one RS, with a detailed explanation on the metrics that have to be added in order to improve the desired recommendation at the end.

Keywords: Recommender Systems · Social Network Analysis · Network trust · Provenance · Metrics

1 Introduction

Recommender Systems (RS) are software techniques and tools that offer suggestions for one particular item that has higher probability to be interesting for a specific user [1]. Suggestions are related with various decision-making processes, such as items to buy, what kind of music to listen to, or what news should one read online on the Internet. RSs are primarily directed toward individuals who lack the sufficient personal experience or competence in order to evaluate the potentially overwhelming number of alternative items that a website, for example, may offer [2]. A prime example is a book recommender system that assists users in selecting a book to read. On the popular website, Amazon.com, the site employs an RS to personalize the online store for each customer [3]. The development of RSs initiated from a rather simple observation: individuals often rely on recommendations provided by others in making routine, daily decisions [4], and it is also important to note that sometimes the user utility for an item is observed to depend on other variables, which we generically call "contextual" [5].

The combination of computers and the users as well, today are considered as nodes within social networks [6]. Hence, calculations and recommendations that are based on

Á. Rocha et al. (Eds.): WorldCIST 2020, AISC 1159, pp. 380–386, 2020.
https://doi.org/10.1007/978-3-030-45688-7_39

Social Network Analysis (SNA) [7, 8] are becoming very popular and worth considering for each and every field of social computing, even for those that did not consider the users' aspects before. Further, motivated by the idea of homophily as the tendency to relate people with similar personal attributes and interests [9], and following the SNA theory of modeling an attribute / interest matrix into a bimodal network as a useful mechanism in cases there are two distinct classes of entities [7].

Additionally, Provenance or network trust, refers to all kind of information that depicts, illustrates and analyze the process of a production for a certain product, which can be anything from a portion of data to a physical object. Hence, provenance information includes meta-data about entities, data, processes, activities, and persons involved in the production process [11]. Basically, provenance can be seen as meta-data that, as an alternative of describing data, describes a production process. The collection and processing of provenance are important in various settings, e.g., to assess quality, to ensure reproducibility, or to reinforce trust in the end product.

2 Background on Relevant Concepts

In line with developments in the fields that we want to correlate - Recommender Systems, Social Network Analysis and Provenance, a background research is conducted, in order to relate the importance and impact of social network analysis global and individual metrics and network provenance on recommender systems.

2.1 Recommender Systems Approach

There are several approaches and methodologies used in RS in order to achieve the most appropriate recommendations for a specific item/s. In [1], among a variety of approaches, 5 different classes of recommendation approaches are distinguished:

- *Content-Based* – where the system learns to recommend items that are similar to the ones that the user liked in the past. The similarity of items is calculated based on the features related with the compared items.
- *Collaborative Filtering (CF)* – the original and most simple application of this approach [12] makes recommendations to the active user based on items that other users with similar tastes liked in the past. The similarity in taste of two users is calculated based on the similarity in the rating history of the users.
- *Demographic* - This type of system recommends items based on the demographic profile of the user [13]. The supposition is that different recommendations should be generated for different demographic domains.
- *Knowledge-Based* - Knowledge-based systems recommend items based on specific domain knowledge about how certain item characteristics meet users' needs and preferences and, eventually, how the item is convenient for the user [14].
- *Hybrid Recommender Systems* - are based on the combination of the above-mentioned techniques. A hybrid system combining techniques A and B tries to use the advantages of A to fix the disadvantages of B. As depicted in the Fig. 1 below, given two (or more) basic RSs techniques, several ways have been proposed and can be used for combining them to create a new hybrid system.

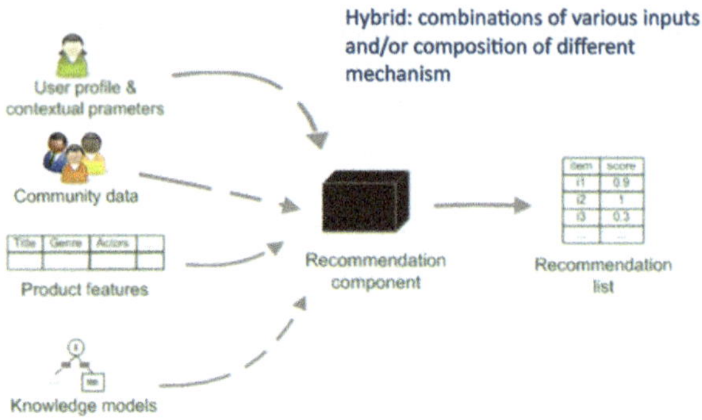

Fig. 1. Hybrid recommender systems

2.2 Social Network Analysis Approach

In terms of SNA analysis, there are certain measures and metrics that help define the behavior of a network and graphs. In this line, the following measures and metrics should be considered when analyzing a network, as elaborated in [15]:

- *Degree centrality* – as the simplest measure of node connectivity, assigns an importance score based purely on the number of links held by each node, and it actually defines how many direct connections each node has within a network.
- *Betweenness centrality* – which is useful for analyzing the dynamics of a communication within a network, measures the number of times a node lies on the shortest path between other nodes, and this measure identifies the nodes that are "bridges" in a network.
- *Closeness centrality* – which helps to identify the "broadcasters" within a network, measures the score of each node based on their "closeness" to all other nodes within the network, and is preferred to be used on cases where we are interested to find the individuals who are best placed to influence the entire network most quickly.
- *Eigen centrality* – which is a very useful measure to understand a network as a propagation use-case, like understanding human social networks. In this way, the Eigen centrality identifies the nodes with influence over the whole network, not just those directly connected to it.

Based on the measures elaborated above, following are some main SNA metrics and their categorization:

1. Global graph metrics: seek to describe the characteristic of a social network as a whole, for example the graphs diameter, mean node distance, the number of components (fully connected subgraphs), cliques, clusters, small-worldness, etc.
2. Individual actor properties: relate to the analysis of the individual properties of network actors, e.g. actor status as central (degree, closeness, or betweenness centrality) or authoritative (eigenvector, PageRank, SALSA, HITS, or weighted HITS), distance, and position in a cluster.

2.3 Network Provenance Metrics Approach

In social media, provenance of information provides a similar value to its users. Social media contains social media data (attributes, links, and contents) which can be used to determine information provenance. Thus, knowing the provenance of a piece of information published in social media—how the piece of information was modified as it was propagated through social media and how an owner of the piece of information is connected to the transmission of the statement—provides additional context to the piece of information [16].

According to [16] following are related provenance attributes related to social media and networks:

- α as a unique identifier - can be constructed to uniquely identify a node. A common example would be a user name that is unique to a particular social media service.
- A is a set of provenance attributes, $(a_1, \ldots, a_n) \in A$, for example name, occupation and education.
- N is the number of provenance attributes sought after for any α.
- W is the set of weights, $(w_1, \ldots, w_n) \in W$, associated with $(a_1, \ldots, a_n) \in A$, where weighting particular provenance attributes allows us to develop strategies for certain criteria.

3 Methodology

From the research conducted regarding the issue of utilization of both: SNA metrics together with provenance metrics, in yielding to better recommendation, it is seen that each of the approaches is treated separately.

In this rationale, our approach was to define, merge and correlate SNA metrics and network provenance in order to develop and enhance the baselines of certain RS.

So, in terms of SNA, while narrowing the findings, we have reached to the point, where we have to consider which of the two general types of SNA categories: individual or global metrics will be more suitable and will perform better in combination with network provenance. From 'global graph metrics' the focus will be on distinguishing between cliques and cascades, and from 'individual graph metrics' the focus is on determining the best metrics to utilize among betweenness, and closeness centrality, as depicted in Fig. 2 below.

Form the network provenance point of view, as elaborated in [17], network trust models are classified in five major categories based on the techniques the use: (i) statistical technique, (ii) heuristics-based, (iii) graph based, (iv) semantic based and (v) fuzzy. Thus, the aim during this research will be to determine the best technique to be used, which in combination with SNA metrics will enhance the recommendation. From the research done until now, a machine learning category of algorithms – named classification, with its naïve Bayes approach seems more feasible as elaborated in [18, 19].

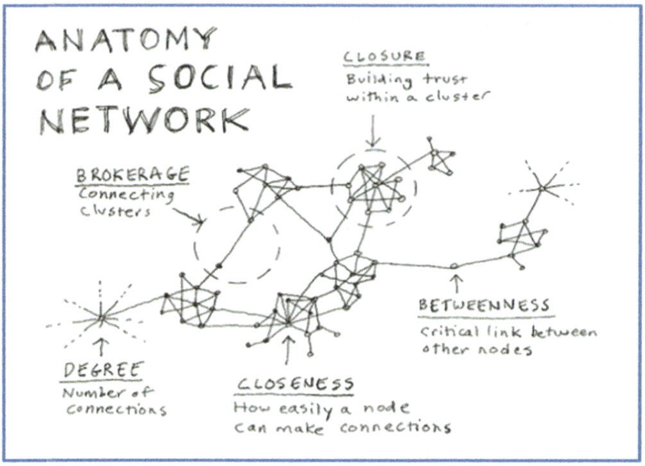

Fig. 2. An anatomy of a social network analyzed on deciding the metrics to be used

4 Evaluation Overview

Since the evaluation of the recommendation remains one of the most important factors on deciding the appropriate metrics to be considered, the evaluation approach which will used are precision and recall. Precision and recall are the measures we use for evaluation, and their definition when applied to our problem domain is following:

$$precision = \frac{|\{\text{relevant items}\} \cap \{\text{recommended items}\}|}{|\text{recommended items}|} \tag{1}$$

$$recall = \frac{|\{\text{relevant items}\} \cap \{\text{recommended items}\}|}{|\text{relevant items}|} \tag{2}$$

Due to the fact that these definitions are adopted for a specific domain, such as the touristic domain that we have used in the first part of experiments [20], we aim to further expand the domains of application such as the usage of datasets of any social network (SN), such as Twitter, Facebook, Instagram, etc. Considering various domains where recommendation can be applied, we have defined precision and recall in terms of satisfaction factor - SF (which is the factor and the feedback that we get from the correspondents and the reviewers which have already received the recommendation (thus they can be considered as testing set) for our evaluation as follows):

$$precision = number\ of\ recommended\ items$$

$$recall = SF/number\ of\ testing\ set\ items$$

Precision and recall calculation that we have received in [20], are depicted in the Fig. 3 below:

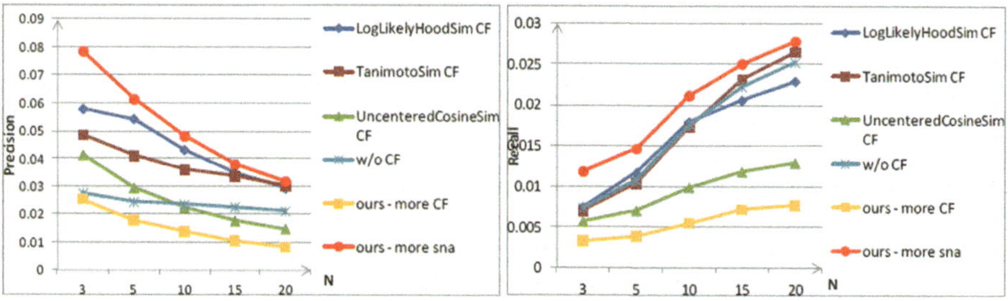

Fig. 3. Precision and recall @N (number of desired recommendation): our algorithms vs. baseline [Source: 20]

It is evident that our SNA-based recommendation algorithm "ours - more sna" outperforms the baseline RSs. This strengthens the conclusion for adopting SNA metrics to items recommender systems. It is to be emphasized that the domain analyzed in [20] deals with data, questioners, interviews and satisfaction factor in the domain of tourism, respectively with different Points of interest in various cities and countries.

5 Conclusion and Future Work

Several studies, including [14, 15, 21] have treated the involvement of social network analysis factors into a recommender system, and to further expand this approach, our intention and conclusion in this paper include the following:

- since it is evident from the experiments that including a variety of individual metrics on SNA performed better than some of the baseline algorithms, it is worth expanding the inclusion of other factors such as global metrics of SNA as betweenness and closeness centrality,
- considering even probabilistic (statistical) techniques such as Bayesian systems, combined with Supervised Machine learning classification algorithm of Naïve Bayes, would strengthen and clearly define the trust of a source within a network, thus making it more convenient for the user that beside that the recommendation is accurate, it is even coming from sources that are trusted.

References

1. Ricci, F., Rokach, L., Shapira, B.: Recommender Systems Handbook. Springer, Boston (2015)
2. Williams, C., Mobasher, B., Burke, R., Bhaumik, R., Sandvig, J.: Detection of obfuscated attacks in collaborative recommender systems. In: Proceedings of the 17th European Conference on Artificial Intelligence, ECAI 2006 (2006)

3. Pampín, H.J.C., Jerbi, H., O'Mahony, M.P.: Evaluating the relative performance of collaborative filtering recommender systems. J. Univ. Comput. Sci. **21**(13), 1849–1868 (2015)
4. Tang, T., Tang, Y.: An effective recommender attack detection method based on time SFM factors. In: 2011 IEEE 3rd International Conference on Communication Software and Networks (ICCSN), pp. 78–81 (2011). https://doi.org/10.1109/iccsn.2011.6013780
5. Wu, Z., Wu, J., Cao, J., Tao, D.: HySAD: a semi-supervised hybrid shilling attack detector for trustworthy product recommendation. In: Proceedings of the 18th ACMSIGKDD International Conference on Knowledge Discovery and Data Mining, pp. 985–993. ACM (2012)
6. Hendler, J., Berners-Lee, T.: From the semantic web to social machines: a research challenge for AI. Artif. Intell. **174**(2)
7. Wasserman, S., Faust, K.: Social network analysis: methods and applications. In: Structural Analysis in the Social Sciences, New York, USA, No. 8. Cambridge University Press (1994)
8. Scott, J.: Social Network Analysis: A Handbook. SAGE Publications, London (2000)
9. Breiger, R.: Duality of Persons and Groups. Soc. Forces **53**, 181–190 (1974)
10. Moreno, J.L.: Emotions mapped by new geography. New York Times (1933)
11. Groth, P., Moreau, L.: PROV-overview: an overview of the PROV family of documents (2013)
12. Trattner, C., Kowald, D., Lacic, E.: TagRec: towards a toolkit for reproducible evaluation and development of tag-based recommender algorithms. SIGWEB Newsl., 1–10 (2015)
13. Domingos, P., Richardson, M.: Mining the network value of customers. In: Proceedings of the Seventh ACM SIGKDD International Conference on Knowledge Discovery and Data Mining, KDD 2001, pp. 57–66. ACM, New York (2001). https://doi.org/10.1145/502512.502525
14. Hofmann, T.: Collaborative filtering via gaussian probabilistic latent semantic analysis. In: Proceedings of the 26th Annual International ACM SIGIR Conference on Research and Development in Information Retrieval, SIGIR 2003, pp. 259–266. ACM, New York (2003). http://doi.acm.org/10.1145/860435.860483
15. Wasserman, S., Faust, K.: Social Network Analysis – Methods and Applications. Cambridge University Press (1994)
16. Barbier, G., Feng, Z., Gundecha, P., Liu, H.: Provenance data in social media. In: Synthesis Lectures on Data Mining and Knowledge Discovery #7. MC Morgan & cLaypool publishers (2013)
17. Tselenti, P., Danas, K.: A review of trust-aware recommender systems based on graph theory. In: Proceedings of the International Conference on Computer Science, Computer Engineering, and Social Media, Thessaloniki, Greece (2014). ISBN: 978-1-941968-04-8 ©2014 SDIWC
18. Mui, L., Mohtashemi, M., Halberstadt, A.: A Computational Model of Trust and Reputation for E-Businesses, Washington, DC, USA, p. 188 (2002)
19. Jøsang, A., Lo Presti, S.: Analysing the relationship between risk and trust. Trust Manag. **2995**, 135–145 (2004)
20. Ahmedi, L., Rrmoku, K., Sylejmani, K., et al.: A bimodal social network analysis to recommend points of interest to tourists. Soc. Netw. Anal. Min. **7**, 14 (2017)
21. Liu, F., Lee, H.J.: Use of social network information to enhance collaborative filtering performance. Expert Syst. Appl. **37**(7), 4772–4778 (2010)

Logic-Based Smart Contracts

Adriana Stancu(✉) and Mihaita Dragan

Faculty of Mathematics and Computer Science, University of Bucharest, Bucharest, Romania
{astancu,mihaita.dragan}@fmi.unibuc.ro

Abstract. With the increasing popularity and diversity of Blockchain systems, smart contracts were introduced as a necessity to automatically execute certain operation depending on the occurred events. The programming languages used in defining the triggering events and their consequent actions depends on the Blockchain implementation. In this article, we will investigate the advantages and suitable scenarios for using a formal approach in defining smart contracts. We will use a particular python implementation of a Blockchain and add the interface with a Prolog component for defining and querying smart contracts.

Keywords: Smart contract · Logic programming · Distributed database

1 Introduction

Blockchain systems were first described in 2008 [1], by the first functional definition of an electronic cash, the Bitcoin. The article was signed pseudonymously by Satoshi Nakamoto and described the Blockchain as "an ongoing chain of hash-based proof-of-work" [1]. The ideas of synchronization of distributed knowledge and obtaining a consensus in peer-to-peer networks were long before deeply studied [2]. The inheritance of previous research was further developed by using the consensus protocols for obtaining a unified view over the transactions added to the Blockchain.

The most developed cash system regarding the diversity of programming paradigms for smart contract is Ethereum [3]. Solidity [4], Vyper [5], or Serpent [6] are all programming languages for Ethereum smart contracts and all of them are compiled in Ethereum byte code - the language executed by the Ethereum virtual machine [4] (EVM).

With the increasing support for various operations, smart contract's code is exposed to programming and design bugs, logical errors or inherent vulnerabilities. Our idea is to use a stable and powerful programming language that already has a logic baseline in order to define the smart contract verification part. We chose SWI Prolog as a starting point to build a Blockchain system. Our goal was to model a scenario in which to analyze the integration of a logic framework to build smart contracts.

2 The Logic Programming Approach of Smart Contracts

Regardless of the programming language, smart contracts have to meet some architectural criteria:

Á. Rocha et al. (Eds.): WorldCIST 2020, AISC 1159, pp. 387–394, 2020.
https://doi.org/10.1007/978-3-030-45688-7_40

- Smart contacts have to be short pieces of code. With the increasing length or complexity of a smart contract, the time required for committing the associated transactions is higher and such is the gas consumption (measure of calculus complexity).
- The immutability is a required characteristic. This is acquired by adding the smart contracts to the Blockchain as a valid transaction.
- If the smart contract is supposed to modify the state of the Blockchain and generate other transactions, these have to be carefully ordered in order to avoid reentrancy attacks [7] or unwanted transfers.

Smart contracts are "automated autonomous programs that reside on the blockchain and encapsulate business logic and code in order to execute a required function when certain conditions are met" [8].

For example, a smart contract can take care of a transaction for renting a house. The contract will be called monthly at a certain due date and verify if the Blockchain contains a transaction from the tenant to the house owner labeled with rent information. If the transaction does not exists it can increase the rent amount or add penalties, depending on the agreement between the tenant and the house owner. The Blockchain system is suitable to this situation because the smart contract is set only once when it is added to the Blockchain. The two parties does not have to trust each other as the contract cannot be altered afterwards and the consequent actions if the rent is not paid will be taken automatically.

The motivation was developed starting with the article *"Evaluation of Logic-Based Smart Contracts for Blockchain Systems"* [9]. The main idea of the article is to promote the combination of logic frameworks into Blockchain systems in order to make easier to work with for communication between customers and developers. A logic approach of programming smart contracts has technical advantages over procedural coding of the contracts by including inherently self verification of the smart contracts logic.

Comparing to smart contracts written in Solidity, Prolog smart contracts are not written in a procedural language, thus they are not vulnerable to out of order execution (reentrancy attacks by breaking the sequence of functions or cross calling of functions). Integer representation is some particularity of Prolog that can support numbers with more than 64 bits by chains of words. This feature ensures protection against integer overflow attacks.

The main reason for using Prolog for smart contracts is the default recursive search for a matching solution. The conditions and their consequences chosen by the user via web interface are automatically translated in Prolog code. The general approach is mainly: explaining the business logic to developers who then translate the requirements into contract's code whilst being prone to misinterpretations. Our approach consists of minimizing the developer's contribution by letting the user create it's contracts via an interface.

3 Implementation Considerations

The Blockchain is implemented in Python backed with the storage functionalities of BigchainDB [10] and synchronization mechanisms of Tendermint [8]. We used Python as the interface between the back-end database and SWI prolog component because this language allows native integration with different frameworks.

The application is distributed over different nodes. The client front-end connects to one of the nodes and send all the database requests to it. The separation between a client and server instance was done because of the necessity the client should be as simple as possible and offer the possibility to connect to any of the accessible nodes on the network.

We added the functionality to change the connection with the address of a new node. Before changing the current node, we check if it is accessible and if there is an instance of the application running on the specified port.

The server component has all the logic and database parts of the application. There is an instance of BigchainDB running on the server and a configurable number of concurrent processes that will be used to consult Prolog contract before making a transaction.

3.1 BigchainDB

BigchainDB 2.0 is an open-source software built on a MongoDB database and embedding Blockchain properties: decentralization, immutability, ownership of assets, sibyl tolerance. BigchainDB driver is build for integration with Python and JavaScript applications. We used the Python driver in order to access and manage the database transactions.

In BigchainDB data is structured as assets. An asset can characterize any object or instance, can be implemented as a file, a class or a dictionary in our case.

In our case, we have two types of assets:

- licenses owned by users;
- user accounts that are signed with their own key.

An asset can have multiple owners, in this case the transaction with this asset must be signed by all of its owners in order to be valid. The process of signing the transaction is called *"fulfillment"* and requires the owner or creator of the asset to sign the data using Ed25519 public-key signature system [9]. The output of the transaction will contain the public key of the owner. When the asset will be spent, the owner has to sign with his/her private key associated to the public key contained by the asset.

The assets can be used in BigchainDB in two ways:

- By users in CREATE transactions.
- Transfer (or update) to other users in TRANSFER transactions.

There is no delete or update method because all the data recorded into the Blockchain is immutable. The well known CRUD database operations (Create, Read, Update, Delete) are replaced with a different approach, namely CRAB [11] (Create, Retrieve, Append, Burn). The entire lifecycle of an asset can be visualized using these operations.

For example, a user account asset will be created by its owner and signed with his/her private key. When the user requires logging in, we will search the user account assets for his/her account. The operation can retrieve data, but cannot commit any modifications on the asset. If the user requires resetting his/her password, he will create a transfer transaction to append some new information to the asset and transfer it to himself/herself.

Finally, a burn operation consists of transferring an asset to an artificial public key. The asset will be accessible for retrieving data, but its state will be locked for any other future transactions because the owner's public key does not have a private key pair.

3.2 Tendermint

Tendermint is an open-source software that ensures the replication of an application's data on multiple machines. The core functionality is to maintain a unified view over the application managed (in our case it is BigchainDB application) even if 1/3 of machines may have arbitrary behavior.

Each machine that is not defective sees the same transaction log and calculates the same state. A fundamental issue in distributed systems is safe and consistent replication. Tendermint plays an important role in the application ensuring a consistent state of the BigchainDB database and tolerance at random machine's behavior by keeping a full replication of the data between the nodes.

Tendermint consists of two core components: an engine that applies the consensus algorithm and an interface for interconnection with various applications.

The consensus engine, called Tendermint Core, ensures that the same transactions are recorded on each machine in the same order. The application interface, named Interface Application Blockchain (ABCI), allows processing of transactions in any programming language.

In this application Tendermint is responsible for sharing blocks and establishing a unique order of transactions. Each BigchainDB node exposes the local database to Tendermint interface and accepts any updates coming from the other nodes via ABCI.

Figure 1 illustrates the interdependence between MongoDB database, BigchainDB application to manage the database and Tendermint used on top of them to synchronize data.

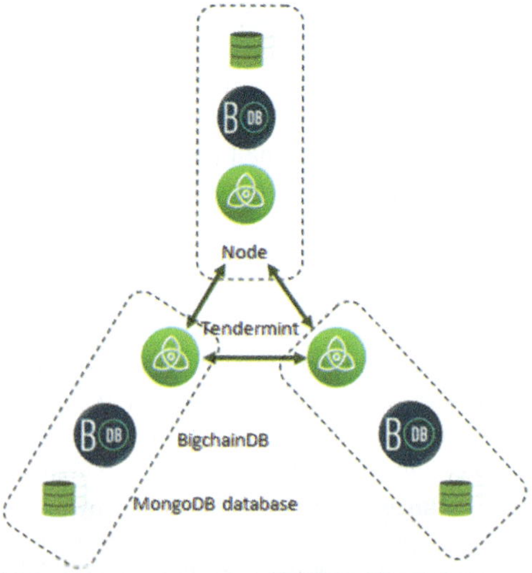

Fig. 1. BigchainDB and Tendermint architecture.

3.3 SWI Prolog

The interface between Python and SWI Prolog is realized by using the PySwipl Python library that allows calling a Prolog process and consulting .pl files. In order to verify a Prolog contract we can consult its clauses using two methods:

- Creating a knowledge database file and consult the file before making a transaction;
- Parsing the Prolog clauses and asserting each one of them into a newly created Prolog process;

We used the second method, as the read/write operations would add unnecessary delay, while we can load the clauses directly into the running memory.

For each of the server applications, we can span multiple processes that will listen for a specific event, receive the contract that has to be verified and return the validation of the transaction.

4 License Management

The scenario analyzed by this implementation is managing a collection of application licenses and controlling their lifecycle: creation of licenses, administrating rights to use and rights to transfer and invalidating them when this is the case.

The application owner can generate a license for its product and transfer it to the client with an associated contract. Depending on the license, the client can be a reseller and transfer it to a end-user or just use it for himself/herself. The reselling process is controlled indirectly by the owner of the application by generating the smart contract along with the license. The smart contract is composed from a set of Prolog clauses that states if the license can or cannot be transferred in certain situations. In our case, we implemented a few rules to check the effectiveness of logic verification:

- There are two types of licenses: full and evaluation. The evaluation licenses are offered by the owner directly to the end-user and cannot be transferred.
- The license cannot be transferred if already expired. The creation date is embedded into the contract and it cannot be modified.
- There are special countries where a certain license can be used. The product owner states the list of valid countries into the contract and any future transfer of the license will be checked to remain in this area.

These rules are our contract statements and any transfer request will be checked against them. The advantage of logic programming is that the translation into Prolog rules will model these requirements even if they can become more complex. The logic paradigm is closer to the natural language used to describe business logic than procedural languages used to implement smart contracts.

Transferring a license requires identification of the license by the identification key of the create transaction and the key of the transfer transaction that gave the ownership to the current user. In other words, a user can transfer a license if he can prove that either he is the creator of the application, or the license is valid and he received it through a legitimate transfer. For a complete view over the application, following there are a diagram of main building blocks in Fig. 2, an overview of a software producer dashboard in Fig. 3 which expose the main options available (generating, transferring and identifying a license) and a raw view of the Prolog in Fig. 4.

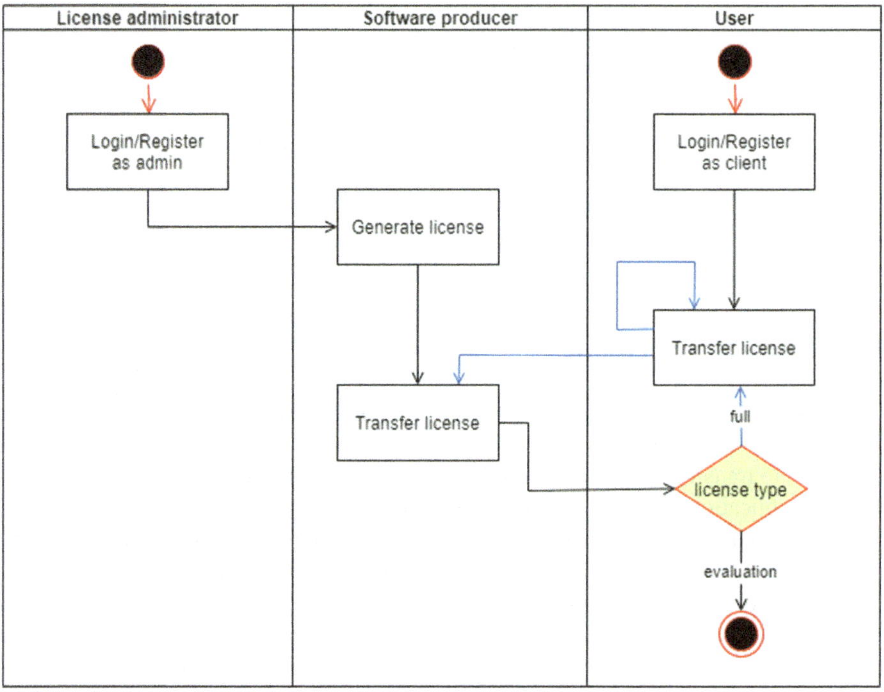

Fig. 2. Diagram of main application components.

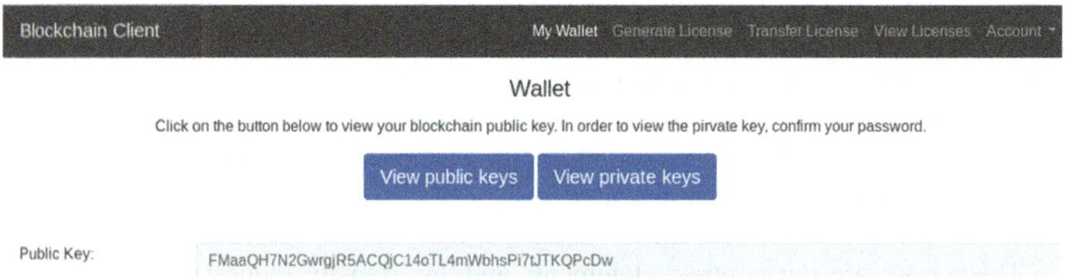

Fig. 3. Dashboard of client application.

```
_id: ObjectId("5ce9197ae3a8be24f2147e97")
data: Object
   type: "evaluation"
   product: "MyAnotherApp"
   valid_from: "2019-05-25 13:31"
   valid_to: "2019-06-14 13:31"
   contract: "transferlicenta(X,Y,T):-cantransfer(T),.
id: "5106eeb8876d1ccc749290cca7834156300b7091327fb3b6
```

Fig. 4. Raw view of license asset.

5 Conclusions

There are multiple use cases where the reliability and immutability of data is important. Contract management always required an additional party to ensure the validity and impartiality of the contract. Using a fully distributed database, we have guaranteed that if a commit is made to BigchainDB, the contract will remain unchanged. Any attempt to modify the data will be visible by changing the hash of the transaction containing this data.

Contract verification and validation using Prolog has the advantage that simplifies the translation of contract clauses into a programming language. A drawback of our implementation is the necessity to run multiple instances of Prolog processes on the server machines in order to allow multiple contracts to be verified simultaneously. The inter-process communication could add a delay in the overall transaction validation process. A possible solution is to implement the Blockchain system directly into a logic framework that would allow both transaction creation and contract verification.

Another advantage of using a logical approach in programming smart contracts is that the specific clauses of the contract can be adjusted to any other use case. A future development of the application is to register data about cars and allow the customers to view the entire chain of previous owners of the car they want to buy. In this scenario, the contract clauses can refer to a selling process or just lending the car for a short period. We can compute the value of the tax from the car details, we have the certitude that the car data is not modified outside the application and we can dynamically compute additional taxes given the history of the car. Also, if the contract conditions are more complex, we just have to add the corresponding Prolog clauses into the application and allow the users to use them while composing their contracts.

References

1. Nakamoto, S.: Bitcoin: A Peer-to-Peer Electronic Cash System, 31 October 2008. https://git.dhimmel.com/bitcoin-whitepaper/
2. Clake, I.: A distributed decentralized Information Storage and Retrieval System. University of Edinburgh (1999)
3. Antonopoulos, A.M., Wood, G.: Mastering Ethereum: Building Smart Contracts and Dapps, 1st edn. O'Reilly Media, Sebastopol (2018)
4. Dannen, C.: Introducing Ethereum and Solidity. Apress Media, New York (2017)
5. Buterin, V.: Vyper Documentation. In: Vyper by Example, p. 13, 4 October 2018. https://readthedocs.org/projects/viper/downloads/pdf/latest/

6. Delmolino, K., Arnett, M., Kosba, A., Miller, A., Shi, E.: A Programmer's Guide to Ethereum and Serpent, p. 6, 21 May 2015. https://mc2-umd.github.io/ethereumlab/docs/serpent_tutorial.pdf
7. Li, J., Yang, C.: How to exploit blockchain public chain and smart contract vulnerability. In: HITBSecConf 2018 (2018)
8. Bashir, I.: Mastering Blockchain, 2nd edn. Packt Publishing, Birmingham (2018)
9. Idelberger, F., Governatori, G., Riveret, R., Sartor, G.: Evaluation of logic-based smart contracts for blockchain systems. In: Rule Technologies. Research, Tools, and Applications: 10th International Symposium, RuleML 2016, Stony Brook, NY, USA, 6–9 July 2016
10. BigchainDB GmbH: BigchainDB 2.0 The Blockchain Database, May 2018
11. Pregelj, J.: CRAB — Create. Retrieve. Append. Burn., 18 October 2017

An Architecture for Opinion Mining on Journalistic Comments: Case of the Senegalese Online Press

Lamine Faty[1](✉), Marie Ndiaye[1], Khadim Dramé[1], Ibrahima Diop[1], Alassane Diédhiou[1], and Ousmane Sall[2]

[1] Ecole Doctorale ED-STI de l'Université Assane SECK de Ziguinchor, Ziguinchor, Senegal
{lamine.faty,marie.ndiaye,khadim.drame,ibrahima.diop, alassane.diedhiou}@univ-zig.sn

[2] Ecole Doctorale Développement Durable et Société (ED2DS), Université de Thiès, Thies, Senegal
osall@univ-thies.sn

Abstract. Comments from the Senegalese online press can create important opportunities for the socio-economic and political actors of our country. These are potentially promising data and useful sources of information. However, the complexity of these data sets no longer allows current methods of opinion mining to exploit this type of comments. This complexity is caused by ambiguous sentences, out-of-context comments and the use of terms borrowed from national languages. To avoid the risk of not reflecting the collective opinion of Senegalese readers, we are interested in proposing an architecture solely for the purpose of valorizing journalistic comments. The architecture will highlight a new solution to solving these types of problems.

Keywords: Online presses · Comments complexity · Opinion mining · Architecture

1 Introduction

The advent of web 2.0 technologies has revolutionized the journalism domain. The new paradigm of communication, called web 2.0 journalism, offers readers the opportunity to read publications and react after by giving their viewpoints. This communication form is widely developed and promotes the expression of a feeling often controlled or repressed during face-to-face interviews. Comments associated with journalism articles can be considered as potentially promising data, therefore an unprecedented opportunity for the socio-economic and political actors of our country. We can discover useful information therein, for example: 1) web comments' potential influence on decision making; 2) web users' agreement or disagreement regarding a specific proposal; 3) overview of public opinion on the country's economic, political and social situation; 4) web users' expectations about public policies. From that moment on, the comments from Senegalese

Á. Rocha et al. (Eds.): WorldCIST 2020, AISC 1159, pp. 395–403, 2020.
https://doi.org/10.1007/978-3-030-45688-7_41

online presses can be considered to be a main source to determine the Senegalese opinion. For this purpose, the new methods of detecting opinions have emerged. This new trend is called opinion mining [1, 2]. The term "opinion mining" is often used when we want to categorize a point of view on positive or negative polarity. Thus, opinion mining consists in treating text content from the web in order to discover the majority of internet users' viewpoints.

In the domain of opinion mining, there are approaches and methods that have been proposed. Some authors have proposed solutions that allow trends to be determined in an independent context of language and domain taking into account sentence disambiguation. Proksch et al. [3] have proposed a semi-automatic solution to analyze critics of deputies. Next to these authors, Satapathy et al. [4] conducted their research on short text (microtext) normalization. In their approach, these authors used Sorensen's similarity algorithm to calculate the phonetic distance between concepts outside the vocabulary and those standard in the form of vocabulary. In the same vein, Gambino et al. [5] worked on the analysis of emotional sentiment associated with journalistic articles published on Twitter. These authors are interested in the comments written less language errors in Spanish. In addition, Kandé et al. [6] proposed a lexicon that can tag words according to positive or negative polarity in two languages, namely French and Wolof. Unfortunately this lexicon is not open source.

However, the complexity of these data sets no longer allows current methods of opinion mining to exploit this type of comments [7]. This complexity is caused by ambiguous sentences, out-of-context comments and the use of terms borrowed from national languages such as Wolof, Pular, Mandinka, etc. Nowadays, Senegalese national languages have a prominent place in the social comments of Internet users. These languages are widely used by a local community in daily conversations, both in the real and virtual worlds. In addition, the presence of national languages in Senegalese comments continues to increase (see Table 1).

Table 1. Language identification.

Number of articles	Number of comments	Terms	Terms in French	Terms in English	Unknown terms
1	*264*	*666*	*49%*	*5%*	*46%*

This leads to a dazzling appearance of unknown tokens in these comments. Sagot et al. [8] proposed a typology of unknown tokens, namely invalid tokens induced by tokenization errors; tokens with inappropriate spelling, borrowed tokens, unknown lexicons. To overcome this issue, two research perspectives are raised:

- Ignore the national languages expressions in favor of French and English;
- Analyze the whole text by taking into account all language.

To avoid the risk of not reflecting the collective opinion of Senegalese readers, it would be important to take into account all comments because the relevant unit for the opinion vocabulary is not necessarily a term in French, but can be in national languages. In situation, the search for opinions based on such data requires a readjustment of current methods. The objective of this article is to propose an architecture that describes a new solution to solving these types of problems. In the rest of this document, we will develop our proposal.

2 Our Proposition

Our architecture is based on journalistic commentaries that reveal complex methods of opinion mining. To enhance the value of this type of data, we propose a process that integrates the collection, analysis and visualization of results. This process is composed of four (4) modules: data acquisition, semantic indexing, opinion analysis and visualization of results (Fig. 1).

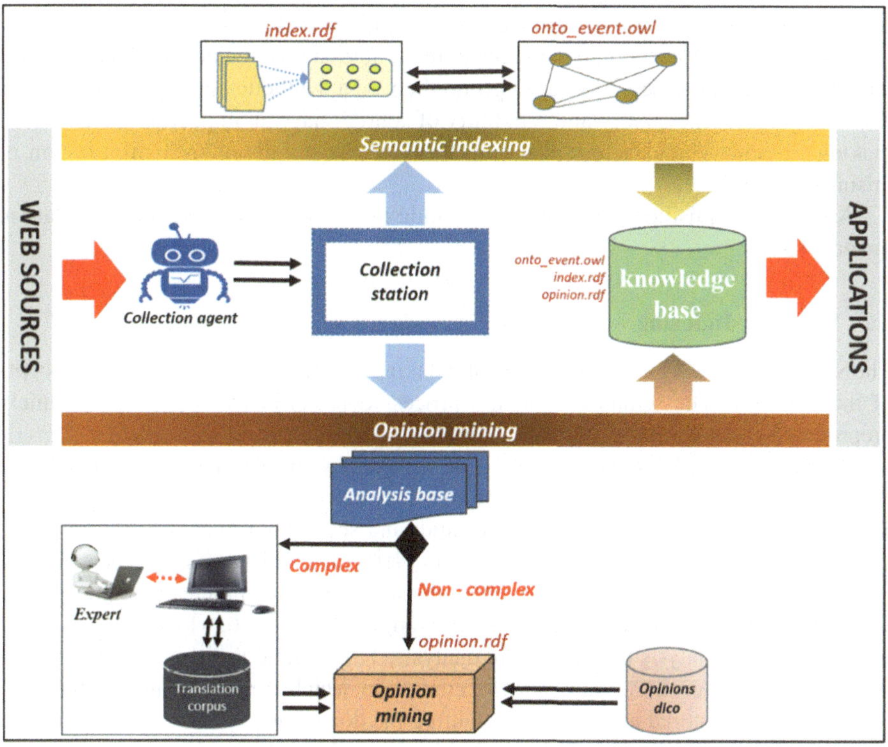

Fig. 1. Architecture of our solution for opinion mining

2.1 Database Constitution

This first module is a web scraping [9] module which role is to collect data and transform them in a consistent way in order to store them in a database.

Data collection is an essential phase for the analysis of data from the web. This operation consists in querying web sources in order to extract useful information from the identified sources. Source identification is the identification of useful, accessible, reliable and sufficiently up-to-date information from legally and technically exploitable sources. It provides relevant data that can provide an answer to our problem. To retrieve the desired content, we use XPath path of the different targeted elements (URL, titles, comments, etc.). However, unstructured data collected sometimes reveal redundancies, noises that can unnecessarily burden the process. The transformation will eliminate them to obtain a clean and exploitable corpus.

Transformation is the process of unifying data from multiple sources into a single format by detecting those that do not comply with certain rules. In practice, it is the operation that merges or aggregates data into a single format by eliminating certain special characters, line breaks or unnecessary spaces, redundancies, case matching, etc.

At the end of the collection, the cleaned data will be classified according to the following variables: source, subject, title, text, comment and metadata. The results obtained will be stored in a database in attribute-value format, notably json.

The interest of this module is to collect the articles and associated comments available in the online presses in order to group them in a synthetic and coherent way in a database. Indeed, the quality and reliability of sources have an impact on the decision-making process. These factors also help to avoid the pitfalls of over information or misinformation.

After data acquisition, we perform semantic indexing to better organize the data in the database.

2.2 Semantic Indexing

The Semantic Indexing module consists of indexing the articles according to the concepts of the ontology. The semantic indexing process consists of two (2) phases, namely preprocessing and similarity measurement.

Preprocessing allows the articles to be represented by key terms called indexes. This operation consists of segmentation, lemmatization and extraction of key terms. Segmentation first takes an article as an input and segments it into several sentences. Then we proceed to identify parts of speech (POS) tagging which consists in determining the grammatical function of each element of the sentence. To complete the segmentation, we use the n-grams extraction method with n ranging from 1 to 6, which allows a much more subtle text analysis than just word analysis. This technique also makes it possible to take into account more generally the context in which the words are written. The result obtained is stored in a matrix containing the documents in lines and the terms in columns (or inverse). Each cell C_{ij} of this matrix is the number of occurrences of the term j in the document $\boldsymbol{i}$.

For the extraction of key terms, the TF * IDF weighting method is more popular for measuring the degree of importance of terms in documents. This approach combines two criteria: the importance of the term for a document (TF) and the power of discrimination of this term (IDF) in the corpus.

$$tfidf(t, d, D) = f_{td} * \log_{10} \frac{N}{n_t}$$

Where

- f_{td} is the occurrence frequency of the term ***t*** in a document ***d***.
- N: number of documents in the corpus
- n_t: number of documents where the term ***t*** appears

The higher the TF-IDF score of a candidate term, the more important it is in the analyzed document.

From there, we can determine the similarities between the documents and the objects in order to create a semantic link between the terms of the documents [10]. Similarity measures consider documents as given vectors. In this article, we determine the distance between two (2) documents thanks to the Cosinus index which is only interested in occurrences. Formula: two vectors $\boldsymbol{d_1}$ and $\boldsymbol{d_2}$

$$\cos(d_1, d_2) = \frac{\langle d_1 | d_2 \rangle}{\|d_1\|.\|d_2\|}$$

$$Dist(d_1, d_2) = 1 - \cos(d_1, d_2)$$

The closer the distance between two documents is to 0, the more semantically similar these documents are. This practice facilitates the categorization of documents. It can be deduced that two documents are close if they have many terms in common. However, this approach does not manage synonyms or similar terms that appear in similar contexts. To overcome this deficiency, we will use Latent Semantic Analysis (or Indexing) [11] (LSA). Resulting from topic modelling approaches, LSA seeks to detect "latent concepts" in documents. A latent concept corresponds to the presence of co-occurrences and correlations between several terms, not two by two but multiple. The fundamental objective of using this method is to bring terms together, but also documents in order to organize them by theme. The result of an indexing is therefore a set of terms that can be either a word, a word root, or a compound term and stored in an ***index.rdf*** file. Finally, these terms are transformed into concepts that create an event ontology that will be stored in an ***onto_event.owl*** file. Following this module, we conduct the opinion mining.

2.3 Opinion Mining

The Opinion Mining module is the module that takes care of the automatic classification of documents according to the positive, negative and sometimes neutral polarities. In this opinion study, we are interested in different aspects of the entity in order to accurately predict the opinions of readers. Aspects are considered as properties on which opinions can be expressed. In other words, it is any clue that can lead to subjectivity in a document. For this purpose, the defined process consists in determining the analysis basis and an automatic classification model.

The basis of analysis is obtained from comments made at regular intervals for a certain duration. It requires an aggregation of comments associated with the selected event. This task is possible thanks to indexing. This data portion is pretreated and is composed of segmentation tasks, parts of speech (POS) tagging. In our approach, we chose the TreeTagger [12] tool which is the most suitable tool for grammatical labeling of a text written in French [13]. Indeed, TreeTagger is an efficient and robust open source tool, developed at the University of Stuttgart that relies on decision trees. To enrich this segmentation, we use the method of extraction of ***n-grams*** with ***n*** which can vary from 1 to 3. This method allows a text analysis much more subtle than the only analysis of the words. And finally, we proceed to the elimination of the elements that have a much less precise meaning. Those elements intervene in particular in the construction of the sentences and often have a uniformed distribution in the majority of the texts. They are prepositions, pronouns, auxiliary verbs, articles, etc. At the end of this step, we propose a matrix representation (the documents in lines and the terms in columns or inverse). Each cell C_{ij} of the matrix is the result obtained thanks to the weighting ***td-idf***. Then we propose a second aspect-based weighting. Finally, we adopt the representation by bag of words where a document is represented by a vector of terms whose component ***i*** indicates the weight of the ith term of the document. From there we apply the classification methods.

For automatic classification, it is a question of modeling the contents of the analysis base in two groups (positive or negative) and the indecidable documents are regarded as neutral. Each group constitutes a probability distribution class of terms. This way of solving the problem of classification of documents is part of supervised learning which could make it possible to easily manage ambiguities. In this article, we will perform a comparative study of the most popular algorithms namely Naïve Bayes classification, Maximum Entropy Classifier, Support Vector Machines Classifier and Logistic Regression to propose a high-performance classifier.

At the same time, we will set up a large multilingual corpus (French-Wolof) to facilitate translation. Beside this corpus, an expert will ensure the verification and validation of the results provided by the algorithms. The results of this module are stored in the ***opinion.rdf*** file which will set us up a knowledge base.

2.4 Knowledge Base Building

A knowledge base consists of synthesizing the expertise of a field generally formalized using an ontology. In our context, it brings together the specific knowledge of Senegalese journalism in a form exploitable by a request from a computer. For this, we will first briefly introduce the events ontology before discussing on search criteria.

An ontology is a structured set of concepts organized in a graph, linked by semantic and logical relations. The first objective of an ontology is to model a set of knowledge in a given domain, which can be real or imaginary in interpretable language, so that they can be used by machines. There are several types of ontologies among which ontologies of domains seem to be adapted for this situation. These types of ontologies describe concepts that depend on a particular domain. In our context, it is a question of creating the domain ontology of Senegalese journalism (event in Senegalese journalism). This task is facilitated by the semantic indexing module. It should be noted that journalistic information is broadcast in a temporal continuity. This allows the refraction of events as constitutive forms of reduction of relevant terms. The retained terms are transformed into concepts of the events ontology. Each concept is translated into multilingual, that is to say in French, English and Wolof. Thus, we can consider defining multilingual and multi-domain search criteria.

The definition of search criteria consists in creating a correspondence between the terms of a user's query and the associated opinion. In practice, it is a question of automatically reformulating the initial request to find the corresponding information in the knowledge base. Search criteria are based on a domain, a subdomain, an event or a keyword in the different languages mentioned above.

Our knowledge base integrates ***index.rdf***, ***opinion.rdf*** and ***onto_event.owl*** files for easy description and information retrieval. The technical logic of our architecture gives our solution the innovative character in that it privileges the dynamic interaction between the information inherited from the information sites and the requests of the users.

3 Discussion

The architecture put in place is a response to the questions of ambiguity, multi-domain and multilingualism.

Ambiguity: An ambiguous text can simply be defined as a text whose meaning can be interpreted in different ways. In Natural Language Processing, Several factors lead methods to ambiguity; among other things we have the mistakes of language (for example grammatical mistakes in French), figures of style, grammatical anaphors, polysemic words and personalized abbreviations in comments. Such a problem has been solved by Part of Speech Tagging, lemmatization and the n-gram method used in the Opinion mining module.

Multi-domain: In opinion mining, the precision of the domain of study is a crucial issue. The orientation of certain vocabularies strongly depends on the domain of interest. Each domain has words or expressions that are intrinsically positive or negative. The question of the multi-domain in opinion mining is a contemporary problem far from being solved. For this purpose, we propose Semantic Indexing which serves to create a relationship between the different domains identified through the contents of the documents. The distance measurement metric used in this architecture is a method that brings together two similar documents.

Multilingualism: We define multilingualism as a text composed of sentences constructed from words or expressions from several languages. Nowadays, national languages are present in online comments. Thus, the majority of Senegalese use words or phrases in Wolof, in French and sometimes even in English in daily conversations on the internet. The establishment of an event ontology whose concepts are translated into French, English and Wolof fully resolves the obstacles brought about by the multilingual. This technique can also enrich the knowledge base with vocabularies of these languages (equivalent terms, translation into other languages).

At the crossroads of obstacles linked to ambiguity, multi-domain and multilingualism, the design of an information system dedicated to opinion mining is an approach that is part of a logic of enhancing journalistic data and knowledge organization for quality and fluidity of access. The architecture put in place formally defines the process of transforming data into information and then information into knowledge. It allows to systematize the exploitation of comments from the Senegalese press online.

4 Conclusion et Perspectives

In the end, we can remind that the architecture put in place constitutes an innovative solution in the field of opinion mining. This solution is particularly interesting for the analysis of opinion of recent textual data, written in languages with little endowment as in the case of comments from the Senegalese press online. Indeed, the ***Data Collection*** module's role is to build the database that will be used to build models. Next to this module, the ***Semantic Indexing*** module makes it possible to link the articles to the events ontology thanks to indexes. From there, we can extract our analysis database from the comments related to an event. This is the beginning of ***Opinion Mining***. Here with the choice of methods and the assistance of an expert, the prediction will be reliable and closer to reality. The adoption of the semantic web approach in an opinion mining platform creates all the intelligence of our solution. For example, the ***Knowledge Base*** module includes ***index.rdf, onto_event.owl and opinion.rdf*** files. This allows for the enrichment of the knowledge base with vocabularies from other languages (equivalent terms, translation into other languages). As a result, the questions related to ambiguity of terms, multi-domain and multilinguism are unequivocally resolved. Our semantic web-based opinion mining platform also offers the following services: 1) Linked open data of web journalist; 2) Statistics; 3) Semantic research; 4) Sharing of information and knowledge; 5) Determination of trends with respect to an event or area of activity, etc.

Our real concern is that the different actors with this need find here a simple and adequate way to improve their decisions. In the future, we intend to extend the solution by taking into account all national languages.

References

1. Turney, P.D.: Thumbs up or thumbs down?: semantic orientation applied to unsupervised classification of reviews. In: Proceedings of the 40th Annual Meeting on Association for Computational Linguistics, pp. 417–424 (2002)
2. Pang, B., Lee, L., Vaithyanathan, S.: Thumbs up?: sentiment classification using machine learning techniques. In: Proceedings of the ACL-02 Conference on Empirical Methods in Natural Language Processing-Volume 10, pp. 79–86 (2002)
3. Proksch, S.-O., Lowe, W., Wäckerle, J., Soroka, S.: Multilingual sentiment analysis: a new approach to measuring conflict in legislative speeches. Legis. Stud. Q. **44**(1), 97–131 (2018)
4. Satapathy, R., Singh, A., Cambria, E.: PhonSenticNet: a cognitive approach to microtext normalization for concept-level sentiment analysis, ArXiv Prepr. ArXiv:1905.01967 (2019)
5. Gambino, O.J., Calvo, H.: Predicting emotional reactions to news articles in social networks. Comput. Speech Lang. **58**, 280–303 (2019)
6. Kandé, D., Camara, F., Ndiaye, S., Guirassy, F.M.: FWLSA-score: French and Wolof Lexicon-based for Sentiment Analysis. In: 2019 5th International Conference on Information Management (ICIM), pp. 215–220 (2019)
7. Faty, L., Ndiaye, M., Diop, I., Drame, K.: The complexity of comments from Senegalese online presses face with opinion mining methods. In: 2019 14th Iberian Conference on Information Systems and Technologies (CISTI), pp. 1–6 (2019)
8. Sagot, B., Nouvel, D., Mouilleron, V., Baranes, M.: Extension dynamique de lexiques morphologiques pour le français à partir d'un flux textuel. In: TALN-Traitement Automatique du Langage Naturel, pp. 407–420 (2013)

9. Sarr, E.N., Ousmane, S., Diallo, A.: FactExtract: automatic collection and aggregation of articles and journalistic factual claims from online newspaper. In: 2018 Fifth International Conference on Social Networks Analysis, Management and Security (SNAMS), pp. 336–341 (2018)
10. Dramé, K., Diop, I., Faty, L., Ndoye, B.: Indexation et appariement de documents cliniques avec le modèle vectoriel. In: DEFT, p. 91 (2019)
11. Deerwester, S., Dumais, S.T., Furnas, G.W., Landauer, T.K., Harshman, R.: Indexing by latent semantic analysis. J. Am. Soc. Inf. Sci. **41**(6), 391–407 (1990)
12. Schmid, H.: Treetaggerl a language independent part-of-speech tagger. Inst. Für Maschinelle Sprachverarbeitung Univ. Stuttg. **43**, 28 (1995)
13. Sarr, E.N., Sall, O., Maiga, A., Faty, L., Marone, R.M.: Automatic Segmentation and tagging of facts in French for automated fact-checking. In: 2018 IEEE International Conference on Big Data (Big Data), pp. 5439–544 (2018)

A Case Study on the Use of Enterprise Models for Business Transformation

Martin Henkel(✉), Georgios Koutsopoulos, and Erik Perjons

Department of Computer and Systems Sciences, Stockholm University, Stockholm, Sweden
{martinh,georgios,perjons}@dsv.su.se

Abstract. Organizations constantly need to change due to their desire to improve, or due to the need to adapt to changes in the environment. The use of enterprise models could be a potential aid in this transformation process - by using models an organization can describe its business, analyze it, and design changes. For an organization, models may be important in order to determine if a transformation should be pursued, and determine the consequences of performing the transformation. In this paper, we report on a case study performed at an organization that uses enterprise models. Based on a set of business transformations that the organization is considering, we examine how the existing models they use can help in expressing the changes needed, and the shortcomings of existing models. We conclude with a set of tentative shortcomings of traditional models and pointers for future research addressing these.

Keywords: Modeling · Business transformation · Enterprise models · Decision-making

1 Introduction

Modern organizations constantly need to adapt to meet new demands from customers and other changes in the environment as well as increase their efficiency. If an organization does not transform to meet these environmental changes it can endeavor meeting organizational goals, or even jeopardizing the organizations existence [1]. There are no signs that the pace at which organizations have to change is slowing down, on the contrary, organization that has a competitive edge tend to keep this advantage for a shorter time due to these environmental changes [2]. Thus, to handle the changes, an organization needs to be able to transform internally and evolve the collaboration with external business partners. However, transforming to meet changes in the environment, or to address internal problems, can be a complex task. About 70% of change initiatives fail [3].

To transform, an organization firstly needs to know its current state, and secondly, there is a need to identify and understand the changes that need to be brought through. Each change needs to be scrutinized and the impact on the organization thought through. There are several ways an organization can be supported when analyzing transformation. One such way that we focus on in this paper is the use of enterprise models. Enterprise

Á. Rocha et al. (Eds.): WorldCIST 2020, AISC 1159, pp. 404–414, 2020.
https://doi.org/10.1007/978-3-030-45688-7_42

models have the potential to provide an overview of an organization and its constituent parts, such as information, process, actors and so on. Among the top benefits of using enterprise models are touted to be its ability to support change management [4].

In this paper, we make use of a case study to examine how enterprise models can help in the transformation effort of an organization. The case study has been performed at a public organization that is responsible for providing healthcare in a Swedish county. In particular, we have examined the healthcare guidance service that is provided by phone to county residents and visitors. The guidance is given by professional nurses specially trained for this job and supported by various information sources incorporated in the software they use. The service is called 1177 that corresponds to the phone number to call. The guidance service is owned by the public organisation but have a number of participants that help deliver it, both private and public, which makes the structure behind the service quite complicated. The complicated structure leads to any proposed change in the service requiring a detailed analysis of which parts of this structure will be affected by the change and how. The organisation has built a number of enterprise models, including detailed process models.

The aim of this paper is to, based on recent transformation request that the organisation got, examine how the consequences of the transformation can be analyzed using the enterprise models found in the case.

The rest of the paper is structured as follows. Section 2 gives a background to the literature in the areas. Section 3 describes the method applied, while Sect. 4 gives an overview of the case. Section 5 describes two example transformations as found in the case, while Sect. 6 extracts requirements on models for analyzing transformation using enterprise models and discusses how the models found in the case complies with the requirements. Section 7 contains a discussion of the implications of the findings and Sect. 8 provides concluding remarks.

2 Background and Related Research

This paper concerns examining the use of enterprise models to manage change. As pointed out in [5], studying change can be done from both a process and a content perspective. The *process perspective* entails studying the steps needed to perform a change, while the *content perspective* deals with the structures that exist before and after the change occurs.

The use of enterprise models is typically done from a *content perspective*. That is, models are used to depict how something is before and after a change. This is evident in the common classification of models as "as-is" or "to-be" [6]. However, models have also been used to design systems that change – adaptable systems. For example, [7] uses models to design how an IT system could adapt when detecting changes in its environment. In essence, the field of enterprise modeling contains a vast type of models, enabling the description of more or less any part of an organization. For example, the Zachman framework [8] contains 60 areas that potentially can be represented with its own type of model. For example, there are ways to represent events, process and data from a number of views. In this paper, we perform a case study and limit the study to the six types of models that were found in the case (this will be described in the next section). In future studies it may be beneficial to study how more types of models can be used to support the analysis of changes in the studied organization.

From a *process perspective,* describing and executing changes to an organization is related to the field of change management. Change management is a set of principles and practices aiming to support the transition of an organizational state to a new one, either being a new implementation or an updated one. A part of change management is to make employees accept changes in their working environments [9]. As pointed out by [10], a part of management of change is also to identify and deal with both intended and unintended consequences. Effort has been put into creating methods for performing changes, such as the [11] unfreeze-change-freeze approach, or the eight guidelines for change [12]. In contrast to these approaches, we focus on aiding the detection of what needs to be changed, rather than how to perform the changes. We do this by looking into what kind of models are useful to analyze the consequences of change in the case study.

From a *content perspective*, it is important to study what is being changed. For example, several authors have described how change impact analysis can be used to discern how a change will impact a software system. Essentially the idea is that it is possible to analyze how much a software system is affected by a change, measured in the number of affected functions. As pointed out by [13], there are two basic approaches to know how a system will be affected. Firstly, a *trace and dependency* approach can be applied by having a comprehensive model of how the IT systems functions are related to organizational elements such as goals and processes, see for example [14]. Secondly, an *experiential* approach can be used where the analysis relies on experts that can perform an analysis based on their tacit knowledge of the organization. The approach that we selected for this paper is the second one, we have relied on experts describing the consequences of the changes as described in the 1177 case. Our aim here is to examine if the enterprise models can be utilized as support to such an expert.

3 Method

When the case study began, 1177 already had created a number of enterprise models, such as goal models and process models. To get information about these models, and what kind of change request the organization need to handle, a number of meetings were held. The following activities were carried out in an iterative fashion:

- *Unstructured group interviews* were used to identify change requests that 1177 had received recently, and describe them in detail. For this, 4 meetings were held, initially one meeting was engaging two experts/strategists at 1177, and the other three meetings engaging one.
- *Workshops* were held to elaborate a value network model. For this, three meetings were held.
- *Document studies* were performed to study how the enterprise models used in the organization covered identified change requests. In total 15 diagrams was examined.

Selecting the change requests to study was done during the four meetings, where we simply asked the experts to identify and describe recent change requests the organization had, and the *potential consequences* of the changes. In total, we identified eight change requests. In this paper we use two of these change request to illustrate our analysis. Table 1 summarizes the types of models and their syntax and contents that were part of the case study.

Table 1. Identified model types

Model type	Syntax	Contents	Number of diagrams
Process	VIA	Tasks, actors, resources	4
Value network	E3 value (simplified)	Actors, resource exchanges	1
Business Model Canvas (BMC)	BMC	Actors, offers, relations, channels	1
Service Design	Service blueprint	Main process, identified problems	1
Goal	BMM compliant	Goals, sub-goals	4
Conceptual	UML Class Diagrams	Concepts	4

4 Overview of the Case

As stated earlier, the 1177 Guidance service is complex. An obvious part of the complexity comes with the wide range of health care issues that the service needs to cater to – ranging from trivial issues and even prank calls, to life-threatening issues. Part of the complexity is also derived from the health guidance being an entry point to the regional health care – thus there is a need not only for the regional resident to know about the service – but the guidance operators also need to know about the region's health care providers in order to advise the patients to the right provision of care.

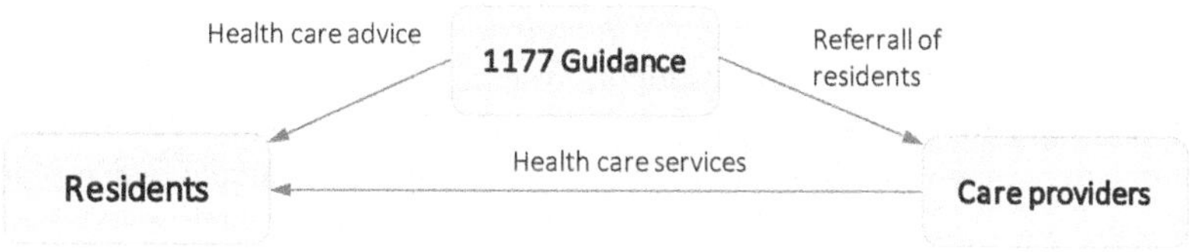

Fig. 1. The 1177 Guidance main actors

The main actors and exchanges of the 1177 Guidance service are shown in Fig. 1. Central to the model is that the 1177 service is located between the care providers and the region's residents. The main task of 1177 is to provide health care guidance, and to help the callers find the appropriate care providers. Carrying out the work requires 1177 to use a number of resources. The main resource is expert nurses, that are specially educated for providing health guidance according to a set of principles. These principles are referred to as LEON principles. Essentially, the LEON principles state that the guidance aims for the most efficient level of care. For example, guidance towards the emergency hospital should only be provided to those in need of emergency care. The guidance is also supported by a number of IT resources, including the telephony and journal systems. There is also a specific Guidance system that the nurses use as a way of guiding the conversation with the callers. This system is developed at a national level. Furthermore, a provider catalogue allows the nurses to find care providers.

5 The Transformations

The organisation and 1177 is in constant transformation. The impetus for transformation comes both via top-down and bottom-up developments. From a top-down perspective, politicians are pushing for reforms to both improve the overall quality, but also to make it easier for the residents in the region to use the service. From a bottom-up perspective changes are also proposed by the staff of the service, and from partners involved.

Whenever a change is proposed, it requires analysis to determine its effect on the service. To examine if enterprise models can help in this analysis, we have selected two recent change requests – a) a proposed change in the guidance support allowing the nurses to guide the patients directly to a care provider based on their symptoms and b) the desire for 1177 to book times at local emergency clinics. We have selected these change requests to illustrate the use of enterprise models for highlighting both internal improvements that the residents may be unaware of (change a) and improvements that affect external partners that the patients have contact with (change b). Both change requests affect external parties and IT systems. Each transformation moves the business from an as-is state to a to-be state. In addition, there may also be a set of *consequences* associated to a change. These consequences are the side-effects of implementing the change. In subsequent sections, we describe the two transformation cases, and the utility of enterprise models for analyzing the changes.

5.1 Transformation A – Guidance Support Improvements

General Description of Business Before Changing: One of the core supports available to the nurses of 1177 is a Guidance support system. It is a national system, developed by a private company, enabling the expert nurses to search for symptoms. Based on the symptoms, the system presents possible sub-symptoms. For example, when searching for "throat pain" the system will list more specific symptoms such as "difficulty to open the mouth" and "difficulties to swallow". No suggested diagnoses are presented in the Guidance support system, however each sub-symptom is connected to an emergency level, ranging from "Immediately" to "Wait and see". To support the nurse, a separate list of health care providers is provided in a catalogue. The Provider catalogue is maintained by a second private company.

The Transformation: A proposed improvement of the support systems is to have better support for recommending health care providers in the region. The novel idea brought forward is to link each provider to the symptoms that they are likely to be able to manage. The novelty in this is that the nurses do not have to reach a diagnosis to select provider, as the identification of a provider may be done directly via the identification of symptoms. The proposed benefit with this change is improved service for the patient, since they will be guided to a provider. Furthermore, a benefit is that the patient can be guided to the providers with the best expertise.

Consequences of the Transformation: In the case, several consequences of the transformation were identified. Firstly, it was clear that several organisations would be affected by the change. Moreover, there was a need to combine data from two ownership domains – the base Guidance support system, and the catalogue of providers. This could be handled by writing contracts such that the data could be used to create the new catalog add-on.

Needed Support for Transformation Analysis: To determine the consequences of this transformation there was a need to understand the interplay among different actors in the case. Initially, there was a general need to know of the actors involved in providing the support systems for the nurses – that is, there was a need to get an overview of the business. We refer to this as the need to know the *actors and their relationships*. There was also a need to know some details of the ownership of the data that the system used. We refer to this as the knowledge of *resource ownership*. Last, since the private organization managing the support system was an external organization, *contracts* were pointed out as being important to manage during the transformation.

5.2 Transformation B – Time Booking at Emergency Clinics

General Description of Business Before: An important part of the health care managed is the local emergency clinics. These clinics are meant to treat acute, but not life-threatening illnesses. For example, the clinics can handle acute allergy reactions, concussions and fractures. The local clinics have set openings hours, and when open, may offload the main emergency units of major hospitals. Currently, the expert nurses at 1177 can recommend the patients to visit an emergency clinic if they deem that the clinic can handle the case. However, even if the 1177 nurses may recommend that a patient visits the emergency clinic, they do not currently put the patient in contact with the clinic in order to book a time slot for an appointment. As a part of the support used by the clinics, they have a journal system.

The Transformation: A proposed change is to extend the capabilities of 1177 to also include the ability to book time slots at the local emergencies. The purpose of doing this is firstly to provide better service to the patients, beside the convenience, having a booked time also makes the patient feel more secure. Secondly, the ability to book times also makes it possible for 1177 to have control of the flow of patients, enabling the booking of timeslots at clinics that are currently having the shortest queue.

Consequences of the Transformation: The first concrete change is that 1177 will now be directly acquiring patients for the clinics. The change entails adding the support infrastructure for 1177 to enable the nurses at 1177 to use an IT system to book times. Practically, this needs to be based on an API provided by the clinics' journal system provider. Moreover, there is a need of a contract that regulates what kind of symptoms the clinics should be capable of handling, and the use of the system API.

Needed Support for Transformation Analysis: In this transformation, just as for transformation A, there is a need to know the affected parties, in this case, the 1177 itself, the clinics involved, and the provider of the journals system. Another similarity is that *contract* management is needed. Compared to Transformation A, there was also a need to know about existing, and design future *IT integrations*. Another issue that was mentioned in the case was that it should be clear what the *overall purpose* with the transformation was. The transformation was deemed useful for two purposes: both to improve the patient experience, but also to have better control of how the patients "flow" through the health care system. By implementing the transformation, the control is improved by being able to pick the clinics that are most competent to handle the symptoms that the patient got, but also by being able to direct the patient to the clinics, given the same competence level, that currently got the lowest load.

6 Transformation Analysis

By comparing the needed analysis support from the requested transformations in the case and the used models in the case it is possible to examine if the models can be used to support the analysis of the transformation. The key question here is: do the models in the case contain the type of information needed to analyse the consequences of the transformation? If the models can convey the needed information, they may provide valuable support.

6.1 Requirement Areas

Based on the transformations identified in the cases (of which two of the cases is described in the previous section), we can summarize that the analysis needs support in the following high-level requirement areas:

a) *Actors and their relations*. The models should help in identify actors, and the relationships that may be affected.
b) *Contracts*. It should be easy to find which existing contracts that exist, and which parties that are in the contract.
c) *Resource ownership*. The ownership of resource such as IT systems, information, and physical items should be clear.
d) *IT systems and integration*. A transformation may impact several systems, and may have a ripple-effect in integrated systems, thus there is a need to know the IT systems being used.
e) *Overall purpose* of the change. As seen in transformation B, there could be multiple purposes with a transformation. When discerning between different implementation

options, it is useful to keep in mind the purpose of the transformation – to ensure that all implementation options fulfil the purposes. Furthermore, it is important to compare the purpose of the transformation with the goal of the organization being transformed.

The two areas, a) actors and relation, and b) contracts, may seem to overlap. The assumption is that if there is a relationship with a business partner there should also be a contract regulating the relationship. However, there are different kinds of business relationships. For example, [15] distinguish between two kinds of relationships: the transactional, relying on formal contracts, and the relational style, which is informal. Therefore, we have separated contracts and relationships in two areas.

6.2 Requirements Fulfilment by Models Type

Based on areas above, it is possible to examine how the enterprise model types support the areas. Table 2 gives an overview of how the used model types support the areas.

Table 2. Model types in the case, and their support for requirement areas

Model (and syntax)	Actors and relationships	Contracts	Resource ownership	IT systems integration	Overall purpose
Process model (VIA)	Actors, control and information flow to other actors			May be shown	
SVN Service value network (E3 value)	Actors, resource exchanges	Implicit	Shown as exchanges		With extension
BMC Business Model Canvas (BMC)	Actors, no detailed relationships	Implicit	Only for the main organisation		
Service design (service blueprint)	Actors, control flow			Shown as "scores"	
Goals (BMM)	Goals of main actors				Main goals
Conceptual (UML class diagram)			Information structures	IT system information content	

The *process models* may show the actors carrying out the task of the 1177 Guidance process. The used model syntax, "VIA", is similar to the more common BPMN syntax used for process modelling. Just as BPMN, it allows including both control and information flow between tasks. This makes it possible to depict some of the relationships between actors. The model may also show where information is stored, in terms of information flows going into IT systems. However the process model has no way of expressing contracts, resource ownership or the purpose/goal of the organization.

The *service value network* model (SVN) has the benefit of giving a good high-level overview of the business by showing the involved actors and their exchanges in term soft resources. Since a resource can be anything that is of value (such as money, goods or services) the model can also be used to depict the source of resources, and thereby the resource ownership. A contract may also be considered as a valuable resource as thus be shown in the model. There are also extensions to SVN that enable the model to show the purpose with performing exchanges, however this was not used in the case.

The *Business Model Canvas* (BMC) is a popular light-way of describing the business offering that an organization provides as well as needed resources supporting such offering. The model lists actors in terms of partners and customers segments, but provide no details on how they are related. Likewise, the BMC list resources used by the main organization but no details linking the resources to value offerings. Just as for SVN, contracts may be considered as a resource and may be included in a BMC.

The *service design model*, following a service blueprint syntax, used in the case includes a main flow of activities that involves interaction with the patients. It also included the IT systems linked to each activity.

In the case *goal models* were used to link the goals of the 1177 Guidance service to the health care regions main goals. The goal model is thus useful when examining if a proposed transformation is aligned with existing goals.

The *conceptual model*, following a UML class diagram syntax, included the information structures needed to guide the patient and record the guidance. However, the model type is not suited for depicting actors. By partitioning the conceptual model it is potentially possible to depict in which IT system the information was stored. However, this was not done is the case.

7 Discussion

Based on the requirements for supporting business transformation as identified in the case, we can see no single enterprise model used in the case supports all requirements (Table 2). However, this is no surprise since each model is tailored to depict certain aspects of the business, such as its goals, processes or information. It is then quite natural that there is a need of a combination of models to perform transformation analysis. Based on the case, we can see that the service value network models (SVN) provides a good overview of the actors involved and their relationships. This, in combination with models showing resource and IT system ownership could make a good foundation for an analysis. Furthermore, without a goal model, or corresponding information in written form, it is difficult to see if a transformation request is in line with the current goal of an organization.

The examination of the model types also led to some not so obvious results. Most notable, the Business Model Canvas did fulfil some of the requirements. However, the Achilles' heel of the BMC is that it does not specify relations between the included concepts. The BMC is simply made for analyzing the value propositions of one organization, and does not fit well when having a case that relies on a network of actors that are collaborating. Since the case was in the area of health care, the focus is on collaboration across organizational boundaries, and transformation of the business is likely to affect several parties as shown in the example transformations.

8 Conclusion

In this paper, we have examined how enterprise models may support the analysis of business transformations. Based on a case study a set of transformation requests were identified, these requests were then used to identify key requirements areas where models could support the transformation analysis. In the case studied, six different types of models were used. The examination pointed towards that none of the models solely could support all requirements, but that the combination of models supported most of the key requirements.

Acknowledgment. We are grateful to the employees who took their time in letting us interview them in order to identify and describe the case presented in the paper.

References

1. Audia, P.G., Locke, E.A., Smith, K.G.: The paradox of success: an archival and a laboratory study of strategic persistence following radical environmental change. Acad. Manag. J. **43**(5), 837–853 (2000)
2. Wiggins, R.R., Ruefli, T.W.: Schumpeter's ghost: is hypercompetition making the best of times shorter? Strateg. Manag. J. **26**(10), 887–911 (2005)
3. Burnes, B.: Managing Change. Trans-Atlantic Publications, Pearson Education, New York (2014)
4. Niemi, E.: Enterprise architecture benefits: perceptions from literature and practice. In: The 7th IBIMA Conference Internet & Information Systems in the Digital Age, Brescia, Italy, 14–16 December (2006)
5. Barnett, W.P., Carroll, G.R.: Modeling internal organizational change. Ann. Rev. Sociol. **21**(1), 217–236 (1995)
6. Sankuhl, K., Stirna, J., Persson, A., Wißotzki, M.: Enterprise Modeling: Tackling Business Challenges with the 4EM Method. The Enterprise Engineering Series. Springer, Cham (2014)
7. Henkel, M., Stratigaki, C., Stirna, J., Loucopoulos, P., Zorgios, Y., Migiakis, A.: Extending capabilities with context awareness. In: International Conference on Advanced Information Systems Engineering, pp. 40–51. Springer (2016)
8. Zachman, J.A.. The Zachman Framework. Zachman International (2003)
9. Nograšek, J.: Change management as a critical success factor in e-government implementation. Bus. Syst. Res. **2**, 13–24 (2011)
10. Huerta Melchor, O.: Managing Change in OECD Governments: An Introductory Framework. OECD Publishing (2008)
11. Burnes, B.: Kurt Lewin and the planned approach to change: a re-appraisal. J. Manag. Studies. **41**, 977–1002 (2004)
12. Fernandez, S., Rainey, H.G.: Managing successful organizational change in the public sector. Public Adm. Rev. **66**, 168–176 (2006)

13. Kilpinen, M.S.: The emergence of change at the systems engineering and software design interface (2008)
14. de Boer, F.S., Bonsangue, M.M., Groenewegen, L.P.J., Stam, A.W., Stevens, S., van der Torre, L.: Change impact analysis of enterprise architectures. In: International Conference on Information Reuse and Integration, Conference 2005, pp. 177–181. IEEE, Las Vegas, NV, USA (2005)
15. Lee, J.-N., Kim, Y.-G.: Effect of partnership quality on IS outsourcing success: conceptual framework and empirical validation. J. Manag. Inf. Syst. **15**, 29–61 (1999)

Credible Information Foraging on Social Media

Yassine Drias[1(✉)] and Gabriella Pasi[2]

[1] LRIA - USTHB, University of Algiers, Algiers, Algeria
y.drias@univ-alger.dz
[2] IKR3 Lab, University of Milano-Bicocca, Milan, Italy
pasi@disco.unimib.it

Abstract. In this paper, an Information Foraging based approach that offers social media's users the ability to get relevant and credible information is proposed. A Social Media Information Foraging System is developed in order to operate on social graphs taking into account the users' interests and their social relations and interactions. To evaluate the performance of the system, a dataset was built using real data extracted from the information sharing network *Twitter*. The results consist in surfing paths leading to relevant information taking into consideration the user's information needs and the credibility of the information.

Keywords: Information foraging · Social media · Information credibility

1 Introduction

Sixty seconds might seem like an insignificant amount of time for us, however when we look at it in terms of how much of data is created on the Web, there is a huge difference. According to the *IBM 2017 Marketing Trends Report*, 2.5 quintillion bytes of data were created every day in 2017 and most of it comes from social media platforms [1]. This big amount of user-generated content represent a rich source of potential information. Nevertheless, one major concern that results from it is to find a way to explore this source and get useful and credible information from it.

The *Information Foraging Theory* aims at discovering paths leading to relevant information on the Web. It was first developed in [2], where the authors established their study on the analogy between the animal food foraging behavior and the behavior of humans while seeking information on the Web. A survey on the advances of the information foraging theory is presented in [3], where the author claims that applying the information foraging theory on social media in order to allow users to get credible information in an efficient way is one of the most interesting new directions.

Á. Rocha et al. (Eds.): WorldCIST 2020, AISC 1159, pp. 415–425, 2020.
https://doi.org/10.1007/978-3-030-45688-7_43

During the past years a few efforts have been deployed in the domain of information foraging on the Web [4,5] and some of them were combined with different techniques and approaches such as deep learning [6], bio-inspired computing [7], ontologies [8], game theory [9] and recommender systems [10]. As social media are gaining an increasing importance and usability, it becomes interesting to perform information foraging on such platforms. In this article we propose a new information foraging model and describe a system based on it, which is capable of automatically defining the topical interests of a user in an implicit way and then use these interests in order to find relevant information to that user on social media. Once the foraging phase is accomplished, the results are ranked based on both their credibility and their relevance to the user.

The rest of the paper is structured as follows. A model for information foraging on social media is proposed in Sect. 2. Section 3 gives details on the implementation of our system. The experiments using real-world data are presented in Sect. 4. Finally, we conclude in Sect. 5 and discuss potential future research.

2 The Foraging Model

Social networks are generally represented using oriented graphs that model interconnections and social relations among people, groups and organizations within the network. Like any other oriented graph, a "social graph" is an ordered pair $G = (V, E)$ composed by a set V of vertices and a set E of directed edges (arrows). Each element $e \in E$ represents a relation of incidence that associates with each edge two vertices. Figure 1 illustrates an example of the social graph structure we consider, where the labels on the vertices represent users and the labels on the edges specify the type of relationship that exists between them.

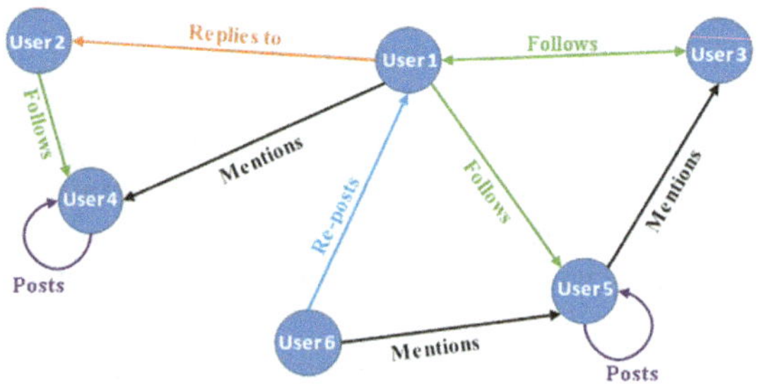

Fig. 1. An example of the considered social graph structure.

The social relations in Fig. 1 are defined as follows:

- *Follows*: is a relationship between two users;
- *Posts*: each post is represented by a self-looping edge on the graph since it only involves one user (the author of the post);

- *Mentions*: the edge starts from the vertex of the author of the post and goes towards the vertex of the user who is being mentioned in that post;
- *Replies to*: the edge goes from the vertex of the user who is writing the reply towards the vertex of the user who published the original post;
- *Re-posts*: like in the previous relation, the edge goes from the user who is re-posting towards the user who published the original post.

It is important to notice that each of the edges representing the relations *posts, mentions, replies to* and *re-posts* have an associated post (message) generated by a user. We denote these edges by *content-sharing-edges.*

Modeling the User's Interests: we consider the case where the user's interests are automatically inferred by the system based on the user's social data, i.e. data related to or explicitly generated by the user in the considered social media platform. The idea is to collect frequent keywords from the user's biography (if available), her/his recent posts and her/his interactions with other users in order to determine the user's interests. We formally represent the information needs (topical interests) of a user as a vector of terms V that is built based on the previously mentioned information and activities on the social network. Algorithm 1 presents the pseudo-code of the process aimed at generating the formal user's interests representation.

Algorithm 1. Generate the user's interests vector

Input: User's ID, h: crawling period in hours, n: size of the vector V;
Output: User's interest vector V;
1. Create an empty text file D;
2. Extract the user's biography from her/his social profile and put it in D;
3. Crawl the user's posts generated during the past h hours and put them in D;
4. Eliminate the non meaningful words from D using a dedicated stop list;
5. Convert D into a list of terms and compute the term frequency of each of them using the following formula: $tf(T_i) = \frac{frequency(T_i)}{\Sigma_{T_j \in D} frequency(T_j)}$;
6. Create a vector of terms V of size n;
7. Put in V the n most frequent terms along with their corresponding tf respectively;

Modeling the Information Scent: the goal of an information foraging system is to get the highest amount of relevant information whilst spending the lowest amount of time. This can be achieved thanks to the *information scent* concept, which aims to guide the foraging process towards sources containing relevant information based on the user's interests. In our case, the goal of the system is to reach social posts that are relevant to a certain user based on her/his interests formally represented by the vector of interests V.

Since each content-sharing-edge is associated with a social post, the information foraging process starts from a content-sharing-edge, then at each step it tries to reach an edge containing more relevant information compared to its predecessor. In other words, at each step of the foraging process the system should select one content-sharing-edge to visit among the reachable edges from the current vertex. This decision is made thanks to the information scent measure. We propose Formula (1) to estimate the information scent the system would get by moving from the current content-sharing-edge e_i to a new one e_j.

$$InfoScent(e_j) = InfoScent(e_i) + Sim(E_j, V) \tag{1}$$

2.1 Foraging Strategy

We propose Formula (2) to define the probability $P(e_i, e_j)$ of moving from a content-sharing-edge e_i to one of its adjacent content-sharing-edges e_j.

$$P(e_i, e_j) = \frac{InfoScent(e_j)^{\alpha} \times sim(E_i, E_j)^{\beta}}{\Sigma_{l \in A_i}(InfoScent(e_l)^{\alpha} \times sim(E_i, E_l)^{\beta})} \tag{2}$$

where α and β are parameters that control the relative weight of the *InfoScent* and the content-based similarity between the social posts vectors on edges e_i and e_j. A_i is the set of the adjacent content-sharing-edges of the edge e_i.

The outcome of our system consists in a set of surfing paths. Each path starts from an initial post (edge) and ends with a target post which is supposed to contain information relevant to the user's interests. The number of edges contained in the surfing path is called the surfing depth. A relevance score can be associated with a surfing path, by assessing the similarity of the last post in the path and the user's interest vector V using Formula (3).

$$f(s) = sim(V, E_j) = \frac{\sum_{k=1}^{m} a_k b_k}{\sqrt{\sum_{k=1}^{m} a_k^2}\sqrt{\sum_{k=1}^{m} b_k^2}} \tag{3}$$

Where:

- s : a surfing path.
- $V = (v_1, v_2, ..., v_u)$ is the vector representing the user's interests.
- $E_j = (t_1, t_2, ..., t_u)$ is the last content-sharing-edge vector on the surfing path.
- $sim(V, E_j)$ is the cosine similarity between the vectors representing respectively the user's interests and the content-sharing-edge e_j.
- a_k and b_k are components of vector V and vector E_j respectively.

This way of computing the score of a social post is based exclusively on the concept of semantic similarity. While it ensures to offer a good ranking based on the relevance, it doesn't take into account the reliability of the information. In the following subsection, we explain how we tackle this issue.

2.2 Assessing the Information Credibility

In order to adress the concern of information credibility on social media, we use the MCDM approach [11] to estimate the credibility of the social posts. For this purpose, we consider several social features related to the social post and to its author (user). We define the social features as follows:

- n_r: number of re-posts of the current post.
- n_w: number of words in the post.
- n_f: number of social connections of the author of the post (followers, friends, ...).
- n_p: total number of posts of the author.
- p_v: whether the profile of the author is verified or not.
- p_p: whether the author is using the default profile picture or a custom one.

We associate with each feature a binary evaluation function ϕ such as, for a given post e_i:

- $\phi_{n_r}(e_i) = 0$ if $n_r = 0$, $\phi_{n_r}(e_i) = 1$ otherwise
- $\phi_{n_w}(e_i) = 0$ if $n_w < 5$, $\phi_{n_w}(e_i) = 1$ otherwise
- $\phi_{n_f}(e_i) = 0$ if $n_f \leq 10$, $\phi_{n_f}(e_i) = 1$ otherwise
- $\phi_{n_p}(e_i) = 0$ if $n_p \leq 20$, $\phi_{n_p}(e_i) = 1$ otherwise
- $\phi_{p_v}(e_i) = 0$ if $p_v = \textit{not verified}$, $\phi_{p_v}(e_i) = 1$ if $p_v = verified$
- $\phi_{p_p}(e_i) = 0$ if $p_p = Default$, $\phi_{p_p}(e_i) = 1$ if $p_p = Custom$

We also define a parameter that represents the importance of each feature in giving a good estimate of the credibility of a social post. We set the values as follows: $I_{n_r} = 0.25$, $I_{n_w} = 0.8$, $I_{n_f} = 1$, $I_{n_p} = 0.9$, $I_{p_v} = 0.1$, $I_{p_p} = 0.95$.

We aim at assessing a credibility score for each surfing path. Only the last social post on the surfing path is considered since it should be the most relevant one on that path. For this purpose, Formula 4 is proposed:

$$g(s) = \frac{\sum_{k \in F} I_k * \phi_k(e_i)}{\sum_{k \in F} I_k} \tag{4}$$

Where:

- $g(s)$: is the overall credibility score associated with the surfing path s;
- e_i: is the last social post on the surfing path s;
- F: is the set of the considered social features, F $= \{n_r, n_w, n_f, n_p, p_v, p_p\}$;
- $\phi_k(e_i)$: is the binary evaluation function of the feature k associated with the social post e_i;
- I_k: is the weight parameter of the feature k.

The assessment of the overall credibility of each surfing path is performed using Formula 4, and therefore for each surfing path a credibility value in the interval [0, 1] is computed. We set a credibility threshold c below which a surfing path won't be considered in the final results list presented to the user.

2.3 Ranking the Results Based on Their Relevance and Credibility

As previously seen in Subsects. 2.1 and 2.2, the system computes the relevance score and assesses the credibility of each surfing path. In order to take advantage of both criteria in the final ranking of the results, we propose Formula 5, which consists in a linear combination between the relevance score and the credibility score of a surfing path.

$$h(s) = \omega * f(s) + (1 - \omega) * g(s) \quad (5)$$

Where:

- h: is the ranking function based on the relevance and the credibility;
- $f(s)$: is the function calculating the relevance score of the surfing pas s;
- $g(s)$: is the function calculating the credibility of the surfing path s;
- ω: is the weight balancing between the relevance and the credibility functions.

The final list of results, which is a list of surfing paths, is ranked using function h. This will ensure that the system takes into account both the relevance and the credibility of the social posts before presenting the final foraging results to the user. The weight ω is an empirical parameter that can be set during the experiments phase, as it may depend on the considered social platform.

3 A Multi-agent Based Social Media IF System

In this section we propose a multi-agent based Social Media Information Foraging System, which uses the information foraging model described in the previous section. By this approach the foraging activity motivated by an information need is carried out by several agents. We consider the social-graph representation of social networks as previously explained, and we assume that each agent browses a part of the social graph with the aim of finding relevant information. Figure 2 illustrates the global architecture of the multi-agent system.

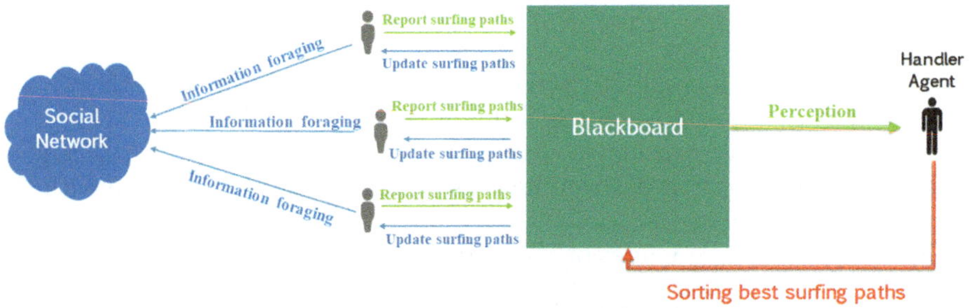

Fig. 2. Architecture of the Social Media Information Foraging Multi-agent System.

The system is composed of two kinds of agents that can work simultaneously and in a collaborative way to address the foraging issue:

a) A group of foraging agents that have the task of foraging relevant information on the social graph. Starting with the same user's interests but from different locations (edges), the agents explore the graph simultaneously with the goal of reaching relevant social posts. Algorithm 2 presents the pseudo-code of the behavior of a foraging agent.

Algorithm 2. Foraging Agent

Input: a social graph structure;
Output: surfing paths ending with relevant social posts;
1: Select an edge (social post) at random;
2: Perceive the environment and check the neighborhood of the current edge;
3: Move to a new social post using Formula 2;
4: Add the new social post to the surfing path;
5: Compute the surfing path score using Formula 5;
6: Communicate the surfing path to the other agents by writing it on the blackboard;
7: Wait for the blackboard-handler agent to sort the solutions;
8: Check the blackboard;
9: **if** the surfing path belongs to the top solutions **then** go to 2
10: **else**
11: Abandon the current surfing path and Go to 1;
12: **end if**

b) A blackboard-Handler agent, which plays the role of the coordinator who takes in charge the management of the system and the sorting of the partial solutions (surfing paths) proposed by the foraging agents based on their relevance and credibility. Algorithme 3 shows the pseudo-code of the agent.

Algorithm 3. Blackboard-Handler Agent

Input: a surfing path *sp*
Output: a sorted list of surfing paths ranked based on their relevance and credibility
1. Check the blackboard;
2. Sort the surfing paths *sp* proposed by the foraging agents according to their score using a Binary Search Tree Insertion;
3. As soon as a new solution is reported by a foraging agent insert it in the right position in the *Binary Search Tree* containing the top solutions;
4. Update the top solutions list on the blackboard;
5. Go to 3;

The blackboard module is an interaction model that allows information sharing between the agents in our system. More concretely, it is an area of shared memory accessible in read and write mode by all the agents, so that it facilitates the collaboration between them.

4 Experiments and Evaluations

To evaluate the performance of our system, we used a dataset that we contructed back in 2015 by crawling data from *Twitter* using *NodeXL* [12]. The keyword "*Paris*" was used to gather tweets during the period when the *United Nations*

Climate Change Conference, COP 21 was being held in the french capital. The resulting social graph we worked on contains 16 896 nodes and 17 706 connections.

The tuning of the values of the empirical parameters used during the experiments took place following an extensive phase of tests, which consisted in varying the values of the different parameters and comparing the yielded outcomes corresponding to these values. The combination of 100 foraging agents, a list of results containing 30 surfing paths, a weight of $\alpha = 2$ associated to the information scent measure, a weight of $\beta = 1$ associated to the similarity between social posts, a credibility threshold $c = 0.35$ and a relevance weight $\omega = 0.9$ led the system to produce the best results on the considered dataset.

4.1 Generating the User's Interests Vector

As mentioned in Sect. 2, one of the new contributions of this work is the automatic generation of the user's interests based on her/his social profile and behavior. Table 1 and Table 2 show an example of different users' information that we extracted from the nodes of the social graph structure. The two tables were split for display purpose. In order to generate the users' interests vector we use their biography labeled as *"v-description"* in Table 1 along with their latest tweets. Table 3 is an example of a user's interests vector generated implicitly using Algorithm 1 for the Google's official Twitter account during the period from September 28th to October 3rd 2019. The terms contained in the vector including "information", "question" and "learn" showcase the company's interest in the domain of information access. The rest of the terms are related to some of the company's products such as its own mobile operating system "android". Note that a term frequency is associated to each term on the vector.

Table 1. Example of users' personal information extracted from Twitter - a

Id	v-followed	v-followers	v-tweets	v-favorites	v-description	v-location
Androidheadline	231	231306	131909	113	All the latest breaking android news & Rumors covering phones, ...	Los Angeles, CA
Google	440	13260154	8086	420	News and updates from Google	Mountain View, CA

Table 2. Example of users' personal information extracted from Twitter - b

Id	v-joined twitter date (utc)	v-default profile	v-default profile image	v-geo enabled	v-language	v-listed count	v-verified
Androidheadline	25/06/2009 20:55	False	False	True	en	3538	False
Google	10/02/2009 19:14	False	False	True	en	92125	True

Table 3. Example of a users' interests vector

Username	User's interests						
Google	Google	Parent	Learn	Question	Android	Information	Treat	Connect

4.2 Defining the Information Credibility

We consider the 5 following features to assess the credibility of the tweets: ***e-retweet:*** number of retweets; ***e-tweet:*** number of words in the tweet; ***v-followers:*** number of followers of the author of the tweet; ***v-tweets:*** total number of tweets of the author; ***v-default-profile-image:*** whether the author is using the default profile picture or a custom one; ***v-verified:*** whether the author's Twitter profile is verified or not. The credibility of a tweet is assessed as described in Subsect. 2.2 and the credibility threshold is set to $c = 0.35$.

4.3 Foraging Results

Table 4 shows the information foraging results on the graph for different users' interests. The first column of the table presents different user's interests generated automatically from random twitter accounts belonging to the social graph. The interests vectors were defined during a period of $h = 72$ h and their sizes n vary between 6 and 8 terms. The second column of the table gives details about the preferred language of each user. Note that the system takes the user's preferred language into account in the foraging process if it is specified. The best result for each user's interests is shown on the last three columns. Column 3 contains the most relevant tweet found by our system, column 4 precises the source of this tweet, while column 5 specifies the length of the surfing path that led to this tweet.

The results reported in the table indicate that our approach is capable of finding relevant tweets with respect to the user's interest and her/his preferred language. Also, the content of the tweets comes from reliable Twitter accounts with a minimum degree of credibility. We notice that the user's interests focusing on the *COP 21* tend to produce longer surfing paths compared to other user's interests. This might be explained by the fact that an important number of users in the network were tweeting about the *COP 21* during that period, which resulted in a high number of connections in the network's graph related to this topic and thus allowing to construct longer surfing paths. Line number 2 on Table 4 is a perfect example of this aspect.

Table 4. Information foraging results on Twitter

User's Interests	Preferred Language	Best Information Foraging Result		
		Tweet	Source Account	Surfing depth
Planet, Climate, Futur, Paris, COP21, Deal	English	New draft climate deal emerges as Paris talks near end #COP21 https://t.co/GLVUyGzTK1	@cop21_news	4
COP21, Pollution, Climate, Nature, Life, Earth, Better	English	As #COP21 continues, Earth breaches a feared pollution threshold. Does it really matter? #Climate https://t.co/RxJq8E1krM https://t.co	@business	5
Translate, French, Leader, Paris, Planet, Climate, Guide	English	Paris Conference Leaders Want You to Think the Planet Is Facing a Climate Change Crisis. https://t.co/S2Zyz1OE0R @DailySi	@bob4gov_now	3
Tourist, Attraction, Relax, Tour, Visit, Vacation, Paris, Beach	English	Gpe_image: #Guadeloupe 🌴🌼#Antilles #Martinique #Paris #Sensation #Gi #Fwi #Westindies #Beach #Caraï... https://t.co/ZbP1Wj4Olu	@relaxincanada	2
Machine, Learning, Engineering, Startup, Etudiants, Python	French	Développeur C++/Python/R dans un contexte "Machine Learning" @SAP #Paris https://t.co/ZKyWZH7KTq #machinelearning #Python	@SiliconArmada	2

5 Conclusion and Future Axes

We proposed in this paper a novel approach dedicated to information foraging on social media. We developed a Social Media Information Foraging System capable of getting credible and relevant information, the first of it's kind in our knowledge. We also focused on generating the users' interests implicitly based on their data on social networks, which consists another novelty of this work. To evaluate our system, experiments were conducted on Twitter. We presented some preliminary results, which we think are really promising. This encourages us to deepen the investigation of information foraging on social media.

As perspectives, we plan to undertake massive experiments during different periods on various topics in order to highlight the usefulness of the developed system for different domains. We also plan to use different approaches including bio-inspired computing and classification in order to enhance the performance of our system.

References

1. IBM Marketing Cloud: 10 Key Marketing Trends for 2017 and Ideas for Exceeding Customer Expectations. IBM (2017). https://www.ibm.com/downloads/cas/XKBEABLN
2. Pirolli, P., Card, S.: Information foraging. Psychol. Rev. **106**(4), 643–675 (1999)
3. Winerman, L.: Monit. Psychol. **43**(3), 44 (2012)
4. Liu, J., Zhang, S.W.: Characterizing web usage regularities with information foraging agents. IEEE Trans. Knowl. Data Eng. **40**, 7478–7491 (2004)

5. Arbelaitz, O., Gurrutxaga, I., Lojo, A., Muguerza, J., Pérez, J.M., Perona, I.: Web usage and content mining to extract knowledge for modelling the users of the Bidasoa Turismo website and to adapt it. J. Expert Syst. Appl. **40**, 7478–7491 (2013)
6. Niu, X., Fan, X.: Deep learning of human information foraging behavior with a search engine. In: International Conference on Theory of Information Retrieval (ICTIR), pp. 185–192 (2019)
7. Drias, Y., Kechid, S., Pasi, G.: A novel framework for medical web information foraging using hybrid ACO and tabu search. J. Med. Syst. **40**(1), 5:1–5:18 (2016)
8. Lamprecht, D., Strohmaier, M., Helic, D., Nyulas, C., Tudorache, T., Noy, N.F., Musen, Ma.A.: Using ontologies to model human navigation behavior in information networks: a study based on Wikipedia. Semant. Web J. **6**(4), 403–422 (2015)
9. Drias, Y., Kechid, S.: Dynamic web information foraging using self-interested agents: application to scientific citations network. J. Concurr. Comput.: Pract. Exp. **31**(22), e4342 (2019)
10. Schnabel, T., Bennett, P., Joachims, T.: Shaping feedback data in recommender systems with interventions based on information foraging theory. In: International Conference on Web Search and Data Mining (WSDM), pp. 546–554 (2019)
11. Viviani, M., Pasi, G.: A multi-criteria decision making approach for the assessment of information credibility in social media. In: Fuzzy Logic and Soft Computing Applications - 11th International Workshop. WILF, Naples, pp. 197–207 (2016)
12. Smith, M., Ceni, A., Milic-Fraylin, N., Shneiderman, B., Mendes Rodrigues, E., Leskovec, J., Dunne, C.: NodeXL: a free and open network overview discovery and exploration add-in for Excel 2007/2010/2013/2016. The Social Media Research Foundation (2010)

Online Geocoding of Millions of Economic Operators

Tiago Santos[1(✉)], Daniel Castro Silva[1], Ana Paula Rocha[1], Henrique Lopes Cardoso[1], Luís Paulo Reis[1], Cristina Caldeira[2], and Ana Oliveira[2]

[1] Laboratório de Inteligência Artificial e Ciência de Computadores (LIACC), Faculdade de Engenharia da Universidade do Porto, Rua Dr. Roberto Frias, s/n, 4200-465 Porto, Portugal
{tiagosantos,dcs,arocha,hlc,lpreis}@fe.up.pt

[2] Autoridade de Segurança Alimentar e Económica (ASAE), Rua Rodrigo da Fonseca, 73, 1269-274 Lisbon, Portugal
{accaldeira,amoliveira}@asae.pt

Abstract. Geocoding is the process of converting an address or a place name into geographic coordinates. This conversion process has become a fundamental subject in many scientific domains and real world applications, from health and crime analysis to route optimization. In this paper, we present a conversion process of over 4.5 million entities, mostly Portuguese Economic Operators, through their addresses or names. We also describe how this information can be useful to detect and remove duplicate information in databases. The results demonstrate the power, flexibility and accuracy of many of today's online geocoding services.

Keywords: Geocoding · Geodesic · Duplication · Economic operator

1 Introduction

From public institutions, including hospitals and tax services, to online stores such as *Amazon* and *eBay*, databases in which physical addresses are stored are prevalent. Without accurate geographic locations, data analysis or real-time tracking is, in many instances, limited to a coarse-grained analysis at a district or even city level. The usage of geographic coordinates, instead of addresses, allows for a wide variety of spatially fine-grained data analysis and applications, such as route optimization, targeted marketing and risk management. Therefore, this fine-grained data has obvious economic advantages, for any organization.

With this in mind, we explore a database of a Portuguese public institution, containing millions of records of entities with physical addresses and proceed to geocode them. One of the main activity of this institution is to supervise and prevent non-compliance with regulatory legislation in the area of economic and food safety. For that purpose, generating and executing inspection routes is an

Á. Rocha et al. (Eds.): WorldCIST 2020, AISC 1159, pp. 426–435, 2020.
https://doi.org/10.1007/978-3-030-45688-7_44

important part of this institution's operational procedures, for which geocoding is crucial to optimize resource usage efficiency.

In Sect. 2 we present a literature review on geocoding-related works. In Sect. 3 we present the approach for efficiently geocoding millions of physical Economic Operators. In Sect. 4 the results of this process are presented and discussed. An application example is presented in Sect. 5, where coordinates are used to identify and eliminate duplicate Economic Operators. Finally, Sect. 6 concludes and points to future work.

2 Related Work

Any geographic point on Earth's surface can be specified by a pair containing two floating numbers: latitude and longitude. Gecoding and reverse geocoding are critical in many of today's applications. Simply speaking, while geocoding is the process of transforming an address into geographic coordinates, specified by the pair $\langle latitude, longitude \rangle$, reverse geocoding is the opposite operation, *i.e.*, the translation of geographic coordinates into an address. These pairs of coordinates are an indispensable piece of data for many applications, such as route planing and optimization [1,2], spatial analysis of epidemiological data [3], crime analysis [4], and political science [5], among others.

Goldberg *et al.* [6] present a survey of the state of the art in geocoding practices. They start as early as the 1960's with the geocoding systems used by the U.S. Census. Crosier [7] addresses the grocoding process, from data preparation to management, in the context of the ArcGIS software package. Murray *et al.* [8] provide a user's guide for indexing and storing spatial data and for developing spatial applications using Oracle Spatial and Oracle Locator. Other works, such as Cayo and Talbot [9], center their study on automated geocoding and on the positional errors that arise from the geometric methods used to interpolate addresses. On their survey, Goldberg *et al.* [6] also discuss sources of error and uncertainties of geocoding processes. Karimi *et al.* [10] focus on understanding uncertainties associated with interpolation techniques used by three geocoding algorithms. Zeiler [11] focuses more on geodatabase structural elements and data modeling, but not on the geocoding process. Asavasuthirakul and Karimi [12] provide an examination on the quality of five online geocoding services: *Geocoder.us*, *Google*, *MapPoint*, *MapQuest*, and *Yahoo*. They used three metrics for evaluation: match rate, positional accuracy, and similarity. They concluded that *Google*, *MapPoint* (currently discontinued) and *Yahoo* (currently deprecated) were the services with higher match rates. Ward *et al.* [13] also compared the accuracy of two geocoding methods: an in-house method, using ArcView 3.2 software and the U.S. Census database; and an automated geocoding service provided by a commercial firm. The authors concluded that, despite the fact that both methods present similar degrees of precision when dealing with urban addresses, the precision of their in-house method is twice more accurate than the commercial service when geocoding addresses in rural areas.

To the best of our knowledge, there are no works that examine the feasibility and usefulness of geocoding a large number of addresses by exploiting, in a combined fashion, several online geocoding services.

3 Approach

In this section we introduce our approach, first presenting the data available, then the selected online geocoding services and finally the process itself.

3.1 Data

Our database contain records on 4,517,174 Economic Operators, which for the rest of this paper, will be referred to as entities. Also, each record consists of 78 fields, such as the entity's name, its address and its economic activity.

The geocoding process success depends in many cases on the degree of details available to represent the input addresses: entity's name, street name, house number, city, and country. Therefore, all entities were divided into four groups according to the available information (see Table 1).

Table 1. Clustering of entities according to the available data

Group	Number of entities
Address + city	4,199,048
Address	127,525
Name + address	39,164
Name	151,437

The resulting information depends on each service. In general, we tried to obtain the following information: *formatted address*, *latitude*, *longitude*, *precision*, *place id*, *type of place*, *postal code*, *locality/city*, *state*, *municipality*, and *parish*.

The value of the *precision* field also depends on the service and there is no uniformity both in value and type. For example, while *Google* provides up to four possible values - ROOFTOP, RANGE_INTERPOLATED, GEOMETRIC_CENTER, and APPROXIMATE - *TomTom* returns a decimal value. The same goes for others fields such as *type of place* and all country subdivisions.

One of the main issues that any work related with geospacial data has to deal with is the inconsistencies and incorrect formats of addresses. Therefore, a very important preliminary task consists in the standardization of the addresses. To correct this issue we could have used libraries such as *libpostal*[1], which parses/normalizes street addresses around the world using statistical natural language processing and open data. However, all modern geocoding services already

[1] https://github.com/openvenues/libpostal.

have a similar system running in the background, which is why they are able to return not only the above information, but also all the different parts that constitute an address.

3.2 Services

There are many online and offline geocoding services and tools available. We have selected five popular online geocoding services that provide a REST API: Google[2], Bing[3], Azure[4], TomTom[5], and Here[6]. By choosing online geocoding services we avoided the need to prepare reference databases and other complex procedures.

Since all of these services offer some sort of free limit or credit, which allows users and developers to test their services without any sort of compromise (one API call $\approx$ one geocodification), we were able to geocode all addresses without the need of using their paid services.

Some of these services, such as *TomTom*, also provide the capacity of batch geocoding. Because of its speed, using this type of service is more effective when needing to process hundreds or even thousands of addresses.

3.3 Geocoding Process

The normal process of geocoding an entity, which is represented in Fig. 1, can be summarized as follows:

1. Get the entity's name and/or full address;
2. Build the service's request *URL*, e.g., for google, the *URL* should take the form: https://maps.googleapis.com/maps/api/geocode/outputFormat?parameters;
3. Make the *http* request, which should return either a *JSON* or an *XML* structure;
4. Parse the response to extract the data and save the results.

For the geocoding process, we could either have followed the above process, where we would have to build a specific *URL* for each service, or use a library capable of standardizing these requests, such as *GeoPy*[7]. In order to simplify the process of building a specific URL for each service, the *GeoPy* library was used, which is capable of abstracting up to twenty four online service APIs. Hence, there is no need to build a *URL* for each service, and instead, we initialize the services and proceed to geocode any number of addresses.

[2] https://cloud.google.com/maps-platform/.
[3] https://www.bingmapsportal.com/.
[4] https://portal.azure.com/.
[5] https://developer.tomtom.com/.
[6] https://developer.here.com/.
[7] https://geopy.readthedocs.io/.

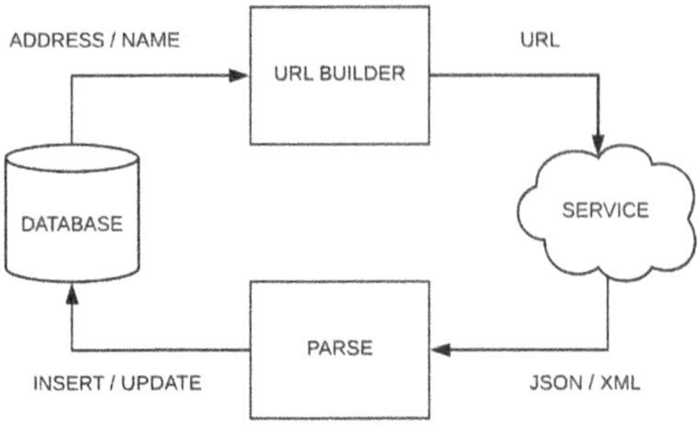

Fig. 1. Typical geocoding process

In order to quickly finalize the geocoding process, we sorted the entities to be processed by service and proceeded in running all services simultaneously. This sorting was made according to the limits of free usage of each service and were also adjusted, later on, according to each service's geocoding speed. Also, considering the groups defined in Table 1, the entities of the last group were geocoded only through *Google*, as it is the service with the highest number of registered entities. Furthermore, every time a service failed to geocode an address, we proceed to select another service, until we reached either a successful geocoding or the end of available services.

4 Results and Analysis

In this section we present the results of the geocoding process, starting with the average speed for each service, followed by how much of each service was employed in geocoding the entire database. Then we explain how we proceed to automatically validate the coordinates. Finally, a simple test to verify the speed of our platform against an expensive geographical query.

4.1 Geocoding Speed

In order to evaluate the efficiency of each of the explored services, we decided to benchmark their use. For that, we randomly selected one hundred complete addresses (containing, at least, the street name, door number, city and country). We then geocoded them twice, with at least four hours interval, so as to prevent possible cache-related issues from the service side. In Table 2 we present the average speed that each service took per each geocode. As we can observe, despite *Here* being the fastest, it was also the only one with less than 100% success rate.

4.2 Geocoding per Service

A total of 4,517,174 addresses were geocoded, with the distribution per service observable in Fig. 2. As expected from the used methodology, while *Here* was the

Table 2. Average speed and success rate of geocoding 100 randomly selected addresses

Service	Time per geocode [s]	Success rate
Here	0,28	96%
Bing	0,43	100%
Azure	0,53	100%
Google	0,56	100%
TomTom	0,90	100%

most used service, due to its speed and also for its geocoding limits, *TomTom* was the least one because of its limit of 2,500 API calls per day.

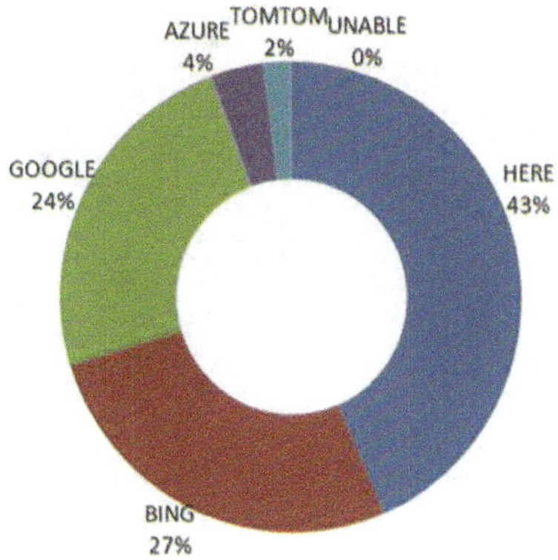

Fig. 2. Database geocoding per service

Despite all the advances of the distinct services in recent years, it was impossible to geocode 777 addresses. After a quick manual inspection of these addresses, it was verified that albeit some of them were non-existent, mainly because they were too incomplete, others were real and complete addresses that were tested directly in *Google Maps* with a positive correspondence. This lead us to believe that, at least in the case of *Google*, the underlying processes behind the available API and the *Google Maps* harbor are not the same.

4.3 Validation

Albeit the results of the used services being usually trustworthy, we decided to not fully trust them. However, proceeding with a manual validation of 4.5 million coordinates is not possible. For now, we are not interested in small uncertainties (street level) that may be introduced by the different algorithms used by the services, but more concerned with large errors, most likely caused either by a bad address interpretation or by an incorrect address. Hence, we have created a quick validation process. Since we have the state/municipality/parish codes of most entities, we used them to verify if their coordinates are actually inside the polygons defined for each area. However, it should be noted that even these codes are not 100% credible, since they were inserted manually into the database, by a human operator, without any type of automatic validation.

Figure 3 depicts every parish of Lisbon and some entities. From a simple observation, it is clear that at least one (the farthest left) should have its coordinates invalidated, since they are clearly outside the boundaries of every parish.

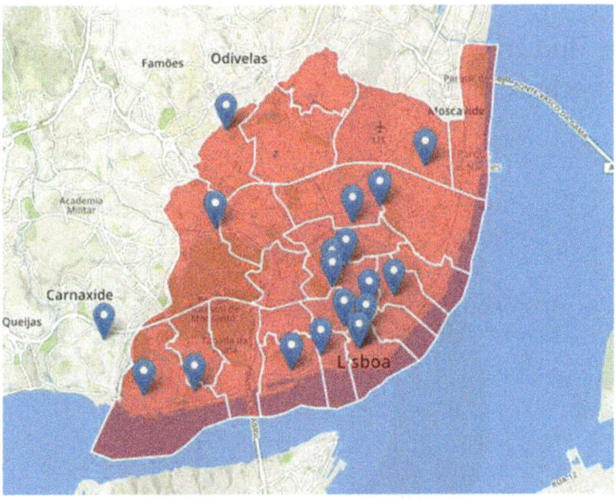

Fig. 3. Parishes of Lisbon and a few entities.

The percentage of valid geocodification per service and per country's subdivision regions can be seen in Table 3. From its analysis, we verified that while *Azure* was the service that gave the highest percentage of invalid coordinates, *Here* was, on the other hand, the one that gave the highest percentage of valid coordinates.

Table 3. Percentages of valid geocodings per country's subdivisions

Service	State	Municipality	Parish
Here	92,9%	92,4%	82,6%
Google	88,8%	87,0%	77,8%
Bing	86,0%	82,2%	70,1%
TomTom	85,8%	84,0%	73,4%
Azure	81,3%	76,5%	66,8%

4.4 Geographical Queries

Because several geographical queries are to be used by the organization, we made a test to verify the query's speed against our platform. We used a MySql 5.6 database, which despite possessing many geospatial capabilities does not offer any function to calculate the geodesic distance between two coordinates. The query is meant to return n entities inside a circumference of radius r meters. For that purpose, we implemented Function 1, which approximately calculates the geodesic distance between two points.

$$\begin{aligned} distance = {} & 6371392.896 \times \arccos(\cos(Radians(pnt\_lat)) \\ & \times \cos(Radians(lat)) \times \cos(Radians(lng) \\ & - Radians(pnt\_lng)) + \sin(Radians(pnt\_lat)) \\ & \times \sin(Radians(lat))) \end{aligned} \tag{1}$$

with *(lat, lng)* and *(pnt_lat, pnt_lng)* as the coordinates of the two points whose distance we want to calculate. *Radians* simply transforms degrees into radians.

We then proceed with testing the query's response time using r = 1,000 m and different values of n. The results, depicted in Fig. 4, show a clear linearity between query speed and number of entities.

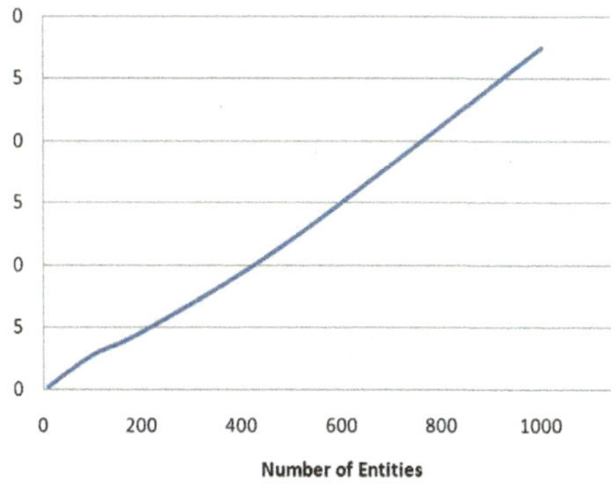

Fig. 4. Query speed vs number of entities.

5 Detecting Duplicate Entities

The usage of sub-optimally designed information systems and databases frequently leads to problems in data, such as inconsistent or duplicate information. The identification and removal of duplicate information is important to enable proper business intelligence and thus optimize decision making.

In this section we describe how the process of geocoding employed as described above has enabled us to mitigate this problem. Albeit the results of the geocoding process are being used in more directions, we describe this particular use case. In fact, the usage of geographical coordinates in finding duplicate data in a database is not common.

Because entries are done manually by different people, it is very difficult to automatically determine if a new entity already exists in a database based solely on fields such as name and address, since it is possible to write this type of information in more than one way. As a consequence, an entity can have more than one entry in the database, under names that differ, often significantly enough to hinder the usage of automatic duplicate resolution mechanisms based on lexical operations. Algorithms such as *Levenshtein distance* are worthless in these cases, since their values would be too low; false positives could also become a problem, as names of actually different entities may in fact be quite similar.

We therefore used the geodesic distance between entities as the main condition to identify whether or not two entities are the same. To differentiate cases where entities share the same pair of coordinates, caused perhaps for being in the same building, we considered others factors such as their economic activities. Their names and addresses were not taken into account. The variation between distance (in meters) and the identified number of duplicate entities can be seen in Table 4. The numbers in the right column represent the potential number of entities that might be removed from the database, given the distance thresholds in the left column.

Table 4. Number of duplicate entities vs geodesic distance.

Distance (meters)	Number of duplicated entities
1	9236
5	9658
25	10501
50	11108
100	11943
250	13823
1000	18943

We can explain this variation by the usage of different geocoding services, meaning that different geometric algorithms were used to calculate the coordinates of similar addresses.

6 Conclusions and Future Work

Because entities were geocoded by individual API calls, it took several weeks to complete the task, which is only reasonable in an academic or scientific environment but not in a production setting. The results show that, for now, *Google* provides the best online service, not only for geocoding through addresses but also through entity names.

In the future, we expect to test other online geocoding services, such as GeocodeFarm[8] and MapQuest[9]. We also plan to address administrative and postal code inaccuracies by using the results of this work. Other than cleaning the database of duplicate information, the results of this work are also being used in routes generation and optimization.

Acknowledgements. This work is supported by project IA.SAE, funded by Fundação para a Ciência e a Tecnologia (FCT) through program INCoDe.2030. This research was partially supported by LIACC (FCT/UID/CEC/0027/2020).

References

1. Amal, L., Son, L.H., Chabchoub, H.: SGA: spatial GIS-based genetic algorithm for route optimization of municipal solid waste collection. Environ. Sci. Pollut. Res. **25**(27), 27569–27582 (2018)
2. Rasmussen, S., Talla, M., Valverde, R.: Case study on geocoding based scheduling optimization in supply chain operations management. WSEAS Trans. Comput. Res. **7**, 29–35 (2019)
3. Mendes, J., Ferreira, M.: Avaliação de métodos de geocodificação para conversão de agravos localizados em endereços, para mapas de pontos em sistemas de coordenadas espaciais. In: A cartografia na geografia da saúde: metodologias e técnicas, chap. 5, pp. 70–82 (2019)
4. Olligschlaeger, A.M.: Artificial neural networks and crime mapping. In: Crime Mapping and Crime Prevention, pp. 313–347. Criminal Justice Press, Monsey (1998)
5. Haspel, M., Knotts, H.G.: Location, location, location: precinct placement and the costs of voting. J. Polit. **67**(2), 560–573 (2005)
6. Goldberg, D., Wilson, J., Knoblock, C.: From text to geographic coordinates: the current state of geocoding. Urisa J. **19**, 33–46 (2007)
7. Crosier, S.: Geocoding in ArcGIS: ArcGIS 9. Esri Press, Redlands (2005)
8. Murray, C.: Oracle spatial user's guide and reference, 10g release 2 (10.2). Oracle Corporation (2006)
9. Cayo, M., Talbot, T.: Positional error in automated geocoding of residential addresses. Int. J. Health Geogr. **2**, 10 (2003)
10. Karimi, H., Durcik, M., Rasdorf, W.: Evaluation of uncertainties associated with geocoding techniques. Comput.-Aided Civ. Infrastr. Eng. **19**, 170–185 (2004)
11. Zeiler, M.: Modeling Our World: The ESRI Guide to Geodatabase Concepts. ESRI Press, Redlands (2010)
12. Asavasuthirakul, D., Karimi, H.: Comparative evaluation and analysis of online geocoding services. Int. J. Geogr. Inf. Sci. **24**, 1081–1100 (2010)
13. Ward, M., Nuckols, J., Giglierano, J., Bonner, M., Wolter, C., Airola, M., Mix, W., Colt, J., Hartge, P.: Positional accuracy of two methods of geocoding. Epidemiology **16**, 542–547 (2005)

[8] https://geocode.farm/.

[9] https://developer.mapquest.com/.

Closed Against Open Innovation: A Comparison Between Apple and Xiaomi

João Lajoso[1], André Sousa[2], João Albuquerque[2], Ricardo Mineiro[3], and Manuel Au-Yong-Oliveira[1,4(✉)]

[1] Department of Economics, Management, Industrial Engineering and Tourism, University of Aveiro, 3810-193 Aveiro, Portugal
{joaolajoso,mao}@ua.pt
[2] Department of Electronics, Telecommunications and Informatics, University of Aveiro, 3810-193 Aveiro, Portugal
{ac.sousa,jp.albuquerque}@ua.pt
[3] Department of Ceramics and Materials Engineering, University of Aveiro, 3810-193 Aveiro, Portugal
mineiro@ua.pt
[4] GOVCOPP, Aveiro, Portugal

Abstract. One of the biggest, most competitive and quicker shifting markets is the smartphone industry. Evaluating significant opportunities and being sustainable in the market is entirely up to a firm's innovation strategy. In this article, we perform an in-depth analysis by available means to study trends and tendencies relative to Apple and Xiaomi, two completely different companies with completely different approaches. We tested the clients' needs and perceptions, in an attempt to see if they are part of the enterprises' strategy (in a survey, with 193 responses). Indeed, customers are interested in powerful cameras, good storage capacity and a good duration of the battery of their smartphones. The market is evaluated above 1.5 billion units sold per year and demonstrates high demand for every new model to be different, more powerful, more capable of fulfilling our lifestyle. Apple uses an incremental innovation sales strategy, based on the billions of euros used in secret research and development. On the other hand, Xiaomi performs open innovation, taking customers' needs and ideas to the world. What is the best approach? Will it be enough to stay on top? An analysis using Google Trends shows how Nokia has faded away over time, how Apple had a spike of interest concerning its latest iPhone 11 launch; while Xiaomi is ever more popular, especially in Portugal.

Keywords: Open innovation · Closed innovation · Xiaomi · Apple · Technology · Google Trends · Survey

1 Introduction

Humans use technology to travel, to communicate, to learn, to work, essentially, throughout their daily lives. As a consequence of that, society demands new gadgets and/or new

Á. Rocha et al. (Eds.): WorldCIST 2020, AISC 1159, pp. 436–448, 2020.
https://doi.org/10.1007/978-3-030-45688-7_45

features in a short space of time and companies try their best to deliver on these requests. The best example that demonstrates this behavior is, without a doubt, the smartphone industry. In Portugal, with data retrieved from PORDATA [1], the total number of mobile subscribers reaches almost 20 million, which is almost an average of two smartphones per person, considering that each mobile plan corresponds to only one smartphone.

Companies started to notice the increase in the demand for innovation to the point that it was too significant to be ignored and could not be left to chance. Therefore, strategies regarding innovation started to appear. Open innovation and closed innovation are two different and opposite methods, and each one has their own "rules", but both can end up with a product that is innovative to the market. There is no clear evidence (no academic references were found) about the impact of the strategy used in the remaining aspects of the companies analysed (Apple and Xiaomi), being these the appearance of the company to the public, public opinion about them, influence of the company on the people and their preferences for the company products and/or services.

The two brands being studied are examples that follow the strategies mentioned above: Apple works in a closed innovation environment while Xiaomi uses an open innovation approach. Although using different strategies, the sales of both brands are pretty similar. According to Gartner [2], in the second quarter of 2019, Apple had a market share of 10.5% with almost 38.5 million units sold worldwide against a market share of 9% for Xiaomi with 33 million units sold worldwide.

This study starts with a literature review about the two strategy concepts being analysed and on the background of Apple and Xiaomi. The methodology used relies on three different sources: surveys distributed on the internet (using Google Forms), focused on the young (18–30 years of age) in the Portuguese population; search queries done in Google Trends that analyse the growth/decrease of interest, not only in Apple and Xiaomi, but also concerning some other mobile phone brands in Portugal and worldwide over the last 15 years; and a search in online databases that study the total number of mobile subscribers in Portugal.

Our main research questions are the following:

- How does the new technology arrive in our hands?
- How are we in touch with innovation?
- Do we (in Portugal) stand out from world trends?

2 Literature Review

2.1 Open and Closed Innovation

"The innovation strategy, innovation, and innovation processes play a decisive role in the matter of acquiring and maintaining the strengths of a company in a battle with the competition" [3]. Innovation came from the necessity of a company to distinguish itself from its competitors through using its knowledge, competencies and/or technologies in the production of a product that would stand out in a market with a constant offering increase of similar products by the competition, and thus being defined as an innovator. One idea that is widely spread is that a company that does not innovate is destined to fail [4]. However, several strategies have been and can be adopted by a company so that it may be considered innovative.

Open innovation is the position where there is a flux of knowledge, competencies and technologies between a company and one or more external subjects. The objective is that anyone who has an idea can share it and work it more efficiently by using the resources of the company to obtain a product which benefits the company and the final consumer [5].

Closed innovation "is based on the concept that innovation in organizations requires the strongest type of control. It is a type of logic that is focused internally within the organization and spread among the employees to reach the sufficient level of quality" [6]. Closed innovation is about self-reliance and about having the best resources in-house to innovate with.

2.2 Background on Apple

Long before making smartphones, Apple already was an established brand. First with computers, then with iPods. In fact, in the first quarter of 2007, Apple sold over 21 million iPods, corresponding to 48% of its revenues [7]. Albeit, in January 2007, Steve Jobs presented the first iPhone at the Macworld Expo in California. As Phil Schiller (senior vice president of worldwide marketing) said, there were some reasons that led to the appearance of the iPhone: "First, Apple had been known for years for being the creator of the Mac, the computer, and it was great, but it had small market share" (…) "And then we had a big hit called the iPod. It was the iPod hardware and the iTunes software. And this really changed everybody's view of Apple, both inside and outside the company" [8]. At the time of the presentation, the iPhone was a revolutionary product because it combined the functionalities from the iPod into a small mobile phone with a touch screen, and it was able to access the internet like a computer. The development took over two years in total secretism, and the most exciting fact is that the hardware team (code-named P1, responsible for the iPod phone) worked utterly separated from the software team (code-named P2, responsible for the still-experimental (at the time) hybrid of multitouch technology and Mac software) [8]. Because the engineers that worked on the original prototype would not have a single clue concerning what the final product would eventually look like, the workers only had access to "special prototype development boards that contained nearly all of the iPhone's parts, spread out across a large circuit board" (Fig. 1) [9].

Because of its natural closed innovation, in the more recent iPhone models, the company moved on to big and bulky security shields. This solution allows the developers to work on the final hardware form in a case while the design secret is maintained [9]. Other phone companies that use closed innovation also adopted this solution to keep the final development product under wraps.

From then on, Apple continues to make small increments to each new generation of the iPhone, with the exception of the iPhone 8 and the X model which were launched at the same time to test if the market was receptive to a phone with face authentication and gestures-based as a major improvement instead of the home button. Last year iPhones made up 62% of Apple's total net revenue, which is between $61 billion and $64 billion (fourth quarter 2019) [10].

Fig. 1. iPhone M68 prototype [9]

The corporate value of the brand has reached the level of being the most valuable brand worldwide, in 2015, being then worth over $170 billion [26]; a position it has maintained into 2018, where it was valued at close to $215 billion [27].

2.3 Background on Xiaomi

Xiaomi was founded in April 2010 with the mission to become an innovative technology enterprise using the internet to develop intelligent hardware products. Since its inception, the company has already been awarded 4,043 patents (1,887 of them international). In August 2010, Xiaomi launched its first product, the operating system MIUI and a few years later, in August 2011, its first mobile phone, the M1. Since then, Xiaomi has been investing in startup companies that develop hardware ("eco-chain products") and co-creating various products with them ("eco-chain companies"). The mobile phone series, the MIUI operating system and the ecological chain products constitute the basis of Xiaomi's open innovation ecology. From this, we identify three main characteristics of innovation carried out by core enterprises in open innovation technology: iterative, social and joint innovation [11].

Iterative innovation "refers to the innovative way that the core enterprise improves its products repeatedly according to user suggestions in order to better meet users' needs" [11]. Applying that to Xiaomi, the iterative innovation occurs with the MIUI, with the creation of the MIUI Forum, where the users can be fully involved in the R&D process of MIUI.

Social innovation "means that core enterprises maintain good communication with users through online social media and offline activities, actively solicit users' innovative ideas on products, and carry out targeted improvement through interactive communication" [11]. From this, Xiaomi expands the scope of users' participation in innovation through social media, including WeChat and Weibo, regarding its mobile phone. Therefore, the innovation of the Xiaomi mobile phone is social innovation.

Regarding joint innovation, it "means that core enterprises jointly develop products through cooperation with partners based on complementary advantages and mutual benefits" [11]. "Therefore Xiaomi introduces partners through a new and more open model. Xiaomi implements the system of "the US Pentagon-special forces". First of all, Xiaomi as "the US Pentagon", provides the product definition and industrial design innovation support to the ecological chain company. In terms of product definition, Xiaomi is good at industrial design, so it guides the ecological chain companies to innovate in product design. (…) Secondly, ecological chain companies, as "special forces", only need to focus on the familiar intelligent hardware field, and innovate and develop new ecological chain products." [11, p. 8]. Herewith, we consider the innovation of the ecological chain products as joint innovation.

At Xiaomi, the customer relationship goes beyond product development, and the Xiaomi phenomenon has been tremendously successful, including in China, where it is an exceptional success story [28].

3 Methodology

We approached the study via three perspectives:

- **How does the new technology arrive in our hands?** An in-depth search of PORDATA and other platforms in order to gather what we think are essential milestones for the success of Apple and Xiaomi;
- **How are we in touch with innovation?** The second approach was through a survey with 235 respondents from the Portuguese population;
- **Do we stand out from world trends?** For the third and last approach we used Google Trends databases to get an insight of reach tendencies, since 2004.

Our first research perspective was gained via research involving INE, Instituto Nacional de Estatística, PORDATA and other Portuguese databases, with the intent of gathering any information on the Portuguese smartphone market. Unfortunately, nothing came up until we founded a table of Mobile Plans, which gave us a tendency of the growing market since 1990. We united the 4th and previous generations of the telecommunications technology market giving us better and affordable plans that encourage the purchase and need to be connected. At the same time, we got to know the strategic devices launched by the brands we are studying.

A second research perspective aimed at getting to know better the reality of Portuguese choices and opinions about innovation, and thus we did a survey (with 193 valid responses) that aimed to compare with worldwide trends and cultural signs. The Portuguese population had passed quite roughly the 2009 economic crisis, and it impacted on our younger generations in a way that only ten years have passed and the oldest respondent to the survey was 20 years old at that time, probably being a university student at the time. The evolution of the internet and international brands makes us more aware of our surroundings and creates more valuable hard skills.

Our third research approach involved the following: since 2004, Google has given to anyone interested search databases so we can get to know the new trends. In this case, we used them to validate the story of the brands we chose to analyse and to provide some insights into the near future. There is no disclosure of the total numbers but they give us the percentage of the interest in particular themes. We can choose the location, the subject, the channels of the search, or the time window we need.

4 Data Analysis and Discussion

In this section, we intend to display our research in order to merge information and obtain conclusions about Closed and Open Innovation.

4.1 How Does the New Technology Arrive in Our Hands?

In order to answer this question, we went through Portuguese statistical databases, and, as a result, we found a timetable with the mobile plan service contracts.

For further inspection, we searched for mobile communication technologies, when they appeared in the market and the company associated with them. Therefore, we can see that a significant impact was with the introduction of 2G and 3G, that resulted in a significant growth of 1100 to 6 million and 10 million to 20 million contracts,

Table 1. Introduction in Portugal of mobile service and brands [1, 12–19]

				Cell phones on mass-market			
Years	Mobile Service Plans Signatures (1000)	Mobile Comunication Technology Penetration	First Company	Nokia	Apple	Samsung	Xiaomi
1990	7						
1991	13						
1992	37			**Nokia 1000/1011 GSM**			
1993	101	2G-GSM	TMN				
1994	174			Nokia 2010/2110			
1995	341						
1996	664			Nokia 1610		**SGH-250**	
1997	1507			Nokia 6110			
1998	4583			Nokia 5110		SGH-500	
1999	4671			Nokia 3210		SGH-2200	
2000	6665			Nokia 3310		A100	
2001	8356	2.5G -GPRS	TMN	Nokia 6610		N400	
2002	9202			Nokia 2300		T400	
2003	10003			Nokia 6600		D410	
2004	10571	3G	VODAFONE	Nokia 7610		D428	
2005	11368			Nokia N70		D820	
2006	12236			Nokia N73		E250	
2007	13477			Nokia N95	**iPhone 1° geração**	E898	
2008	14953			Nokia E63	iPhone 3G	i780	
2009	16051			Nokia 5230	iPhone 3GS	E1100	
2010	19686			Nokia N8	iPhone 4	Galaxy S	
2011	20034			Nokia N9	iPhone 4S	Galaxy Note	
2012	19861	4G	VODAFONE	Nokia 808 PureView	iPhone 5	Galaxy Note 2	**Mi 1S**
2013	19076			Nokia Lumia 520	iPhone 5S	Galaxy Note 3	Mi 2S
2014	18974			Nokia Lumia 930	iPhone 6	Galaxy Note 4	Mi 3S
2015	19352			Nokia 230	iPhone 6S	Galaxy Note 5	Mi Note Pro
2016	19928			Nokia 216	iPhone 7	Galaxy S7 EDGE	Mi Max
2017	19420			Nokia 6	iPhone X	Galaxy S8	Mi Mix 2
			Models Total 2019	**484**	**74**	**1206**	**115**

respectively. Therefore, this represents two mobile phones per Portuguese citizen. Of course, these contracts are divided into private and company plans.

From the communication organizations that were a pioneer at the time, just Vodafone has remained because TMN (Telecomunicações Móveis Nacionais) was merged with MEO that is currently a sub-brand of Altice. TMN was the first company in the world to present pre-charged plans for mobile phones.

We went a step further to include a focus on the brands Xiaomi and Apple, as well as on older competitors (Nokia and Samsung), for a better analysis and contextualization. Table 1 consolidates all of the information.

4.2 How Are We in Touch with Innovation?

We did a survey which had 193 respondents, focusing on young people from 18 to 30 years old. Using an infinite sample calculation with 50% proportion and an error of 10%, we estimated that 193 persons represent 2.5 million of the young generation, with 7.05% of error [24]. Therefore, our sample is reliable and robust. The formula we used was the following [25]:

$$n = \frac{Z\alpha^2 \times N \times p \times q}{i^2 \times (N-1) + Z\alpha^2 \times p \times q} = \frac{1,96^2 \times 2.500.000 \times 0.5 \times 0.5}{0,0705^2 \times (2500000 - 1) + 1,96^2 \times 0.5 \times 0.5}$$
$$\cong 193\, respondents$$

In our survey, the results are as follows:

- 55.4% of the survey respondents are female and 44.6% are male;
- Concerning the question about the smartphone brand that they own, the results are in Table 2.
- 28.5% think that Apple is the most innovative brand, followed by Samsung with 28.0% and 16.6% for Xiaomi;
- From the total survey takers, only 28.0% want to change their current smartphone;
- For those who want to change their current smartphone, 64.8% intend to stay with the same brand, being willing to spend around 750€ for a new device. The remaining 35.2% want to change the brand, being prepared to invest around 500€.

Table 2. Smartphone brands owned

Brand	Percentage (%)
Apple	31.1
Huawei	25.4
Samsung	20.2
Xiaomi	13.0
Nokia	1.5
Others	8.8

- In terms of qualities/functionalities, the subjects surveyed prefer a good smartphone camera (33.9%) as the main feature, followed by storage capacity and the duration of the battery, with 23.5% and 14.8% respectively;
- Using the total desired average price of 509€, we estimate the current market value of the survey takers at around 98 thousand euros. Extrapolating this number for the Portuguese population aged between 18–30 years, it gives a total market worth of 1.2 billion euros.

4.3 Do We Stand Out from World Trends?

Using Digital Marketing tools such as Google Trends (which explores what the [Google] world is searching for and shows how frequently a given search term is entered into Google's search engine relative to the site's total search volume), we did a close comparison to the Worldwide as well as Portugal's relevance in a search involving the brands Xiaomi, Apple, Nokia, Samsung and Huawei. The time window was 2004 until the most recent data available, October 2019. In Fig. 2 and Fig. 3 we can see that Nokia predominates for eight years until Samsung launched a wide variety of models and the Galaxy big screen in particular. Earlier, Apple began its smartphone launch in 2007 with its first-generation iPhone. The world was just thrilled with touch gestures and there being only one big button. The beginning of the end of "dumb phones" began. Although the Portuguese population was afraid of the uncertainty like everyone else, they saw the versatility of the IOS and Android operating systems. The App era launched the interest not only to be online but also to be connected to everyone.

There is some seasonality and frequency, meaning that twice each year we see an increase in searches, from Black-Friday until Christmas Eve and for every launch of a Galaxy or an iPhone, in February or in September, respectively. Worldwide, we recently saw the best performance ever of Apple at the time of the iPhone 11 launch, but it is just a peak. The Xiaomi rising started in 2014, and ever since the enterprise has continued to grow. For the companies that we studied they are the only one to use Open Innovation.

After a closer look, in the past 15-year plots (Fig. 4 and Fig. 5), we saw the significant tendency of Xiaomi's growth, and we think the stores opening worldwide may be one of the reasons. In Portugal, Xiaomi opened three stores this year with a meaningful impact in the local news and there are already plans to open more next year [20], and the population seems to like their variety of products. They had stood apart from the other four brands in Portugal and increased towards Christmas. This implies that they could be in a leadership position although Samsung is also on the rising.

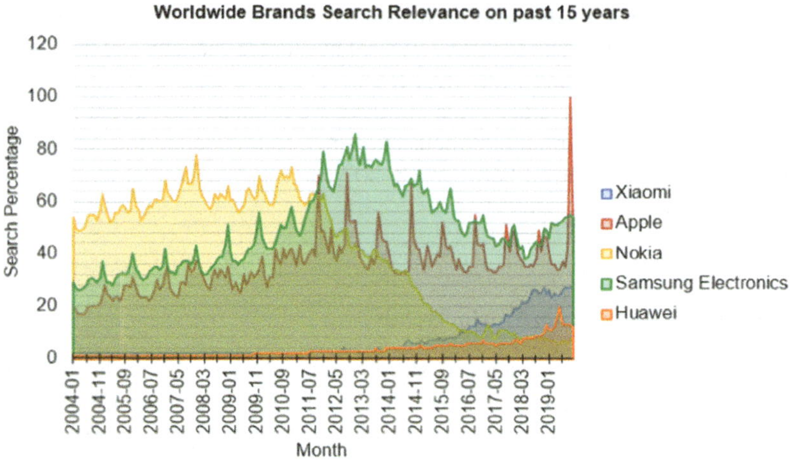

Fig. 2. Worldwide brands search relevance in the past 15 years [based on 21]

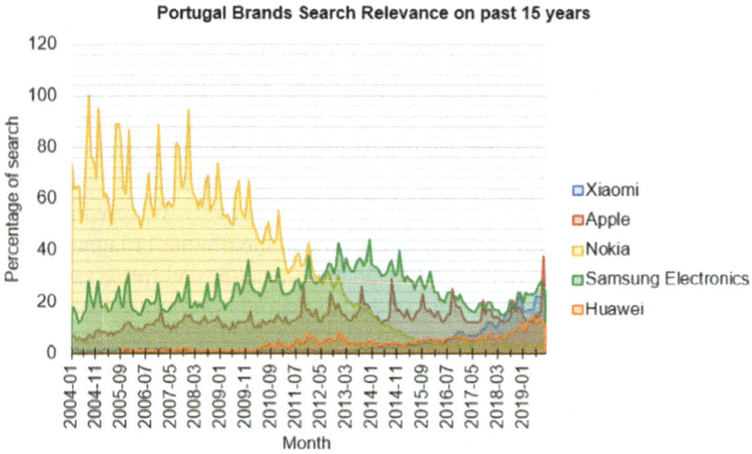

Fig. 3. Portugal brands search relevance in the past 15 years [based on 22]

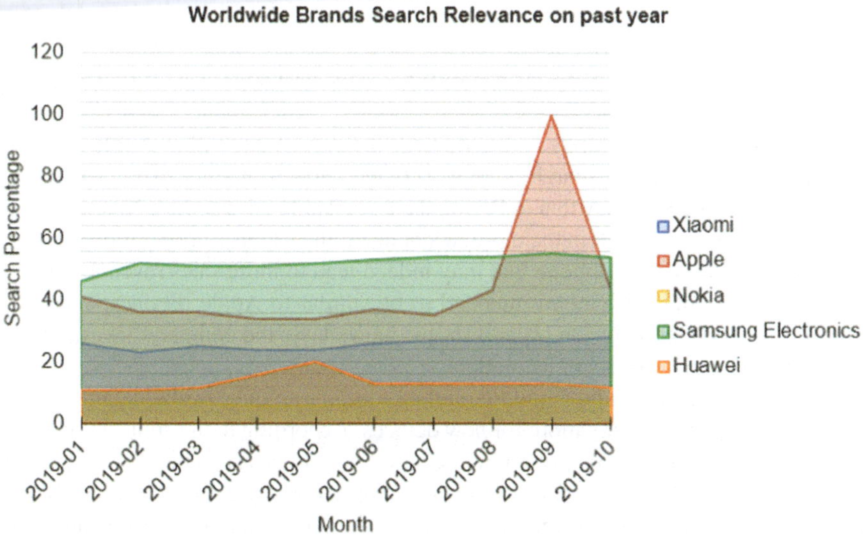

Fig. 4. Worldwide brands search relevance in the past year [based on 21]

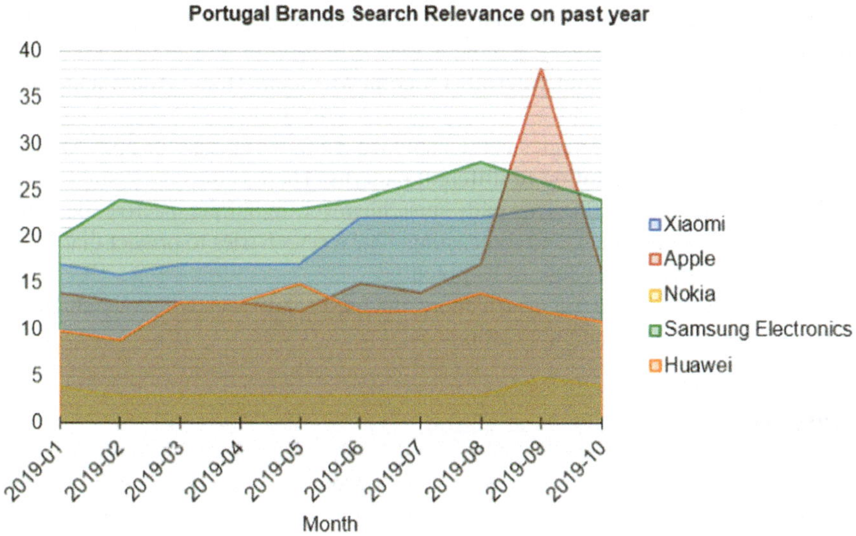

Fig. 5. Portugal brands search relevance in the past year [based on 22]

5 Conclusions

The main questions are yet to be answered; What is the best approach? Will it be enough to stay on top?

Considering the two brands being studied, the results show that Apple overtakes Xiaomi on every sales and search done worldwide. However, since 2012, Xiaomi has exhibited a growing tendency that in Portugal passes Apple on searches. Xiaomi operate in a middle price market but, in the more recent news, the Chinese corporation presented the first highest value smartphone that may indeed lead to more sales revenue [23].

The initial models of Xiaomi intended to be similar to Apple iPhones, and that is a well-known strategy between organizations. Nowadays, the company invests in new design and technology shifts in its models.

In Portugal, we saw that the mobile market is similar to the worldwide interests and that people are willing to pay more for new devices. According to the subjects surveyed, the main features that must be present in the new smartphones are a better camera, more storage capacity and a longer battery lifecycle. With an open innovation strategy, Xiaomi Mi Mix Alpha brings together the most requested features with a 108 MP camera and 512 GB of storage capacity. Apple, following a different approach (closed innovation), presented last September the iPhone 11 Pro Max with the same 12 MP camera and 64 GB ROM of the base model (iPhone 11), keeping almost the same specifications of previous models and trying to sell the iCloud service as an alternative for storage capacity.

It is important to highlight that the smartphone market is not the only sales revenue source of these brands. There are other products such as smartwatches (Mi Band and Apple Watch), the MacBook and the iCloud subscription service (Apple), scooters (Xiaomi M365) and even home utensils (Xiaomi Mi Smart Home Aqara and Xiaomi Mijia aspirator robot).

To conclude, the two strategies being analyzed led to the success of both Apple and Xiaomi, even though they followed different patterns. There is thus not a single unique path in order to be successful. This path must, however, in our perspective, be entirely focused on database feedback and forecasting. Our research tries to predict that Xiaomi should be under analysis because of its potential.

For further studies, we suggest a more intensive and extended research of both companies' environments and interactions.

References

1. PORDATA - Assinantes/equipamentos de utilizadores do serviço móvel. https://www.pordata.pt/Portugal/Assinantes+++equipamentos+de+utilizadores+do+serviço+móvel-1180. Accessed 17 Oct 2019
2. Gartner Says Global Smartphone Sales Continued to Decline in Second Quarter of 2019. https://www.gartner.com/en/newsroom/press-releases/2019-08-27-gartner-says-global-smartphone-sales-continued-to-dec. Accessed 28 Oct 2019
3. Rolik, Y.A.: A complex approach to evaluating the innovation strategy of a company to determine its investment attractiveness. Procedia – Soc. Behav. Sci. **99**, 562–571 (2013)
4. Le, H.T.T., Dao, Q.T.M., Pham, V.-C., Tran, D.T.: Global trend of open innovation research: a bibliometric analysis. Cogent Bus. Manag. **6**(1), 1–20 (2019)

5. Greco, M., Grimaldi, M., Cricelli, L.: Benefits and costs of open innovation: the BeCO framework. Technol. Anal. Strateg. Manag. **31**(1), 53–66 (2019)
6. Alawamleh, M., Bani Ismail, L., Aladwan, K., Saleh, A.: The influence of open/closed innovation on employees' performance. Int. J. Organ. Anal. **26**(1), 75–90 (2018)
7. Apple Reports First Quarter Results. https://www.apple.com/newsroom/2007/01/17Apple-Reports-First-Quarter-Results/. Accessed 30 Oct 2019
8. The secret origin story of the iPhone. https://www.theverge.com/2017/6/13/15782200/one-device-secret-history-iphone-brian-merchant-book-excerpt. Accessed 30 Oct 2019
9. Warren, T.: An exclusive look at an original iPhone prototype - Apple's red iPhone M68 in all its glory. The Verge, 19 May 2019. https://www.theverge.com/2019/3/19/18263844/apple-iphone-prototype-m68-original-development-board-red. Accessed 08 Jan 2020
10. Apple Reports Third Quarter Results. https://www.apple.com/newsroom/2019/07/apple-reports-third-quarter-results/. Accessed 30 Oct 2019
11. Ortiz, J., Ren, H., Li, K., Zhang, A.: Construction of open innovation ecology on the internet: a case study of Xiaomi (China) using institutional logic. Sustainability **11**(3225), 1–17 (2019)
12. All Xiaomi phones - page 3. https://www.gsmarena.com/xiaomi-phones-f-80-0-p3.php. Accessed 04 Nov 2019
13. All Samsung phones - page 15. https://www.gsmarena.com/samsung-phones-f-9-0-p15.php. Accessed 02 Nov 2019
14. anos de iPhone: veja a evolução dos modelos e os preços de lançamento. https://canaltech.com.br/smartphone/10-anos-de-iphone-veja-a-evolucao-dos-modelos-e-os-precos-de-lancamento-96070/l. Accessed 31 Oct 2019
15. The History of Mobile Phones From 1973 to 2008: The Phones That Made it All Happen. https://www.knowyourmobile.com/retro/the-history-of-mobile-phones-from-1973-to-2008-the-handsets-that-made-it-all-happen-d58/. Accessed 31 Oct 2019
16. Vodafone lançou serviços 4G em Portugal – Pplware. https://pplware.sapo.pt/informacao/vodafone-lancou-servicos-4g-em-portugal/. Accessed 31 Oct 2019
17. Mobile Connect Card introduz primeiro serviço 3G de dados em Portugal - Telecomunicações - SAPO Tek. https://tek.sapo.pt/noticias/telecomunicacoes/artigos/mobile-connect-card-introduz-primeiro-servico-3g-de-dados-em-portugal. Accessed 06 Nov 2019
18. Telemoveis.com - Como ter GPRS em Portugal. https://www.telemoveis.com/tecnologia/como-ter-gprs-em-portugal.html. Accessed 31 Oct 2019
19. Queluz, P., Pereira, F.: Introdução às Comunicações Móveis. Sistemas de Telecomunicações, pp. 1–99. Instituto Superior Técnico
20. Xiaomi. 'Portugal é um mercado-chave para expansão na Europa'. https://www.dinheirovivo.pt/empresas/xiaomi-portugal-e-chave-para-expansao-na-europa/. Accessed 04 Nov 2019
21. Xiaomi, Apple, Nokia, Samsung Electronics, Huawei - Explorar - Google Trends. https://trends.google.pt/trends/explore?q=%2Fm%2F0h6955b,%2Fm%2F0k8z,%2Fm%2F05b5c,%2Fm%2F01nn79,%2Fm%2F01qkl1. Accessed 03 Nov 2019
22. Xiaomi, Apple, Nokia, Samsung Electronics, Huawei - Explorar - Google Trends. https://trends.google.pt/trends/explore?date=all&geo=PT&q=%2Fm%2F0h6955b,%2Fm%2F0k8z,%2Fm%2F05b5c,%2Fm%2F01nn79,%2Fm%2F01qkl1. Accessed 03 Nov 2019
23. Xiaomi's Mi Mix Alpha is almost entirely made of screen - The Verge. https://www.theverge.com/circuitbreaker/2019/9/24/20881378/xiaomi-mi-mix-alpha-announced-wrap-around-screen. Accessed 04 Nov 2019
24. PORDATA - População residente, média anual: total e por grupo etário. https://www.pordata.pt/Portugal/Popula%C3%A7%C3%A3o+residente++m%C3%A9dia+anual+total+e+por+grupo+et%C3%A1rio-10. Accessed 13 Nov 2019
25. Vilelas, J.: Investigação - O processo de construção do conhecimento. Edições Sílabo, Lisbon (2009)

26. Gehani, R.R.: Corporate brand value shifting from identity to innovation capability: from Coca-Cola to Apple. J. Technol. Manag. Innov. **11**(3), 11–20 (2016)
27. Best global brands 2018 rankings. https://www.interbrand.com/best-brands/best-global-brands/2018/ranking/. Accessed 08 Jan 2020
28. Shih, C.C., Lin, T.M.Y., Luarn, P.: Fan-centric social media: the Xiaomi phenomenon in China. Bus. Horiz. **57**(3), 349–358 (2014)

Testing the Causal Map Builder on Amazon Alexa

Thrishma Reddy[1], Gautam Srivastava[2(✉)], and Vijay Mago[1]

[1] Department of Computer Science, Lakehead University, Thunder Bay, Canada
{tshivara,vmago}@lakeheadu.ca
[2] Department of Mathematics and Computer Science, Brandon University, Brandon R7A6A9, Canada
SRIVASTAVAG@BrandonU.CA

Abstract. When it comes to any problem, their causes, and solutions, people often have very different perspectives. Agreeing on the same course of action can sometimes be difficult. A causal map is a way to capture different perspectives people have about any situation. In this paper, we study the use of conversational artificial intelligence to capture and store the thought process of a particular problem. Our experiments consisted of developing a model for a voice-activated personal assistant (Alexa) that interacts, captures, and converts the responses of the participants into causal maps. Next, a detailed pre-test and post-test questionnaire that focuses on assessing interactions and willingness of the participants to collaborate with the developed model was conducted. Our results show that we were able to build an Alexa *skill* that could successfully capture participants thought process and transform it into a causal map that could be analyzed along with data from other participants.

Keywords: Causal maps · Intelligent systems · Artificial intelligence · NLP

1 Introduction

Problems come in all forms, easy or difficult. The difficult ones such as ecological management or obesity are tedious to work with and are often labeled as *complex* problems. While the natural sciences provide many definitions and tools to measure complexity, complex problems often share at least two traits which are central to this research. First, these complex problems are *multi-factorial*. The traditional reductionist approach that attempts to fix the 'root' cause does not lend itself well to a complex problem [12], and may even cause harm through unintended consequences [5]. Instead, the importance is often on mapping [7] and navigating [8] the *complex system of interactions* between factors that contribute to a problem of interest and factors affected by it. Second, dissemination and implementation research emphasizes that solutions to complex problems often

Á. Rocha et al. (Eds.): WorldCIST 2020, AISC 1159, pp. 449–461, 2020.
https://doi.org/10.1007/978-3-030-45688-7_46

require *coordinated actions between stakeholders* from multiple sectors (i.e., a multi-actor view [16]). For instance, actions regarding population and obesity involve sectors as varied as food production, city infrastructure (e.g., to promote walkable cities and access to fresh food), mental and physical well-being of people [19] and so on. Coordinated actions should produce a coherent policy, which implies that stakeholders work together at least by sharing a mission [4].

Causal maps are a widely used form of systems maps, in which concepts are represented as nodes, and their causal connections are captured through directed edges. Causal maps usually have a core concept, edges connecting them, polarity on the edges (+/−) and in some cases weight as well [3]. In Fig. 1 the core concept node is highlighted in red and the rest are nodes connected to it through causal connections.

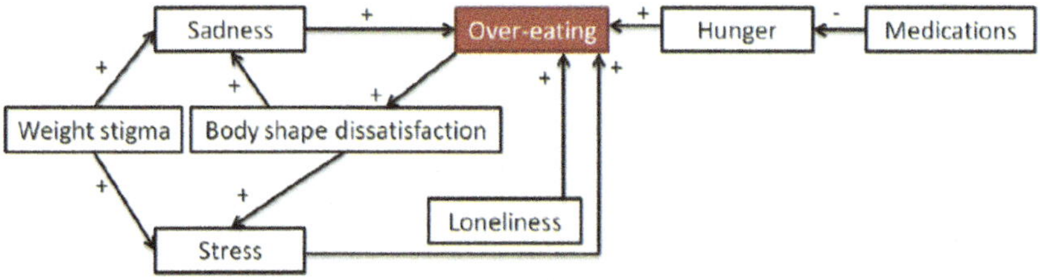

Fig. 1. Sample causal map where *over-eating* is the problem of interest [9].

2 Background and Motivation

Creating explanations of human learning experiences is an essential aspect of everyday life. These explanations, in turn, depend on their understanding of the given situation or circumstances where the learning takes place. This understanding is gained by individual experiences which are acquired through induction or deduction of a topic. Induction is drawing a conclusion based on previous experience [1]. Whereas a deduction is drawing a conclusion based on the models we have created in our minds based on our experience. By constructing abstract models or systems that abridge and summarize their features, we understand the different characteristics of everyday life. We tend to call these models or systems as ideas in the purest form. For instance, our abstract concept of a bird is a model or system for thinking about actual birds to make sense of their behaviour, as opposed to, say, the behaviour of cats, dogs, tortoises, beetles, and people. In short, our concepts provide our minds with systems for experiencing and thinking; our minds operate (reason) within them to investigate the world we experience with implications and consequences that are rich in meanings. Of course, a lot of this is done entirely automatically and subconsciously. During the formation of policy, either in a company or government, solutions to problems are shaped and debated. These problems can vary from *climate change* to *obesity*.

Instead of looking at its symptoms, it is essential to get to the root cause of the problem; because if not handled, the root cause will likely reoccur and make it challenging to handle. Hence to address this problem the policy must emphasize on finding the root cause of the problem first. The root cause can be determined by finding the causal factors of a problem. Causal factors can be determined by asking "why" to the problem statement. Once the causal factors are recognized the next step is to ask "why" questions again to the above-identified causal factors. Such as "why are there no jobs?" or "why is there inadequate access to clean water and nutritious food". The process of asking "why" needs to be continued until all the responses have been exhausted. You can refer to Fig. 2 for a clear understanding.

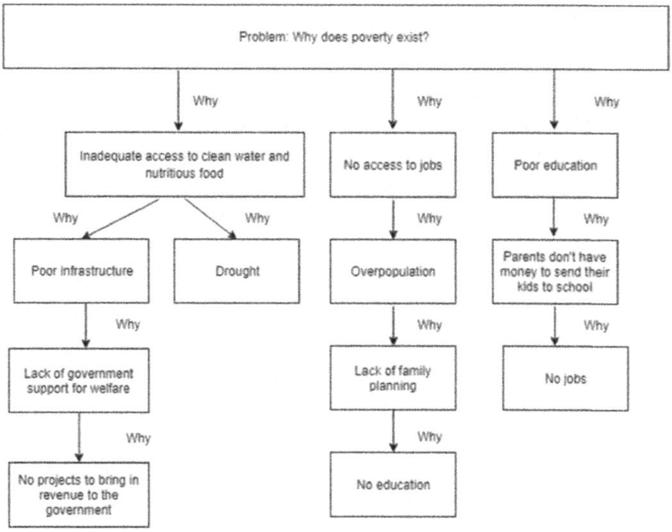

Fig. 2. Using the Why-tree process to uncover the underlying reasons for poverty

The process to produce a map as shown in Fig. 1 is relatively simple: participants create concept nodes and link them by indicating the causal relationship to be an increase ('+') or a decrease ('−') [13,14]. However, at least three issues may arise if the facilitator does not provide further guidance. First, participants need to choose node labels that have an unambiguous quantification: having 'more' or 'less' of this concept should be a straightforward notion. Second, users may forget about concepts that they already have, and add one with a similar name. Facilitators thus continuously monitor the maps to either avoid creating a redundant concept or merge them once they are discovered. Given the tremendous potential for (subtle) variations in language, discovering similar concepts is a difficult problem, particularly as the number of concepts increases [10]. Third, case studies have shown that cognitive limitations make it difficult for participants to think of structures such as loops and disjoint paths [2]. In particular, Ross observed how peculiar it was that "those who set policy think only acyclically, especially since the cyclical nature of causal chains in the real world has been amply demonstrated" [17]. Without paying particular attention to loops, participants may produce star diagrams with the one central problem at the core, and every other factor directly connecting to it. Facilitators may thus prompt participants extensively for relationships, to minimize the risk of missing loops or additional paths [11]. Participants interested in developing causal models have often done it with the support of a trained facilitator, who elicits concepts and causal relations [6]. Alternatively, tech-savvy participants may receive training and independently develop causal models using software such as `cMap` (common in education research), `MentalModeler` (most used in socio-ecological systems), or `Vensim` (typical in health and systems engineering). However, both approaches(trained facilitator and available software) have limitations. A trained facilitator can provide ample guidance but may be costly or unavailable. The software may be free and accessible anytime, but it does not guide the participant through the process of building a causal map. Also, both approaches rely on a visual inspection of the map as it is built, which does not easily scale as participants start to have many concepts and interrelationships. For instance, a participant may add a concept that is synonymous with an existing concept. To notice this redundant concept, all other concepts should be examined manually by the facilitator and participant, which becomes prohibitive as the number of concepts increases.

Thus, there is a need for an approach to causal model building that can be available at any time, without costs, and scales easily. In this research, we address this need by leveraging voice-activated virtual assistants (`Amazon Alexa`) to design and implement a *virtual facilitator* and justify the acceptance among human participants. Our solution guides participants in developing a model through a conversation (like a human facilitator), but is available at any time without cost (as software) and continuously examines the map to avoid typical issues such as synonymy of concepts.

3 Methodology

Our study was conducted with a small number of participants. Since the objective was to show Voice activated Personal Assistant (VAPA) that we developed [15] can capture the thinking process of a person regardless of their English dialect. We did not foresee any specific requirement for selecting participants apart from the standard preconditions such as no psychological or neurological disorders and normal English speaking capabilities. Two surveys pre-test and post-test surveys were designed. Pre-test survey was to understand each participant's previous experience with VAPA, and their knowledge about entities and causal map. The post-test survey was designed to capture the participants experience with using our *Build a Model* skill. We observed that during development phase of the tests a person found the questions asked by Alexa annoying when they had no idea regarding what information it was trying to capture from them. With this in mind, a user guide was designed which included – the workings of the skill, instructions on how to use and answer to the questions asked by Alexa, and an example conversation.

3.1 Hypothesis

The central goal of this research is to examine if it is possible to use VAPA to guide a participant in capturing their thought process of any problem without any guidance from a human facilitator. Based on this requirement, three null hypothesis was developed with respect to their experience with *Build a Model*:
$H_0{}^1$: There is no difference between participants with or without prior VAPA experience.
$H_0{}^2$: There is no difference between participants with or without experience with 3rd party skills/actions (Amazon/Google Home).
$H_0{}^3$: There is no difference between participants with or without the knowledge of causal maps and entities.

3.2 Materials

The pre-test survey consisted of 15 questions which were aimed at gathering the participant's demographic information as well as their experience with VAPA on a scale from 1 (not likely to use) to 10 (extremely likely to use). These ratings were expected to be correlated with participants experience of the skill's usefulness in real life asked in post-test survey. The survey also required the user to rate if they have read the user guide and understood the workings of the skill, its questioning pattern and the way it communicated its results. These results were compared with the result of the post-test survey to find whether participants found the skill's to be useful. This was needed as the skill's success rate depended on the participant's understanding of the skill. Once the user has read the user guide and taken the pre-test survey, they will then be able to proceed to the post-test survey. The post-test survey aims to capture the participant's experience with the skill to find out if they find Alexa skill to be a useful voice activated tool for converting the user's thoughts into causal maps. The results of the post-test survey will be compared with pre-test results to see the conditions which affected their reactions. Both surveys had a question that asked the participant to give their concerns or feedback, if any, could be used to enhance future skills. It was expected that each participant would select a problem of their choice to form causal maps of their thinking about the chosen issue.

3.3 User Study

Data from 13 participants who agreed to participate in this study have been included. Also as mentioned before participants were researchers and students studying at a Canadian University and few of them had completed their education. The skill was not deployed on the Amazon skill store and could be used only on one testing device which was handled by the student researcher. All the participants tested the skill and completed the survey under the supervision of the researchers at the Canadian University. The participants choose voluntarily to test the skill.

The participant's pools were roughly split into gender, with 38.46% of them being Female and 61.54% of them Male. Out of the female participants, 60% were native English speakers, and 40% were non-native English speakers. Out of the male participants, 25% were English speakers and 75% were non-native English speakers. A majority of 77% of the participants were pursuing their master's degree and aged between 22–27 years of age.

4 Results

All 13 participants responded to all the questions. What we have noticed during testing with all these participants is that Alexa has been unable to correctly identify a few of the words spoken by the participants as each person has a different way of pronouncing words when using English of different dialects. This created unintended entities to be captured, creating questions that could not be answered. For example, a non-native English speaker from India wanted to form a causal map for *hate* but Alexa consistently identified the word as *hit*, because it is an entity identified by Google Natural Language API. The next question from Alexa to the participant was "can you name the causes of hit?". This question made no sense since the word *hate* was incorrectly identified as *hit*. Build a Model has been designed in five different English dialects, including USA, Canada, India, Australia, and the UK. During testing the corresponding English language of the skill was enabled depending on the English dialect of the person. These dialects in Alexa represent that the pronunciation of each person with different dialects can be correctly identified by Alexa when the correct dialect in Alexa is used. We tried the skill with the English (Indian) dialect with an Indian participant hoping that *hate* would be identified as *hate* and not *hit*, but it was misidentified on each trail. We also tested the same with a native English speaker with English (CN) and noticed that *hate* was identified as *hate* but more complicated words were misidentified such as cardiovascular disease, lymphoma and so on. What we have understood from this is that words in different dialects can be identified by Alexa as long as we have created *intent* slots in Alexa skill during development, which we did not. The intent slots act as Alexa's reference guide that says these are the words that can be said by participants. Since we don't give Alexa any idea of what to capture, this affects her word identification ability.

4.1 Hypothesis Testing

To test our three hypothesis, we employee a 2×2 cross tabulation analysis along with Chi-square and Fisher's exact test. The cross tabulation is a descriptive analysis of two given variables with percentage description of the row, column and the total.

H_0^1 Null Hypothesis: Let us consider the $H_0{}^1$ null hypothesis: There is no difference between participants with or without prior VAPA experience in their experience with "Build a Model". To analyze this hypothesis, we need to cross-tabulate the participants previous VAPA experience and their experience with the skill after testing (Table 1). We may notice that having previous experience with a VAPA was irrelevant to finding full use of the skill or not. Since each participant had to read the user guide, we can say that the instructions provided in it helped guide the participant with the skill. We carried out a Chi-square and a Fisher's test statistically to further test the null hypothesis. As you can see in Table 5, the p-value of Chi-square test is 0.522 which is above the significance value (0.05) considered. Similarly, p-value from Fisher's test is greater than 0.05. Therefore we can accept the $H_0{}^1$ null hypothesis and say there is no significant relationship between the experience of a participant with VAPA and their experience with Build a Model. Which means if a person wants to have a positive experience with the skill, they don't need to have any VAPA experience.

Table 1. Cross tabulation of previous VAPA experience of participants with their response to if they find *Build a model skill* useful

			Post-test Q8: Do you think "Build a model" is a useful Alexa skill that can be used in the future? [Usefulness_Alexa_skill]			
			Yes	No	Maybe	Total
Pre-test Q6: Have you used voice-activated personal assistant before (such as Alexa, Google Home/mini, Apple Siri)? [VAPA_experience]	No	Count	1	0	1	2
		% within VAPA_experience	50.0%	0%	50.0%	100.0%
		% within Usefulness_Alexa_skill	11.1%	0%	25.0%	15.4%
		% of Total	7.7%	0%	7.7%	15.4%
	Yes	Count	8	0	3	11
		% within VAPA_experience	72.7%	0%	27.3%	100.0%
		% within Usefulness_Alexa_skill	88.9%	0%	75.0%	84.6%
		% of Total	61.5%	0	23.1%	84.6%
Total		Count	9	0	4	13
		% within VAPA_experience	69.2%	0%	30.8%	100.0%
		% within Usefulness_Alexa_skill	100.0%	0%	100.0%	100.0%
		% of Total	69.2%	0%	30.8%	100.0%

H_0^2 Null Hypothesis: Similar to the first null hypothesis test, we wanted to see if any experience with installing skills from the store affected the experience of the participants with Build a Model. Experience from installing skills from the skills store tells us about if a person knows how to invoke a skill, follow the given instructions and give Alexa proper commands accurately.

We performed a cross-tabulation of VAPA experience with skills and their Build a Model experience after testing (Table 2). What we could notice from the table is that most of the participants had never enabled a skill from the Alexa skill store which meant they did not know how to invoke a skill through an invocation name. For example: In Build a Model the invocation name is *Model Creator*. However, with the user guide provided they could follow along with the instructions properly. Most participants who have never enabled skills or participants who have enabled skills found Build a Model to be a useful skill in converting thoughts through voice input into causal maps. We can see in Table 5 that both Chi-square and Fisher's exact test p-value is more than 0.05. Which says there is no correlation between finding Build a Model useful and having any experience with installing and using skills from the Alexa skills store. $H_0{}^2$ has been supported but with the condition that the participants be provided with a user guide which gives proper instructions and explanation of Build a Model working.

Table 2. Cross tabulation of previous *VAPA experience* of participants with skills from the Amazon skill store and their response to if they find *Build a model skill* useful

			Post-test Q8: Do you think *Build a model*is a useful Alexa skill that can be used in the future? [Usefulness_Alexa_skill]			
			Yes	No	Maybe	Total
Pre-test Q7: If you have answered yes for the above question. Have you installed any skill for Alexa, or any action for Google home/mini? [VAPA_experience_with_skills]	No	Count	6	0	6	2
		% within VAPA_experience_with_skills	66.7%	0%	66.7%	100.0%
		% within Usefulness_Alexa_skill	66.7%	0%	66.7%	15.4%
		% of Total	46.2%	0%	46.2%	15.4%
	Yes	Count	3	0	1	4
		% within VAPA_experience_with_skills	75.0%	0%	25.0%	100.0%
		% within Usefulness_Alexa_skill	33.3%	0%	25.0%	30.8%
		% of Total	23.1%	0%	7.7%	30.8%
Total		Count	9	0	4	13
		% within VAPA_experience_with_skills	69.2%	0%	30.8%	100.0%
		% within Usefulness_Alexa_skill	100.0%	0%	100.0%	100.0%
		% of Total	69.2%	0%	30.8%	100.0%

H_0^3 Null Hypothesis: We wanted to see if there is a relation between "knowing what a causal map or an entity is" and "participants finding Build a Model helpful or not in converting their thoughts into causal maps just the way they wanted to". This was necessary because if the participant did not know what an entity is, he would rate the skill as not helpful as it depends on the entities being captured in the response of the participant. One of the participants, for example, responded *intrusive parents* as the cause of *stress* and only parents were captured from their response. If they did not know the skill captures only entities, then they will assume that the skill is not working accurately, which is not correct. During development testing of Build a Model, we came across this problem and added what an entity and a causal map is in the user guide for the participants to get an idea regarding them before testing.

We performed cross tabulations with "if they knew what an entity and a causal map is" against "if they found Build a Model useful", see Table 3 and Table 4. In these cross tabulations, we can see that the majority of the participants knew what a causal map (75.0%) and entity (66.7%) was and rated the skill as a helpful skill. 25% of participants who knew what a causal map is and 33.3% who knew what an entity is rated that the skill may be helpful to convert the thoughts into causal maps. These participants left comments such as - entity capturing needs to improve and does not capture every entity they indented Alexa to capture. Participants who lacked entity and causal maps knowledge also rated it as somewhat useful Alexa skill. Statistically, we can see in Table 5 that p-value of both Chi-square (0.488 for entity * usefulness and 0.118 for causal map * usefulness) and Fisher's exact test(1.00 for entity * usefulness and 0.308 for causal map * usefulness) is greater than our assumed significance value 0.05. Which supports our $H_0{}^3$ that states that there is no relationship or dependency of knowing what a causal map or entity is to have a positive experience with Build a Model.

Table 3. Cross tabulation of whether the participant knew what an *entity* was with if *Build a Model* skills were found to be useful

			Post-test Q8: Do you think *Build a model* is a useful Alexa skill that can be used in the future? [Usefulness_Alexa_skill]			
			Yes	No	Maybe	Total
Pre-test Q11: Do you know what an *entity* is? [Entity]	No	Count	1	0	0	1
		% within Entity	100.0%	0.0%	0.0%	100.0%
		% within Usefulness_Alexa_skill	11.1%	0.0%	0.0%	7.7%
		% of Total	7.7%	0.0%	0.0%	7.7%
	Yes	Count	8	0	4	12
		% within Entity	66.7%	0.0%	33.3%	100.0%
		% within Usefulness_Alexa_skill	88.9%	0.0%	100.0%	92.3%
		% of Total	61.5%	0.0%	30.8%	92.3%
Total		Count	9	0	4	13
		% within Entity	69.2%	0	30.8%	100.0%
		% within Usefulness_Alexa_skill	100.0%	0	100.0%	100.0%
		% of Total	69.2%	0	30.8%	100.0%

4.2 Summary of the Results

All three null hypotheses were conceptualized to see that "there is no significant relationship with any factors to have a positive relationship with Build a Model". It is confirmed to be true with statistical tests. We must note here, however, that this depends on the user's first reading the user guide, to gain an understanding of the working of the skill. 11 out of the 13 participants rated between 8–10 for their overall experience with Build a Model which was a positive experience. Only one participant gave a low rating of 3, but we noticed that this participant had not gone through the provided user guide even though it was mandatory. This may be one of the reasons why he gave such a low rating. This participant did not provide any comments to support his rating.

4.3 Implementation of Causal Maps Using Build a Model

When Alexa completes asking questions or if the user stops the conversation in the middle. The causal map generated during the conversation will be emailed, and the conversation will be stored in DynamoDb.

Figures 3a and 3b are few of the maps created by participants during testing. Participants also formed maps for obesity, climate change, hate, water scarcity, fashion, stress, smoking, depression and headache. In Fig. 3a there were two layers of questions asked by Alexa. *Drinking alcohol* has been sent to Google Natural Language API and *alcohol* has been identified as the entity, so Alexa has developed a causal map for *alcohol* even though the map was for *drinking alcohol.* There are two layers of questions asked by Alexa which is one for *alcohol* and then questions for the causes answered. Then the participant has requested to stop the questions once the second layer of questions were answered. Even though Build a Model identified *alcohol* as the core problem and proceeded to ask what causes alcohol? The participant knew why it was asking such a question and proceed to answer without stopping the skill thinking it was not working properly. The causal map data is stored in JSON format in AWS DynamoDb table. Figure 3 shows a few causal maps generated by the participants during testing.

Table 4. Cross-tabulation of whether the participant knew what a *causal map* was with if *Build a model* skill were found to be useful

			Post-test Q8: Do you think *Build a model* is a useful Alexa skill that can be used in the future? [Usefulness_Alexa_skill]			
			Yes	No	Maybe	Total
Pre-test Q11: Do you know what a *causal map* is? [Causal_map]	No	Count	0	0	1	1
		% within Causal_map	0.0%	0.0%	100.0%	100.0%
		% within Usefulness_Alexa_skill	0.0%	0.0%	25.0%	7.7%
		% of Total	0.0%	0.0%	7.7%	7.7%
	Yes	Count	8	0	3	12
		% within Causal_map	66.7%	0.0%	25.0%	100.0%
		% within Usefulness_Alexa_skill	88.9%	0.0%	75.0%	92.3%
		% of Total	61.5%	0.0%	23.1%	92.3%
Total		Count	9	0	4	13
		% within Causal_map	69.2%	0.0%	30.8%	100.0%
		% within Usefulness_Alexa_skill	100.0%	0.0%	100.0%	100.0%
		% of Total	69.2%	0.0%	30.8%	100.0%

Table 5. Chi-square and Fisher's exact test using SPSS software

	Chi-square statistic value	p-value (Chi-square)	p-value (Fisher's Exact test)
VAPA experience * Usefulness	0.410	0.522	1.0
VAPA experience with skills * Usefulness	0.090	0.764	1.0
Knowledge of Entity * Usefulness	0.481	0.488	1.0
Knowledge of Causal Map * Usefulness	2.438	0.118	0.308

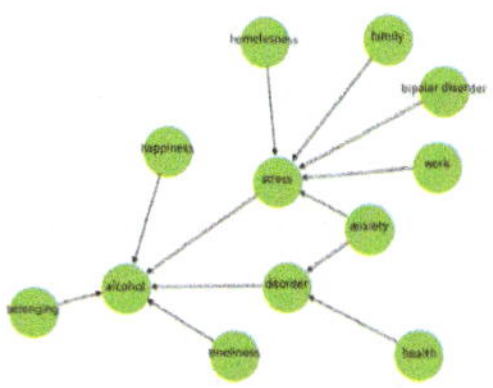

(a) *Drinking Alcohol*

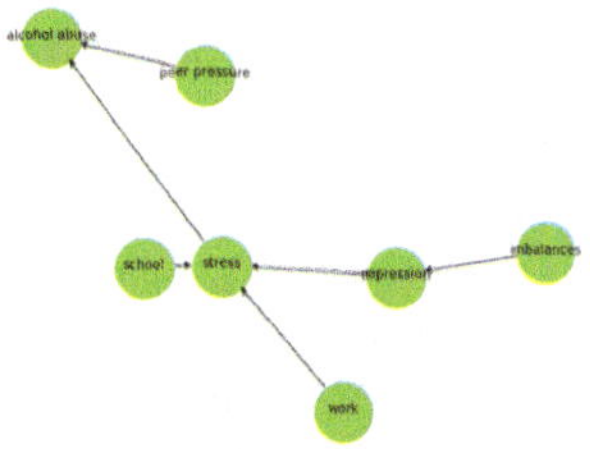

(b) *Alcohol Abuse*

(c) *Hate*

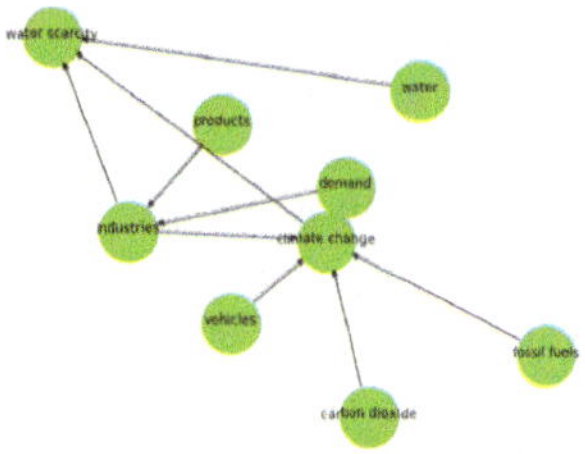

(d) *Climate Change*

Fig. 3. Causal maps created by *Build a model*

5 Conclusion and Future Directions

Build a Model can be used by the government/any company as a survey tool to extract reasoning for a specific problem. Data from a specific location can produce causal maps of an issue that can be used to understand what people at that location think about a particular problem [18]. For example, Anti-vaxxers movement can be analyzed using this tool if people participate in answering Alexa's questions. We can look at questions like - What if we were able to capture what goes on in people's head? and then aim at such people to educate them if they have a wrong understanding of a particular problem's causes. Creating causal maps using voice conversation has never been done before. We have used the available technologies to create *Build a Model* which can replace a human facilitator in creating a causal map for a particular problem. In short we use

Alexa to capture the user's response, converts it to text and this text is converted into causal map. We performed a test with the participants to see if what we have created is a useful skill that can be used in real life to convert a human's thoughts into causal maps. With the test data we could confidently say that participants who tested the skill found it a useful skill.

References

1. How to conduct a root cause analysis. https://www.thecompassforsbc.org/how-to-guides/how-conduct-root-cause-analysis. Accessed 01 Apr 2019
2. Axelrod, R.: Decision for neoimperialism: the deliberations of the British Eastern Committee in 1918. In: Structure of Decisions: The Cognitive Maps of Political Elites, pp. 77–95 (1976)
3. Dikopoulou, Z., Papageorgiou, E., Mago, V., Vanhoof, K.: A new approach using mixed graphical model for automatic design of fuzzy cognitive maps from ordinal data. In: 2017 IEEE International Conference on Fuzzy Systems (FUZZ-IEEE), pp. 1–6. IEEE (2017)
4. Dube, L., et al.: From policy coherence to 21st century convergence: a whole-of-society paradigm of human and economic development. Ann. N. Y. Acad. Sci. **1331**(1), 201–215 (2014)
5. Fink, D.S., et al.: Wrong answers: when simple interpretations create complex problems. In: Systems Science and Population Health, pp. 25–36 (2017)
6. Frerichs, L., et al.: Mind maps and network analysis to evaluate conceptualization of complex issues: a case example evaluating systems science workshops for childhood obesity prevention. Eval. Program Plan. **68**, 135–147 (2018)
7. Giabbanelli, P.J.: Analyzing the complexity of behavioural factors influencing weight in adults. In: Giabbanelli, P., Mago, V., Papageorgiou, E. (eds.) Advanced Data Analytics in Health, pp. 163–181. Springer, Cham (2018)
8. Giabbanelli, P.J., Baniukiewicz, M.: Navigating complex systems for policymaking using simple software tools. In: Giabbanelli, P., Mago, V., Papageorgiou, E. (eds.) Advanced Data Analytics in Health, pp. 21–40. Springer, Cham (2018)
9. Giabbanelli, P.J., et al.: Creating groups with similar expected behavioural response in randomized controlled trials: a fuzzy cognitive map approach. BMC Med. Res. Methodol. **14**(1), 130 (2014)
10. Giabbanelli, P.J., et al.: Overcoming the PBL assessment challenge: design and development of the incremental thesaurus for assessing causal maps (ITACM). Technol. Knowl. Learn. **24**, 1–8 (2017)
11. Giabbanelli, P., et al.: Developing technology to support policymakers in taking a systems science approach to obesity and well-being. Obes. Rev. **17**, 194–195 (2016)
12. Heitman, K.: Reductionism at the dawn of population health. In: Systems Science and Population Health, p. 9 (2017)
13. Mago, V.K., Bakker, L., Papageorgiou, E.I., Alimadad, A., Borwein, P., Dabbaghian, V.: Fuzzy cognitive maps and cellular automata: an evolutionary approach for social systems modelling. Appl. Soft Comput. **12**(12), 3771–3784 (2012)
14. Mago, V.K., Morden, H.K., Fritz, C., Wu, T., Namazi, S., Geranmayeh, P., Chattopadhyay, R., Dabbaghian, V.: Analyzing the impact of social factors on homelessness: a fuzzy cognitive map approach. BMC Med. Inform. Decis. Mak. **13**(1), 94 (2013)

15. Reddy, T., Giabbanelli, P.J., Mago, V.K.: The artificial facilitator: guiding participants in developing causal maps using voice-activated technologies. In: International Conference on Human-Computer Interaction, pp. 111–129. Springer (2019)
16. Riley, B., et al.: Systems thinking and dissemination and implementation research. In: Dissemination and Implementation Research in Health: Translating Science to Practice, p. 143 (2017)
17. Ross, S.: Complexity and the presidency. In: Axelrod, R. (ed.) The Structure of Decision: Cognitive Maps of Political Elites, pp. 96–112. Princeton University Press, Princeton (1976)
18. Sharma, G., Srivastava, G., Mago, V.: A framework for automatic categorization of social data into medical domains. IEEE Trans. Comput. Soc. Syst. **7**, 129–140 (2019)
19. Verigin, T., et al.: Supporting a systems approach to healthy weight interventions in British Columbia by modeling weight and well-being. In: Proceedings of the 49th Annual Simulation Symposium, p. 9. Society for Computer Simulation International (2016)

Assessing the Communicative Effectiveness of Websites

Antonio Sarasa, Ana Fernández-Pampillón, Asunción Álvarez, and José-Luis Sierra(✉)

Universidad Complutense de Madrid, 28040 Madrid, Spain
{asarasa,apampi,jlsierra}@ucm.es, asuncion.alvarez@gmail.com

Abstract. Websites are the main mechanisms used by companies and institutions to communicate their activities to the world. For this reason, it is critical to ensure that the message being conveyed by the website is exactly the one intended by the institution. However, analyzing what is being communicated is complex given the semantic and contextual factors that intervene. To face this concern, this paper describes a method that makes it easier for a human expert to evaluate the communicative effectiveness of a website. Since, in most websites, textual content is used to explain the institution's main mission and activities to visitors, the method heavily relies on a quantitative analysis of the textual web content. Once this analysis finishes, the result is represented by a semantic network that takes account of the conceptual structure underlying the website. In addition to presenting the method, we describe a web application prototype that implements it, as well as we evaluate it with a meaningful case study.

Keywords: Content text processing · Communicative effectiveness · Website analysis · Text mining

1 Introduction

The popularization of the Internet has had a significant impact on the way in which institutions and companies communicate with their stakeholders. In this respect, websites are a basic tool to convey institutional messages and corporate images [7, 14, 18]. Through websites, organizations can convey different semantic messages, such as showing that a company is dynamic, committed to the environment, or aimed at a specific type of audience, among others. Therefore, organizations must take special care with the content displayed on their websites, as it will be the main entry point for any visitor approaching them and, therefore, it will generate a first, but still decisive, impression about the organization and its activity.

Establishing exactly how the contents of a website are perceived and interpreted is a complex task due to the semantic and contextual aspects involved. The combination of these aspects shapes and influences the message that is conveyed [29]. In this sense, although the content formats of a website can be diverse (videos, downloadable documents, iconographies, images, audio, text, etc. [21]), it is textual content that typically

Á. Rocha et al. (Eds.): WorldCIST 2020, AISC 1159, pp. 462–471, 2020.
https://doi.org/10.1007/978-3-030-45688-7_47

defines the messages that should be conveyed to readers, while the rest of the content serves to complement or clarify this message [26]. For this reason, one way to analyze the communicative effectiveness of a website is to analyze the communicative effectiveness of the text it contains [2, 20, 28]. Therefore, this paper proposes an approach that is aimed at helping a human expert (e.g., a linguist) evaluate the communicative effectiveness of the contents of a website by integrating a well-established set of text processing techniques within a semi-automated analysis workflow. By the "communicative effectiveness" of a text, we mean the ability of a text to communicate the message that its source (institution, company, etc.) actually wants to convey. For this purpose, a basic linguistic analysis followed by a quantitative analysis of the website text content are carried out, and the results are represented as a *semantic network*: a graph of concepts connected by co-occurrence distances.

The paper is structured as follows. Section 2 describes the proposed approach. Section 3 describes a web application that implements this approach. Section 4 provides some evaluation results. Section 5 outlines some works related to ours. Finally, Sect. 6 outlines some conclusions and lines for future work.

2 The Communicative Effectiveness Assessment Method

The main goal of the approach presented in this paper is to assist a human expert in the evaluation of the communicative effectiveness of a website. For this purpose, the expert is assisted during the semantic analysis of the textual content of the website with the aim of determining its communicative effectiveness. The approach is domain-neutral, in the sense that it should be applicable to any website, in any informational domain. In addition, while we have tested the approach with websites in Spanish, it should also be easily adaptable to other languages.

In our approach, the semantic analysis of the text begins with a quantitative analysis of both the whole text and the parts into which the text is divided, which identifies the Most Frequent Concepts (MFCs) in the text. This enables the expert, usually a linguist, to select those concepts that can facilitate the representation and interpretation of the results. Then, a second quantitative analysis is carried out to determine the most frequent contextual collocations or co-occurrences among the MFCs selected by the expert and the other concepts of the text. As a result, a semantic network is generated, where nodes represent concepts and arcs represent proximity relationships between the connected concepts. The expert can use this semantic network to determine: (i) the underlying discursive threads in the whole text and its parts; and (ii) the topic or topics of each thread. This text patterns of threads and topics are used to determine where the company message is effectively conveyed (i.e., where it is communicated in an accurate and consistent way).

Figure 1 outlines the workflow of the approach. In this workflow, the proposed textual semantic analysis method consists of the following steps:

- *Text extraction.* In this step the web pages in the *website* to be analyzed are formatted to remove all structural, descriptive or format tags, and are converted into plain *text*. The rationale here is that if the text is not consistent (for example, because there is no

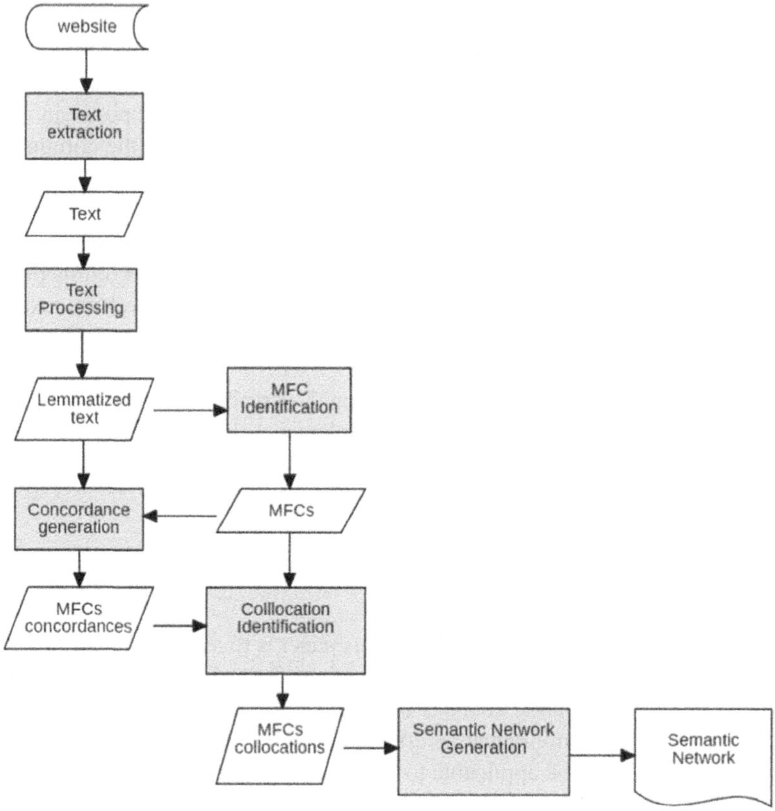

Fig. 1. Processing workflow

coherence among the concepts in a title and those in the body of the text), this will be easier to detect, because concepts are not "masked" by the stronger communicative effect of prominent textual elements (e.g. the title).

- *Text processing*. In this step, the extracted text is fed into a standard natural language processing pipeline, in which: (i) words are filtered in accordance with a stop list; and (ii) the resulting words are disambiguated and lemmatized. As a result, the raw text is converted into a *lemmatized text*, consisting of a sequence of lemmas (i.e., canonical forms of words) associated with the representative words for the text.
- *MFC identification*. In order to identify MFCs, the frequencies (i.e., number of occurrences in the text) of simple lemmas, as well as of *bigrams* (groups of two consecutive lemmas) and *trigrams* (groups of three consecutive lemmas) are computed. This frequency list is then sorted in descending order by frequency. The ordered list is presented to the expert, who selects the most relevant concepts. These concepts are the MFCs.
- *Concordance generation*. In this step, the *concordances* of each MFC (i.e., the contexts in which this MFC occurs) are established. For this purpose, each occurrence of a MFC *c* in the lemmatized text determines a potential concordance, which is given by the

text fragment $\alpha\ c\ \beta$ containing c (the exact number of lemmas included in α and β is a parameter of the process).

- *Collocation identification*. In this step, the MFC concordances are used to compute the *collocations* of each pair of MFCs c_1 and c_2 in each concordance γ. Such collocations are summarized by tuples (c_1, c_2, γ, n), where n is the number of times that c_1 and c_2 occur together in the concordance γ (of all the possible tuples of this form, only those with $n > 0$ make sense). In addition to generating these MFC *collocations*, the process gives the expert the opportunity of reviewing them to select the most significant ones.
- *Semantic network generation*. Finally, during this step a *semantic network* is generated from the MFC *collocations* identified. This network is an undirected graph in which: (i) nodes are given by MFCs; (ii) for each tuple (c_1, c_2, γ, n) summarizing the collocations of c_1 and c_2 in γ, an edge is generated that connects c_1 and c_2; this edge is labelled by γ:n, and n is called the *semantic weight* of the edge. This network provides valuable information to the expert. In particular, it makes it possible to determine the *semantic degree* of a concept c (i.e., the number of edges connected to c), as well as the *semantic weight* of the connection between two concepts c_1 and c_2 (i.e., the sum of the semantic weights of all the edges connecting c_1 and c_2).

3 Implementation

We have implemented the method described in the previous section as a web application. The functionalities of this application can be summarized as follows (see Fig. 2):

- The application makes it possible to either select a website to be analyzed or input a plain text for testing purposes (Fig. 2a). The website or the text entered is then processed to generate the list of potential MFCs, ordered by frequency. For informative purposes, the user can access the result of the *text processing* step (Fig. 2b). In addition, to achieve maximum accuracy, the expert performs word disambiguation.
- Once the list of candidate MFCs is available, the expert can select the more relevant ones to establish the final MFCs (Fig. 2c). For this purpose, the frequency of the MFCs is displayed (see Fig. 2d). Notice that, as indicated in the previous section, these MFCs can be both single lemmas, bigrams, or trigrams.
- Once the MFCs have been established, the application generates the information regarding the MFC collocations, and it gives the expert the opportunity to review and adjust it (Fig. 2e). This process can be customized by setting the maximum length of the concordances, as well as the minimum frequency from which collocations of two concepts are considered significant.
- Finally, the collocations are used to generate the semantic network (Fig. 2f).

From a technical point of view, the application was implemented using a typical Model-View-Controller scheme with PHP-Apache-MySQL technology, in which: (i) requests made in the client are transferred in the server, via PHP scripts, to a MySQL database using SQL queries; (ii) the returned data is processed in the server; and (iii) results are sent back to the client in the form of HTML pages.

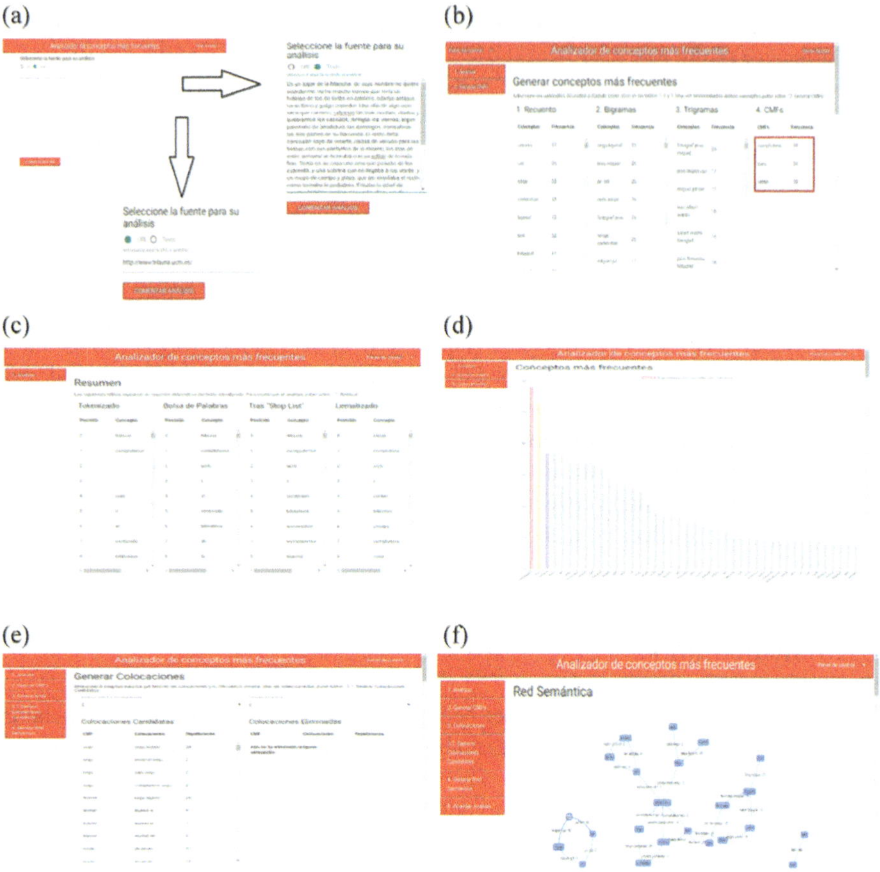

Fig. 2. Screenshots of the application (labels are in Spanish): (a) source selection; (b) summary of the text processing step; (c) MFCs selection; (d) histogram of candidate MFCs; (e) summary of collocations; (f) semantic network

4 Evaluation

For the evaluation of the effectiveness and usability of the approach, a real case study was carried out, specifically the semantic analysis of a Spanish-based multinational company's website (the identity of the company and the details of the specific content –e.g., the specific lemmas of the selected MFCs and concordances- are omitted by confidentiality reasons). The analysis was carried out by a linguist who had user-level technical knowledge.

The website analyzed was structured in nine sections. The distribution of words per section is shown in Fig. 3. Once processed, 20 MFCs were selected. Figure 4 summarizes the frequencies of these MFCs. As regards the collocation analysis, MFC collocations were identified in 19 different concordances. This information is summarized in the semantic network of Fig. 5. In this semantic network, the linguist realized how relationships among MFCs represent the dominant discourse of the text.

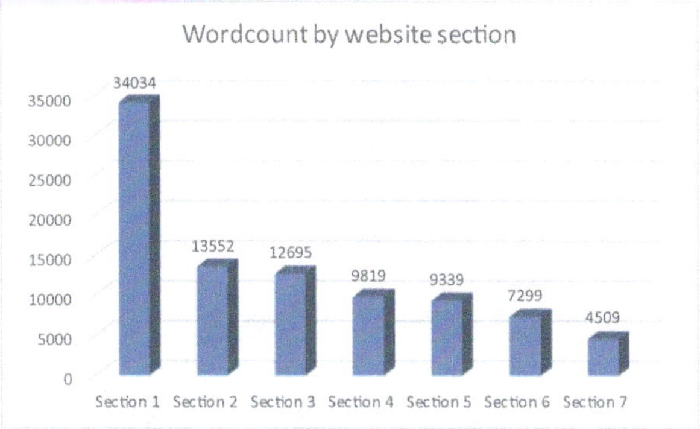

Fig. 3. Number of words per section in the case-study website

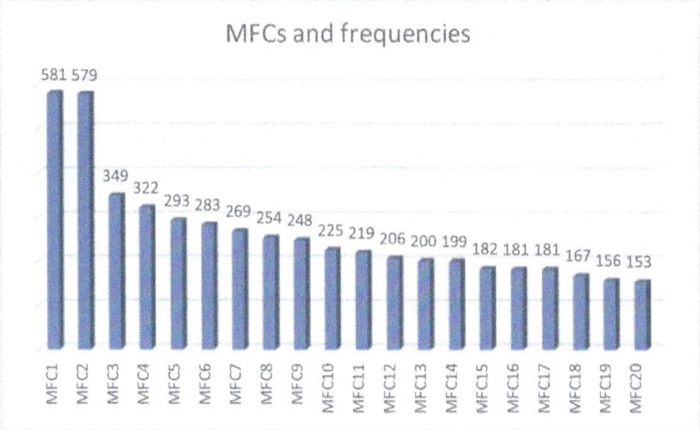

Fig. 4. Distribution of the selected MFCs (for confidentiality reasons actual names are hidden for sections and MFCs: Sect. 1, Sect. 2, MFC1, MFC2 … etc. are used instead)

This case-study, carried out on a main company website, shows how the linguist was able to locate and analyze the various discursive lines, their themes and the integration, and the coherence among them and with the global website discourse. It is also worth noting that the linguist was able to detect, in the semantic network, some discursive elements that did not "fit" the general discourse of the corporation. These elements constitute the so-called subtext or latent discourse of the web which are critical because the latent discourse may not be perceived but the underlying message it conveys is not the desired one and, consequently, it could reduce the website's communicative effectiveness. In this regard, the approach makes it possible for the expert to make recommendations to improve communication by integrating the website's latent manifest discourses.

Regarding the usability of the web application implementing the approach, even though the application was successfully used by the linguist (who is a non-computer specialist user) with no technical support, several weak points which might be improved were identified. Firstly, the need to improve the lemmatization stage in order to optimize

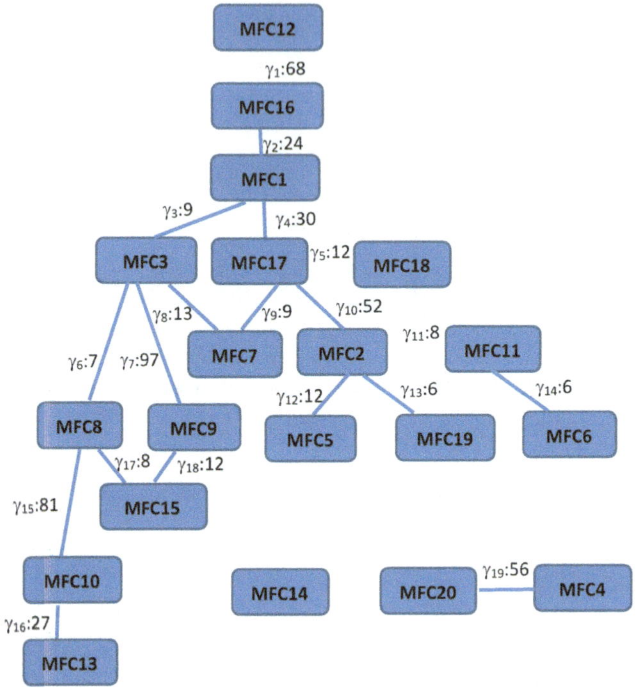

Fig. 5. Semantic network for the case-study website (actual MFCs and concordances intentionally not shown to maintain confidentiality)

the selection of concept candidates was clear. Also, the expert pointed out the convenience of visualizing the way in which the web pages were preprocessed, as well as the possibility of working with text in different file formats, not only html and plain text (e.g., to be able of processing PDF or other text format files). The expert also remarked on the convenience of generating reports as PDF files, including both the semantic network and other analyses (e.g., a collocation report). Finally, the expert also highlighted the convenience of specifying a maximum level of analysis in website navigation trees, as well as the option of restricting the analysis to specific sections of the website in order to carry out comparisons between the communicative messages of those sections.

5 Related Work

Various types of textual analysis have been proposed to extract semantic information from texts which try to find certain linguistic characteristics or patterns in the text analyzed, and which constitute the field of *text mining* [19]. Text mining includes, among others, the following operations [4]: (i) creation and preprocessing of the text corpus that will serve as the basis for the analysis [15, 17]; (ii) recognition of linguistic patterns, mainly lexical patterns [25]; (iii) identification and extraction of the concepts underlying the text [1]; (iv) disambiguation of the concepts when necessary [12]; (v) identification and extraction of relationships between the identified concepts [2]; and (vi) quantitative analysis of the concepts found, to establish their frequency distributions [3]. For instance, a popular application of text mining is sentiment analysis [22], which consists in establishing the proximity of the semantics of a text with respect to a previously fixed concept (or concepts) based on the frequency of words related to the fixed concept [6, 24]. The work proposed in this paper can be described as falling within the field of text mining. However, rather than fully automating the analysis of a text source, our aim is to provide assistance during the analysis of a website by a knowledgeable human expert (typically a linguist).

A way to depict graphic representations between concepts is a *context graph*. Such a graph represents the relationships between a set of concepts (e.g. countries) as reflected in a corpus with respect to a given context (e.g. crude oil) [10]. In order to obtain the semantic representation of a text as a context graph, two main approaches have been proposed: graph-theoretical techniques for web content mining [23] and conceptual graphs [27]. Web content mining [13] proposes six methods that are based on the adjacency of terms in a webpage [8]: standard, simple, n-distance, n-simple distance, absolute frequency, and relative frequency. These methods provide representations of texts only as words and simple relations (unknown relations) between words [16]. Conceptual graphs, on the other hand, provide a more descriptive representation of the semantics of a text [9], based on concepts (not words) and well-defined semantic relationships (such as synonymy, hyperon/hyponymy etc.). However, the complexity and cost of linguistic analysis is considerable, which makes it less effective when it comes to analyzing large quantities of text. In addition, the resulting graphs may be difficult for humans to read due to the detail of the representations [17]. In this regard, the approach proposed in this paper lies between web content mining and conceptual graphs. This approach is a simple, but still efficient, way to assist human experts in the difficult task of analyzing the communicative effectiveness of websites.

6 Conclusions and Future Work

This work provides a methodology to facilitate the analysis of the communicative effectiveness of a website based on creating, by means of a quantitative analysis, a semantic network that represents the most frequent concepts of the text and their collocations. It has been possible to prove, in a case study, the effectiveness of the semantic network in identifying and analyzing the discursive elements of the text in order to establish its communicative effectiveness and identify potential points for improvement. Specifically, the network made it possible to identify the conceptual "skeleton" of the dominant discourse in the website. It was also possible to analyze the thematization of certain concepts (i.e., whether a concept becomes the main subject of a text and, therefore, if it is consistent with the message to be conveyed). The approach has been implemented in a web application, which has been successfully used by an expert linguist in the analysis of a non-trivial, real-world, website.

We are currently working on incorporating of the improvements suggested by the expert in the web application. We are also refining our approach by addressing the relationships between MFCs and other, non-frequent, concepts in the text. The study of these relationships can be useful to identify the secondary or marginalized discourses that provide additional information about the background discursive lines of the website, which can reduce the communicative effectiveness of the text. We are also addressing different strategies to compare the analysis of different parts of a website in order to verify the degree of coherence and overlap between the discourses of the different sections of a website, as well as with respect to the overall discourse. This analysis can be useful to identify any semantic "gaps" between some sections and the rest of a website. Finally, we plan to apply the techniques developed in this paper to help experts to formulate cataloguing schemata in the framework of our *Clavy* tool for managing reconfigurable digital collections [5, 11].

Acknowledgment. This work has been supported by the Research Project TIN2017–88092–R.

References

1. Amensisa, A.D., Patil, S., Agrawal, P.: A survey on text document categorization using enhanced sentence vector space model and bi-gram text representation model based on novel fusion techniques. In: 2018 2nd International Conference on Inventive Systems and Control (ICISC), pp. 218–225. IEEE (2018)
2. Arendt, F., Karadas, N.: Content analysis of mediated associations: an automated text-analytic approach. Commun. Methods Meas. **11**(2), 105–120 (2017)
3. Bauer, C., Scharl, A.: Quantitive evaluation of web site content and structure. Internet Res. **10**(1), 31–44 (2000)
4. Berendt, B., Hotho, A., Stumme, G.: Towards semantic web mining. In: International Semantic Web Conference, pp. 264–278. Springer, Heidelberg (2002)
5. Buendía, F., Gayoso, J., Sierra, J.L.: Generation of standardized e-learning content from digital medical collections. J. Med. Syst. **43**(188), 8 (2019)
6. Cambria, E., Poria, S., Hazarika, D., Kwok, K.: SenticNet 5: discovering conceptual primitives for sentiment analysis by means of context embeddings. In: Thirty-Second AAAI Conference on Artificial Intelligence (2018)
7. Capriotti, P., Moreno, A.: Communicating corporate responsibility through corporate web sites in Spain. Corp. Commun. **12**(3), 221–237 (2007)
8. Cela, K.L., Sicilia, M.Á., Sánchez, S.: Social network analysis in e-learning environments: a preliminary systematic review. Educ. Psychol. Rev. **27**(1), 219–246 (2015)
9. De Diego, I.M., Fernández-Isabel, A., Ortega, F., Moguerza, J.M.: A visual framework for dynamic emotional web analysis. Knowl.-Based Syst. **145**, 264–273 (2018)
10. Feldman, R., Sanger, J.: The Text Mining Handbook: Advanced Approaches in Analyzing Unstructured Data. Cambridge University Press, Cambridge (2007)
11. Gayoso, J., Rodríguez, C., Sierra, J.L.: Browsing digital collections with reconfigurable faceted thesauri. In: Gołuchowski, J., et al. (eds.) Complexity in Information Systems Development. LNISO, vol. 22, pp. 69–86. Springer, Cham (2017)
12. Gómez, P.C., Sánchez, M.A. (eds.): Lexical Collocation Analysis: Advances and Applications. Springer, Cham (2018)
13. Herring, S.C.: Web content analysis: expanding the paradigm. In: Hunsinger, J., Klastrup, L., Allen, M. (eds.) International Handbook of Internet Research, pp. 233–249. Springer, Dordrecht (2009)
14. Hwang, J.-S., McMillan, S.J., Lee, G.: Corporate web sites as advertising. J. Interact. Advert. **3**(2), 10–23 (2003)

15. Kilgarriff, A., Kosem, I.: Corpus tools for lexicographers. In: Granger, S., Paquot, M. (eds.) Electronic Lexicography. Oxford University Press, Oxford (2012)
16. Kosala, R., Blockeel, H.: Web mining research: a survey. ACM SIGKDD Explor. Newslett. **2**(1), 1–15 (2000)
17. Lin, Y., Michel, J.B., Aiden, E.L., Orwant, J., Brockman, W., Petrov, S.: Syntactic annotations for the google books ngram corpus. In: Proceedings of the ACL 2012 System Demonstrations, pp. 169–174. Association for Computational Linguistics (2012)
18. Llopis, J., González, R., Gasco, J.: Web pages as a tool for a strategic description of the Spanish largest firms. Inf. Process. Manage. **46**(3), 320–330 (2010)
19. Miner, G., Elder IV, J., Fast, A., Hill, T., Nisbet, R., Delen, D.: Practical Text Mining and Statistical Analysis for Non-structured Text Data Applications. Academic Press, Cambridge (2012)
20. Mostafa, M.M.: More than words: social networks' text mining for consumer brand sentiments. Expert Syst. Appl. **40**(10), 4241–4251 (2013)
21. Powers, S.: HTML5 Media: Integrating Audio and Video with the Web. O'Reilly, Sebastopol (2011)
22. Redhu, S., Srivastava, S., Bansal, B., Gupta, G.: Sentiment analysis using text mining: a review. Int. J. Data Sci. Technol. **4**(2), 49 (2018)
23. Schenker, A.: Graph-theoretic techniques for web content mining (2003)
24. Schouten, K., Baas, F., Bus, O., Osinga, A., van de Ven, N., van Loenhout, S., Vrolijk, L., Frasincar, F.: Aspect-based sentiment analysis using lexico-semantic patterns. In: International Conference on Web Information Systems Engineering, pp. 35–42. Springer, Cham (2016)
25. Shi, F., Chen, L., Han, J., Childs, P.: A data-driven text mining and semantic network analysis for design information retrieval. J. Mech. Des. **139**(11), 111402 (2017)
26. Snell, S.: Clear and effective communication in web design. Smashing Mag. (2009). https://www.smashingmagazine.com/2009/02/clear-and-effective-communication-in-web-design/
27. Sowa, J.F.: Conceptual Structures: Information Processing in Mind and Machine. Addison-Wesley, Boston (1983)
28. Thorleuchter, D., Van Den Poel, D.: Predicting e-commerce company success by mining the text of its publicly-accessible website. Expert Syst. Appl. **39**(17), 13026–13034 (2012)
29. Wells, J., Valacich, J., Hess, T.: What signal are you sending? How website quality influences perceptions of product quality and purchase intentions. MIS Q. **35**(2), 373–396 (2011)

Teaching Pedigree Analysis and Risk Calculation for Diagnosis Purposes of Genetic Disease

Noureddine Kerzazi[1](✉), Mariam Tajir[2,4], Redouane Boulouiz[2,4], Mohammed Bellaoui[2,4], and Mostafa Azizi[3]

[1] ENSIAS, University Mohammed V in Rabat, Rabat, Morocco
n.kerzazi@um5s.net.ma
[2] Genetics Unit, Medical Sciences Research Laboratory, Faculty of Medicine and Pharmacy, University Mohammed Premier, Oujda, Morocco
mariam_tajir@yahoo.fr, bmbellaoui@gmail.com
[3] MATSI Lab, ESTO, University Mohammed First, Oujda, Morocco
azizi.mos@ump.ac.ma
[4] Mohammed V University in Rabat, Avenue Mohammed Ben Abdallah Regragui, Madinat Al Irfane BP 713, Agdal, Rabat, Morocco

Abstract. A Faculty of Medicine needs to respond appropriately to the rapid changes in medical education, to ensure that future geneticists are well trained to meet the challenges of medical practice. This paper presents five core requirements for a tool integrating new methodologies for teaching graduate-level a course of medical genetics. Our methodology presented here focuses on exploiting pedigree analysis in the field of medical genetics, particularly to explore them in the diagnosis of genetic diseases. Because of its important relevance as one of the skills that future medical practitioners must have, we designed this approach as a learning process supported by a tool. We also discuss an ongoing supported effort at utilizing common tools and IT resources to make it easier for the learners to reach their expected skill levels and provide them with a rich learning experience. We find that not only do our tool prototype has a positive impact on the learning process, practitioners also have expectations to feed a bio-bank of medical cases as inputs for future empirical studies.

Keywords: Pedigree · Genetic disease · Pedigree analysis · eLearning · Genotype elimination · Medical informatics

1 Introduction

Working with human genomes and techniques of genetics is becoming a daily medical practice. A medical pedigree is a graphic representation of a family's health history and genetic relationships (see Fig. 1). A pedigree model is used as a key tool in practice to diagnose genetic diseases. To this end, health professionals

Á. Rocha et al. (Eds.): WorldCIST 2020, AISC 1159, pp. 472–485, 2020.
https://doi.org/10.1007/978-3-030-45688-7_48

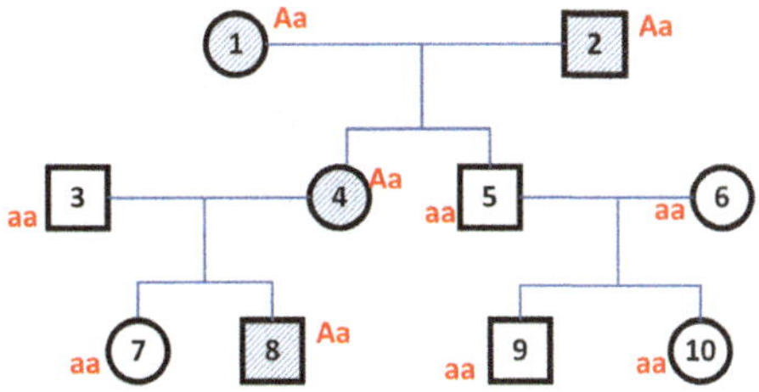

Fig. 1. A real example of pedigree

and genealogists must develop their ability to design pedigrees and to draw reliable diagnoses. However, students are introduced to pedigree modeling briefly through the exploration of few exercises on drawing and interpreting hand made pedigrees on papers. Other courses attempted to teach pedigrees by the mean of a drawing software but not usually a dedicated software for pedigrees. There is no assurance at the end of the course that the learner has really gained the required skills and competencies on pedigree modeling and analysis. This lack in mastering theoretical and practical concepts related to genes inheritance affect negatively the ability of students to perform reliable and accurate diagnoses of genetic diseases. Moreover, drawing large or complex pedigrees by hand is a challenging, time-consuming, and error-prone task.

Despite the use of automated tools for drawing pedigrees, significant questions remain about analysis and risk computation. While pedigree models provide substantial information about genes inheritance, the analysis of these models creates a huge space of variability that can be misleading [6]. Since it is impossible to explore all variant cases, a weak analysis could lead to an invalid diagnosis or even inconsistencies when interpreting pedigree information. Such situations are referred to as "Diagnose Failures".

1.1 Problem Statement

Previous study [1] showed that the related problem of genotype consistency checking is NP-complete if we consider at least three alleles (i.e., each of two or more alternative forms of a gene that arise by mutation and are found at the same place on a chromosome). The problem of genotype consistency checking considers pedigrees with typed and untyped individuals as shown in Fig. 1 and asks the extent to which all untyped individuals can be assigned a genotype so that the resulting pedigree is consistent with the laws of Mendelian inheritance.

To alleviate this problem we introduce a tool that could help students in reducing checking problem to polynomial time by applying techniques such as genotype elimination.

1.2 Objective

In this paper, we aim to support teaching pedigrees by proposing a methodology helping learners, both students and professionals, to design, analyze and understand easily a pedigree model while respecting reliability, accuracy, and consistency. At the end of the course, the learners will be more familiar with pedigrees, thankful for the information that this tool could give them, and aware of the hypothesis to further explore for efficient diagnosis.

1.3 Contributions

The paper makes the following contributions. First, it makes a focus on the importance of pedigrees as a genealogical tool for the diagnosis of genetic diseases. Secondly, we introduce and discuss five core requirements for a tool that should support pedigree modeling and analysis. Third, we propose a prototype that should support acquiring knowledge used such as a material course for teaching pedigrees to students and professionals acting in the medical field.

1.4 Outline

The remainder of this paper is organized as follows. The second section gives an overview of pedigrees, its different nodes, its vertices, and colors. The third section discusses the five core requirements for a tool that supports pedigree modeling and analysis. Related works are summarized in the fourth section. Before concluding, we give some results obtained through an evaluation of our prototype usability performed by students in Genetics Lab.

2 Foundations

In this section, we first introduce the terminologies and concepts used in the medical course at hand, including Pedigree Modeling and Analytics. We subsequently discuss the challenges with the actual teaching formats. Finally, we present our pedagogical approach (i.e., active learning) underpinning our new course format.

2.1 Basic Concepts of Pedigree

A medical pedigree is a graphic representation of genetic disorders that are inherited in a family [3]. Doctors can use pedigrees to analyze the probability that someone in a family will inherit a condition. It is used as a key tool in practice to diagnose genetic diseases [16], and therefore it is a daily medical practice in genetic counseling. Accordingly, health professionals (medical doctors and genetic counselors) must develop their competencies in designing and interpreting pedigrees. Health professionals could draw pedigree according to the classic way by handwriting on a paper, or instead by the mean of dedicated software. Sometimes one pedigree could be so large to cover more than one page. If there is a mistake somewhere on it, it should be corrected by making some partial changes or redrawing the model entirely from scratch. Therefore, it is obvious

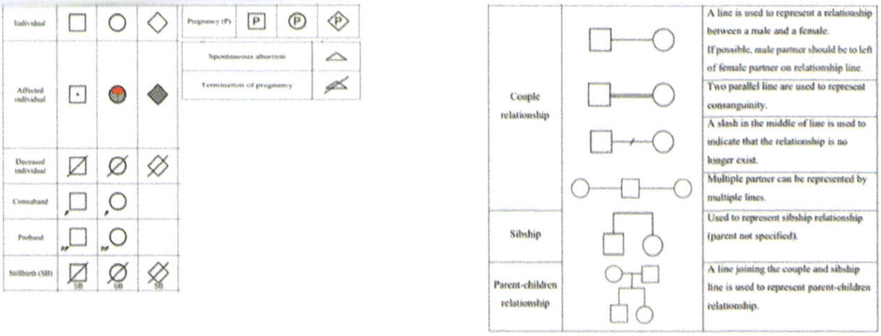

Fig. 2. Symbols and relationships used to design a pedigree model [3,4]

Table 1. Overview of the 4 types of disease transmission.

Type	Description
Autosomal Dominant	1. Don't skip generations 2. Affected parents can have unaffected children
Autosomal Recessive	1. Skip generation 2. Unaffected parents can have affected children
X Linked Dominant	1. Disease never transfers from father to son 2. All daughters of an affected father will be affected
X Linked Recessive	1. Males are more affected 2. The disease tends to transfer from Mother to son and father to daughter 3. Disease never transfers from Father to son

that in the case of large or complex pedigrees, drawing them by hand becomes a challenging task, which is time-consuming and error-prone.

Following this understanding, drawing a pedigree requires symbols and relationships [6,8,12,16,18]. Medical students are introduced to the symbols and colors compliant with standards as well as extended pedigrees, as shown in Fig. 2. These symbols represent females, males and their relationships of marriage and inheritance. Each level is equivalent to one generation of the family. The colors are used for tracking the genetic traits, both dominant and recessive. For example, circles, squares, and diamonds represent respectively females, males, and individuals with no identified gender. All added information to these symbols, including colors and arrows, are attributes with specific meanings useful for the task of pedigrees analysis. Relationships (see Fig. 2) are lines connecting these symbols. We distinguish different kinds of relationships: a couple (marriage or union), a divorced couple, sibship (no specified parent), kin-ship (parents-children relationship). Most of the attributes associated with symbols and relationships are set via questions directly to the patient or his companions, by filling given forms or extracted from his health records or IDs. All these data are stored in a database.

Pedigree analysis can lead to four types of disease transmission as described in Table 1.

2.2 Challenges of Teaching Pedigrees

Previous literature about medical education has raised needs for developing novel curricula and competencies which integrate information technology as a cornerstone [10]. Moreover, it has been explored how existing teaching approaches from traditional course formats can be extended and adapted to meet the new requirements of medical practices [11]. However, according to Ozuah [11], there is still a need to incorporate technological advancements into the delivery of teaching and to incorporate new and emergent domains of knowledge into the existing curriculum.

To respond appropriately to the rapid changes in medical education and to ensure that future geneticists are well trained to meet the requirements of the medical practice, universities face the challenge to train medical students on various technological and clinical skills. A major challenge for teaching medical courses is to prepare students adequately for a diversity of problems, which they will encounter in their daily practice and to overcome the gap between real professional scenarios and those used in academia.

What Sort of a Geneticist Is Needed? Huge advances have taken place in information technology (IT) applied to medical knowledge. IT can also be added to curricula as medical training tools. Our goal is to produce a geneticist who is fully competent and prepared to start a basic medical practice by the time of graduation. A medical graduate should not be wasting time in learning too many new concepts - a debt of knowledge - after putting that many years into graduation. In other words, our goal should be to have medical graduates who have less knowledge but more abilities to be efficient in their daily practice. These abilities will include the use of information technology, problem-solving skills, and appreciation of a wider variety of possible solutions to genetics problems.

Workload and Time Management - The most challenging part of being a medical student is to manage time. some topics should be set aside for self-study. The precious time set aside can be used to cover other important topics.

Memorizing - It is difficult to understand and memorize medical concepts and syntax. There are whole syntax and semantic related to pedigree modeling, which needs to be memorized by a medical student before they step into practical life.

Beginning a Clinical Exposure - The faculty is engaged to ensure that medical students will be transformed into the most effective clinicians. When medical students start getting their clinical exposure, they ...

Establishing an Educational Culture - In which trainees and physicians examine their performance and measure patient outcomes, with the ultimate aim of continually improving the quality of care they provide. For instance, Developing a system to support continuing medical education that increases diagnosis reliability. Moreover, a good teacher is one who teaches not by words but by actions. Pedigrees are clinical activities within practical topics that should not be repeated in theory classes. Students reported that too much theory is boring, sometimes leading to a loosening of self-motivation. Consequently, medical teachers need support in their quest to provide a high-quality education to medical students.

Changes in Students' Expectations - Students have seemingly become aware and well-informed about health, disease and treatment options thanks to open access to medical information via the Internet. Students request the clinical skills required for performing the role of a competent and effective medical geneticist.

Table 2. Mapping requirements to challenges: R1 (Usability & Design), R2 (Automatique recommendations to support the analysis), R3 (Support for case based approach & Bio-Bank), R4 (Automatic Validation), and R5 (Integrated Algorithms).

Challenges	Task	R1	R2	R3	R4	R5
Usable modeling tool		+				+
Lack of debugging tools	6	+				
Lack of modeling documentation	1			+	+	
Model review	8	+	+			+
Lack of automatic validation	7				+	+

To address the challenges described above and foster active learning, we are: *(1)* revisiting the content of the course, *(2)* building a tool, *(3)* and adjusting the pedagogical approach.

2.3 Pedagogical Approach

Due to its interdisciplinary and complex nature, teaching genetics challenges both medical students and teachers. Therefore, traditional lectures based on theoretical content should be supplemented by *(1)* active learning approach [14] as well as by *(2)* real-world case studies such as those encountered in the real practice [5]. Active learning allows students to learn through interaction with peers and by working on hands-on tutorials and real-world case studies [2]. In order to organize such practice-oriented exercises, we need software support and open educational resources. Different kinds of learning objectives could be achieved by varying teaching formats to ensure an in-depth knowledge transferred to the students.

The underlying learning theory on which we based our pedagogical approach is a revised Blooms Taxonomy that was introduced by Krathwohl [9]. Krathwohl distinguishes between two dimensions: knowledge dimension and cognitive process [9], as illustrated in Fig. 3. Knowledge dimension distinguishes four levels: *Factual*, *Conceptual*, *Procedural*, and *Metacognitive*. While cognitive process dimension distinguishes six levels: *Remember*, *Understand*, *Apply*, *Analyze*, *Evaluate*, and *Create*. Our pedagogical actions take place at the intersection between those two dimensions. For instance, students need factual knowledge (FK) to use terminologies and basic elements to build a pedigree and conceptual knowledge to draw relationship between those elements. On the other dimension, i.e. cognitive process, they need skills to analyze pedigree' models and to drive diagnoses.

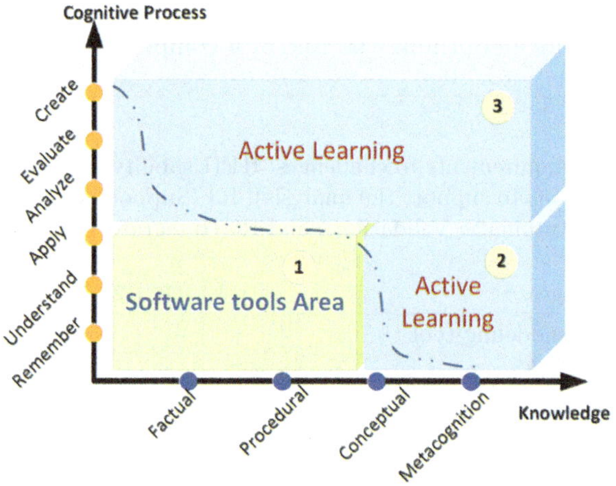

Fig. 3. Revised blooms taxonomy Krathwohl [9]. 1) Activities set aside for the tool; 2&3) Active learning.

3 Requirements for Pedigree Analysis Course

This section discusses the prototyping approach used to identify five core requirements for a tool which support pedigrees' modeling, followed by a detailed discussion of each of the five requirements.

3.1 Identification of Requirements

The major trigger for this work occurred during a discussion session aiming to improve the materials of a course involving concepts of pedigree design when one of the participants exclaimed that the course needs interactive materials to support the understanding of theoretical concepts. While the request for a tool was a soft goal, the teachers actually meant something different: on the one hand, a tool to ease the modeling of correct (i.e., syntax and semantic) pedigrees, good conceptual analysis is tightly dependent on a good representation of pedigrees. On the other hand, medical students need guidelines and a recommendation system to master pedigree analysis.

Since the extent of these ideas and their impact on pedigree analysis activities was not precise enough to derive concrete requirements for support tools, we decided to adopt a rapid prototyping process [17]. Such a process typically is used when the requirements of a software product are unclear. The end goal of such a process is not the prototype itself, but rather a specific list of requirements that can then be used to implement a real product (e.g., using an iterative process). Similarly, our aim was to obtain a list of core requirements for a tool that support teaching pedigrees modeling and analysis (see section challenges). In addition, the prototype would serve as a vehicle for a user study that empirically evaluates the impact of the requirements on the ability to perform typical pedigree analysis activities.

Our rapid prototyping process basically consisted of the following activities:

1. extracting new ideas by analysing the document of challenges
2. deriving requirements from these ideas
3. comparing these requirements with those already prototyped
4. adapting the prototype's existing features and adding new ones according to the new requirements
5. testing and stabilizing the prototype on example code snippets
6. at the end of each iteration, we would solicit ideas for tool support from the teachers and students
7. once all ideas have been noted down, we would demo the prototype
8. obtaining feedback about the prototype and its relation to the teachers' own ideas
9. iterate back to step 1

Once the final prototype was obtained, we extracted all its features, then used open coding to group related features into more abstract requirements. Open coding [19] aims at representing key characteristics (by adding codes) of textual data and categorizing the elicited concepts from the text in a hierarchy of categories as an abstraction of a set of codes. The process is repeatedly performed until reaching a "state of saturation".

For each such requirement, we also identified all modeling challenges of Table 2 that it impacts. Apart from a number of utility and convenience features, we ended up with 5 core requirements that, together, cover all challenges, as shown in Table 2. The rest of this section discusses these 5 requirements.

R1. Usability & Design

The first requirement states that **pedigree design tool support should physically integrate the syntax and semantics of modeling pedigrees and its related metadata (e.g., type, attributes, description of concepts and constraints) into a representation of problems related to genotype** [20]. For example, we seek the usability and intuitive design of pedigree models such as represented in Fig. 4. This requirement allows students to remain compliant with the nomenclature of pedigrees.

Apart from reducing the pain of handmade pedigrees, this first requirement also brings other benefits to tool support. For example, reliable abstraction of genome inheritance. Furthermore, determining a diagnosis is now possible and supported by the tool.

R2. Automatique recommendations to support the analysis
The second requirement focuses on a component of recommendation to support students analyzing their pedigree models. Based on common software engineering sense [17], this requirement states that **pedigree modeling tool should provide a well-encapsulated component for recommandation**, reducing time and errors of this complex and error-prone task. Note that it does not suffice to have a good pedigree model, we still need to analyze it and drive sound medical diagnose.

Apart from reducing the effort of analysis, again making it easier help to reduce the pain of learning. That said, we seek increasing the ability to be guided by the system. While requirement 1 has made an abstract representation of pedigree explicit, the usage of these models could be supported through a system component devoted to the recommendation, further encouraging systematic review of the process of analysis.

Moreover, supporting the exploration of individuals' genotypes using a tool will open a plethora of new avenues such as proposing new methods or implementing existing ones. For instance, implementing Markov chain Monte Carlo analyses of genetic data on pedigrees [13].

R3. Support for case based approach & Bio-Bank
While requirement 1 brought a framework for pedigree modeling and requirement 2 discussed a system of recommendation in an articulated way, requirement 3 closes the loop by stating that **we should learn from an empirical database of pedigree models**. This approach presents a good opportunity for reusing knowledge from a similar case (i.e., a case-based approach to support the process of learning).

Requirement 3 basically ensures that students can learn from previous case studies. As such, a database of pedigree models representing typical cases provides not only a learning opportunity but also a way to get more skills as well as good practices. This automatic extraction of related cases also allows both students and practitioners to easily compare the previous analysis to the model at hand.

Furthermore, the requirement also targets the achievement of the increase of empirical body of knowledge, especially when combined within the e-learning process.

R4. Automatic Validation
Requirement 4 states that **pedigree modeling tool support should automatically validate the syntax of models as well as the value of the attribute of each element within the pedigree model**:

This validation should be performed automatically, either during the modeling of a new pedigree (by students) or during the process of evaluation (carried out by teachers). If a violation of a syntactical constraint is detected, the system should either halt or fall back to options that are known to be good. Automatic constraint validation improves comprehension and enforces naming conventions and the presence of documentation.

As also mentioned in previous studies [11], students should have the opportunity to learn from typical cases and hence apply good practices to pedigree representation and analysis as well, including empirical evaluation. The main idea behind our tool is to bring the body of knowledge close to pedigree analysis activities and automatically suggest similar cases from the empirical database. For example, to analyze a pedigree model, students should only focus on pedigree modeling, where the tool can automatically generate a wizard to support the process of analysis step by step.

R5. Integrated Algorithms
While algorithms such as those for machine learning are frequently used in modern applications because they can support decision making, the speed, and simplicity of algorithms learning from existing information mean that it is still important to integrate them within the tool. For instance, algorithms can help to compute risks and probabilities for the propagation of specific diseases. Algorithms can also help to identify disease patterns within specific communities by looking into empirical data.

4 Related Work

Different concepts have been proposed to teach clinical data analysis. Besides traditional classroom teaching, we distinguish between approaches based on practical exercises, experiments, similar cases, and simulation [15].

The most prominent teaching approach for pedigrees focuses on theoretical aspects instead of practical exercises [15]. The underlying idea is to apply pedigree principles at a conceptual level of students as well as practitioners. Studies have shown that applying such processes of modeling at the individual level can lead to significant performance improvements [7,21]. In contrast to the approach presented here, the pedigree design exercises deal with pen and paper.

Empirical approaches focus on teaching by conducting experiments, typically as part of a regular course. The students take the role of experimental subjects to explore the effects of selected processes or techniques themselves. Typical objectives of such experiments consider comparisons of different analyses of pedigrees. In contrast to the approach presented here, the experimental treatments are usually medical level processes and not process modeling or analysis activities.

Approaches for teaching pedigrees practices are often based on educational approaches [13]. Analyzing a pedigree, for instance, is often demonstrated with the means of theoretical concepts to impart knowledge about genomic inheritance principles. In contrast to the approach presented in this paper, this kind of teaching typically aims at a better understanding of a specific philosophy rather than a better understanding of the challenges of modeling and analyzing large pedigrees.

Finally, simulation is sometimes used to support teaching in the area of pedigrees [20]. Here, students can make local decisions and see their global effects. Simulation is also suited for playing "What-if Analysis" and, thus, help to better understand pedigrees. In contrast to our approach, simulation has a more limited scope that consists of understanding pedigree analysis and risk computation of the probability of diseases.

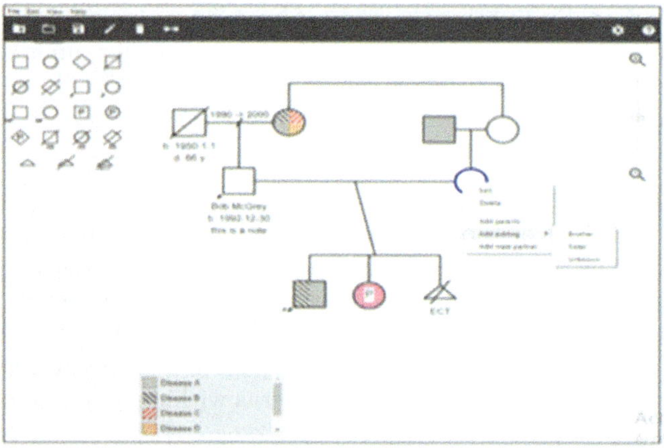

Fig. 4. Demo of the tool

All these different approaches have their specific strengths and drawbacks. We build on these valuable approaches to introduce requirements for a new course design integrated into our medical school curricula.

5 Tool Implementation

This section discusses the main components of the final prototype obtained at the end of requirement analysis. This prototype has been used subsequently to empirically evaluate the impact of the five core requirements on pedigree modeling and teaching activities. This prototype, which we named *GenDis*, is a modeling framework implemented as a standalone web application.

Syntax - The core component of *GenDis* is a modeling framework as shown in Fig. 4. It uses the defacto syntax described by Bennett et al. [3,4]. This allows to design and annotate pedigree models in a harmonized way in terms of predefined symbols and shapes.

GenDis - Has a number of configuration options by itself, which basically allow to enable or disable features like the consistency validation, risk calculation probability, recommendations, generation of reports file in different formats. It's worth noticing that so far the implementation effort is focussed on designing correct (syntactically) pedigrees and that the other features prototype are not yet implemented.

Constraint Checker - This component checks the correctness of pedigree models designed by students. This is mainly by verifying syntax against predefined symbols and allowed relationships [3,4]. While this works to some extent even for complex pedigrees, dedicated constraint checkers would improve usability. We plan to extend it to consider other types of constraints by implementing a rule engine to be able to import/export rules and turn on/off levels of security for each rule.

6 Conclusion and Future Work

This paper presents and empirically evaluates 5 requirements related to a tool support dedicated to teaching pedigree modeling and analysis. Our findings show how the requirements contribute to improving the correctness, the reliability of analysis, and the integration of system recommendations. Furthermore, the improvements of correctness and reliability supported by our tool *GenDis* will help even novice students to get more insights in the field of medical genetics at the faculty of medicine and pharmacy.

While we prototyped the five requirements in the *GenDis* framework, they are in fact independent. For example, students can focus only on mastering the syntax of pedigrees, while others will focus on pedigree analysis by looking at similar cases in the pedigree models'database, and more importantly, researchers could integrate new algorithms to explore new avenues. On the other hand, requirements 3 (case-based approach) is less common in today's tool and our results suggest that it is worth for students to learn using similar case-based models.

Where do we go next? Can excellence in teaching be fostered? an old inspirational teacher quote said "A good teacher is one who teaches not by words but by actions".[1] Use of information technologies for medical education has become increasingly prevalent. IT provides medical students with the tools to continue to access the medical knowledge necessary not only to perform quality diagnozes, but to be a life-long learners. We continue to evaluate knowledge acquisition as a measurable learning outcomes of our new educational environment, adding content towards a bio-bank of case-based pedigrees. Also, challenges of preparing the future doctor involve emphasis and standardization of competencies and learning outcomes, integration of formal knowledge and clinical experience, patient-centered care, population health, cost-conscious high value care, and understanding the organization of health services.

We plan to evaluate our tool along with the new format of the course lecture against our requirements by conducting a formal evaluation that follows the faculty standards in place. Also, we plan to conduct an informal evaluation in which the students directly rate the course and materials used to teach them pedigree concepts. The direct comparison of our new teaching format with the previous sessions will lead our effort for the improvement of the *GenDis* tool.

---

[1] https://www.wow4u.com/teacher/.

References

1. Aceto, L., Hansen, J., Ingolfsdottir, A., Johnsen, J., Knudsen, J.: The complexity of checking consistency of pedigree information and related problems. J. Comput. Sci. Technol. **19**(1), 42 (2004). Special Issue on Bioinformatics
2. Asamoah, D., Sharda, R., Zadeh, A.H., Kalgotra, P.: Preparing a data scientist: a pedagogic experience in designing a big data analytics course. Decis. Sci. Innov. Educ. **15**(2), 161–190 (2017)
3. Bennett, R.L., French, K.S., Resta, R.G., Doyle, D.L.: Standardized human pedigree nomenclature: update and assessment of the recommendations of the national society of genetic counselors. J. Genet. Couns. **17**(5), 424–433 (2008)
4. Bennett, R.L.: The family medical history as a tool in preconception consultation. J. Commun. Genet. **3**(3), 175–183 (2012)
5. Biggs, J., Tang, C.: Teaching For Quality Learning at University (Society for Research Into Higher Education), 4th edn. Open University Press, London (2011)
6. Facio, F.M., Feero, W.G., Linn, A., Oden, N., Manickam, K., Biesecker, L.G.: Validation of my family health portrait for six common heritable conditions. Genet. Med. Official J. Am. Coll. Med. Genet. **12**(6), 175–183 (2010)
7. He, D., Wang, Z., Parida, L., Eskin, E.: IPED2: inheritance path based pedigree reconstruction algorithm for complicated pedigrees. IEEE/ACM Trans. Comput. Biol. Bioinform. **14**(5), 1094–1103 (2017)
8. Jouali, F., Laarabi, F.-Z., Marchoudi, N., Ratbi, I., Elalaoui, S., Rhaissi, H., Fekkak, J., Sefiani, A.: First application of next-generation sequencing in moroccan breast/ovarian cancer families and report of a novel frameshift mutation of the brca1 gene. Oncol. Lett. **12**(2), 1192–1196 (2016)
9. Krathwohl, D.: A revision of bloom's taxonomy: an overview. Theory Into Pract. **41**(4), 212–264 (2004)
10. Majumder, M.A., D'Souza, U., Rahman, S.: Trends in medical education: challenges and directions for need-based reforms of medical training in South-East Asia. Indian J. Med. Sci. **58**, 369–380 (2004)
11. Ozuah, P.O.: Undergraduate medical education: thoughts on future challenges. BMC Med. Educ. **2**(1), 8 (2002)
12. Paterson, T., Graham, M., Kennedy, J., Law, A.: VIPER: a visualisation tool for exploring inheritance inconsistencies in genotyped pedigrees. J. BMC Bioinform. **3**(3), 175–183 (2012)
13. Pirinen, M., Gasbarra, D.: Finding consistent gene transmission patterns on large and complex pedigrees. IEEE/ACM Trans. Comput. Biol. Bioinform. **3**(3), 252–262 (2006)
14. Prince, M.: Does active learning work? A review of the research. J. Eng. Educ. **93**(3), 223–231 (2004)
15. Ranaweera, T., Makalic, E., Hopper, J.L., Bickerstaffe, A.: An open-source, integrated pedigree data management and visualization tool for genetic epidemiology. J. Genet. Couns. **47**(4), 1034–1039 (2018)
16. Solomon, B.D., Muenke, M.: When to suspect a genetic syndrome. Am. Fam. Physician **86**(9), 826–833 (2012)
17. Sommerville, I.: Software Engineering, 9th edn. Addison-Wesley Publishing Company, Boston (2010)
18. Tokutomi, T., Fukushima, A., Yamamoto, K., Bansho, Y., Hachiya, T., Shimizu, A.: f-treeGC: a questionnaire-based family tree-creation software for genetic counseling and genome cohort studies. BMC Med. Genet. **18**(1), 175–183 (2017)

19. Wagner, S., Fernández, D.M.: Analysing text in software projects. CoRR, abs/1612.00164 (2015)
20. Yang, L., Cheng, E., Özsoyoğlu, Z.M.: Using compact encodings for path-based computations on pedigree graphs. In: Proceedings of the 2nd ACM Conference on Bioinformatics, Computational Biology and Biomedicine, pp. 235–244 (2011)
21. Yuan, S.-M., Zhang, Y.-N., Du, J., Li, W., Tu, C.-F., Meng, L.-L., Lin, G., Lu, G.-X., Tan, Y.-Q.: Phenotypic and molecular characteristics of androgen insensitivity syndrome patients. Asian J. Androl. **20**(5), 473–478 (2018)

Multi-label Classifier to Deal with Misclassification in Non-functional Requirements

Maliha Sabir(✉), Christos Chrysoulas(✉), and Ebad Banissi(✉)

London South Bank University, 103 Borough Road, London SE1 0AA, UK
{sabirm3,chrysouc,banisse}@lsbu.ac.uk
http://www.lsbu.ac.uk

Abstract. Automatic classification of software requirements is an active research area; it can alleviate the tedious task of manual labeling and improves transparency in the requirements engineering process. Several attempts have been made towards the identification and classification by type of functional requirements (FRs) as well as non-functional requirements (NFRs). Previous work in this area suffers from misclassification. This study investigates issues with NFRs in particular the limitations of existing methods in the classification of NFRs. The goal of this work is to minimize misclassification and help stakeholders consider NFRs in early phases of development through automatically classifying requirements. In this study, we have proposed an improved requirement detection and classification technique. The following summarizes the proposed approach:

(a) A newly created labelled corpus,

(b) Textual semantics to augment user requirements by word2vec for automatically extracting features, and

(c) A convolution neural network-based multi-label requirement classifier that classifies NFRs into five classes: reliability, efficiency, portability, usability, and maintainability.

Keywords: Non-functional requirements · Requirement engineering · Machine learning approaches · Natural language processing · Neural networks

1 Introduction

Requirements originate from stakeholders in natural language [1]. Non-functional requirements (NFRs) are always associated with functional requirements (FRs) [2]. While FRs describe the functions, tasks or behavior that a system must support [3], NFRs represent the particular qualities the systems must have. For a successful implementation, it is crucial to have comprehensive and transparent requirements for the identification of relevant constraints, assumptions, risks, and dependencies [4]. Fulfillment of a single FR can lead to the achievement of one or more NFRs [5]. NFRs are significant for the success of a software system [4]. However, literature suffers from both terminological and theoretical conflict [6]. Given the diversity of definitions of NFRs, stakeholders

Á. Rocha et al. (Eds.): WorldCIST 2020, AISC 1159, pp. 486–493, 2020.
https://doi.org/10.1007/978-3-030-45688-7_49

tend to disagree on important NFR attributes [7]. Each NFR attribute is treated differently and requires specialized expertise [8], but due to lack of resources and engineering skills, these have not been effectively addressed, resulting in higher structural defects and project failure rates [9, 10]. The complexity of NFR concepts has led to the development of numerous automated and semi-automated classification methods based on natural language processing (NLP), rule-based (RB) approaches, and machine learning (ML) approaches [11]. However, these tools are naïve and exhibit misclassification [12]. The literature highlights concerns in this domain, described below.

Limited in Generalizability: The RB approach has limited generalizability [13]. Rules need to be manually crafted and enhanced all the time. Moreover, the system can become overly complicated with some rules beginning to contradict each other [14]. The use of keywords or certain vocabulary is correlated with specific NFR attributes [15]. Furthermore, the RB approach is limited to defining rules for handling security, usability, and maintainability [14].

Limitation of Feature Selection Techniques: The fit criteria for NFR identification and ML training lacks clarity [16]. Mostly, syntactic feature part-of-speech (POS) tagging is used [13, 17–19]. Abad found that pre-processing improved the performance of an existing classification method [16]. The second most common technique, Bag of Word (BOW) [17, 18, 20, 21], fails to maintain the sentence order and cannot deal with polysemy [22]. In comparison, n-grams feature can consider the word order in a close contest to its neighbouring words [20, 23], but it also suffers from data scarcity [24]. However, it is observed that the semantic knowledge of sentences based on word2vec reaches better classification performance [5, 25]. In ML approach reliability, portability [26] and usability [16] have not been adequately addressed.

Limited NFR Ontology: Most of the classification refers to functionality, availability, fault tolerance, legal, look and feel, maintainability, operational, performance, portability, scalability, security and usability. The NFR attributes chosen to be a part of existing studies are not the critical NFR attributes to be representative of its whole domain [27], leading to misclassification [5, 12].

Lack of Corpus: To a large extent, ML techniques are dependent on training data [28]. They require a representative and balanced corpus which is currently lacking [17, 26]. The PROMISE dataset used in previous studies has misconceptions about requirements. A corpus with an unbalanced number of examples of a specific linguistic construction can cause an algorithm to apply a specific label more frequently than others, resulting in a bias in the classification results [29].

Limitation to Handle Multi-label Classification: Peng [25] argues that one requirement could exhibit the characteristics of more than one attribute. In such a case, the classification should be made based on more than one label [29]. However, this has not been addressed.

This study proposes a novel approach for handling the misclassification of NFRs based on multiple labels using deep learning techniques. The paper is organized as follows: Sect. 2 describes work related to NFRs and their classification; Sect. 3 illustrates a proposed conceptual framework; and finally, Sect. 4 details recommendations for further research.

2 The Classification of NFRs

This section presents work related to the detection and classification of NFRs. Cleland designed an NFR classifIer which uses a fixed set of keywords to identify NFRs [30]. In another study by Cleland, he describes the classification algorithm and then evaluates its effectiveness through a series of experiments and against a large requirement document obtained from Siemens Logistics and Automotive Organization [21].

Hussain proposes a supervised automatic classifier with the use of syntactic and keyword features for information retrieval for classifying NFRs. The experiment was performed for binary as well as multi-class NFR classification with the hybrid training set [19].

Vlas and Robinson developed RCNL, a multilevel ontology in which the lower levels apply generic English grammar-based concepts while the upper, abstract levels apply OSS requirements-based concepts [13].

Ormandjieva proposes a supervised learning approach based on a support vector machine (SVM) for the classification of the requirement into ISO/IEC 9126 ontology classes using Wen ontology language (OWL). The ontology contains dependency relationship between FRs and NFRs, further an association of NFRs with subcategories has been highlighted. They also proposed gold standard annotated NFR corpus [31].

Slankas proposes an NFR locator to locate and classify 14 NFRs using k-nearest neighbors (K-nn), SVM and naïve Bayes algorithms. The tool resulted in various misclassification and suffered from generalizability issues. Slankas found that certain features are associated with specific NFR attributes [15].

Sharma suggests an approach to identify NFRs based on rules extracting multiple syntactic and semantic features and relationship among them through a domain-specific language. The approach was practiced on a publicly available requirement corpus [14].

Singh proposes automated identification and classification of NFRs, and subclasses based on the ISO 9126 quality model. The study proposes an RB classifier using the thematic role as well as identified the priority of the extracted NFRs according to their occurrence. The results were analyzed on two corpora. The PROMISE corpus contains higher-ranked sentences than the Concordia Corpus [32].

Casamayor suggests a semi-supervised approach based on user feedback, iteratively collecting data using an expectation maximization strategy to enable learning an initial classifier for NFRs. It uses a multinomial naïve Bayes classifier with expectation maximization (EM) for the initial training of requirement documents labeled manually to train the binary classifier [18].

Zhang uses three kinds of index terms at different levels of linguistic semantics, as n-grams, individual words, and multi-word expressions (MWE) are used in the representation of NFRs. Then, an SVM with the linear kernel is used as the classifier [23].

Sunner proposes a supervised learning-based cluster technique requirement mining and classification of FR and NFRs in agile software development. In the study, k-means was used with the UPGMA classifier model, SVM was used with RBF kernel, and a neural network used with a genetic algorithm. The results show that a cluster neuro-genetic approach provides better results than the SVM and RBF kernel [17].

Mahmoud A. proposes an unsupervised approach that exploits the textual semantics of FRs to identify potential NFRs with a clustering algorithm. The experiment was performed on datasets from SafeDrink, SmartTrip, and BlueWallet [33].

Kurtanovic develops and evaluates a supervised machine learning approach employing meta-data, lexical, and syntactical features—in particular, usability, security, operational, and performance requirements [26].

Abad uses various feature extraction and feature selection techniques to maximize the accuracy of classification algorithms. This study was performed on the tera-PROMISE repository and shows that pre-processing improved the performance of an existing classification method [16].

Lu proposes an approach that uses textual semantics by word2vec for automatically classifying user reviews from two popular apps, iBooks and WhatsApp. The approach is used for classifying user reviews into NFRs with each user review sentence labeled as one type [25].

Younas came up with an automated semi-supervised semantic similarity-based approach that uses an application of the word2vec model and popular keywords for the identification of NFRs. The study as performed on the PROMISE-NFR dataset [5].

Baker made a first attempt to use a convolution neural network (CNN) and artificial neural network (ANN) for the classification of NFRs into five classes—maintainability, operability, performance, security, and usability—using an unsupervised approach. The experiment was performed on the PROMISE dataset. The resulting tool shows inexperience and lacks practicality [27].

3 Handling the Misclassification of NFRs

We have adopted a multi-label classification approach for handling misclassification. The proposed approach uses a CNN, which has proven to be successful for single-label classification [27]. A CNN has a clear advantage over traditional machine learning algorithms as it can learn and generate dense vectors for word representation. Traditional machine learning algorithms tend to divide the multi-label problem into multiple binary classifications, whereas a CNN has the ability to learn multiple labels in one classifier. The experiment is a two-step procedure which is described in Fig. 1.

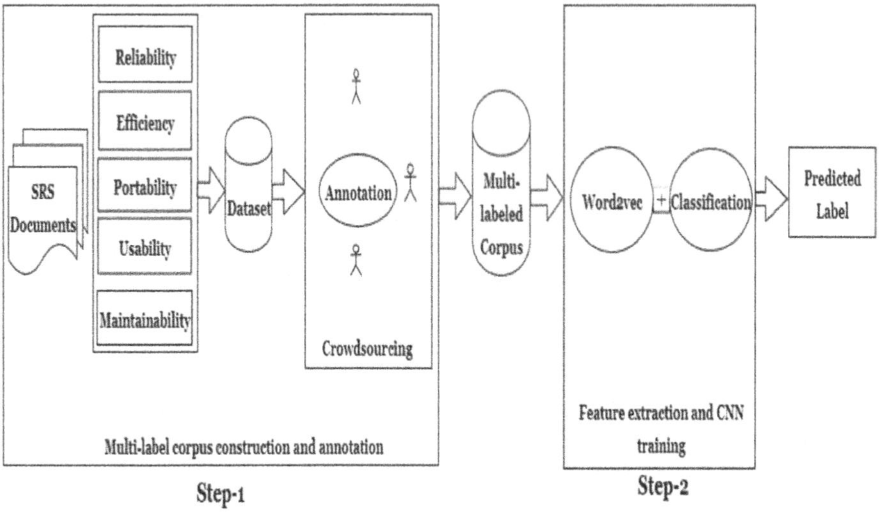

Fig. 1. Conceptual framework for multi-label classification

3.1 Step 1: Corpus Construction and Annotation

In Fig. 1, the first step describes the multi-label corpus construction and annotation. The procedure starts with the identification of the critical NFR attributes—reliability, efficiency, portability, usability, and maintainability—based on the software quality model: McCall's, Boehm's, Dromey's, ISO 9125/25010 and FURPS/FURPS+ [34]. Various Software requirement specification (SRS) documents has been collected. The researcher manually extracts requirements related to the selected NFR attributes and records them in the form of a dataset.

To overcome the limitations of the existing corpus and construction of a new corpus, we have followed a set of guidelines through the MATTER-MAMA framework proposed by Stubs [35]. We aim to have a representative and balanced corpus which makes the corpus to be generalized to the domain it belongs to and contains roughly equal numbers of training examples of all attributes selected to be a part of this ontology.

To have a labeled corpus, all the requirements related to the selected categories will be re-assessed by a crowd. For instance, a requirement related to efficiency may also be referred to usability. So, It will be labeled for both NFR classes. This minimizes the chances of mislabeling. Three annotators perform specification to the given dataset based on the guidelines and instruction [36].

The selection of annotators, annotation tool, and annotation scheme have all been performed through the MATER-MAMA framework under the supervision of the researcher. Step 1 delivers the outcome in the form of the multi-labeled corpus.

3.2 Step 2: Feature Extraction and CNN Training

Step 2 involves designing a CNN for feature extraction and classification. This step takes the labeled corpus from the previous step, extracts features on word2vec, and trains a word-level CNN for multi-label text classification that uses libraries like TensorFlow and Keras. These libraries provide a set of tools for building neural network architecture.

Whereas, the use of word2vec for word embeddings finds the semantic dependency and relatedness of the words in a requirement to determine class label(s). The CNN learns and generates dense vectors for word representation [37], which assumes that words with similar meanings occur in similar contexts [38]. Hence, predicts the labels with improved fit criteria for classification.

4 Conclusion and Future Work

This study provides a good starting point for handling misclassification with regards to NFRs. In this work, we have proposed a CNN based multi-label classifier to automatically classify stakeholder requirements into five classes of NFRs: reliability, usability, portability, maintainability, and efficiency.

The adoption of the proposed tool will aid in manual effort and time invested in the management of NFRs. It will also result in practical benefits to stakeholders. The identification of hidden requirements, the reduction of misclassification, and timely detection of NFRs dependencies and conflicts will benefit stakeholders in the prioritization of requirements and management of resources.

In a prospective course of action, we will bring the design and implementation details of the following: (1) a labeled corpus for NFRs and (2) a multi-label CNN based classifier trained on a representative and balanced corpus for improved training of the classifier on the basis of a broad ontology of NFR domain.

References

1. Ibrahim, N., Wan Kadir, W.M.N., Deris, S.: Documenting requirements specifications using natural language requirements boilerplates. In: 2014 8th Malaysian Software Engineering Conference (MySEC), Langkawi, Malaysia, pp. 19–24. IEEE (2014)
2. Khatter, K., Kalia, A.: Impact of non-functional requirements on requirements evolution. In: 2013 6th International Conference on Emerging Trends in Engineering and Technology, Nagpur, India, pp. 61–68. IEEE (2013)
3. Mijanur Rahman, Md., Ripon, S.: Elicitation and modeling non-functional requirements – a POS case study. Int. J. Future Comput. Commun. 485–489 (2013). https://doi.org/10.7763/IJFCC.2013.V2.211
4. Hussain, A., Mkpojiogu, E.O.C., Kamal, F.M.: The role of requirements in the success or failure of software projects. Int. Rev. Manage. Mark. **6**(7), 306–311 (2016)
5. Younas, M., Wakil, K.N.D., Arif, M., Mustafa, A.: An automated approach for identification of non-functional requirements using Word2Vec model. Int. J. Adv. Comput. Sci. Appl. **10** (2019). https://doi.org/10.14569/IJACSA.2019.0100871
6. Glinz, M.: On non-functional requirements. In: 15th IEEE International Requirements Engineering Conference (RE 2007), Delhi, pp. 21–26. IEEE (2007)
7. Berntsson Svensson, R., Gorschek, T., Regnell, B.: Quality requirements in practice: an interview study in requirements engineering for embedded systems. In: Proceedings of the 15th International Working Conference on Requirements Engineering: Foundation for Software Quality. pp. 218–232. Springer, Heidelberg (2009)

8. Zhang, M.-L., Zhou, Z.-H.: A review on multi-label learning algorithms. IEEE Trans. Knowl. Data Eng. **26**, 1819–1837 (2014). https://doi.org/10.1109/TKDE.2013.39
9. Helmy, W., Kamel, A., Hegazy, O.: Requirements engineering methodology in agile. Environment **9**, 8 (2012)
10. Khan, F., Jan, S.R., Tahir, M., Khan, S., Ullah, F.: Survey: dealing non-functional requirements at architecture level. VFAST Trans. Softw. Eng. **9**, 7 (2016). https://doi.org/10.21015/vtse.v9i2.410
11. Binkhonain, M., Zhao, L.: A review of machine learning algorithms for identification and classification of non-functional requirements. Expert Syst. Appl. X **1**, 100001 (2019). https://doi.org/10.1016/j.eswax.2019.100001
12. Gries, S.Th., Berez, A.L.: Linguistic annotation in/for corpus linguistics. In: Ide, N., Pustejovsky, J. (eds.) Handbook of Linguistic Annotation, pp. 379–409. Springer, Netherlands, Dordrecht (2017)
13. Robinson, W.N.: Two rule-based natural language strategies for requirements discovery and classification in open source software development projects (2012). https://doi.org/10.2753/MIS0742-1222280402
14. Sharma, V.S., Ramnani, R.R., Sengupta, S.: A framework for identifying and analyzing non-functional requirements from text (2014). https://doi.org/10.1145/2593861.2593862
15. Slankas, J., Williams, L.: Automated extraction of non-functional requirements in available documentation. In: 2013 1st International Workshop on Natural Language Analysis in Software Engineering, NaturaLiSE 2013 – Proceedings, pp. 9–16 (2013). https://doi.org/10.1109/NAturaLiSE.2013.6611715
16. Abad, Z.S.H., Karras, O., Ghazi, P., Glinz, M., Ruhe, G., Schneider, K.: What works better? A study of classifying requirements. In: 2017 IEEE 25th International Requirements Engineering Conference (RE), pp. Lisbon, Portugal. pp. 496–501. IEEE (2017)
17. Sunner, D., Bajaj, H.: Classification of functional and non-functional requirements in agile by cluster neuro-genetic approach. Int. J. Softw. Eng. Appl. **10**, 129–138 (2016). https://doi.org/10.14257/ijseia.2016.10.10.13
18. Casamayor, A., Godoy, D., Campo, M.: Identification of non-functional requirements in textual specifications: A semi-supervised learning approach. Inf. Softw. Technol. **52**, 436–445 (2010). https://doi.org/10.1016/j.infsof.2009.10.010
19. Hussain, I., Kosseim, L., Ormandjieva, O.: Using linguistic knowledge to classify non-functional requirements in SRS documents. In: Kapetanios, E., Sugumaran, V., Spiliopoulou, M. (eds.) Natural Language and Information Systems, pp. 287–298. Springer, Heidelberg (2008)
20. Mahmoud, M.: Software requirements classification using natural language processing and SVD. Int. J. Comput. Appl. **164**, 7–12 (2017). https://doi.org/10.5120/ijca2017913555
21. Cleland-Huang, J., Settimi, R., Zou, X., Solc, P.: Automated classification of non-functional requirements. Requirements Eng. **12**, 103–120 (2007). https://doi.org/10.1007/s00766-007-0045-1
22. Tsai, C.-F.: Bag-of-words representation in image annotation: a review. ISRN Artif. Intell. **2012**, 1–19 (2012). https://doi.org/10.5402/2012/376804
23. Zhang, W., Yang, Y., Wang, Q., Shu, F.: An empirical study on classification of non-functional requirements. In: Proceedings of the 23rd International Conference on Software Engineering & Knowledge Engineering (SEKE 2011) (2011)
24. Chong, T.Y., Banchs, R.E., Chng, E.S., Li, H.: Modeling of term-distance and term-occurrence information for improving n-gram language model performance, p. 5 (2013)
25. Lu, M., Liang, P.: Automatic classification of non-functional requirements from augmented app user reviews. In: Proceedings of the 21st International Conference on Evaluation and Assessment in Software Engineering – EASE 2017, Karlskrona, Sweden, pp. 344–353. ACM Press (2017)

26. Kurtanovic, Z., Maalej, W.: Automatically classifying functional and non-functional requirements using supervised machine learning. In: 2017 IEEE 25th International Requirements Engineering Conference (RE), Lisbon, Portugal, pp. 490–495. IEEE (2017)
27. Baker, C., Deng, L., Chakraborty, S., Dehlinger, J.: Automatic multi-class non-functional software requirements classification using neural networks. In: 2019 IEEE 43rd Annual Computer Software and Applications Conference (COMPSAC), Milwaukee, WI, USA, pp. 610–615. IEEE (2019)
28. Roh, Y., Heo, G., Whang, S.E.: A survey on data collection for machine learning: a big data - AI integration perspective. arXiv:1811.03402 [cs, stat] (2018)
29. Tsoumakas, G., Katakis, I.: Multi-label classification: an overview. Int. J. Data Warehouse. Min. **3**, 1–13 (2007). https://doi.org/10.4018/jdwm.2007070101
30. Cleland-Huang, J., Settimi, R., Xuchang, Z., Solc, P.: The detection and classification of non-functional requirements with application to early aspects. In: 14th IEEE International Requirements Engineering Conference (RE 2006), Minneapolis/St. Paul, MN, pp. 39–48. IEEE (2006)
31. Ormandjieva, O.: Ontology-based classification of non-functional requirements in software specifications: a new corpus and svm-based classifier (2013). https://doi.org/10.1109/COMPSAC.2013.64
32. Singh, P., Singh, D., Sharma, A.: Rule-based system for automated classification of non-functional requirements from requirement specifications. In: 2016 International Conference on Advances in Computing, Communications and Informatics (ICACCI), Jaipur, India, pp. 620–626. IEEE (2016)
33. Mahmoud, A., Williams, G.: Detecting, classifying, and tracing non-functional software requirements. Requirements Eng. **21**, 357–381 (2016). https://doi.org/10.1007/s00766-016-0252-8
34. Miguel, J.P., Mauricio, D., Rodríguez, G.: A review of software quality models for the evaluation of software products. Int. J. Softw. Eng. Appl. **5**, 31–53 (2014). https://doi.org/10.5121/ijsea.2014.5603
35. Pustejovsky, J., Stubbs, A.: Natural Language Annotation for Machine Learning. O'Reilly Media, Sebastopol (2013)
36. Hinze, A., Heese, R., Luczak-Rösch, M., Paschke, A.: Semantic enrichment by non-experts: usability of manual annotation tools. In: Cudré-Mauroux, P., Heflin, J., Sirin, E., Tudorache, T., Euzenat, J., Hauswirth, M., Parreira, J.X., Hendler, J., Schreiber, G., Bernstein, A., Blomqvist, E. (eds.) The Semantic Web – ISWC 2012, pp. 165–181. Springer, Heidelberg (2012)
37. Nematzadeh, A., Meylan, S.C., Griffiths, T.L.: Evaluating vector-space models of word representation, or, the unreasonable effectiveness of counting words near other words. p. 6 (2017)
38. Mikolov, T., Chen, K., Corrado, G., Dean, J.: Efficient estimation of word representations in vector space. arXiv:1301.3781 [cs] (2013)

A Game Logic Specification Proposal for 2D Video Games

Carlos Marín-Lora(✉), Miguel Chover, and Jose M. Sotoca

Institute of New Imaging Technologies, UJI, Castellón, Spain
cmarin@uji.es

Abstract. The game engines are one of the essential and daily used applications on game development. These applications are designed to assist in the creation of games' contents. However, the games definition and specification is still complex. The mechanisms involved in the behaviour definition are not easy to standardise, given their dependency with the subjective game logic design and the selected programming language features. In this sense, this work presents the design and development of a game logic system for 2D video games. It draws from the study and analysis of the behaviour and game logic definition literature. From this, a game logic system has been developed from a first-order logic semantics and a reduced set of actions and conditions. The model has been tested with several games to determine its potential. Finally, one of these games is described with the proposed semantics, and further on it is also used as a reference for a user test against other game logic systems.

Keywords: Computer games · Game logic · Behaviour specification

1 Introduction

The development of video games is a complex process where multidisciplinary skills are required to complete it [10,12,13]. The appealing of the games depends on the designer's skills to complete the game requirements, either for character design, animation, sound composition, narrative or game-play design. To address this problem, game engines provide the required tools to develop a game. In fact, professional game studios have their own one, although there is a trend towards the usage of commercial engines such as Unity [15] or Unreal [8]. However this fact has two issues, on the one hand, there is some obscurantism about the proprietary game engines, while on the other hand, the commercial engines implement a huge number of features turning them into really complex applications to use for the non-expert public. Additionally, from a theoretical point of view, there are claims to expand the game engines research field, since the formal requirements are unknown and there is little literature [1,2]. Specifically, Anderson et al. [2] highlights the lack of literature in this regard and proposes several research lines that should be explored in the future: a list of common software components, a

Á. Rocha et al. (Eds.): WorldCIST 2020, AISC 1159, pp. 494–504, 2020.
https://doi.org/10.1007/978-3-030-45688-7_50

unified language, generic content creation tools, a generic architecture and the identification of the best practices.

Conversely to the general trend towards the democratisation of content creation, it is still very difficult to work with game engines. Especially on the game logic definition, since they imply knowledge of scripting and game development APIs. The game logic definition is arguably the most subjective technical process in the video game development. In a script-based game engine such as Unity, two different programmers could solve the same mechanic differently depending on their initial approach. This is one of the reasons why the game logic standardisation is complicated, which hinders the game logic optimisation.

Currently, there are alternatives such as visual scripting systems. These systems transfer traditional programming to more comfortable visual elements for non-expert users. Indeed, commercial game engines, as mighty as Unreal, use a Blueprints system based on message passing. However, these systems are still not easy to use and they still rely on huge APIs. It is delicate to determine which is the proper alternative to save these issues since the simplicity perception or ease of use brings forward subjective experience variables.

In this sense, this work proposes a 2D game logic specification along with a game engine requirements to run it. In order to verify its proper operation, a minimal game engine has been implemented meeting essential requirements. However, the game logic specification method proposed in this work could be implemented in any game engine. The proposal aims to generate a theoretical knowledge that serves as a reference for the generation and optimisation of game engines, providing knowledge on the essential requirements for the creation of 2D games. As a starting hypothesis, it is considered that it is possible to define complex behaviours from a reduced set of logical elements. During its development, a large number of 2D games have been implemented to check the capabilities of the system and, as far as it has been tested, the response has been positive since until now all the mechanics have been completed. As an example for this paper, a 2D arcade platform game is described down below. In addition, that game has also been used as a reference for a user test where its development is compared with two game engines and their game logic systems.

The rest of the article is organised as follows: Sect. 2 lists the game engine requirements, in Sect. 3 the game logic specification system is stated and in Sect. 4 the functions that encapsulate it are presented. Next, in Sect. 5, the 2D arcade game is described and then Sect. 6 presents the experiment carried out to determine if it is perceived as easier to use. Finally, Sect. 7 shows the conclusions gained from this development and the possible future work lines.

2 Game Engine Overview

Before detailing the game logic system, it is necessary to define some functional requirements for the minimal game engine. For simplicity, the game engine is conceived for 2D games and it must include the following features:

- The game is composed of a certain number of scenes.
- The game has a set of properties, available for the entire system.
- A scene is composed of a set of game objects with no hierarchies.
- A game object has different components: physics, sound, rendering, etc.
- Each component has a set of properties.
- Each game object has a set of behaviour rules.

Besides that, the game logic system requires a series of characteristics for its specification. These must allow game objects to support the game logic completeness, providing them with the following capabilities:

- Create and delete game objects.
- Read and update game properties or game objects properties.
- Execute changes and queries about any property.
- Check execution properties such as collisions, interaction events and timers.

In such a way, the system shall ensure a complete operation of the game state and its elements. Where the run of the game loop works with its properties. These properties define the game and its game objects features. In the proposed game engine, a set of generic properties have been selected for the game and the game objects, and both can handle new knowledge as new properties.

Game

- **General**: Name, resolution, active scene, and camera properties.
- **Sound**: Sound files, start, volume, pan and loop.
- **Physics**: Gravity and air friction.
- **New**: Custom shared properties for the game, admitting boolean or numeric.

Game Object

- **General**: Name, position, angle, dimensions, collision shape and tags.
- **Render**: Texture, opacity, flip, scroll and tile.
- **Text**: Text, font, size, color, style and offset.
- **Sound**: Sound files, start, volume, pan and loop.
- **Physics**: Velocity, rotation, density, friction, restitution and damping.
- **New**: As for the game, custom properties to complete functionalities.

3 Game Logic Specification

The game logic assigns to the game objects the tasks to perform based on the game state, following a predefined semantics. Its development is based on first-order logic (FOL), widely used in other areas such as robotics and automation [9,14,17]. The semantics are defined by a syntax of logical and non-logical symbols. Logical symbols include logical connectives and variables while non-logical symbols include predicates and functions [4]. The domain D for this interpretation I is the set of the game objects and the game. The predicates Ψ, defined in

FOL, specify the behaviour rules. In such a way, the tasks that a game object must perform are organised in predicate formulas

$$\{\Psi_0 \wedge \Psi_1 \wedge \Psi_2 \wedge \ldots\}$$

where its elements can have the following structures based on just two predicates: the *If* condition and the *Do* action.

- **Action structure**: Composed by an atomic element that includes a single literal predicate *Do*.
- **Conditional structure**: Modelled by a predicate sequence based on the IF-THEN-ELSE rule structure [11] where *If* is a conditional literal, and both ϕ and θ are new predicate sequences to be evaluated if the condition is met or if it is not, respectively.

$$(\mathit{If} \rightarrow \phi) \wedge (\neg \mathit{If} \rightarrow \theta)$$

For the predicates, expressions are essential since they evaluate the game state based on parameters, either as a boolean or as an arithmetic expressions.

- **Arithmetic expression**: Performs mathematical operations either with numerical constants, game or game objects properties, or mathematical functions such as *sin*, *cos*, *tan*, *asin*, *acos*, *atan*, *sqrt*, *random* and *abs*.
- **Boolean expression**: Evaluates logical relationships between game elements and arithmetic expressions using relational operators such as *greater*, *greater-or-equal*, *equal*, *less-or-equal* and *less*.

Action Predicate *Do*

An action is defined as a behaviour to be performed by a game object. In this specification, the actions are formalised as the non-logical function elements which can handle parameters as arithmetic expressions. Where the set of actions is based on the *create*, *read*, *update* and *delete* (CRUD) operations for information persistence on databases [6] applied to game and game objects properties. From these operations, the system can produce more complex actions raising the system's level of abstraction. In this way, the available operations are:

- **Create**: Creates a new element, as a copy of an existing game object.
- **Read**: It allows to read the information of a game or a game object property. The syntax *gameObject.property* is used to read this information.
- **Update**: Modifies the value of a game or a game object property. The new value is determined from the evaluation of an arithmetic expression.
- **Delete**: Removes a game object from the game.

Conditional Predicate *If*

The conditional predicate represents the evaluation element for a condition in this decision-making process. Where the condition evaluation is based on the result of a boolean expression that assesses the relationship between properties.

$$\mathit{If}(booleanExpression)$$

Conceptually, this boolean expression evaluates the relationship between properties with an arithmetic expression. For this, it has an established structure composed of a *read* operation, a relational operator and an arithmetic expression.

$$property\,[\geq, >, ==, <, \leq]\;arithmeticExpression$$

Following this theoretical framework, the proposed game logic specification is able to define behaviours for 2D games.

4 Functions

From the previous section, the game logic system could compose game mechanics. However, this low-level semantics is not the most comfortable to use. This section presents the function encapsulation of the low-level operations presented above aiming to increase the level of abstraction without giving up on its potential. For this purpose, a reduced set of actions and conditions are defined, in order to create high-level functions to ease the behaviours definition.

Actions

- **Spawn**: Directly derived from the *create* operation, it spawns a game object in a position and in an angle as a copy of an existing one. This action is normally used to create enemies or projectiles.
- **Edit**: Execution of an *update* operation to modify a property from the game or a game object. The value comes from an arithmetic expression evaluation.
- **Destroy**: Implements the *delete* operation over a game object.
- **Move**: Specialisation of the *update* operation that applies a displacement on the game object by an angle and a speed as arithmetic expressions.
- **Rotate**: Equivalent to the *Move* action for angular displacements depending on a pivot and a speed, where both parameters are arithmetic expressions.
- **Push**: Physics-based alternative to the *Move*, where a force is applied to the game object in a given angle. It relies on the object's physics component.
- **Animate**: Texture swapping to produce a key-frame animation controlled by an arithmetic expression for the frames-per-second rate.
- **Play Sound**: Audio playback of a sound stored in the game data.
- **Change Scene**: Scene swapping to change the information from a scene to another. This action could be used to switch between game levels.

Conditions

- **Check**: Evaluation of a boolean property from the game or a game object.
- **Compare**: Relational condition between a property with an arithmetic expressions in modes *greater*, *greater-equal*, *equal*, *less-equal* or *less*.
- **Timer**: Temporary condition for a specific time where a system timer is compared with a cut-off time. It will always be false as long as the timer is lower than the cut-off time. When true, the timer needs to be restarted.
- **Collision**: Collision detection between game objects based on tags.
- **Keyboard**: Keystrokes condition for predetermined keys in a key mode.
- **Pointer**: Pointer controller managed by the game engine's input system.

5 Use Case

During the development of this work, several 2D games have been implemented in order to test its performance and its capabilities. In this section, a 2D platformer inspired on the *London's Tower Bridge Defense* game [16] is presented as an example. The selection of this game is underpinned by the reliability of its design, since it is one of the advanced 2D tutorial games of the Unity learning platform and it is assumed that its mechanics and procedures are optimised and well structured. Figure 1 shows a capture of the outcoming game in its execution.

Fig. 1. Capture of the 2D platformer game.

For copyright reasons, the assets employed are from a public library with no usage restrictions [3]. Which has just affected the game's aesthetics aspect. Next, the game is described by the semantics established throughout this work.

Game Objects

- **Player**: The game's character controlled by the user. It has to accumulate as many points as possible by destroying *Enemy* game objects while avoids their contact. It can move to the right, to the left, jump and shoot *Bullet* game objects over the enemies. Points and lives are stored in custom properties.
- **Enemy**: The antagonist character who falls into the scene and randomly chooses a movement direction. It bounces when collides with the scene horizontal limits until it falls through the drain and is eliminated. It can harm the *Player* and can be destroyed with *Bullet* game objects.
- **Ground**: Static object setting the *Player* and the *Enemy* walkable zones.
- **Bullet**: Spawned by the *Player* to destroy *Enemys*. It moves in the *Player*'s direction until it reaches an *Enemy* or the scene limits, then it is destroyed.
- **Aid**: *Player*'s health recovery, randomly spawned by the *Aid Spawner*. When they interact, the *Player*'s lives are increased and it is destroyed.

- **Score**: Scoring display for the *Player*'s points property. It is increased if a *Enemy* is destroyed by a *Bullet.*
- **Stopwatch**: Time controller for the sixty seconds available for the game. When it reaches zero, the game ends.
- **Enemy Spawner**: Auxiliary object who randomly drops *Enemy* objects from the screen top.
- **Aid Spawner**: *Enemy Spawner* akin, but dropping *Aid* in a lesser frequency.

Each game object has a behaviour rules set to perform their tasks. Since present all the behaviour rules is unrealistic given its extension, the rules associated with the *Player* and the *Enemy* movements are described next. For its correct operation, some of their properties have been balanced. Generally, in a platform game, the *Player* is restricted to shifts from left to right and to jumps. In this case, the jumps are constrained by a ground-contact condition. Additionally, the user interaction works with keyboard events, but the same structure would be valid with pointer events.

- **Player movement to the left**:
  $If(keyboard(keyLeft, pressed)) \rightarrow$
  $If(compare(maxSpeed, greater, abs(player.velocityX))) \rightarrow$
  $If(collision(groundTag)) \rightarrow$
  $Do(push(player.thrust, 180)) \wedge Do(edit(player.horizontalFlip, true)) \wedge$
  $Do(animate(player.walkAnimationTextureList, 24))$
- **Player movement to the right**:
  $If(keyboard(keyRight, pressed)) \rightarrow$
  $If(compare(maxSpeed, greater, abs(player.velocityX))) \rightarrow$
  $If(collision(groundTag)) \rightarrow$
  $Do(push(player.thrust, 0)) \wedge Do(edit(player.horizontalFlip, false)) \wedge$
  $Do(animate(player.walkAnimationTextureList, 24))$
- **Player jump**:
  $If(keyboard(keySpace, pressed)) \rightarrow$
  $If(collision(groundTag)) \rightarrow$
  $Do(push(10 * player.thrust, 90)) \wedge$
  $Do(animate(player.jumpAnimationTextureList, 24))$

Conversely, the *Enemy* movement is completely different since its movements are autonomous and are determined by two behaviour rules dependent on collision events. The first happens with its first contact with the central ground, when it will randomly select a movement direction (left or right). The second is triggered if it reaches the scene horizontal limits, when it will shift its direction.

- **Enemy movement initialization**:
  $If(collision(centralPlatformTag)) \rightarrow$
  $If(check(enemy.first)) \rightarrow$
  $Do(edit(enemy.velocityX, thrust * (1 - 2 * random(0, 1)))) \wedge$
  $Do(edit(enemy.first, false))$

- **Enemy boundaries**:
  $If(collision(boundariesTag)) \rightarrow$
  $Do(edit(enemy.velocityX, enemy.velocityX * -1))$

6 Experiment

From the previous section, an experiment has been carried out in order to check if it works in an easier way than other methods. During the activity, the participants developed the *Player* movement (left and right shifts, and jumps) in Unity, Unreal and in a minimal game engine known as GLS arised from this work. For Unity, the scripting system was C# while for Unreal was its Blueprints system. Besides that, the GLS method relies on a decision tree visual editor based on the Sect. 4's theory. Upon completion, the participants filled a usability test in order to check if the proposed system is perceived as easier to use than the other commercial methods. It draws from the hypothesis that the game logic specification can be stated in a easier way than other traditional methods. The method selected to measure the degree of usability perceived is the System Usability Scale (SUS) [5]. This method provides a quick metric through a 10-item questionnaire with five response options, from the most absolute disagreement (1) to the most absolute agreement (5). The questionnaire items are the following:

Q1. I think that I would like to use this system frequently.
Q2. I found the system unnecessarily complex.
Q3. I thought the system was easy to use.
Q4. I think that I would need the support of a technical person to be able to use this system.
Q5. I found the various functions in this system were well integrated.
Q6. I thought there was too much inconsistency in this system.
Q7. I would imagine that most people would learn to use this system quickly.
Q8. I found the system very cumbersome to use.
Q9. I felt very confident using the system.
Q10. I needed to learn a lot of things before I could get going with this system.

6.1 Results

The experiment was carried out with 27 participants. Three of them had never used any of the evaluated systems, while the rest knew at least one and only two of them had worked with the three systems. In order to measure the users' previous experience, they have rated their previous knowledge in a range from 1 to 5, from nonexistent to expert respectively, where the average result obtained was 2.85. Besides that, game comprehension, a practical demonstration was carried out in all three cases with the *Player*'s movement specification. The results obtained are shown in Table 1. The three central columns show the results average obtained by the methods in each item of the questionnaire along with its standard deviation shown in parentheses. In order to check the questionnaire

Table 1. SUS test results and Friedman significance evaluation

Items	Unreal	GLS	Unity	Friedman
Q1	3.04 (±0.98)	3.96 (±1.13)	3.04 (±1.43)	10.9 (+)
Q2	3.52 (±1.25)	1.81 (±1.04)	3.93 (±0.96)	24.5 (+)
Q3	2.67 (±1.21)	4.56 (±1.09)	2.41 (±1.39)	22.2 (+)
Q4	3.81 (±1.14)	1.70 (±0.82)	4.00 (±1.18)	32.1 (+)
Q5	3.11 (±1.15)	4.22 (±1.01)	2.93 (±1.33)	9.9 (+)
Q6	3.04 (±1.06)	1.93 (±0.87)	3.37 (±1.60)	9.2 (+)
Q7	2.67 (±1.18)	4.81 (±0.48)	2.04 (±1.06)	35.6 (+)
Q8	3.85 (±0.99)	1.37 (±0.63)	3.89 (±1.22)	33.9 (+)
Q9	2.25 (±0.94)	4.44 (±0.70)	2.59 (±1.45)	25.7 (+)
Q10	3.93 (±0.96)	1.59 (±0.75)	4.26 (±0.98)	38.7 (+)

statistical significance, a Friedman's non-parametric test has been applied [7]. This is a non-parametric technique to measure the significance of the statistical difference between the applied methods. In this case, the Friedman test considers all ratings with the same importance and to discern the statistical significance of the results the Chi-square distribution was used. For all three methods, the Chi-square value is 5.99 with a confidence level of $p = 0.05$, which would imply that the resulting values lesser than this reference value do not have statistical significance. The values of the statistical significance are shown in the right column of Table 1, where the positive statistical significance is indicated by (+) when the value is greater than 5.99, and by (−) in the opposite case.

7 Conclusions and Future Work

This work proposes a game logic specification to define 2D games mechanics based on a reduced set of behaviour elements. The analysis carried out on the requirements and on game engines features have shown that there is a research need to determine the games' essential requirements and the correct procedures for their design. From this information, a data model has been built supporting the functional requirements raised at the beginning of the paper. Besides that, a semantics based on first-order logic has been defined with the aim of generate logic rules from a reduced set of actions and conditions. All this without hierarchies, loops or complex data structures such as matrices or vectors. Additionally, the proposed system has been tested against two other commercial systems such as Unity and Unreal. The first one based on traditional scripting and the second one based on message passing. The results show that the users perceived GLS as more usable than the other systems analyzed. The Unity and Unreal methods were virtually tied for second place, although there is a certain advantage in favour of Unity, probably due to some users were previously experienced on this one. This fact is stated in Q1, where GLS beats Unity and Unreal in preference of use with a difference of almost one point. Nevertheless, this has to be carefully considered due to its relationship with subjective preferences, motivations and previous experience. The questions Q2 and Q3 show that GLS is perceived as easier to use against the other two. In addition, it seems that the Unity's scripting language is a bit more complex for the users than a message-passing language as the Unreal Blueprints. In this way, it follows that users establish as unnecessary the use of complex methods to complete these game development tasks. In fact, question Q4 supports this conclusion and confirms that it is necessary to reduce learning curves in order to democratise game development. On questions Q5 and Q6 regarding the integration of functions and the inconsistencies detected, the results show a clear advantage of GLS. Since it uses a reduced set of actions and conditions while Unreal and Unity methods are penalised for the large number of functions they contain and their huge amount of parameters. In addition, the users considered GLS as much faster to learn than the others systems and that they also find it much less cumbersome to use than Unity and Unreal, with a difference of almost two points in questions Q7 and Q8. Finally, Q9 and Q10 show that GLS seems safer for upstarters, again a two points difference in their averages. As future work, it is intended to keep exploring the capabilities of this design in three ways: its extension to a 3D game engine, its application on a multi-user point of view and its potential implementations in parallel.

Acknowledgments. This work has been supported by the Ministry of Science and Technology (TIN2016- 75866-C3-1-R, TI2018-098651-B-C54) and the Universitat Jaume I research projects (UJI-B2018-56, UJI-B2018-44).

References

1. Ampatzoglou, A., Stamelos, I.: Software engineering research for computer games: a systematic review. Inf. Soft Tech **52**(9), 888–901 (2010)
2. Anderson, E.F., Engel, S., McLoughlin, L., Comninos, P.: The case for research in game engine architecture (2008)
3. Art, K.: Abstract platformer (2019)
4. Brachman, R., Levesque, H., Reiter, R.: Knowledge Representation. MIT (1992)
5. Brooke, J., et al.: SUS-A quick and dirty usability scale. In: Usability Evaluation in Industry, vol. 189, no. 194, pp. 4–7 (1996)
6. Daissaoui, A.: Applying the MDA approach for the automatic generation of an MVC2 web application. In: 2010 Fourth International Conference on Research Challenges in Information Science (RCIS), pp. 681–688. IEEE (2010)
7. Friedman, M.: The use of ranks to avoid the assumption of normality implicit in the analysis of variance. J. Amer. Stat. Assoc. **32**(200), 675–701 (1937)
8. Games, E.: Unreal engine 4 (2019)
9. Geller, J.: Knowledge representation: Logical, philosophical, and computational foundations, brooks/cole, 2000, 512 pp. Mind. Mach. **13**(3), 441–444 (2003)
10. Gregory, J.: Game Engine Architecture. AK Peters/CRC Press (2017)
11. Karplus, K.: Using if-then-else DAGs for multi-level logic minimization. University of California, Santa Cruz, Computer Research Laboratory (1988)
12. Millington, I.: AI for Games. CRC Press (2019)

13. Nystrom, R.: Game programming patterns. Genever Benning (2014)
14. Patel-Schneider, P.F.: A decidable first-order logic for knowledge representation. IJCAI **9**, 455–458 (1985)
15. Unity Technologies: Unity 3D engine (2019)
16. Unity Technologies: Unity tower bridge defense tutorial (2019)
17. Wang, C., Joshi, S., Khardon, R.: First order decision diagrams for relational MDPs. J. Artif. Intell. Res. **31**, 431–472 (2008)

Organizational Models and Information Systems

Measuring Consumer Behavioural Intention to Accept Technology: Towards Autonomous Vehicles Technology Acceptance Model (AVTAM)

Patrice Seuwou[1,2](✉), Christos Chrysoulas[2], Ebad Banissi[2], and George Ubakanma[2]

[1] Northampton University, Northampton NN1 5PH, UK
patrice.seuwou@northampton.ac.uk, seuwoup@lsbu.ac.uk
[2] London South Bank University, London SE1 0AA, UK
{chrysouc,banisse,george.ubakanma}@lsbu.ac.uk

Abstract. The work presented in the paper aims at exploring information technology acceptance in the context of Autonomous Vehicles (AV) with the objectives of identifying and testing the constructs that will influence future AVs acceptance. Most models of technology acceptance focus on barriers of successful information technology implementation in organisations and technologies that have already been deployed. There is only a small number of studies conducted on emerging and disruptive technologies such as AV. We address this issue by deriving context-related determinants from an extensive literature analysis and further describing a technology acceptance modeling process to provide an explanation for drivers' and factors influencing people behavioural intention to accept AV technology. Based on our evaluation we take the determinants self-efficacy, perceived safety, trust, anxiety and legal regulation into consideration. The outcome and main contribution of this paper is the proposal of a theoretical AV technology acceptance model (AVTAM). This study concluded that the performance of these AV technologies, their safety on the road and consumer's trust for the AV manufacturers will play a very important role for mass AV adoption.

Keywords: Unified Theory of Acceptance and Use of Technology (UTAUT) · UTAUT2 · Autonomous vehicles · Technology acceptance model · AVTAM · Context

1 Introduction

For decades, the user acceptance of information technology has been a vital field of study for psychologists and sociologists investigating new insights into the acceptance of behaviour at individual and organisational levels [1]. As the application of computer technology continues to proliferate and diversify, vehicles are becoming increasingly intelligent and it is expected that soon they will be equipped with autonomous technologies allowing them to drive passengers from one location to another without any

Á. Rocha et al. (Eds.): WorldCIST 2020, AISC 1159, pp. 507–516, 2020.
https://doi.org/10.1007/978-3-030-45688-7_51

human intervention. AV also known as self-driving cars or driverless cars are vehicles equipped with several technologies including radar, laser light, GPS, odometry, computer vision and are capable of sensing their environment and navigating without human input. Nowadays, many car drivers do not realise that their cars are already equipped with various levels of assistance. Indeed, the current generation of vehicles can already detect, anticipate and correct autonomously other road users' maneuvers. Certainly, AV is a disruptive technology. Not all technologies predicted by popular media are immediately welcomed into society, and AV is one such technology. Despite numerous models being proposed to predict consumer use behaviour of technology, there is a gap in the literature as the latest models and theories are still not able to fully capture the complexity of the factors influencing people behavioural intention to adopt AV [1, 2, 7]. The Unified Theory of Acceptance and Use of Technology (UTAUT) and UTAUT2 are models that have been used for years to help us understand the drivers for acceptance of technologies by its users [4]. It appears that AV adoption cannot be measured only using existing models. Furthermore, there are several important factors also not represented on the latest car technology acceptance research model, therefore, the absolute need to develop a model specifically for AV is required. The paper proceeds as follows: First, technology acceptance concept is thoroughly reviewed to provide insights in the development process and the predictive character of acceptance models, second, determinants that effect acceptance of AV are examined. We perform a literature review as well as an analysis of qualitative results from a technology acceptance evaluation to derive and define the constructs "self-efficacy", "perceived safety", "trust", "anxiety" and "legal regulation". They form the basis for a theoretical AV Technology Acceptance Model (AVTAM) which is presented based on the UTAUT2 structure. Finally, the implications followed by future work are discussed. The rest of the paper is organised as follows: Sect. 2 gives an overview of the literature Review on AV acceptance, Sect. 3 discusses the main constructs identified for the proposed AVTAM model with the hypothesis, Sect. 4 provides the methods and techniques used to test the proposed research model, the results are presented in Sect. 5 followed by a brief conclusion with future work.

2 Literature Review

Automation of vehicles and driverless technology has attracted a lot of people from the stakeholders to the media. Stakeholders have been yearning to develop and take the automated driving technology a notch higher and meet the increasing demands of consumers. Public exposure to the different forms of electric vehicles has slowly gained momentum in recent years with various manufacturers carrying out test runs and pilot programs in Europe. Such programs help towards examining the user acceptance of these modern vehicles and act as prerequisites to ensure such investment pay off. Many authors believe that customer acceptance is likely to be the biggest obstacle to autonomous vehicle penetration [1–3]. At first, many consumers may be reluctant to put their lives in the hands of a robot. Recent studies and surveys have shown a split in opinion as to whether people would like autonomous capability to be available in their vehicles or not. User acceptance is a key ingredient to the successful adoption of autonomous vehicles. In the last three decades, the global car market has gone through some rapid changes especially in driver experiences, thanks to the technological advancement. For instance, a typical middle-class classic car has standard features such as automatic gearbox, an electronic stability program and power windows [4].

For the high-end vehicles, they can have intelligent features such as an automatic startup framework, self-parking technology and active lane assistance. The innovations have led to comfortable driving. However, the automobiles still need human interaction to foster their movement. Today's innovations focus on automobiles that are fully autonomous and free from direct human interaction. Driverless vehicles are those that operate without direct input by the driver to control the acceleration, steering or braking and are designed in such a way that the controller does not continuously observe the roadway when the car is moving in self-driving mode. The driver-less vehicle technology has been too advanced; opening new opportunities for the automotive dealers, individual users and also the potential manufacturers [5].

There are diverse factors that influence vehicle-purchasing behaviour amongst the people. There are the situational factors that include regulatory environments. Besides the situational factors, psychological factors also play an essential role in the purchase of autonomous vehicles and include the personal attitudes. The extensive adoption of electric cars can be understood through the examination of the different factors that influence consumer acceptance. The acceptance is critical towards the commercial success or failure of the AVs.

According to [4], there are different factors that influence and motivate the environmentally conscious behaviours. The elements encompass the personal values; environmental concerns as well as the belief that one can make a difference in the use of such vehicles. Also, these factors contribute towards an individual's purchase intentions for the alternative fuel vehicles. Gas prices significantly impact the individual's interest in buying hybrid electric cars. There are other customer preferences such as reduced air pollution, reduced maintenance and better handling. There is no doubt that environmentalists and people that have strong preferences for energy security will automatically adopt the hybrid electric cars. In their research, [6] concluded that among the top attributes considered before a new vehicle purchase include price, reliability and the fuel economy.

Different societies have distinct attributes relating to their preferences for the autonomous vehicles. According to the unified theory of acceptance, the degree of technology acceptance varies. People in Hong Kong prioritize the environmental benefits of these innovations followed by the economic and social benefits. Negative factors have also been identified as barriers towards the purchase of the electric vehicles. These include lack of AV infrastructure, limited selection of the vehicles and potential increases in the electrical rates. Results of the study by [7] highlighted four determinant factors affecting purchasing behaviours of the potential buyers: high costs, charge inconvenience, psychological factors and short battery range. A study in China also revealed some deterministic factors or concerns when purchasing autonomous vehicles such as the cost of operations, overall buying cost, possible speeds and battery capacity. In the USA, the potential buyers considered the charge time and range as concerns other than the value of the hybrid vehicles. Findings by [8] also show that the attributes, personal norms, the perceived behavioural control as well as the planning abilities influenced the decision to buy an autonomous vehicle. Customers will always be concerned about the potential advantages as well as the promises such as safety benefits, times savings, fuel savings.

3 Formulation of the Autonomous Vehicles Technology Acceptance Model (AVTAM)

Our study was conducted in the UK and our target population was potential AV users which represented any car users. Several interviews were conducted with academics and practitioner from the fields of Psychology, Sociology and Computer Science. Numerous constructs were proposed. The following research model was created (see Fig. 1).

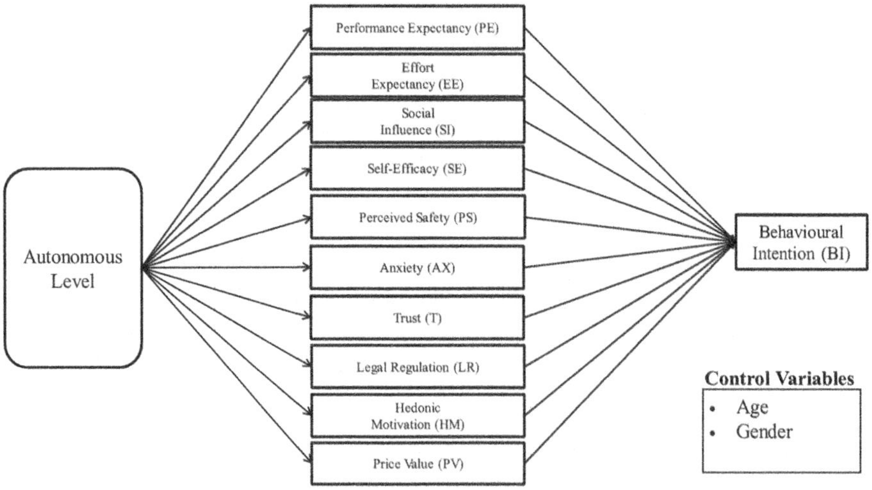

Fig. 1. Research model for measuring consumers' behavioural intention to adopt autonomous vehicles: Autonomous Vehicle Technology Acceptance Model (AVTAM)

In order to maximise understanding of different viewpoints and explore other potentials factors not previously identified by existing models, an initial survey was conducted at the beginning of the study. Our target population was mainly in England, the sample size was 408 participants from all age groups, race and cultural backgrounds with a minimum of 18 years of age. In this study, a convenience sampling technique was used data was collected from staffs and students at various universities in UK, mainly due to the challenges of having access to existing AV users and technology manufacturers. Furthermore, due to the complexity of the questionnaires more appropriate for a certain demographic. The following factors and hypotheses were identified.

Performance Expectancy (PE)

Performance expectancy is defined as the degree to which an individual believes that using autonomous vehicles will help him or her to attain gains in daily life activities, increase productivity, decrease possibilities of accidents on the road and makes driving more secure. PE also takes into consideration security features of the car; provide some level of satisfaction about trust, and privacy protection. Gender and age is theorised to play a moderating role. The influence of performance expectancy on behavioural intention will be moderated by gender and age, such that the effect will be stronger for men and in particular for younger men.

H1: *Performance Expectancy will be positively related to behavioural intention of using autonomous vehicles.*

Effort Expectancy (EE)

Effort Expectancy is defined as the degree of ease associated with the use of autonomous vehicles. The ease is associated with learning how to use autonomous vehicles and how clear and understandable the interaction with the technology is. The influence of effort expectancy on behavioural intention will be moderated by gender, age and experience. It is suggested that effort expectancy is more salient for women than for men. The gender differences predicted here could be driven by cognitions related to gender roles. Increased age has been shown to be associated with difficulty in processing complex stimuli and allocating attention to information. Thus, the study proposes that effort expectancy will be most salient for women, particularly those who are older and with relatively little experience with autonomous vehicles. The influence of effort expectancy on behavioural intention will be moderated by gender, age and experience, such that the effect will be stronger for women, particularly younger women and particularly at early stages of experience.

H2: *Effort Expectancy will be positively related to behavioural intention of using autonomous vehicles.*

Social Influence (SI)

Social Influence is defined as the degree to which an individual perceives that important others believe he or she should use the technology. Social influence occurs when one's emotions, opinions, or behaviours are affected by others. Social influence takes many forms and can be seen in conformity, socialisation, peer pressure obedience, persuasion, sales and marketing, and review of information. Furthermore, social influence as a direct determinant of behavioural intention contains the explicit or implicit notion that the individual's behaviour is influenced by the way in which they believe others will view them as a result of having used autonomous vehicles. The impact of social influence on behavioural intention will be moderated by gender, age and experience, such that the effect will be stronger for women, particularly older women, particularly older women in mandatory stages of experience. The following elements are included in this factor: Journalists, Expert reviewers, Friends, Family, media.

H3: *Social Influence will be positively related to behavioural intention of using autonomous vehicles.*

Self-Efficacy (SE)

The determinant self-efficacy is defined as a person's belief in his/her ability and competence to use a technology (e.g. a radio) to accomplish a particular task. It can play a major role in how one approaches a goal or a task as it is developed from external experiences and self-perception. In the Autonomous car context, people with high self-efficacy - those who believe they can perform well - are more likely to view a difficult task as something to be mastered rather that something to be avoided. Aligning results from this construct with results from the perceived safety construct, the researcher believes that this will gives us an interesting insight into how an information system-related task is judged based on users' personality characteristics.

H4: *Self-Efficacy will be positively related to behavioural intention of using autonomous vehicles.*

Perceived Safety (PS)

Perceived safety is defined as the degree to which an individual believes that using a system will affect his or her well-being. The researcher named the construct perceived safety considering the self-reflective character of perceiving a situation hazardous. Within the autonomous car, this also comprises the judgment of one's own driving skills and safety feeling in relation to other drivers. The impact of perceived safety is assumed as critical in the process of predicting the behavioural intention to use, as the user will estimate the potential effect of safety-related consequences through using the technology on public road and dangerous environments. The following elements are included in this factor: Vehicle security and safety, cyber security concerns

H5*: Perceived Safety will be positively related to behavioural intention of using autonomous vehicles.*

Anxiety (AX)

Anxiety is defined in the autonomous car context as the degree to which a person responds to a situation with apprehension, uneasiness or feelings of arousal. The factors Anxiety and Behavioural Anxiety differ with regards to their origin, as Anxiety was derived from a computer anxiety construct described within the SCT and was used earlier in the UTAUT validation. Behavioural Anxiety otherwise reflects anxiety in a more general understanding towards the autonomous vehicle or system behaviour which addresses e.g. the fear to lose control of the car.

H6*: Anxiety will be negatively related to behavioural intention of using autonomous vehicles.*

Trust (T)

Acceptance of the truth of statements from AV manufacturers without evidence or investigation. The belief in the reliability, truth, or ability of the car. The belief that the user can trust the car and the algorithm to function as advertised while protecting consumers' privacy.

H7*: Trust will be positively related to behavioural intention of using autonomous vehicles.*

Legal Regulation (LR)

The rules or directives made and maintained by an authority and transportation regulatory bodies. The following elements are included in this factor:

Legislation, policies, law, liability, regulations, effort by the government and car manufacturers to better support users.

H8: *Legal Regulation will be positively related to behavioural intention of using autonomous vehicles.*

Hedonic Motivation (HM)

Hedonic Motivation is defined as the fun or pleasure derived from using a technology, and it has been shown to play an important role in determining technology acceptance and use. Thus, the influence of hedonic motivation on behavioural intention will be moderated by gender, age and experience.

H9: *Hedonic Motivation will be positively related to behavioural intention of using autonomous vehicles.*

Price Value (PV)

The cost and pricing structure may have a significant impact on consumers' technology use. In marketing research, the monetary cost/price is usually conceptualized together with the quality of products or services to determine the perceived value of products or services. We followed these ideas and define price value as consumers' cognitive tradeoff between the perceived benefits of the applications and the monetary cost for using them. The price value is positive when the benefits of using a technology are perceived to be greater than the monetary cost and such price value has a positive impact on intention. Thus, price value has been added as a predictor of behavioural intention to use autonomous vehicles. The following elements are included in this factor: Cost of the vehicle, operational cost and maintenance cost, cost of network protocols being used LTE/WAVE/Wifi.

H10: *Price Value will have a significant influence on behavioural intention of using autonomous vehicles.*

Behavioural Intention (BI)

Behavioural intention is defined as an individual's intention to perform a given act which can predict corresponding behaviours when an individual acts voluntarily. Besides that, behavioural intention is the subjective probability of carrying out behaviour and also the cause of certain usage behaviour [9]. Thus, intentions show the motivational factors that influence behaviour and are indicators of how hard people are willing to try and the effort they put in to engage in a behaviour. Also, it was found that behavioural intention is to be the main factor of individual mobile services usage and that usage intentions are rational indicators of future system use.

4 Method

In this study, we first examined the basic UTAUT model with the fundamental constructs in the original UTAUT model [10] and its extension, the UTAUT2 together with the Car Technology Acceptance Model (CTAM). To test and validate the proposed model, our target population was the current vehicles drivers and general transportation consumers.

Our study was conducted in the UK. All items of the model were measured using a seven-point Likert scale, with the anchors being "strongly disagree" and "strongly agree." There were 482 valid responses for the second survey. Data analysis for the final conceptual model was performed by Structured Equation Modelling (SEM) using the R Studio programming and the Statistics Package for Social Science (SPSS) system to analyse the data using reliability analysis, correlation analysis, and regression analysis.

5 Results

In this section, we use Cronbach Alpha to measure the internal consistency that is, how closely related set of items are as a group. All the values are greater than 0.7, which clearly shows the reliability of the scale/research instrument. Dillon-Goldstein rho should be higher than 0.8, as the minimum is 0.91 this is satisfactory (Table 1).

Table 1. Cronbach's alpha

#	Measure	Cronbach's alpha
1	Performance expectancy	0.91
2	Effort expectancy	0.93
3	Social influence	0.82
4	Self-efficacy	0.86
5	Perceived safety	0.80
6	Anxiety	0.88
7	Trust	0.85
8	Legal regulation	0.90
9	Hedonic motivation	0.96
10	Price value	0.87
11	Behavioral intention	0.92

Table 2 below presents the correlation matrix showing the correlation coefficients between different constructs part of the proposed model. The most correlated variables are PE & BI (0.81), so with the higher performance expectancy there is higher behavioural intentions, and PE & PS (0.73) and the least correlated variable is the Age. The anxiety is negatively correlated and that is good, the maximum is for Perceived safety (−0.74), which corresponds to what could be expected, as higher Perceived Safety implies less Anxiety.

Both the goodness of fit indices and parameter estimates coefficients were examined to check whether the hypothesised structural model fitted the data and to test the hypotheses. The fit indices indicated that the hypothesised structural model provided the good fit to the data. All hypotheses were supported (Table 3).

Table 2. Correlations matrix

Measure	PE	EE	SI	SE	PS	AX	T	LR	HM	PV	BI
PE	1										
EE	0.66	1									
SI	0.73	0.62	1								
SE	0.55	0.62	0.6	1							
PS	0.68	0.6	0.48	0.42	1						
AX	0.51	−0.5	0.38	0.33	0.74	1					
T	0.64	0.57	0.55	0.53	0.6	0.45	1				
LR	0.55	0.48	0.54	0.46	0.43	−0.3	0.72	1			
HM	0.58	0.45	0.53	0.42	0.36	−0.3	0.5	0.45	1		
PV	0.56	0.49	0.5	0.4	0.51	0.43	0.54	0.5	0.42	1	
BI	0.81	0.63	0.72	0.55	0.7	0.57	0.66	0.55	0.64	0.66	1

Table 3. Results of the original structural model

Construct	Code name	Hypotheses	Relationship (Positive)	Standardized regression weights (β)	Supported
Performance expectancy	PE	H1	PE → BI	0.256	YES***
Effort expectancy	EE	H2	EE → BI	0.038	YES
Social influence	SI	H3	SI → BI	0.199	YES***
Self-efficacy	SE	H4	SE → BI	0.020	YES
Perceived safety	PS	H5	PS → BI	0.176	YES***
Anxiety	AX	H6	AX → BI	−0.077	YES**
Trust	T	H7	T → BI	0.084	YES
Legal regulation	LR	H8	LR → BI	0.043	YES**
Hedonic motivation	HM	H9	HM → BI	0.184	YES***
Price value	PV	H10	PV → BI	0.193	YES***

*** Significant at 0.001 level (two tailed), **Significant at 0.01 level (two tailed)

6 Conclusion and Future Work

Transportation has immense implications for social welfare, economic development, and environmental sustainability. Congestion, environmental degradation, social inequity, and public health issues are problems that sustainable transport policies urgently need to resolve. In this article, we explored AVs and their potential as a solution for smart and sustainable development, we have also explored the factors that will influence consumers' adoption. The results suggest that the proposed model of the Autonomous Vehicles acceptance demonstrates a considerable explanatory and predictive power. Thus, the integration of the internal and external factors with the UTAUT2 is both theoretically appealing as well empirically significant. Future studies could focus in exploring the effect of factors such as cultural influence and socio-economic status on technology adoption.

References

1. Becker, F., Axhausen, K.W.: Literature review on surveys investigating the acceptance of autonomous vehicles (2017)
2. Morgan Stanley: Autonomous Cars: Self-Driving the New Auto Industry Paradigm, s.l.: Morgan Stanley Blue Paper (2013)
3. RTRAC: Automated Driving Roadmap: Status 3rd Draft for public consultation (2015)
4. Fraedrich, E., Lenz, B.: Societal and individual acceptance of autonomous driving. J. Auton. Driving, 621–640 (2016)
5. 2025ad.com: 2025AD|2025AD - The Automated Driving Community. https://www.2025ad.com/categories/latest/technology/. Accessed 24 Oct 2019
6. Rödel, C., Stadler, S., Meschtscherjakov, A., Tscheligi, M.: Towards autonomous cars: the effect of autonomy levels on acceptance and user experience. In: Proceedings of the 6th International Conference on Automotive User Interfaces and Interactive Vehicular Applications, pp. 1–8. ACM (2014)
7. Cho, Y., Park, J., Park, S., Jung, E.S.: Technology acceptance modeling based on user experience for autonomous vehicles. J. Ergon. Soc. Korea **36**(2) (2017)
8. Niculescu, A.I., Dix, A., Yeo, K.H.: Are you ready for a drive?: User perspectives on autonomous vehicles. In Proceedings of the 2017 CHI Conference Extended Abstracts on Human Factors in Computing Systems, pp. 2810–2817. ACM (2017)
9. Yi, M.Y., Jackson, J.D., Park, J.S., Probst, J.C.: Understanding information technology acceptance by individual professionals: Toward an integrative view. Inf. Manag. **43**, 350–363 (2006)
10. Venkatesh, V., Morris, M.G., Davis, G.B., Davis, F.D.: User acceptance of information technology: toward a unified view. MIS Q. **27**(3), 425–478 (2003)

Sustaining E-Government Website Services: An Investigation of Dynamic Relationships of Organisational Factors in a Government Agency

Gunadi Gunadi(✉)

Gajayana University, Malang, East Java 65144, Indonesia
gunadi@unigamalang.ac.id

Abstract. The availability of e-government website services can deteriorate over time. Accordingly, persistent quality maintenance is necessary to sustain high availability levels of website services. The government of Surabaya, Indonesia has delivered and sustained website services with a high-level of availability over time. This research aims to reveal important organisational factors and their feedback relationships that influence availability levels. Guided by a previously proposed model, a causal loop diagram on e-government website availability, data was collected by observing the Surabaya website and interviewing the government staff unit that maintains the website. A revised causal loop diagram was constructed and analysed. Some leveraging factors have been identified, such as staff satisfaction over their tasks and rewards, communication and coordination among staff based on trust and legitimate procedures, sufficient staff competence, informal training and communication with leadership. Through several reinforcing loops, these factors both influence and are influenced by the availability of website services.

Keywords: Website availability · Maintenance · E-government of surabaya

1 Introduction

The use of a website to deliver governmental services is ubiquitous. In many countries, most citizens and businesses rely on websites to provide governmental services [1–3]. They access e-government websites (eGW) to obtain information, communicate, participate, or complete transactional services. Therefore, to maximise service benefits, e-government management must sustain consistent availability [4] of the services that are accessible, functional and useful.

However, the availability of eGW services can deteriorate over time due to new requirements, programming errors, invalid hyperlinks, inadequate website maintenance [5], and dynamic changes in laws and regulations [6]. Accordingly, persistent and quality website maintenance is necessary to sustain high availability levels of website services.

Much research has studied website maintenance and eGW services delivery. For example, Ivan and Despa [7] studied web application cost, Nagy et al. [8] investigated a

Á. Rocha et al. (Eds.): WorldCIST 2020, AISC 1159, pp. 517–527, 2020.
https://doi.org/10.1007/978-3-030-45688-7_52

decision-making process to maintain or replace an organisation's website, Alsmadi and Abu-Shanab [9] investigated the security of the website, and Al-Jamal and Abu-Shanab [10] researched the involvement of citizens in government policy formulation and government data transparency programs. However, the existing literature in eGW maintenance focuses less on the sustained availability of services. Specifically, they do not examine the importance of organisation-related factors and their feedback-relationships with availability levels.

In Indonesia, most government agencies have used a website to deliver some of their services. This is especially true for Surabaya, which has achieved a significant level of eGW service delivery and several national and international e-government awards over the past years. The role of its eGW has been maintained such that the services delivered highly benefit the city's citizens and business. Therefore, it is important to understand and learn how continuous quality maintenance leads to such an achievement.

Distinct from previous research, and as part of a broader scope of a study, this research aims to identify factors and model feedback relationships that influence eGW service availability over-time in Surabaya based on the previously proposed conceptual model on e-government website availability [11]. This model, which is a qualitative system dynamics model [12], argues that sustaining a high-level of availability of the eGW services over time involves dynamic feedback relationships between the availability level, complexity of eGW maintenance and factors associated with organisational effort, competence, communication and coordination, trust, procedure and resource sharing.

The resulting model could improve understanding of the importance of eGW maintenance, how its success can be achieved, and assist e-government management in understanding the feedback relationships and other factors that sustain the eGW service delivery over time.

The rest of the paper is organised as follows: a review of the existing literature, followed by an adopted method consisting of qualitative system dynamics modelling, the case and data collection, a detailed discussion of the findings, and finally, the conclusion.

2 Literature Review

2.1 E-Government Website Service Availability and Maintenance

An e-government website can be defined as an e-government service delivery media consisting of a collection of government related documents or files that are organised and published on the World Wide Web. It is a virtual space that allows citizens and businesses to interact with the government to obtain information, communicate, or accomplish transactions [13]. Currently, eGW is an important gateway for service deliveries [14] due to its capacity to provide highly efficient processes and effective services [15].

Considering its critical role, and in order to sustain its benefits, an eGW service must be consistently available [16]. The eGW services are available if it is always accessible by citizens and businesses, adaptable to new requirements [15], contain appropriate content and features, function properly and align with e-government processes [13]. However, this level of availability could be affected by additional factors, such as the quality of the website development [17], organisational and external pressures [5], and the reliability of website infrastructure [18]. Accordingly, e-government managers must undertake continuous eGW maintenance.

Maintenance is defined as the activities undertaken, after the website development and launch, that ensure that the information accessible to the user is current, accurate, relevant [19], legally reliable, free of programming errors and demonstrates a perfected website design to suit new requirements [5, 20]. Maintenance activities may consist of 'exploration, development, testing and deployment of web features', [21, p. 5] which are continuous activities unless a complete redevelopment takes place. Lack of quality maintenance may decrease the dynamic availability [21, 22] of the eGW.

In general, the literature indicates that successful maintenance is influenced by the motivation, competence and collaboration of e-government staff [23, 24]. By considering the complex nature of e-government, for which a systems thinking approach [25] is deemed appropriate [26, 27], understanding the complexity of the eGW maintenance and its level of availability requires recording dynamic feedback relationships of these factors [24]. Especially, much research suggested the use of causal loop diagram (CLD) of System Dynamics method to capture and reveal the complexity of these factors and their feedback relationships structure [24, 28, 29] in influencing the level of eGW services availability over time. The resulted CLD indicates the leveraging factors and how a factor eventually influences and is influenced by the availability level.

3 Method

3.1 Qualitative System Dynamics

Per the model, which adopts a qualitative system dynamics (SD) method, a causal loop diagram (CLD) was constructed based on the data collected from the case study. The model was built with the following steps: formulating a dynamic problem, identifying the main factors, constructing a conceptual CLD, collecting field data, revising the CLD and analysing the functionality of the CLD [25, 30]. In this research, the first three steps have been undertaken and presented in [11].

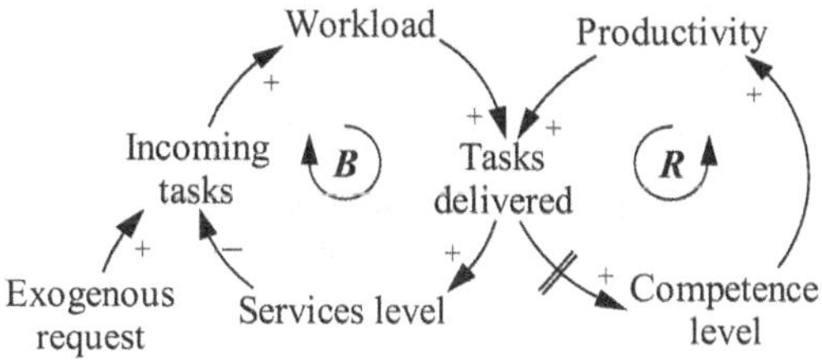

Fig. 1. An example of a causal loop diagram

A CLD consists of words or phrases that are linked by curved arrows and attached to each of which are polarity and time delay symbols [29, 31] (see Fig. 1). The arrow denotes a causal relationship between two factors. The polarity (+/–) indicates the direction, either in the same direction or in different directions, of two related variables. The time delay (//) is shown by crossing the arrow. Tracing a chain of causality arrows can form two types of loops: reinforcing (R) or balancing (B) loops. Reinforcing loops express exponential growth or decaying behaviour, and balancing loops indicate a goal-seeking behaviour of a system.

3.2 The Case

This research investigates the e-government system of Surabaya (eGoS), the capital city of the East Java Province of the Republic of Indonesia. Its website address is http://www.surabaya.go.id. This case was selected as a result of its performance achievements, the eGoS has received several national and international awards, and its ability to meet the case study selection criteria [32].

The eGoS website is an important gateway for government services by providing public information, facilitating communication between the citizens and government and allowing citizens and businesses to make use of government services.

The eGoS website is managed by the Management of Public Communication Section (MPCS). This unit is headed by a head of section and supported by information technology (IT) and non-IT staff. The staff, in many cases, must collaborate within the MPCS as well as with staff from other units. The organisational structure of the management of the eGoS is shown in Fig. 2.

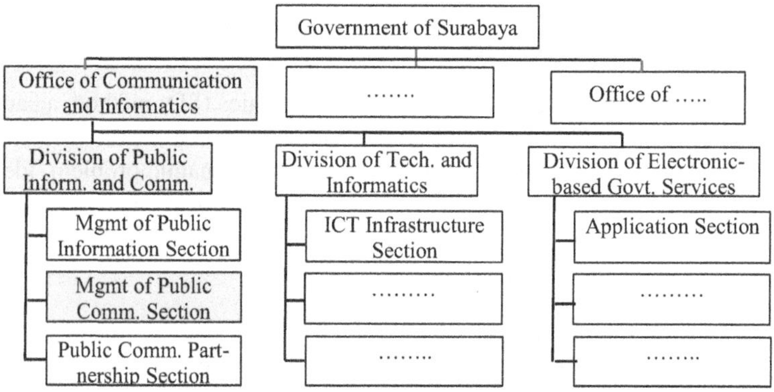

Fig. 2. The organisational structure of the management of the eGoS

3.3 Data Collection

To answer the research question, data was collected by observing the eGoS website and interviewing the eGoS staff. The eGoS website observation sought to obtain a level of the eGW services availability over time. The interviews sought to collect qualitative data to understand and identify possible causal loops among key factors that could explain the eGW services availability level.

Various e-government related awards have been received by the eGoS for more than seven consecutive years until 2018 [33] indicating that the eGW service availability always at a high-level over time. The observation took place for approximately 12 consecutive weeks in 2018 to obtain the actual quantitative data of the availability level. Observers were required to open the website and click through the available menus, sub-menus, sub-sub-menus and down to the lowest level of hyperlinks, buttons and application names. The observers then noted whether the facilities were functional.

Each week, each observer calculated the percentage of the working website services out of the total available services. The plot of the sum of these percentages over the observation periods defines the level of the eGoS website service availability over time.

Qualitative data was collected by interviewing six staff members and two managers from the Division of Public Information and Communication of the eGoS, who were responsible for maintaining the availability of website services. The staff responsible included a news content contributor, a website and communication designer, a content administrator, a website programmer, a website database administrator and an e-government infrastructure employee. These respondents were deliberately chosen to obtain data that reflected the true operation of the website maintenance system. These interviews were guided by a semi-structured open-ended questionnaire which was created based on the model in [11]. Each of these on-site interviews lasted about one hour.

4 Findings

4.1 Availability of eGW Services Over Time

The observations indicated that the government of Surabaya has been utilising its eGW to deliver various quality services to its citizens and businesses. External users can access information, communication, and transactional services via the eGW. They can obtain government-related news and announcements as well as various downloadable documents. Furthermore, they can communicate, report or request solutions about city-related problems and receive a response directly. They may complete business permit applications, access online procurements systems or obtain tickets for public health services.

While the website is complex in the type and number of services delivered, its availability level over time is relatively high as shown in Fig. 3. Its overall availability level was above 90%.

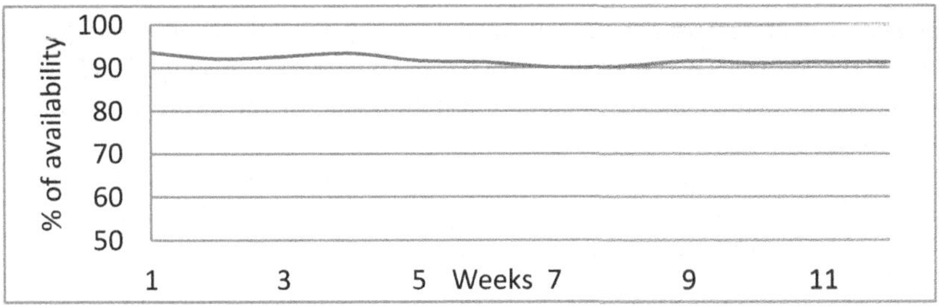

Fig. 3. eGoS website services availability level over time

4.2 Factors and Their Feedback Relationships

Figure 3 indicates the ability of the management of the eGoS to sustain the eGW over time. This section explores feedback relationships of important factors using a CLD from the organisational dimension as shown in Figs. 4 and 5. The exploration is based on a proposed CLD in [11].

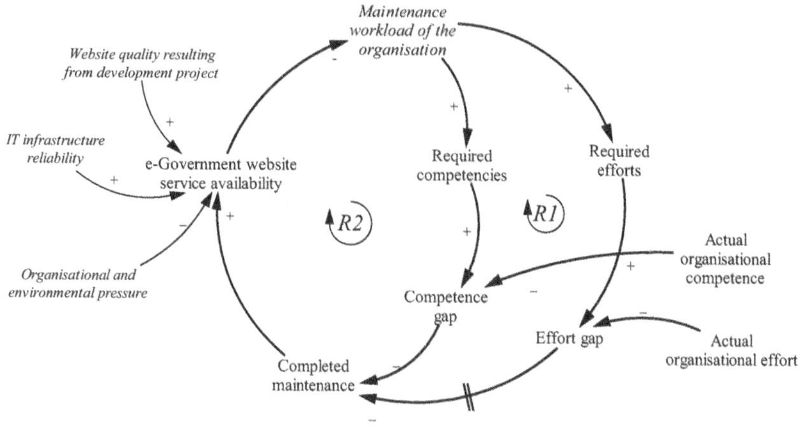

Fig. 4. Maintenance model

Organisational Effort (*R1*) and Competence (*R2*). The data supports the CLD that to sustain the level of *eGW services availability* the eGoS management has to carry out continual website maintenance due to the persistence of *organisational and environmental pressures*, the *website quality resulting from development project* and the *IT infrastructure reliability* (see Fig. 4). Website maintenance included creating new service features, overcoming errors, and improving website content and design. Any maintenance requests became *maintenance workloads of the organisation*, which increased the level of both *required competencies* and *required efforts* of the organisation.

In response to this need, the eGoS management has provided levels of *actual organisational competence* and *actual organisational effort* in such a way that they can close the *competence gap* and *effort gap,* respectively. This improved the quality and quantity of completed maintenance and kept the *eGW service availability* at a high level over time. These feedback relationships explained the importance of the effort and competence factors through two reinforcing loops: *R1* and *R2*. These loops persuaded the management to maintain these factors at sufficient levels. The eGoS management employs and organises staff who, simultaneously, are highly motivated toward effort and are equipped with the required and appropriate competencies (see Fig. 5).

The *actual organisational effort* Factor. The level of the *actual organisational effort* factor over time is positively influenced by *staff effort*, *total staff collaboration*, and *communication and coordination* between them. The *staff effort* factor represents the

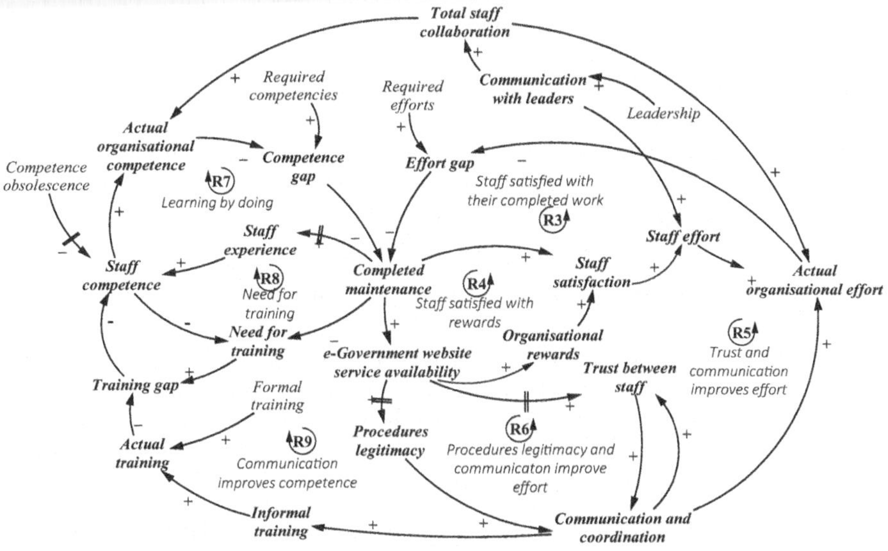

Fig. 5. Feedback relationships of organisational factors

effort exerted by staff members of the eGoS in response to the maintenance workload. The time required to complete the maintenance tasks depended on the complexity of the tasks. In some cases, staff were willing to work overtime. Normally, completing a maintenance task required collaboration from several staff members, and they needed to communicate and coordinate their responsibilities.

Staff Satisfied with Their Completed Work (*R3*). At the eGoS, successfully *completed maintenance* improved *staff satisfaction*. They wanted to do more tasks, and therefore, they completed jobs quickly. This satisfaction improved *staff effort* and, consequently, the *actual organisational effort*. Ultimately, this upheld the high-level of *eGW service availability*.

Staff Satisfied with Organisational Rewards (*R4*). The top leaders of the eGoS provided *organisational rewards* by promoting the website to a broader audience. This improved *staff satisfaction*. Staff members were highly motivated as a result of the rewards and were aware that it was their responsibility to complete the maintenance tasks. Therefore, they were willing to exert more effort when maintenance was urgent. This kept the *actual organisation effort* at a high level.

Trust and Communication Improved Effort (*R5*). Website maintenance of the eGoS invariably involved more than one staff member. A high-level of *eGW services availability* meant that staff collaboration over time was functional. As a result, *trust between staff* gradually grew, which meant that they believed their colleagues will commit to and complete their responsibilities. Improving trust eased *communication and coordination* between them, which was adequate, quick and positive, and it was often informal, direct and flexible. This eased the way they collaborated, therefore improving the level of *actual organisational effort*.

Procedures Legitimacy and Communication Improved Effort (*R6*). At the eGoS, the staff had to understand and closely adhere to organisational procedures to coordinate effort. This adherence made communication easier, although many of the operational procedures were not formally written. A high-level of *eGW services availability* meant that the procedures they followed were proven to be workable and, therefore, legitimate. This gradually increased the *procedure legitimacy* and made staff communicate seamlessly using the same procedure, therefore, improving *communication and coordination* and, accordingly, improving *actual organisational effort.*

The *actual organisational competence* Factor. Several factors influenced the levels of *actual organisational competence*, such as *staff competence* and *total staff who collaborate*. The eGoS management employed staff that met a specific level of *staff competence*, such as in a database, programming, design, front-end website programming and graphic design. Their competence levels were sufficiently high for undertaking maintenance so that any required maintenance tasks could always be completed. Effective collaboration between staff improved *actual organisational competence*.

Learning by Doing (*R7*). As the *completed maintenance* increased in number and quality over time, *staff experience* also gradually increased. In some cases, this experience improved *staff competence* of up to 50%. Experience obtained by staff caused maintenance work to become easier and faster.

Need for Training (*R8*). In some cases, the advancement of IT is still beyond the achieved experience level. Therefore, *competence obsolescence* decreased *staff competence* level. This happened when the staff was assigned a new maintenance job but they did not have the required knowledge and skill to complete it. Eventually, this caused the staff of the eGoS to become unable to complete their maintenance tasks properly. This reduced *completed maintenance,* which led to low quality or late completion. This, in turn, increased the *need for training* and the widening of the *training gap*. Consequently, this can further lower the *staff competence level.* Fortunately, the staff of the eGoS were aware that they needed to renew their competencies regularly due to the persistent advancement of IT.

Communication Improves Competence (*R9*). At the eGoS, the *training gap* could be closed by the *actual training* factor, therefore, improving *staff competence*. The training was predominantly informal as *informal training* was always available from other staff members. They were available to share their knowledge skill, which included web design, server troubleshooting, programming language, database synchronisation and so on. The impact of this *informal training* was quite significant. For example, training on a website could improve a skill by up to 50%. This kind of training was realised as a result of seamless *communication and coordination* among staff members, and, ultimately, let to a high-level of *eGW services availability* over time.

Leadership Factors. Leadership at the eGoS played an important role in improving *communication with leaders*, especially operational managers. Good communication both motivated staff to further exert effort and encouraged them to fruitfully collaborate. This factor was critical to the eGoS as any required maintenance needed approval from the leadership.

5 Concluding Discussion

Findings showed that the e-government of Surabaya city was able to sustain the e-government website services at a high and valuable level of availability over time, but they did not fully support the proposed CLD model [11]. For example, the resulting model indicated that the staff tended to submit a task once it was completed. Moreover, reinforcing loops dominated the model as was predicted. This suggests that management should focus on the leveraging factors in the loops to prevent them from decreasing.

Some of the leveraging factors can be determined from the model. First, the organisation should provide rewards to the staff so that they have a sense of pride in their completed maintenance tasks. Second, communication and coordination among staff should be seamless by developing trust between them and establishing reliable procedures. This enables various staff members with varied competencies to fruitfully collaborate and perform informal training.

Overall, the resulting CLD can improve understanding of important factors and how they causally relate to each other. It suggests strategies for e-government management on how to sustain a high-level of availability of the eGW services. This CLD will be used as a basis for developing a simulation model of quantitative system dynamics.

Acknowledgment. This research is funded by the Directorate of Research and Community Services, Ministry of Research, Technology, and Higher Education of Indonesia. [Grant no. 7/E/KPT/2019, 19-02-2019].

References

1. Cataldo, A., et al.: Design of a single window system for e-government services: the chilean case. J. Ind. Manag. Optim. **14**(2), 561–582 (2017)
2. United Nations: United Nations E-Government Survey 2016. In: E-Government in Support of Sustainable Dvelopemnt. United Nations, New York (2016)
3. Gil-Garcia, J.R., Vargas-Marin, E., Gascó, M.: Understanding the success of government portals: the role of political leadership, standards, and a powerful centralized IT agency. In: Proceedings of the 51st Hawaii International Conference on System Sciences (2018)
4. Sá, F., Rocha, Á., Cota, M.P.: Potential dimensions for a local e-Government services quality model. Telematics Inform. **33**(2), 270–276 (2016)
5. Peters, J.C.: Web information systems: a study of maintenance, change and flexibility. Thesis. Brunel University, United Kingdom (2010)
6. Manoj, A.T., Jaffar, A.A.: Designing a semantic tool to evaluate web content of government websites. Int. J. Public Adm. Digit. Age (IJPADA) **3**(2), 19–36 (2016)
7. Ivan, I., Despa, M.L.: Estimating maintenance cost for web applications. Inform. Econ. **20**(4), 34 (2016)
8. Nagy, D., Schultz, L., Wiederker, T.: Replace or revise? A Case Study Investigating the Replacement of an Organizational Website, AMCIS 2017 (2017)
9. Alsmadi, I., Abu-Shanab, E.: E-government website security concerns and citizens' adoption. Electron. Gov. Int. J. **12**(3), 243–255 (2016)
10. Al-Jamal, M., Abu-Shanab, E.: The influence of open government on e-government website: the case of Jordan. Int. J. Electron. Gov. **8**(2), 159–179 (2016)

11. Gunadi, G.: Qualitative system dynamics modelling of the impacts of maintenance, effort, competence and collaboration on e-government website availability. Electron. Gov. Int. J. **15**(2), 189–212 (2019)
12. Walters, J.P., et al.: Exploring agricultural production systems and their fundamental components with system dynamics modelling. Ecol. Model. **333**, 51–65 (2016)
13. Rorissa, A., Demissie, D., Gharawi, M.: A descriptive analysis of contents of Asian E-Government websites. In: E-Government Website Development: Future Trends and Strategic Models: Future Trends and Strategic Models, vol. 3, no. 56.35, p. 102 (2010)
14. Chen, Y.-C.: Managing Digital Governance: Issues, Challenges, and Solutions. Routledge, London (2017)
15. Bhattacharya, D., Gulla, U., Gupta, M.: An assessment study of Indian state government portals. In: Downey, E., Ekstrom, C.D., Jones, M.A. (eds.) E-Government Website Development: Future Trends and Strategic Models, pp. 130–152. IGI Global, Hershey (2010)
16. Kienle, H.M., Distante, D.: Evolution of web systems. In: Mens, T., Serebrenik, A., Cleve, A. (eds.) Evolving Software Systems, pp. 201–228. Springer (2014)
17. Melin, U., Axelsson, K.: Managing e-service development–comparing two e-government case studies. Transforming Gov. People Process Policy **3**(3), 248–270 (2009)
18. Roy, C., et al.: IT infrastructure downtime preemption using hybrid machine learning and NLP. In: FedCSIS Position Papers (2015)
19. Vasconcelos, W.W., Cavalcanti, J.: An agent-based approach to web site maintenance. In: Koch, N., Fraternali, P., Wirsing, M. (eds.) Web Engineering, pp. 271–286. Springer (2004)
20. Bailey, J.O.: Web accessibility diagnosis, improvement and maintenance. Thesis. Durham University, United Kingdom (2007)
21. Kong, X., Liu, L., Lowe, D.: Modeling an agile web maintenance process using system dynamics. In: 11th ANZSYS Conference/Managing the Complex V. Christchurch, New Zealand (2005)
22. Black, L.J., et al.: A Dynamic Theory Of Collaboration: A Structural Approach To Facilitating Intergovernmental Use Of Information Technology. In: The 36th Annual Hawaii International Conference on System Sciences, HICSS 2003. IEEE, Hawaii (2003)
23. Gunadi, G., McGrath, G.M., Sandy, G.: A model of feedback relationships between software maintenance and information systems staff management: a case of an e-government system. In: ISICO 2013, Bali (2013)
24. Luna-Reyes, L.F., Gil-García, J.R.: Using institutional theory and dynamic simulation to understand complex e-Government phenomena. Gov. Inf. Q. **28**(3), 329–345 (2011)
25. Maani, K.E., Cavana, R.Y.: Introduction to Systems Thinking. Pearson Education New Zealand, Auckland (2009)
26. Scholl, H.J.: Central research questions in e-government, or which trajectory should the study domain take? Transforming Gov. People Process Policy **1**(1), 22 (2007)
27. Shareef, M.A., et al.: e-Government Adoption Model (GAM): differing service maturity levels. Gov. Inf. Q. **28**(1), 17–35 (2011)
28. Abdelgawad, A., Snaprud, M., Krogstie, J.: Accessibility of Norwegian municipalities websites: a qualitative system dynamics approach. In: Proceedings of the 28th International Conference of the System Dynamics Society, Seoul, Korea (2010)
29. Fang, Y., et al.: System dynamics modeling for information systems research: theory development and practical application. MIS Q. **42**(4) (2018)
30. Tegegne, W.A., Moyle, B.D., Becken, S.: A qualitative system dynamics approach to understanding destination image. J. Destination Mark. Manag. **8**, 14–22 (2018)

31. Sterman, J.D.: Business Dynamics: System Thinking and Modeling for a Complex World. The McGraw-Hill Companies, Inc., Boston (2000)
32. Yin, R.K.: Case Study Research, Design and Methods. Third ed. Applied Social Research Methods Series. Sage Publishing, Inc., California (2003)
33. Anonymous: Penghargaan: Daftar Prestasi Surabaya Tiap Tahun. https://surabaya.go.id/id/page/0/49215/penghargaan. Accessed 7 July 2018

Framework for Bridge Management System in Montenegro

Željka Radovanivić(✉)

Građevinski fakultet, Univerzitet Crne Gore, Džordža Vašingtona b.b, 81000 Podgorica, Montenegro
zeljkar@ucg.ac.me

Abstract. The problem of construction road infrastructure and bridge management system, as one of the segments of road infrastructure, is extremely complex and requires a multidisciplinary approach. In this paper the problems are presented from the standpoint of the civil engineering profession. About forty bridges, which are part of the Bar-Boljare highway will be included in the road network of Montenegro, very soon. Within the Main Design of Highway Bridges, a project Program of Bridge Control During the Exploitation is provided, also. The bridge control program should be included in the database for bridge maintenance on the highway. Highway bridges should be only one segment of the database of bridges on the road network of Montenegro. The existing bridges in Montenegro are mostly over thirty years old, traffic loads and traffic density are increasing very rapidly, floods are becoming more frequent, and the effect of salt, water and air pollution affect the accelerated decaying of bridges. Therefore, there is a need to build bridge management information system for complete territory of Montenegro. Before establishing a database for bridges, it is necessary to establish a database for roads, per municipalities or regions. In the aim precise and easy bridge identification, it is necessary to use a geodetic system, GIS. The paper proposes method for setting priorities for the maintenance of bridges and method for assessing the bearing capacity of bridge during the exploitation. The purpose of this research was to provide a framework for the development Bridge Management System (BMS) in Montenegro.

Keywords: Bridges · Database · Information system

1 Introduction

The problems of object's maintenance can be grouped into: Problems of maintenance of infrastructural objects, Problems of maintenance of hydro-technical objects and Problems of maintenance of residential buildings, administrative buildings and public buildings. Each of these groups are subdivided into subgroups. Maintenance of infrastructure objects can be divided into maintenance of railway infrastructure and maintenance of road infrastructure. Each of these subgroups is then subdivided into the maintenance of specific elements. Road infrastructure maintenance issues can be divided into monitoring of the following road infrastructure elements: roads, bridges, tunnels, retaining

Á. Rocha et al. (Eds.): WorldCIST 2020, AISC 1159, pp. 528–538, 2020.
https://doi.org/10.1007/978-3-030-45688-7_53

walls, slope stability and so on. In this paper the emphases will be on the problem of monitoring, observation and maintenance of bridges.

The procedures and methods of control and maintenance of bridges, which are prescribed in the currently valid building regulations of Montenegro, have not been implemented in practice. Some of methods of monitoring, bridge inspection, condition assessment and estimation of the maintenance cost, applied in the countries of the region have only been partially implemented in Montenegro. As a consequence, we have poor maintenance of bridge with numerous damage to non-structural and structural elements. Such condition, over a long period of time, may endanger people and property or result in a significant increase in cost of rehabilitation and/or strengthening of the bridges.

Issues related to the maintenance of infrastructure objects will be particularly emphasize in the coming period because most bridges in Montenegro are older than thirty years. Traffic loads is on the rise and many bridges are not dimensioned for such increased traffic loads. Aggressive action of salts for defrosting pavement's areas, flood effects, combined with irregular maintenance of bridges can lead to catastrophic consequences.

The Ponte Morandi Bridge, Genoa-Italy, which collapsed in August 2018, warns us of the poor bridge infrastructure management. The bridge was put into service in 1967 and repaired in 2016. The bridge was maintained by a private company, Autostrada, which maintains most of Italy's highways (Fig. 1).

Fig. 1. Bridge Ponte Morandi, Genoa: a) state of the bridge before, [1] and b) after collapse, [2]

In the Montenegro's roads network the 27 bridges, 2 overpasses, 6 underpasses and three loops will be included very soon. These constructions are part of the Smokovac-Uvač section on the Bar-Boljare highway, which is currently under construction. The Bar-Boljare highway is part of the Trans-European Transport Network, TEN, from Gdansk to Istanbul. The Chinese company CRBC is the designer of the all technical documentation and contractor for the first section of the highway, which is 41 km long. The highway will be equipment for continuous monitoring of vehicle movement and control of road signalization. Program of Bridge Control During the Exploitation is done as a part of the Main Designs of Highway Bridges. This part of the technical documentation will need to be included in the bridge maintenance database, which will be integrated in the Bridge Management System, BMS.

"The Strategy of Transport Development in Montenegro for period 2019–2035", [3], defines the strategic and operational goals of development transport sector in order the socioeconomic needs of Montenegro, but also to align with European Union policy and Trans-European Transport Network (TEN-T guidelines and standards).

The problem of marking, monitoring and maintenance of highway bridges will need to be based very wide, with the task of integration the highway bridge management system into a much wider system, that is monitoring, controlling and maintaining the all roads networks in Montenegro.

The problem of building bridge management system is extremely complex and requires a multidisciplinary approach. A successful modern information system, with tendency to install artificial intelligence systems, can only be built with the joint work of civil engineers, programmers, geodesy engineers, electrical engineers, geologists, meteorologists, and other professions involved in the construction and maintenance of bridges.

The system should be simple, easy to understand and adapted to civil engineers, who should make decision about of priority investment in the revitalization of particular bridge or multiple bridges in the roads network. At the same time, the system should enable the implementation of the latest knowledge in the field of bridges monitoring in order to make promptly and timely decisions.

2 Problems of Maintenance of Infrastructure Objects

Partial or total exclusion of bridges from operations can lead to bigger traffic and economic disorder. In the last time significance of this problem was observed and some countries require that a register be done for all bridges on the main roads and for all works performed on them. In addition, computerized data processing is performed to evaluate rational maintenance of bridges. In the first decade of the 21st century, in the United States, one billion dollars was set to create a global network for monitoring infrastructure objects and bridges. The goal was to create a centres where not only the register of bridges will be guide, in these centres live signal regarding the stresses and deformations in large span bridges will be transfer, also. The programs are developed that account degree of security for particular action and enable faster decision making in the event of emergency.

In Montenegro building a bridge database is at the beginning. The Directorate for Transportation purchased software for the processing data about the bridges on the main roads fifteen years ago, but the has not yet been created and the software is not used.

In municipalities, part of the municipal administration takes care of local roads, city roads, bridges and other objects on the roads. The method of managing the bridges register, carrying out the prescribed inspections and monitoring of bridges is not harmonized with the Directorate for Transport.

Obviously, we need a new, modern and widely deployed BMS, as well as the Roads Infrastructure Management System, RIMS.

3 Bridges Database

Well organized and regularly filling with data, bridges database is the basis for good BMS and for assessment of rating condition of a bridge or bridges generally.

The data, that can be obtained from the bridges database, can be used to estimate the load of vehicles on the roads, timely planning and works execution on the bridge rehabilitation or strengthening, improvement of design and construction practice, decision making on priority investment and assessment of actual bearing capacity of the bridge.

The establishment of bridge database begins with composing a list of bridges, which is accompanied by all technical documentation and with file where is put data about registered condition of the bridge. The bridge should be get an identification mark. Croatia has developed the HRMOS Bridge Management System [4], which is based on the Danish Bridge Management System DANBRO, developed by the Danish Roads Directorate. In this system, prior to the commencement of the collection of bridge data, it is necessary to carry out the identification of the bridge in the Road Inventory Object Identification. In order to allow a unique marking of bridges, it is necessary to establish a unique register of the road network. The road section on which the bridge is located should also be entered in the register. The system should be set up so that it is always possible to import road and bridge databases, with the possibility of continuous expansion of both databases.

So, in Montenegro, before establishing BMS (or in parallel with the establishment of BMS), it is necessary to establish a road database per municipalities, with made road maps and marked road objects. In defining object identification, it is necessary to use a geodetic identification system, GIS, and give each object a code mark. This means that each bridge will need to have a start and end mark on the bridge in GIS (Fig. 2).

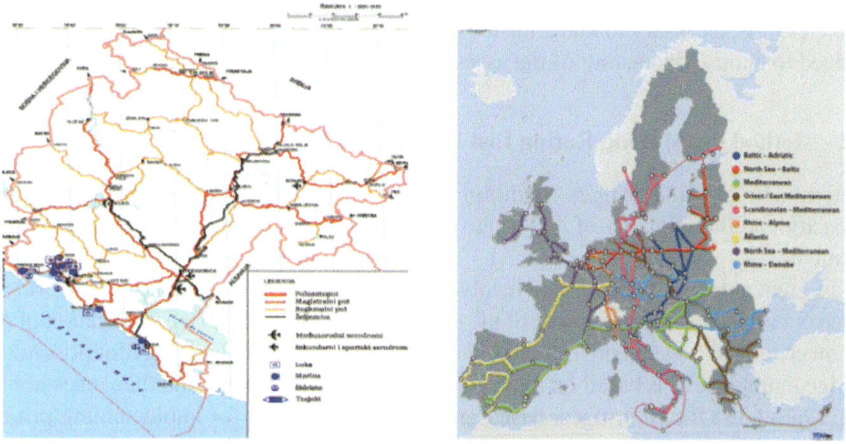

Fig. 2. a) Traffic map of Montenegro; b) Trans-European Transport Network, [3]

The bridge database consists of: bridge identification sheet, bridge drawing (longitudinal and cross-sections with dimensions of essential structural elements), assessment of the bridge condition determined after the inspection, data on bridge interventions and data about the passage of special loads over the bridge, rating of the individual bridge according to a specific method, photo documentation of the bridge sorted by date of origin. The database has to be updated every two years during the regular bridge inspection.

3.1 Bridges Controls

The current regulation has prescribed bridge inspections in order to contribute to road safety, usability and durability of the bridges. The prescribed inspections of the bridges are: technical inspection (required before commissioning of the bridge), ongoing inspection (performed twice a year, before and after the winter season), regular inspection (performed every two years), main inspection (every six years and prior to the expiry of the warranty period), an extraordinary inspection (due to emergency events, earthquakes, fires, floods or war events) and a detailed inspection (performed if there is any doubt about the bearing capacity and safety of the bridge or if special loads were crossed over the bridge).

The regulations define the content of each of these inspections, and the tables in which the status is recorded are part of the Program of the Bridge Control in the Exploitation, which is a mandatory part of the Main Design of each bridge. Most of the prescribed bridge controls are visual controls. As part of the regular inspection of the bridges, a geodetic survey of the bridge should be carried out every two years. The survey report consist measurement of deformation registered on the of the rappers that are installed during the bridge construction. After survey the results should be compared with results of the previous geodetic survey, with the tolerance defined in the Main Design of the Bridge for the exploitation phase.

Based on the procedure explained above it is clear that effective bridge management is not possible without the establishment of BMS.

Why should the system be set up at the state level? In the case of the passes special vehicles, it is necessary to know the corridors for the movement of special loads, which is need to choose depending of the assessment of the bridges bearing capacity.

3.2 Method for Making Rating List for Bridges Maintenance

Determination of the bridges condition consists of assessing the state of load-bearing capacity and functionality of the bridge and the degree of damage to the non-structural elements.

A great number of persons are involved in the assessment of the condition of bridges, so errors in the visual assessment of the bridge are possible due to the application of different criteria and a subjective approach to assessment. The methodology for prioritizing bridge maintenance is based on the concept of rating each bridge, and then on establishing a rating list for all bridges, based on which the criterion for implementing priority measures for revitalization or strengthening is set. The overall rating of the bridge can be divided into partial ratings, as shown in Tables 1, 2 and 3.

One bridge has structural and non-structural elements with different levels of influence on stability and functionality of the bridge. For the assessment of the rating of bridges in Bosnia and Herzegovina, [5], 28 elements were selected. For the assessment of bridges in Serbia the rating is made on 26 elements, [6]. The numerical rating of the bridge is obtained by evaluating each element by importance factor a_i (varying from 2 to 12) and points b_i which depend from the current state of the element, [5].

The partial ratings serve to indicate which element groups have been given priority by the bridge. R_I rating, refers to a group of elements that affect the load-bearing capacity and stability of a bridge, Table 1. The group of elements on the basis of which a R_{II} rating is obtained are elements that affect the functionality and durability of a bridge, and R_{III} is the rating of the service provided by the bridge.

Table 2 shows the method for determining the R_{II} functionality rating, and Table 3 presents the method for determining the R_{III} rating. The total rating of the bridge R, is the sum of all three partial ratings. Summary rating R is:

$$R = RI + RII + RIII \quad R = \sum_{i=1}^{28} \text{ai bi} \tag{1}$$

Table 1. Rating list of bridge capacity

	Elements	Importance factor, a_i	Elements state	Points, b_i	Rating, Ri = ai x bi
1	Foundations	12	Dangerous	100	Rating of capacity
2	Abutments		Not satisfy	20	$RI = \sum_{i=1}^{10}$ ai bi
3	Middle piers		Bad	15	
4	Bearings		Not favourable	10	
5	Main girder		Acceptable	5	
6	Transverse girder		Good	1	
7	Slab		There is not	0	
8	Bracings, ties, hangings		Unknown	9	
9	Corrosion of steel structures	10	Dangerous	80	
			Not satisfy	10	
	Corrosion of reinforcement and cables		Bad	7	
			Not favourable	4	
	Decline of concrete		Acceptable	2	
			Good	1	
10	Extraordinary transport (loads bigger than regular loads)	8	Frequent	10	
			Rarely	3	
			No	0	

In the database for bridges, the rating of bridges is formed on the basis of a falling rating. Bridges with higher ratings have priority in maintenance.

Table 2. Rating list of bridge maintenance

	Elements	Importance factor, a_i	Elements state	Points, b_i	Rating, Ri = ai x bi
11	Waterproofing	6	There is no	20	Rating of maintenance RII = $\sum_{i=11}^{24}$ ai bi
			Not satisfy	15	
			Bad	10	
			Not favourable	5	
12	Drainage system		Acceptable	3	
			Well or it is not necessary	1	
			Unknown	4	
13	Pavement construction	6	Necessary replacement	10	
			Not satisfy	7	
			Bad	5	
			Not favourable	4	
14	Dilatation device		Acceptable	2	
			Good	1	
			There is not	0	
15	Climatic condition	6	Heavy	4	
			Not favourable	3	
16	Characteristics of river and terrain		Acceptable	2	
			Good	1	
17	Quality of former bridge exploitation	6	Neglect	10	
			Not satisfy	5	
			Bad	3	
			Acceptable	2	
			Good	1	
18	Bridge importance in traffic net	6	L > 100 m	5	
			L < 50 m	3	
			Less signific.	1	
19	Possibilities of bridge inspection	4	Not possibly	3	
			Not favourable	2	
			Favourable	1	
20	Transition slabs	4	There is not	10	
21	Retaining walls		Not satisfy	5	

(*continued*)

Table 2. (*continued*)

	Elements	Importance factor, a_i	Elements state	Points, b_i	Rating, Ri = ai x bi
22	Railing		Bad	4	
23	Curbs		Not favourable	3	
24	Pedestrian paths		Acceptable	2	
			Not necessary	1	
			Good	1	
			There is not	0	

Table 3. Rating list of bridge service

	Elements	Importance factor, a_i	Elements state	Points, b_i	Rating, Ri = ai x bi
25	Geometry of the bridge	4	Discontinuity	5	Rating of services RIII = $\sum_{i=25}^{28}$ ai bi
			Width of pavement < 6.0 m	4	
			Width of pavement > 6.0 m	3	
			There is no pedestr. paths	2	
			In accordance with norms	0	
26	Traffic loads	4	More than 5000	5	
			3000–5000	3	
			1000–3000	2	
			To 1000	1	
27	Installation	2	Not satisfy	5	
			Bad	4	
			Not favourable	3	
28	Signalling		Acceptable	2	
			Good	1	
			There is not	0	

3.3 Method for Assessment of Bearing Capacity of the Bridge During Exploitation

Assessment of the load-bearing capacity and level of safety of bridges is on relation with real vehicle load and the real resistance of the material achieved during the construction of the bridge. Assessment of the load-bearing capacity of the bridge is a more complex procedure than the design of the cross section of the bridge. There are two ways of proving that a structure has sufficient bearing capacity, deterministic and probabilistic. The safety problem of load-bearing structures is understood as a probability problem and most of the sizes introduced into the load-bearing proof are treated as random sizes. The load-bearing control of existing bridges refers to the span structure, which is the load-bearing structure above the piers. Control of the load-bearing capacity of existing steel bridges, structural fatigue must also be taken into account.

It is proposed that the USA, National Bridge Inspection Standard and AASHTO Manuel for Bridge Evaluation be applied to assess the bearing capacity of bridges in operation, in the absence of our applicable standards, [7]. These standards define a method for determining the rating factors of new bridges and bridges that have been in operation for a long time. The standard specifies that the rating factor be determined for all bridges that have been designed or rehabilitated after 2010, according to the procedure shown in the standard, [7]. Probabilistic assessment of the existing structure safety is based on: the actual carrying capacity of the critical element of the span structure, the estimation of the real load in relation to the normative (applied in the project), and the calculation of the load-bearing factor through the introduction of a reliability index, which defines the probability of exceeding the carrying capacity.

The traffic load on the all bridges on the section of the highway Smokovac – Uvač are designed in accordance with the standard in which the standard vehicle is 600 + 300 kN. This is lower than the traffic load prescribed by European standards. Many bridges in Montenegro are built according to old standards that have even less normative traffic load. In order to monitor current trends in the field of the bridges management a probabilistic principle for determining rating factors for the both new and old bridges should be applied.

4 Monitoring the State of Construction of Important Bridges

The bridges with large span or high piers are very expensive investments and disconnecting them from the road network produce economic losses. This is reason why this bridges should be specially monitored and controlled.

The bridge Moračica, on the highway, deserves this approach. The bridge has pier height 160 m, a total length of 960 m, with spans 95 + 170 + 3 × 190 + 125 m (Fig. 3).

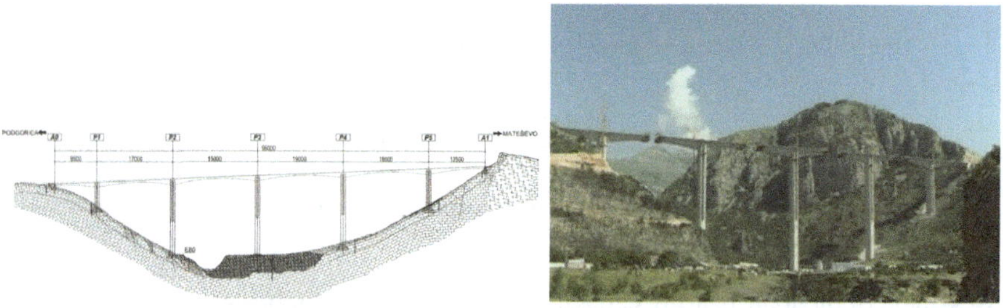

Fig. 3. Moračica bridge: a) Longitudinal section, [11]; b) Photo, august 2019, [12]

The development of measurement technology, in the field of sensor technology and wireless signal transmission, has enabled the monitoring of structures in real time, Structural Health Monitoring, SHM. The essence of SHM is development of autonomous systems for continuous monitoring, inspection and damage detection with minimal involvement of the workforce. Typically, an SHM system consists of a sensors network that measures parameters that are relevant to the design state and record the effects of the environment. For bridges, the following parameters are relevant: position in the GIS, deformation, rotations, stresses, forces, vibrations, accelerations, temperature in the structure, humidity of the structure, chlorine concentration (values higher than prescribed lead to accelerated corrosion of reinforcement), as well as environmental parameters (air temperature, wind speed, snow level, water level and flow rate, CO_2 concentration in the air). All of these sensors had to be connected to the power supply and data logging unit almost through the cable network, [8].

The SHM implementation capability is limited by the available budget, and the number of sensors that can be set is usually limited. Therefore, there are proposals to use the Wireless Smart Sensor Network (WSSN) for cost benefit, [9, 10].

The possibility of accurate surveying of bridges has been significantly increased. Researching the collapse of the Morandi Bridge, a geodetic deformation observation method based on Synthetic Aperture Radar (SAR) observation, with an accuracy of 1 mm, was applied in order to determine the condition of the bridge before collapse, [13]. This innovative method will be used much more frequently in the future.

5 Conclusion

There is a need to build BMS for complete territory of Montenegro. The paper proposes method for setting priorities for the maintenance of bridges and method for assessing the bearing capacity of bridge during the exploitation. The purpose of this research was to provide a framework for the development BMS in Montenegro.

References

1. Mail Online: Shocking photo shows Genoa bridge ‘crumbling’ and ‘caving in’ a few weeks before it collapsed. https://www.dailymail.co.uk/news/article-6062911/Shocking-photo-shows-Genoa-bridge-crumbling-weeks
2. Irish times: Genoa motorway collapse. https://www.irishtimes.com/news/world/europe/genoa-motorway-collapse
3. Projekat Evropske komisije za Ministarstvo saobraćaja i pomorstva Crne Gore: Strategija razvoja saobraćaja Crne Gore 2019–2035, p. 411 (2019)
4. Hrvatske ceste: HRMOS Sustav za upravljanje i gospodarenje mostovima – Popis, p. 57
5. Pržulj, M.: Mostovi, Udruženje izgradnja (2014)
6. Matić, B.: Sistem upravljanja mostovima u Srbiji. Građevinsko arhitektonski fakultet u Nišu. Nauka+praksa 12.1/2009
7. WSDOT Bridge Design Manual (2019). https://www.wsdot.wa.gov/publications/manuals/fulltext/M23-50/Chapter13.pdf
8. Radovanović, Ž., Ulićević, M.: Influence of temperature on box girder bridges. Građevinar **60**, 109–121 (2008)

9. Shachu, P., Manjunatha, S.: Automatic bridge health monitoring system using wireless sensors. Int. J. Sci. Res. (IJSR) **6**(6), 2472–2475 (2017)
10. ChccKian, T.: Structural health monitoring of a bridge structure using wireless sensor network. Master's theses on Western Michigan University, p. 144 (2012)
11. Main Design nof the Moračica Bridge, Company CRBC, (2016)
12. Photo of the Moračica bridge. https://kodex.me/clanak/193137/najskuplji-objekat-autoputa-most-moracica-vrijedan-74-5-milona-eura
13. Milillo, P., Giardina, G., Perissin, D., Milillo, G., Coletta, A., Terranova, C.: Pre-collapse space geodetic observations of critical infrastructure: the Morandi bridge, Genoa, Italy. Remote Sens., 1–14 (2019)

An IoT Approach to Consumer Involvement in Smart Grid Services: A Green Perspective

Miloš Radenković[1], Zorica Bogdanović[2(✉)], Marijana Despotović-Zrakić[2], Aleksandra Labus[2], Dušan Barać[2], and Tamara Naumović[2]

[1] School of Computing, Union University, Knez Mihailova 6, Belgrade, Serbia
milos@elab.rs
[2] Faculty of Organizational Sciences, University of Belgrade, Jove Ilića 154, Belgrade, Serbia
{zorica,maja,aleksandra,dusan,tamara}@elab.rs

Abstract. The paper presents an innovative model for consumer participation in electricity market based on IoT. The goal is to develop a new business model and the underlying IoT infrastructure to enable the participation of individual household devices in smart grid services. The purpose of the proposed model is to allow for higher inclusion of renewable energy sources by providing cheap and flexible balancing services. The approach is based on the premise that consumers' attitudes towards green energy and environmental protection are important incentives for acceptance of the proposed model, and as such should be closely monitored during its development. The proposed model has been evaluated through the study of consumer' attitudes in the context of a smart grid in Serbia. The results show that consumers are interested in environmental conservation and are willing to participate in these new smart grid services.

Keywords: Smart grid · IoT · Consumer participation · Business model

1 Introduction

Energy sector worldwide is traditionally regarded as one of the most regulated, and has naturally drifted towards a vertically integrated model. With the sudden rise in information technologies and new economic concepts, new ways of doing business and industry branches have emerged. In order to support these new business environments the energy sector has to adopt alongside them. This adaptation comes in the form of deregulation and restructuring of the energy sector as a whole, where more and more segments are organized into open markets, exchanges, and services. One of these services that seek to modernize both the grid and the business model is demand-response. These emerging services often rely on new technologies and ways of doing business, and energy grids that seek to support them both technologically and in the business sense are called the "Smart Grid" [1, 2].

The focus of this paper is on affordable and widely available Internet of Things technologies as a backbone for designing new smart grid services [3]. With the recent advances in IoT and the utilization of widely available LAN-enabled microcontrollers

Á. Rocha et al. (Eds.): WorldCIST 2020, AISC 1159, pp. 539–548, 2020.
https://doi.org/10.1007/978-3-030-45688-7_54

connected to the power outlets, we can control and coordinate thousands of consumer appliances at a time. These "smart-plugs" will be capable of turning any individual appliance, no matter how old, into a smart device capable of basic decision-making and adapting its use to be most beneficial to the energy grid. In addition, by utilizing open source hardware as a base for our "smart-plugs" we will be removing any monetary and technological restrictions giving the consumers to opportunity to focus purely on the ecological incentives and social responsibility.

The rest of the paper is organized as follows. Section 2 describes the theoretical aspects of consumer participation in smart grid, basics of demand response, and the ecological aspects of demand response. In the Sect. 3, we present the original business model for consumer participation in demand response based on IoT. The proposed IoT infrastructure is described in details. In the Sect. 4, we present the study of Serbian consumers' readiness to participate in the proposed business model. Section 5 presents the concluding remarks.

2 Theoretical Background

2.1 Consumer Participation in Smart Grid Services and Demand Response

Demand response as a technology entails management of energy use through manipulation of end user facilities away from normal consumption patterns in response to certain events [4]. These events are often either energy prices or distinct states of the energy grid. The manipulation of user facilities often takes form of turning on or off certain appliances, in order to ensure a more stable energy grid. Any participation in the demand response process is compensated to the appliance and loss of comfort, either through direct monetary compensation or through energy bills. There are two standard demand response models: curtailment and shifting. With curtailment, the appliance lowers its energy consumption, without any plans to move it to another time period. Shifting, on the other hand, requires the appliance to be turned on at a different time. The shifting offers more flexibility to the operators, allowing them to shift load from a peak period to a non-peak period. Additionally, demand response services can be used to maintain the stability of the grid, and participate in the frequency regulating activities taking place in the balancing energy market [5].

Demand response is recognized as a key technology of the future by ENSTO-E [6], which is best seen in 2030 and 2050 European energy policies and decarbonisation targets that are deeply reliant on demand response as a technology. Despite its future and current importance, demand response is still a relatively new technology, with only a few larger European countries recognizing demand response as a market resource.

In order to introduce demand response into an energy market, many steps need to be taken, especially in countries where there are not that many smart devices, or any household smart meters. These two facts alone would make the implementation of a conventional demand response system if not impossible than an extremely costly undertaking. To overcome these problems, we have turned to the Internet of Things technologies, by working with individual household devices - smart plugs, which make any plugged device into a demand-response able device. In this way, we can overcome the need for household smart meters, and the general lack of intelligent devices.

As energy market participation can be considered a complex process, requiring accordance to many regulations and rules, it requires a well-defined business model. A demand response model which has become an industry standard is the model of demand response aggregator. Aggregator as a market participant performs many activities such as metering, verification, base calculations, and end-user billing and compensation.

2.2 Ecological Aspects of Demand Response

The ecological aspect of demand response services lies in the new role of the balancing market. Due to solar plants and wind farms constantly deviating from their "goal" energy production, the stability of the grid can be negatively affected. In order to offset these deviations other types of generators must make up for their lacking or overproducing energy, by selling it on the balancing market. Due to this, it is often the case that in order to cover for the wind deviations, a coal plant must be utilized for balancing purposes.

The approach that we propose hopes to change this fact by making the majority of the balancing market into controllable load that can be shed or utilized in order to balance the grid. In this way, renewable plants can cover their deviations with household devices, rather than conventional power plants, the end result being a completely green system, where there is little need for conventional plants in the balancing market.

In order to make any notable change to the environment and the energy sector as a whole, we have recognized that end-users are fast becoming a driving factor in many environmental programs [7, 8]. By focusing on the end-users, and conducting a thorough acceptance study, a better understanding of the driving factors behind the end-users decisions can be obtained, and their eventual stance towards demand response as a technology.

Previous researches have shown that consumers' attitudes towards green energy and environmental protection are becoming important incentives for acceptance of new consumer-centric business models in energy sector. However, the 'green gap' between consumer attitudes to the environment and the adoption of green behaviour has been identified [9] and should be taken into account during the models development in order to achieve a better acceptance.

3 A Model of Household Participation in Demand Response

3.1 Business Model

The business model that we propose offers demand response services to the consumers in the Serbian electricity market. The model is based on a system for distributed control of IoT-enabled household devices, through the use of "smart-plugs". The developed system should be capable of delivering the required controllable demand-side energy for the purpose of participating in the Serbian Balancing Energy Market.

Table 1 shows the business model canvas for the proposed approach. The concept is based on the participation of consumers [10, 11] and aggregation of their individual devices. By aggregating the individual devices into a single controllable load, we can effectively predict the effective capacity with statistical models in the same manner that every day energy predictions are made by grid operators. This aggregated load of all individual devices will be made available on the Serbian balancing market for activation. Upon the activation of our bid, the individual devices would coordinate their operation in order to curtail their collective energy usage, while seeking to maintain user comfort. The user comfort in this case can be defined as minimizing the time the device is unavailable to the user. This comfort can be increased by improving the coordination of individual devices as well as the reducing the potential capacity of the aggregated load. Upon the completion of the demand response action, any earnings made on the balancing market are to be used to compensate end-users for the brief loss of comfort, in addition to system maintenance costs.

Table 1. Business model canvas

<table>
<tr><td colspan="5">Business idea: Household participation in Serbian balancing market Services: Demand response</td></tr>
<tr><td rowspan="2">Key partners
JSC Elektromreža Srbije
PE Elektroprivreda Srbije</td><td>Key activities
Development of platform, services and infrastructure
Marketing and CRM</td><td rowspan="2">Value prepositions
Social responsibility
Support to green energy
Support to environmental protection</td><td>Customer relationships.
Recognition of the most valued customers by assigning green badges</td><td rowspan="2">Customer segments
Households</td></tr>
<tr><td>Key resources
Network and IoT infrastructure
Smart plugs</td><td>Channals
WiFi, LAN, mobile network</td></tr>
<tr><td colspan="3">Cost structure
Software development
Development of smart plugs</td><td colspan="2">Revenue streams
Income by participation in energy market.
Subventions</td></tr>
</table>

3.2 IoT Infrastructure

The IoT infrastructure for the proposed model is based on connecting individual household devices using “smart-plugs”. The developed system should be capable of delivering the required controllable demand-side energy for the purpose of participating in the Serbian Balancing Energy Market [12]. In order to be suitable for the Serbian market, this approach focuses on affordable and widely available technologies as a backbone for its demand response approach. This approach is suitable and sustainable because it is not expected that the complex metering infrastructures will be implemented in the Serbia in the foreseeable future. Through the utilization of widely available LAN-enabled microcontrollers connected to the power outlets, we can control and coordinate the consumption of thousands of consumer appliances at a time [13]. These “smart-plugs” will be capable of turning any individual appliance, no matter how old, into a smart device capable of basic decision-making and adapting its use to be most beneficial to the energy grid. In addition, by utilizing open source hardware as a base for our "smart-plugs" we will be effectively removing any monetary and technological restrictions for any future researchers and applications.

The concept of smart plug is shown in Fig. 1. The smart plug is realized using the Arduino microcontroller wirelessly connected to the aggregator's infrastructure. Communication and coordination of hardware components are done over the internet through the use of local end-user networks. The use of TCP/IP in this case would allow for the use of RESTful services as the basic building block of the model, which in turn allows for high scalability and flexibility of the distributed network. Software components developed in this way are in accordance with international standards on interoperability of smart grids while still being capable of supporting large number of users.

Fig. 1. The concept of smart plug

By focusing on the individual device, rather than the smart meter for the household, we can simplify the calculations and the business logic entirely. The calculations regarding the "base load" of a single device and its usage patterns are easily isolated, and as such can be grouped together with similar neighbouring devices. These neighbouring devices would then coordinate with each other and make up a distributed computing network. A network that works in such a fashion, would be easily expandable, and could function with minimal investment, which is a stark contrast to conventional smart grid projects that require a sizable investment.

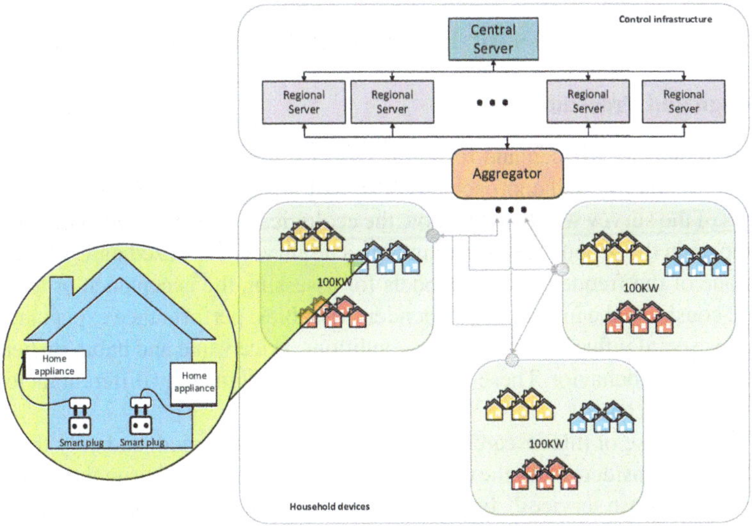

Fig. 2. Control infrastructure for aggregation of smart plugs

The aggregation of the distributed network is done by the aggregator's servers. A server contains most business logic pertaining to the communication with all external agents. In the sense of demand response, this external communication involves B2B energy market participation. The architecture of the aggregation aspect of the proposed model is shown in Fig. 2.

Each household is supplied with a web application, where reports for the consumption and the participation of each device in demand response can be seen (Fig. 3).

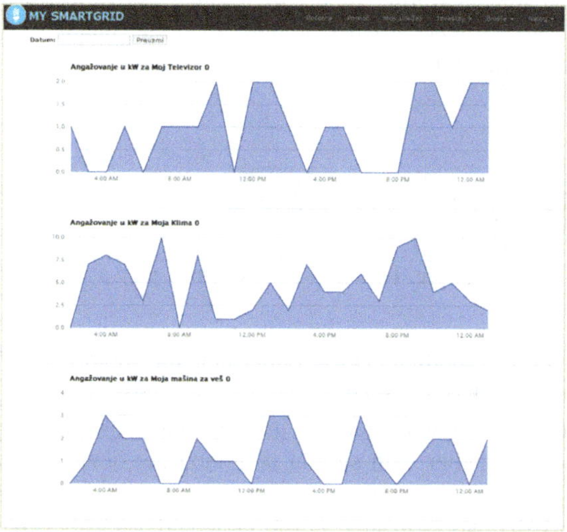

Fig. 3. Report on the consumption of smart plugged devices

4 Assessing Readiness for Participation in Demand Response

4.1 Design and Procedure

Having in mind that the demand response services are not available in Serbia yet, the evaluation of the proposed approach has been realized by surveying the potential users. The focus of the survey was to assess how the ecological incentives affect their readiness to participate in the proposed business model. The survey was based on UTAUT2 model [14], as one of the frequently used models for assessing the acceptance of technology. UTAUT2 considers the impact of independent variables: performance expectancy, effort expectancy, social influence, facilitating conditions, price value and habit, on behavioral intention, and use behavior. These influences can be moderated by different factors, such as age, education, etc.

For the purpose of this research, UTAUT2 has been modified, and only the selected variables were considered. As the dependent variable, we have chosen the "readiness to use demand response services", based on Behavioural intention variable from UTAUT2. As independent variables, we have chosen "the attitude towards green energy" and "the attitude towards environmental protection", both based on Social influence from UTAUT2. The impact of each variable was analysed depending on age, income and employment status of respondents.

The following hypotheses were set:

- H1.1: The attitude of consumers towards green energy has impact on their readiness to use demand response services.
- H1.2: The attitude of consumers towards environmental protection has impact on their readiness to use demand response services.

Besides demographic questions, the questionnaire included questions for measuring readiness to participate in the proposed demand response services:

- To what extent would you be interested to use demand response services if you would contribute to green energy development/environmental protection in this way?
- For how long would you be willing to let the aggregator company turn off/on your air condition/storage heater in the summer/winter period if you would contribute to green energy development/environmental protection in this way? (regulation up/regulation down, offered options in minutes: 60, 15, 10, 5)
- Do you intend to use demand response services when they become available in Serbia?

Before completing the survey, each respondent was given an explanation of the proposed demand response services. There were 244 participants in the research; 42,6% were male, 57,4% female. Most respondents had high school as their highest education level (56%), 35% had completed undergraduate studies, only 9% had a MSc or a PhD degree. Demographic data is shown in Table 2.

Table 2. Demographic data

Age		Employment status		Income	
<30	39%	Employed	56%	Up to 250€	10%
31–40	16%	Retired	8%	250€–500€	35%
41–50	18%	Student	34%	>500€	17%
>50	27%	Unemployed	2%	No answer	38%

4.2 Results

Readiness of consumers for participation in *demand response* services depending on the considered incentives is shown in Table 3. The results show that the respondents generally have more support for environmental protection (4.33) but the support for green energy is also high (4.12).

Table 3. Incentives for participation in demand response

	Incentive	Mean	Std. Dev.
General attitude	Attitude towards environmental protection	4.33	0.89
	Attitude towards green energy	4.12	0.97
Regulation up	Contribution to environmental protection	3.54	1.35
	Contribution to green energy	3.58	1.32
Regulation down	Contribution to environmental protection	3.57	1.37
	Contribution to green energy	3.58	1.32

As an example of demand-response service in regulation up we have considered turning the AC off in the summer period, and as an example of demand-response service in regulation down we have considered turning the storage heaters on in the winter period. Table 3 shows that mean values and deviations in both cases are similar, and that both considered incentives are equally valued. Still, it can be noticed that the consumers have a generally high awareness on environmental protection and green energy, but they are not entirely ready to accept discomfort in order to contribute to these causes. Most of the respondents answered that they would accept their device to be turned off for up to 60 min, but lower times are preferred. We can conclude that in the proposed scenario, AC or TA furnace should not be affected longer than 15 min.

Figure 4 shows the distribution of the attitudes toward green energy and environmental protection by age. It can be seen that consumers value environmental protection more that green energy support, and that age groups 31–50 value both these aspects more than older and younger population.

Similarly, the results indicate that the respondents with the highest incomes value green energy and environmental protection more than consumers with lower incomes. Respondents from all employment statuses value the environmental protection, but the green energy is a higher incentive to the employed and retired, comparing to student population. The obtained results can be used for adjusting the business model with respect to users' preferences, and more precise market segmentation.

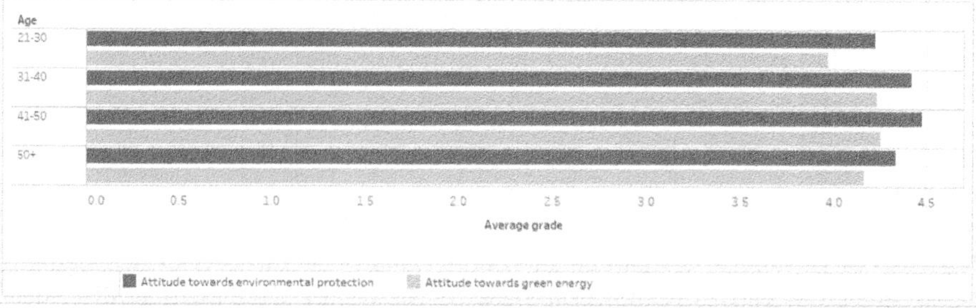

Fig. 4. Distribution of attitudes toward green energy and environmental protection by age

5 Conclusion

The implementation of the proposed approach should result in a novel system for an advanced smart grid technology, and is expected to be the first of its kind in the region. The proposed solution offers flexible demand-side energy as an alternative to coal plants the tertiary and secondary balancing ancillary services. Any solution adoptions resulting from this approach would result in less coal use in the energy sector. Additionally, the flexibility of demand response allows for finer balancing when renewable generators, such as wind farms, are involved making them less of a burden for the energy grid.

Economically speaking, the results of the application of the proposed approach would have the most impact in the industry, where its results could be immediately translated into value for the demand response company. The greatest impact would be to the energy market and the energy grid as a whole, as the implementation of this technology is bound to improve the quality of the grid, modernize it, and offer potential new business practices to both the market and the grid operator. Additionally, by informing the public of demand response as a technology, as well as by promoting the consumers who participated the most, we can hopefully change the stance of the end-users into a pro-active one in regards to energy consumption and environment.

As the results of the survey show, consumers are generally motivated to participate in new smart grid services by their attitudes towards environmental protection and green energy. In addition, there are differences towards these incentives among different age groups. Higher interest shown by participants in age groups 31–50 may be due to the fact that these groups are more aware of environmental problems. However, further analysis is necessary in order to fully understand the perceptions of different consumer groups, and to gather enough information to be able to make the personalized demand response services.

Acknowledgement. Authors are thankful to Ministry of education, science and technological development, Republic of Serbia, grant 174031.

References

1. Ipakchi, A., Albuyeh, F.: Grid of the future. IEEE Power Energy Mag. **7**(2), 52–62 (2009)
2. Farhangi, H.: The path of the smart grid. IEEE Power Energy Mag. **8**(1), 18–28 (2010)
3. Mashima, D., Chen, W.P.: Residential demand response system framework leveraging IoT devices. In: 2016 IEEE International Conference on Smart Grid Communications, SmartGridComm 2016, pp. 514–520 (2016)
4. Rahimi, F., Ipakchi, A.: Demand response as a market resource under the smart grid paradigm. IEEE Trans. Smart Grid **1**(1), 82–88 (2010)
5. Albadi, M.H., El-Saadany, E.F.: A summary of demand response in electricity markets. Electr. Power Syst. Res. **78**(11), 1989–1996 (2008)
6. ENTSO-E: Guideline on Electricity Balancing (2017). https://www.entsoe.eu/Documents/Network codes documents/NC EB/Informal_Service_Level_EBGL_16-03-2017_Final.pdf. Accessed 08 Jan 2020
7. Shuqin, W., Ang, T., Jancenelle, V.E.: Willingness to pay more for green products: the interplay of consumer characteristics and customer participation. J. Retail. Consum. Serv. **45**, 230–238 (2018)

8. Vand, B., Hast, A., Bozorg, S., Li, Z., Syri, S., Deng, S.: Consumers' attitudes to support green energy: a case study in Shanghai. Energies **12**(12), 2379 (2019)
9. Gupta, S., Ogden, D.T.: To buy or not to buy? A social dilemma perspective on green buying. J. Consum. Mark. **26**(6), 378–393 (2009)
10. Schuitema, G., Ryan, L., Aravena, C.: The consumer's role in flexible energy systems: an interdisciplinary approach to changing consumers' behavior. IEEE Power Energy Mag. **15**(1), 53–60 (2017)
11. Abi Ghanem, D., Mander, S.: Designing consumer engagement with the smart grids of the future: bringing active demand technology to everyday life. Technol. Anal. Strateg. Manag. **26**(10), 1163–1175 (2014)
12. Kiviluoma, J., et al.: Harnessing flexibility from hot and cold: heat storage and hybrid systems can play a major role. IEEE Power Energy Mag. **15**(1), 25–33 (2017)
13. Pipattanasomporn, M., Feroze, H., Rahman, S.: Multi-agent systems in a distributed smart grid: design and implementation. In: IEEE/PES Power Systems Conference and Exposition 2009, PSCE 2009, pp. 1–8 (2009)
14. Venkatesh, V., Thong, J.Y., Xu, X.: Consumer acceptance and use of information technology: extending the unified theory of acceptance and use of technology. MIS Q. 157–178 (2012)

A Reference Model for Smart Home Environment: Functions, Semantics and Deployment

Wenbin Li[1(✉)], Matthieu Liewig[1], and Fano Ramparany[2]

[1] Orange Labs, 35510 Cesson-Sévigné, France
{wenbin.li,matthieu.liewig}@orange.com
[2] Orange Labs, 38240 Meylan, France
fano.ramparany@orange.com

Abstract. Along with the proliferation of smart home solutions, Smart Home Environment (SHE) has been constantly evolving with diverse functions and services to improve households' living experiences. Consequently, a global view of SHE is desired to represent the characteristics of the ever-changing domain for both solution adoption and innovation purposes. In this paper, we present a reference model of SHE aiming at capturing the environmental characteristics by leveraging the quick evolution pace and the model abstraction level. The reference model consists of three views i.e., functional view, deployment view and ontological view organized following the middle-out methodology. The functional view firstly presents the hardware components and software features necessary to build modern SHE, and then the deployment view and ontological view respectively describe the deployment structure in lower level and the SHE semantics in higher level. The objective is to provide, from a service provider perspective, a common understanding of SHE for household adoption, better industry positioning and research innovation.

Keywords: Smart home · Smart home environment · Reference model · Functional component · Ontology · Deployment · Service provider

1 Introduction

Driven by the latest advance in Information and Communications Technology (ICT), Internet of Things (IoT) and Artificial Intelligence (AI), large number of smart home devices have arisen in recent years, varying from traditional home infrastructure devices such as resident gateway, to emerging IoT sensors and actuator, as well intelligent terminal devices such as smart speakers. According to the 2019 Global Smart Home Forecast [1], consumer spending on smart home related hardware, services and installation fees will reach $157B by 2023, and 15% of all households will have at least one type of smart system installed. The Smart Home Environment (SHE) has been constantly evolving along with the proliferation of smart home devices, changing domestic behaviors and improving households' living experiences.

Á. Rocha et al. (Eds.): WorldCIST 2020, AISC 1159, pp. 549–559, 2020.
https://doi.org/10.1007/978-3-030-45688-7_55

The concept of smart home is defined in [2] as "a residence equipped with computing and information technology which anticipates and responds to the needs of the occupants, working to promote their comfort, convenience, security and entertainment through the management of technology within the home and connections to the world beyond", and is then expanded to the idea of SHE [3] as being able to exhibit various forms of artificial intelligence by enhancing traditional home automation systems with new smart functions addressing diverse high-level goals of well-being like increasing comfort, reducing operational costs, and guaranteeing safety and security for the house holders. Nevertheless, along with the ongoing mutualization of SHE and its enabling technologies (mainly ICT, IoT and AI), the SHE concept is extended by devices and functions to offer advanced services of security, automation, entertainment and assisted living. It is fundamental for both academia and industry to capture the latest SHE essence as well as future trends, and therefore a global view of SHE is needed reflecting the integration of enabling technologies. In this paper, we present a SHE reference model consisting of functional, deployment and ontological views to capture overall features of this ever-changing domain.

Several works defined and modelled the SHE: An early work in [4] presented an overall description of smart home technologies, products and services, which is a helpful reference to structure later works. The work in [3] then introduced a SHE overview of architectures and application areas, and described the SHE as the integration of control system, home automation system and network. Benefits and risks of SHE technologies are discussed in [5] with highlight on EU policy challenge with respect to the inconsistencies between industry, prospective users and policymakers. The software aspect of SHE is detailed in [6] to manage smart home resources. At last, the SHE is considered in [7] as a real-time interactive response between the power grid and users, with illustrated technologies for smart grid system.

As complementary of the SHE technologies, our paper emphasizes on the SHE functions with reference domains instead of detailed technologies, and presents the lower deployment and higher semantics of SHE. The objectives are threefold: to provide a common definition of SHE from service provider perspective, to attract research and industry attention for solution positioning and innovation, as well as to promote the household adoption. The remainder of the paper is organized as follows: Sect. 2 presents an overview of the SHE reference model; Sect. 3 introduces the SHE functional view; Sect. 4 and Sect. 5 respectively illustrate the SHE deployment structure and ontologies for semantic solutions. Section 6 concludes our work.

2 Reference Model Overview

In this paper, the SHE is regarded as the application and integration of pertinent technologies to home and the enhancement of home services. We firstly identify a number of common SHE features as follows:

- **Intelligence:** The SHE behaves at a certain intelligence level with learning, deduction, decision making and interaction capability.
- **Security and Privacy:** The SHE provides security and privacy measures to prevent both physical intrusions and the disclosure of information.

- **Upgradeability and Expandability:** The SHE is regularly updated and upgraded to incorporate additional users, devices and external environments.
- **Energy Efficiency and Low Cost:** The SHE solutions are energy efficient and low cost, comparing to large scale deployed industry solutions.
- **Limited Capability:** The SHE has limited hardware capability (e.g., storage, computing and networking) due to the energy, cost and size constraints.
- **Comfort and Entertainment:** The SHE is comfortable for households with the diversity of services such as multimedia and entertainment.

The reference model presented in this paper consists of the following three views. The three views are respectively introduced in the latter sections following a middle-out methodology, meaning that we start by identifying functional components of SHE, and then taking the functional view as the focus point, the introduction descends to a lower deployment view supporting the SHE functions and ascends to a higher ontological view empowering the SHE functions. The three views incorporate each other with intersection to build the global SHE reference model.

- **Functional View.** The functional view presents the functional features of the SHE, and emphasizes on the SHE system-of-systems feature with three-layer modelling.
- **Deployment View.** The deployment view illustrates the inner- and inter-connection nature of SHE following the functional view.
- **Ontological View.** The ontological view highlights the semantic technology for SHE to enable interoperability, federation and knowledge management.

3 Functional View

The functional view of SHE is presented in Fig. 1 and modelled as the union of three layers, i.e., individual layer, interconnection layer and interaction layer. The rectangles with solid fill represent common existing features in SHE, while the rectangles with pattern fill represent the emerging features being researched.

Individual Layer: In this layer, the SHE is device-centric and modelled as the individually deployed home devices which function in an intelligent and independent manner. The following types of home devices are considered:

- Terminal devices, such as smart speaker and smart TV.
- Home infrastructure devices, such as residential gateway and Wifi repeater.
- IoT devices, including home sensors, actuators and legacy home objects powered by sensors and actuators.
- Other Microcontroller (MCU) based devices, such as home automation devices.

Interconnection Layer: In this layer, the SHE is group-centric and modelled as the communication and interworking of individual devices, and the interconnection layer describes the SHE features from the group of home individuals' perspective.

Interaction Layer: In this layer, the SHE is user-centric and modelled as the interaction between users and home individuals, and correspondingly the interaction layer describes the SHE features from the user service perspective.

To indicate the priority and necessity of each layer features, the keywords "SHALL", "SHOULD" and "MAY" are used in this paper following the RFC2119 specification [8].

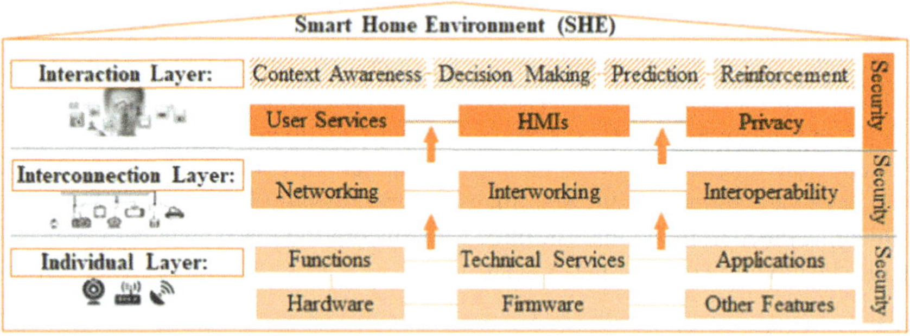

Fig. 1. Functional view of SHE

3.1 Individual Layer

The individual layer describes the SHE features from six aspects: hardware, firmware, function and technical services, applications, security and other features.

Hardware: From hardware aspect, the individual holds the following features.

- Processors. The individual shall have at least one Generic Purpose Processor (GPP) (e.g., CPU, microprocessor) to work in various application contexts, should have one or more Application Specific Instruction-Set Processor (ASIP) (e.g., Graphics Processing Unit (GPU), Digital Processing Unit (DPU)) to accelerate specific tasks, and may have one or more Single Purpose Processor (SPP) (e.g., timer, counter, Analog-to-Digital Converter (ADC)) to realize particular tasks.
- Memory. The individual shall have a Non-Volatile Memory (NVM) for general storage of applications and data, should have one or more secured ROMs for security features (e.g., secure boot, cryptographic key storage), may have one or more specific NVMs (e.g., NAND Flash [9], NAND for Not-AND) to support specific services (e.g., video recording or time shifting) for better performance and data isolation, and shall have at least one Random-Access Memory (RAM).
- Input/Output (I/O). According to the intended functions and user services, the individual shall have corresponding I/O interfaces to be controlled by processor(s) and to communicate with internal and external components, as well as I/O ports for physical connection with other devices.

– Microcontroller (MCU) or System on Chip (SoC). Depending on the complexity, the individual shall use either MCUs or SoCs as the substrate to integrate processors, memory controllers and optionally along with other components (e.g., memory and I/O peripherals) for size improvement and power efficiency purposes.

Firmware: The term firmware is used here to refer to the program set tied to the particular hardware components of a device (including boot loader, kernel/OS, drivers and middleware). The individual shall have a firmware to provide an operating environment for the developed functions, technical services and applications.

Functions and Technical Services: The individual shall realize certain functions (e.g., device management, sensing and actuating, content delivery) and exposes technical services. Different from the user service, the technical service refers to the representation of one or more functions to a network that makes the functions discoverable, registerable and remotely controllable by other devices.

Applications: An application is defined as the developed program to demonstrate the functions and consume the technical services by users or devices. In the case of infrastructure devices and terminal devices, the individual shall have one or more applications deployed in the device; in the case of IoT devices and other MCU devices, the individual may have one or more applications deployed in the same device, and one or more applications deployed in other terminal or infrastructure devices to consume the technical services provided by this device.

Security: Generally, the infrastructure and terminal devices shall support all following security features, while IoT and other MCU devices should support as well if environment permits. From hardware perspective, the individual supports the secure boot [10] allowing establishing a trust chain from a hardware root of trust to the firmware itself. The individual has a secure element (e.g., cryptographic coprocessor or Trusted Execution Environment) for sensitive code execution, and a one-time programmable memory allowing hiding key and secrets in hardware. From software perspective, the SHE shall be developed according to the least privilege paradigm: no superfluous rights or accesses might be granted to any application, and has hardening mechanisms [11], such as memory protection and mandatory access control.

Other Features: The individual should have simplified housing to include all necessary components with internal or external power supply. Moreover, the individual shall meet the existing standards to assure home safety, environment protection, noise and energy consumption, such as the standards specified in [12–15].

3.2 Interconnection Layer

In this layer, the SHE is the collection of all devices with the following features.

Networking: The SHE shall have one or more network access points to enable both of the Local Area Network (LAN) connection for all internal devices and the Wide Area

Network (WAN) connection between the SHE and external environments. While the deployment is detailed in Sect. 4, the networking is generally managed by the residential gateway in a centralized home deployment, and in a decentralized deployment one or more devices may have independent network access capabilities.

Interworking: The SHE should support the information exchange among internal devices and more importantly the interworking of home devices to orchestrate or compose individual technical services [16] together to enhance home functionality.

Interoperability: During the networking and interworking of home devices, standards protocols and models (if any) should be used in each SHE individual and cover as many layers as possible of the OSI model (i.e., from the bottom physical layer to the top application layer) to achieve the SHE interoperability [17].

Security: The security feature in the interconnection layer groups all security features from its constituting individuals. Especially the networking and interworking should be protected by access control mechanisms and firewall, and enable secured communication protocol such as SSL and HTTPS.

In the case that the individual collection is deployed following a decentralized architecture, each individual shall apply its own security features to interwork with others, while in the case of centralized deployment architecture, the centralized control point (e.g., residential gateway) shall apply the security features of itself to all individuals in addition to the individual security features themselves.

3.3 Interaction Layer

The interaction layer describes the features of the interaction between home users and the collection of home individuals.

User Services: The smart home environment shall eventually provide one or more user services to household through the individuals, the individual collection the interworking with external environment such as cloud servers. Different from the technical service, the user service is the operations of the entire SHE that benefit home users, such as multimedia, telecommunication and entertainment services. Taking the user service of TV as an example, the delivery of television content to home involves the running of backend set in a cloud server, the running of frontend set in the home set-top box, the execution of receiving and displaying content functions in TV and optionally the execution of a remote control application in a mobile phone.

Human-Machine Interface (HMI): The SHE shall have at least one HMI for users to interact directly with the individual layer and the interconnection layer both locally (at home) and remotely (outside home). The HMI is either in the form of a software interface from an application such as presented in [18], or a physical device for I/O purpose such as introduced in [19].

Security: The interaction layer regroups all security features from the individual layer and interconnection layer. Additionally, the SHE shall have user authentication and authorization mechanisms to ensure the correct use of devices, applications and user services by home users. However, a compromise has to be made between the SHE security level and the usability for common household, as effective security measures come with high resource consumption and the cost of usability.

Privacy: The SHE shall provide a secured and protected environment for user's privacy and shall be standard compliant (e.g., General Data Protection Regulation (GDPR) [20]). Home data, services and discovered context shall be only available and accessible to legitimate devices and users via privacy management framework [21].

Besides the above common features of the interaction layer, a number of SHE features are being studied in both academics and industry with promising prototype results, and some representative features are identified below.

Context-Awareness: The SHE should be able to automatically identify home contextual information [22] regarding the environment and users (e.g., activity, situation and preference), and correspondingly provide adapted services to users according to the identified context and empower the functionality of existing services.

Decision Making Support: The smart home environment should be able to assist the decision making of users under specific situations [23], which does not only include management and coordination decision, but also intelligent interaction.

Prediction: The SHE should be able to anticipate users' requirements via behavioral histories and provide suggestions and recommended services accordingly without explicit requirements.

Reinforcement: The SHE should continuously enrich its knowledge and improve user service quality via iterative reinforcement learning [24] for comfort and autonomy at home and customize services according to the obtained SHE context.

4 Deployment View

The deployment view illustrates the lower organization structure supporting the SHE functions presented in the previous section. As shown in Fig. 2, the SHE is the collection of all devices deployed following either centralized organization (e.g., star network topology) or decentralized organization (e.g., mesh network topology) with Personal Area Network (PAN) or LAN connection. The SHE is then further connected via WAN to cloud environment which provides both public Internet services and ISP managed services (e.g., VoIP and IPTV). The home PAN is often realized by protocols such as Zigbee, Zwave and Bluetooth; the LAN is commonly based on computer networks; and the WAN is achieved by computer networks, mobile networks or low-power WAN, such as LoRa and Sigfox based ones.

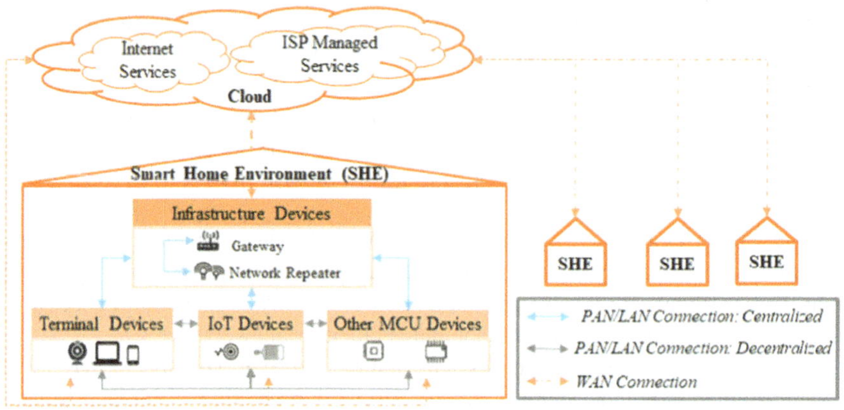

Fig. 2. Deployment view of SHE

Each device in the deployment view holds the features described in Sect. 3. In the centralized home deployment, the end nodes (i.e., terminal devices, IoT devices and other MCU based devices) are connected to a central control point (e.g., residential gateway) either directly or via network (e.g., Wifi and Zigbee) repeaters. The residential gateway is responsible for receiving, controlling and forwarding the communications among devices or from devices to cloud infrastructure and platforms. In a decentralized deployment, no central control point exists and the sourcing device directly communicates with the targeting device via PAN/LAN, and the end node devices also directly connect to cloud via WAN connection (e.g., 4G).

5 Ontological View

The ontological view presents the semantic modeling of SHEs based on ontologies [25], which is a higher level abstraction and a SHE enabling technology to bring semantic interoperability, automatic federation and knowledge discovery capabilities. The ontological view is placed above the deployment view and functional view and further serves as an complementary view, as the ontologies on one hand capture the semantic meaning, logics and rules behind SHE, and on the other hand improve the interoperability and knowledge discovery capability at a global scale of all SHE individuals and potentially among SHEs, by breaking the existing silos [26] resulting from the heterogeneous devices and diversified user services of SHEs.

A number of SHE ontologies have been developed, while few have been standardized with wide adoption as the SAREF ontology [27] and the W3C SSN and SOSA [28]. Due to the page limitation, we here briefly introduce the SAREF ontology offering reasonable model and concept coverage.

The SAREF (Smart Appliances REFerence) ontology was developed and standardized by the European Commission in close cooperation with ETSI (European Telecommunications Standards Institute) to provide a modular and domain-independent semantic layer for smart appliances. As shown in Fig. 3, the starting point of the SAREF ontology

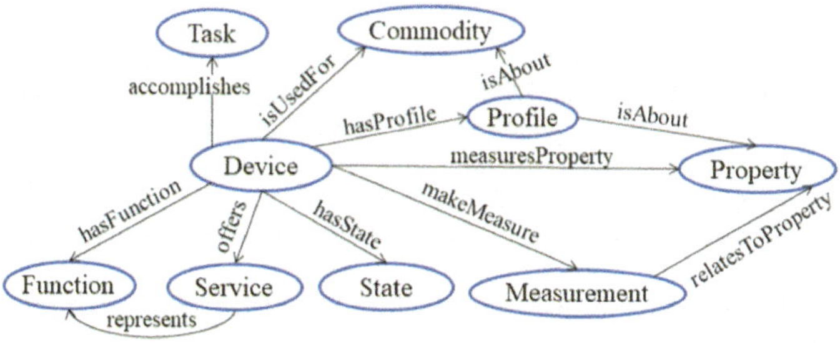

Fig. 3. SAREF core concepts

is the concept of *Device* representing tangible objects designed to accomplish one or more *Tasks* in diverse types of locations and associated with States. The SAREF ontology offers a list of basic *Functions* that can be combined towards more complex functions in a single device. A *service* specifies the device that is offering the service, the functions to be represented, and the input and output parameters necessary to operate the service. Furthermore, a device can be characterized by a Profile used to collect information about a certain Property or Commodity (e.g. energy or water) for optimizing its usage in the home/building in which the device is located. Together with the Measurement, Property and UnitOfMeasure, the ontology allows relating different measurements from a given device for different properties measured in different units. While SAREF details can be found in [27], SAREF ontology is defined essentially as a device-centric ontology for device physical and data aspects. An application of ontologies to SHE with examples is presented in [29].

6 Conclusion

Three views of Smart Home Environment are presented in this paper targeting a reference for SHE functions, deployment structure and ontologies. With a high coverage of the SHE functions on both fundamental and cutting-edge aspects, this work also illustrates the semantic enabling technology potential to further link and evolve the SHEs. In the future work, we will enrich each view presented in the paper with technology details and examples, as well as include and characterize the SHE use cases.

References

1. Global Smart Home Forecast. Strategy Analytics (2019). https://www.strategyanalytics.com/access-services/devices/connected-home/smart-home/reports/report-detail/2019-global-smart-home-forecast---september-2019?slid=959383&spg=1. Accessed 18 Nov 2019
2. Aldrich, F.K.: Smart homes: past, present and future. In: Harper, R. (ed.) Inside the Smart Home, pp. 17–39. Springer, London (2003). https://doi.org/10.1007/1-85233-854-7_2
3. Costin, B., Marius, B., Amelia, B.: An overview of smart home environments: architectures, technologies and applications. In: Christos, K.G., Petros, K., Demosthenes, S. (eds.) Local Proceedings of the Sixth Balkan Conference in Informatics, vol. 1036, pp. 78–85 (2013). http://ceur-ws.org/Vol-1036/p78-Badica.pdf

4. Bierhoff, I., et al.: Smart home environment. COST Brussels (2007)
5. Wilson, C., Hargreaves, T., Hauxwell-Baldwin, R.: Benefits and risks of smart home technologies. Energy Policy **103**, 72–83 (2017). https://doi.org/10.1016/j.enpol.2016.12.047
6. Xu, K., Wang, X., Wei, W., Song, H., Mao, B.: Toward software defined smart home. IEEE Commun. Mag. **54**, 116–122 (2016). https://doi.org/10.1109/MCOM.2016.7470945
7. Li, M., Gu, W., Chen, W., He, Y., Wu, Y., Zhang, Y.: Smart home: architecture, technologies and systems. Procedia Comput. Sci. **131**, 393–400 (2018). https://doi.org/10.1016/j.procs.2018.04.219
8. Bradner, S.: Key words for use in RFCs to Indicate Requirement Levels. https://tools.ietf.org/html/rfc2119. Accessed 18 Nov 2019
9. Micheloni, R., Crippa, L., Marelli, A.: Inside NAND Flash Memories. Springer, Dordrecht (2010)
10. Dietrich, K., Winter, J.: Secure boot revisited. In: 2008 the 9th International Conference for Young Computer Scientists, pp. 2360–2365 (2008). https://doi.org/10.1109/ICYCS.2008.535
11. Turnbull, J.: Hardening Linux. Apress, New York (2006)
12. IEC 62474 - Material Declaration for Products of and for the Electrotechnical Industry. http://std.iec.ch/iec62474. Accessed 18 Nov 2019
13. Electromagnetic compatibility (EMC). https://ec.europa.eu/growth/single-market/european-standards/harmonised-standards/electromagnetic-compatibility_en. Accessed 18 Nov 2019
14. EUR-Lex - 32009L0125 - EN - EUR-Lex. https://eur-lex.europa.eu/legal-content/EN/ALL/?uri=CELEX%3A32009L0125. Accessed 18 Nov 2019
15. Smith, R.: Directive 2006/54/EC of the European Parliament and of the Council of 5 July 2006. In: Core EU Legislation, pp. 333–345. Macmillan Education UK, London (2015). https://doi.org/10.1007/978-1-137-54482-7_31
16. Ruta, M., Scioscia, F., Loseto, G., Di Sciascio, E.: Semantic-based resource discovery and orchestration in home and building automation: a multi-agent approach. IEEE Trans. Industr. Inf. **10**, 730–741 (2014). https://doi.org/10.1109/TII.2013.2273433
17. Ganck, A.D.: The New European Interoperability Framework. https://ec.europa.eu/isa2/eif_en. Accessed 18 Nov 2019
18. Song, G., Zhou, Y., Zhang, W., Song, A.: A multi-interface gateway architecture for home automation networks. IEEE Trans. Consum. Electron. **54**, 1110–1113 (2008). https://doi.org/10.1109/TCE.2008.4637595
19. Sellers, E.W., Vaughan, T.M., Wolpaw, J.R.: A brain-computer interface for long-term independent home use. Amyotroph Lateral Scler. **11**, 449–455 (2010). https://doi.org/10.3109/17482961003777470
20. General Data Protection Regulation (GDPR) – Official Legal Text. https://gdpr-info.eu/. Accessed 18 Nov 2019
21. Townsend, D., Knoefel, F., Goubran, R.: Privacy versus autonomy: a tradeoff model for smart home monitoring technologies. In: 2011 Annual International Conference of the IEEE Engineering in Medicine and Biology Society, pp. 4749–4752 (2011). https://doi.org/10.1109/IEMBS.2011.6091176
22. Wongpatikaseree, K., Ikeda, M., Buranarach, M., Supnithi, T., Lim, A.O., Tan, Y.: Activity recognition using context-aware infrastructure ontology in smart home domain. In: 2012 Seventh International Conference on Knowledge, Information and Creativity Support Systems, pp. 50–57 (2012). https://doi.org/10.1109/KICSS.2012.26
23. Chahuara, P., Portet, F., Vacher, M.: Context-aware decision making under uncertainty for voice-based control of smart home. Expert Syst. Appl. **75**, 63–79 (2017). https://doi.org/10.1016/j.eswa.2017.01.014
24. Li, D., Jayaweera, S.K.: Reinforcement learning aided smart-home decision-making in an interactive smart grid. In: 2014 IEEE Green Energy and Systems Conference (IGESC), pp. 1–6 (2014). https://doi.org/10.1109/IGESC.2014.7018632

25. Bauer, M., et al.: Towards semantic interoperability standards based on ontologies. AIOTI White paper (2019)
26. Li, W., Privat, G., Cantera, J.M., Bauer, M., Gall, F.L.: Graph-based semantic evolution for context information management platforms. In: 2018 Global Internet of Things Summit (GIoTS), pp. 1–6 (2018). https://doi.org/10.1109/GIOTS.2018.8534538
27. Smart Appliances Reference Ontology and oneM2M Mapping. https://www.etsi.org/deliver/etsi_ts/103200_103299/103264/02.01.01_60/ts_103264v020101p.pdf
28. Semantic Sensor Network Ontology. https://www.w3.org/TR/vocab-ssn/. Accessed 18 Nov 2019
29. Li, W., Privat, G.: Cross-fertilizing data through web of things APIs with JSON-LD. In: Maria, M., Ruben, V., Felix, L.K. (eds.) Proceedings of the 4th Workshop on Services and Applications over Linked APIs and Data Co-located with the 13th Extended Semantic Web Conference (ESWC 2016), vol. 1629 (2016). http://ceur-ws.org/Vol-1629/paper2.pdf

The Digital Transformation at Organizations – The Case of Retail Sector

Maria João Ferreira[1,3,6], Fernando Moreira[1,2,3,4](✉), Carla Santos Pereira[1,3,5], and Natércia Durão[1,3]

[1] DCT, Universidade Portucalense, Rua Dr. António Bernardino de Almeida, 541, 4200-070 Porto, Portugal
{mjoao,fmoreira,carlasantos,natercia}@upt.pt
[2] IJP, Universidade Portucalense, Porto, Portugal
[3] REMIT, Universidade Portucalense, Porto, Portugal
[4] IEETA, Universidade de Aveiro, Aveiro, Portugal
[5] CEMAT, Instituto Superior Técnico, Universidade Técnica de Lisboa, Lisbon, Portugal
[6] Centro Algoritmi, Universidade do Minho, Braga, Portugal

Abstract. Digital transformation has changed the way we do business in all industries, from health to education. In this scenario, the retail sector is no exception. It is continually affected by advances in digital technologies, which contribute to significant, sometimes drastic and/or disruptive changes in the competitive landscape. Retail is now increasingly digital as multi-sided marketplaces are bringing the online and offline markets together to create better shopping experiences for consumers. In this context, consumers expect to find technology-enriched retail environments. Retailers are looking for advantages, e.g. to grow and create new market opportunities by using technological tools, creating new business models, optimizing and modernizing their practices in a consumer-centric approach. The research presented aims to provide an overview of the critical areas of the sector, perform a trend analysis of the key technologies used and contribute to the prediction of future retail trends.

Keywords: Retail · Digital transformation · Consumer · Multi-sided marketplaces · Technologies

1 Introduction

There are new types of consumers, especially those who are digital natives. These consumers are more connected to the Internet, and have transformed the way they select, buy and consume the products and services offered [1]. According to the same authors [1] market volatility increased further with the arrival of disruptive new companies. These companies proposing new offerings (products and services) through web and mobile applications, leading to a shared or platform economy (Uber, Airbnb, among others). Additionally, with the emergence of "born digital" companies (Amazon, Facebook and Google), which become dominant, while companies that dominate their industries are

Á. Rocha et al. (Eds.): WorldCIST 2020, AISC 1159, pp. 560–567, 2020.
https://doi.org/10.1007/978-3-030-45688-7_56

considered threatened [2]. In this context, the success of purely digital organizations, according to Riasanow et al. [3] such as Netflix, Spotify or Amazon, as well as the bankruptcy of traditional companies like Kodak or Blockbuster, are examples of these changes and turbulence, as well as the digital transformation (DT) that is going on [4].

However, according to Loonam et al. [5], organizations implement DT initiatives in a piecemeal manner, as noted in [6] "*recent work in academia has been largely concerned with guiding certain aspects of digital transformation.*" The need for guidelines for the implementation of DT is presented in the study conducted in [7], where the authors "*surveyed 391 large companies with revenues of $500 million or more across 30 countries, found that organizations with successful digital strategies were 26% more profitable than their in-industry peers and generated 9% higher revenue from their physical assets.*"

One sector that is undergoing profound transformation is the retail sector due to the innovation of its business model through the use of digital platforms [8]. The emergence of new players in the market, such as Alibaba, without their inventory and hybrids like Amazon, combining their stock, opening the platform to independent suppliers as well [9]. According to [8] and [10], digital platform business models have transformed the nature of retail trading as platforms link consumers to the independent supplier base. In this new concept, the retailer only intermediates transactions between buyers and sellers, transferring the risk of storing the retailer's products to the supplier. This new concept made it possible, and according to [11], to create multifaceted markets that use shopping mall-like principles, functioning as networks that facilitate social interactions and co-creation of value, only now in digital. However, the effects of platform economics and platform-based businesses in the retail sector have received little attention from the academic community, as stated in [8].

In this study, our focus is on the DT in the retail sector, in particular, it is intended to give an overview of the sector critical areas, conduct a trend analysis of the most commonly used technologies and contribute to the forecast of future retail trends.

The paper is organized as follows: In Sect. 2, we explore the current context of the digital transformation. Section 3 introduces the digital transformation in the retail sector, followed by successful retail experiences through digital transformation in Sect. 4. The paper is concluded in Sect. 5.

2 Current Context of the Digital Transformation

The companies that today argue to compete and dominate the global market (Amazon, Alibaba, among others) are mostly based almost entirely on technology, and have been undergoing an unprecedented digital transformation. The most representative technologies that allow companies to be digitally empowered, according to [5], can be grouped as follows: (i) virtualization systems (cloud computing); (ii) mobility systems (social media, internet of things, smartphones and tablets); (iii) big data analytics systems. These technologies, combined appropriately with business processes and aligned with external opportunities, will provide superior competitive advantages [12].

2.1 Digital Transformation

In the literature, there are already some studies [3] and [13] concerned with finding the definition of digital transformation (DT). In both studies, there is a concern to present as many conceptual definitions as possible to find some common points or to organize the descriptions by specific characteristics, to propose a definition that combines all existing ones. In the study presented, for example, in [13] 23 definitions were found, which led to its classification in three groups in relation to the definition of DT: (i) refer mainly to organizations; (ii) there are essential differences between the definitions regarding the types of technologies involved [14] as well as the nature of the transformation taking place [15]; (iii) there are similarities between the definitions, for example using common terms such as "*digital technologies*" [16].

To illustrate this diversity, and due to the limited number of pages, only two definitions are presented, chronologically ordered: "*the use of new digital technologies (social media, mobile, analytics or embedded devices) to enable major business improvements (such as enhancing customer experience, streamlining operations or creating new business models)*" [17], and "*goes beyond merely digitizing resources and involves the transformation of key business operations, products, and processes, culminating in revised or entirely new business models*" [14].

Based on the previous definitions and the study presented in [13], the author proposes the following conceptual definition "*a process that aims to improve an entity by triggering significant changes to its properties through combinations of information, computing, communication, and connectivity technologies*".

2.2 The Importance of Digital Transformation

As mentioned, recent advances in technology have had a sometimes disruptive transformative effect on different sectors of activity. Today, organizations to maintain their competitiveness have to optimize their business processes to become faster, more efficient and more competitive [18].

In [19, 20], a study is presented that assessed the situation of Portuguese organizations concerning DT. From the study, the authors' highlight that there is the awareness of the importance of DT in organizations begins to be noticeable, both the importance of organizational awareness and the adoption of digital technologies. However, the DT in Portugal is still at a relatively immature stage.

According to [21], with the emergence of DT, organizations are thus faced with two options: (1) adapt to changing market conditions or (2) be outdated, bound by legacy software and obsolete and biased business strategies many of them even disappear.

2.3 How to Lead Digital Transformation?

Given the shift to the digital world, organizations cannot focus their reflection and behavior solely on the physical/traditional world. Managers of any organization must recognize and anticipate technology-enabled change, estimate its potential and impact, as well as understand how to leverage digital technology to create and capture value for their organizations. Managing organizations in a digital world for success requires rethinking organizational strategies, business models, and key business drivers [22].

The DT implementation in an organization depends, both the changes in processes and technologies and how these changes come together and articulate. For DT in an organization to succeed, a whole new approach is required where the mission is to transform the organization with the involvement of all employees in the process. Thus, for DT to succeed, a clear and coherent digital strategy is needed to drive transformation within the organization. Honestly, digital organizations know that digital technologies must be used to achieve the organization's strategic goals. Digital strategy can be defined as "*A business strategy, inspired by the capabilities of powerful, readily accessible technologies, intent on delivering unique, integrated business capabilities in ways that are responsive to constantly changing market conditions* [23]. Managers must use corporate digital strategy to create competitive advantage, value and customer satisfaction by combining existing technology with the capabilities of other digital technologies [24].

In [22] is presented a framework with six DT corporate dimensions that, according to the authors, can position an organization for a successful competitive posture due to DT. These dimensions are (1) Strategic vision (for a digital world); (2) Culture of innovation, (3) Know-how assets and intellectual property; (4) digital resources (talent); (5) Strategic alignment and, (6) technology assets. The proposed framework allows, according to the authors, (1) the benchmarking of one organization with others within a sector of activity or "against" organizations that are in the same state of progress towards DT, (2) executives/managers measure their organization's growth over time, and (3) help to diagnose resource gaps in an organization by identifying its performance in each of the six dimensions.

3 Digital Transformation in the Retail Sector

The retail sector "*is the part of a country's economy that is made up of businesses that sell goods through stores, on the internet, etc. to the public share prices in the retail sector have been driven up by takeover activity*". In this section it is analyzed the DT in the retail sector [25].

3.1 Importance and Potential in Retail Sector

DT in the retail sector aims to reshape the business and provide innovative experiences for consumers. Just as in other business sectors, it aims to increase competitiveness by digitizing purchasing processes and means of payment. According to [7], the organizations are divided into four categories of digital maturity: fashionistas, digital masters, conservatives, and beginners. The authors emphasized that levels of expertise vary according to the area of activity. To the authors the retail sector is classified as a digital master sector, i.e., "*Strong over-arching digital vision/Excellent governance across silos/Many digital initiatives generating business value in measurable ways/Strong digital culture*".

With DT, the retail sector is increasing customer-oriented concerning customer service, where mechanisms can be adapted to make processes more flexible and tailored to the needs of the business target audience, wherever he is, i.e. at home, on the street or even on public transport [26].

Existing retail business models are challenged by new competition, and evolving consumer expectations as retail increasingly move online. In this scenario, new business models emerge taking advantage of digitization and changing consumer expectations, thus requiring the industry to react to the ever-changing situation. An example of this retail business model innovation is digital platforms [8, 27].

According to [26] the evolution of the retail sector presents three distinct moments over time (Retail 1.0 (1900's); Retail 1.5 (mid 1900's); Retail 2.0 (early 2000's)) as shown in Fig. 1.

Fig. 1. The transformations of retail [26]

3.2 Consumer Behavior

Today, with the so-called digital age, there is a change in consumer behavior. Consumers are currently "*more informed, communicate more with other customers and are forming ever higher expectations regarding digital service provision that spans across all channels and industries*" [28]. In this context, today's consumers expect organizations to anticipate their current and future needs; needs that are still mostly unknown to organizations. Organizations adopting these new trends will succeed, while others will no longer be competitive and may even disappear.

As mentioned, the retail sector is one of the business sectors that has great digital innovation and organizational change. In this sector, consumers are attracted by the ease and convenience of having at their disposal a range of indicators (evaluations by other users, comparability of prices, etc.) that make them trust online and mobile shopping [29]. In this context, it can easily be seen that the so-called "traditional" retailers are joining and making their presence via these digital channels.

4 Successful Retail Experiences Through Digital Transformation

As discussed earlier, DT has changed the lives of people and organizations. The transformation from "traditional" retail to digital retail that is taking place is a complete business transformation [29].

The four retail organizations presented in this section use, as shown in [13] the "combinations of information, computing, communication, and connectivity technologies", as the basis of their business and are cited in the literature as success stories. Table 1 presents a characterization and comparison of the four organizations indicated.

However, as noted, the success of the organizations described is significantly due to the use of digital technologies. The organizations presented viewed digital as an opportunity to create a long-term sustainable business model, i.e., moved away from

Table 1. Table comparison of Amazon.com, Alibaba Group, eBay and Farfetch (2019).

	Alibaba[b]	Amazon[b]	eBay[b]	Farfetch[a,c,d]
Type of platform(s)	B2B, B2C, C2C platforms	B2C marketplace	C2C, B2C platforms	B2C platforms
Market value	$480.8B	$916.1B	$31.5B	$2.4B
Year founded	1999	1994	1995	2007
Headquarters	Hangzhou, China	United States	United States	London
N. of employees	66,421	647,500	14,000	3,232
Revenue (2019)	$51.9B	$232.9B	$17.9B	$837M
Profits (2019)	$10.3B	$10.1B	$2.5B	$89 million in losses

[a]https://www.macrotrends.net/stocks/charts/FTCH/farfetch/market-cap; [b]https://www.forbes.com/companies/; [c]https://www.forbes.com/sites/glendatoma/2019/08/08/farfetch-earnings-china/; [d]https://www.owler.com/company/farfetch

traditional business paradigms. In the digital approach is the vision of the retail ecosystem as a network of suppliers and franchisees that support the central actor (retailers) in harmonizing a business model with long-term goals to maximize the value of customer loyalty in the long term, not just focused on the transactional approach.

With preservation being the overriding goal of the digital approach, customer relationships are achieved by supporting the combination of information, computing skills, communications and connectivity technologies. In this context, the technologies mentioned above, together with the use of DT's four pillars, combine information gathering to understand consumer needs (e.g. cognitive technologies), provide a more extensive range (e.g. visual merchandising), help consumers decide (e.g. Virtual Trial Rooms), and improve customer service (e.g. AI based self-learning systems) [30].

5 Conclusions

DT that is currently being analyzed and investigated should be seen as preparing for any organization to survive and compete in the future by continually adapting to a changing environment. The need for continuous transformation will not diminish. DT involves continual study/analysis of the environment to recognize trends in technological evolution, constant experimentation to determine how to respond to those trends effectively, and how to propagate successful experiences by organizations.

As regards the retail sector, mentioned above, it should not be discouraged. It should be accompanied by analysis, not only from a technology perspective but also from consumer trends and preferences, i.e., the study of current and future profiles consumers in the digital age.

The research presented makes it possible to gauge the positioning of the retail sector against DT. In this research are also analyzed organizations that have in their genesis the digital, such as eBay or hybrids as is the case of Amazon.

As future work, it is intended to apply the framework, described in Sect. 2.3, and check the status regarding the DT, for each of the organizations presented. It is also designed to explore in more detail consumer behavior/profiles and trends in the retail sector in the coming years.

References

1. Henriette, E., Feki, M., Boughzala, I.: Digital transformation challenges. In: MCIS, p. 33 (2016)
2. Sebastian, I., Ross, J., Beath, C., Mocker, M., Moloney, K., Fonstad, N.: How big old companies navigate digital transformation (2017)
3. Riasanow, T., Setzke, D.S., Böhm, M., Krcmar, H.: Clarifying the notion of digital transformation: a transdisciplinary review of literature. J. Competences Strateg. Manag. **10**, 5–31 (2019)
4. Goh, J., Gao, G., Agarwal, R.: Evolving work routines: adaptive routinization of information technology in healthcare. Inf. Syst. Res. **22**(3), 565–585 (2011)
5. Loonam, J., Eaves, S., Kumar, V., Parry, G.: Towards digital transformation: lessons learned from traditional organizations. Strateg. Change **27**(2), 101–109 (2018)
6. Hess, T., Matt, C., Benlian, A., Wiesböck, F.: Options for formulating a digital transformation strategy. MIS Q. Executive **15**(2), 123–139 (2016)
7. Westerman, G., Bonnet, D., McAfee, A.: Leading Digital: Turning Technology into Business Transformation. Harvard Business Review Press, Boston (2014)
8. Hänninen, M., Smedlund, A., Mitronen, L.: Digitalization in retailing: multi-sided platforms as drivers of industry transformation. Baltic J. Manag. **13**(2), 152–168 (2018)
9. Hagiu, A., Wright, J.: Multi-sided platforms. IJIO **43**(11), 162–174 (2015)
10. van Alstyne, M.W., Parker, G.G., Choudary, S.P.: Pipelines, platforms, and the new rules of strategy. Harvard Bus. Rev. **94**(4), 54–60 (2016)
11. Teller, C., Wood, S., Floh, A.: Adaptive resilience and the competition between retail and service agglomeration formats: an international perspective. J. Mark. Manag. **32**(17–18), 1537–1561 (2016)
12. Tucto-Mechán, L., Bazán-Martínez, J.: The digital transformation triangle. A framework to set the foundations for a successful digital journey. In: I Congreso Internacional de Ingeniería de Sistemas, pp. 225–231 (2019)
13. Vial, G.: Understanding digital transformation: a review and a research agenda. J. Strateg. Inf. Syst. (2019)
14. Horlacher, A., Klarner, P., Hess, T.: Crossing boundaries: organization design parameters surrounding CDOs and their digital transformation activities. In: Americas Conference of Information Systems, San Diego, CA (2016)
15. Andriole, S.J.: Five myths about digital transformation. MIT Sloan Manage. Rev. **58**(3), 20–22 (2017)
16. Singh, A., Hess, T.: How chief digital officers promote the digital transformation of their companies. MIS Q. Executive **16**(1), 1–17 (2017)
17. Fitzgerald, M., Kruschwitz, N., Bonnet, D., Welch, M.: Embracing digital technology: a new strategic imperative. Sloan Manag. Rev. **55**(2), 1–13 (2013)
18. Chanias, S., Myers, M., Hess, T.: Digital transformation strategy making in pre-digital organizations: the case of a financial services provider. J. Strateg. Inf. Syst. **28**(1), 17–33 (2019)
19. Pereira, C.S., Moreira, F., Durão, N., Ferreira, M.J.: Towards the digital transformation: are Portuguese organizations in this way? In: New Knowledge in Information Systems and Technologies. AISC, vol. 930, pp. 326–336 (2019)

20. Durão, N., Ferreira, M.J., Pereira, C.S., Moreira, F.: Current and future state of Portuguese organizations towards digital transformation. Procedia Comput. Sci. (2019, accepted for publication)
21. Ho, J.C., Chen, H.: Managing the disruptive and sustaining the disrupted: the case of Kodak and Fujifilm in the face of digital disruption. Rev. Policy Res. **35**(3), 352–371 (2018)
22. Gurbaxani, V., Dunkle, D.: Gearing up for successful digital transformation. MIS Q. Executive **18**(3), 209–220 (2019)
23. Sebastian, I.M., Ross, J.W., Beath, C., et al.: How big old companies navigate digital transformation. MIS Q. Executive **16**(3), 197–213 (2017)
24. van Dyk, R., Van Belle, J.: Factors influencing the intended adoption of digital transformation: a South African case study. In: Federated Conference on Computer Science and Information Systems, p. 519 (2019)
25. Cambridge Dictionary (2019). https://dictionary.cambridge.org/dictionary/english/retail-sector
26. Grewal, D., Roggeveen, A.L., Nordfält, J.: The future of retailing. J. Retail. **93**(1), 1–6 (2017)
27. Hänninen, M., Mitronen, L., Kwan, S.K.: Multi-sided marketplaces and the transformation of retail: a service systems perspective. JRCS **49**, 380–388 (2019)
28. von Leipzig, T., Gamp, M., Manz, D., Schöttle, K., et al.: Initialising customer-orientated digital transformation in enterprises. Procedia Manuf. **8**, 517–524 (2017)
29. Aperion, EnsembleIQ: Top Technologies Driving Retail Digital Transformation. Convenience Store News, 2–3 (2018)
30. Deloitte: Disruptions in Retail through Digital Transformation - Reimagining the Store of the Future (2017). https://www2.deloitte.com/content/dam/Deloitte/in/Documents/CIP/in-cip-disruptions-in-retail-noexp.pdf

Towards a Business Model for Post-industrial Tourism Development in Jiu Valley, Romania

Ionela Samuil[1], Andreea Cristina Ionica[1](✉), Monica Leba[1], Sorin Noaghi[1], and Alvaro Rocha[2]

[1] University of Petrosani, 332006 Petrosani, Romania
andreeaionica2000@yahoo.com
[2] University of Coimbra, Coimbra, Portugal

Abstract. The research is carried out in Petrila, a small town in Jiu Valley, Romania, a mono-industrial area, which was formed and developed on the basis of mining activity, an area that declined after 1990 and which now seeks a sustainable recovery formula, tourism development being one of the alternatives. The research was conducted to establish the premises for a Business Model for Petrila Theme Park. For this, firstly, the potential effects generated by the development of tourism, economic, social and environmental activities from the point of view of the inhabitants were determined. Thus, the model and hypotheses were tested to determine the relationship between the types of effects generated and the general community satisfaction. Also, are presented the results of evaluating the attitude of the inhabitants toward the development of a theme park and the use in this context of modern technologies such as augmented reality. The perspective opened by the presented research is to build an industrial tourist area, as an alternative to the closed mining activity.

Keywords: Industrial tourism · Petrila · Tourism development · SPSS

1 Introduction

In order to diminish the negative effects generated by the restructuring of the mining activity as well as other factors of economic and social nature and to obtain a revitalization of the area, a series of actions and projects have been started for the tourism development and the integration of the resorts into the national and international tourist circuits. These take into consideration that the Jiu Valley has a rich and varied tourism potential, suitable for both adventurers and those who want to discover the world at a glance. Besides the anthropic potential, the Jiu Valley also has a multitude of underground coal mining industrial structures, testimony of past times, which can be transformed, exploited and integrated into the local tourism ecosystem, thus offering a particular approach. In these conditions, it is desired to develop a strategy of reconversion of the area by reusing the industrial heritage, one of the directions being industrial tourism, a relatively new branch in the tourism field that promotes sites and events that come from the field of industrial

Á. Rocha et al. (Eds.): WorldCIST 2020, AISC 1159, pp. 568–577, 2020.
https://doi.org/10.1007/978-3-030-45688-7_57

production. It is closely linked to the notion of industrial culture as a dynamic socio-cultural concept that evolves along with the economic environment, undergoing a series of transformations. The paper addresses this very important topic for all areas affected by industrial activities, but especially for the Jiu Valley. The results of an empirical study testing a number of nine hypotheses is then presented. Also, it is explored the potential of building a theme park and the use of augmented reality based on the results regarding the attitude of the inhabitants. Based on these findings, the elements needed to develop a Business Model with community involvement are established. The research was carried out in a former mining community where, although there is potential for natural tourism, organized tourism activities have only recently been considered as a strategy for economic reorganization.

2 Postindustrial Tourism: From Concept to Business Model

Over time, a number of terms have been used to explain the concept of industrial tourism, some authors using the term to describe aspects of industrial tourism. Each of these terms has limitations in that they do not cover the full range of possible sites. Based on the definitions stated by different authors, "… the process of presenting contemporary manufacturing processes" [1], "…tourists' visits to operational sites where basic activity is not tourism-oriented" [2], "… site visits that allow visitors to learn about past, present and future economic activities" [3], we believe that industrial tourism can be defined as making visits to sites that allow visitors to learn about past activities, observe the present and project the future. Potential tourists can opt to visit a site, operational or post-exploitation, to find out both about the economic activities carried out in the past but also of other types of activities carried out in time, to better understand the evolution so far, based on these elements trying to make a projection in the future.

To successfully implement this concept, it is necessary to develop a detailed Business Model (BM). The tourism sector, more than other sectors, is continually undergoing a series of profound transformations, mainly due to the globalization of services but also to the changing dynamics of international competition. The enrichment of the tourism offers with new destinations but also the development of new tourism niches such as industrial tourism requires the increasingly complex use of new technologies and is based on the transformation of huge amounts of data received through applications into more valuable proposals for the customer [4]. These elements involve all stages of tourism production, but the introduction of new technologies in information and communication management is particularly relevant [5, 6]. These are some of the reasons why different approaches of the BM in tourism were released, for instance, recently the notion of smart tourism [7] integrated in an BM so that based on the data regarding the preferences of tourists you can get a personalized product to be transmitted using an efficient communication channel, or, the creative BM in tourism [8], with differences from a conventional BM, in terms of using a different set of resources, such as dances, shows, festivals, art, craft fairs for achieving authentic experiences by the tourists.

3 Context of the Research. Petrila's World – Between History and Perspectives

After 1990, following the massive restructuring, effect of the mine closure process, both the number of inhabitants and the number of year-employees in the extractive industry decreased from year to year. The number of these employees reduced from 60,679 in 1989 to only 3,767 in 2017.

Petrila is a town located in the Jiu Valley, Hunedoara County - Romania, with an area of 308.7 km^2 located on the banks of the Transylvanian Jiu River. The locality is certified, in different documents, both in 1493 and in 1499, the inhabitants being shepherds. Currently, the local infrastructure does not have an adequate structure for tourism development, this being on the priority list of the town development strategy.

The massive layoffs in the mining sector have led to a decrease in the standard of living in this area and a significant increase in the unemployment rate at the local level compared to the unemployment rate at the national level. At the local level the unemployment rate is 26%, but the real rate is around approximative 50.8%, [9] because there are many people who have come out of the employment records.

The Jiu Valley is defined not only by mining and mines, by cities and people with a precarious financial situation but also by cities and hospitable people, eager to share their traditions and culture. This intra-mountain basin is known both for its natural riches but especially for its mountain potential, being one of the most tourist-friendly. Normally, in the transition to a service-based economy, tourism is considered the main, if not the only, substitute for industrial activity. Although some areas may resort to other tourism resources – e.g. natural landscapes, gastronomy, architecture or history - paradoxically, most consider that industrial remains are the only resources that can be transformed into heritage elements. So far, a study has been carried out on the conversion of the former Petrila Mining Assembly into a Thematic Park [10], aiming at the restoration of the affected buildings, technologies and landscapes and their recovery in a dedicated tourist circuit, integrating also the heritage objects. In the Jiu Valley there is no specific industrial touristic establishment, so such a project can generate an increase in the number of tourists but only in the conditions in which it would be included in touristic circuits, next to the other attractions of the area. Also, the role of the community, evaluated throughout the perceptions that inhabitants have regarding this project, is part of the problem statement.

Although the opportunities for socio-economic development of the area are reduced, Petrila has a number of advantages: areas of great tourist attraction (wild, quasi-virgin areas, the existence of several cottages in the surroundings of the city, rich and varied forest vegetation with potential for hunting and fishing); organizing traditional folk holidays. Like the other cities of the Jiu Valley and, in fact, all the mono-industrial cities, Petrila is also looking for a survival solution, the most handy being the recovery and reconversion of the industrial heritage from the former Petrila Mining Assembly, under the conditions where, due to passion and work sustained by the recently established organization Petrila Planet important objectives were listed in the National Heritage.

4 Methodology

The methodology ensures the achievement of the identification of the attitude of the inhabitants toward tourism development in the area but, especially, to what extent they are willing to get involved in this process. The results will allow developing the research in the direction of designing a BM for the case of post-industrial tourism with the involvement of the community.

It was used a quantitative method – a questionnaires-based survey. So, a questionnaire-based instrument, drawn from a sum of variables, most using Likert scale as evaluation method, with 5 ranks (1 = Strongly disagree; 5 = Strongly agree) was applied to the inhabitants of Petrila. A number of factual questions were also used, with pre-formulated answers related to the individual characteristics of the respondent.

The sampling was carried out, in several stages, by age and gender quotas. During the period May - September, a number of 973 questionnaires were sent, both online and offline. Of the questionnaires submitted, a total of 694 questionnaires were completed, returned and validated, the respondents being 355 female (51.2%) and 339 males (48.8%) aged over 18 years. Most of the respondents are employed in the public sector 46.6%, the others being distributed as follows: 16.5% in the private sector, 5.3% have their own business, 2.3% unemployed, 0.8% households, 0.8% retirees and 27.8% students, most (77.4%) living here for over 20 years. This sampling is statistically representative of the population of the town, with a marginal error of ±5%.

Data analysis was performed in two phases, using SPSS.20 and AMOS.18. The first part of the reliability analysis considered on one hand the scales to determine the perception of the inhabitants regarding the effects generated by the tourism development and on the other hand the scales to determine the general satisfaction of the community as a result of tourism development. Cronbach alpha coefficients were determined for each item separately and then a series of Spearman correlations were performed to test the issued hypotheses. The reliability analysis was performed on the scales used for identifying the perception regarding the positive impact and also the negative impact of tourism development in the area. The Cronbach alpha index values exceeded the level of 0.85 for each item analysed, so both scales are consistent.

The questionnaire was structured on 5 sections, aimed to determine the impact of tourism development on the local and individual economy, the social and cultural influence as well as the impact on the environment. It was also estimated of the impact on the general satisfaction of the community as well as the attitude of the inhabitants toward additional tourism development in the studied area. The items used in this study were obtained following a comprehensive review of the existing literature [11–14] containing results of empirical studies on the models, respectively the constructs mentioned in the 5 sections, customized on the specific conditions of the researched area. So, the model is developed on nine hypotheses (Table 1), represented by the relationships between the five latent constructs: the individual benefits obtained from tourism development (IBTD), the positive effects generated by tourism development (PETD), the negative effects generated by tourism development (NETD), the general satisfaction of the community as a result of tourism development (GSC) and the attitude of the inhabitants toward tourism development (IATD). In each hypothesized relationship an effect can be identified either positive (+) or negative (−).

Table 1. Statistical hypotheses

H01 => The idea of obtaining individual benefits does not influence the perception regarding the positive effects generated by the tourism development H11 => The idea of obtaining individual benefits is positively correlated with the perception regarding the positive effects generated by the tourism development
H02 => The idea of obtaining individual benefits does not influence the general satisfaction of the community as a result of tourism development H12 => The idea of obtaining individual benefits is positively correlated with the general satisfaction of the community as a result of tourism development
H03 => The idea of obtaining individual benefits does not influence the perception regarding the negative effects generated by the tourism development H13 => The idea of obtaining individual benefits is negatively correlated with the perception regarding the negative effects generated by the tourism development
H04 => The idea of obtaining individual benefits does not influence the attitude of the inhabitants toward tourism development H14 => The idea of obtaining individual benefits is positively correlated with the attitude of the inhabitants toward tourism development
H05 => The perception of positive effects generated by the tourism development does not influence the attitude of the inhabitants towards tourism development H15 => The perception of positive effects generated by the tourism development is positively correlated with the attitude of the inhabitants toward tourism development
H06 => The perception of positive effects generated by the tourism development does not influence the general satisfaction of the community as a result of tourism development H16 => The perception of positive effects generated by the tourism development is positively correlated with the general satisfaction of the community as a result of tourism development
H07 => The general satisfaction of the community as a result of tourism development does not influence the attitude of the inhabitants toward tourism development H17 => The general satisfaction of the community as a result of tourism development is positively correlated with the attitude of the inhabitants toward tourism development
H08 => The perception of negative effects generated by the tourism development does not influence the general satisfaction of the community as a result of tourism development H18 => The perception of negative effects generated by the tourism development is negatively correlated with the general satisfaction of the community as a result of tourism development
H09 => The perception of negative effects generated by the tourism development does not influence the attitude of the inhabitants toward tourism development H19 => The perception of negative effects generated by the tourism development is negatively correlated the attitude of the inhabitants toward tourism development

5 Findings and Results

Making the correlations and determining the Spearman coefficients allowed the presentation of the results regarding the hypothesis testing. Thus, H01 ("The idea of obtaining individual benefits is positively correlated with the perception of positive effects generated by the tourism development") was confirmed with $\rho = 0.817$, with $p < 0.01$ having a very strong association between variables; H02 ("The idea of obtaining individual benefits is positively correlated with the general satisfaction of the community as a result of tourism development") is confirmed with $\rho = 0.644$, $p < 0.01$ having a strong association between the two variables. Also, H03, H04, H05, H07 and H08 were confirmed with optimum values but with statistically insignificant associations. H06 ("The perception of positive effects generated by the tourism development is positively correlated with the general satisfaction of the community as a result of tourism development") is confirmed by demonstrating a high intensity association with values of $\rho = 0.609$, $p < 0.01$. For H09 ("The perception of negative effects generated by the tourism development is negatively correlated with the attitude of the inhabitants toward tourism development."), the null hypothesis is accepted, having $\rho = -0.035$ and $p = 0.258$.

In the second phase of the analysis, using AMOS was performed the standardized equation model, determining chi square for the proposed model. The indicators of global fit have the following values: CMIN/DF = 0.697; CFI = 1.000; RMSEA = 0.000.

We can see that, regarding the positive effects to be generated by the development of tourism activities in the area, most of the respondents have the opinion that they will certainly appear. In relation to the positive effects to be registered at the economic level (PEETD) 42% of the respondents place the answers in the Agreed area and 42% in the Strongly agreed area, the situation being similar for the positive effects that will appear in the social sectors (PSETD) 43% of the respondents place the answers in the Agreed area and 33% in the Strongly agreed area.

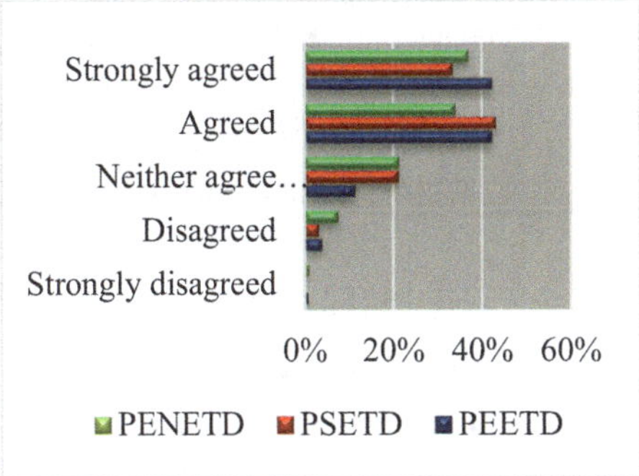

Fig. 1. The positive effects generated by the tourism development

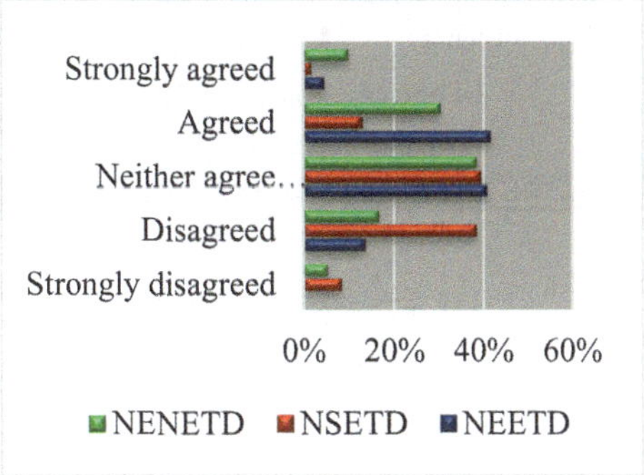

Fig. 2. The negative effects generated by the tourism development

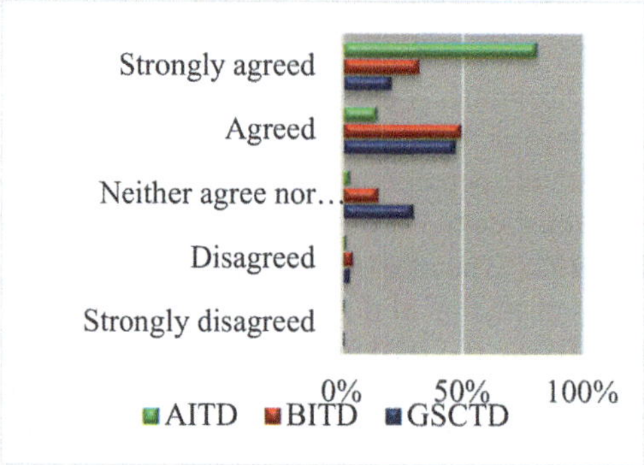

Fig. 3. The perception of the inhabitants toward tourism development

At the same time, the respondents consider that there will be beneficial policies for the protection of the environment (PENETD) 34% being in the Agreed area and 37% in the Strongly agreed area (Fig. 1). Analyzing the possibility of the occurrence of negative effects following the tourism development in the area, we observe that the population is quite retained in the estimates, most placing the answers in the area Neither agree nor disagree, such as: 41% in related of the occurrence of possible negative economic effects (NEETD), 39% regarding the possibility of the occurrence of negative effects in social sectors (NSETD) and 38% talking about the negative effects on the environment (NENTD) (Fig. 2).

The results of the analysis show, without a doubt, that the questioned population has the opinion that by placing the area on the tourist map there will be benefits in all sectors, a general satisfaction at the community level (GSCTD) with 47% of the answers in the Agreed area and 20% in the Strongly agreed area and also a series of benefits at the individual level (BITD) with 49% in the Agreed area and 32% in the Strongly agreed area. In conclusion, it can be observed that the attitude of the inhabitants toward tourism development (AITD) is placed in the Strongly agreed area with 81% (Fig. 3).

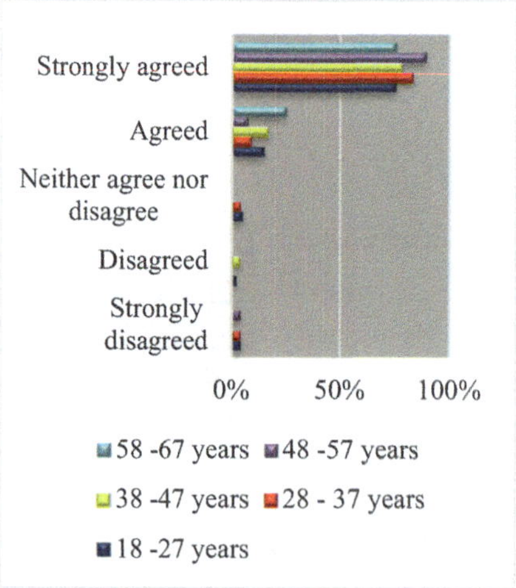

Fig. 4. The agreement for the development Petrila Theme Park

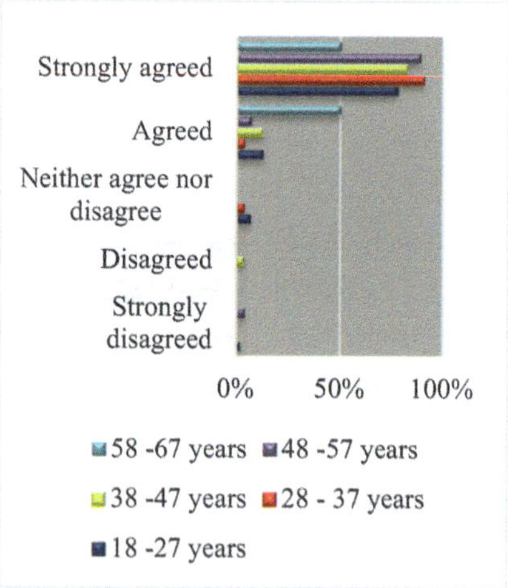

Fig. 5. The agreement for including Thematic Park, within a touristic circuit

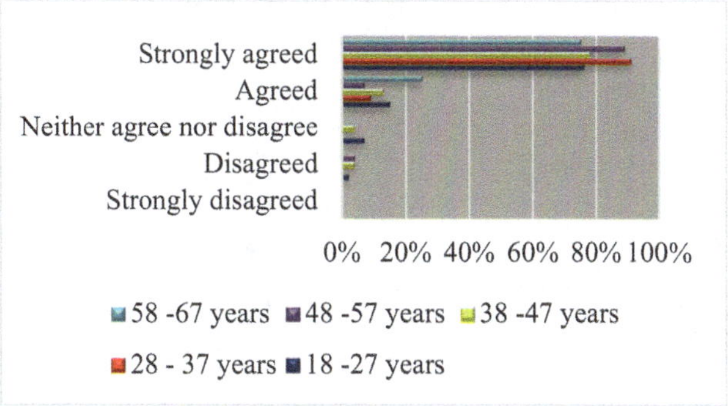

Fig. 6. The agreement for using modern technologies to promote touristic potential

By presenting to the locals the idea regarding the enrichment of the local tourism offer by building a Theme Park in the former Petrila mining complex (Fig. 4) we got positive feedback; the most of the answers are found in the Strongly agreed area, the respondents most excited about this idea are in the age ranges 28–37 years with 83% and 48–57 years with 89%. As well as its integration into a touristic circuit along with the other local tourist destinations (Fig. 5) the feedback was positive, showing the same age ranges, both 28–37 years with 91% and 48–57 years with 89% of answers in the Strongly agreed area. Also, the possibility of using modern technologies to increase the attractiveness (Fig. 6), has aroused interest being registered in the Strongly agreed area for both age ranges 28–37 years with 91% and for 48–57 years 89%.

All these results create the premises for the proposal of a business model with the involvement of the community.

6 Influencing Factors for the Realization of a Business Model for Petrila Theme Park

In order to achieve, in the future, a BM for the Petrila Theme Park, organized on the ruins of a former mining assembly, is presented a simplified view on how a community-based redevelopment can be achieved (Fig. 7), by stimulating other fields and improving the local economy taking into account the attitudes of the inhabitants (the current stage of the research) and the others stakeholders (the next stage of the research).

The development and exploitation of the Petrila Theme Park, included in a touristic circuit will generate a series of positive effects materialized in the increase of the number of work places, the improvement of living conditions, the change of the image on the area of both locals and visitors, a higher concern for improving environmental protection policies, etc. Although this activity will not be able to compensate for lost jobs, it is difficult to replace an entire industry, it may be the starting point for investments, small businesses, new jobs, but without allowing for imbalances due to economic uncontrolled growth. Given that the role of data analysis in tourism has become increasingly important for understanding travel practices and for designing and providing efficient services, digitizing tourism in general and the intensive use of digital tools by participants in tourism activities can give us an advantage in placing the project in the smart tourism area [15, 16].

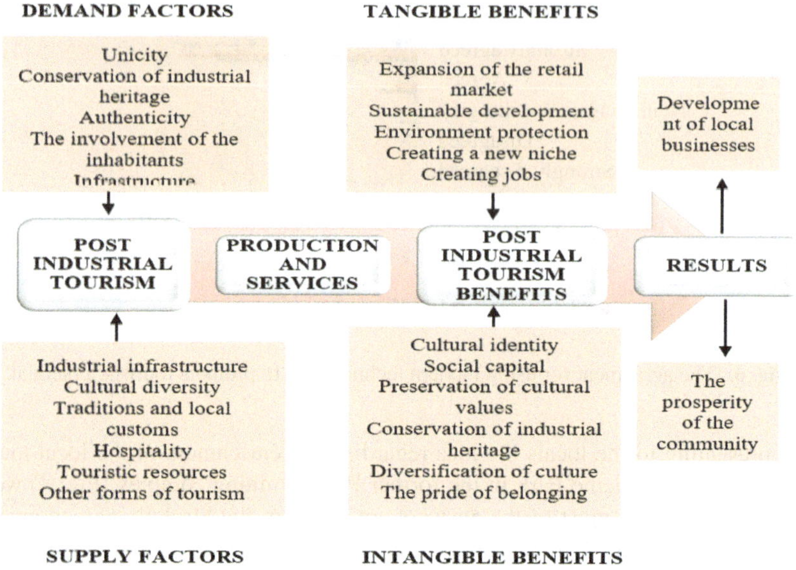

Fig. 7. Influencing factors in tourism development

7 Conclusions and Further Enhancements

The research carried out and presented in this paper has focused mainly on the identification of the key elements necessary for designing a BM applicable in industrial tourism, taking into account the perception of the effects (economic, socio-cultural but also environmental) generated by the development of tourism activities in the proposed area. It was also followed within the analysis, to what extent the obtaining of personal benefits can influence the attitude of the inhabitants towards tourism development, a confirmed hypothesis, there is a very strong correlation between these variables, so that the possibility of increasing the income and, in general, an improvement in the standard of living can be a factor influencing the perception of the locals. The study shows that the population believes that tourism will bring to the town more advantages than disadvantages and the predominantly positive perception on the effects that the development of tourism activities will generate leads to a favourable attitude in this regard. It must be taken into account that Petrila is a small town, for a long time dependent on the mining activity, at the moment looking for strategies of economic conversion, the most feasible being the use of the tourist potential. This element can be considered a factor of influence in the perception of the locals in the conditions under which this idea will increase the economic opportunities and tourism will act as a dynamic factor for the whole area.

Based on the results obtained, we have also drawn up an outline that will allow us as a further enhancement to achieve a Business Model with the help of which the needs of the locals and the community in general can be identified and satisfied. At the same time, planning for the closure of industrial activities should be carried out by municipalities in a continuous and consultative process, taking into account the real needs of the stakeholders. Both the mine closure planning process and subsequent processes for identifying needs and solutions should be mediated by NGOs, as buffer institutions between businesses and individuals, thus ensuring transparency in decision-making for all stakeholders. NGOs are important actors in the process of building an active civil society and play a complex role in society and through them civil society can draw the state's attention to the neglect of some of its responsibilities, regarding the violation with or without intention of legitimate rights of its citizens, especially with regard to the vulnerable ones. A partnership between companies and physical persons, mediated by NGOs, formalized and legalized, can be established as a management council in project management.

References

1. Yale, P.: From Tourist Attractions to Heritage Tourism, Huntingdon. Elm Publications, Huntingdon (1991)
2. Frew, E.A.: Industrial tourism: a conceptual and empirical analysis. Ph.D. thesis, Victoria University, Melbourne, Australia (2000)
3. Otgaar, A.H.J.: Industrial tourism: where the public meets the private. Ph.D. thesis, Erasmus University Rotterdam, Rotterdam, Netherlands (2010)
4. Gravari-Barbas, M.: Tourism as a heritage producing machine. Tour. Manag. Perspect. **26**, 5–8 (2018)
5. Dibra, M.: Rogers theory on diffusion of innovation - the most appropriate theoretical model in the study of factors influencing the integration of sustainability in tourism businesses. In: World Conference on Technology, Innovation and Entrepreneurship. Elsevier B.V., Netherlands (2015). Procedia - Social and Behavioral Sciences 195, 1453–1462
6. Shafiee, S., Ghatari, A.R., Hasanzadeh, A., Jahanyan, S.: Developing a model for sustainable smart tourism destinations: a systematic review. Tour. Manag. Perspect. **31**, 287–300 (2019)
7. Szromek, A.R., Herman, K.: A business creation in post-industrial tourism objects: case of the industrial monuments route. Sustainability **11**(5), 1451 (2019)
8. Ohridska-Olson, R.V., Ivanov, S.H.: Creative tourism business model and its application in Bulgaria. In: Proceedings of the Black Sea Tourism Forum 'Cultural Tourism, The Future of Bulgaria, Varna, Bulgaria, pp. 23–39 (2010)
9. http://www.orasulpetrila.ro/wp-content/uploads/2014/07/strategie-2016-2020.pdf. Accessed 3 Dec 2018
10. Toderaş, M., Samuil, I., Ionică, A.C., Olar, M.L., Militaru, S.: Aspects regarding a mining area rehabilitation for post-industrial tourism. In: The International Conference on Manufacturing Science and Education, Sibiu, Romania (2019)
11. Ko, D.W., Stewart, W.P.: A structural equation model of residents' attitudes for tourism development. Tour. Manag. **23**, 521–530 (2002)
12. Perdue, R.R., Long, P.T., Allen, L.R.: Resident support for tourism development. Ann. Tour. Res. **17**(4), 586–599 (1990)
13. Gursoy, D., Rutherford, D.G.: Host attitudes toward tourism: an improved structural model. Ann. Tour. Res. **31**, 495–516 (2004)
14. Vargas-Sanchez, A., Porras-Bueno, N.: Understanding residents' attitudes toward the development of industrial tourism in a former mining community. J. Travel Res. **47**(3), 373–387 (2008)
15. Koo, C., Cantoni, L.: Special issue on informatics/data analytics in smart tourism. Inf. Process. Manage. https://www.sciencedirect.com/science/article/pii/S0306457319312129
16. Yoo, C.W., Goo, J., Huang, C.D., Nam, K., Woo, M.: Improving travel decision support satisfaction with smart tourism technologies: a framework of tourist elaboration likelihood and self-efficacy. Technol. Forecast. Soc. Change **123**, 330–341 (2017)

Application of Industry 4.0 Methods in Russian Industrial Companies: A Qualitative Approach

Sergei Smirnov[1], Ekaterina Mochalina[2(✉)], Galina Ivankova[2], Oleg Tatarnikov[2], and Alena Esareva[1]

[1] St. Petersburg State University, 7/9 Universitetskaya nab., St. Petersburg 199034, Russia
[2] Plekhanov Russian University of Economics, Stremyanny lane, 36, Moscow 117997, Russia
mochalina77@gmail.com

Abstract. For the past 10 years implementation of Industry 4.0 technologies and methods at manufacturing companies has shown a dynamic growth. Russian companies are significantly behind the leading countries in terms of speed and mass application of these methods. At the same time, there are some companies in Russia that have been already actively using methods and technologies of Industry 4.0 for several years. Paper focuses on the study of their experience in Industry 4.0 methods implementation. It is a high-quality descriptive study of application the Industry 4.0 methods in 9 large industrial Russian companies. The authors used secondary sources, namely: interviews with managers and project managers, conferences and exhibitions speeches, press releases, information presented on corporate websites. Using the Grounded Theory methodology, and the open coding procedure, the authors propose their own organizational model of significant factors affecting the realization of projects of Industry 4.0 methods implementation. Further, based on qualitative analysis of the available secondary data sources, conclusions are drawn regarding the patterns and features of the implementation of production digitalization methods, as well as assumptions are made about the prospects for using some Industry 4.0 methods in Russian companies.

Keywords: Russian industry companies · Industry 4.0 methods · Grounded theory · Organizational systems

1 Introduction

Nowadays the world is undergoing a transition to the digital economy. It is associated mostly with the concept of the manufacturing processes automatization, which got the name "smart factory" as a part of the more general approach, Industry 4.0. There are government support and development programs of this route in different countries: Industry 4.0 (Germany), American Consortium of Industrial Internet (USA), Smart Factory (Netherlands), Usine du Futur (France), High-Value Manufacturing Catapult (Great Britain), Fabbrica del Futuro (Italy), Made Different (Belgium), "Made in China-2025" (China), Digital Economy of the Russian Federation (Russia). The qualitatively new approach brings changes to the automatization of manufacturing processes and practical

Á. Rocha et al. (Eds.): WorldCIST 2020, AISC 1159, pp. 578–588, 2020.
https://doi.org/10.1007/978-3-030-45688-7_58

application is often ahead of the theoretical understanding of them. However, today many works have appeared discussing the principles and the content of Industry 4.0 concept [1, 2].

The Smart Factory concept involves the transition to automated digital production, controlled by intelligent systems in real-time in constant interaction with the external environment, going beyond the boundaries of one company, with the prospect of combining things and services into a global industrial network. These networks are information technology systems that combine the virtual and real worlds, creating chains of interconnections where machines with intelligence can interact with each other, often without human intervention. The Smart Factory key principles are interoperability, virtualization, decentralization, real-time capability, service orientation, and modularity. Such production systems can provide more flexibility, reduce lead times, customize with small batch sizes, and reduce costs [3]. Modern manufacturing companies must produce complex, often personalized products. An additional requirement is a quick and flexible response to customer needs. All this creates a new vision of a contemporary manufacturing company [1]. Paper [4] notes that the rapid development of Industry 4.0 contributed to the fact that engineers began to adapt quickly to new opportunities in production that give digital transformation. Paper [5] considers various configurations of assembly processes built on the Smart Factory concept. Chong et al. [6] note as a key issue the opportunity to collect a variety of data allowing to make quick and rational decisions. All these advantages are not bounded by the framework of one production division (subdivision) since it becomes possible to link different departments and optimize the company's whole production process or even a group of companies. Most authors agree that the opportunity to deploy digital technologies will become a significant competitive advantage for companies in the nearest future [7–9]. Another popular trend in literature nowadays is the discussion of methods [10–14]. Basing on these works we consider the following 9 key digital technologies contribute to the Industry 4.0 concept: autonomous robots, 3D-simulation, horizontal and vertical integration, industry Internet of things, information security, cloud software, additional 3D-printing, augmented reality (AR), big data.

The main objective of this paper is to conduct an exploratory consolidation of the practice learnt from the Industry 4.0 methods application in several large Russian companies. We intend to identify the most significant factors of industry 4.0 project that lead to success. Among secondary goals, we are aimed to recognize the most promising and easily implementable methods of the industry 4.0. Basing on grounded experience, we also strive to identify certain patterns of successful project implementation.

2 Methodology

This qualitative study uses the methodology of the grounded theory by Strauss and Korbin [15]. The theory is built and verified by collecting and analyzing data about the encoding object, thus a mutual connection between the qualitative analysis data is achieved. Each stage of the analysis affects the next, thereby determining it. Also, the "labels" (codes), categories and data identified in the interview are constantly compared with each other. The companies' selection was carried out according to the following characteristics: large industrial companies that actively implementing industry 4.0 methods, and at the same time practicing a high level of corporate openness. Secondary external sources from open access were used as initial data, namely, interviews with directors, managers, project managers, their speeches at conferences and exhibitions, company press releases and information presented on corporate websites. Interviews with top and digital project managers were used as open secondary sources of analyzed data. Interviews were conducted by professional journalists for industry and business media, all interviews dates from 2018 and 2019. An observation method was also used to collect data. In combination with other methods, it allows to ensure the richness and completeness of the information. In this paper, we study data of nine major Russian companies, representing different industries (metallurgy, oil refining, mechanical engineering). Table 1 below shows the companies participated in the study.

Table 1. Russian industry companies (participated the study)

Company	Description	Turnover (USD millions)	Employees
MMK	The global steel producer, a leader in the iron and steel industry in Russia, provides full production cycle	8 197	17 887
SIBUR	An integrated petrochemical company, has own raw materials and infrastructure	9 058	26 164
TAT-NEFT	The vertically integrated oil company, has complex manufacturing and service capacities	14 512	48 080
KAMAZ	The largest truck manufacturer in Russia, covers the entire technological cycle	2 969	34 109
Norilsk Nickel	The leader of the mining and metallurgical industry of Russia, the largest producer of nickel and palladium	11 624	75 900
OMK	One of the largest manufacturers of metal products for energy, transport and industrial Russian companies	2 775	22 800
SEVER-STAL	A vertically integrated mining and metallurgical company with major assets in Russia and an insignificant part abroad	8 566	36 257
NLMK	A leading international steel producer with a vertically integrated business model	12 059	53 300
UEC-SATURN	A company specializes in the development, production, and after-sales service of gas turbine engines	649	12 425

As one of the company's selection criteria, we accept the secondary sources' data availability about the Smart Factory and 4.0 Industry methods implementing projects. The large size of companies indirectly indicates the demand for the implementation of such methods and technologies. Paper studies the nature and patterns of projects and using the Industry 4.0 methods at large Russian industrial enterprises based on the collected data, identifies implementation projects difficulties and suggests ways of solutions.

Data analysis in the framework of the grounded theory includes the coding procedure. Line-by-line coding is carried out, the text is analyzed, line by line and a short designation of "what is happening" in this line is put in correspondence with each line. Coding is the procedure through which data is divided, conceptualized and reconnected in a new way, thereby creating a new theory based on the received data.

There are three types of coding that correspond to several stages of reading the text: open, axial and selective. The focus of this study will be open coding, which is carried out in the process of a detailed analysis of the interview - according to words, sentences, lines and paragraphs. Thus, we will be able to create some basis for a future grounded theory. Figure 1 shows the general logic for constructing the theory.

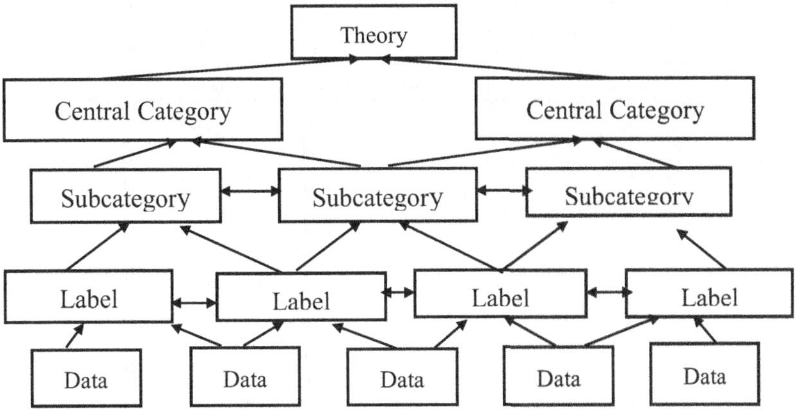

Fig. 1. The grounded theory building principal logic scheme)

Writing notes about each of the formed analytical concepts or categories. Writing notes begins with the selection of the first analytical concept and continues in the process of analyzing new data and constructing the theory.

Integration, refinement and formulation of the theory. As categories are formed from codes, they are interconnected into a theoretical model that is built around one central category. As a result, a created model explains the studied phenomenon, from the very beginning of data collection. Then, its adequacy is tested in the process of further data collection and analysis. In the process of data analysis, the text was divided into separate parts, "labels were glued" (codes), which were subsequently assigned to certain sub categories (according to words, phrases, text). Codes similar in meaning were assigned predefined names (concepts) that are logically associated with "coding labels". Then the concepts were grouped into certain categories (for convenience in the subsequent analysis).

The last part of the study based on open coding is the analysis of the number of categories, identifying the reasons for this correlation of the results and the interconnection between them based on their characteristics and measurements. The process of theorizing the data after coding and categorization are necessary to determine and systematize the interconnections that arose in the analysis process.

3 Results

In the coding process, the following codes or "labels" were determined: dates, cooperation, external assistance, nine Smart Factory technologies, tasks, results, implementation problems, prerequisites for development and the implementation projects' budget (see Fig. 2).

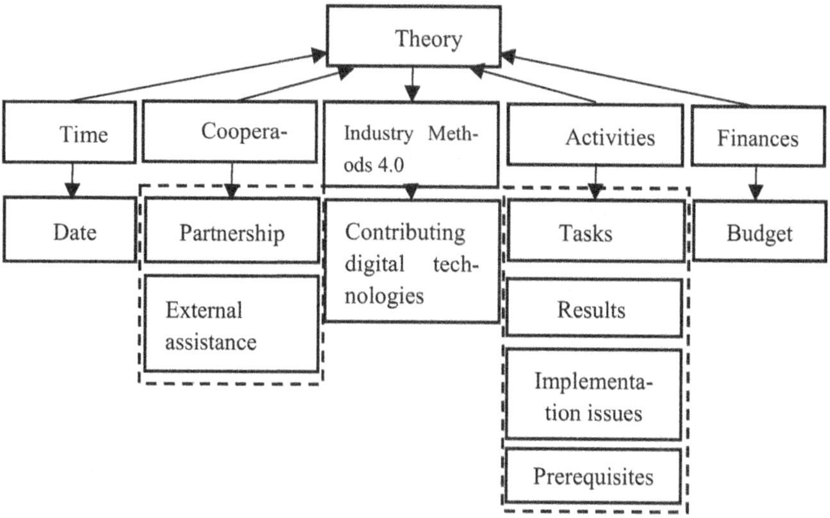

Fig. 2. Interconnection between data elements, "codes" and categories

The selected codes formed new categories of analysis, including time, cooperation, methods of Industry 4.0, activity and finances. These categories became the basis for the theoretical explanation of the reasons for using digital methods in production. One of the problems in our analysis was that most companies report not wholly the information concerning the implementation of Industry 4.0 elements. Quite often (in interviews) company representatives, talking about technologies being introduced, report on partnerships between them and foreign technology companies that help to develop and implement specific solutions. It was also found that cooperation with various Universities of the country where the development and training of future specialists take place is a common practice. Thus, companies train personnel in advance so that employees are immediately involved in the company's work process. Several companies reported the opening of scientific and technical centers, various laboratories at the territory of the special economic zone "Skolkovo Innovation Center". Table 2 below provides examples of typical extracts from company information materials.

Table 2. Examples of data elements from the categories "time" and "cooperation"

Category		
Example	Time	Cooperation
	• PJSC MMK - Oracle. The cooperation has begun since 2005, the contract is signed for the next 5 years • SIBUR. The implementation of digital production methods has been going during the last two years • TATNEFT. By 2020-2021, digital copies are planned to create for all main facilities of the company • PJSC KAMAZ. The strategy for the transition to digital production in the company was outlined in 2006 • JSC OMK. In 2017, a unified system of operational production management was implemented (the basis for creating a unified MES system)	• New strategic partner of PJSC MMK - Oracle • SIBUR Digital Technology Center (Tomsk) • SIBUR cooperation with Teradata to optimize production lines • TATNEFT cooperates with Schneider Electric, SAP, Siemens, Emerson • KAMAZ signed an agreement on cooperation and partnership with Siemens to create a united platform • KAMAZ and KUKA Robotics RUS signed an agreement on partnership (09/2018)

The code "date" (category "time") was created to estimate the time intervals for certain events (the signing of a cooperation agreement, the date of implementation and implementation of the project or the achievement the key goals of the company, etc.) which had to happen at the company.

The main tasks that companies set themselves are to increase economic efficiency, labor productivity and product quality. However, in most cases, companies emphasize the desire to retain personnel who are not able to work with new technologies. Such workers are sent for retraining or training, which is carried out using virtual or augmented reality methods. Table 3 below shows typical excerpts from company information materials on the remaining three categories.

It is interesting to note that not much information is available in open sources about the size of budgets of projects for the digitalization of production. Only a few companies in the interview gave approximate funds that are spent on the development and modernization of production using the methods of Industry 4.0. Company representatives note a shortage or lack of funds for the development of the digitalization process [16]. Even large companies cannot always afford technology implementation. The digitalization process has not yet gained proper dynamics, therefore, now, all the companies reviewed are experiencing a shortage of qualified personnel. This leads to the construction of partnerships with universities. Several sources note that there is still no certainty that a positive financial effect will be obtained as a result of the implementation of digitalization technologies.

To summarize the results, a general scheme was created (see Fig. 3 below), the basis of which is the methods of Industry 4.0, which combine all the categories for

Table 3. Examples of data elements from categories "Methods of Industry 4.0", "Activities" and "Budget"

Category		
Methods of Industry 4.0	Activities	Budget
• PJSC MMK. The use of 3D-technical vision, the implementation of digital copies, the use of neurotechnology, VR/AR, chat bots, cybersecurity, mobile applications • SIBUR. Implementation Big Data Technologies • PJSC TATNEFT. The implementation of digital copies and artificial intelligence, the concept of a virtual company • SEVERSTAL. Using Big Data, VR, and Cybersecurity Technologies • Norilsk Nickel. The implementation of artificial intelligence, neural networks and unmanned underground dump trucks	• PJSC SIBUR: the implementation of new technologies and expansion of production have set the problem of an acute shortage of personnel • TATNEFT. Building an expert team based on the Moscow Scientific and Technical Center of the company in Skolkovo • KAMAZ. 2012–2020 - the period of development of digital production, the development of integrated PLM-, ERP-, MES-systems • PJSC SEVERSTAL. The task of carrying out outreach among employees (who see a threat to their workplace in digitalization of business)	• MMK. More than $2 billion allocated for production modernization program • TATNEFT. More than 2-fold increase in investment in research and development and experimental development • SEVERSTAL. Investments of 3.4 billion RUB for building the IT infrastructure • NLMK. The economic effect of the digital projects implementation is hundreds of millions (in rubles)

which the analysis was conducted. All information about the methods and processes of implementation is described directly in these "labels" (categories) of encoding.

Having analyzed the elements of Industry 4.0 (which are used in nine Russian companies) it is required to assess the frequency of application of a particular method and analyze the reasons for their implementation in companies (see Table 4 below).

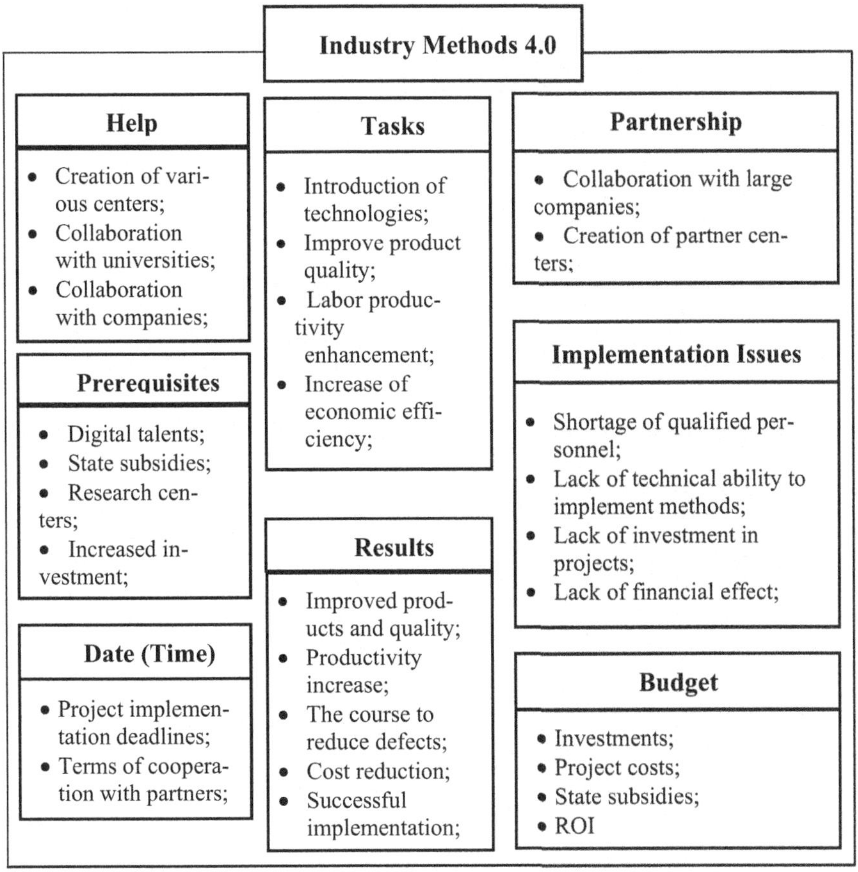

Fig. 3. Open coding results

4 Conclusion and Discussion

Based on the results that we gained from the analysis of secondary sources data for nine large Russian companies in the oil, automobile and metallurgical industries the conclusions can be drawn about the frequency, nature and problems of using Industry 4.0 methods. The most digitalized companies are MMK, SIBUR, TATNEFT, KAMAZ, Norilsk Nickel and UEC-SATURN. These companies invest significant funds in the development and implementation of digital production technologies, have a certain scientific base for the development and application of Industry 4.0 elements. Among the most frequently deployed technologies are: digital simulation, vertical and horizontal integration, Internet of things (IoT) and Big Data). These technologies allow companies to have a unified computer network with the ability to remotely control and automate technological process control, which allows now to increase economic efficiency, improve the quality and productivity of employees. The next most popular factor (among the reviewed companies) is the use of cloud technologies in production. The development

Table 4. Comparative table of the application of Industry 4.0 technologies

Company name/Industry 4.0 methods	MMK	SIBUR	TATNEFT	KAMAZ	Norilsk Nickel	OMK	SEVER-STAL	NLMK	UEC-SATURN	TOTAL
Autonomous robots	+	+		+	+				+	5
3D-simulation	+	+	+	+	+			+	+	**8**
Horizontal and vertical integration	+	+	+	+	+			+	+	**8**
Industry Internet of Things	+	+	+	+	+	+	+	+		**8**
Information Security	+	+	+		+		+		+	6
Cloud software	+	+	+	+		+	+	+		7
Additional 3D-printing	+	+		+	+				+	5
Augmented Reality (AR)/VR	+	+	+				+		+	5
Big Data	+	+	+	+	+	+	+		+	8
Total	9	9	7	7	7	3	5	4	7	

and implementation of technologies take time, and many companies are currently implementing such projects. Autonomous robots, information security, additive production, augmented and virtual reality (AR/VR) are the methods that only half of the companies we have analyzed use it. This may be because in these sectors the use of additive production and 3D printing technologies is only partially applicable. The use of robots is also not yet widespread, although some companies have successful projects based on this technology. The number of robots in the enterprise is increasing nowadays, but this does not always help to reduce personnel. Speaking about augmented and virtual reality, we conclude that companies use already these technologies in the process of staff training. The common line is the idea that companies are ready to try and continue to implement digital technologies that will contribute to the development of production but would like to save jobs for people. A typical argument for this position is that people can change their position within the company itself, without changing their place of work. It is noted that with the implementation of digital technology, the demand for qualified personnel only grows.

The limitations of our study include the unrepresentative cases for the qualitative analyses. SME's also have capabilities and realize industry 4.0 projects. Further, a significant limitation of our research is the use of secondary open sources, which always imply limited information with an inability to clarify it. Future research should be based on in-depth interviews with top and project managers directly involved in digital projects.

The digital economy and the digitalization process embrace more and more companies every day. All companies strive to keep up with the implementation of digital production methods and are ready to use various components of Industry 4.0. Most companies will continue to implement them with increasing activity. Elements of Industry 4.0 should be applied as a complex system. It will allow to organize the production process into integrated and optimized process flow, providing a high level of efficiency at all stages of production, as well as provide a stable connection between producers, customers and partners. It can be predicted that in the process of implementing the methods of Industry 4.0, companies will also begin to transform their business models.

References

1. Veza, I., Mladineo, M., Gjeldum, N.: 15th IFAC Symposium on Information Control Problems in Manufacturing, INCOM 2015 (2015). Managing innovative production network of smart factories. IFAC-PapersOnLine 28, 3, 555–560
2. Lu, Y.: Industry 4.0: a survey on technologies, applications and open research issues. J. Ind. Inf. Integr. **6** (2017). https://doi.org/10.1016/j.jii.2017.04.005
3. Shafiq, S., Sanin, C., Toro, C., Szczerbicki, E.: Virtual engineering object (VEO): toward experience-based design and manufacturing for Industry 4.0. Cybern. Syst. **46**(1–2), 35–50 (2015)
4. Shafiq, S., Sanin, C., Szczerbicki, E., Toro, C.: Virtual engineering factory: creating experience base for Industry 4.0. Cybern. Syst. **47**(1–2), 32–47 (2016)
5. Cohen, Y., Faccio, M., Galizia, F., Mora, C., Pilati, F.: Assembly system configuration through Industry 4.0 principles: the expected change in the actual paradigms. IFAC-PapersOnLine **50**, 14958–14963 (2017). https://doi.org/10.1016/j.ifacol.2017.08.2550

6. Chong, S., Pan, G.-T., Chin, J., Show, P.L., Yang, T.C.K., Huang, C.-M.: Integration of 3D printing and Industry 4.0 into engineering teaching. Sustainability **10**, 3960 (2018). https://doi.org/10.3390/su10113960
7. Mrugalska, B., Wyrwicka, M.: Towards lean production in Industry 4.0. Procedia Eng. **182**, 466–473 (2017). https://doi.org/10.1016/j.proeng.2017.03.135
8. Vaidya, S., Ambad, P., Bhosle, S.: Industry 4.0 - a glimpse. Procedia Manuf. **20**, 233–238 (2018)
9. Kobara, K.: Cyber physical security for industrial control systems and IoT. IEICE Trans. Inf. Syst. **99**(4), 787–795 (2016)
10. Thoben, K., Wiesner, S., Wuest, T.: Industrie 4.0 and smart manufacturing - a review of research issues and application examples. Int. J. Autom. Technol. **11**(1), 4–16 (2017)
11. Monostori, L., Kadar, B., Bauernhansl, T., Kondoh, S., Kumara, S., Reinhart, G., Ueda, K.: Cyber-physical systems in manufacturing. CIRP Ann.-Manuf. Technol. **65**(2), 621–641 (2016)
12. Siegert, J., Schlegel, T., Bauernhansl, T., Siegert, J., Schlegel, T., Bauernhansl, T.: Matrix fusion factory. In: 8th International CIRP Conference on Learning Factories, CLF 2018, 01 January 2018. Procedia Manufacturing 23, 177–182
13. Petrushenskaya, A., Korshunov, G., Smirnov, S.: Digital production management methods of radio-electronic industry, vol. 537, no. 3 (2019)
14. Khadra, J.B., Goncharova, N.L., Radwan, Y.: Regional aspects the small and medium enterprises and their impact on the social and economic development. In: Proceedings of the 33rd IBIMA Conference (2019)
15. Strauss, A.L., Corbin, J.M.: Basics of Qualitative Research: Grounded Theory Procedures and Techniques. Sage Publications, Newbury Park (1990)
16. Pishchulov, G., Richter, K., Pahomova, N., Tchenzharik, K.: A circular economy perspective on sustainable supply chain management: an updated survey. St Petersburg Univ. J. Econ. Stud. **34**(2), 267–297 (2018). https://doi.org/10.21638/11701/spbu05.2018.204

Efficiency and Productivity of Communication Companies: Empirical Evidence from Ecuador Using Panel Data and DEA

Angel Higuerey[1,2](✉), Reinaldo Armas[1], and Miguel Peñarreta[1]

[1] Ciencias Empresariales, Universidad Técnica Particular de Loja, San Cayetano Alto, 110150 Loja, Ecuador
aahiguerey@utpl.edu.ec

[2] Instituto Experimental de Investigaciones Humanísticas, Económicas y Sociales (IEXIHES), Universidad de Los Andes, Carmona, Trujillo 3150, Venezuela

Abstract. This study analyzes the efficiency and productivity of communication companies in Ecuador. The authors use the technique of data enveloping analysis (DEA) to measure the efficiencies of companies and the Tobit regression to determine the incidence of working capital factors on efficiency. The data corresponding to the financial information of the 122 companies investigated between 2015 and 2018. The results determined a low efficiency in the companies of the sector and that the squared cash cycle is significant, but with a negative sign. These findings mean the first attempt to contribute empirical evidence about this little-studied sector, but with a lot of impact in the world.

Keywords: Efficiency · Productivity · DEA, communication companies · Ecuador

1 Introduction

For management, the efficiency and productivity of companies represent the ability to design and implement strategies that improve company performance [1]. From the point of view of public policy, it represents the opportunity to improve the productivity of companies in strategic sectors such as communication [2].

The purpose of this study is to analyze the efficiency and productivity of 122 media companies in Ecuador, a country where this type of industry has had an impact on the dissemination and generation of news. It also seeks to provide new empirical evidence considering that it hasn't been extensively researched.

Data enveloping analysis (DEA) has been the most used programming technique for this type of research. This study, continuing with this trend, applies a CRS model with an input orientation.

The document is organized by sections. Section 2 presents a review of the relevant literature. Section 3 presents the data and methodology used. Section 4 shows the results of the study. Finally, the conclusions and the bibliography used are shown.

Á. Rocha et al. (Eds.): WorldCIST 2020, AISC 1159, pp. 589–597, 2020.
https://doi.org/10.1007/978-3-030-45688-7_59

2 Literature Review

The efficiency and productivity of large, small and medium-sized enterprises have been of interest for a long time. The modern literature determines that these aspects in the industry 4.0 still remain. The data enveloping analysis (DEA) technique is a linear programming tool based on the concepts of measuring productive efficiency proposed by [3] and notably improved by [4], with high applicability to productivity and efficiency analyses in both public and private sector activities, in areas like agriculture, banking, transportation or supply chains [1].

In the context of Spanish media companies, it was determined that not all firms presented technical efficiency and much less efficiency on scale [5, 6]. These results are similar to the findings in the telecommunications industry. For example, [7] measured the efficiency of branches of this type of companies in Korea, at the level of small and medium enterprises (SMEs). It was identified that efficiency is conditioned by the size of the companies as opposed to productivity levels where the gap is reduced in part to public policy interventionism [8]. In India, the results showed that private firms have better levels of efficiency than public ones [9]. In Greece, only 41% of the analyzed companies showed results close to the efficiency frontier [10]. In China and Taiwan, studies have focused on the field of competitiveness analysis after liberalizing the telecommunications market and establishing privatization processes of state enterprises. It seems that the market leaders find opportunities in mergers and acquisitions of small companies to improve scale efficiency but with less technical efficiency [11, 12]. In the U.S., after the regulation to increase incentives in telecommunications [13], improvements weren't found in the technical efficiency of the sector. An empirical analysis for local telephony in Brazil showed low efficiency in the sector due to stable competition and growing government compensation [14]. It appears that companies operating in countries with a higher penetration rate are more likely to have total asset efficiency [15].

In this sense, the importance of the communication and telecommunication industry in the global economy is highlighted. US, Japan, China, and the EU are trying, through public policy, to boost efficiency, productivity, and competitiveness of the companies [2].

In perspective, the scope of communication companies as objects of research is still limited. In this line our study aims to provide more empirical evidence because companies based on the results of the DEA can develop strategic plans to improve efficiency and performance.

3 Data and Methodology

The companies were selected on the basis of the International Standard Industrial Classification (ISIC), media sector, identified by the letter "J", referring to "Information and Communication" activities. Not all classifications have been considered in this paper; only those dedicated to media activities have been considered.

In order to form a balanced data panel, 122 companies have been considered whose financial information between 2015 and 2018 has been audited and is publicly available. Figure 1 shows the composition of the panel made up mainly of 58% for radio, 15%

for promotion and advertising, and 13% for TV production. The financial data used to determine the efficiency of the companies were collected from the Superintendence of Companies, Securities and Insurance on the website (www.supercias.gob.ec).

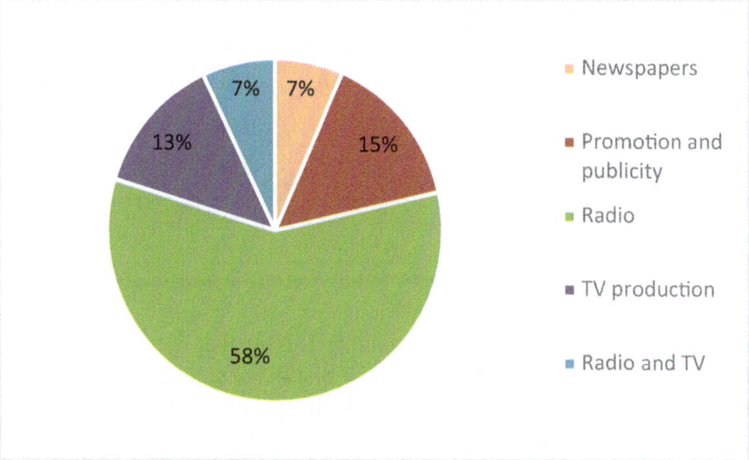

Fig. 1. Composition of communication companies in Ecuador

In determining efficiency, there hasn't been consensus on the variables that measure the outputs and inputs to be used. Some authors have suggested the use of variables that are measured physically [16]. Other authors recommend the use of variables in monetary units to measure economic efficiency [17].

As output, Income has been selected, a variable that measures the efficiency of companies and represents all the revenues, expressed in US dollars. A similar variable has been used in the telecommunications sector by [2, 5, 15]. Revenues correspond to the sum of revenues from the main activities plus other revenues.

The inputs used are total workers (Employees), non-current assets (Act_no_corr) and materials and resources (Mat_recur). Employees represent the total number of people working in media enterprises [18, 19]. Non-current assets (Act_no_corr) is represented by money invested in property, plant, and equipment, and other non-current assets necessary for the provision of services, expressed in US dollars; and Mat_recur, represents the necessary expenses incurred to provide its services, excluding personnel expenses. It was determined by adding operating costs and expenses and subtracting total personnel expenses.

However, in order to measure the management of working capital, the variable Liquidity, Indebtedness and the Cash Cycle (CC) have been used. The Liquidity (ratio between current assets and current liabilities) was determined to see if you can back your short-term debts. On the other hand, Indebtedness (ratio of total liabilities vs. total assets) represents the proportion of liabilities in terms of all the company assets. While the CC (the difference between the average collection period and the average payment period) is the time it takes for the company to recover the cash.

Table 1. Descriptive statistics of the variables

Variable	Obs	Mean	Std, Dev,	Min	Max
Income	488	1.107.954,66	3.786.950,59	2.000,00	28.347.310,00
Employees	488	18,34	47,56	1,00	471,00
Act_no_corr	488	443.465,00	1.919.537,00	1,00	16.700.000,00
Mat_recur	488	646.886,90	2.414.256,00	860,19	20.600.000,00
Liquidity	473	3,71	7,76	0,00	95,94
Indebtedness	488	0,60	0,46	0,00	7,12
CC	488	25,88	72,98	-301,54	312,68
CC2	488	5.985,49	11.828,75	0,00	97.767,67

The descriptive statistics are shown in Table 1. Note that there is a great difference in income and Act_no_corr and due to the fact that in the sample are all the companies without distinction of size, and in this sector, there are different levels of coverage.

The methodology to be applied to determine efficiency is Data Envelopment Analysis (DEA). This technique uses linear programming algorithms to estimate the frontier; its precursor was [20]. A company is efficient if there isn't another or a combination that produces more product given the inputs; or uses less input given the outputs. It doesn't impose any a priori functional form on the data, it can accommodate multiple outputs and inputs, and it produces information on the "reference companies" for each one of the inefficient companies.

With respect to the methodology to be used to determine the efficiency of the companies, DEA and an econometric Tobit regression were used. This means that the variables are classified into outputs (income) and inputs (employees, non-current assets and materials and resources). In the Tobit regression, the dependent variable is efficiency, which is a variable that is established between 0 and 1, the independent variable is CC, being the control variables Liquidity and Indebtedness.

Mainly, there are two DEA models, constant returns to scale (CCR) and variable returns to scale (BCC) [21]. In this work, the CRS model is used with an input orientation. The formulation is as follows:

$$Min\, ef,\ \lambda\, efi \tag{1}$$

Subject to:

$$-yi + Y\lambda \geq 0 \tag{2}$$

$$ef\, xi - X\lambda \geq 0 \tag{3}$$

$$\lambda \geq 0 \tag{4}$$

Where:

ef is efficiency, is a constant vector, *X* is a matrix with all the inputs of all the companies, *Y* is a matrix of all the outputs of the companies, *xi* is a vector of the inputs of the company *i*, *yi* is a vector of the outputs of the company *i* and *i* represents the company *i*-th. With respect to the Tobit model, the econometric specification is as follows [22]:

$$efij* = \alpha xij + \mu ij \;\; \mu ij \; \sim N(0, \sigma 2) \tag{5}$$

Where *ef*ij* is a latent variable that takes a value between one and zero.

4 Results

This section presents the results of determining technical efficiency, productivity and the impact of working capital variables on efficiency.

4.1 Technical Efficiency

Table 2 shows the efficiency results by service and years, using the proposal of [4]; it also shows the average efficiency per year. It should be noted that the average efficiency per year increased until 2016, followed by a significant decrease in recent years. However, it can be seen that, in the last year, average efficiency has increased.

Table 2. Average efficiency per service and years

Service	2015	2016	2017	2018	Mean
Newspaper	0,340	0,503	0,547	0,593	0,496
Promotion and advertising	0,366	0,512	0,455	0,493	0,456
Radio	0,395	0,548	0,441	0,473	0,464
Radio stations	0,325	0,575	0,363	0,315	0,394
TV production	0,428	0,597	0,574	0,509	0,527
TV	0,340	0,520	0,477	0,453	0,447
Mean	0,387	0,545	0,469	0,485	

In this sense, it can be observed that during the first three years, TV production occupied the first place in average efficiency and the second place in 2018. While newspapers occupy first place in 2018, but in 2016 it was ranked the last in terms of efficiency.

On the other hand, the radio stations sector has remained in the last position in most of the years, and with an average distance from the others. It is necessary to point out that the general average is influenced by the radio sector, which has the largest number of observations.

It is interesting to detail the behavior of the companies in the annual ranking in order to contrast what has been expressed in the preceding paragraphs. A table has been

Table 3. Positioning of the companies of the sector that have occupied the first place.

N_emp	Sector	Position			
		2015	2016	2017	2018
37	Radio	1	1	1	1
41	Radio	1	1	1	1
108	TV Production	1	1	1	4
15	Radio	1	5	10	58
71	Radio	1	58	59	48
102	TV Production	17	1	1	1
78	Radio	3	1	2	2
39	Radio	12	1	4	1
44	TV Production	36	1	12	44
122	Promotion and advertising	84	1	26	1
99	Radio	7	1	28	49
84	Radio	58	1	40	38
90	TV Production	73	1	95	70
34	Newspaper	13	12	1	1
98	Radio	44	2	1	1
107	Newspaper	78	7	1	1
69	TV Production	2	48	1	3
97	TV Production	15	20	1	24
101	Promotion and advertising	26	16	5	1
110	Newspaper	37	37	7	1
85	Radio	60	17	38	1
103	Radio	75	74	77	1

constructed in which all the companies that at some time occupied the first place are included. They were sorted by sector in each year.

The results show that only two companies were able to maintain the first place in the ranking of efficiency during the years and these belong to the radio sector. As mentioned above, the radio sector didn't have the highest average efficiency, due to it has the largest number of observations and most of them have low-efficiency indicators.

Table 3 also shows in more than two consecutive years the leadership position, but this is lost in the following years, located in much lower scales, reaching in some cases the places that are at the end of the ranking (N_emp 71 and 90). It is observed the recovery of some companies, who have improved to be located, at least one year in the first place (N_emp 98, 107, 85 and 103).

In general, it could be said that the media sector hasn't been led in terms of efficiency by a group of companies, these have been altering the positioning of the overall ranking, which also affects the average efficiency of the sector.

4.2 Change in Total Factor Productivity (TPF)

Change in total factor productivity (TPF) is composed of changes in technical efficiency (CTE), technological change (TC), change in pure technical efficiency (CPTE) and change in scale efficiency (CSE). This indicator is important because it allows appreciating if the companies in the media sector are improving or not their total productivity.

It can be seen that the greatest change in TPF was obtained by the newspapers. Table 4 shows which sector has the greatest change in CTE and TC. Newspapers are followed by the TV and promotion and advertising sectors; it should be remembered that the TV sector requires adaptation and changes in order to be able to be in an informative world. The radio sector, which ranks fourth, has the largest number of companies in the sample and is widely used to communicate news and entertainment.

Table 4. Change in Total Productivity Factor (TPF)

Service	CTE	TC	CPTE	CSE	TPF	N° of Companies	Ranking
Newspaper	1,146	1,155	1,060	1,082	1,343	8	1
Promotion and advertising	1,122	0,905	1,024	1,102	1,002	18	3
Radio	1,129	0,900	1,050	1,082	1,001	71	4
Radio stations	1,035	0,831	1,066	0,965	0,827	2	6
TV Production	1,069	0,935	1,028	1,044	0,997	16	5
TV	1,091	0,913	1,024	1,065	0,996	7	2
Mean	1,091	0,913	1,024	1,065	0,996		

On the other hand, it is also appreciated that the sectors of TV production and radio station occupy the last places due to their low change in technical efficiency and technological change. In these sectors, innovations must be up to date, since they are necessary to improve their efficiency and increase their productivity.

4.3 Incidence of Working Capital Variables on Technical Efficiency

Table 5 shows the Tobit regression with efficiency as a dependent variable, the CC variable in its different forms as an independent variable and the control variables.

When analyzing the Tobit regression, it can be seen that liquidity is not significant to explain the efficiency of the company, as well as indebtedness (see Table 5). The cash cycle squared is significant and with a negative sign, which indicates that efficiency as a function of CC is a concave function, that is, it improves to a certain point from which it worsens.

Table 5. Tobit regression

Efficiency	Coef.	Robust Std. Err.	t	P > t	[95% Conf. Interval]	
Liquidity	−.0027457	.0015018	−1.83	0.068	−.0056969	.0002054
Indebtedness	.0360677	.0291679	1.24	0.217	−.0212484	.0933837
CC	.0001822	.0001356	1.34	0.180	−.0000843	.0004488
CC2	−2.30e-06	9.45e-07	−.44	0.015	−4.16e-06	−4.46e-07
_cons	.4718492	.0229345	20.57	0.000	.426782	.5169163
var(e.Efficiency)	.0563281	.0036529			.0495885	.0639837

5 Conclusions

The present work has measured the technical efficiency and the change in the productivity of companies in the media sector in Ecuador in the period 2015 - 2108. On the other hand, it's analyzed if technical efficiency is influenced by working capital factors such as liquidity, indebtedness and the cash cycle. In order to measure technical efficiency, income was used as output while the inputs used were employees, non-current assets (Act_no_corr) and other materials (Mat_recur).

The average technical efficiency of the media increased in the second year of study and then decreased in the following years. That decline is deeper in some communication sectors which make the overall average tend to decrease.

With regard to productivity, two of the most important companies have the biggest changes in the total productivity of the factors and are positively influenced by changes in technical and technological efficiency. It can also be seen that technical efficiency is influenced by the cash cycle, but in a concave manner; technical efficiency improves to a certain point, after that it worsens.

According to the results shown, it's important that the managers improve the procedures in order to provide incentives to increase their technical efficiency if their companies are located in the sectors with inefficiency results.

From an academic point of view, it is important to continue studies in the media sector. Efficiency studies show weaknesses in the management of factors and allowing them to search for improvements.

References

1. Emrouznejad, A., Yang, G.L.: A survey and analysis of the first 40 years of scholarly literature in DEA: 1978–2016. Socio-Econ. Plann. Sci. **61**, 4–8 (2018)
2. Halkos, G.E., Tzeremes, N.G.: International competitiveness in the ICT industry: evaluating the performance of the top 50 companies. Glob. Econ. Rev. **36**(2), 167–182 (2007)
3. Farrell, M.J.: The measurement of productive efficiency. J. Roy. Stat. Soc. **35**, 1578 (1957)
4. Banker, R.D., Charnes, A., Cooper, W.W.: Some models for estimating technical and scale inefficiencies in data envelopment analysis. Manag. Sci. **30**(9), 1078–1092 (1984)

5. Campos Lucena, M.S., Velasco Morente, F.: La eficiencia de las televisiones públicas en españa: La aplicación del dea como modelo de medición. Revista de Ciencias Sociales **19**(2), 362–378 (2013)
6. Fernández-Menéndez, J., López-Sánchez, J.I., Rodríguez-Duarte, A., Sandulli, F.D.: Technical efficiency and use of information and communication technology in Spanish firms. Telecommun. Policy **33**(7), 348–359 (2009)
7. Cooper, W.W., Park, K.S., Yu, G.: An illustrative application of idea (imprecise data envelopment analysis) to a Korean mobile telecommunication company. Oper. Res. **49**(6), 807–820 (2001)
8. Yang, J.C.: The efficiency of SMEs in the global market: measuring the Korean performance. J. Policy Model. **28**(8), 861–876 (2006)
9. Nigam, V., Thakur, T., Sethi, V.K., Singh, R.P.: Benchmarking of Indian mobile telecom operators using DEA with sensitivity analysis. Benchmark. Int. J. **19**(2), 219–238 (2012)
10. Giokas, D.I., Pentzaropoulos, G.C.: Evaluating productive efficiency in telecommunications: evidence from Greece. Telecommun. Policy **24**(8), 781–794 (2000)
11. Wang, Q., Geng, C.X.: Research on financing efficiencies of strategic emerging listed companies by six-stage DEA model. Math. Prob. Eng. (2017)
12. Yang, H.H., Chang, C.Y.: Using DEA window analysis to measure efficiencies of Taiwan's integrated telecommunication firms. Telecommun. Policy **33**(1–2), 98–108 (2009)
13. Uri, N.D.: Measuring the impact of price caps on productive efficiency in telecommunications in the United States. Eng. Econ. **46**(2), 81–113 (2001)
14. Façanha, L.O., Resende, M.: Price cap regulation, incentives and quality: The case of Brazilian telecommunications. Int. J. Prod. Econ. **92**(2), 133–144 (2004)
15. Hu, J.L., Hsu, H.H., Hsiao, C., Tsao, H.Y.: Is mobile jumping more efficient? Evidence from major Asia-Pacific telecommunications firms. Asia Pac. Manag. Rev. **24**(2), 190–199 (2019)
16. Bucklin, L.P.: Research in productivity measurement for marketing decisions. Res. Mark. **1**, 1–22 (1978)
17. Duhan, D. F. A taxonomy of marketing productivity measures. In: Proceedings of American Marketing Association, Chicago, pp. 229–232 (1985)
18. Higuerey, A., Trujillo, L., González, M.M.: Has efficiency improved after the decentralization in the water industry in Venezuela? Util. Policy **49**, 127–136 (2017)
19. De Jorge, J., Díaz, J.: Análisis de la productividad, eficiencia y sus factores explicativos: el caso de las empresas colombianas, 2005–2010. Revista de Métodos Cuantitativos Para La Economía y La Empresa **26**(26), 315–343 (2018)
20. Boles, J.N.: Efficiency squared–efficient computation of efficiency indexes. In: Proceedings of the Annual Meeting (Western Farm Economics Association), vol. 39, pp. 137–142 (1966)
21. Grmanová, E., Strunz, H.: Efficiency of insurance companies: application of DEA and tobit analyses. J. Int. Stud. **10**(3), 250–263 (2017)
22. Sağlam, Ü.: A two-stage performance assessment of utility-scale wind farms in Texas using data envelopment analysis and Tobit models. J. Clean. Prod. **201**, 580–598 (2018)

Researches Regarding the Burnout State Evaluation: The Case of Principals from Arab Schools from South Israel

Yunnis Nassar[1], Andreea Cristina Ionica[1], Monica Leba[1], Simona Riurean[1](✉), and Álvaro Rocha[2]

[1] University of Petrosani, 332006 Petrosani, Romania
andreeaionica2000@yahoo.com, Simona@yahoo.com
[2] University of Coimbra, Coimbra, Portugal

Abstract. It is a reality that wearable devices are part of our lives. We find them in different forms and answering different utilities and needs. The present research responds to an expressed necessity, namely the awareness on the psychological stress work-related syndrome of burnout, increasingly present in different fields of activity, education being among them. The research is carried out on two directions: (1) The analysis of the relationship between the organizational climate and the burnout state using a quantitative research. The results are concerning the identification of the organizational climate dimensions with the highest influence on the state of burnout and also, the fact that the studied demographic variables have no influence on the state of burnout. (2) The development, prototyping and testing of a burnout state evaluation and prediction device based on a burnout detection algorithm that allows the creation of a user profile. The findings of the research will further allow the testing of the prototype for evaluation and prediction of the burnout state in response to the stressors related to the identified high influence dimensions of the organizational climate. The research is characterized by a high degree of specificity due to the identified premises and the research context focused on the principals from Arab schools from South Israel.

Keywords: Wearable device · Prototyping · Organizational climate

1 Introduction

1.1 Burnout - Concept

According to the well-known definition provided by Maslach [1], burnout is a psychological syndrome described as a specific response to prolonged exposure to stressors in the workplace and has three components: exhaustion, depersonalization and reduced self-efficacy. Exhaustion is represented by feelings of over-exposure to stressors and depletion of emotional and physical resources; depersonalization (cynicism, isolation) refers to the indifference or attitudes of indifference regarding colleagues or tasks at work; and reduced self-efficacy (reduced personal achievement, dissatisfaction with

Á. Rocha et al. (Eds.): WorldCIST 2020, AISC 1159, pp. 598–608, 2020.
https://doi.org/10.1007/978-3-030-45688-7_60

others) refers to a feeling of incompetence or lack of achievement and productivity in the workplace.

A general definition of burnout in relation to the diagnostic criteria is still under discussion, with the enablers stirring up controversy [2]. Bianchi et al. [3] listed four reasons why burnout should not become a nosologically structure: the insufficient scientific basis, the tendency to overlap burnout with depression, the three-dimensional structure of burnout syndrome considered as unrealistic, the mere definition of the syndrome as being related to the workplace cannot be considered nosologically discriminant.

However, today the stress-related phenomenon of burnout is increasingly present in different fields of activity and has become an important issue not just for the affected people, but also for the organization and society. Even though this concept has been analyzed by many specialists, at present, there is a need for means and methods that allow the early detection of the tendency of burnout.

1.2 Context of the Research

The education sector is among the most affected areas of activity of burnout. As the research regarding the relationship between school principals' burnout and organizational climate is conducted in Israel's Arab education sector, the special situation of teachers and school principals in this sector will be considered. So, are noted both the existence of the Arab tradition and the characteristics of modern Western organizations, and consequently, the reality of the fact that modern Arab organizations often have a mix of values.

Thus, the principal's role in determining the school climate in Arab schools in Israel is considered central, a conclusion that coincides with findings from other regions with traditional influences, non-Western societies such as Japan, but differs significantly from schools in Israel's Jewish education system, with larger western influences, in which the role of the principal is no longer so important and is rather correlated with other factors, such as the autonomy of the teachers and the prestige of the school, the tendency towards innovation.

Also, must be considered for the contextualization of the research the characteristics of the tribal family and the Arab tradition, in which the power and responsibility are at the leader, and the subordinates have a strong need for dependence on their superiors, and the fact that the teachers in Arab schools from Israel expect the principal to be "obliged" to provide them a model of a hard-working man, who will practice an effective management style and support and encourage teachers. This is in line with the Arab tradition of the paternalist, in which the leading figure, the existence of authority is essential to ensure purpose and motivation, and to determine the social relations of the group [4]. The research context would not be well defined if are omitted the school characteristics, like the very crowded classes with students from very poor families, the high dropout rates due to tribal practices, the low level of education in the Bedouin schools, the lack of modernization of the educational field and not achieving the desired results by the students which leads to dissatisfaction in the work of teachers who have also, a low sense of security due to the high rate of violence [5, 6].

The above characteristics are also factors of the organizational climate. All of these, burden and overload the school principal, enhancing the appearance of the burnout state.

The climate is different from one school to another, it changes due to the perceptions of teachers, students and parents. These changes determine different reactions of the principals to stress. The analysis of the dimensions of the organizational climate given by the socio-professional context are input elements in the evaluation system, which will be able to signal the emergence of the burnout state, and according to each individual the appropriate coping strategies will be chosen (centered on problem or on emotions).

1.3 The Premises and the Objective of the Research

Previous research supports the predominant role of the principal in the schools from Arab educational system from Israel. Even though the most of the researches on burnout has been focused on teachers, the phenomenon of school principal's burnout has been addressed by Friedman [7, 8] and Kremer-Hayon et al. [9] leading to the premises' settings of the present research.

The premises are also based on a preliminary study on the aspects: the coordinates of the Arab educational system in Israel, the complex environment characterized by the Bedouin-tribal-Hamula spirit and the position of the school principal in the patriarchal society, and the importance of communication and interpersonal relations in the context of the principals' burnout and organizational climate relationship [10].

The main objective of the research is to analyze the relationship between the organizational climate and the level of burnout of school principals. This will be achieved, in the first phase (the first direction of the research) by evaluating the influences of the demographic variables on the perception of the organizational climate and the level of burnout at the principals level and also, the influences of the dimensions of the organizational climate on the burnout state of the principals of the Arabic schools from Bedouin area in southern Israel; and in the next phase (the second direction of the research) by designing a wearable device integrated in the conceptual framework of the burnout evaluation system [11], based on the results concerning the organizational climate analyze, and the results of the evaluation of the level of the principals' burnout state after the questionnaires-based survey and after the collection of physiological parameters of the principals.

2 Research Area. Methods and Materials

For reaching the objective it was used a quantitative method, a questionnaire-based survey. From a total of 35 schools from the Bedouin area of Southern Israel, agreed to participate a number of 30. The questionnaire was applied to the total number of employees of the 30 schools, namely 705, being validated a number of 686 questionnaires (filled by 30 principals and 656 teachers). The research was conducted during the period June-August 2019. In the initial phase of the research, the distribution of the questionnaires and the analysis and interpretation of the results of the questionnaires were carried out at the level of each school. However, there were no significant differences between the results obtained at each school level regarding the studied variables. This allowed the generalization of the interpretation of the results and moving on to the next phase of the study in order to achieve the research objective. The second phase of the research involved combining the results in a single report based on the coding of the questionnaires.

The tools used in the questionnaire-based survey were:

(1) Survey among teachers' staff – school as an environment supporting pedagogic teaching – learning - evaluating processes (included in Meitzav evaluation, School Growth and Efficiency Measures). The questionnaire was used in order to assess the perception regarding the school organizational climate. The questionnaire consisted of 38 items grouped into eight major dimensions: work/job satisfaction, satisfaction on management, sense of security, conditions for prolonged stay in school, professional development, teamwork, teacher autonomy, pedagogical leadership. The organizational climate questionnaire contains items to which the study participants provided answers on a scale from 1 to 5 (1-total disagreement, 5-total agreement).
(2) Principals' burnout scale [8]. The questionnaire was used in order to assess the burnout level of the principal applied on 30 subjects. The questionnaire consisted of 22 items grouped into 3 dimensions: exhaustion, segregation, dissatisfaction with the others. The questionnaire contains items to which the study participants provided answers on a scale from 1 to 6 (1-never, 6-always) (Table 1).

Table 1. Values of the internal reliability index (Cronbach's Alpha)

Variable	Statement	Scale	O. statement	Alpha V.
Managerial Erosion	1-22	1-6		0.917
Exhaustion	1-9	1-6	6,8,9	0.915
Segregation	10-16	1-6		0.795
Dissatisfaction with the others	17-22	1-6		0.826
Organizational climate	1-38	1-5		0.959
Work satisfaction	1-3	1-5	1	0.817
Satisfaction on management	4-8	1-5		0.828
Sense of security	9-10	1-5		0.971
Conditions for prolonged stay in school	11-14	1-5		0.827
Professional development	15-19	1-5		0.838
Team work	20-27	1-5		0.910
Autonomy of Teachers	28-29	1-5		0.731
Leadership	30-38	1-5		0.931

The internal reliability indices for the research are clear. Reliability is considered good if it is higher than 0.7, and so the reliability for the Burnout variable, including the three sub-variables is good. Also, the reliability is good for the Organizational Climate variable including the eight sub-variables and indicates that the items examine the same content for each variable.

The data analysis tools and procedures contained the statistical analysis and processing that was performed with the SPSS 20 program and included descriptive statistics (dispersion, averages, standard deviations, etc.), Alpha Cronbach internal consistency, correlation and regression analyzes (Pearson correlation, paired T-test, multiple regression test) (Tables 2 and 3).

Table 2. Socio-demographic characteristics of the principals

Variable	N	%	Mean (±S.D)	Min.	Max.
Male	25	83.3%			
Female	5	16.7%			
Age (years)			(8.020 ±) 48.13	30	64
Work in current school (years)			(9.353 ±) 18.20	1	35
Work in any school (years)			(7.360 ±) 25.20	14	45
BA	4	13.3%			
MA	26	86.7%			

Table 3. Socio-demographic characteristics of the teachers

Variables	N	%	Media (±S.D)	Min	Max.
Male	231	35.2%			
Female	425	64.7%			
Age (years)			(± 9.294) 35.57	20	63
Work in current school (years)			(± 6.991) 8.59	1	43
Work in any school (years)			(± 8.607) 11.27	1	43
BA	4	13.3%			
MA	21	.86.7%			
PhD	1	0.2%			

For the wearable device was used the prototyping approach based on Arduino Hardware and software platform. This platform allows rapid prototyping and testing of the hardware design allowing also software implementation and testing at each step.

3 The Research Hypotheses

The research hypotheses address the analysis of the relationship between the principals' burnout state and the perception regarding the organizational climate from the perspective of the demographic variables: age, seniority at work in current school and seniority in work in all schools, education level, gender, position.

Testing the Research Hypotheses
H1. There is a negative correlation between the principals' burnout state and the perception regarding the organizational climate. The findings show that there is a significant negative correlation between principals' burnout state and the organizational climate perception ($p < .01$, $r = -0.637$). The hypothesis was accepted (Table 4).

Table 4. The relationship between the organizational climate and principals' burnout state

Principals' Burnout				
		Exhaustion	Segregation	Dissatisfaction with others
Organizational Climate	Work Satisfaction	−0.572**	−0.418*	−0.450*
	Satisfaction on management	−0.485**	−0.492**	−0.457*-
	Sense of security	−0.389*	−0.552**	−0.558**
	Conditions for prolonged stay	−0.223	−0.251	−0.322
	Professional develop.	−0.349	−0.568**	−0.500**
	Team work	−0.245	−0.671**	−0.560**
	Autonomy of teachers	−0.184	−0.313	−0.447*
	Leadership	−0.200	−0.458*	−0.379*

p > .05, *p < .05, **p < .01

H2: The sub-variables of organizational climate: work satisfaction, teachers' sense of security affects the overall change of the principals' burnout state. The hypothesis was accepted (Table 5).

Table 5. Predicting values for the organizational climate influence on principals' burnout state

Predicting value	B	β	t	R^2
Work satisfaction	−0.440	−0.463	−2.783*	0.384
Satisfaction on management	−0.317	−0.227	−0.957	0.464
Sense of security	−0.620	−0.471	−2.604*	0.611
Conditions for prolonged stay	0.217	0.198	1.052	0.633
Professional development	−0.164	−0.122	−0.422	0.634
Team work	−0.002	−0.001	−0.005	0.637
Autonomy of teachers	−0.112	−0.106	−0.519	0.645
Leadership	0.317	0.221	1.020	0.662

*p < .01

The findings indicate that the variables *work satisfaction* and *sense of security of teachers* clearly explain the principals' burnout state. Teachers' sense of security has a slightly greater effect on the principals' burnout (−0.471- | β) than the work satisfaction (0.463 = β). The other variables were found to be statistically insignificant and without significant contribution.

H3: There are influences of the demographic variables (gender and level of education) regarding the principals' burnout state and the perception of the organizational climate. The hypothesis was rejected.

Table 6. T Test for principals' burnout and organizational climate by gender

		Men N = 25	Female N = 5	*t*(28)
Principals' Burnout	Mean	2.213	2.073	0.402
	S.D.	0.684	0.856	
Organizational Climate	Mean	4.333	4.342	−0.044
	S.D.	0.455	0.335	

p > .05

According to Table 6 there is no significant difference in the principals' burnout state between men and women [t (28) = 0.402, p > .05]. In addition, there is no significant difference in the organizational climate between men and women [t (28) = −0.044, p > .05].

Table 7. T Test for principals' burnout and organizational climate by level of education

		BA N = 4	MA N = 26	$t(28)$
Principals' Burnout	Mean	2.193	2.189	0.011
	S.D.	0.727	0.712	
Organizational Climate	Mean	4.421	4.321	0.425
	S.D.	0.364	0.447	

$p > .05$

According to Table 7 there is no significant difference in the principals' burnout with Bachelor degree (BA) (M = 2.193, S.D. = 0.727) and with Master degree (MA) (M = 2.189, S.D. = 0.712). In addition, there is no significant difference in the organizational climate [t (28) = 0.425, p > .05]. The mean value regarding the perception of the organizational climate for the principals with BA (M = 4.421, S.D. = 0.364) was similar to the organizational climate mean for the principals with MA (M = 4.321, S.D. = 0.447).

H4: There is a positive correlation between the principals' burnout state and the age, the seniority at work in current school and the seniority in work in all schools. The hypothesis was rejected.

Table 8. Correlation between principals' burnout, age and seniority

	Principals' burnout	Age	Seniority in current school	Seniority in all schools
Principals' Burnout	1	−0.159	−0.171	−0.047
Age		1	0.647**	0.882**
Seniority in current school			1	0.650**
Seniority in all schools				1

**p < .01, p > 0.05

The findings (Table 8) show that there is no significant correlation between *Principals' Burnout* and *Age* (p < .05, r = −0.159), *Seniority in the current school* (p > .05, r = 0.171), and *Seniority in all schools* (p .05, r = −0.047).

The results obtained in the first research direction are: the identification of the dimensions of the organizational climate with the greatest influence on the appearance of the burnout state (work satisfaction and feeling of security), with no influence from the demographic variables. Also, based on these results and the results presented by Friedman [8] a synthesis of the relationship between the dimensions of burnout, the dimensions of the organizational climate and the risk factors was made pointing out the fact that the risk factors are taken into account through their influence on the climate dimensions (Fig. 1). In the next stage of the research will be pursued using a designed wearable device (prototype stage) predicting the occurrence of the burnout state in the school directors in the Bedouin area of southern Israel under the influence of the organizational climate related risk factors, the device is calibrated using both the directors' answers to questionnaires and the values of physiological parameters.

4 Burnout State Evaluation and Prediction Device

The necessary hardware components for the device are: oximetry and heart rate sensor, skin conductivity sensor, Bluetooth module, and microcontroller. For measuring the oximetry and heart rate it is used MAX30102 Sensor (Fig. 2).

A Bluetooth module is used to transmit data obtained after the measurements to the mobile device. In order to prototype the software, Arduino IDE and Arduino MEGA 2560 were used as the main platform. Before the program was implemented, there were made the connections between the Arduino board and the sensors (Fig. 3).

After the final hardware build, the serial port was monitored and the data was acquired. Testing the program was afterwards done by connecting the Bluetooth module to a mobile phone and building a minimalistic Android application in M.I.T. App Inventor IDE. The algorithm (Fig. 4) for burnout state estimation will run in a loop,

Burnout dimensions	Exhaustion		Segregation		Dissatisfaction with others	
Organizational climate dimensions	Work satisfaction Satisfaction on management Sense of security					
	Professional development		Team work Leadership		Professional development Team work Teachers' autonomy Leadership	
Stressors	Work and responsibility overload	Lack of support and cooperation	Lack of support and cooperation	Difficulties with teachers	Difficulties with teachers	Difficulties with parents, students and administrative staff

Fig. 1. Burnout dimensions-Organizational climate dimensions-Stressors/Risk factors

Fig. 2. MAX30102 sensor module.

Fig. 3. Arduino Mega hardware prototype

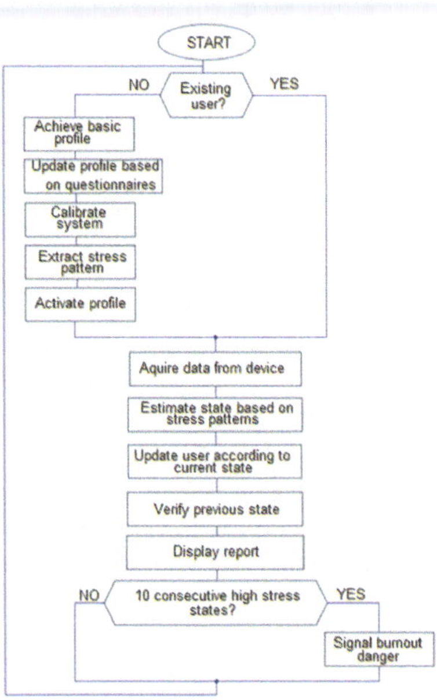

Fig. 4. Algorithm logic diagram

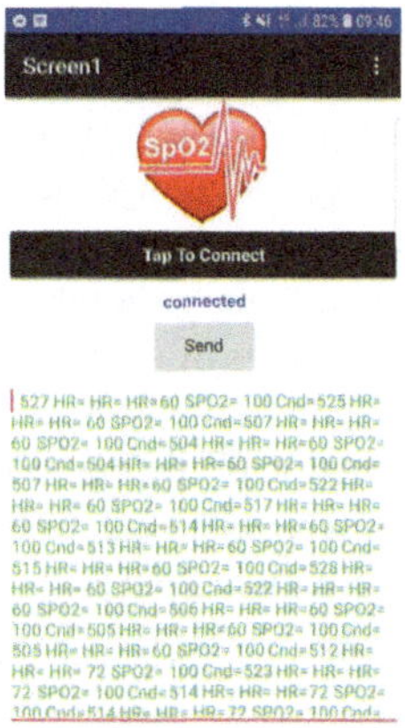

Fig. 5. Android application

starting with the verification of the user's profile. In case the profile is not registered, a new profile will be created, containing the initial data, like name, age, etc. that also includes results obtained from questionnaires.

After the profile data is filled, the connection to the measurement device, its calibration and the application of an acquisition protocol for the user's state is completed. The obtained data is processed, to extract the estimation blueprints of the general state, but mostly of the high stress state. The blueprints are included in the user profile. Once the user profile is completed or if it already exists, the real-time acquisition of the physiological data from the user can be started. These are processed and analyzed together with existing templates in the profile to estimate the current state of the user and will be stored in the user profile. The last consecutive states stored in the profile are checked in order to display a user report, and if there are consecutive high stress states, the warning about the danger of the burnout state is signaled. To test the algorithm and to see the functionality of the sensors and Bluetooth module, a minimal smartphone application has been created that shows the transmission of values to the smartphone. The results from running the program can be seen in Fig. 5.

5 Conclusions

The burnout state is a real problem, and the sooner it's being predicted, the easier it is to deal with. Stress is an everyday life factor and has to become recognizable.

The first research direction highlighted that the level of burnout of school principals depends on the organizational climate mainly expressed through the work satisfaction and sense of security dimensions. The level of burnout of the principals does not depend on the demographic variables studied (sex, age, seniority in school, seniority in work and level of education) and these will not be further taken into account.

The second research direction led to building a prototype of wearable device both hardware and software parts for the estimation of current state and prediction of burnout state. The prototype embeds a dedicated prediction algorithm, having as burnout state estimation an artificial neural network, that creates a user profile based on the results of organizational climate analysis, questionnaires-based survey and physiological parameters acquisition. The instant results of the burnout state estimation done by this algorithm is displayed as color-coded alarms directly on the wearable device and the acquired physiological parameters together with the state estimation are stored in the device and transmitted to the smartphone for a graphical representation.

The findings of the two directions of the research are integrated by using the organizational climate analysis and currently used as inputs in a pilot study case for evaluation and prediction of the burnout state using the developed prototype. The further expected results are related to the validation and enhancement of the wearable device that is subject to an international patent.

References

1. Maslach, C., Jackson, S.E.: The measurement of experienced burnout. J. Organ. Behav. **2**(2), 99–113 (1981)
2. Kaschka, W., Korczak, D., Broich, K.: Burnout: a fashionable diagnosis. Dtsch Arztebl Int **108**, 781–787 (2011)
3. Bianchi, R., Schonfeld, I.S., Laurent, E.: Is it time to consider the "Burnout syndrome" a distinct illness. Front Public Health **3**, 158 (2015)
4. Al Asad, S., Danaiata, D., Natase, M.: The influence of leadership in teachers; practice in bedouin high schools. Rev. Int. Comp. Manag. **18**(4), 362–375 (2017)
5. Abu-Bader, S., Gottlieb, D.: Poverty, education, and employment among the Arab Bedouin in Israel. In: Poverty and Social Exclusion Around Mediterranean Sea, pp. 213–245 (2013)
6. Al Baqain, R., Valle-Zaerate, A.: Economic analysis of bedouin sheep farming in Jordan and Palestinian Territories. Livestock Res. Rural Dev. **23**, 249 (2011)
7. Friedman, I.A.: Multipathways to burnout: cognitive and emotional scenarios in teacher burnout. Anxiety Stress Coping **9**, 245–259 (1996)
8. Friedman, I.A.: Burnout in school principals: role related antecedents. Soc. Psychol. Educ. **5**, 229–251 (2002)
9. Kremer-Hayon, L., Faraj, H., Wubbels, T.: Burn-out among Israeli Arab school principals as a function of professional identity and interpersonal relationships with teachers. Int. J. Leadersh. Educ. **5**(2), 149–162 (2002)
10. Ionica, A.C., Nassar, Y., Mangu, S.: Organizational climate aspects and principal's burnout in Southern Israel schools. In: MATEC Web of Conferences, vol. 290(3), p. 07009 (2019)
11. Nassar, Y., Marcus, R., Ionica, A.C., Leba, M.: Integrated system for burnout state assessment. Int. J. Econ. Stat. **6**, 95 (2018)

An Evaluation of How Big-Data and Data Warehouses Improve Business Intelligence Decision Making

Anthony Martins[1], Pedro Martins[1], Filipe Caldeira[1], and Filipe Sá[1,2](✉)

[1] Department of Computer Sciences, Polytechnic Institute of Viseu, Viseu, Portugal
martinsanthony_arm@outlook.com, {pedromom,caldeira,filipe.sa}@estgv.ipv.pt

[2] Artificial Intelligence and Computer Science Laboratory, University of Porto, Porto, Portugal

Abstract. Analyze and understand how to combine data warehouse with business intelligence tools, and other useful information or tools to visualize KPIs are critical factors in achieving the goal of raising competencies and business results of an organization. This article reviews data warehouse concepts and their appropriate use in business intelligence projects with a focus on large amounts of information. Nowadays, data volume is more significant and critical, and proper data analysis is essential for a successful project. From importing data to displaying results, there are crucial tasks such as extracting information, transforming it analyzing, and storing data for later querying. This work contributes with the proposition of a Big Data Business Intelligence architecture for an efficiently BI platform and the explanation of each step in creating a Data Warehouse and how data transformation is designed to provide useful and valuable information. To make valuable information useful, Business Intelligence tools are presented and evaluates, contributing to the continuous improvement of business results.

Keywords: Data Warehouse · Big Data · Data Mart · Business Intelligence · OLAP · ETL · Power BI · Analytical methods

1 Introduction

Information Systems (IS) present themselves as the engine that drives companies to success by using vast amounts of information collected and stored in databases (DB). On the IS field, several authors try to create a definition for IS. In [1] authors define Information System as any system that has information as input to generate output information. In [2] an Information System is a combination of procedures, data, people and IT to achieve the organisation's goals.

For the current economic ecosystem, exploring and learning from raw data has become a necessary core competence. It is clear that by focusing on organisational learning and competitive advantage, many organisations are achieving successful results and becoming more independent in today's business market. As organisations learn how

Á. Rocha et al. (Eds.): WorldCIST 2020, AISC 1159, pp. 609–619, 2020.
https://doi.org/10.1007/978-3-030-45688-7_61

to use information, it enables quick answers to demand and fast adaptation to a turbulent environment [3]. In addition to active business adaptation, it became necessary to decide what to do, how to do it, and finally, how to perform the required operations [4].

This article arises from a need to find solutions for the efficient and useful management of indicators within companies, a reality for many companies in the news. For this, and according to the information retained during research and reading of different themes and ideas, as a contribution comes the proposal of a solution through the demonstration and explanation of a Big Data BI architecture adapted for the management of indicators and remaining information focused on the presentation of optimised results, the effective management of indicators, as well as better decision making.

Several concepts, such as Big Data, Data Warehouse (DW) and Business Intelligence (BI), can be combined to provide an information analysis and management tool, putting more emphasis on results and subsequent decisions.

This article is organised in five chapters, with the presentation of the themes: Data Warehouse and Business Intelligence, as well as the introduction of a Big Data BI architecture proposal for information management, indicators and decision-making assistance. Subsequently, the chapter presented main rules of the GDPR, to show the main risks on privacy when analysing information and the conclusion which determines the importance of BI and its architecture for the analysis of data and the management of indicators within a company.

Resuming, the real importance of BI includes architectures, tools, DB, applications and methodologies, and expression of content for different people, which will have different meanings. For the information to have real benefit for the business, it is necessary to provide accurate information on-time, when needed, and it must be vital for decision making, strategic planning and even business triumph [5, 6].

Having up-to-date information allows for improving processes and support strategies. BI systems allow combining data from different environments, locations, operating systems, and DB's, in a speedy time [7, 8]. Most of the times, these massive amounts of data are unstructured or stored inside databases with different schemas, requiring special treatment before integration (Big Data). Two stages are proposed to reach the success goals in BI [9]. First is to identify, collect, store and maintain data. Second, is to retrieve information, process, and present data in a way which is useful for decision-making.

2 Data Warehouse

More than just storing data, a Data Warehouse is a repository where data is organised in a central data deposit. Stored data is extracted in the form of operational data, subject-oriented, non-volatile and historical nature [10]. The Data Warehouse should meet specific operational objectives, each adapted to the organisational business requirements. Important to note that, the preparation, design and storage of data need to be carried out appropriately in data warehouses, because, it is not enough to merely copy data from a database to a data warehouse [11].

Data can be differentiated into two types. The first type consists of operational data, where applications process data based on transactions (OLTP). This operation mode is commonly applied in production systems and executive applications. The design of the functional and operational data structure is following the business rules, and the design of OLTP systems are compliant to meet the daily business needs of the enterprise but do not support integrated data analysis [12]. The second type is focused on decision support, OLAP, the goal is to enable competitive advantage by exploiting the increasing amount of data collected and stored in files and DB for faster and more accurate decision making [13]. Figure 1 shows the underlying architecture of a DW.

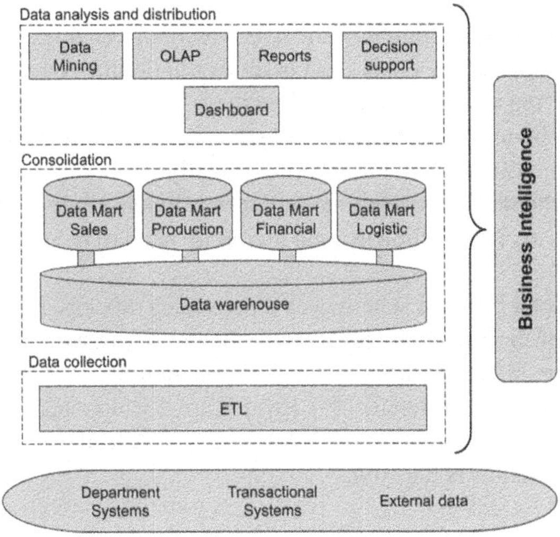

Fig. 1. Architecture of BI, Data Warehouse [14]

Figure 1 illustrates the following DW and business intelligence processes:

- Data collection: Composed of ETL (Extraction, Transforming and Loading) processes.
  - Extraction: Data extraction or collection consists of obtaining relevant data from various sources, internal or external to the company.
  - Transformation: Data is adjusted and prepared during conversion and uniformly adapted to its own DW target format.
  - Loading: The loading phase involves logging or updating collected data and transformed into the DW.
- Consolidation/Storage Processes: After collecting information, using ETL techniques, data is loaded into DW, as well as later into Data Marts. The process of loading information to DW or DM is critical, so it should be well planned, and then well implemented, so that there is no possibility of anomalies or loss of information.
- Data analysis and distribution: Data mining, OLAP, Dashboards, Decision support, and Reports, are created for better representation and visualization of the data. Some of the crucial features of DW are:

- Have a subject orientation, storing information related to the business core to meet the needs of decision-makers.
- Define a unique way of representing data by converting data from various enterprise systems, thus establishing an efficient integration.
- The value of the entered data is not lost over time, as it does not show volatility.
- Owner of an efficient granularity, being characteristic concerning the level of detail or summary of the data contained in a DW.

After the ETL process is complete, the data warehouse enables queries and analysis, transforming sparse data, previously inaccessible or underutilized information, into useful information for the business strategy. To achieve better data warehouse performance and results, star schemas and snowflake schemas are the most popular for implementations in multidimensional data models (not directly supported for the relational data model). In a star schema, all the information for one dimension is store in one denormalized table, and this avoids the dimensional hierarchy need to be re-assembled via costly join operations.

The definition of a DW star schema is a three stages process, in a loop cycle. First is the identification of the facts that characterize the problem under analysis. Second, the dimensions that can influence these facts. Third the definition of the granularity of the stored data (i.e., the level of detail) [15, 16]. Figure 2 illustrates a star-schema with a central fact table and many dimensions surrounding it. A traditional dimension present in almost all star-schemas is the "time".

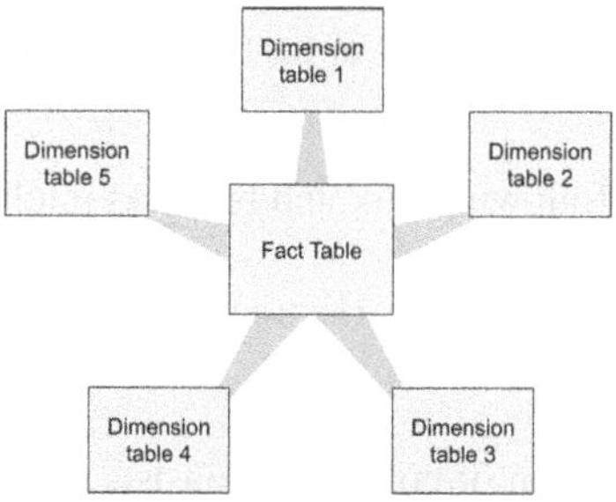

Fig. 2. Shows the basic star schema

Once the star-schema is identified, it is also important to understand the snowflake concept. The snowflake schema represents a dimensional model that is composed of a central fact table and a set of constituent dimension tables which can be further broken up into sub-dimension. Snowflake schemas adhere to the normalization principles of relational design theory for the dimension tables, and, as a consequence, there is no redundancy in tables, which decreases the storage overhead. Additionally, there are no anomalies, and the schema is more natural to extend. However, there is also a downside, during query processing, the different hierarchy levels of a dimension have to be joined together with surrogate keys, which introduces a computational overhead [17].

Figure 3 shows a star-schema alongside with several snowflakes. For instance, the dimension time can be decomposed on a snowflake, with the sub-dimensions (DT): week, day, month, year, weekday.

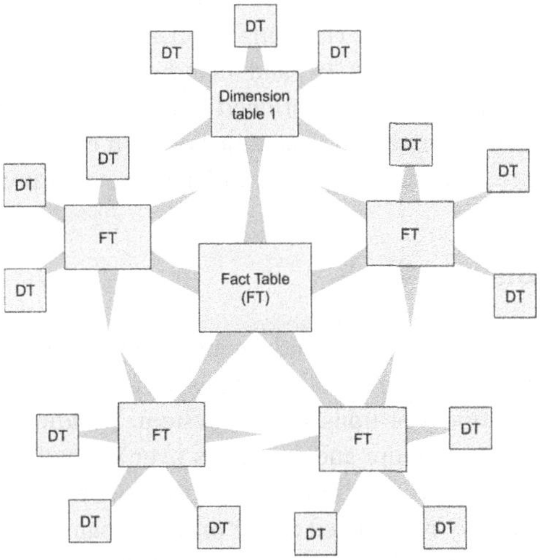

Fig. 3. Snowflake basic schema

Big Data refers mostly to unstructured data, in addition to the challenges posed by the volume and variety of data, and velocity about collecting, analyzing and using data. According to [18], "Big Data does not mean a large volume of data, but how hard its analysis is, to reach a result in the desired time". However, at the same time, a DW includes the E (extraction) and T (transformation), which is c the conversion of unstructured data to some structure. The DW also contains large data volumes. Finally, the DW also includes complex data analysis which requires time to process. Accounting all the similarities between the DW processes and the definition of Big Data, it is fair to assume that a DW dedicated to business intelligence also relates to Big Data [19].

A DW must be prepared to accept all changes and possible challenges regarding up to 10 or more Vs: Velocity, Veracity, Volume, Variability, Validity, Vulnerability, Volatility, Visualization, Value. When carefully analyzing the V's, it is clear the also Business Intelligence data, and the DW, govern themselves by the same V's, therefore, once again, Big Data. Both Big Data and Business Intelligence DW's challenge the massive increase of data volume by adopting and managing the technology evolution, thereby improving and changing data analysis methods. In this way, it is necessary to evolve and adapt to the constant challenges posed by Big Data [20, 21].

The concept of Big Data Analytics (BDA) focuses on improving the capabilities of an organisation to create more value given information collected over, days, months, years. By finding patterns in sparse details and translating it to useful business knowledge, it allows to leverage investment direction, and its impact on corporate value [22, 23]. On another positive side, BDA is becoming a trend in organisations that for several years, collected data without knowing if it was going to be useful someday. The analytics process, including deploying and using BDA tools, is seen by organisations as a tool for improving operational efficiency, generate new revenue, and increase competitiveness advantage over rivals [24, 25].

Databases with raw data, Big Data (i.e. unstructured and untreated data), presents excellent potential once data is explored and analysed. Advances in technologies such as NoSQL, BigQuery, MapReduce, Hadoop, WibiData and Skytree, make it possible for a wide range of business sectors to expand their data analysis capacities. Without the advance of technologies as the ones mentioned before, many business sectors did not have the means to explore stored data. Examples of such expansion include the public areas of finances, justice, healthcare, energy, and nature events prediction [26].

3 Business Intelligence Tools

The core of most large business is currently widely spoken in the area of Information Technologies. The implementation of a BI system is not a conventional application-based project, such as an operational or transactional system. Building a BI tool is complex, requiring appropriate infrastructure and resources over a lengthy period, preferably as short as possible. Many projects that introduce BI, the importance of the DW for quality analysis is key, subsequently recognized as the principal cause of the project's success. In BI, an integrated set of tools, technologies and products are used to collect, integrate, analyze and make available data. This includes intelligent exploration, integration, and aggregation, as well as multidimensional analysis of data from various resources [27].

In a traditional BI system, power users serve less experienced casual users. They analyze and collect the data requested by casual users and produce the reports and visualizations on which casual users base their decisions. The Balanced Scorecard (BSC) concept is essential in the BI Project, and it facilitates a multidimensional overview of an enterprise. The BSC method highlights the management of intellectual resources as an essential factor and integrator in the realization of the business's strategy.

Business Intelligent systems/techniques such as Online Analytical Processing (OLAP) and also Data Mining, represent decision support tools which allow obtaining more value from data. Data value gain is achieved by exploring stored data, aggregating and revealing unknown trends [28]. Many statistical and quantitative techniques explain, and predictable models allow to visualise data in a way that before was not possible, leading intelligent business system to be a priority for successfully manage and extract data knowledge. However, information extraction is essential, so it is the commuting time required to do so, in this context near-real-time data extraction becomes a challenge in terms of distributed processing method [9, 29, 30].

Some case studies show that BI is used mainly for: strategic planning; improving relations with customers; analyzing the profitability of products and services; analyzing internal processes and operational efficiency of an organization; controlling and management accounting [31]. Additionally, some examples of BI tools are IBM Cognos Business Intelligence [32]; Microsoft Power BI [33]; Oracle BI Foundation Suite [34]; SAP Business Objects [35]; MicroStrategy Analytics Platform [36]; Qlik Sense [37]; Tableau [38]; Tibco Spotfire [39].

4 Big Data Solution for the Success

For delivering a sustained competitive advantage in the organization, a Big Data Architecture is a plan to deploy an agile, state-of-the-art BI solution. Adapting organizational processes to the reality of the Big Data era and assessing the effectiveness of data analysis methods is halfway to achieving prosperity in any BI Project, and success in creating innovative solutions and ideas within the organization. The other halfway is the efficiency of ETL methods for integration into the data collection process. Figure 4 shows the proposed layout solution for Big Data Architecture of a Business Intelligence Project to achieve success in different business areas within organizations. The visible three processes:

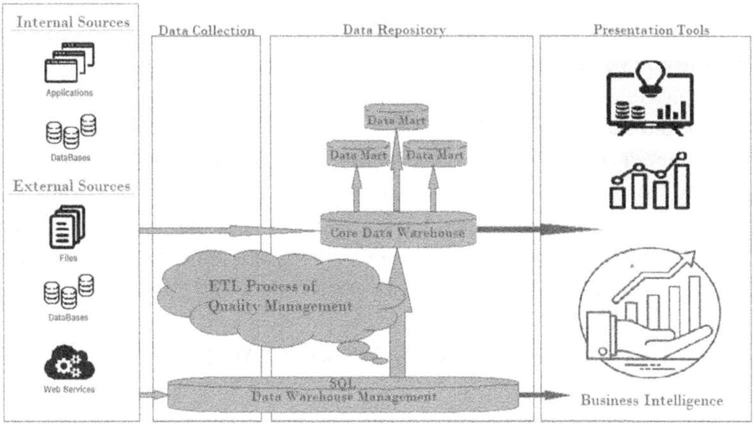

Fig. 4. Big Data architecture of Business Intelligence project

- Data Collection: Analysis of data from different sources, which can be of different formats, and using appropriate tools to increase data collection efficiently and quickly.
- Data Repository: Processing and transforming of data into useful information using an ETL performant process and store of this information into DW.
- Presentation Tools: Ideas, solutions and actions are presented using BI tools and provided from the information handled and stored in DW.

In the recent past, the BI solutions focused on only structured and internal data. Consequently, much valuable unstructured and external data is hidden and unused, potentially leading to an incomplete view of reality and biased business decision making. Nowadays BI primarily uses transactional database data from multiple sources, as well as many sources other than relational DBs [11, 40].

Advances in technology make it easier to continuously collect large volumes of heterogeneous data, bringing new challenges and opportunities for BI. Any BI project must combine internal and external data from the organisation, coming from different data sources, focused on gathering reliable information and finding appropriate solutions.

With this large amount of data, optimized management of large data can help companies improve customer satisfaction, manage risk and even generate competitive intelligence. The ability to deliver real-time results and help make important decisions is now simple and manageable using BI tools [8].

Data collection is performed on a specific set of data and serves to obtain results, assessments and other important information through BI reports. These reports serve to present the results with a simple and intuitive structure. After the collected data, is the exploration and analysis of data to enable the creation of reports with valuable information. The usefulness of reporting focuses on the need to improve and assist decision making by managers and administrators in each area. BI has its limitations: it is necessary to interpret essential and relevant data, always maintaining common sense and intelligence, to obtain satisfactory results.

Currently, BI systems are essential for organisations in terms of monetary gains and reduced task time. The most relevant advantage of BI is to obtain and disseminate useful information from large amounts of data, with a primary focus on assessing future decisions more quickly and credibly, through the help and use of reports and dashboards. The structure and aesthetics of these reports are also considered a quality advantage, and there are numerous possibilities to structure these reports and make each result and interpreted information visible, stimulating and more concise: graphs, tables, cartography, infographics, images and others. Giving users an easy and clear view of results is a goal, thus making each professional area of the organization more competitive in its field of activity. Without quality data, the BI project is not an intelligent system.

5 General Data Protection Regulation (GDPR)

Knowing the benefits of analyzing this data for decision making, it is also relevant to note that there are significant risks directly applied to privacy data analysis. Each country has its own rules regarding the exploration, storage, and protection of data. These standards are now strength at the European level through the GDPR. In this context, the following are the most interesting and important rules to retain:

- Fairness, Lawfulness, and Transparency: The data subject must know when processing will occur (transparency), the processing must match the initial description (fairness), and the processing must a purpose and specificity described under Regulation (lawfulness) [41, 42].
- Finality: Personal data should be collected for specific, explicit and legitimate purposes, and should not be processed outside of those specified purposes.
- Confidentiality: A first security objective following the controller's security obligation is to maintain the confidentiality of information, so, privacy is a security objective described as keeping the content of the information secret from all parties except those authorized to access it.
- Integrity and Authenticity: Integrity, as a security objective, must ensure that data are not altered by unauthorized or unknown means, and authenticity, as a security objective, must ensure that data are transmitted by an entity and not modified by unauthorized persons or unknown means (the authenticity of data also implies data integrity).

- Availability: Data controllers have an obligation to protect personal data against destruction or possible accidental loss.

6 Conclusions

One of the critical factors for the success of any BI solution is the performance of the ETL process. The relation of BI with Big Data and DW opens the door to create an intelligent platform, which can present relevant information to organizations through detailed reports and graphs. Based on the various reviewed research works, BI methods are the best ways to focus on the business reality and to discover useful information for decision-making. Despite the recognized importance attributed to the optimization of ETL, it is rare to find Information Technology teams that develop metrics and indicators that allow the accurate monitoring of their performance. The concept of Organizational Big Data Warehouse is a solution for analyzing big data and selecting only value-added data to store the essentials information in historical DW. The architecture platform presented above, in chapter four, is a solution adapted for monitoring metrics and indicators of the organization.

Big Data includes structured, semi-structured and unstructured data. Due to increasing data volume, relational DB needs to be adapted and prepared to support storage scalability and flexibility.

According to this research study, adopting Big Data technology with BI tools has immense benefits to companies, both in terms of information organization, structure, and management, as well as in the speed of data analysis and further storage. However, the focus and most significant benefit are on obtaining reliable information, a fundamental and central issue about defining and choosing the right strategy for a complete BI project, with the focus on an efficient Big Data BI architecture.

Nowadays, BI reports are the most talked-about topics at the business level, being the most suitable method to gain competitiveness and reinforce competencies towards competitors.

Some interesting topics for future in-depth analysis:

- Analyze and evaluate the differences between the use of OLAP techniques and Data Mining methods for data analysis in a DW.
- What are the advantages of integrating AI and ML in a BI project?

Acknowledgements. This work is financed by national funds through FCT – Fundação para a Ciência e Tecnologia, IP, under the project UID/Multi/04016/2019. Furthermore, we would like to thank the Instituto Politécnico de Viseu and CI&DETS for their support.

References

1. Casagrande, N.G.: Metodologia para Modelagem de Arquitetura de Informação estratégica para pequenas empresas: Uma aplicação no setor de turismo rural (2005)

2. Amaral, L., Varajão, J.: Planeamento de Sistemas de Informação, p. 228. FCA-Editora de Informática, Lda, Lisboa (2007)
3. Namada, J.M.: Organizational learning and competitive advantage. In: Handbook of Research on Knowledge Management for Contemporary Business Environments, pp. 86–104. IGI Global (2018)
4. Velloso, F.: Informática: conceitos básicos, vol. 9. Elsevier, Rio de Janeiro (2014)
5. Turban, E., et al.: Business intelligence: um enfoque gerencial para a inteligência do negócio. Bookman Editora, Taipei City (2009)
6. Martins, P., Abbasi, M., Furtado, P.: A scale: auto-scale in and out ETL + Q framework. In: Kozielski, S., Mrozek, D., Kasprowski, P., Małysiak-Mrozek, B., Kostrzewa, D. (eds.) Beyond Databases. Architectures and Structures. Advanced Technologies for Data Mining and Knowledge Discovery, pp. 303–314. Cham, Springer (2015)
7. Cody, W.F., et al.: The integration of business intelligence and knowledge management. IBM Syst. J. **41**(4), 697–713 (2002)
8. Olszak, C.M., Ziemba, E.: Business intelligence as a key to management of an enterprise. In: Proceedings of Informing Science and IT Education Conference (2003)
9. Wieder, B., Ossimitz, M.-L.: The impact of business intelligence on the quality of decision making–a mediation model. Procedia Comput. Sci. **64**, 1163–1171 (2015)
10. Corrêa, Â., Sferra, H.H.: Conceitos e aplicações de data mining. Revista de ciência tecnologia **11**, 19–34 (2003)
11. Bele, D., Weis, L., Maher, N.: Sustainable Development Under the Conditions of European Integration, Part II. MISC (2019)
12. Caldas, M.P.K., Scandelari, L., Luiz Kovaleski, J.: Aplicações sobre uma Data Warehouse no ambiente das organizações e suas vantagens. In: Simpósio de Engenharia de Produção, XIII SIMPEP (2006)
13. Chaudhuri, S., Dayal, U.: Data warehousing and OLAP for decision support. In: ACM Sigmod Record, vol. 26. no. 2. ACM (1997)
14. Meirelles, F.: Fundação Getulio Vargas-Escola De Administração De Empresas De São Paulo. Pesquisa Anual De Tecnologia De Informação 13 (2000)
15. Taurion, C.: A Busca da Qualidade para Alavancar os Negócios. Developers' Mag. 10–11 (1997)
16. Garani, G., Helmer, S.: Integrating star and snowflake schemas in data warehouses. Int. J. Data Warehous. Min. (IJDWM) **8**(4), 22–40 (2012)
17. Levene, M., Loizou, G.: Why is the snowflake schema a good data warehouse design? Inf. Syst. **28**(3), 225–240 (2003)
18. Abbasi, M., et al.: SINGLE vs. MapReduce vs. Relational: Predicting Query Execution Time. In: International Conference: Beyond Databases, Architectures and Structures. Springer, Cham (2018)
19. Marquesone, R.: Big Data: Técnicas e tecnologias para extração de valor dos dados. Editora Casa do Código, São Paulo (2016)
20. Anuradha, G., Roy, B.: Suggested techniques for clustering and mining of data streams. In: International Conference on Circuits, Systems, Communication and Information Technology Applications (CSCITA). IEEE (2014)
21. Yin, S., Kaynak, O.: Big data for modern industry: challenges and trends (point of view). Proc. IEEE **103**(2), 143–146 (2015)
22. Grover, V., et al.: Creating strategic business value from big data analytics: a research framework. J. Manag. Inf. Syst. **35**(2), 388–423 (2018)
23. Ji-fan Ren, S., et al.: Modelling quality dynamics, business value and firm performance in a big data analytics environment. Int. J. Prod. Res. **55**(17), 5011–5026 (2017)
24. Sivarajah, U., et al.: Critical analysis of Big Data challenges and analytical methods. J. Bus. Res. **70**, 263–286 (2017)

25. Chaturvedi, A., Lone, F.A.: Analysis of Big Data security schemes for detection and prevention from intruder attacks in cloud computing. Int. J. Comput. Appl. **158**(5) (2017)
26. Yi, X., et al.: Building a network highway for big data: architecture and challenges. IEEE Netw. **28**(4), 5–13 (2014)
27. Yeoh, W., Koronios, A.: Critical success factors for business intelligence systems. J. Comput. Inf. Syst. **50**(3), 23–32 (2010)
28. Bimonte, S., et al.: Design and implementation of active stream data warehouses. Int. J. Data Warehous. Min. (IJDWM) **15**(2), 1–21 (2019)
29. Martins, P., Abbasi, M., Furtado, P.: A scale: big/small data ETL and real-time data freshness. In: Kozielski, S., Mrozek, D., Kasprowski, P., Małysiak-Mrozek, B., Kostrzewa, D. (eds.) Beyond Databases. Architectures and Structures. Advanced Technologies for Data Mining and Knowledge Discovery, pp. 315–327. Springer, Cham (2015)
30. Naidoo, S.S.: Business intelligence systems input: effects on organizational decision-making. Diss. Capella University (2019)
31. Lennerholt, C., van Laere, J., Söderström, E.: Implementation challenges of self-service business intelligence: a literature review. In: 51st Hawaii International Conference on System Sciences, 3–6 January 2018, Hilton Waikoloa Village, Hawaii, USA, Vol. 51. IEEE Computer Society (2018)
32. Volitich, D.: IBM Cognos 8 BI: The Official Guide. McGraw-Hill, New York (2008)
33. Lachev, T., Price, E.: Applied Microsoft Power BI: Bring Your Data to Life!. Prologika Press (2018)
34. Gligor, G., Teodoru, S.: Oracle exalytics: engineered for speed-of-thought analytics. Database Syst. J. **2**(4), 3–8 (2011)
35. Färber, F., et al.: SAP HANA database: data management for modern business applications. ACM Sigmod Rec. **40**(4), 45–51 (2012)
36. Halper, F., Stodder, D.: TDWI analytics maturity model guide. TDWI research, pp. 1–20 (2014)
37. Ilacqua, C., Cronstrom, H., Richardson, J.: Learning Qlik Sense®: The Official Guide. Packt Publishing Ltd, Birmingham (2015)
38. Murray, D.G.: Tableau Your Data! Fast and Easy Visual Analysis with Tableau Software. Wiley, New York (2013)
39. Choo, A., Saeger, T.: Data analysis for yield improvement using TIBCO's spotfire data analysis software. In: CS Mantech Conference (2011)
40. Ram, J., Zhang, C., Koronios, A.: The implications of Big Data analytics on business intelligence: a qualitative study in China. Procedia Comput. Sci. **87**, 221–226 (2016)
41. ITGP Privacy Team. EU General Data Protection Regulation (GDPR): An Implementation and Compliance Guide - Second edition. IT Governance Ltd (2017)
42. Van Alsenoy, B.: Regulating data protection: The allocation of responsibility and risk among actors involved in personal data processing (2016)

Multi-model Environment Generation and Tailoring Model for Software Process Improvement

Gloria Piedad Gasca-Hurtado[1], Jesús Andrés Hincapié Londoño[1(✉)], and Mirna Muñoz[2]

[1] Universidad de Medellín, Carrera 87 N° 30-65, Medellín, Colombia
{gpgasca,jehincapie}@udem.edu.co
[2] CIMAT – Unidad Zacatecas, Av. Universidad No. 222, 98068 Zacatecas, Mexico
mirna.munoz@cimat.mx

Abstract. Software development organizations take great risks when are faced with process improvement projects and the inclusion of best practices. Such risks include the selection of a standard, framework or model for the improvement process. Some organizations decide to integrate best practices from different sources to be non-dependent on a particular model, framework or standard. However, the integration of different models is an additional challenge since the complexity of the implementation increases as best practices come from different natures. Hence, we propose a model to generate and customize a multi-model environment for software process improvement. We intend to reduce the complexity of combining agile and traditional frameworks by generating a frameworks integration catalog, which, from heuristics, graphically represents the possible paths to follow for the process improvement project. We use design-based science for this research because of its focus on problem analysis in real-world environments. In this way, we can consolidate the model to customize and generate a multi-model environment for the implementation of best practices of different nature during an improvement process. We also define a proposal of organizational profiles, which will allow us to validate the model in a business case in future work.

Keywords: Software process improvement · Multi-model environment · Software engineering

1 Introduction

The implementation of process improvement proposals and best practices in software development companies has a high and variable complexity that depends on the type of standard, framework or model used. The integration of best practices from different sources is used to achieve concrete improvement that is independent of the model, framework or standard. However, the complexity of the implementation increases when the sources of such best practices have a different nature [1].

Á. Rocha et al. (Eds.): WorldCIST 2020, AISC 1159, pp. 620–630, 2020.
https://doi.org/10.1007/978-3-030-45688-7_62

Therefore, we intend to define a methodology to generate multi-model catalogs of best practices for the improvement of software processes that include agile and traditional practices. This proposal seeks to reduce the complexity of combining agile and traditional frameworks in a multi-model environment [2–4], through the design of a frameworks integration catalog, including a strategy for the design of the graphic representation of the catalog with the use of heuristics.

We select Design-Based Science as the research methodology for this work, because of its approach to the analysis of unsolved problems in a real-world environment. By implementing the selected methodology, it is possible to consolidate all the components that make up the model to generate and tailor a multi-model catalog for software process improvement. Such components include the proposal of organizational profiles that facilitate the multi-model environment personalization.

In the Configuration Engine Component, we present the structural model of the software tool to graphically represent the environment. Such a tool is under development and it will allow the use of technology to validate the model in a later stage. The rest of the model components are described in this work.

This paper is structured as follows: Sect. 2 presents a background of the topics related to this work. Sect. 3 shows how the research methodology is applied to this work. Section 4 explains the model for defining a multi-model catalog and its four components. Finally, Sect. 5 presents conclusions and future work.

2 Background

2.1 Software Process Improvement

The main goals of software process improvement (SPI) are to develop better-quality software, increase customer satisfaction, and to increase returns on investment. SPI is a strong approach used by software organizations to improve their competitive market position [5]. SPI was developed to manage and improve the quality of software development [6]. Through SPI it is possible to resolve issues related to ad-hoc software processes because SPI aims to obtain optimal solutions considering the planning, development, and production cycles, as well as to resolve organizational issues.

SPI can provide benefits for software development organizations. SPI is adequate when organizations need to improve product quality, shorten the time-to-market, increase productivity, reduce costs, and more. To realize these benefits, the effective implementation of SPI requires time, careful scheduling, resources and knowledge management [5, 7, 8].

In conclusion, SPI is the sequence of activities that involves checking the current development process, then planning to change it until achieving a specific goal, sometimes repeatedly and constantly [9]. This sequence is related to an awareness of self-reflection and pursues a better process and result. Specifically, in the field of software engineering, the SPI is an important area, taking into account that the software process refers to a collection of activities, actions, and tasks performed while the product is being built [10].

2.2 Agile and Traditional Methodologies

Agile Software development is just one of the methodologies used in software development. Agile is a word used to describe a process model concept that is different from the existing process model concepts [11]. Agile software development concepts were coined by Kent Beck and her team by stating that Agile Software Development is a way to build software by doing it and helping others to build it [12]. However, just like other process models, Agile Software Development has its advantages and is not suitable for all types of projects, products, people and situations. Agile Software Development enables a process model that is tolerant of changes in the requirements so the response to the changes can be done faster [13].

For its part, the traditional software development process is considered as a set of activities, methods, practices, and transformations that people employ to develop and maintain software and associated products (for example, project plans, project documents, design, code, test cases, user manual).

Authors like Pressman [9] consider the software process as the set of partially ordered activities that a project must follow to perform some task. This task should aim to achieve a goal and is associated with the development of one or more products [14].

In general, when referring to traditional methodologies, it is essential to relate to the System Development Life Cycle (SDLC). A definition associated with SDLC is one that considers a whole process in building a system through several steps [15]. There are several models of the SDLC, the model which is quite popular and widely used is the waterfall. Some other models of SDLC, for example, are fountain, spiral, rapid, prototyping, incremental, build and fix, and synchronize and stabilize. With the SDLC cycle, the process of building the system is divided into several steps and on large systems, each step is done by different teams [16].

3 Design Science in Information System Research

Information Systems (IS) Research lies at the intersection of people, organizations, and technology [17]. It relies on and contributes to cognitive science, organizational theory, management sciences, and computer science. It is both an organizational and a technical discipline that is concerned with the analysis, construction, deployment, use, evaluation, evolution, and management of information system artifacts in organizational settings [18, 19].

Within this setting, the design science research paradigm is proactive concerning technology. It focuses on creating and evaluating innovative IT artifacts that enable organizations to address important information-related tasks.

Nowadays, IS research must address the interplay among business strategy, IT strategy, organizational infrastructure, and IS infrastructure. This interplay is typical in this research project; thus, we select the Framework for IS Research as Fig. 1 shows.

4 Model to Generate a Multi-model Catalog

Considering the difficulties that organizations experience when trying to harmonize different frameworks, models, and standards of SPI best practices, we propose a model to create multi-model environments, generating a catalog of best practices.

This proposal intends to address the following difficulties:

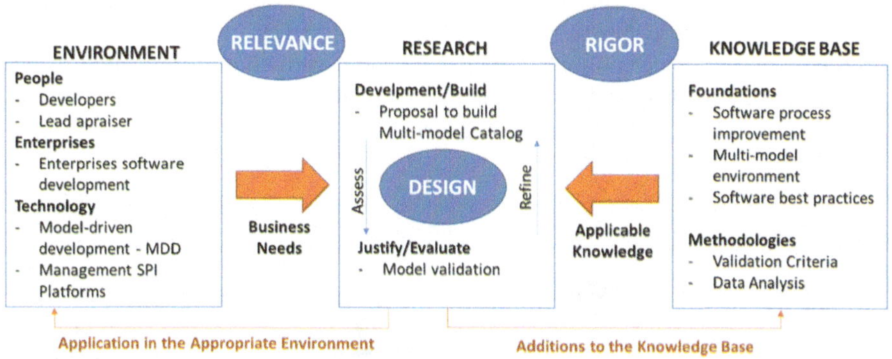

Fig. 1. Framework for IS research

- the need to classify good practices and their associated activities [2];
- the lack of criteria for the selection of tasks that generate value to the project and the exclusion of those that do not generate value [3]; and
- the need to divide activities and correlate them between different models or frameworks that constitute a multi-model environment [4].

These difficulties mean that the implementation of a multi-model environment has a complexity inherent in integrating best practices from different sources such as CMMI-DEV, ISO 29110 and other standards; or of different natures such as SCRUM.

In this sense, the problem lies in the complexity of selecting and relating key elements of each of these sources, to generate a best practice model adapted to the needs of the company. This makes organizations give up in their attempt to solve the problem of implementing frameworks, standards or process improvement models under a multi-model approach, which results in increased costs, low quality of the products developed, teams' demotivation and processes inefficiency in general.

Studies have been developed where integration between agile and traditional frameworks [20, 21] and some applications to the industry [4, 22] are proposed. Abstract solutions models have also been found to achieve integration [2, 3, 23] and sets of best practices from the union of both agile and traditional frameworks [24–26]. However, these proposals focus on specific models and do not consider a generic strategy to integrate any model. Furthermore, they do not include a tool to support the multi-model environment definition.

A proposal that includes the integration between frameworks is the multi-model environment, which is an approach that allows harmonizing process improvement models [1].

However, the definition of a multi-model environment requires a detailed analysis of the best practices of each model, which leads practitioners to face the complexity of harmonizing them. This analysis allows identifying similarities and eliminating redundancies to achieve greater benefits of harmonization [1].

Therefore, this paper presents a model for the construction of a catalog of best practices from a multi-model environment. This model is designed from components shown in Fig. 2. Next, each of these components is described.

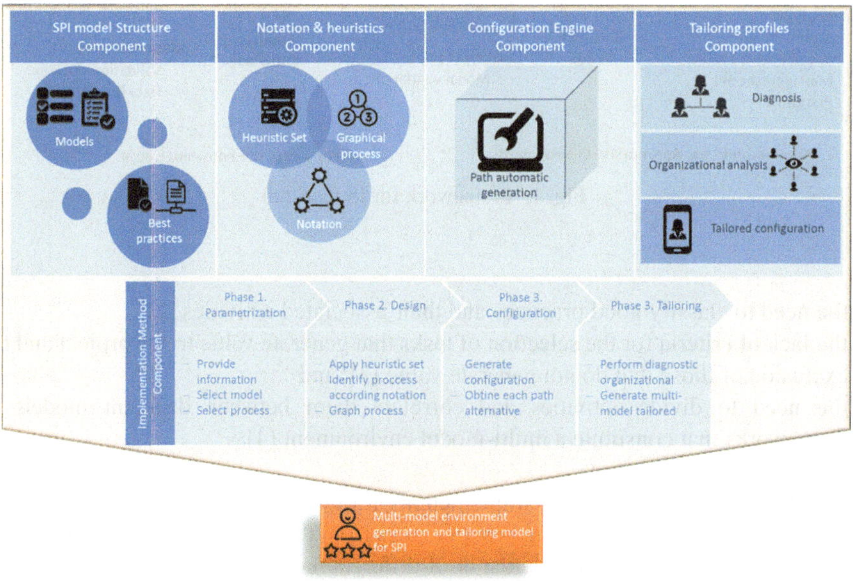

Fig. 2. Model to build multi-model catalog components

4.1 SPI Model Structure Component

This component represents the conceptual model of the issues related to our problem. It describes the different entities, their attributes, roles, and relationships, as well as the restrictions that govern the problem domain. The domain model for the SPI Model Structure Component is the one shown in Fig. 3, which is required for the automatic generation of the multi-model environment.

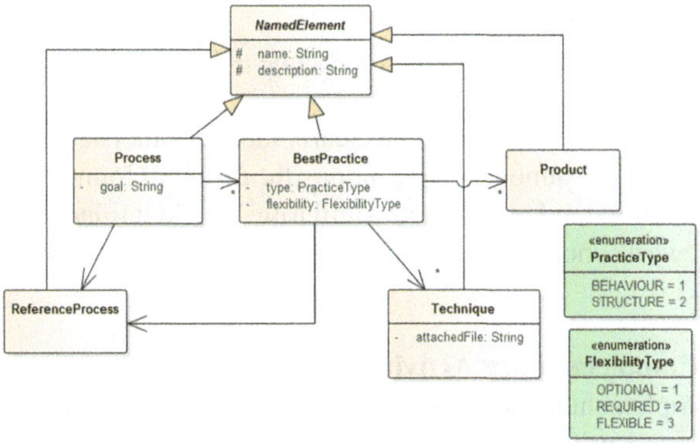

Fig. 3. Domain model of the multi-model environment software tool

4.2 Notation and Heuristics Component

Set of heuristics conceived from four dimensions that allow defining the catalog of best practices. To represent these heuristics, it was necessary to define a proper notation. This notation is equivalent to a system of conventional signs that is adopted to express the use of the dimensions established in the set of heuristics.

The set of heuristics were organized in a process representation structure in terms of four dimensions. This structure is described in the Multi-model Catalog for Software Projects Management including Agile and Traditional Practices [27]. Next, each of the dimensions of the catalog is summarized.

Dimension 1 – Activity Type

This dimension allows cataloging activities in structural and behavioral and is defined in terms of the following rules:

1. Define the level of detail for the comparison of practices between frameworks.
2. An activity is structural when it generates an artifact after its execution.
3. An activity is behavioral when it indicates recommendations to solve a specific situation but does not imply the generation of an artifact.

Dimension 2 – Activity Flexibility

It allows determining if an activity is optional, required or flexible. For its implementation, the following heuristics are defined:

1. An activity is optional when, in the opinion of an expert, according to the experience in developed projects or using a metric of the size of the project, it does not generate value for its development.
2. An activity is required when, according to an expert, to the experience in projects developed or using a metric of the size of the project, it is key to the project success.
3. An activity is flexible when, according to an expert, to the experience in projects developed or using a metric of the size of the project, it can be adapted to the needs or conditions of the project without affecting its success.

Dimension 3 – Redundant Activities
It allows identifying activities focused on the same objective.

1. An activity is redundant when you can clearly identify other activities that are geared towards achieving the same objective among the different frameworks being studied.
2. An activity is not redundant when no similarities are identified with other activities among frameworks, nor are similar objectives that direct such activity identified.

Dimension 4 – Complementary Activities
An activity is complementary when a dependency between the different activities of the same framework is identified.

4.3 Configuration Engine Component

In this component, the configurations of multi-model environments are automatically generated. This component is associated with the implementation of the heuristics defined from the dimensions described in the Notation and Heuristic Component. For its implementation, it is necessary to develop a regulated technological tool so that the different configurations are automatically generated. An example of the implementation of heuristics is presented in [28].

4.4 Tailoring Profile Component

It facilitates the customization of automatically generated configurations, based on an organizational diagnosis. This diagnosis is defined under the analysis of the maturity of the organization and its ability to start a software process improvement initiative.

According to Petersen and Wohlin [29], the software development process should be in harmony with the actual environment in which the software is developed and delivered. This means that to have success in the establishment of a multimodel environment, it is necessary to achieve the characterization of the organizational context as complete and accurate as possible.

To achieve the characterization of the actual organizational environment, this proposal selected a set of conceptual aspects proposed by Petersen and Wohlin and by Clark and O'Connor [30]. The set of the conceptual aspects are listed below:

1. *Product*: the actual type of products that the organization develops. Besides, the time in which the organization delivers its products. The aspects covered in this element are the type of systems and project time.
2. *Process*: identification of the actual practices that the organization performs. The aspects covered in this element are tasks, work products and the methodology used to develop the software.
3. *Human resources*: experience and training of the actual organization's employees. The aspects covered in this element are experience, roles, and staff turnover.
4. *Organization*: the actual structure of the organization. This element covers the certification obtained, the organization distribution, and the number of employees.

5. *Knowledge*: communication channels of the organization and how the knowledge is shared and disseminated. The aspects covered in this element are the way to disseminate organizational assets, the communication channels established within a project and the way to share project assets.

Having characterized the organizational environment, it will be possible to establish a multi-model environment tailored to the organizational culture and to provide the best practices that could be implemented in an easy way within the organization.

4.5 Implementation Method Component

Next, each of the phases of the method that were listed in Fig. 2 is detailed.

Phase 1. Parametrization
The objective of this phase is to declare the values that characterize the models. In this phase, the basic parameters of interaction with the different models are defined based on the structure and its rules. Activities defined for this phase are as follows:

1. Provide information for the structure of the models.
2. Select software process improvement (SPI) models: the organization must define a reference model that will facilitate the comparison of best practices. This reference model will be the guide on which the application of the different heuristics is based.
3. Select the improvement process to be structured: this activity is related to the definition of the organization's improvement interest. It is the starting point for the execution of the process improvement initiative.

Phase 2. Design
The objective of this phase is to obtain the graphic representation of the different alternatives that an organization can achieve to guide its improvement initiative. It has a direct dependence on the parameterization values established in the parameterization phase. The activities associated with this phase are:

1. Apply each of the heuristics defined in the Notation and Heuristic Component.
2. Identify the best practices that comprise the process selected in the parameterization phase according to the defined notation.
3. Generate the graphical representation of the multi-model environment using the graphical representation tool.

Phase 3. Configuration
The objective of this phase is to automatically generate all possible configurations of the multi-model environment. Each configuration is defined based on a possible path to follow, resulting from the application of the different heuristics described in the Notation

and Heuristic Component, in order to implement a process improvement initiative in an organization independent of a model, standard or framework.

Phase 4. Tailoring
The objective of this phase is to adapt the generated multi-model environment to the organization profile. This profile is defined from a diagnosis where the organization's maturity and its ability to initiate an SPI initiative are analyzed. The interests of the organization are considered, even from the parameterization phase, when the reference model and the process are selected. However, it is at this stage that the customization process is completed with activities such as:

1. Perform company diagnosis: select questions associated with the organization and answer a survey. The results of this survey are analyzed to define the profile of the organization according to the Tailoring Profile Component.
2. Generate recommendations according to the organization's diagnosis and deliver the multi-model environment and custom configurations.

5 Conclusions

In this research, we proposed a model to create multi-model environments to cope with the difficulties of harmonizing different frameworks, models and standards of SPI. This model comprises four components: SPI Model Structure, Notation and Heuristics, Configuration Engine, Tailoring Profile, and Implementation Method.

The main contribution of this paper is the model structure to generate a multi-model catalog, especially the SPI model Structure, Tailoring profiles, and Implementation Method components since the other two were already designed in previous works. These components and the model in general still lack proper validation, which will be carried out as future work once the software tool is completely developed and we can put the model to the test in a case study in a software development organization.

References

1. Marino, L., Morley, J.: Process improvement in a multi-model environment builds resilient organizations. In: NEWS SEI, Software Engineering Institute (2009)
2. Salinas C.J.T., Escalona, M.J., Mejías, M.: A scrum-based approach to CMMI maturity level 2 in web development environments. In: Proceedings of the 14th International Conference on Information Integration and Web-based Applications & Services - IIWAS 2012, p. 282. ACM Press, New York (2012)
3. Monteiro, P., Borges, P., Machado, R.J., Ribeiro, P.: A reduced set of RUP roles to small software development teams. In: International Conference on Software and System Process, ICSSP 2012 – Proceedings, pp. 190–199 (2012)
4. Tuan, N., Thang, H.: Combining maturity with agility: lessons learnt from a case study. In: Proceedings of the Fourth Symposium on Information and Communication Technology, pp. 267–274 (2013)

5. Niazi, M., Babar, M.A., Verner, J.M.: Software process improvement barriers: a cross-cultural comparison. Inf. Softw. Technol. **52**, 1204–1216 (2010)
6. Samalikova, J., Kusters, R.J., Trienekens, J.J.M., Weijters, A.J.M.M.: Process mining support for capability maturity model integration-based software process assessment, in principle and in practice. J. Softw. Evol. Process **26**, 714–728 (2014). https://doi.org/10.1002/smr.1645
7. Lars, M.: Managing knowledge in a software organization. J. Knowl. Manag. **7**, 63–80 (2003). https://doi.org/10.1108/13673270310477298
8. Meehan, B., Richardson, I.: Identification of software process knowledge management. Softw. Process Improv. Pract. **7**, 47–55 (2002). https://doi.org/10.1002/spip.154
9. Pressman, R.S.: Software Engineering: A Practitioner's Approach. Palgrave Macmillan, London (2005)
10. Conradi, H., Fuggetta, A.: Improving software process improvement. IEEE Softw. **19**, 92–99 (2002). https://doi.org/10.1109/MS.2002.1020295
11. Martin, R.C.: Agile Software Development: Principles, Patterns, and Practices. Prentice Hall PTR, Upper Saddle River (2003)
12. Dingsøyr, T., Dybå, T., Moe, N.B.: Agile Software Development Current Research and Future Directions. Springer, Heidelberg (2014)
13. Dhir, S., Kumar, D., Singh, V.B.: Success and failure factors that impact on project implementation using agile software development methodology. In: Hoda, M., Chauhan, N., Quadri, S., Srivastava, P. (eds.) Software Engineering, pp. 647–654. Springer, Singapore (2019)
14. Adi, P.: Scrum method implementation in a software development project management. Int. J. Adv. Comput. Sci. Appl. **6**, 198–204 (2015). https://doi.org/10.14569/IJACSA.2015.060927
15. Langer, A.M.: Guide to Software Development: Designing and Managing the Life Cycle. Springer, London (2012)
16. Gonçalves, E.F., Drumond, G.M., Méxas, M.P.: Evaluation of PMBOK and scrum practices for software development in the vision of specialists. Indep. J. Manag. Prod. **8**, 569–582 (2017). https://doi.org/10.14807/ijmp.v8i5.598
17. Silver, M.S., Markus, M.L., Beath, C.M.: The information technology interaction model: a foundation for the MBA core course. MIS Q. **19**, 361 (1995). https://doi.org/10.2307/249600
18. Madnick, S.E.: The challenge–to be part of the solution instead of being the problem (1993)
19. Orlikowski, W.J., Barley, S.R.: Technology and institutions: what can research on information technology and research on organizations learn from each other? MIS Q. **25**, 145 (2001). https://doi.org/10.2307/3250927
20. Špundak, M.: Mixed agile/traditional project management methodology – reality or illusion? Procedia – Soc. Behav. Sci. **119**, 939–948 (2014). https://doi.org/10.1016/J.SBSPRO.2014.03.105
21. Hornstein, H.A.: The integration of project management and organizational change management is now a necessity. Int. J. Proj. Manag. **33**, 291–298 (2015). https://doi.org/10.1016/J.IJPROMAN.2014.08.005
22. Pinheiro, P.R., Machado, T.C.S., Tamanini, I.: Dealing the selection of project management through hybrid model of verbal decision analysis. Procedia Comput. Sci. **17**, 332–339 (2013). https://doi.org/10.1016/j.procs.2013.05.043
23. Buglione, L.: Light maturity models (LMM). In: Proceedings of the 12th International Conference on Product Focused Software Development and Process Improvement - Profes 2011, p. 57. ACM Press, New York (2011)
24. Brown, A.W., Ambler, S., Royce, W.: Agility at scale: economic governance, measured improvement, and disciplined delivery. In: 35th International Conference on Software Engineering (ICSE), pp. 873–881. IEEE (2013)
25. Ng, P.-W.: Theory based software engineering with the SEMAT kernel: preliminary investigation and experiences. In: Proceedings of the 3rd SEMAT Workshop on General Theories of Software Engineering - GTSE 2014, pp. 13–20. ACM Press, New York (2014)

26. Van Hilst, M., Fernandez, E.B.: A pattern system of underlying theories for process improvement. In: Proceedings of the 17th Conference on Pattern Languages of Programs - PLOP 2010, pp. 1–24. ACM Press, New York (2010)
27. Bustamante, A.F., Hincapié, J.A., Gasca-Hurtado, G.P.: Structure of a multi-model catalog for software projects management including agile and traditional practices. In: Mejia, J., Munoz, M., Rocha, Á., Calvo-Manzano, J. (eds.) Trends and Applications in Software Engineering, pp. 87–97. Springer, Cham (2016)
28. Hincapie, J.A., Gasca-Hurtado, G.P., Bustamante, A.F.: Multimodel catalogue heuristics for software project managemet. In: 11th Iberian Conference on Information Systems and Technologies (CISTI), pp. 1–6 (2016)
29. Petersen, K., Wohlin, C.: Context in industrial software engineering research. In: 3rd International Symposium on Empirical Software Engineering and Measurement, pp. 401–404. IEEE (2009)
30. Clarke, P., O'Connor, R.V.: The situational factors that affect the software development process: Towards a comprehensive reference framework. Inf. Softw. Technol. **54**, 433–447 (2012). https://doi.org/10.1016/J.INFSOF.2011.12.003

Using IoT and Blockchain for Healthcare Enhancement

Mohamed A. El-dosuky[1] (✉) and Gamal H. Eladl[2]

[1] Computer Sciences Department, Faculty of Computer and Information, Mansoura University, Mansoura, Egypt
mouh_sal_010@mans.edu.eg
[2] Information Systems Department, Faculty of Computer and Information, Mansoura University, Mansoura, Egypt
gamalhelmy@mans.edu.eg

Abstract. Modern technologies such as Internet of Things (IoT) and blockchain have a valuable contribution in improving healthcare services. This paper aims at achieving and democratizing healthcare services by providing healthcare-as-a-service. It was achieved by developing medical devices with sensors for healthcare. It connects the medical devices such as temperature sensor to the network of medical physicians and nurses through the cloud. This paper presented integration between IoT and Blockchain as a secured platform to mitigate nurses' shortage. Blockchain was used in the proposed operational framework to store and validate patients' records. Remarkable results were found in decreasing the gap in the number of nurses for large scale of patients. Prototypical implementation of this proposed healthcare service was presented with all technical requirements to make it applicable easily.

Keywords: Blockchain · Cloud · E-Healthcare · IoT

1 Introduction

Healthcare is an important field that requires a continuous enhancement. One of its components is nursing human resource. Skilled nurses are considered as a primary pillar of healthcare success. Healthcare faces daily issues. Nursing shortage is a crucial problem in the globe [1]. This issue has been acknowledged by the World Health Organization and other health centers and organizations in all over the world. The shortage is caused by an increased demand for nurses, while fewer people are choosing nursing as a job.

As technology advances, so does its influence in healthcare too. Thus, technology may provide a solution to the problem of nursing shortage through new trends such as cloud computing, internet of things, and the emergence of blockchain technology that could be a technical solution to such common problems. Moreover, blockchain technology can support transparency and accountability of stored data.

Therefore, this paper aims at democratizing healthcare service by providing healthcare as a service, which can be achieved by developing medical devices with sensors

Á. Rocha et al. (Eds.): WorldCIST 2020, AISC 1159, pp. 631–639, 2020.
https://doi.org/10.1007/978-3-030-45688-7_63

for healthcare and connecting them to the network of medical physicians and nurses through a cloud is utilized. Blockchain is used to store patients' records.

The following sections present a synthetic literature review on the use of blockchain in healthcare, the proposed methodology, and conclusion and future work.

2 Previous Work

Blockchain can be considered as a distributed and secured database in which each block has a hash to the previous block. An enormous body of literature embraces the use of blockchain in healthcare.

In 2014, bitcoin was cleverly proposed in medical research [2]. In 2015, blockchain popularity was increased to be a new economic model [3] and decentralizing privacy [4].

In 2016, blockchain evolved [5], Electronic Patient Record systems (EPRs) [6], to empower the patient-physician relationship [7]. A notable application is Medrec that utilizes blockchain for managing permission in medical domain [8]. Blockchain can be also useful for Interoperability [9]. Blockchain can provide protocols for medical trustworthiness [10] and for transparency [11] as well.

In 2017, blockchain has been evolved rapidly [12]. It has contributed for health care applications [13–15]. That contribution was proven to be very vital for health care [16], thus empowering e-health [17]. Many previous works address challenges and opportunities of blockchain in e-healthcare such as [18–20].

Metrics for blockchain-based healthcare decentralization are [21]:

M1. Compliance of workflow with Health Insurance Portability and Accountability Act (HIPAA)
M2. Scalability through huge populations of participants
M3. Cost-effectiveness
M4. Supporting for Turing-complete operations
M5. Supporting for user identification
M6. Supporting for structural interoperability
M7. Supporting of patient-centered healthcare

In 2018, blockchain gained its fame as a remedy for security and privacy of e-healthcare [22]. Many systems were proposed such as Blochie [23], FHIRchain [24] and Mistore [25].

We notice that most researchers are rushing to build blockchain-based health systems without integrating IoT. Others worked with a different perspective using IoT without using blockchain. Therefore, this paper is proposed to handle this issue.

This research paper integrates these two technologies to improve healthcare in general, and the process of equipping patients in particular.

3 Proposed Methodology

In this section, we present the stakeholders of proposed system. And move towards a mathematical model with operational framework. Furthermore, a brief discussion of how to construct temperature sensing circuit is presented.

The stakeholders are the patients with sensors, their relatives, nurses, physicians and ambulances. Figure 1 illustrates the stakeholders of the proposed system.

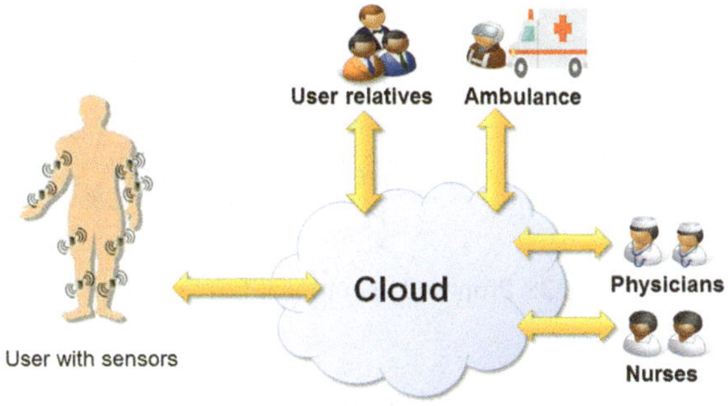

Fig. 1. Stakeholders of the proposed system

The proposed system proposes that the patient is equipped with sensors. Physicians and nurses are linked to monitors to track the status of the patient. In emergency cases, the system sends a message to the ambulance and/or patient's relatives.

To make is easy for pondering the communications among sets of stockholders, Table 1 shows cardinality of stakeholders.

Table 1. Nomenclature

Stakeholder set	Cardinality of:
B	Beds
D	Physicians/Doctors
N	Nurses
P	Patients

Assume that there are P patients, N nurses, and B beds in hospital. There are many scalability dimensions for example from beds to patients (BP Scalability, the B÷P ratio), and from patients to nurse (PN Scalability, the P÷N ratio).

BP Scalability does not concern us, as it defines the physical capacity of hospitals. PN Scalability, however, can scale up and be enhanced with the technology presented in this paper.

Figure 2 shows a proposed framework, inspired by [26]. At the beginning, CARE can be redefined to stand for medical Countermeasure (C), based on Analysis (A) and Repository (R) of Events (E). Nurses are tracking the status of enrolled patients. All events are stored in the blockchain repository. Events undergo an analysis process which is recursive by nature. This means that most operational time is attributed to the analysis phase. Analysis log data are stored in the blockchain repository too. Countermeasure corresponds to the action to be taken.

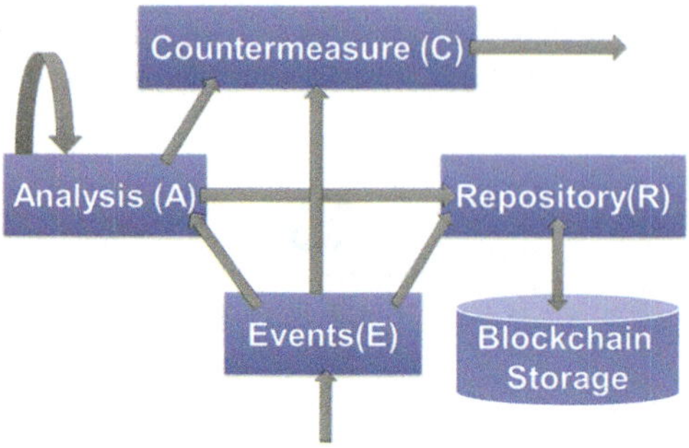

Fig. 2. Proposed operational framework

Figure 3 shows temperature sensing circuitry. The components are ARDUINO Uno board, breadboard, temperature sensor, Wi-Fi, wires and resistors, Light Emitting Diode (LED), and Micro-USB cable. To build the circuit, there is a good tutorial [27].

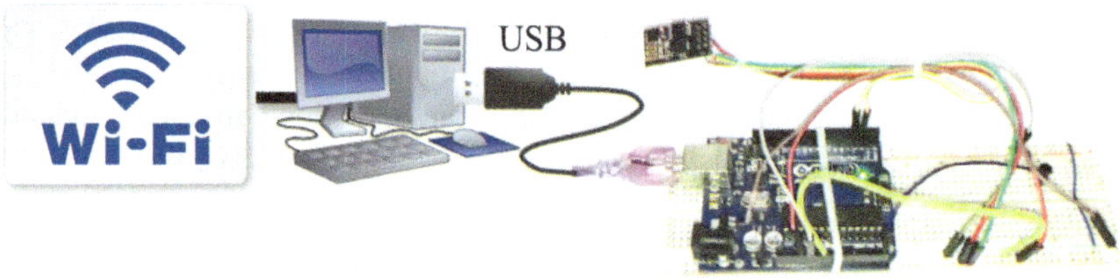

Fig. 3. Temperature sensing circuit

The IoT circuit for temperature measurement is connected to internet. A channel in thingspeak.com is created and linked to the circuit to get readings from the ARDUINO device. ARDUINO device is more available and durable in the market rather other devices according to regional market.

Figure 4 shows the architecture of the proposed system. The first run of blockchain will be mining the Genesis block. Then the proposed system displays the hash of the Genesis block.

After that, the system interactively asks about some data and encapsulates the answers in the blockchain. These data are:

a) Patient's Full Name
b) Patient's Address
c) Hospital
d) Disease
e) Medication given
f) Final Bill Amount

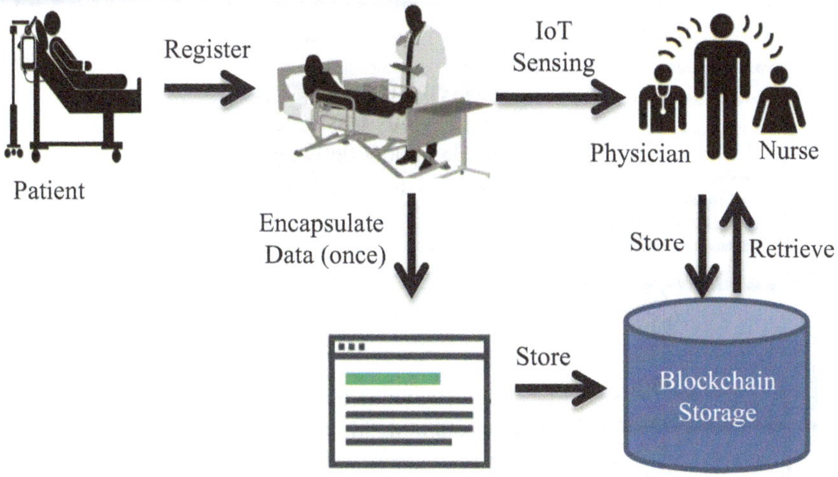

Fig. 4. Architecture of the proposed system

4 Experiments and Results

The implementation of the private blockchain is achieved in Java as listed in the appendix. Figure 5 shows the blockchain structure.

The first run of blockchain will be mining the Genesis block. Then the proposed system displays the hash of the Genesis block, which happens to be:

00000d99d5b74ae4083b0dc72eb037911c7e4fc3b57cdf96

It has 5 leading zeros (nonce).

After that, the system interactively asks about some data and stores the answers/patient records in the blockchain.

For the first patient, a block is mined, with hash:

000007e79730cbf13d1d35cf7138b6248b2df3e6093583e6

It has 5 leading zeros. For the second patient, a block is mined, with hash:

00000b632ec77b8a4eae6c78b08932624e84ab3814e13dc9

It has 5 leading zeros. After each data entry, the blockchain can be validated.

Table 2 shows computational time in nanoseconds for creating blocks on Acer Extensa laptop, running 32-bit Windows 8.

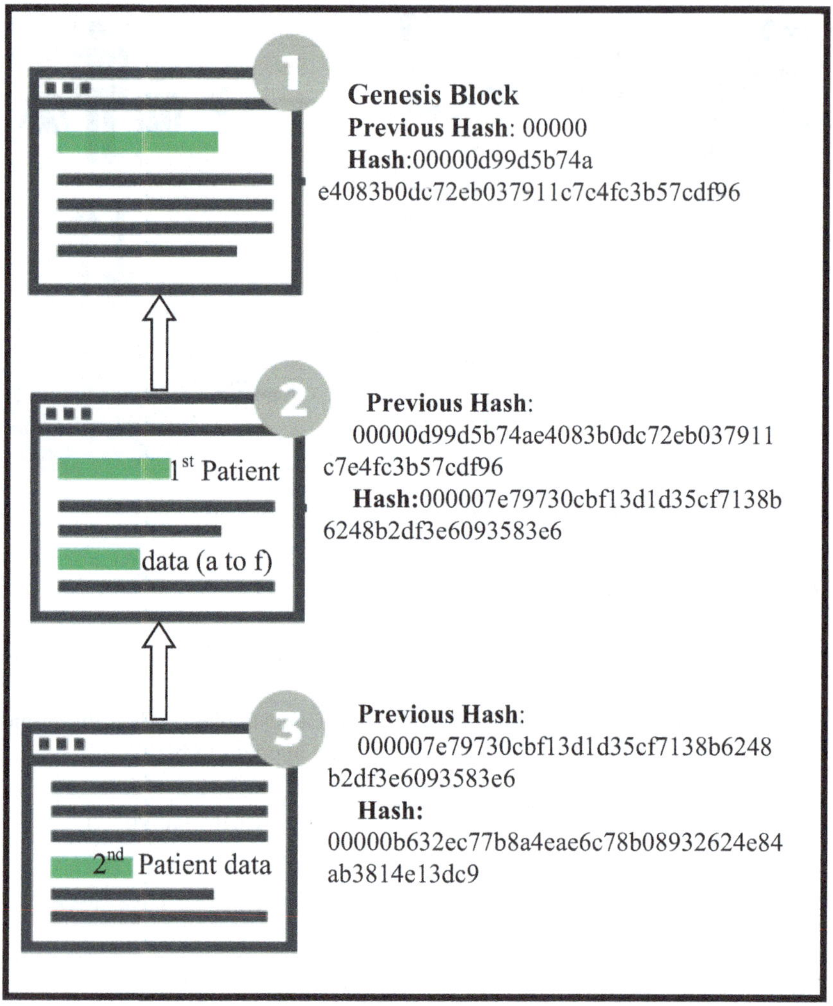

Fig. 5. Blockchain structure

Table 2. Creation time of blocks

Block	Time in seconds
Genesis block	0.747965977
1st patient block	5.210931889
2nd patient block	4.2750578509
3rd patient block	12.061223465
4th patient block	8.542168901

For testing a sensor-based data for randomly generated 100 patients, the results show that the average of creating blocks is about 6 s per patient. This is compared to manually temperature measurement time which is 24 s per patient. With large scale of patients, this accumulative difference will make a noticeable significant. This proposed system does not validate its scalability by minimizing the elapsed time for measuring patient temperature but also handling the nurse shortage noticeable.

5 Conclusion and Future Works

This paper proposed healthcare-as-a-service using IoT and blockchain. A temperature sensing circuit is designed and linked to thingspeak.com by using wi-fi. Nurses are tracking the sensing signal on monitors. The paper also presented how the blockchain is used to store the health records. The proposed system is enhanced by time reduction and minimizing nurse human resource. It is very easy to use integration of other medical data such as pressure and, vital signs in general.

A possible future direction may integrate big data and mobile based application to monitor the patient's status with the IoT equipment to come.

Other possible directions are to secure the blockchain [28] and to investigate the integration between the IoT and the blockchain [29].

Appendix A. Blockchain Implementation in Java

```
import java.util.Date;
import java.time.format.DateTimeFormatter;
import java.time.LocalDateTime;
import java.security.MessageDigest;
public class Block
{
    public String hash;
    private String data;
    private String timeStamp;
    public String previousHash;
    private int nonce;

    public Block(String data,String previousHash ){
        this.data = data;
        this.timeStamp = CarePlus.time;
        this.previousHash = previousHash;
        this.hash = calHash();
    }
```

```
    public String calHash() {
        return Sha256(previousHash +timeStamp +
            Integer.toString(nonce) +data);
    }
    public void mineBlock(int difficulty) {
          String difstr =getDificultyString(difficulty);
    while(!hash.substring( 0,
              difficulty).equals(difstr)) {
            nonce ++;
            hash = calHash();
        }
        System.out.println("Block Mined Successfully!  Hash: " + hash);
    }

    public static String getDificultyString(int difficult){
        return new String(new char[difficult]).replace('\0', '0');
    }

    public static String Sha256 (String data){
        // return java.security.MessageDigest of the data
    }
}
```

References

1. Buchan, J., Aiken, L.: Solving nursing shortages: a common priority. J. Clin. Nurs. **17**(24), 3262–3268 (2008)
2. Carlisle, B.G.: Proof of prespecified endpoints in medical research with the bitcoin blockchain. The Grey Literature 25 (2014)
3. Swan, M.: Blockchain: Blueprint for a New Economy. O'Reilly Media Inc, Sebastopol (2015)
4. Zyskind, G., et al.: Decentralizing privacy: using blockchain to protect personal data. In: 2015 IEEE Security and Privacy Workshops, pp. 180–184. IEEE (2015)
5. Baliga, A.: The blockchain landscape. Persistent Systems (2016)
6. Baxendale, G.: Can blockchain revolutionise EPRs? ITNow **58**(1), 38–39 (2016)
7. Gropper, A.: Powering the physician-patient relationship with hie of one blockchain health it. In: ONC/NIST Use of Blockchain for Healthcare and Research Workshop. ONC/NIST, Gaithersburg. ONC/NIST (2016)
8. Azaria, A., et al.: Medrec: using blockchain for medical data access and permission management. In: 2016 2nd International Conference on Open and Big Data (OBD), pp. 25–30. IEEE (2016)
9. Brodersen, C., et al.: Blockchain: Securing a New Health Interoperability Experience. Accenture LLP (2016)
10. Irving, G., Holden, J.: How Blockchain-Time Stamped Protocols Could Improve the Trustworthiness of Medical Science, p. 5. F1000Research (2016)

11. Nugent, T., Upton, D., Cimpoesu, M.: Improving Data Transparency in Clinical Trials Using Blockchain Smart Contracts, p. 5. F1000Research (2016)
12. Dai, F., et al.: From bitcoin to cybersecurity: a comparative study of blockchain application and security issues. In: 2017 4th International Conference on Systems and Informatics (ICSAI), pp. 975–979. IEEE (2017)
13. Angraal, S., Krumholz, H.M., Schulz, W.L.: Blockchain technology: applications in health care. Circ. Cardiovasc. Qual. Outcomes **10**(9), e003800 (2017)
14. Benchoufi, M., Ravaud, P.: Blockchain technology for improving clinical research quality. Trials **18**(1), 335 (2017)
15. Dhillon, V., Metcalf, D., Hooper, M.: Blockchain in health care. In: Blockchain Enabled Applications, pp. 125–138. Apress, Berkeley (2017)
16. Heston, T.: Why Blockchain Technology Is Important for Healthcare Professionals. Available at SSRN 3006389 (2017)
17. Dubovitskaya, A., et al.: How blockchain could empower e-health: an application for radiation oncology. In: VLDB Workshop on Data Management and Analytics for Medicine and Healthcare, pp. 3–6. Springer, Cham (2017)
18. Rabah, K.: Challenges & opportunities for blockchain powered healthcare systems: a review. Mara Res. J. Med. Health Sci. **1**(1), 45–52 (2017)
19. Karafiloski, E., Mishev, A.: Blockchain solutions for big data challenges: a literature review. In: IEEE EUROCON 2017-17th International Conference on Smart Technologies, pp. 763–768. IEEE (2017)
20. Tama, B.A., et al.: A critical review of blockchain and its current applications. In: 2017 International Conference on Electrical Engineering and Computer Science (ICECOS), pp. 109–113. IEEE (2017)
21. Zhang, P., et al.: Metrics for assessing blockchain-based healthcare decentralized apps. In: 2017 IEEE 19th International Conference on e-Health Networking, Applications and Services (Healthcom), pp. 1–4. IEEE (2017)
22. Esposito, C., et al.: Blockchain: a panacea for healthcare cloud-based data security and privacy? IEEE Cloud Comput. **5**(1), 31–37 (2018)
23. Jiang, S., et al.: Blochie: a blockchain-based platform for healthcare information exchange. In: 2018 IEEE International Conference on Smart Computing (SMARTCOMP), pp. 49–56. IEEE (2018)
24. Zhang, P., et al.: Fhirchain: applying blockchain to securely and scalably share clinical data. Comput. Struct. Biotechnol. J. **16**, 267–278 (2018)
25. Zhou, L., Wang, L., Sun, Y.: Mistore: a blockchain-based medical insurance storage system. J. Med. Syst. **42**(8), 149 (2018)
26. Pfleeger, C.P., Pfleeger, S.L.: Security in Computing. Prentice Hall Professional Technical Reference (2002)
27. http://forum.arduino.cc/index.php?action=dlattach;topic=206943.0;attach=63910. Accessed 7 Nov 2019
28. El-Dosuky, M.A., Eladl, G.H.: DOORchain: deep ontology-based operation research to detect malicious smart contracts. In: World Conference on Information Systems and Technologies. Springer, Cham (2019)
29. El-dosuky, M.A., Eladl, G.H.: SPAINChain: security, privacy, and ambient intelligence in negotiation between IoT and blockchain. In: World Conference on Information Systems and Technologies. Springer, Cham (2019)

Interactive Inspection Routes Application for Economic and Food Safety

Telmo Barros[1], Tiago Santos[1], Alexandra Oliveira[1,2], Henrique Lopes Cardoso[1(✉)], Luís Paulo Reis[1], Cristina Caldeira[3], and João Pedro Machado[3]

[1] Laboratório de Inteligência Artificial e Ciência de Computadores (LIACC), Faculdade de Engenharia da Universidade do Porto, Rua Dr. Roberto Frias, s/n, 4200-465 Porto, Portugal
{up201405840,tiagosantos,aao,hlc,lpreis}@fe.up.pt

[2] Escola Superior de Saúde do Instituto Politécnico do Porto (ESS-IPP), Rua Dr. António Bernardino de Almeida 400, 4200-072 Porto, Portugal

[3] Autoridade de Segurança Alimentar e Económica (ASAE), Rua Rodrigo da Fonseca, 73, 1269-274 Lisbon, Portugal
{accaldeira,jpmachado}@asae.pt

Abstract. This paper describes an application aimed at improving the current state of enforcement in the areas of food safety and economic activities in Portugal. More specifically, the application focuses on a flexible and interactive approach to generate inspection routes, to be followed by surveillance brigades with the aim of verifying Economic Operators' compliance to national and European legislation on economic and food safety. The problem is modeled as a Multi-Depot Periodic Vehicle Routing Problem with Time Windows, and the algorithmic approaches employed seek to maximize either the number of inspected Economic Operators or a utility function that takes into account the utility gained from inspecting each Economic Operator. The generated solutions are shown in an intuitive platform, where human operators can visualize the solutions details (including georeferenced locations in a map) and fully customize them on time by manually removing or adding Economic Operators to be targeted.

Keywords: Decision support · Vehicle Routing Problem · Geographic information system · Web applications

1 Introduction

A considerable amount of work has been done in order to put to good use the huge quantity of data produced worldwide. From business to government, any sector, public or private, can benefit from the treatment of such big data. This information can serve to assess key indicators, support decisions and improve the quality of existing systems in general. However, due to the speed and volume that this data is being generated, several challenges need to be overcome such as the processing of the referred data or its visualization [1].

Á. Rocha et al. (Eds.): WorldCIST 2020, AISC 1159, pp. 640–649, 2020.
https://doi.org/10.1007/978-3-030-45688-7_64

The system presented in this paper is part of a more general project that aims to develop models for semi-automated analysis of economic and food related risks as well as for identification and planning of optimized inspection routes to Economic Operators that present a higher risk. This is accomplished by employing techniques such as information extraction, machine learning, data and text mining, and interactive information visualization. In the end, the project aims at improving the monitoring and inspection processes of an Economic and Food Safety Authority in Portugal. There are more than 3,500,000 Economic Operators registered in the Authority's internal system. These Economic Operators are subject to customer complaints, and one of the Authority's role is to detect infraction indications in such complaints, along with their severity. This will affect the usefulness of inspecting a complaint-targeted Economic Operator, its inspection utility. Regarding the territorial organization of the Authority, there are three regional units, composed of a total of twelve operational units nationwide with a variable number of inspectors and vehicles at their disposal. The information system in use is very disperse, depending on multiple data sources and platforms. All the referred factors make it very difficult to automate some processes, such as the identification of the Economic Operators that must be inspected (a time-consuming manual procedure), the assignment of these Economic Operators to brigades of inspectors, or the determination of optimal inspection routes, taking into consideration the minimization of travel distances or time.

This paper focuses on an interactive Web application to generate and visualize flexible inspection routes plans for the Authority inspectors. These routes are created with the purpose of maximizing the number of inspections to Economic Operators that present a higher risk (namely to public health or food safety), while minimizing the specific resources to such operation.

Regarding the mentioned optimization problem, it is a variant of the Vehicle Routing Problem (VRP) which is a complex combinatorial optimization problem that, *given a fleet of vehicles with uniform capacity, a common depot, and several costumer demands, find the set of routes with overall minimum route cost which service all the demands* [2]. Previous research has shown that vehicle routing optimization can promote significant economic savings [3–5].

In Sect. 2, it is discussed the related work. The details about the selection of Economic Operators to be inspected, data size and problem constraints are described in Sect. 3. Section 4 explains the architecture of the web application and related components, its capabilities and major details of implementation. A brief discussion of the main achievements achieved with the developed solution is included in Sect. 5. In Sect. 6 it is presented the conclusions and pointed some directions for future work.

2 Related Work

This type of problems is not new to the scientific community, the discussion focuses on how to best solve them due to the degree of complexity inherent to its resolution.

A VRP can be modeled as a graph composed of a set of points, representing cities or customers and a distribution center, a set of arcs joining them, and a set of available vehicles (at the distribution center). Each arc has associated a non-negative value which may represent a distance, a travel time or other travel cost. The purpose of the problem is to determine the lowest cost of a vehicle route set subject to the following restrictions [6]: (i) each city is visited exactly once by exactly one vehicle; and (ii) all vehicle routes starts and ends at the distribution center.

Problems modeled as VRP include supplying products to customers, collecting waste from a locality, or supplying supermarket chains. One of the first problems modeled as a VRP was the *Truck Dispatching Problem* [7], considered as a generalization of the *Traveling Salesman Problem* [8,9].

Another factor that keeps the problem alive is the many existing variants [10]. For instance, in the Periodic VRP [11] the routing planning period is extended to multiple days, not requiring neither all the cities to be visited in one day nor the vehicles to return to the distribution center in the same day they depart. Other variants include the Open VRP [12], the VRP with Time Windows [13], and the Multi Depot VRP [14].

3 Problem

Concerning the mentioned vehicle route optimization, the described problem is a Multi-Depot Periodic Vehicle Routing Problem with Time Windows (MDPVRPTW) [15]. Some of its characteristics are shown in Fig. 1, which portraits a solution composed of 4 inspection routes, given 3 operational units and 13 Economic Operators.

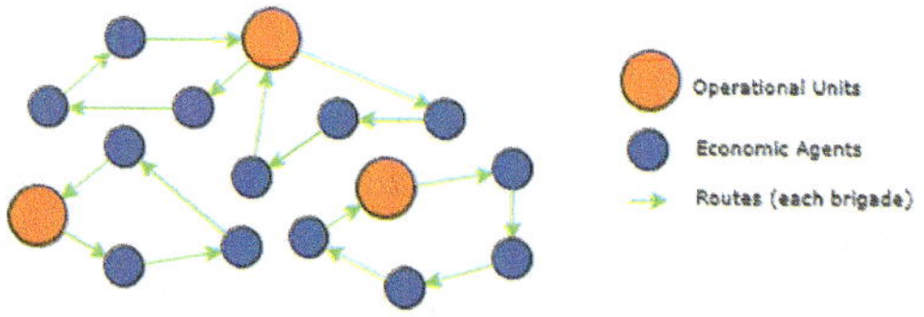

Fig. 1. MDPVRPTW applied representation.

This family of route generation problems, MDPVRPTW, encompasses the main constraints of the problem at hand. The problem constraints are:

- The inspection actions planning can be carried out in multiple levels, the two addressed here are the **Operational** level – which focuses on the generation of inspection plans for the brigades, through the selection of specific Economic Operators to be inspected and the definition of starting and finishing points for each brigade – and the **Brigade** level – which is responsible for making minor changes in the inspection plan in real-time, due to unexpected reasons (such as the closure of an Economic Operator);

- Each inspection plan is carried out by a group of one or more brigades/vehicles. Each brigade is composed of two or more inspectors;
- An inspection plan is targeted to a specific economic activity or a set of economic activities, which directly affects the Economic Operators selection for inspection;
- Inspections can only take place during the period of operation of the Economic Operators;
- Each operational unit is associated with a set of municipalities and/or parishes that make up the geographical boundary on which the brigades of this operational unit can operate. This limit is well defined and must be respected without exception;
- Interruptions are foreseen during the working period of the brigades;
- The inspections plans may have a duration of multiple days;
- The departure and arrival points of the brigade vehicles are the headquarters of the operational unit by default, but they can change and be two different points, since each operational unit has a number of parking spaces (such as fire stations). These locations are used both as a starting or ending point for inspections lasting more than one day or taking place far away from the headquarters of the selected operational unit.

These conditions turn the generation of flexible routes into a very challenging problem. In the following section it is described the approach to solve this problem, in which we relax some of the constraints in order to obtain valid solutions in a short period of time.

4 Approach

It was designed an intuitive Web application to interact with a logic layer (where the methods to solve the problem in the most efficient way are embedded) and to visualize the solutions from a geographical and chronological points of view.

4.1 Architecture

The whole project is composed by multiple layers which appear illustrated in Fig. 2. It shows a simplified view of the entire system. The *Inspections Planning Assistant*, from the Web application, is supported by the Routes Generation API (RGAPI), responsible for the execution of the routing algorithms. The RGAPI itself relies on the Open Source Routing Machine (OSRM) [16],[1] a high-performance open-source C++ routing engine used to calculate travel time and/or distances between two or more geographic points [17]. This architecture allows for a good overall performance, reliability and the ability to scale easily.

[1] http://project-osrm.org/.

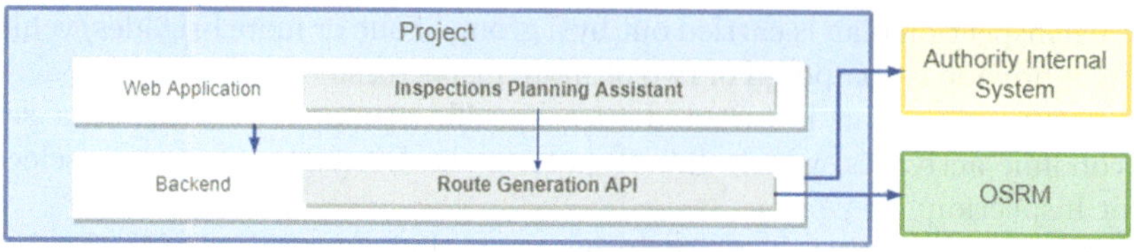

Fig. 2. System's architecture overview.

4.2 Routes Generation

Some aspects concerning the RGAPI need to be taken into consideration before addressing the visualization of the solutions themselves, such as the determination of the inspection utilities and the opening hours for each Economic Operator. The utility of inspecting an Economic Operator, $u_{eo}(\in [0,1])$, is the basis for determining optimal inspection routes, since it is the element that is intended to be maximized when seeking for an optimal inspection route. The value of u_{eo} is determined by a function of the number of complaints NC regarding a particular Economic Operator eo, registered in the system and can be defined as shown in Eq. 1:

$$u_{eo}(NC) = \begin{cases} 0.05 & \text{if } NC = 0 \\ \frac{NC}{10} & \text{if } 1 \leq NC < 10 \\ 0.9 + \frac{NC-9}{100} & \text{if } 10 \leq NC < 19 \\ 1 & \text{if } NC \geq 19 \end{cases} \tag{1}$$

Figure 3 provides a visual representation of this function.

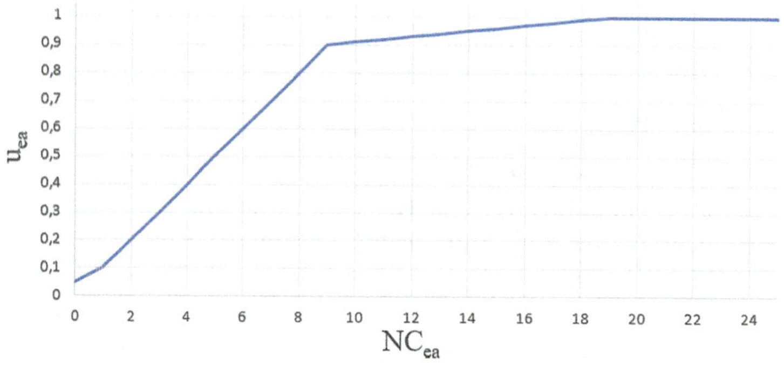

Fig. 3. Utility transformation function chart.

Given that the inspections may only occur while the Economic Operator is operating, it was implemented a strategy to get this information from the

Economic Operators using the Google Places[2] and Yelp[3] APIs. These services allowed to retrieve the most updated opening hours of the Economic Operators that use their services.

Regarding the implemented methodologies to solve the vehicle routing problem itself, four different methods were developed: one exact approach based on the Branch and Bound algorithm, and three meta-heuristic approaches – Hill Climbing, Simulated Annealing and Genetic Algorithms. The variety of methods allowed to explore different behaviors given the input constraints and to select the most adequate algorithm to solve the problem.

4.3 Web Application

The Web application is the entry point to the visualization and interaction with the routing algorithms. The *Inspections Planning Assistant* component has two main views or pages: (i) the *Inspection Planning Form*, shown in Fig. 4, which can be used to customize route generation, and (ii) the *Inspection Planning Overview*, shown in Figs. 5 and 6, where it is possible to see the solution and its details, also enabling editing and storing operations.

The *Inspection Planning Form* allows to input the custom conditions for a new inspection plan generation. These constraints are used to execute the algorithms in the RGAPI and are directly correlated with the ones specified in Sect. 3. The input variables are the following:

- *Objective function:* It is either the maximization of u_{eo} or the maximization of agents to be inspected;
- *Starting date:* Starting date and time for the inspections route;
- *Meal break:* Time interval where a 1 hour break may occur;
- *Economic activities:* List of economic activities to include in the solution (may be empty to consider all registered Economic Operators);
- *Brigade:* List of brigades composed by:
  - *Vehicle:* Vehicle to be used;
  - *Duration:* Maximum amount of time to be spent on the field preforming the inspections;
  - *Inspectors:* List of inspectors (minimum of 2 per brigade);
  - *Starting/finishing points:* Geographic location of the starting and finishing points (the operational unit headquarter is used by default).

The second view, portrayed in Fig. 5, presents the solutions found and is composed of three subviews. In the **map** subview, the Economic Operators used as input for the algorithms are visible. The ones that are part of one of the generated inspection routes are displayed with a colored marker, which can be green if $0 \leq u_{eo} \leq 0.5$, orange if $0.5 < u_{eo} \leq 0.8$ or red if $0.8 < u_{eo} \leq 1$ and blue if it is excluded from the solution. By clicking on a marker, more information about the agent is displayed, together with an option button to either add or remove

[2] https://cloud.google.com/maps-platform/places/.
[3] https://www.yelp.com.

the Economic Operator from the solution. A new solution is then recalculated given the new changes.

The routes are outlined in the map and can be selected or toggled from the left sidebar. This sidebar also contains main information about each brigade and global details of the solution. It is also possible to generate a new solution by clicking on the button "New Route".

Fig. 4. Input conditions page (multiple views).

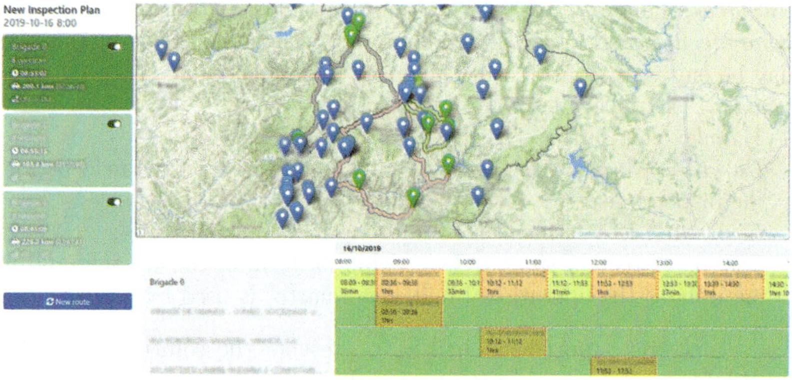

Fig. 5. Web application overview (map and timetable subviews).

The **timetable** subview, visible in the bottom part of Fig. 5 is a chronological view of the events of the selected brigade on left sidebar. The yellow blocks in the first row represent travels from one point to another, the orange ones are the inspections (currently fixed with a 60 min duration). The opening hours of each

Fig. 6. Web application overview (instructions subview).

Economic Operator appear alongside the orange blocks, in the rows below. For the example in the Fig. 5 the three visible agents are open 24 h.

The **instructions** subview, visible in Fig. 6, displays a table with the selected agents to be supervised, their inspection utility and the times of departure and arrival of each point in the route. It may also display textual instructions of the whole route or parts of the route (between waypoints). It is also possible in this view to browse directly to a third party application that offers a real-time directions feature (based on Google Maps).

5 Achievements

As a decision support system, the developed application is capable of creating a daily planning for one or more brigades of inspectors. These routes minimize the total waste time (time difference between the estimated time of the route and the maximum duration input), and always find the maximum number of agents inspected in a day or the maximum utility value for the set of inspected agents. This is the main goal of the application.

The generated solutions can be consulted and the system is amenable to the evaluation of human operators, who apply changes in the route (additions, exchanges or removals of Economic Operators from the inspection plan).

The routing algorithms supporting the application were subject to multiple tests, in a simulated environment, whose results are described elsewhere. The analyzed assessment metrics were the execution times, the global utility values and the global waste time. An important point to stress out is the capability of returning a satisfactory solution within 1 s for inputs of 100 Economic Operators and within 60 s without input size limitations. Given that the duration currently assigned to every inspection is 60 min, a satisfactory solution is a solution with a waste time below 60 min.

The presented system represents a big step with respect to the integration of a support decision system in the Economic and Food Safety Authority current procedures. It will allow to automate the process of selecting the Economic Operators to be inspected as well as the complex process of brigade assignment and respective route generation. Facilitating the execution of these actions may substantially reduce the time spent on inspections planning.

6 Conclusions and Future Work

In this paper it is present an interactive platform capable of generating flexible inspection routes to an Economic and Food Safety Authority inspectors. It offers functionalities that facilitate the whole process of selecting and assigning Economic Operators to the multiple brigades, respects the main constraints from the original problem and allows platform users to customize the desired inspection plan before and after the solution has been generated, through the addition and removal of Economic Operators to be inspected.

The solutions of generation and optimization of routes analyzed always aim at determining the most effective routes for a fleet of vehicles to visit a set of points through the application of various strategies, from Branch and Bound to Genetic Algorithms. The modeling of sub-problems involves varying the original problem by introducing or removing restrictive variables. The application in the final solution has thus gone through the creation of several problem-solving algorithms that have linked a type of MDPVRPTW to a function of attribution of utilities to each Economic Operator using the number of associated complaints.

Given the nature and scope of this paper's problematic, the development prospects are varied. It is considered the following directions:

- **Routing algorithms:** Expansion of the number of implemented methods and/or parameters in order to obtain better solutions;
- **Utility function:** Definition of an utility transformation function that takes into consideration more input variables, such as the Economic Operator's reputation, based on previous inspections, or associated economic activities;
- **Directions view:** Targeted for the inspectors to follow up real-time driving and driving directions to the next stop while driving;
- **Real-time location:** Direct connection with the brigade vehicle to allow a real-time view of the geographical positions of all vehicles on the field.

Acknowledgements. This work is supported by project IA.SAE, funded by Fundação para a Ciência e a Tecnologia (FCT) through program INCoDe.2030. This research was partially supported by LIACC (FCT/UID/CEC/0027/2020).

References

1. Marvin, H.J.P., Janssen, E.M., Bouzembrak, Y., Hendriksen, P.J.M., Staats, M.: Big data in food safety: an overview. Crit. Rev. Food Sci. Nutr. **57**(11), 2286–2295 (2017)
2. Machado, P., Tavares, J., Pereira, F.B., Costa, E.: Vehicle routing problem: doing it the evolutionary way. In: Proceedings of the 4th Annual Conference on Genetic and Evolutionary Computation, pp. 690–690. Morgan Kaufmann Publishers Inc. (2002)
3. Hasle, G., Lie, K.-A., Quak, E.: Geometric modelling, numerical simulation, and optimization. Springer, Heidelberg (2007)
4. Toth, P., Vigo, D. (eds): The Vehicle Routing Problem. Society for Industrial and Applied Mathematics (2002)

5. Cattaruzza, D., Absi, N., Feillet, D., González-Feliu, J.: Vehicle routing problems for city logistics. EURO J. Transp. Logist. **6**(1), 51–79 (2017)
6. Laporte, G.: The vehicle routing problem: an overview of exact and approximate algorithms. Eur. J. Oper. Res. **59**(3), 345–358 (1992)
7. Dantzig, G.B., Ramser, J.H.: The truck dispatching problem. Manag. Sci. **6**(1), 80–91 (1959)
8. Flood, M.M.: The traveling-salesman problem. Oper. Res. **4**(1), 61–75 (1956)
9. Kruskal, J.B.: On the shortest spanning subtree of a graph and the traveling salesman problem. Proc. Am. Math. Soc. **7**(1), 48 (1956)
10. Braekers, K., Ramaekers, K., Van Nieuwenhuyse, I.: The vehicle routing problem: state of the art classification and review. Comput. Ind. Eng. **99**, 300–313 (2016)
11. Campbell, A.M., Wilson, J.H.: Forty years of periodic vehicle routing. Networks **63**(1), 2–15 (2014)
12. Li, F., Golden, B., Wasil, E.: The open vehicle routing problem: algorithms, large-scale test problems, and computational results. Comput. Oper. Res. **34**(10), 2918–2930 (2007)
13. Caric, T., Gold, H.: Vehicle Routing Problem. In-Teh, Vienna (2008)
14. Montoya-Torres, J.R., López Franco, J., Isaza, S.N., Jiménez, H.F., Herazo-Padilla, N.: A literature review on the vehicle routing problem with multiple depots. Comput. Ind. Eng. **79**, 115–129 (2015)
15. Cordeau, J.-F., Gendreau, M., Laporte, G., Potvin, J.-Y., Semet, F.: A guide to vehicle routing heuristics. J. Oper. Res. Soc. **53**(5), 512–522 (2002)
16. Luxen, D., Vetter, C.: Real-time routing with OpenStreetMap data. In: Proceedings of the 19th ACM SIGSPATIAL International Conference on Advances in Geographic Information Systems, GIS 2011, pp. 513–516. ACM (2011)
17. Huber, S., Rust, C.: Calculate travel time and distance with Openstreetmap data using the open source routing machine (OSRM). Stata J. **16**(2), 416–423 (2016)

Prediction of Mobility Patterns in Smart Cities: A Systematic Review of the Literature

Nelson Pacheco Rocha[1](✉), Ana Dias[2], Gonçalo Santinha[3], Mário Rodrigues[4], Alexandra Queirós[5], and Carlos Rodrigues[3]

[1] Medical Sciences Department & Institute of Electronics and Informatics Engineering of Aveiro, University of Aveiro, Aveiro, Portugal
npr@ua.pt

[2] Department of Economics, Industrial Engineering, Management and Tourism & GOVCOPP - Governance, Competitiveness and Public Policies, University of Aveiro, Aveiro, Portugal
anadias@ua.pt

[3] Department of Social, Political and Territorial Sciences & GOVCOPP - Governance, Competitiveness and Public Policies, University of Aveiro, Aveiro, Portugal
{g.santinha,cjose}@ua.pt

[4] Águeda School of Technology and Management & Institute of Electronics and Informatics Engineering of Aveiro, University of Aveiro, Aveiro, Portugal
mjfr@ua.pt

[5] Health Sciences School & Institute of Electronics and Informatics Engineering of Aveiro, University of Aveiro, Aveiro, Portugal
alexandra@ua.pt

Abstract. This study aimed to identify the current approaches to determine mobility patterns using smart cities' infrastructures, which might be useful to disseminate good practices. Therefore, a systematic review was performed based on a search of the literature. From an initial search of 8207 articles, 25 articles were retrieved for the systematic review. These articles reported different approaches to predict mobility patterns using smart city data sources, namely data from mobile carrier networks, from social networks or from transit agencies' smart cards.

Keywords: Smart cities · Smart mobility · Mobility patterns · Systematic review

1 Introduction

The continuum of human spatial mobility, which can be analyzed according to different geographic and temporal scales, is a real challenge for urban governance since it affects the cities development, namely in terms of transportation flows [1]. Therefore, smart mobility is a key aspect of smart cities [2, 3].

Considering the different types of sensing infrastructures of smart cities, heterogeneous data sources might be aggregate by big data algorithms [4] to infer mobility patterns (i.e., patterns that describe the mobility behaviors of the inhabitants) [2, 5]. In turn, mobility patterns can enrich the experiences of how cities function by providing

Á. Rocha et al. (Eds.): WorldCIST 2020, AISC 1159, pp. 650–659, 2020.
https://doi.org/10.1007/978-3-030-45688-7_65

information that might be used, for example, by traffic management systems to compute the best routes or by urban planners to enforce policies or to anticipate the dimension of the congestion of certain zones of the cities [2].

The study reported in the present was performed to identify the current approaches to determine mobility patterns using smart cities' infrastructures. This is useful to inform smart cities' stakeholders about state-of-the-art solutions and researchers about the gaps of the current research.

2 Methods

The research study report in this article was informed by the following research questions:

- RQ1: What approaches are being used to determine mobility patterns based on smart cities' infrastructures?
- RQ2: What are the maturity levels of the solutions being reported?

Boolean queries were prepared to include all the articles that have in their titles, abstract or keywords one of the following expressions: 'Smart City', 'Smart Cities' 'Smartcity', 'Smartcities', 'Smart-city' or 'Smart-cities'. The resources considered to be searched were two general databases, Web of Science and Scopus, and one specific technological database, IEEE Xplore. The literature search was performed in January 2018.

As inclusion criteria, the authors aimed to include all the articles that report evidence of explicit use of smart cities' infrastructures to determine mobility patterns.

Considering the exclusion criteria, the authors aimed to exclude all the articles not published in English, without abstract or without access to full text. Furthermore, the authors also aimed to exclude all the articles that report overviews, reviews, applications that do not explicitly require smart cities' infrastructures or that were not relevant for the specific objective of this study.

After the removal of duplicates and articles without abstract, the selection of the remainder articles according to the outline inclusion and exclusion criteria was performed in following steps: i) the authors assessed all abstracts for relevance and those clearly outside the scope of the present systematic review were removed; ii) the abstracts of the retrieved articles were assessed to verify if they were related to smart mobility, which includes mobility patterns, and those articles reporting studies not related to smart mobility were excluded; iii) the abstracts of the remaining articles were assessed to select the articles related to mobility patterns; and iv) the full text of the retrieved articles were analyzed and classified. In all these four steps the articles were analyzed by at least two authors and any disagreement was discussed and resolved by consensus.

3 Results

The initial step of the screening phase (i.e., step 1 of the abstracts screening) yielded 8130 articles by removing the duplicates (67 articles) or the articles without abstracts (ten articles).

Based on abstracts, 2676 articles were removed due to the following reasons: i) not published in English; ii) overviews or systematic revisions; iii) editorials, prefaces, and announcements of special issues, workshops or books; and iv) not related to applications for smart cities.

The abstracts of the remaining 5454 articles were analyzed and it was concluded that 696 articles were related to smart mobility. However, of these 696 articles, 658 articles were excluded because although they are related to smart mobility (e.g., parking's management or drivers assistance), they are not specifically related to mobility patterns.

Finally, the full texts of remainder 38 articles were assessed (i.e., full texts screening) and one article was excluded since it is a commentary, nine articles were excluded because they were not considered relevant for the purpose of this study and three articles were excluded due to the impossibility to access the respective full texts. Therefore, 25 articles were considered eligible for the systematic review.

From those 25 articles, 19 were published in conference proceedings [1, 2, 6–22] and only six were published in scientific journals [23–28]. All the retrieved articles intend to contribute to predict mobility patterns, aiming to enable proactivity in planning and managing transport systems as well as other cities' infrastructures with impact on mobility. However, different specific approaches were considered, namely using data from mobile carrier networks that implicitly represent operations with personal mobile devices, data from social networks, data from transit agencies' smart cards, or data from other sources.

3.1 Personal Mobile Devices

Modern urban systems require monitoring different types of data originate from various sources. In this respect, data from mobile carrier networks present the advantage of being possible to infer inhabitants' behavioral fingerprints based on the use of their mobile devices [6]. Therefore, several articles report the use of mobile devices to implicitly collect data from inhabitants to predict mobility patterns [2, 6–11, 23, 24].

However, the data related to mobile carrier networks, as other types of data, have validity problems. Considering this aspect, article [23] focuses two validity aspects of the data collected from mobile carrier networks: internal time discrepancy and data loss. Therefore, records of transit agencies' smart cards and records related to the location of vehicles from a Chinese urban transportation system were used along with manual results investigation to explore and verify the data validity of mobile carrier networks.

In terms of prediction of mobility patterns based on data from mobile carrier networks, [6] presents a conceptual method of analyzing selected elements of the collected datasets (i.e., events describing the presence of the devices in particular areas to determine inhabitants behaviors, namely static or mobile behaviors) to obtain patterns such as people traveling by public transport or people traveling by private cars. However, [6] do not report any implementation, which means it is not possible to conclude about the adequacy of the approach.

Also, data from mobile carrier networks, namely the caller details from mobile devices, were used in the study reported by [7]. The study is related to the implementation of a public transportation crowd prediction. Using neural network models and computation power of parallel streaming analytic engines, the proposed system aims to track activities performed by mobile phone users in their devices to predict the number of passengers entering public transports stations, the number of waiting passengers on transport platforms and other metrics on the crowd density related to public transportation systems. The feasibility of the proposed systems was demonstrated in an interchange station of the Singapore's subway system.

Similarly, [8] explores whether travel time prediction can be performed from mobile carrier networks data. More precisely, the features being used were cell dwell times, which describe the amount of time that a mobile device stays associated with a specific telecommunications base station before performing a handover. The model predictions were evaluated using a floating car data set contained vehicle's geographical position, bearing and velocity annotated with a timestamp and a vehicle identifier car ground truth data. This dataset have approximately 40 million data points from a total of 27124 trips that were gathered during a traffic monitoring campaign performed in Luxembourg in 2015. In terms of results, it was observed that the distribution of the predictions based on the dwell times reflect the ground truth data. Based on the achieved results, the authors conclude that mobile phone signaling data is, under some assumptions, a sufficient predictor for travel times.

In turn, the study reported by [9] used mobile carrier networks data to identify transportation corridors (i.e., corridors that reveal the spatial travel patterns of urban residents), which help the understanding of the city transportation structure and the travel demand features. According to the reported experience, the transportation eigenlines were identified from a mobile carrier network dataset of Shanghai. Then, the obtained eigenlines were testified by comparing with real road network and population flow maps. Finally, the transportation corridors were drawn according to the determined eigenlines and priori knowledge.

Article [10] proposes a method to understand the mobility directions of certain groups of people within a city, by using the internal logs of a 3G mobile carrier network. The authors propose a mathematical model that has been experimentally validated by using a dataset extracted from an Italian telecom operator. The results confirm a good match between the flux estimated by the model, and the real flux across the roads.

In [11] the position information of the individuals was retrieved from activities performed in their mobile device (e.g., voice call, any instant message like Twitter or any other data from the Internet like Facebook). The focus of the study was the identification of the mobile users' flows applied to tourism, by associating the flows to the position of the monuments and other important attractions or events. This can help local administrations to envisage high tourist density in certain areas, to drive in other attractions with less affluence and to adapt the tourism structures accordingly. In terms of validation, the authors could identify the mobility of the tourists in the city of Rome from the data provided by an Italian telecommunications operator. Results show the possibility to detect tourist flows in different zones of the city.

Similarly, [2] also propose to use data coming from mobile carrier networks to compute real-time mobility patterns related to point of interests of a city. Based on the monitoring of 3G signaling collected from an Italian telecom operator, the authors tested different scenarios, namely vehicular traffic estimations, speed profile estimations, origin/destination matrices or subway flows. The results were positive, since the quality of the predictions and estimations are often better than well-adopted competitors [2].

Finally, article [24] presents UrbanMobilitySense, a smartphone-based mobility sensing system to conduct transportation activity surveys. The system is based not only on data from mobile carrier networks but also in other human mobility data that can be automatically collected by mobile devices (e.g., WiFi or 3-axes accelerometer data). The collected raw data is transformed into high-level mobility information (e.g., stops and transportation modes), which are presented by interactive interfaces to represent specific knowledge so that the respective users might provide feedback. In terms of validation, the system has been deployed in Singapore to support surveys of transportation authorities and the authors conduct experiments to study how sensing scheme affects the quality and quantity of collected data and place information.

3.2 Social Sensing

When looking at mobility patterns, sensing based in social networks also has significant relevance and is complementary to other infrastructure sensors, as it was reported in [12–17].

Based on a literature study, [12] proposes an architecture of an information system to estimate the value of reliability of the information extracted from social networks to monitor traffic conditions. Moreover, [13] presents an unsupervised method to extract real world geo-referenced events from Twitter streams in order to decompose the complex mobility flows of a bike sharing network and assess its impact on city services such as public transport and public safety. Likewise, [14] proposes a method to filter the tweets related to traffic congestion in Bandung and to extract information (e.g., location, date and time). In turn, [15] presents a social sensing data analysis framework to gather, cleanse, analyze and show social sensing data to support smart mobility. Moreover, article [16] aims to translate big data into adaptive urban designs, namely by analyzing multi-source geo-tagged data from Instagram and Twitter for the detection of frequently used places and the activities being performed there. Finally, [17] applies a machine learning method for semantic enrichment of mobility data records (e.g., trip counts by using geo-referenced social media data) of bike sharing network.

Concerning the type of technology adopted within the references analysed, the Twitter datasets used in [12–17] were collected using the Twitter streaming Application Programme Interface (API). In [13] an OpenNLP machine learning tool was used as well as Google Maps Geocoding API. Similarly, in [15] data from Twitter streaming API were integrated with Geographical Information System.

In terms of validation, [12] reports an experiment with 1000 tweets collected in Bekasi City, which consist of training data (500 tweets) and testing data (500 tweets). Also, in [14] 700 tweets were used as the training data and 100 tweets were used as the testing data. In [15], the social sensing data analysis framework was validated by a mobility analyzer, a subsystem of a pilot project in Italy. Moreover, more than 30000 geo-tagged tweets from the city of London were collected to validate the algorithm presented by [13]. In turn, [17] reports a proof of concept comparing a Twitter dataset containing 37335 geo-tagged tweets with a subset of a bike-share dataset containing 43636 unique bike rentals records collected during a one-week period. Finally, the algorithms presented in [16] were tested with 0.8 million raw records collected in Singapore, which proved they were able to effectively and efficiently detect hot places in a given area and classify discussing topic with high accuracy.

3.3 Smart Cards

The need to progress towards a more sustainable transport informed several studies aiming to examine the relevance of public transit infrastructure to many travelers by using data from transit agencies' smart cards [1, 18, 19, 25, 26].

The efforts reported in [25] aim to understand the diurnal mobility across Brisbane, a city where most trips are completed by automobile, by using smart cards transit data together with Google's General Transit Feed Specifications. The use of these two data sources made it easier to disclose the way in which public transit remains relevant, revealed daily mobility, and underlined the way in which different locals across a metropolitan area are connected by public transportation. In terms of validation, the authors retrieved 205560 distinct transit riders' trip trajectories, which allowed to determine how local public transit infrastructure were used and how this usage was linked to various locals around the city.

Zone discovery is considered important to urban system designers, public service providers and commercial marketers, who need to understand people's behaviors and functional regions in a city. In this respect, [26] report a study that aims to discover significant zones of a city based on smart cards transit data that were processed by special purpose algorithms. For the validation of the algorithms, smart cards transit data were collected from the bus network in Seoul during a week. Urban zones were discovered with different levels of abstraction and the authors concluded that such zones can be used to understand inhabitant's behaviors.

For improving the operations of inter-zonal buses in the cities, the study reported in [18] defines specific trip patterns (i.e., frequent bus passenger trip patterns for bus lines) and mine the passenger trajectories from bus smart card data to recommend inter-zonal bus lines. The authors verified the validity of the proposed method by conducting extensive experiments on a real data from Beijing public transports.

In turn, for the study reported by article [19] one-week smart card transit data were analyzed together with a control passenger flow survey in two subway stations of Beijing to determine the travel time and the passenger flow distribution in the multi-mode public transport. The authors conclude that the results might provide a quantitative basis for the development of a smart city.

Finally, in [1] the focus was the evolution extreme transit behaviors of travelers in urban area by using smart card transit data. Based on several aspects of descriptive statistics of the collected data from Beijing, the authors propose a concept to depict the extreme level of the passengers' travel patterns.

3.4 Other Data Sources

Additional sources of information, different from the ones previously reported (i.e., data from mobile carrier networks, data from social networks or smart card transit data) were used in the studies reported by five articles [20–22, 27, 28].

In [20], the project MOBISEC is presented. The project was focused on improving road safety for inhabitants as cyclists and pedestrians of a medium size city and included the collection of information from volunteers (using smartphones or tablet PCs equipped with global positioning and Wi-Fi systems). The article only presents a conceptual view of the project, since an application to gather information from volunteers was designed but not implemented.

In turn, article [21] presents the UnCrowdTPG project aiming to develop a research platform for understanding the mobility patterns of the inhabitants using public transportation in Geneva, as well as their experience with respect to the crowds perceived aboard the vehicles. The research platform allow the inhabitants to provide their assessment of traveler experiences, being the solicitation provided by an application that asked its users to rate the crowd volumes at particular stops and in particular vehicles. To do so, the application detects automatically when the user is at a relevant stop or in a vehicle. Based on these data, together with the open data related to the public transportation, the platform aimed to predict future experience for an individual in a given vehicle [21]. The UnCrowdTPG is planned to be deployed in Geneva (Switzerland), but the algorithms were not evaluated since the authors were still waiting from a set of objective data [21].

Additionally, article [22] proposes an approach that exploits knowledge discovery technique to extract information (e.g., travel time, average speed during the day or fuel consumption per km during the day) from car sharing data available online from major providers in Milan. The data analysis confirmed that during the day there are a lot of cars in the city center while during the evening the cars move to the suburbs. As expected, findings showed that the speed is lower during the day than during the night and the fuel consumption has the opposite trend [22].

Moreover, article [27] proposes an approach to forecast passenger boarding choices and public transit passenger flows based on mining common inhabitant's behaviors and using knowledge from geographic data (e.g., public bus information, point of interest or weather data). All the experimental data comes from the Ridge Nantong Limited Bus Company and Alibaba platform which is also open to the public. Experimental results show that the proposal performs better than baselines in the prediction of passenger boarding choices and public transit passenger flows.

Finally, [28] presents a vector machine-based model to predict future mobility behavior from crowd sourced data. The data of 8303 inhabitants were collected through a dedicated smartphone app to track mobility behavior. The overall success rate of the forecasting model was 82%. The most challenging part was to differentiate between trips made by personal car, bike and public bus transportation. The authors conclude that a forecasting model of this type can facilitate the management of a smart city mobility system while simultaneously ensuring the timely provision of relevant pre-travel information to its inhabitants.

4 Discussion and Conclusion

Concerning the approaches being used to determine mobility patterns (i.e. the first research question) within the articles analyzed, the results are diverse and various types of data sources are being used.

Several articles [2, 6–11, 23, 24] use data related to mobile carrier networks logs (e.g., voice call, any instant message like Twitter or any other data from the Internet like Facebook), namely to determine the position and movements of the mobile devices' users. In some circumstances the mobile carrier networks are complemented with data from other sources, such as data from Wi-Fi networks [24], 3-axes accelerometers [24], geographical positioning systems [8], surveys [24] or data related to traffic monitoring [8].

Other group of articles [12–17] use social sensing datasets. To retrieve the data, APIs such as the Streaming APIs provided by Twitter were used. Moreover, as complementary information, the study reported in [12] also gathered weather data.

Solutions based on the data provided by smart cards transit data were proposed by five articles [1, 18, 19, 25, 26]. Complementary data sources include Google's General Transit Feed Specifications [25] and a control passenger flow survey [15].

Finally, several articles [20–22, 27, 28] consider different data sources, namely information provided by volunteers [20], assessment of traveler experiences provided by the inhabitants together with open data from public transports network [21], car sharing data available online [22], geographic data such as public bus information or weather data [27], or crowd sourced data collected through a dedicated smartphone app to track mobility behavior [28].

Regarding the proposed solutions maturity level (i.e., the second research question), article [6] presents a conceptual method of analyzing selected elements of the collected datasets but does not report any implementation. Similarly, [20] presents a conceptual view of a project, the MOBISEC project, but the article does not present experimental results. Moreover, in the study reported by [21], the developed algorithms were not validated since the author were still waiting from a set of objective data.

The algorithms reported in [2, 7, 9–11, 23] have been tested using mobile carrier networks data belonging to different telecom operators. In [8] a comparative analysis was performed between the results of the proposed algorithms when applied to a telecom dataset and the data points from a total of trips that were gathered during a traffic monitoring campaign. In turn, six articles report the development of algorithms to extract mobility information from real social sensing data [12–17]. Moreover, five articles report algorithms to process data from transit agencies' smart cards [1, 18, 19, 25, 26] and three articles [22, 27, 28] considered other data sources (e.g., information reported by the inhabitants or car sharing data) to validate the developed algorithms.

Finally, article [24] presents a smartphone-based mobility sensing system to conduct transportation activity surveys, the UrbanMobilitySense. However, the authors did not report how the deployed system is being used by the transportation authorities.

After this revision is possible to state that, in the case of mobility patterns, relevant arguments were made regarding the importance of smart cities' infrastructures. Furthermore, in some cases, the authors tried to redefine the role of smart applications with impact on the cities of the future. Nevertheless, it is evident the lack of robust systems.

Although the authors tried, in methodological terms, to guarantee that the review selection and the data extraction of this systematic review were rigorous, it should be acknowledged that this study has limitations, namely the dependency on the keywords and the selected databases and the fact that both grey literature and publications written in other languages than English were excluded.

Acknowledgements. This work was financially supported by National Funds through FCT – Fundação para a Ciência e a Tecnologia, I.P., under the project UI IEETA: UID/CEC/00127/2019.

References

1. Cui, Z., Long, Y., Ke, R., Wang, Y.: Characterizing evolution of extreme public transit behavior using smart card data. In: IEEE First International Smart Cities Conference (ISC2), pp. 1–6. IEEE (2015)
2. Tosi, D., Marzorati, S.: Big data from cellular networks: real mobility scenarios for future smart cities. In: IEEE Second International Conference on Big Data Computing Service and Applications (BigDataService), pp. 131–141. IEEE (2016)
3. Bellini, P., Cenni, D., Nesi, P.: AP positioning for estimating people flow as origin destination matrix for smart cities. In: 22nd International Conference on Distributed Multimedia Systems, pp. 202–209. KSI Research Inc (2016)
4. Batty, M.: Big data, smart cities and city planning. Dialogues Hum. Geogr. **3**(3), 274–279 (2013)
5. Calabrese, F., Colonna, M., Lovisolo, P., Parata, D., Ratti, C.: Real-time urban monitoring using cell phones: a case study in Rome. IEEE Trans. Intell. Transp. Syst. **12**(1), 141–151 (2010)
6. Klimek, R., Kotulski, L.: Towards a better understanding and behavior recognition of inhabitants in smart cities. A public transport case. In: International Conference on Artificial Intelligence and Soft Computing, pp. 237–246. Springer, Cham (2015)
7. Liang, V.C., Ma, R.T.B., Ng, W.S., Wang, L., Winslett, M., Zhang, Z.: Mercury: metro density prediction with recurrent neural network on streaming CDR data. In: IEEE 32nd International Conference on Data Engineering, ICDE 2016, pp. 1374–1377. IEEE (2016)
8. Derrmann, T., Frank, R., Faye, S., Castignani, G., Engel, T.: Towards privacy-neutral travel time estimation from mobile phone signalling data. In: IEEE International Smart Cities Conference (ISC2), pp. 1–6. IEEE (2016)
9. Wang, Y., Li, L., Yu, J., Li, Z., Wang, S., Ke, R.: Identifying the urban transportation corridor based on mobile phone data. In: IEEE First International Smart Cities Conference (ISC2), pp. 1–6. IEEE (2015)
10. Scaloni, A., Micheli, D.: Estimation of mobility direction of a people flux by using a live 3G radio access network and smartphones in non-connected mode. In: IEEE 15th International Conference on Environment and Electrical Engineering (EEEIC), 1869–1873. IEEE (2015)
11. Caruso, M. C., Giuliano, R., Pompei, F., Mazzenga, F.: Mobility management for smart sightseeing. In: International Conference of Electrical and Electronic Technologies for Automotive, pp. 1–6. IEEE (2017)
12. Alifi, M. R., Supangkat, S. H.: Information extraction for traffic congestion in social network: case study: Bekasi city. In: International Conference on ICT for Smart Society, ICISS 2016, pp. 53–59. IEEE (2016)
13. Zhou, Y., De, S., Moessner, K.: Real world city event extraction from Twitter data streams. Procedia Comput. Sci. **58**, 443–448 (2016)
14. Hanifah, R., Supangkat, S. H., Purwarianti, A.: Twitter information extraction for smart city. In: International Conference on ICT for Smart Society (ICISS), pp. 295–299. IEEE (2014)
15. You, L., Motta, G., Sacco, D., Ma, T.: Social data analysis framework in cloud and mobility analyzer for smarter cities. In: IEEE International Conference on Service Operations and Logistics, and Informatics, SOLI 2014, pp. 96–101 (2014)

16. You, L., Tunçer, B.: Exploring the utilization of places through a scalable "activities in places" analysis mechanism. In: IEEE International Conference on Big Data (Big Data), pp. 3563–3572. IEEE (2016)
17. Coffey, C., Pozdnoukhov, A.: Temporal decomposition and semantic enrichment of mobility flows. In: 6th ACM SIGSPATIAL International Workshop on Location-Based Social Networks, LBSN 2013, pp. 34–43. ACM (2013)
18. Wang, Z., Jin, B., Zhang, F., Yang, R., Ji, Q.: Discovering trip patterns from incomplete passenger trajectories for inter-zonal bus line planning. In: IFIP International Conference on Network and Parallel Computing, pp. 160–171. Springer, Cham (2016)
19. Zhou, Y., Yao, L., Jiang, Y., Chen, Y., Gong, Y.: GIS-based commute analysis using smart card data: a case study of multi-mode public transport for smart city. In: Bian, F., Xie, Y. (eds.) Geo-Informatics in Resource Management and Sustainable Ecosystem, pp. 83–94. Springer, Heidelberg (2015)
20. Melendreras-Ruiz, R., García-Collado, A.J.: MOBISEC: an european experience directed towards improving cities through citizen participation. In: International Conference on New Concepts in Smart Cities: Fostering Public and Private Alliances, pp. 1–5. IEEE (2013)
21. Gustarini, M., Marchanoff, J., Fanourakis, M., Tsiourti, C., Wac, K.: UnCrowdTPG: assuring the experience of public transportation users. In: 10th International Conference on Wireless and Mobile Computing, Networking and Communications, pp. 1–7. IEEE (2014)
22. Pagani, A., Bruschi, F., Rana, V.: Knowledge discovery from car sharing data for traffic flows estimation. In: Smart City Symposium Prague, pp. 1–6. IEEE (2017)
23. Liu, Y., Weng, X., Wan, J., Yue, X., Song, H., Vasilakos, A.V.: Exploring data validity in transportation systems for smart cities. IEEE Commun. Mag. **55**(5), 26–33 (2017)
24. Wu, F.J., Lim, H.B.: UrbanMobilitySense: A user-centric participatory sensing system for transportation activity surveys. IEEE Sens. J. **14**(12), 4165–4174 (2014)
25. Zhou, J., Corcoran, J., Borsellino, R.: Mapping cities by transit riders' trajectories: the case of Brisbane, Australia. Environ. Plann. A **49**(8), 1707–1709 (2017)
26. Jung, J.Y., Heo, G., Oh, K.: Urban zone discovery from smart card-based transit logs. IEICE Trans. Inf. Syst. **100**(10), 2465–2469 (2017)
27. Zhang, N., Chen, H., Chen, X., Chen, J.: Forecasting public transit use by crowdsensing and semantic trajectory mining: case studies. ISPRS Int. J. Geo-Inf **5**(10), 180 (2016)
28. Semanjski, I., Lopez, A., Gautama, S.: Forecasting transport mode use with support vector machines based approach. Trans. Marit. Sci. **5**(02), 111–120 (2016)

Four Enterprise Modeling Perspectives and Impact on Enterprise Information Systems

Boris Shishkov[1,2,3(✉)], Aglika Bogomilova[3], and Magdalena Garvanova[1]

[1] Faculty of Information Sciences, University of Library Studies and Information Technologies, Sofia, Bulgaria
m.garvanova@unibit.bg
[2] Institute of Mathematics and Informatics, Bulgarian Academy of Sciences, Sofia, Bulgaria
[3] Institute IICREST, Sofia, Bulgaria
b.b.shishkov@iicrest.org, abogomilova@gmail.com

Abstract. The alignment between Enterprise Modeling (EM) and Software Specification (SS) is still uncertain, this leading to enterprise information systems of low quality. Hence, only the EM-driven software generation could help aligning software functionalities to domain requirements. This inspires the emergence of innovative approaches, such as the SDBC (Software Derived from Business Components) approach, considered by us. It steps on a conceptual invariance (embracing concepts whose essence goes beyond the barriers between social and technical disciplines), while SDBC also builds upon this, to accommodate a modeling duality featuring (1) technology-independent EM rooted in social theories; (2) SS rooted in computing paradigms. The proposed EM-SS alignment is component-based, featuring a potential re-use of modeling constructs, such that the modeling effectiveness and efficiency are stimulated. We consider particularly *(1)*, observing insufficient EM maturity in general: many analysts conduct intuitive EM (not scientifically grounded); they often fail to be exhaustive (some mainly focus on behavior, others – on data, and so on); some analysts mix up essential business things with information exchange that is not featuring essential business things; other analysts are unaware of the importance of communicative acts; many analysts overlook regulations and values; and so on. We address 4 EM perspectives, namely language acts, regulations, public values, and energy – each of them is a theory/paradigm on its own and studying them in isolation is important. It is also important considering them in combination, identifying possibilities for bringing them together, in order to achieve a more exhaustive EM foundation with regard to corresponding SS. We argue that the 4 perspectives make our EM vision usefully broad but we do not claim exhaustiveness. We have studied each of them, providing accordingly theoretical justification and partially demonstrating their practical applicability (by means of an example). Thus, the contribution of our paper is two-fold: (i) We make a small contribution to the development of the SDBC approach; (ii) We analyze different EM perspectives.

Keywords: Enterprise modeling · Language acts · Norms · Public values · Energy

Á. Rocha et al. (Eds.): WorldCIST 2020, AISC 1159, pp. 660–677, 2020.
https://doi.org/10.1007/978-3-030-45688-7_66

1 Introduction

The alignment between enterprise modeling and software specification is still uncertain, this leading to enterprise information systems of low quality; hence, only the enterprise-modeling-driven software generation could help aligning software functionalities to domain requirements [1]. This has inspired the emergence of innovative approaches, such as the SDBC (Software Derived from Business Components) approach [2], referred to in the current paper.

On the one hand, the approach steps on a conceptual invariance (embracing concepts whose essence goes beyond the barriers between social and technical disciplines), while on the other hand, the approach builds upon that "common ground" to accommodate a modeling duality featuring (1) technology-independent enterprise modeling that is rooted in social theories and (2) software specifications that are rooted in computing paradigms. Further, the proposed alignment between enterprise modeling and software specification is component-based, featuring a potential re-use of modeling constructs, such that the modeling effectiveness and efficiency are stimulated [3].

In the current paper, we consider especially *(1)*, observing insufficient maturity as it concerns enterprise modeling in general: many analysts conduct intuitive enterprise modeling that is not scientifically justified; they often fail to be exhaustive in their modeling (some of them would only focus on modeling behavior and others would only focus on modeling data, and so on); some analysts would mix up essential business things (e.g., John paid for his service subscription) with information exchange that is not featuring essential business things (e.g., John entered his PIN incorrectly while using an ATM); other analysts would be unaware of the importance of communicative acts in real-life communication, through which commitments are generated, that are in turn crucially important with respect to the processes within an organization; many analysts would overlook regulations and public values as key restrictions with regard to the functioning of an organization [1].

In particular, we address **four enterprise modeling perspectives**, namely *language acts*, *regulations*, *public values*, and *energy* – each of them is a theory/paradigm on its own. Hence, studying them in isolation is important. Nevertheless, it is also important considering them in combination, such that possibilities are identified for bringing them together, in order to achieve a more exhaustive enterprise modeling foundation with regard to corresponding software specifications.

Even though we argue that those *four perspectives* make our enterprise modeling vision usefully broad, we do not claim exhaustiveness. What we have done in this paper is to study each of them, providing accordingly theoretical justification and partially demonstrating their practical applicability (by means of an illustrative example). Thus, the contribution of the paper is two-fold: (i) We make a small contribution to the development of the SDBC approach; (ii) We analyze different enterprise modeling perspectives.

Finally, we are to elaborate each of those perspectives:

- As studied in [1], one way of modeling an enterprise is to capture the entity-to-entity communications and related actions, as featured in [4]. For example, at a pizza desk, we observe customers going for pizzas, sandwiches, and so on. Imagine that John is staying at the desk and Richard is a customer, and Richard would like to have for lunch a piece of Pizza Margherita. By asking for this, Richard is stating a REQUEST and in turn, John could either DECLINE this request (if they have no Pizza Margherita) or PROMISE to deliver a piece of Pizza Margherita to Richard. Then if promised, the pizza is to be delivered. If this is done, the mere act of delivering the piece of Pizza Margherita is a STATEMENT featuring the result of what was done, triggered by the request. Nevertheless, the statement is not "completing" the interaction because Richard may ACCEPT the result (if the piece of pizza was delivered in time and looks OK) or NOT (if the piece of pizza does not look OK and/or was delivered with a huge delay). Hence, all those communicative acts ("request", "promise", "state", and so on) are straightforwardly related to corresponding ACTIONS (for example: "a piece of Pizza Margherita is being delivered by John to Richard", "Richard is paying for the piece of pizza", and so on). Those actions in turn represent "building blocks" as it concerns the business processes "flowing" among enterprise entities. We therefore argue that by capturing communicative acts and corresponding actions, one is capable of identifying and modeling business processes. This in turn allows for delivering enterprise models accordingly.

- As studied in [1], another way of modeling an enterprise is to establish what may, may not or must happen in a particular situation, as featured in [5]. For example, in case George is a VISA credit card holder, then: (i) If George has not reached his credit limit, he MAY pay (up to the limit) using his VISA credit card; (ii) If George has reached his credit card limit, he MAY NOT pay using his VISA credit card; (iii) If VISA has billed George for a minimal monthly payment, George MUST do the payment to VISA. Those are examples of norms (or rules) that in turn allow for bringing forward the REGULATIONS governing an enterprise. We therefore argue that by capturing norms, one is capable of modeling the potential processes as it concerns an enterprise.
- As studied in [1], public values, such as PRIVACY, TRANSPARENCY, and ACCOUNTABILITY are important as it concerns the CONTEXT-AWARENESS of enterprise information systems [6, 7]. In the current paper, we would also like to address the potential for usefully considering public values in the enterprise modeling process, inspired by [8, 9]. In this, we refer to previous work [10] where we superimpose such public values (labelled "atomic values") to what we call "composite values" that are featured in three bipolar dimensions, namely: hierarchy - egalitarianism, autonomy - conservatism, and harmony – mastery. We argue that by identifying the relevant bipolar dimension and positioning an enterprise accordingly, one is capable of modeling the general structure of business processes. Then this could be usefully related to atomic values that would be adequately weaved in as requirements with regard to the enterprise being modeled.
- And in the end, we argue that considering energy-related issues could also be useful in modeling enterprises. By this we mean intangible issues that are not "visible" but are of importance with regard to what the modeled enterprise actually is. Let us take an example featuring Tom who is a human being. One would agree that what Tom actually is extends far beyond what the eyes can perceive, far beyond the flesh, blood, and bones. There are also other things, both gross and subtle, which are claimed to constitute in combination what the person is. Inspired by Hindu Philosophy [11], we see all this as related to five essential elements, namely: earth, water, fire, air, and ether.

Through specific features related to each of those elements, we are able to describe a human being much better than a purely visual description. We hence argue that this also holds for enterprises: two enterprises (for instance) could be identical in terms of structure, business goals, turnover, and so on. Still, they could differ a lot and those differences are not always trivial to see. Through "five-elements glasses" we would be able to "see more": one of the enterprises could be managed in an "ether" style (space) and this means well-established teams, managers who delegate, equal importance as it concerns everyone in the hierarchy, and so on; the other enterprise could be managed in a "fire" style (for example) and this means much responsibility for the leader with less delegation, and also push for control of the leader as it concerns all hierarchical levels.

We study the appropriateness and strengths of modeling enterprises, driven by each of the above perspectives, and the remaining of the current paper is organized as follows: We consider related work in Sect. 2 and in particular – we present the SDBC modeling foundations and relevant works touching upon each of the abovementioned enterprise modeling perspectives. Our views concerning those perspectives are presented in Sect. 3, especially with regard to the SDBC-driven modeling. We provide partial exemplification in Sect. 4. Finally, we conclude the paper in Sect. 5.

2 Related Work

As mentioned already, in the current section we consider firstly (in Sect. 2.1) our way of modeling (driven by the SDBC approach) and secondly (in Sect. 2.2) – relevant works with regard to each of the four enterprise modeling perspectives, discussed in the previous section.

2.1 Way of Modeling

Our way of modeling is driven by the **SDBC approach** ("**SDBC**" stands for "*Software Derived from Business Components*"); *SDBC* is a software specification approach (consistent with MDA [12]) that covers the *early phases* of the software development life cycle and is particularly focused on the `derivation of software specification models on the basis of corresponding (re-usable) enterprise models` [1, 2]. *SDBC* is based on three key ideas: (i) The software system under development is considered in its enterprise context, which not only means that the software specification models are to stem from corresponding enterprise models but means also that a deep understanding is needed on real-life (enterprise-level) processes, corresponding roles, behavior patterns, and so on. (ii) By bringing together two disciplines, namely `enterprise engineering` [4] and `software engineering` [13], *SDBC* pushes for applying `social theories` in addressing enterprise-engineering-related tasks and for applying `computing paradigms` in addressing software-engineering-related tasks, and also for integrating the two, by means of sound methodological guidelines. (iii) Acknowledging the essential value of re-use in current software development, *SDBC* pushes for the identification of re-usable (generic) enterprise engineering building blocks whose models could be reflected accordingly in corresponding software specification models. We refer to [1] for information on *SDBC* and we are reflecting the *SDBC* outline in Fig. 1.

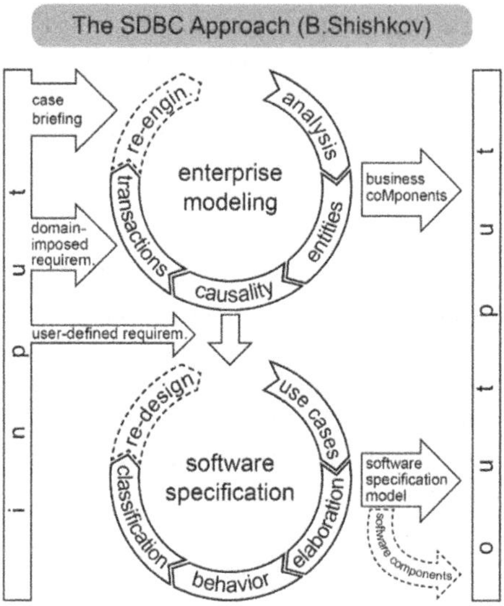

Fig. 1. Outlining the *SDBC* approach (Source: [14], p. 48)

As the figure suggests, there are two *SDBC* modeling milestones, namely `enterprise modeling` (*first milestone*) and `software specification` (*second milestone*). The first milestone has as input a *case briefing* (the initial (textual) information based on which the software development is to start) and the so-called *domain-imposed requirements* (those are the domain regulations to which the software system-to-be should conform).

Based on such an input, an analysis should follow, aiming at structuring the information, identifying missing information, and so on. This is to be followed by the identification (supported by corresponding social theories) of enterprise modeling *entities* and their inter-relations. Then, the *causality* concerning those inter-relations needs to be modeled, such that we know what is required in order for something else to happen [14]. On that basis, the dynamics (the entities' behavior) is to be considered, featured by *transactions* [3]. This all leads to the creation of enterprise models that are elaborated in terms of composition, structure, and dynamics (all this pointing also to corresponding data aspects) – they could either feed further software specifications and/or be "stored" for further use by enterprise engineers. Such enterprise models could possibly be reflected in corresponding *business coMponents* (complete models of business components) [1]. Next to that, re-visiting such models could possibly inspire enterprise re-engineering activities – see Fig. 1.

Furthermore, the second milestone uses as input the enterprise modeling output (see above) and the so-called *user-defined requirements* (those requirements reflect the demands of the (future) users of the software system-to-be towards its functioning) [14].

That input feeds the derivation of a *use case model* featuring the software system-to-be. Such a software specification starting point is not only consistent with the Rational Unified Process - RUP [15] and the Unified Modeling Language – UML [16] but is also considered to be broadly accepted beyond RUP-UML [17]. The use cases are then *elaborated* inspired by studies of Cockburn [17] and Shishkov [1, 3], such that *software behavior models* and *classification* can be derived accordingly. The output is a *software specification model* adequately elaborated in terms of *statics* and *dynamics*. By applying de-composition, such a model can be reflected in corresponding *software components*, as shown in the figure. Such an output could inspire software engineers to propose in a future moment software re-designs, possibly addressing new requirements.

Further, in bringing together the first milestone of *SDBC* and the second one, we need to be aware of possible *granularity mismatches*. The enterprise modeling is featuring business processes and corresponding *business coMponents* (for the sake of brevity we do not provide further elaboration as it concerns the business coMponent concept; for more information, interested readers are referred to [1]) but this is not necessarily the level of granularity concerning the software components of the system-to-be. With this in mind, an ICT (Information and Communication Technology) APPLICATION is considered as matching the granularity level of a business component – an ICT application is an implemented software product realizing a particular functionality for the benefit of entities that are part of the composition of an enterprise system and/or a (corresponding) enterprise information system [1]. Thus, the label "software specification model" as presented in Fig. 1, corresponds to a particular ICT application being specified. Hence, software components are viewed as implemented pieces of software, which represent parts of an ICT application, and which collaborate among each other driven by the goal of realizing the functionality of the application [1] (functionally, a software component is a part of an ICT application, which is self-contained, customizable, and composable, possessing a clearly defined function and interfaces to the other parts of the application, and which can also be deployed independently [13]). Hence, a software coMponent is a conceptual specification model of a software component [1]. Said otherwise, THE SECOND SDBC MILESTONE is about the identification of software coMponents and corresponding software components.

2.2 Language Acts, Regulations, Public Values, and Energy

As already mentioned, in the current Sect. we consider works relevant to the *four enterprise modeling perspectives* discussed in the Introduction.

As it concerns `language acts`, we are mainly considering the works of **Dietz** [4] that touch upon *Enterprise Ontology*, aiming at extracting the *essence* of an enterprise from its actual appearance. The *Organization Theorem* has crucial importance with regard to that and it is in turn backed by four axioms, namely: the *operation axiom*, the *transaction axiom*, the *composition axiom*, and the *distinction axiom*. The operation axiom states that the operation of an enterprise is constituted by actors who perform two kinds of acts, namely *production acts* (that contribute to bringing about the goods/services being delivered) and *coordination acts* (that are about complying with commitments regarding the performance of corresponding production acts). Referring to **LAP** – the *Language-Action Perspective* [18] and to the *operation axiom*, the transaction axiom poses that a coordination act is performed by one actor (called "producer") and directed towards another actor (called "customer"). Hence, the notion of **transaction** refers to the question how production acts and coordination acts are related to each other, and this all points to two "conversations", namely: an *actagenic conversation* (it is about the order) and a *factagenic conversation* (it is about the result). Finally, it is established that the INITIATOR of a *transaction* is the customer while the EXECUTOR of the transaction is the producer. The composition axiom concerns the notion of a *business process* seen as a structure of causally related transactions. Hence, next to the *operation axiom* (that considers the elementary acts) and the *transaction axiom* (that considers putting them together as *transactions*), the *composition axiom* considers structures of transactions driven by causality and this is what is meant by "business process". *Causality* can be illustrated in a simple way: In order to configure a local-area network, one would need to have firstly provided personal computers, a switch, a server, and so on, and in order to configure in turn a personal computer, one would need to have firstly provided a monitor, a hard disk, a motherboard, and so on. In a similar way, *causality* is considered when addressing structures of *transactions*. Finally, the distinction axiom serves to separate the distinct human abilities playing a role with regard to communication, namely: PERFORMA (the actual act of *evoking an attitude*), INFORMA (it is about *conveying semantics*), and FORMA (it is about *conveying information*).

As it concerns **regulations**, we are mainly considering the works of **Liu** [5] that touch upon *Organizational Semiotics* (**OS**), in general, and the *Norm Analysis Method* (**NAM**) – in particular. Actually, *OS* is a branch of **semiotics** while *NAM* is one of the two *OS* methods. *OS* focuses on the nature, characteristics, and behavior of signs – it is claimed that in contrast to the concept of *information*, **signs** offer a more rigorous and solid foundation to understand *information systems*. For example, within a business context, a bank note is much more than a piece of colored paper with digits on it; it stands for the bank note holder's wealth and ability to pay, as well as the issuing bank's authority and credibility, and much more. Next to *signs*, *OS* considers the notion of **affordance** featuring *dependencies* – for example, in the context of a university library, a book *affords* to be borrowed. Finally, it is through *NAM* that *OS* addresses the **norms** based on which behaviors are realized – norms are the rules and patterns of behavior, either *formal* or *informal*, *explicit* or *implicit*, existing within a society, an enterprise, or even a small group of people working together to achieve a common goal; *four types* of norms are considered, namely: evaluative norms, perceptual norms, cognitive norms, and behavioral norms. Each type of *norm* governs human behavior from different aspects. A **norm analysis** is normally carried out on the basis of the results of a prior **semantic analysis** (featuring the concepts under study and their inter-relations). The *semantic* model delineates the area of concern of an enterprise. The patterns of behavior specified in the *semantic* model are part of the fundamental *norms* that retain the ontologically determined relationships between agents and actions without imposing any further constraints. In general, a complete *norm analysis* can be performed in four steps: responsibility analysis (it enables one to identify and assign responsible entities to each action), proto-norm analysis (it helps one to identify relevant types of information for making decisions concerning a certain type of behavior), trigger analysis (it is to consider the actions to be taken in relation to the absolute and relative time), and detailed norm specification (it concerns the actual specification of *norms* in two versions, a *natural language* and a *formal language*).

As it concerns **public values**, we are mainly considering the works of **Schwartz** [9, 19] touching upon *three universal human needs* – the needs of individuals as biological organisms, the needs of coordination of social interaction, and the needs of preservation and well-being of a social group. Through socialization, those *needs* are reflected in as **public values** ("**values**", for short), to give them different significance, and to use culturally shared concepts in the communication process [9]. Schwartz observes consensus in behavioral sciences about some of the leading characteristics of values, namely that they are: 1) beliefs closely related to emotions; 2) motivational construct – refer to the preferred goals; 3) have an abstract nature that distinguished them from norms and attitudes; 4) function as standards or criteria in the selection or evaluation of behaviors, people or events; 5) rank in order of importance and build a hierarchical system of *value* priorities; 6) many relevant *values* are involved in the formation of a particular attitude or behavior [19]. The organization of *values* at the individual level is the result of the psychological dynamics of conflict and compatibility that humans experience in the process of pursuing different goals in daily life. On the contrary, the structure of the *value* system at the societal level reflects in particular the various models that communities use to solve problems, arising from the regulation of human activity, such as: *a)* the relationship between the individual and the group – to what extent people are autonomous or included in their groups, described by embeddedness vs. autonomy value orientations (the undifferentiated vs. the differentiated from the group individual); *b)* ensuring responsible social behavior – how to motivate people to coordinate their actions and respect the other's right, described by hierarchy vs. egalitarianism value orientations (inequality vs. equality), and *c)* the role of the individual in the natural and social environment – whether it is more important to adapt to the outside world and accept it as it is or to constantly strive to change and exploit it, described by harmony vs. mastery value orientations (adaptation to the environment vs. control and change). *Embeddedness*, *egalitarianism* and *harmony* are collectively oriented values, while *autonomy*, *hierarchy* and *mastery* are individually oriented ones [9, 19]. The ways in which those alternatives are solved is reflected in the social value priorities.

As it concerns "**energy**"-related issues, we mainly refer to **Satyasangananda** [11] featuring the **Hindu philosophy**. According to it, all matter is composed of a combination of *five tattwas*, i.e. **elements**. The *Shiva Swarodaya* (an ancient Sanskrit tantric text) explains that creation takes place due to these five elements and by them it is sustained. Further, *Tantraraja Tantra* stipulates that the five elements permeate the entire body and mind. Everything we do and think is under the influence of these *elements*. The *five elements* are known as **ether**, **air**, **fire**, **water** and **earth**. Nonetheless, the *five elements* should not be mistaken for physical or chemical elements. They should rather be regarded as a consequence of emanations which are created by different energies or life-force vibrations. However, according to *Hindu* studies (particularly **Yoga** and **Tantra**) that have examined the elements in detail, we are composed of those *elements* and are continuously subject to their influence; they are not different *energies* but are different aspects of the same *energy* manifesting itself in infinite various combinations. The *elements* constantly interact and when there is a balance between them, we feel *physically healthy*, *mentally calm* and *aware*. As they are constantly changing, their *environment* and their *balance* are disturbed and one or two *energies* may prevail over the rest. As for the *elements* themselves: The **ether** element concerns issues such as *space of mind* and *inspiration*. This *element* governs mental aspects among which are *balance, care, support*, and so on. The **air** element is "connected" to parts of the brain that are responsible for *creativity* and *art*. This element is about *motion, development, rhythm*, and so on. The **fire** element nourishes *growth, change*, and *evolution*. This element governs *conscious actions, dynamics*, and *will*. The **water** element concerns *emotions* that in turn undergo different "phases" and may change unexpectedly. This element is about *directing the focus inwards, possessing the power to overcome obstacles*. Finally, the **earth** element is about *patience, stability*, and *sustainability*. This element is an essential *security* issue.

3 Four Enterprise Modeling Perspectives and an SDBC-Driven Software Specification

As mentioned already, in the current section we will present our modeling proposal featuring the consideration (with regard to the enterprise modeling challenge) of language acts, regulations, public values, and energy, as part of our *SDBC-driven software specification* that is essentially based on *underlying enterprise modeling*. This enterprise modeling stays on many "pillars" and modeling *language acts, regulations, public values*, and *energy* are only some of them. Hence, we do not claim exhaustiveness, "keeping the doors open" for further research, as illustrated in Fig. 2:

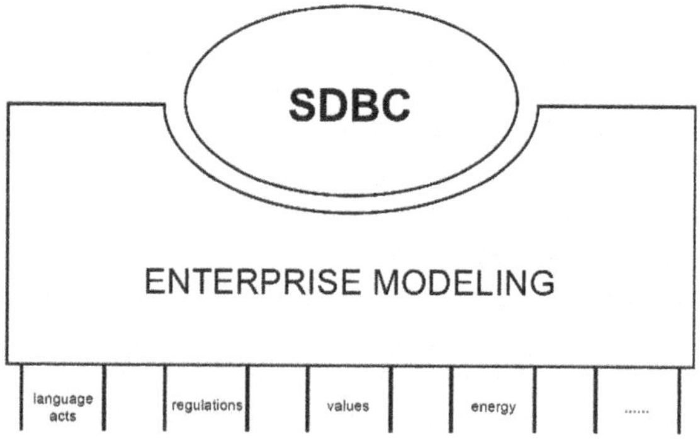

Fig. 2. Addressing four enterprise modeling perspectives

As the figure suggests, we consider each of those perspectives as reflecting corresponding *enterprise modeling foundations*, such that in the end we have a solid enough *enterprise model* that can be reflected in adequate *software specifications*, following the

SDBC guidelines. Nevertheless, for the sake of brevity, in the current paper we only limit ourselves to addressing the enterprise modeling perspectives.

3.1 Language Acts

Referring to the discussion presented in the previous section, we go for particularly considering the Transaction Axiom and especially, the Transaction Pattern as a key modeling element to capture language acts in the context of business process modeling, as studied in [1].

We interpret the transaction concept (see Sect. 2) as centered around a particular *production fact*. The reason is that the actual output of any *enterprise system* represents a set of *production facts* related to each other. They actually bring about the useful value of the business operations to the outside world and the issues connected with their creation are to be properly modeled in terms of *structure*, *dynamics*, and *data*.

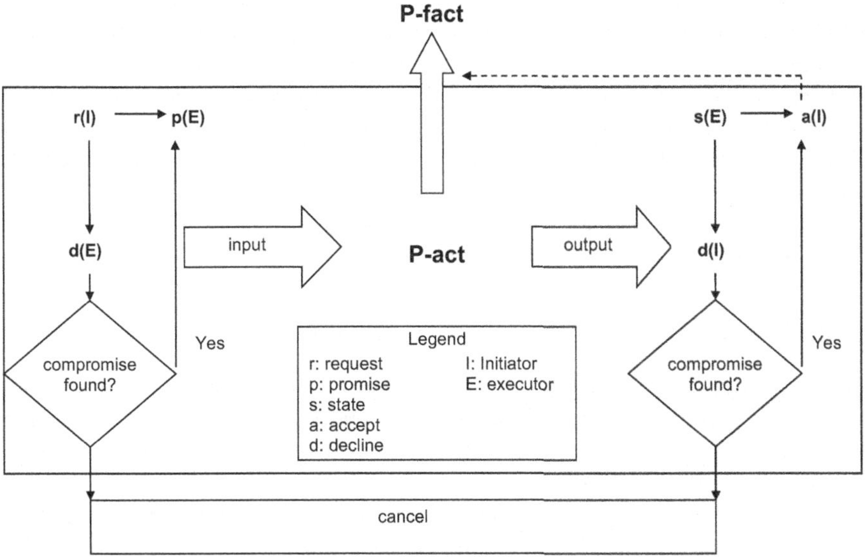

Fig. 3. A proposed interpretation of the transaction concept (Source: [3], p. 70)

However, the already justified necessity of considering also the corresponding *communicative* aspects is important. Although they are indirectly related to the *production facts*, they are to be positioned around them. As already stated, we address this through our interpretation of the *transaction* concept, as depicted in Fig. 3; as seen from the figure, the *transaction* concept has been adopted, with a particular stress on the *transaction's* output – the *production fact*. The *order phase* is looked upon as an input for the *production act*, while the *result phase* is considered to be the *production act*'s output. The dashed line shows that a *transaction* could be successful (which means that a *production fact* has been (successfully) created) only if the *initiator* (the one who is initiating the *transaction*, as presented in Fig. 3) has accepted the *production act* of the other party (called *executor*). As for the (*coordination*) *communicative acts*, grasped by the *transaction*, they are also depicted in the figure. The *initiator* expresses a *request* attitude towards a proposal (any *transaction* should concern a *proposition* – for example, a shoe to be repaired by a particular date and at a particular price, and so on). Such a *request* might trigger either *promise* or *decline* - the *executor* might either *promise* to produce the requested product (or service) or express a *decline* attitude towards the *proposition*. This expressed attitude actually triggers a discussion (negotiation), for example: "I cannot repair the shoe today, is tomorrow fine?… and so on". The discussion might lead to a compromise (this means that the *executor* is going to express a *promise* attitude towards an updated version of the *proposition*) or might lead to the *transaction's* cancellation (this means that no *production fact* will be created). If the *executor* has expressed a *promise* attitude regarding a *proposition*, then (s)he must bring about the realization of the *production act*. Then the *result phase* follows, which starts with a *statement* expression from the *executor* about the requested *proposition* that in his/her opinion has been successfully realized. The *initiator* could either accept this (expressing an *accept* attitude) or reject it (expressing a *decline* attitude). Expressing a *decline* attitude leads to a discussion which might lead to a *compromise* (this means that finally the *initiator* is going to express an *accept* towards the realized *production act*, resulting from *negotiations* that have taken place and *compromise* reached) or might lead to the *transaction's cancellation* (this means that no *production fact* will be created). Once the realized *production act* is *accepted* the corresponding *production fact* is considered to have appeared in the (business) reality.

Hence, we adopt language acts in our enterprise modeling, by considering transactions as the elementary modeling building blocks.

3.2 Regulations

Referring to the previous discussion (see Sect. 2), we essentially count on semiotic norms for capturing, expressing, and establishing regulations because regulations represent sets of rules and rules in turn can be adequately brought forward via semiotic norms, as justified by Liu [5].

Further, in *enterprise modeling*, most *rules* and *regulations* fall into the category of **behavioral norms**. Those *norms* prescribe *what people must, may, and must not do*, which are equivalent to *three deontic operators*: "*is obliged*", "*is permitted*", and "*is prohibited*". Hence, the following format is considered suitable for specifying *behavioral norms*.

```
whenever <condition>
if <state>
then <agent>
is <deontic operator>
to <action>
```

To exemplify the above, we consider a credit card holder - Josh. Imagine that: (i) The credit card limit is 5000.00 EURO; (ii) Josh has used already 4285.58 EURO; (iii) Josh needs to purchase an airplane ticket for the price of 425.19 EURO. Hence, a norm derived based on the above information is:

```
whenever Josh has a valid credit card
if Josh has not reached the credit card limit
then Josh
is permitted
to use the credit card for an amount up to the difference between the
   credit card limit and the amount currently used
```

In our enterprise modeling, we use norms in elaborating transactions.

3.3 Public Values

Values are considered to be desires of the general public (or public institutions/organizations that claim to represent the general public), that are about properties considered societally valuable, such as respecting the privacy of citizens or prohibiting polluting activities [7]. Even though values are to be broadly accepted (that is why they are public), they may concern individuals (for example: considering privacy) [1]. Hence, put broadly, values concern the societal expectations with regard to the way services should be delivered [20]. Further, we argue that "values" become actual "values" only if resources are committed for this (for example, a government finds privacy so important that time and money are invested to regulate and enforce privacy); otherwise things only remain at the level of "hollow" abstract desires (such as for example: "Make the World a better place") that are stated but are never effectively realized.

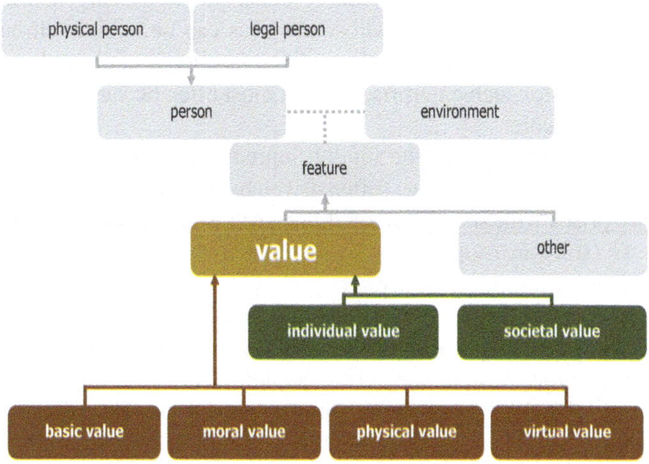

Fig. 4. Categorizing values (Source: [7], p. 404)

We consider a value categorization (Fig. 4) according to which values are desires relevant to particular *persons* (either *physical* or *legal* persons) or their *societal environment*. As such, values may either concern a particular individual or society altogether. Hence, we can distinguish between *individual values* (for example, privacy) and *societal values* (for example, sustainability). We also distinguish between *basic values* (for example, love), *moral values* (for example, justice), *physical values* (for example, nature), and *virtual values* (for example, intelligence).

The way to reflect values in enterprise modeling is to "translate" them in functional requirements – even though values themselves are non-functional in nature, the only way to expose them is by means of functional (software) solutions that are to reflect underlying requirements. More information featuring the SDBC-driven requirements specification can be found in [1].

3.4 Energy

Referring to the previous discussion (see Sect. 2), we explicitly consider the five elements, namely: *earth*, *water*, *fire*, *air*, and *ether*.

We consider each enterprise as an "organism", similarly to a human; we observe similarities between human behavior and organizational (enterprise) behavior, at least as it concerns "influences" from those five elements, and in particular:

- An earth-driven enterprise would be stable and conservative, in the sense that business processes would be evolutionary, changing slowly over time. Examples for this are enterprises driven by narrow expertise and specific business processes, such as hand-made souvenir production, air-co compressor repair, and so on. IT (software) support for such enterprises would most often be aligned with the specific business processes characterizing the enterprise.
- A water-driven enterprise would be solid but unstable and changeable, in the sense that business processes may stay essentially stable but often changeable as realization. Examples for this are enterprises whose business can be realized through different channels, such as consultancy (it can be realized face-to-face, distantly, and so on). IT (software) support for such enterprises would most often be variant-driven, assuming the same software core.
- A fire-driven enterprise would be totally unpredictable, in the sense that business processes can significantly change (often driven by personal decisions). Examples for this are enterprises with strong personal presence, such as art agencies, campaigns, and so on. IT (software) support for such enterprises would most often assume new software instances starting from scratch.
- An air-driven enterprise would be dynamic and fast developing, in the sense that business processes may change even essentially. Examples for this are enterprises whose business processes are subject to technological and/or legal influences, such as road-traffic-related enterprises, e-Businesses, and so on. IT (software) support for such enterprises would most often assume powerful interfaces towards integration with other technologies.
- Finally, an ether-driven enterprise would be balanced and communicative, in the sense that business processes assume integration and coordination with regard to complex

environments. Examples for this are tourist agencies, car rental companies, and so on. IT (software) support for such enterprises would most often assume standardization and component-based solutions, allowing for fast "plug-and-play" replacements.

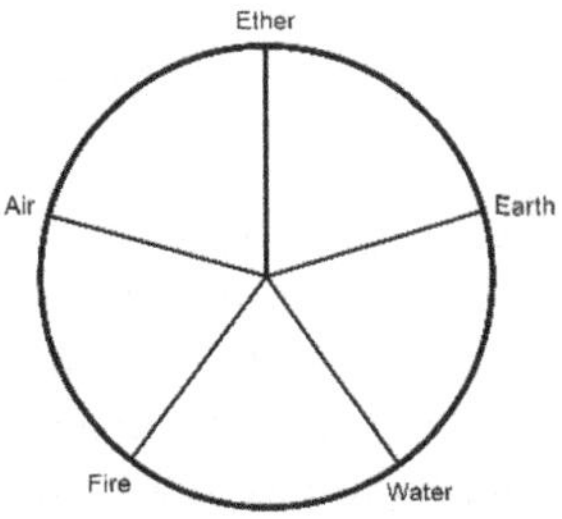

Fig. 5. Considering the 5 elements

All five elements are reflected in Fig. 5 and the circle suggests "continuity" in the sense that none of those elements pops up in isolation – most often it is the case that more than one of them have impact. Hence, our discussion in the current sub-section is slightly simplified, assuming influence from just one element. We do this for the sake of brevity, driven by the purpose to just outline our ideas on weaving the 5-elements-analysis in enterprise modeling.

What we would do in the end with all this is to facilitate our design as it concerns the software specifications and also the enterprise modeling preceding them. The assumption we make is that we are able to "sense" which element is predominantly influencing the enterprise under study. Discussing this further is left beyond the scope of the current paper.

4 Illustrative Example

As already mentioned, we do partial exemplification in order to illustrate our enterprise modeling touching upon: (i) language acts; (ii) regulations; (iii) public values; (iv) energy. For this we use an illustrative example (following guidelines of Yin [21]) running throughout the current section, featuring the challenge of specifying a financial e-Mediator that offers advices for purchasing insurance products. To do this, it is needed to realize match-making between what the customer wants and what products are available.

We represent the *Customer*, *Advisor, Match-maker*, *Request Processing Unit* (we call it "Request Handler", for short) and *Data Search and Processing Unit* (we call it "Data Searcher", for short), as just entities and put them in named boxes, as follows: Customer (**C**); Advisor (**A**); Match-maker (**MM**); Request handler (**R**); Data searcher (**D**) – see Fig. 6.

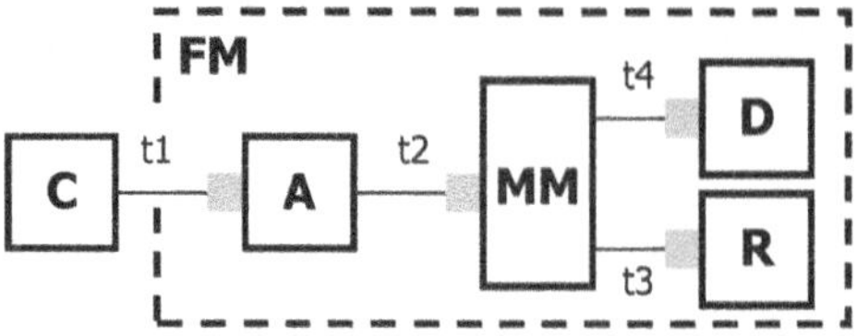

Fig. 6. FM business entity model (Adapted from: [1], p. 223)

On the figure, the connections indicate the need for *interactions* between *entities*, in order to achieve the business objective of financial mediation; with each connection, we associate a single *transaction* (**t**): C-A (**t1**); A-MM (**t2**); MM-R (**t3**); MM-D (**t4**). Further, **C** is positioned in the *environment* of the financial mediation system – **FM**, and **A**, **MM**, **R** and **D** together form the **FM** system. Through **t1**, **FM** is related to its *environment* (represented by **C**). Thus, from the perspective of **C**, there is no difference between **FM** and **A**.

That is how we weave language acts in our enterprise modeling, since behind each transaction "stays" the transaction pattern – see Fig. 3.

Further, we go for straightforwardly elaborating the above model in terms of semiotic norms, by providing (just for illustrative purposes) several norms:

-------**Whenever** C has requested advice
If MM has realized match-making
Then A
Is obliged to formulate and deliver an advice
-------**Whenever** C has requested advice
If R has received submitted customer information
Then R
Is obliged to deliver standardized customer specification
-------**Whenever** C has requested advice
If D has received information about the type of a customer need
Then D
Is obliged to deliver a candidate-matches list

This norm elaboration is partial – we have only identified several norms to demonstrate how transactions could be usefully elaborated in terms of regulations.

Further, as it concerns public values, our exemplification will also be partial, for the sake of brevity. We will consider just one public value, namely: ACCOUNTABILITY.

In the case of automated financial mediation, it is expected that any design/maintenance/operational failure would be easily traceable and reportable, thus leading to corresponding accountabilities. This would concern responsibility for directing the customer to an inappropriate (with regard to his/her requirements) insurance product, violations with regard to his/her privacy-sensitive data, and so on. Accountability requires the curation of software and algorithms, and also failure of components should be traced.

In this regard, we argue that DESIGN should be important and for this reason, we lean towards weaving accountability in the FM system design – this represents a Value-Sensitive Design – VSD [8].

Our VSD-inspired view on accountability's implications with regard to automated financial mediation featuring insurance products, is depicted in Fig. 7:

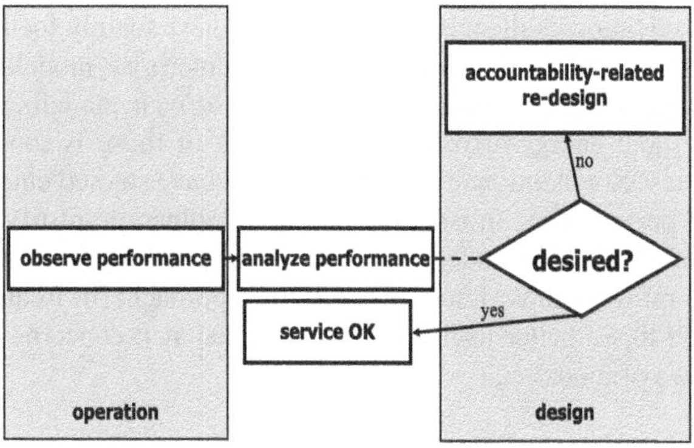

Fig. 7. Accountability implications

As it is seen from the figure: (i) at design time we specify what is DESIRED while at run time it shows up what ACTUALLY HAPPENED – what was observed; (ii) if the observed performance corresponds to what was desired, then the service delivery has been adequate; (iii) otherwise, the desired performance was not achieved and corresponding ACCOUNTABILITY would need to be considered and this would only be possible if the accountability value has been reflected in the design, such that the customer can effectively trace back what happened and identify the responsible "actor"(s).

Next to that, establishing accountability through re-designs could possibly lead to some value "tensions", if tracing back what happened would lead to: (i) disclosing privacy-sensitive information; (ii) making technical data explicit including such data that represents copyright-protected "know-how"; (iii) reducing the system availability (during the traceability-related actions).

Finally, as it concerns energy, the FM case is clearly a WATER-driven one because: (i) The business entity model, depicted in Fig. 6, would look absolutely the same no matter if the advising is delivered by a human (who in turn collaborates with other humans for the match-making, request processing, and so on) or by a software component (that in turn collaborates with other software components for the match-making, request processing, and so on) => The business processes are ESSENTIALLY STABLE. (ii) At the same time, those business processes can be realized through different "channels", such as human-driven and software-driven (see above) and therefore, the BUSINESS OPERATION IS CHANGEABLE.

In summary, through the FM example, we have demonstrated enterprise modeling activities in 4 perspectives, namely: language acts, regulations, public values, and energy, such that we not only demonstrate how this can be done (and how different modeling activities could be considered in combination) but we have also implicitly justified the importance of each of those enterprise modeling perspectives.

5 Conclusions

Building upon previous research of the authors, this paper concerns the SDBC approach that is about the enterprise-modeling-driven specification of software. Abstracting from the software specification challenge, in the paper, we have mainly focused on the modeling of enterprises, considering in particular four enterprise modeling perspectives, namely: language-acts-driven modeling, regulations-driven modeling, public-values-driven modeling, and energy-driven modeling. Each of those is rooted in particular underlying theories. Not claiming exhaustiveness, we have studied enterprise modeling in all those four perspectives, in isolation and in combination, justifying their importance and illustrating possible modeling activities. Our plans for future research include: (i) Better incorporation of those issues in the SDBC approach; (ii) Realization of bigger case studies, such that a better justification is achieved as it concerns the adequacy of our proposed ways of modeling.

Acknowledgement. This work is supported by: (i) Bulgarian National Science Fund, Project: KP-06-N 32/4/2019; (ii) Ministry of Education and Science, Project of the National Scientific Program "Information and Communication Technologies for a Single Digital Market in Science, Education and Security (ICTinSES)".

References

1. Shishkov, B.: Designing Enterprise Information Systems, Merging Enterprise Modeling and Software Specification. Springer, Cham (2020)
2. Shishkov, B., Quartel, D.: Combining SDBC and ISDL in the modeling and refinement of business processes. In: Manolopoulos, Y., Filipe, J., Constantopoulos, P., Cordeiro, J. (eds.) Enterprise Information Systems. ICEIS 2006. Lecture Notes in Business Information Processing, vol. 3. Springer, Heidelberg (2008)
3. Shishkov, B.: Software Specification Based on Re-usable Business Components. Delft University Press - Sieca Repro, Delft (2005)
4. Dietz, J.L.G.: Enterprise Ontology, Theory and Methodology. Springer, Heidelberg (2006)
5. Liu, K.: Semiotics in Information Systems Engineering. Cambridge University Press, Cambridge (2000)
6. Shishkov, B., Larsen, J.B., Warnier, M., Janssen, M.: Three Categories of Context-Aware Systems. In: Shishkov, B. (ed.) BMSD 2018. LNBIP, vol. 319, pp. 185–202. Springer, Cham (2018)
7. Shishkov, B., Mendling, J.: Business process variability and public values. In: Shishkov, B. (ed.) Business Modeling and Software Design. BMSD 2018. Lecture Notes in Business Information Processing, vol. 319. Springer, Cham (2018)
8. Van den Hoven, J.: Value Sensitive Design and Responsible Innovation. In: Owen, R., Bessant, J., Heintz, M. (eds.) Responsible Innovation: Managing the Responsible Emergence of Science and Innovation in Society. Wiley, Hoboken (2013)
9. Schwartz, S.H.: An overview of the Schwartz theory of basic values. Online Read. Psychol. Cult. **2**(1), 11 (2012)
10. Garvanova, M., Shishkov, B., Janssen, M.: Composite public values and software specifications. In: Shishkov, B. (ed.) Business Modeling and Software Design. BMSD 2018. Lecture Notes in Business Information Processing, vol. 319. Springer, Cham (2018)
11. Satyasangananda, S.: Tattwa Shuddhi. Yoga Publications Trust, Munger (2007)
12. MDA: The OMG model driven architecture (2020). http://www.omg.org/mda
13. Szyperski, C.: Component Software, Beyond Object-Oriented Programming. Addison-Wesley, Boston (1998)

14. Shishkov, B., Janssen, M., Yin, Y.: Towards context-aware and privacy-sensitive systems. In: Shishkov, B. (ed.) Business Modeling and Software Design. BMSD 2017. SCITEPRESS (2017)
15. Kruchten, P.: The Rational Unified Process, An Introduction. Addison-Wesley, Boston (2003)
16. UML: The unified modeling language (2020). http://www.uml.org
17. Cockburn, A.: Writing Effective Use Cases. Addison-Wesley, Boston (2000)
18. Searle, J.B.: Speech Acts: An Essay in the Philosophy of Language. Cambridge University Press, Cambridge (1969)
19. Schwartz, S.H.: The refined theory of basic values. In: Roccas, S., Sagiv, L. (eds.) Values and Behavior. Springer, Cham (2017)
20. Friedman, B., Hendry, D.G., Borning, A.: A Survey of Value Sensitive Design Methods, vol. 1, p. 76. Hanover, Now Foundations and Trends (2017)
21. Yin, R.: Case Study Research: Design and Methods. Sage Publications, Thousand Oaks (1994)

Complex Systems Modeling Overview About Techniques and Models and the Evolution of Artificial Intelligence

Wafa Mefteh[1](✉) and Mohamed-Anis Mejri[2]

[1] University of Gabes, Gabès, Tunisia
[2] ISETKb, Kélibia, Tunisia

Abstract. Several natural systems are defined as complex (related to several domains including: physics, biology, social sciences, cognitive sciences, ...) and many artificial and industrial systems fall into this category. These systems are often distributed, open, large-scale and heterogeneous. With their interconnections so complicated, they are beyond the general understanding of a human being. Modelling such systems is not an easy task and so the need of efficient techniques and models. Different techniques and models was proposed to tackle the complexity of modelling such systems. The evolution of artificial intelligence techniques can solve several problems related to complex systems design.

Keywords: System modelling · Complex systems design · Artificial intelligence

1 Introduction

Complicated systems are difficult systems to model but they are deterministic. All of their components react in reliably predictable ways. The laws governing the system are stable, and they are applied the same way every time. Complicated systems don't have a mind of their own (don't have goal, don't have their own agenda that determines how they respond to conditions). Indeed, we can say that, for complicated systems, we have the capability of understanding such systems but it is not easy to do it. To understand complicated systems, we must understand each component of the system and then bring it all together to understand the overall system.

But, the notion of complexity is a concept invented by Man to allow him to describe a set of things that seem related, that we do not find simple and for which it is difficult or impossible for us to anticipate the result, to understand it by understanding the behaviour of its components. A complex system is defined as a system composed of many interacting elements whose overall characteristics cannot be reduced to those of its components. It is a system whose overall behaviour cannot be analysed as a succession or juxtaposition of the subsystems behaviours independently. All components contribute simultaneously to

Á. Rocha et al. (Eds.): WorldCIST 2020, AISC 1159, pp. 678–688, 2020.
https://doi.org/10.1007/978-3-030-45688-7_67

the behaviour of the system. Indeed a complex system involves the following four main factors: The system is composed of interacting elements (or components). Many relations are established between these elements. The system is immersed in a dynamic environment that it must adapt to it. The system adopts a dynamic behavior over time, a non-linear behavior. It can evolve to a growing complexity. In this paper, we begin by presenting the complex systems theory and giving some examples of complex systems. Then, we give an overview about techniques and models proposed to tackle the complexity of modelling such systems. Finally, we discuss the evolution of artificial intelligence and how it can solve problems related to complex systems design.

2 The Complex Systems Theory

The theory of complex systems is recent, its content and vocabulary are still uncertain. However, it is the subject of several research works because it develops an awareness of the omnipresence of complex systems in many disciplines: economic, social, ecological ... and the need to understand their functioning. Several important theories developed during the twentieth century have contributed to the development of the theory of complex systems: *The theory of chaos* which studies systems which are extremely sensitive to small causes and their behavior has a cyclical aspect. *Cybernetics* that studies the operation of slave systems with the concepts of positive or negative feedback. It formalizes the graphs of interactions between the organs of a system. *The systemic* which is based on the concept of holism, antithesis of reductionism, which states that to understand the functioning of a system we must simultaneously study all of its components with their interactions and not study separately each complosant.

Murray Gell-Mann [10] traces the meaning of complexity to the root of the world. Plexus means braided from which is derived complexus meaning braided together. Complexity is therefore associated with the inter-connectivity elements with the system and between a system and its environment [2]. A complex system consists of agents (components) that interact with each other, with their environment and with the emergent phenomena created by these interactions. Agents can be of a varied nature: an animal, a person, a group of people, an institution, an organ, a cell, an enzyme ... . The behavior rules of an agent define the stimuli that it emits towards the other agents based on the stimuli he receives from other agents and its environment. These rules are evolutionary according to the experience of the agent: stimuli that it has received and that it has emitted. In the following, I touch the most important common characteristics of complex systems that have been found so far: *Emergence*, *Self-organisation*, *Non-linearity*, *Evolutionary dynamics*, *Limited predictability*. *Emergence*; is an important property of complex systems, is a process of phenomena creation by the interactions of the agents: between them and with their environment. Emerging phenomena are of various natures, for example the appearance of a new agent, a modification of the environment, a distribution law of events ... These phenomena can be surprising and challenge intuition and common sense. Emergencies are not planned or piloted by an authority that would have an overview of the system. This so-called higher order behaviour cannot simply be derived by aggregating behaviour at the level of the elements. The whole is more than the sum of its parts. This higher order was not intended by the elements. It is a spontaneous order. *Self-organisation*; complex adaptive systems operate without central control. However, they are often characterised by a certain order. They, as it were, organise themselves from the bottom-up. *Non-linearity*; complex systems may suddenly change behaviour or move to another regime. They may move from a high degree of stability to very unstable behaviour. Think for example about revolutions and financial crises. *Evolutionary dynamics*; complex adaptive systems are often shaped by evolutionary dynamics. The mechanism of evolution starts with variation. Then there is selection of elements that are fit for the changed conditions. These elements flourish and multiply in the system. They may also change the external environment of the system, causing new variation. New variation may also come from outside the system. A new cycle of variation-selection-multiplication-variation starts. The system is never at rest. There is no movement to a knowable "end point" or equilibrium. There is constant change and innovation. *Limited predictability*; the behaviour of complex systems cannot be predicted well. Small changes in initial conditions or history can lead to very different dynamics over time. The existence of non-linear behaviour also adds to unpredictability.

Many natural systems (immune systems, ecologies, brains, ...), industrial systems (supply chain, automotive ...) and artificial systems (computing systems, artificial intelligence systems, evolutionary programs ...) are characterized by complex behaviors. In this section we give some examples of natural, industrial and artificial complex systems. *Anthill;* Ants are social insects and workers whose colonies can reach several million individuals. The collective behavior of ants has always fascinated and amazed naturalists. Everything seems to be as if each colony behaved like one super-organism and that there existed virtually within these societies a mysterious force capable of coordinating the activities of several thousand individuals. One of the main characteristics of ants is their ability to collectively solve a variety of problems, often quite complex, that they face daily. These problems are of varied nature: research and selection of food sources, sharing of tasks and organization of work, etc. It should also be noted that ants also adapt very quickly to changes in their environment. Such collective prowess leads us to question the mechanisms by which the individuals composing these societies coordinate their activities. Ants have no representation, or explicit knowledge of the structures they produce. In particular, they do not use predefined plans to build their nests. Their brains, which comprise about one hundred thousand neurons, are not powerful enough to allow a single individual to integrate all the information on the state of the colony at a given moment and then ensure the division of tasks and the good running of society. So there is no conductor among the ants. Although an anthill may seem modest from the outside, it hides an extraordinary network of underground galleries. In an anthill the agents are the ants, the ant interactions are indirect by means of deposits of odorous hormones called pheromones, the emergent phenomenon is the habitat of ants consisting of galleries. There is no centralized coordination, only local interactions between ants that have no overview of the anthill.

Group Flights of Birds or Benches of Fish. We all have in mind these images of birds twirling in the sky. But flying in a group is not everyday use. This is often done on long flights, very high flights. When they fly together, they are less vulnerable than when they are alone, because the predator does not have only one target. It is therefore a technique used by these small birds to try to save their skin. Another hypothesis is that they would seek to save their energy. This is particularly the case when they are flying in a line or a V-shape. The bird in front is more busy because it is facing the wind, there is a friction of the air. When we are behind, we feel less. This is called self-organization. Flying together, this often forms a group of several thousand individuals surprisingly very homogeneous. The mystery of this harmony lies in the fact that birds do not have a leader and yet manage to remain very united, even during rapid changes of direction. Birds are not the only ones to self-organize. We thus find these same behaviors in some fish that move in benches. The synchronized movements of benches of fish are conceptually simple. By observing twirling thousands of fish, one can not help but be fascinated by the complex vision offered to us. How do fish communicate inside a school to turn so synchronized? How does a benche of fish go from one collective behavior to another? While these fish seem randomly distributed, one has the impression of an intentional and centralized control. However, all the results indicate that the movement of the bench results from the individual actions of each animal, acting solely on the basis of the local perception of its environment. Most of the fish that form benches are small. It is generally thought that the main evolutionary advantage of a bank is to protect these small fish from predators. The formation of group flights of birds or benches of fish results from interactions between birds or fish, each of which obeys some very simple rules of behavior such as maintaining a distance from their neighbors and obstacles.

Market. As we knoW, a market is a medium allowing buyers and sellers of a specific good or service to interact in order to facilitate an exchange. The market study is based on several elements including the behavior of the consumer which represents the principal element. "Consumer behavior" refers to all the behaviors that relate to the acquisition of goods and services, but which are not the same as those preceding the acquisition, the choice itself, the use of goods and services purchased and the eventual abandonment of these products. A market can be conceptualized as a complex system whose components are the consumers and the environment is the supply. By his act of buying, a consumer interacts with other consumers by the impact of this purchase on the evolution of the products and by the influence of his act on the behavior of consumers. Emerging phenomena are market prices, changes in supply.

Several industrial systems fall into this category (such as chemical process, power network, sustainable energy systems, manufacturing process, transportation systems, wireless communication network, robotic systems, biomedical systems, ... In a chemical process, chemical compounds are changed with the help of chemical reactions. The chemical process may occur itself when two chemicals (chemical compounds) get in touch with each other or also with the help of any chemical agent. Note that chemical composition of the chemicals or materials changes in a chemical process. The power network is one of the largest, most complicated, and most sophisticated systems in the world. It consists of: generating stations that produce electric power, electrical substations for stepping electrical voltage up for transmission, or down for distribution, high voltage transmission lines that carry power from distant sources to demand-centers, distribution lines that connect individual customers, ... [14]. The complexity of transport systems consists essentially in understanding the spatial organization of mobility.

The realization of such industrial systems makes use of hundreds of engineers from many different specialties. The complexity of industrial systems remains a vague and subjective notion, but it corresponds to a strong industrial reality: it characterizes systems whose control of design, maintenance and evolution poses significant problems, related to their size and the number of technologies used, which make the whole difficult to grasp.

In the automotive industry, a project "vehicle" represents a significant burden man-years of work, spread over years, involves several different trades and involves very large budgets.

The engineering of complex computer systems is all activities pertinent to specifying, designing, prototyping, building, testing, operating, maintaining, and evolving complex computer systems [22]. The new and emerging demands of applications and the evolution of computer architectures and networks now essentially force systems to be complex. Complex computer systems are found in almost every industry. These include industrial process control, aerospace and defence, transportation and communications, energy and utilities, medicine and health, commercial data processing, and others. Indeed, computer systems have to keep pace with the evolution of the industry the requirements of industrial, commercial, and government complex computer systems.

During the last years, speed, capacity, and throughput had exclusively dominated computer development and so developers need to address the increasingly urgent problem of growing complexity by optimizing computer-system design towards mastering complexity.

3 Complex Systems Modelling: Overview About Techniques and Models

To study a complex system, all its components must be considered simultaneously: agents, interactions and the environment. Should consider that "everything is more than the sum of the parts" because we must also consider interactions [16]. Multidisciplinary skills must be brought together to cover all facets of the system under study. The diversity of agents improves the properties of complex adaptive systems: emergence, innovation, self-organization, ... The medium and long-term evolutions of complex systems are unpredictable because it is impossible to define all the variables with precision. It may be futile to want to identify the cause of a situation observed in a system because this situation is often due to multiple causes. It is impossible to identify the first cause among these multiple causes. Complex systems are subject to shifts that are abrupt changes in their scale and speed. For example an economic crisis, a revolution. A tipping point is a state of a system where a small cause can cause a profound and brutal change in the state of the system. In the following, we present some techniques and models used to model some complex systems in order to answer similar questions.

As the transport industry becomes more complex, conventional approaches, focusing on a narrow range of factors, have to be replaced by more nuanced analysis and solutions. The integration of complex networks methods by transport specialists is very recent and comes as a complement to other approaches such as circulation routing and flow optimization which strongly focus on transport costs. Transport networks are mostly studied in a static view. Research in this field is very voluminous and diverse. The work in this field can be classified into two classes: works which describe the entire network (the global level) and works which are interested in describing either groups or individual nodes of within the network [5]. Hu and Zhu [13] present an empirical study of the global (worldwide) maritime transport network (WMN) using the different representations of network topology. To construct the WMN (in which the nodes are ports and links are container liners connecting the ports), they used data obtained from an authoritative container industry data base named CI-online and based on the idea of spaces L and P which is widely used in the study of public bus transportation networks and railway networks.

Biological regulatory mechanisms, including gene expression, are inherently complex systems as defined above. As such, they cannot be understood by mere identification of components, products, ensembles and connections. Many researchers worked on modeling biological systems in order to understand evolutionary systems. This includes understanding the representation and communication of information in living systems, predicting protein function from gene sequence, ... Many algorithms was also proposed to tackle some biological problems, such as the development of efficient DNA sequencing procedures. Several researchers have worked on the modeling of biological systems. Julien Delile et al. [4] propose an agent-based model and simulation platform of animal embryogenesis, called MecaGen, 1 centered on the physico-chemical coupling of cell mechanics with gene expression and molecular signaling. Embryonic development is viewed as an emergent, self-organized phenomenon based on a myriad of cells and their genetically regulated, and regulating, biomechanical behavior. Roberta Bardini et al. [3] review some of the most relevant multi-level and hybrid modelling approaches for biological systems presented in the literature. The review is organized presenting the different models based on their relevance to the different aspects of the modelling process.

The biological immune system is a complex adaptive system. There are lots of benefits for building the model of the immune system. For biological researchers, they can test some hypotheses about the infection process or simulate the responses of some drugs. Osvaldo D. Kim et al. [15] describe approaches developed to undertake issues regarding the mathematical formulation and the underlying optimization algorithms, and that address the phenotype prediction by including available kinetic rate laws of metabolic processes. They discuss how these have been used and combined as the basis to build computational strain optimization methods for metabolic engineering purposes and how they lead to bi-level schemes that can be used in the industry, including a consideration of their limitations.

Hannelore Aerts et al. [1] simulated large-scale brain dynamics in 25 human brain tumor patients and 11 human control participants using the Virtual Brain, an open-source neuroinformatics platform. Local and global model parameters of the Reduced Wong–Wang model were individually optimized and compared between brain tumor patients and control subjects. Thyh present a summary of computational modeling workflow using TVB. A number of works have focused on developing new methods for modeling TBI based on current neuropathology and neuroimaging techniques. Horstemeyer et al. [11] developed a new approach to modeling brain tissue, a mechanics-based brain damage framework that was able to correlate Chronic Traumatic Encephalopathy (CTE) pathology in deceased football players to the damage nucleation, growth, and coalescence mechanisms within the tissue model. Garimella et al. [9] and Wu et al. [23] both developed new techniques for utilizing Diffusion Tensor Imaging (DTI) data to embed axonal tract information into existing finite element brain models. Mark F. Horstemeyer et al. [12] gives an interesting state-of-the-art of modeling and simulation methods related to traumatic brain injuries arising from mechanical loads.

The manufacturing process is one of the important processes in a product's life cycle. The sharing of manufacturing process information among different functional application systems, such as process planning, manufacturing simulation, manufacturing execution and project management, has become difficult to implement due to the growing complexity of the manufacturing information of product, process, resource and plant. Lihong QiaoEmail et al. [20] present a manufacturing process information modelling method which builds definition of manufacturing process data by applying current PSL specifications. Enzo Frazzon et al. [7] propose and apply a data-driven adaptive planning and control approach that uses simulation-based optimization to determine most suitable dispatching rules in real-time under varying conditions. Modelling AM processes is important; because it allows AM practitioners to enhance understanding and will help improve the process at the planning stage of the manufacturing process. Panagiotis Stavropoulos et al. [21] reviewed and classified the literature on modelling the existing AM processes. They classified the AM modelling according to key performance indicator (KPI), process parameters and the modelling approach (analytical, numerical or empirical). For this study a classification according to the process mechanism (ISO 17296-2) has been followed.

Because that failure of robotic software may cause catastrophic damages and in order to establish a higher level of trust in robotic systems, formal methods are often proposed. However, their applicability to the functional layer of robots remains limited because of the informal nature of specifications, their complexity and size. Creating intelligent robots and systems requires skills from a broad range of subjects. It poses fundamental questions about the design, physical modeling and control of complex and highly interactive systems. Wenfu Xu et al. [24] give a survey of modeling, planning, and ground verification of space robotic systems. Sergio García et al. [8] illustrate their experiences from EU, academic, and industrial projects in identifying, modeling, and managing variability in the domain of service robots. Indeed, defining a general reference architecture for robotic applications is elusive: robots can have many purposes, forms, and functions, can operate to accomplish different missions, and often operate in an open-ended (changing) environment. As a consequence, each robotic system has to be equipped with a specific mix of functionalities that strongly depend on several factors. Mohammed Foughali et al. [6] propose the use of formal methods to verify the functional layer of robotic systems. They focus on verification by means of model checking, and use statistical model checking to tackle scalability issues. A particular interest was given to real-time properties, e.g. schedulability and bounded response, crucial in robotics.

Synthesis. To model complex systems, researchers are often using metaphors from nature in Simulation and Scientific Models. In general, complex systems models can be distinguished into mathematical and computational ones. A computational model is a formal model whose primary semantics is operational. The model prescribes a sequence of steps or instructions that can be executed by an abstract machine, which can be implemented on a real computer. However, a mathematical model is a formal model whose primary semantics is denotational. The model describes by equations a relationship between quantities and how they change over time. But it is not a strict separation. During the last years, speed, capacity, and throughput had exclusively dominated computer development and so developers need to address the increasingly urgent problem of growing complexity by optimizing computer-system design towards mastering complexity. Performance prediction, and in general behavior prediction, may exploit simulation based approaches [17] *or analytical techniques to evaluate in advance the effects of design choices, or variability under different workloads, or emerging behavior, of systems, and provide a valuable support in all the phases of the lifecycle of a system by means of proper modeling approaches* [18,19].

4 Discussion: Complex Systems Design and the Evolution of Artificial Intelligence

Recent advances in the field of artificial intelligence have shown that combining reinforcement learning with deep neural networks can produce outstanding results. Reinforcement learning (RL) algorithms learn to attain a complex objective over many steps. RL learn to find the relations between immediate actions and late returns. They operate in an environment where feedback is delayed and where it can be difficult to identify what action leads to what result over time. Deep Learning (DL) has attracted much interest in a wide range of applications. The advent of DL has significantly improved many areas in machine learning such as speech and image recognition. The main advantage of DL is that it can automatically find low-dimensional representations of high-dimensional data. DL has accelerated progress in RL, defining the field of "Deep Reinforcement Learning" (DRL). DL enables RL to scale to decision-making problems involving high-dimensional state and action spaces. One of the driving forces behind DRL is the vision of creating systems that are capable of learning how to adapt in the real-world. Unfortunately, most conventional reinforcement learning approaches depends on stationary dynamics to backpropagate learning throughout time. Some of the technical challenges that are currently faced in this domain are the support of non-stationary environment dynamics and the scalability to large multi-agent systems. Despite the fact that autonomous systems require the development of learning approaches robust to the challenge of dynamic and noisy environments, deep reinforcement learning has been loosely extended to the multi-agent domain. Human intelligence surely didn't arise in isolation but by cooperating and competing in a cumulative cultural evolution. Why should it be possible to create artificial intelligence in a single agent framework? Multi-agent designs are often more robust than monolithic designs because failure can be compensated, they are scalable in the sense that we can add more agents to an architecture without the need to redesign the whole architecture and are able to reuse or reconfigure the constituents in different ways. Multi-Agent thinking can provide a design framework providing robustness, scalability and flexibility. Intuitively, for an object to be referred to as an agent it must possess some degree of autonomy, that is, it must be in some sense distinguishable from its environment by some kind of spatial, temporal, or functional boundary. It must possess some kind of identity to be identifiable in its environment. To make the definition of agent useful, we often further require that agents must have some autonomy of action, that they can engage in tasks in an environment without direct external control. Traditionally, Agent-Based Models draw on examples of phenomena from biology such as social insects and immune systems. These systems are distributed collections of interacting entities (agents) that function without a "leader." From simple agents, who interact locally with simple rules of behavior, merely responding befittingly to environmental cues, and not necessarily striving for an overall goal, we observe a synergy which leads to a higher-level whole with much more intricate behavior than the component agents, e.g. insect colonies and immune responses. The field of Artificial Life (AL) produced a number of models based on simple agent rules capable of producing a higher-level identity, such as the flocking behavior of birds, which were called Swarms. In these models, agents are typically described by state-determined automata: that is, they function by reaction to input and present state using some iterative mapping in a state space.

5 Conclusion

We conclude that Complex systems are distinguished from complicated systems whose design difficulties can be solved by a talented engineer. Complex adaptive systems are everywhere (natural, industrial and artificial). The theory of complex systems has helped to conceptualize properties common to systems of various kinds (biological, ecological, economic ...). Advances in computer science considerably strengthen the methods and tools for studying, modeling and simulating complex systems. Recent advances in the field of artificial intelligence have shown that combining reinforcement learning with deep neural networks can produce outstanding results to design and build complex systems. Multi-Agent thinking can provide a design framework providing robustness, scalability and flexibility. The theory of complex systems is a young science with content still poorly defined but it is booming. In this paper, we presented the complex systems theory, examples of complex systems, an overview about techniques and models to design some complex systems and finally a brief discussion about the evolution of artificial intelligence and how using its techniques can deal with complex problems and systems design.

References

1. Aerts, H., Schirner, M., Jeurissen, B., Van Roost, D., Achten, E., Ritter, P., Marinazzo, D.: Modeling brain dynamics in brain tumor patients using the virtual brain. ENEURO **5**(3), 15 (2018)
2. Amaral, L.A., Ottino, J.M.: Organisation as co-evolving complex adaptive systems. In: British Academy of Management Conference, pp. 8–12 (1997)
3. Bardini, R., Politano, G., Benso, A., Di Carlo, S.: Multi-level and hybrid modelling approaches for systems biology. Comput. Struct. Biotechnol. J. **15**, 396–402 (2017)
4. Delile, J., Doursat, R., Peyriéras, N.: Computational modeling and simulation of animal early embryogenesis with the MecaGen platform. In: Kriete, A., Eils, R. (eds.) Computational Systems Biology, 2nd edn, pp. 359–405. Academic Press, Oxford (2014)
5. Ducruet, C., Lugo, I.: Structure and dynamics of transportation networks: models, methods and applications. In: Rodrigue, J.P., Notteboom, T.E., Shaw, J. (eds.) The SAGE Handbook of Transport Studies, pp. 347–364. SAGE, Thousand Oaks (2013)
6. Foughali, M., Ingrand, F., Seceleanu, C.: Statistical model checking of complex robotic systems (2019)
7. Frazzon, E., Kück, M., Freitag, M.: Data-driven production control for complex and dynamic manufacturing systems. CIRP Ann. - Manuf. Technol. **67**, 515–518 (2018)
8. García, S., Strüber, D., Brugali, D., Di Fava, A., Schillinger, P., Pelliccione, P., Berger, T.: Variability modeling of service robots: experiences and challenges. In: VaMoS (2019)
9. Harsha, T., Garimella, R., Menghani, R., Gerber, J.I., Sridhar, S., Kraft, R.H.: Embedded finite elements for modeling axonal injury. Ann. Biomed. Eng. **47**(9), 1889–1907 (2019)
10. Gell-Mann, M.: Complexity at large. Complexity **1**(5), 3–5 (1996)
11. Horstemeyer, M.F., Berthelson, P.R., Moore, J., Persons, A.K., Dobbins, A., Prabhu, R.K.: A mechanical brain damage framework used to model abnormal brain tau protein accumulations of national football league players. Ann. Biomed. Eng. **47**(9), 1873–1888 (2019)

12. Horstemeyer, M.F., Panzer, M.B., Prabhu, R.K.: State-of-the-art modeling and simulation of the brain's response to mechanical loads. Ann. Biomed. Eng. **47**(9), 1829–1831 (2019)
13. Hu, Y., Zhu, D.: Empirical analysis of the worldwide maritime transportation network. Phys. A **388**, 2061–2071 (2009)
14. Kaplan, S.M.: Electric power transmission: background and policy issues. CRS Report for Congress (2009)
15. Kim, O.D., Rocha, M., Maia, P.: A review of dynamic modeling approaches and their application in computational strain optimization for metabolic engineering. Front. Microbiol. **9**, 1690 (2018)
16. Lepicard, G.: Théorie des systèmes complexes. Contributions des membres du Groupe Emergence è des publications ou présentations (2009)
17. Mefteh, W.: Simulation-based design: overview about related works. Math. Comput. Simul. **152**, 81–97 (2018)
18. Mefteh, W., Migeon, F., Gleizes, M.P., Gargouri, F.: ADELFE 3.0 design, building adaptive multi agent systems based on simulation a case study. In: Computational Collective Intelligence - Proceedings, Part I 7th International Conference, ICCCI 2015, Madrid, Spain, 21-23 September 2015, pp. 19–28 (2015)
19. Mefteh, W., Migeon, F., Gleizes, M.P., Gargouri, F.: Simulation based design for adaptive multi-agent systems with the ADELFE methodology. IJATS **7**(1), 1–16 (2015)
20. Qiao, L., Kao, S., Zhang, Y.: Manufacturing process modelling using process specification language. Int. J. Adv. Manuf. Technol. **55**(5–8), 549–563 (2011)
21. Stavropoulos, P., Foteinopoulos, P.: Modelling of additive manufacturing processes: a review and classification. Manuf. Rev. **5**, 26 (2018)
22. Stoyenko, A.: Engineering complex computer systems: a challenge for computer types everywhere. i. let's agree on what these systems are. Computer **28**(9), 85–86 (1995)
23. Wu, T., Alshareef, A., Sebastian Giudice, J., Panzer, M.B.: Explicit modeling of white matter axonal fiber tracts in a finite element brain model. Ann. Biomed. Eng. **47**, 1908–1922 (2019)
24. Wenfu, X., Liang, B., Yangsheng, X.: Survey of modeling, planning, and ground verification of space robotic systems. Acta Astronaut. **68**(11), 1629–1649 (2011)

Business Process Modelling to Improve Incident Management Process

Rúben Pereira[1(✉)], Isaías Bianchi[2], Ana Lúcia Martins[1],
José Braga de Vasconcelos[3,4], and Álvaro Rocha[5]

[1] Instituto Universitário de Lisboa (ISCTE-IUL), Lisbon, Portugal
{ruben.filipe.pereira,almartins}@iscte-iul.pt
[2] Federal University of Santa Catarina, Florianópolis, Brazil
isaias.bianchi@ufsc.br
[3] Universidade Europeia, Lisbon, Portugal
jose.vasconcelos@universidadeeuropeia.pt
[4] Centro de Administração e Políticas Publicas (CAPP),
Universidade de Lisboa, Lisbon, Portugal
[5] Universidade de Coimbra, Coimbra, Portugal
amrocha@dei.uc.pt

Abstract. Business process management (BPM) is an approach focused on the continuous improvement of business processes, providing for this a collection of best practices. These best practices enable the redesign of business processes to meet the desired performance. IT service management (ITSM) defines the management of IT operations as a service. There are several ITSM frameworks available, consisting in best practices that propose standardizing these processes for the respective operations. By adopting these frameworks, organisations can align IT with their business objectives. The constant need for change, competitiveness and innovation in organisations compels managers to analyse its business processes and find improvement opportunities. Therefore, the objective of this research is to understand how BPM can be used to improve of ITSM processes. An exploratory case study in a multinational company based in Lisbon, Portugal, is conducted for the improvement of the time performance of an incident management process. Data were gained through documentation, archival records, interviews and focus groups with a team involved in IT support service. The results of this study demonstrate how BPM highlighted the current incident management process incongruences. The preliminary evidences of this research indicate that with the application of the remain BPM lifecycle phases may be possible to improve incident management process. During the next months the authors intend to identify the main problems and simulate the appropriate BPM heuristics to understand the impact in the business organisation.

Keywords: Business process management · IT service management · Case study · Process improvement

Á. Rocha et al. (Eds.): WorldCIST 2020, AISC 1159, pp. 689–702, 2020.
https://doi.org/10.1007/978-3-030-45688-7_68

1 Introduction

Change is a constant in business environments. Classic management authors, such as Porter [1], helped to understand that change, whether of the internal or of the external environment, affects organisations and that it is an issue that should be addressed. As Harmon [2] argues, "change and relentless competition call for constant innovation and for constant increases in productivity, and those, in turn, call for an even more intense focus on how work gets done. To focus on how the work gets done is to focus on business processes". This means a continuous call for business process (BP) change and improvement [2]. For this challenge, business process management (BPM) appears as a discipline for the management of BPs focused on continuous improvement [3]. Aalst [4] argues that BPM has the "potential for significantly increasing productivity and saving costs" and also states that "BPM has a broader scope: from process automation and process analysis to operations management and the organization of work", providing different ways to approach change and to improve BPs. Best practices have been proposed for improvement initiatives, consisting in redesign heuristics to modify BPs and align them with business objectives (BO) [5, 6].

With the increasingly pressing developments in communication and technology, IT is seen today as "an integral part and fundamental to support, sustain, and grow a business" [7], being impossible for many organisations to function and succeed without it [8]. IT became a complex and dynamic landscape in organisations [9]. With this, came the need to align IT operations with the BO, which led gradually to the *servitization* of IT operations [10]. Thus, IT service management (ITSM) arose. In order to help organisations to perform ITSM, several IT frameworks were proposed, providing managers and organisations with customer-centred sets of processes and best-practices, for managing IT operations and aligning them with their BO [11]. As the literature shows [12–15], the most used of these frameworks is Information Technology Infrastructure Library (ITIL). A recent report about the current state of ITSM indicates that 47% of the surveyed IT managers employ ITIL or some of its processes in their ITSM strategy, being the most adopted IT framework [16].

The constant need for change, competitiveness and innovation in organisations compels managers to analyse its BPs and find improvement opportunities [2]. Being BPM a discipline that integrates process-oriented improvement initiatives [17], it is relevant to understand if it can be employed to improve IT services, which are based on processual ITSM frameworks. This research focuses on revealing how BPM can be employed for the improvement of the ITIL's incident management process (IM process), one of the most adopted ITSM processes [17–19]. Being an underexplored topic, an exploratory case study is conducted to elicit qualitative insights, following the methodology proposed by Yin [20]. This article has the following objectives. First, to explore the relationship between BPM and ITSM processes. Second, to understand how BPM can improve the IM process. Last one, to produce managerial recommendations for improvement of the IM process.

2 Theoretical Background

Business processes have always been part of the organisations, whether in a formal or informal ways [21]. BPs define the tune and performance of an organisation when delivering a service or a product to a customer [3].

Dumas et al. [3], in a 20-year updated definition, establish "business process as a collection of inter-related events, activities and decision points that involve a number of actors and objects, and that collectively lead to an outcome that is of value to at least one customer".

BPs allow organisations to reach its BOs efficiently and effectively by enabling the coordination and connection of its resources [22]. Thus, the management and improvement of BPs is vital for organisations, as it allows to achieve BOs while also coping with change [2].

BPM, by implying a continuous commitment to manage BPs, requires a lifecycle methodology with structured steps and feedback that establish a managerial practice in organisations, which is the premise for continuous improvement and to meet the BOs [23]. This BPM lifecycle is based in principles such as modelling and documentation of BP, customer-orientation, constant assessment of the performance of BP, a continuous approach to optimization and improvement, following best practices for superior performance, and organisational culture change [24]. Several BPM lifecycle proposals appeared, contributing for the growth of BPM as a concept. The BPM lifecycle proposed by Dumas et al. [3], have these phases: process identification, process discovery, process analysis, process redesign, process implementation and process monitoring and controlling. By applying BPM lifecycles, organisations can address BP change and identify the improvements required to achieve the desired BOs [3, 25].

Galup et al. [12] state "because ITSM is process-focused, it shares a common theme with the process improvement movement (such as, TQM, Six Sigma, Business Process Management, and CMMI)". This section presents the literature review performed to collect existing related work on the improvement of IM process through BPM. The related work found can be divided in two categories: improvement of the IM process and BP improvement through BPM. Table 1 presents the examples found of work developed on the improvement of the IM process.

The related work found for the improvement of the IM process revealed that most of the work developed is technology-oriented, with different methods and techniques being applied for technological improvement solutions. Automation and incident correlation are the most common improvement methods found.

Table 2 presents examples of works developed on BP improvement through BPM.

Several more related works were found concerning BP improvement through BPM, being the most common example located in healthcare services. Although several examples were found in each of the two categories mentioned, there was not a single work found concerning the improvement of IM process through a BPM approach, the topic of this research. Two examples were found that slightly approached this topic:

- The work of Mahy et al. [17], published in the 3rd International Conference on Systems of Collaboration, that only modelled the IM process.

Table 1. Related work on the improvement of the IM process

Authors	Publication	Improvement method
Ghrab et al. [26]	3rd International Conference on Artificial Intelligence and Pattern Recognition	Automation through constraint programming
Goby et al. [27]	24th European Conference on Information Systems	Automation through business intelligence
Trinkenreich et al. [28]	27th International Conference on Software Engineering and Knowledge Engineering	Automation through business intelligence
Salah et al. [29]	Journal of Network and Systems Management	Incident correlation model
Bezerra et al. [30]	9th International Conference on the Quality of Information and Communications Technology	Optimization through the reuse of experiences and natural language processing techniques
Bartolini et al. [31]	19th IFIP/IEEE International Workshop on Distributed Systems	Optimization through discrete event simulator
Gupta et al. [32]	5th IEEE International Conference on Autonomic Computing	Automation through incident correlation

- The work of Bezerra et al. [30], published in the 9th International Conference on the Quality of Information and Communications Technology, that also only modelled the IM process.

The review of related work in the literature did not provided a single reference on the topic approached by this research, which leads to the conclusion that this topic is not explored.

3 Research Methodology

Grounded on the previous section, one may argue that this study is exploratory in nature rather than hypothesis testing. Exploratory analysis should be considered where there are none or few prior works presented on the subject studied [20, 42]. Reinforced by Runeson and Höst [43] as having exploration as primary objective, to Zainal [44] a case study analysis is suitable to analyse a limited number of events or conditions in detail. Thus, this research follows the case study methodology proposed by Yin [45].

Following Yin [45] recommendations, a research question is formulated, to fulfil the purpose of this study. What are the BPM redesign heuristics that best suit IM process improvement? This question will guide the authors in the research for a common ground between BPM and an ITSM process that has not been proper explored.

Table 2. Related work on the BP improvement through BPM

Authors	Publication	Area
Sanka Laar and Seymour [33]	13th European Conference on Management, Leadership & Governance	Small and medium enterprises
Rebuge and Ferreira [34]	Journal of Information Systems	Healthcare services
Netjes et al. [35]	Workshops of 7th International Conference on Business Process Management	Healthcare services
Becker et al. [36]	13th Americas Conference on Information Systems	Healthcare services
Küng and Hagen [37]	Business Process Management Journal	Banking services
Hertz et al. [38]	Supply Chain Management: An International Journal	Automotive industry
Ferretti and Schiavone [39]	Business Process Management Journal	Operations in seaports
Rinaldi et al. [40]	Business Process Management Journal	Public administration
Haddad et al. [41]	Business Process Management Journal	Non-profit organisations

3.1 Research Context Unit of Analysis

The unit of analysis of this case study will consist in a team that belongs to a multinational company in the markets of electrification, digitalization and automation, which is present in at least 190 countries and employs directly more than 350.000 people. The team was formed in 2014 and is composed by 17 members, being characterized in Table 3.

The team is based in Lisbon, Portugal, integrated in a IT support service to the whole Human Resources (HR) department of the company, being the owner of a local IM process. With this process, the organisation provides an IT support service that receives and processes all incidents reported by the users of the HR's IT services, which are the customers of the process. This team is chosen as the unit of analysis for this case study because it is the owner of an IM process, the selected ITSM process for research. Being the goal of the team and of its IM process to provide a fast service, time is considered as the main performance dimension. Thus, time shall be the driving performance dimension for the improvement proposals in this research.

3.2 Data Collection

A plan for collection of evidence was defined following Yin [45] recommendations. In order to achieve the validity of the analysis that will be performed and avoid the

Table 3. Team members information

ID	ITIL experience	Time (in years)		Function
		IT experience	With IM team	
I1	Yes	19	3	Team leader
I2	Yes	16	3	Support expert
I3	No	0,16	0,16	Support expert
I4	No	5	2	Support expert
I5	Yes	7	3	IT specialist
I6	Yes	23	2	Support agent
I7	Yes	22	3	Support expert
I8	No	3	0,16	Support expert
I9	Yes	22	3	IT specialist
I10	Yes	5	2	IT specialist
I11	Yes	13	2	Support expert
I12	Yes	4	2	Support expert
I13	Yes	25	3	Support expert
I14	Yes	14	3	Support expert
I15	Yes	9	1	Support expert
I16	No	0,25	0,25	Support agent
I17	Yes	18	3	IT specialist
	Average	**12,1**	**2,1**	

weaknesses inherent to each source of evidence, it is desirable to collect evidence from difference sources. For this case study, four different sources of evidence are expected to be made available by the organization, as presented in Table 4.

Table 4. Expected sources of evidence

Sources of evidence	Description
Documentation	Internal documentation and private web content about the organisation, the IT support service, the team and the IM process
Archival records	Data records and data reports generated by the daily operation of the IM process
Interviews	Open and focused interviews, with all the team members. Four rounds of interviews
Focus groups	Structured focus groups, with all the team members

Being an IT-nature process, observation and physical artefacts are not expected to be available as sources of evidence. Adding to the enounced sources, it is also expected that informal sources, such as punctual conversations, may occur with team members. The plan for collection of evidence is designed according with the different objectives of each BPM lifecycle phase [3] and also taking into account that the performance dimension to improve in this research is time.

3.3 Process Discovery

The goal of Process Discovery is to document the current state of the process, by producing an as-is model [3]. The first step consisted in the collection of the evidence required, through the conduction of the two rounds of interviews and gathering of documentation. The analysis of the evidence was performed in parallel with its collection. This analysis provided the desired information for this phase and allowed to clear the incongruencies detected in the collection of evidence. The IM process was then documented in its current state, through an as-is model and respective description. Lastly, the 1st focus group was convened to approve the documentation.

The 1st round of interviews followed a script directed to the mapping of the process. The interviewees were asked to map the IM process, from end-to-end, detailing the activities performed, the participants involved, the decisions and exceptions along the process.

The 2nd round of interviews was conducted based on the results of the previous round of interviews. This time, the team members were faced with the initial draft of the as-is model and requested to validate it: to spot and correct inaccuracies, and to address incongruencies and clarify them. They were also asked to detail even more the model with relevant artefacts and information inherent to the IM process.

Documentation was collected, being available administrative documents about the support service, descriptions of the participants in the IM process and respective roles, as well as a depiction of the workflow adopted.

To have an approved as-is model, the 1st focus group was convened. The team was asked to collectively analyse the final as-is model, discuss its validity and point any required adjustments. These methods of evidence collection were employed because they enable a thorough Process Discovery, having as output the documentation of the as-is model, which is the basis for the analysis of the IM process and its issues.

In the two rounds of interviews conducted, four incongruencies were detected, concerning the mapping of thc IM process.

Incongruency A, depicted in Table 5, had its source in the 1st round of interviews, and concerned what of activity the Incident Prioritization in the 1st support level is: a user activity, performed by a team member (scenario 1), or a script activity, automatically performed the IM system (scenario 2).

The 2nd round of interviews cleared this incongruency, by defining Incident Prioritization has a script activity performed automatically by the IM system (scenario 2).

Incongruency B, shown in Table 6, had its source in the 1st round of interviews, and concerned existence of a gateway for the triage of incident validity in the 1st support level (scenario 1) or not (scenario 2).

Table 5. Incongruency A

Incongruency A	
Scenario 1	**Scenario 2**

Table 6. Incongruency B

Incongruency B	
Scenario 1	**Scenario 2**
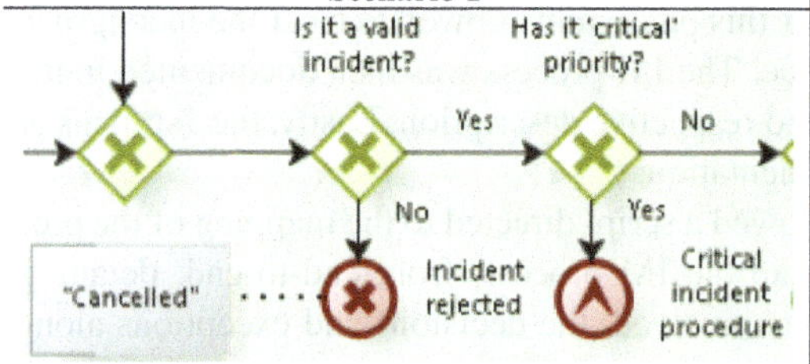	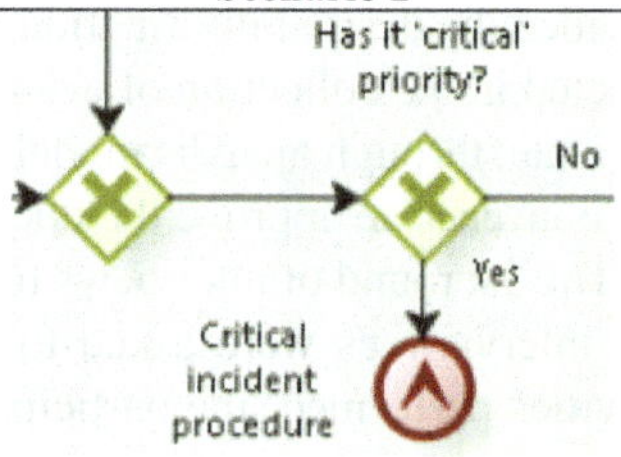

The documentation analysis solved this incongruency, by revealing that there was no gateway for the triage of incident validity in the 1st support level (scenario 2).

Incongruency C, presented in Table 7, also had its source in the 1st round of interviews, and concerned the position of a gateway for the triage of critical priority incidents in the 1st support level: if before the Incident Categorization activity (scenario 1) or after (scenario 2).

The 2nd round of interviews cleared this incongruency, by setting the position of the gateway before Incident Categorization activity (scenario 1).

Incongruency 4, presented in Table 8, was detected in the 2nd round of interviews, and concerned the period of working days required by the IM system for closing an incident: if 14 working days (scenario 1) or 7 working days (scenario 2).

Table 7. Incongruency C

Incongruency C	
Scenario 1	**Scenario 2**
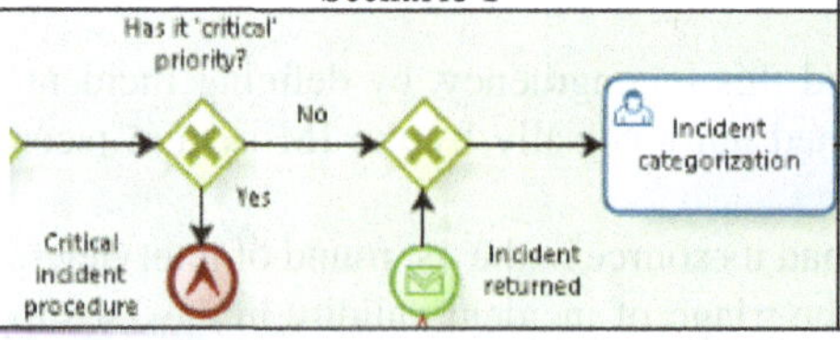	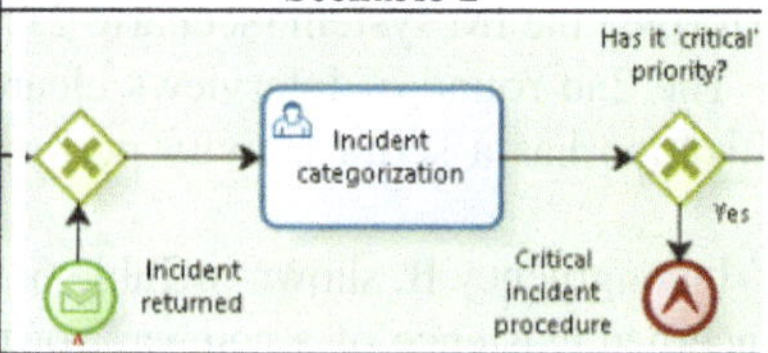

Table 8. Incongruency D

Incongruency D	
Scenario 1	**Scenario 2**
 14 working days	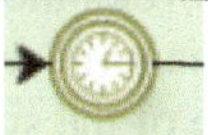 7 working days

The documentation solved this incongruency, by revealing 14 working days as the time period required by the IM system for closing an incident (scenario 1). With all the incongruencies cleared, the final documentation of the IM process was produced, including the final draft of the as-is model.

3.4 As-is Model and Documentation

There are three main identified participants in the IM process:

- The customers, which are the users who report incidents to the support service.
- The support service, composed by the three support levels (SL) that perform the IM process itself, thus being the focus of this research.
- The IT suppliers, which are external to the support service and are the providers the IT services used by the organisation.

The support service is composed by three SLs that have different roles and teams involved, as presented in Table 9.

Table 9. Support levels in the support service

Support level	Description & responsibilities	Staff
1st SL	Support agents that perform the initial reception, triage and forwarding of incidents to the 2nd SL, the so called dispatching activities	2
2nd SL	Support experts that perform a technical diagnosis and resolution of the incidents, being also responsible to contact the customer and to always close incidents. If unable to find the solution for the incidents, it must forward it for the 3rd SL	10
3rd SL	IT specialists that perform an extensive investigation and resolution, being the last resort of the support service to solve the incidents. If unable, it must request for the intervention of the respective IT supplier. With the resolution performed, the 3rd SL must return the incident to the 2nd SL for closure	4 (in the team)

Of the three SL presented in Table 9, the team fully incorporates the 1st SL with 2 support agents, and the 2nd SL with 10 support experts. However, the 3rd SL is partially represented in the team, with only 4 IT specialists. The 3rd SL is composed by several IT specialists from various IT development teams in the organisation, that are called to participate in the IM process whenever required, being the quantity of staff involved in this SL unknown. This structure of the support service is designed to handle and solve the incidents according with their complexity and severity, and with the level of expertise and specialization required, being one of goals of the team to retain and solve as much incidents as possible in the 2nd SL, avoiding a high workload for the 3rd SL.

The IM process starts whenever a customer reports an incident to the 1st SL, either through the email (Incident reported) or through call (Call arrival). From here, the IM process is mainly grounded in the workflow defined by ITIL, being its activities and gateways easily recognized in the as-is model. These activities are performed with the support of a single IM system, which is used for the storage and management of all incidents, through the logging and update of incidents in tickets with all the respective information and actions performed.

There are two rework situations in the middle of the IM process:

- Whenever there is an incident returned from the 2nd SL to the 1st SL, due to mistake of the 1st SL.
- Whenever there is a reopening request from the customer to the 2nd SL, after an incident is labelled as resolved.

The IM process finishes in the Incident solved & service restored, the main end event of the process in the 2nd SL, with the incident solved and the ticket closed. However, two alternative end events may occur:

- Incident rejected, when the 2nd SL determines that there is no valid nature in the received incident that justifies the deployment of the IM process and cancels the respective ticket.
- Critical incident procedure, a special procedure designed to deal with incidents that have a perceived critical priority or critical nature. These procedures are different and customized according with the different IT services.

The documentation of the IM process reveals that, despite being adapted to the organisation, it is mainly grounded on the ITIL standard and that most of the recommendations proposed by the standard are followed. The final documentation of the IM process was submitted for approval in the 1st focus group. This 1st focus group, convened with all the team members, validated the documentation, without any opposition, not being required any adjustments or corrections to the as-is model. With the 1st focus group finished, the Process Discovery phase was completed. The next section presents the discussion and conclusion.

4 Discussion and Conclusion

Business Process Modelling is a broad discipline that offers methods and tools for the control and improvement of business processes. By using BPM, managers can thoroughly analyse processes and discover improvement opportunities, which is increasingly requested in the organisational environment. This research aims to explore how the incident management process can be improved through a BPM approach, a relationship that has not yet been much investigated by the scientific community. For this goal, a case study methodology is being performed in a multinational organization.

So far, the authors were able to prove that using BPM it was possible to find and solve some process incongruencies which is a positive sign and raises our expectations on forthcoming findings.

This research intends to contribute to reduce the gap existent between BPM and ITSM processes. There is a clear relation between both areas, due to their process-oriented nature, but few researches were developed in this specific area.

The authors will continue the case study for the next couple of months where is expected to elicit the main incident management process problems using interviews, observation, document analysis and focus group techniques. Then the proper heuristics will be selected and discussed with the entire team to reach a tuned set to test. Finally, the selected heuristics will be simulated and the impact in the business and daily operations studied.

References

1. Porter, M.: Competitive Strategy: Techniques for Analyzing Industries and Competitors. Free Press, New York (1980)
2. Harmon, P.: Business Process Change. Elsevier, Waltham (2014). https://doi.org/10.1017/CBO9781107415324.004
3. Dumas, M., La Rosa, M., Mendling, J., Reijers, H.A.: Fundamentals of Business Process Management. Springer, Berlin (2013)
4. van der Aalst, W.M.P.: Business process management: a comprehensive survey. ISRN Softw. Eng. **2013**, 1–37 (2013). https://doi.org/10.1155/2013/507984
5. Mansar, S.L., Reijers, H.A.: Best practices in business process redesign: use and impact. Bus. Process Manag. J. **13**, 193–213 (2007). https://doi.org/10.1108/14637150710740455
6. Hanafizadeh, P., Moosakhani, M., Bakhshi, J.: Selecting the best strategic practices for business process redesign. Bus. Process Manag. J. **15**, 609–627 (2009). https://doi.org/10.1108/14637150910975561
7. Mohamed, N., Kaur, J., Singh, G.: A conceptual framework for information technology governance effectiveness in private organizations. Inf. Manag. Comput. Secur. **20**, 88–106 (2012)
8. Sallé, M.: IT service management and IT governance: review, comparative analysis and their impact on utility computing (2004). https://doi.org/10.1007/b115738
9. Jamous, N., Bosse, S., Gorling, C., Hintsch, J., Khan, A., Kramer, F., Muller, H., Turowski, K.: Towards an IT Service Lifecycle Management (ITSLM) concept. In: Proceedings - 4th International Conference on Enterprise Systems: Advances in Enterprise Systems, ES 2016, pp. 29–38 (2017). https://doi.org/10.1109/ES.2016.10
10. Conger, S., Winniford, M., Erickson-Harris, L.: Service management in operations. In: 14th Americas Conference on Information Systems, pp. 3884–3894. AIS Toronto, Canada (2008)

11. Marrone, M., Kolbe, L.M.: Impact of IT service management frameworks on the IT organization. Bus. Inf. Syst. Eng. **3**, 5–18 (2011). https://doi.org/10.1007/s12599-010-0141-5
12. Galup, S.D., Dattero, R., Quan, J.J., Conger, S.: An overview of IT service management. Commun. ACM **52**, 124 (2009). https://doi.org/10.1145/1506409.1506439
13. Pereira, R., da Silva, M.M.: Designing a new integrated IT governance and IT management framework based on both scientific and practitioner viewpoint. Int. J. Enterp. Inf. Syst. **8**, 1–43 (2012)
14. Valiente, M.-C., Garcia-Barriocanal, E., Sicilia, M.-A.: Applying an ontology approach to IT service management for business-IT integration. Knowl.-Based Syst. **28**, 76–87 (2012). https://doi.org/10.1016/j.knosys.2011.12.003
15. Iden, J., Eikebrokk, T.R.: Implementing IT service management: a systematic literature review. Int. J. Inf. Manag. **33**, 512–523 (2013). https://doi.org/10.1016/j.ijinfomgt.2013.01.004
16. BMC, Forbes Insights: Delivering value to today's digital enterprise: the state of IT service management. In: Forbes Insights, Jersey City, NJ, USA (2017)
17. Mahy, Y., Ouzzif, M., Bouragba, K.: Supporting ITIL processes implementation using business process management systems. In: Ouzzif, M., Jarir, Z. (eds.) 3rd International Conference on Systems of Collaboration, pp. 77–80. IEEE, Casablanca (2016). https://doi.org/10.1109/SYSCO.2016.7831338
18. Gupta, R., Prasad, K.H., Mohania, M.: Automating ITSM incident management process. In: 5th International Conference on Autonomic Computing ICAC 2008, vol. 1, pp. 141–150 (2008). https://doi.org/10.1109/ICAC.2008.22
19. AXELOS: ITIL Foundation: ITIL, 4th edn. TSO (The Stationery Office) (2019)
20. Yin, R.K.: Case Study Research and Applications: Design and Methods. SAGE Publications, Thousand Oaks (2009). https://doi.org/10.1097/FCH.0b013e31822dda9e
21. Roeser, T., Kern, E.-M.: Surveys in business process management - a literature review. Bus. Process Manag. J. **21**, 692–718 (2015). https://doi.org/10.1108/BPMJ-07-2014-0065
22. Weske, M.: Business process management. In: Business Process Management: Concepts, Languages, Architectures, pp. 3–23. Springer, Berlin (2012). https://doi.org/10.1007/978-3-642-28616-2
23. de Morais, R.M., Kazan, S., de Pádua, S.I.D., Costa, A.L.: An analysis of BPM lifecycles: from a literature review to a framework proposal. Bus. Process Manag. J. **20**, 412–432 (2014). https://doi.org/10.1108/BPMJ-03-2013-0035
24. Pyon, C.U., Woo, J.Y., Park, S.C.: Service improvement by business process management using customer complaints in financial service industry. Expert Syst. Appl. **38**, 3267–3279 (2011). https://doi.org/10.1016/j.eswa.2010.08.112
25. Harmon, P.: The scope and evolution of business process management. In: Handbook on Business Process Management, vol. 1, pp. 37–81 (2010). https://doi.org/10.1007/978-3-642-00416-2
26. Ghrab, I., Ketata, M., Loukil, Z., Gargouri, F.: Using constraint programming techniques to improve incident management process in ITIL. In: 2016 Third International Conference on Artificial Intelligence and Pattern Recognition (AIPR), pp. 1–6. IEEE (2016). https://doi.org/10.1109/ICAIPR.2016.7585231
27. Goby, N., Brandt, T., Feuerriegel, S., Neumann, D.: Business intelligence for business processes: the case of IT incident management. In: 24th European Conference on Information Systems, pp. 1–15. AIS, Istanbul (2016). https://doi.org/10.13140/RG.2.1.2033.9604
28. Trinkenreich, B., Santos, G., Confort, V., Santoro, F.: Toward using business process intelligence to support incident management metrics selection and service improvement. In: 27th International Conference on Software Engineering and Knowledge Engineering, pp. 522–527. IEEE, Pittsburgh (2015). https://doi.org/10.18293/SEKE2015-110

29. Salah, S., Maciá-Fernández, G., Díaz-Verdejo, J.E., Sánchez-Casado, L.: A model for incident tickets correlation in network management. J. Netw. Syst. Manag. **24**, 57–91 (2015). https://doi.org/10.1007/s10922-014-9340-6
30. Bezerra, G., Pinheiro, V., Bessa, A.: Incident management optimization through the reuse of experiences and natural language processing. In: da Silva, A.R., Silva, A.R., Brito, M.A., Machado, R.J. (eds.) 9th International Conference on the Quality of Information and Communications Technology, pp. 58–65. IEEE, Guimarães (2014). https://doi.org/10.1109/QUATIC.2014.14
31. Bartolini, C., Stefanelli, C., Tortonesi, M.: SYMIAN: a simulation tool for the optimization of the IT incident management process. In: Turck, F., Kellerer, W., Kormentzas, G. (eds.) 19th IFIP/IEEE International Workshop on Distributed Systems: Operations and Management, pp. 83–94. IEEE, Samos Island (2008)
32. Gupta, R., Prasad, K.H., Mohania, M.: Information integration techniques to automate incident management. In: NOMS 2008 - IEEE/IFIP Network Operations and Management Symposium: Pervasive Management for Ubiquitous Networks and Services, pp. 979–982 (2008). https://doi.org/10.1109/NOMS.2008.4575262
33. Sanka Laar, D., Seymour, L.F.: Redesigning business processes for small and medium enterprises in developing countries. In: Mokoaleli-Mokoteli, T., Ndaba, Z. (eds.) 13th European Conference on Management, Leadership & Governance, pp. 512–519. Academic Conferences and Publishing International Ltd., Saint Petersburg (2017)
34. Rebuge, Á., Ferreira, D.: Business process analysis in healthcare environments: a methodology based on process mining (2012). https://doi.org/10.1016/j.is.2011.01.003
35. Netjes, M., Mans, R.S., Reijers, H.A., Aalst, W.M.P., Van Der Vanwersch, R.J.B.: BPR Best Practices for the healthcare domain. In: Rinderle-Ma, S., Sadiq, S., Leymann, F. (eds.) Workshops of 7th International Conference on Business Process Management, pp. 605–616. Springer, Berlin (2010). https://doi.org/10.1007/978-3-642-12186-9
36. Becker, J., Fischer, R., Janiesch, C.: Optimizing U.S. healthcare processes: a case study in business process management. In: 13th Americas Conference on Information Systems, pp. 2236–2247, Keystone, CO, USA (2007)
37. Küng, P., Hagen, C.: The fruits of business process management: an experience report from a swiss bank. Bus. Process Manag. J. **13**, 477–487 (2007). https://doi.org/10.1108/14637150710763522
38. Hertz, S., Johansson, J.K., de Jager, F.: Customer-oriented cost cutting: process management at volvo. Supply Chain Manag. Int. J. **6**, 128–141 (2001). https://doi.org/10.1108/13598540110399174
39. Ferretti, M., Schiavone, F.: Internet of Things and business processes redesign in seaports: the case of Hamburg. Bus. Process Manag. J. **22**, 271–284 (2016). https://doi.org/10.1108/BPMJ-05-2015-0079
40. Rinaldi, M., Montanari, R., Bottani, E.: Improving the efficiency of public administrations through business process reengineering and simulation: a case study. Bus. Process Manag. J. **21**, 419–462 (2015). https://doi.org/10.1108/BPMJ-06-2014-0054
41. Haddad, C.R., Ayala, D.H.F., Maldonado, M.U., Forcellini, F.A., Lezana, Á.G.R.: Process improvement for professionalizing non-profit organizations: BPM approach. Bus. Process Manag. **22**, 634–658 (2016). https://doi.org/10.1108/MRR-09-2015-0216
42. Zaidah, Z.: Case study as a research method. Jurnal Kemanusiaan, 1–6 (2007). https://doi.org/10.1177/15222302004003007
43. Runeson, P., Höst, M.: Guidelines for conducting and reporting case study research in software engineering. Empir. Softw. Eng. **14**, 131–164 (2009). https://doi.org/10.1007/s10664-008-9102-8

44. Zainal, Z.: Case study as a research method. J. Kemanusiaan. **9**, 1–6 (2007). https://doi.org/10.1177/15222302004003007
45. Yin, R.K.: Case Study Research: Design and Methods (2009). https://doi.org/10.1097/FCH.0b013e31822dda9e

Study of the Pattern Recognition Algorithms for the Automation of a Smart Semaphore Through FPGAs

Erik Fernando Méndez(✉), Gabriela Mafla, and José Ortiz

Universidad Regional Autónoma de los Andes, Ambato, Ecuador
us.erikmendez@uniandes.edu.ec, gabriella_1912@hotmail.com,
jose_ortize89@hotmail.com

Abstract. The following document presents the development of an intelligent traffic light, based on FPGA technology, which assessing the amount of traffic gives priority to the lane with the highest number of cars. The comparative study of pattern recognition algorithms was performed: color matching algorithm, cross correlation algorithm and optical character recognition algorithm (OCR). The methodology for the development of the system is based on a matrix of programmable logic gates in the field or FPGA, a camera that acquires images and sends them for digital processing programmed in LabView intelligently with pattern recognition algorithms for decision making in the area of vehicular traffic control. The density of vehicular traffic is determined and the card changes the duration of the green light given for each lane according to the number of existing vehicles. As a result, it was obtained that the most appropriate algorithm to implement an intelligent traffic light prototype using FPGAs was the color matching algorithm that has an accuracy of 100% and a response time of 3 ms.

Keywords: Algorithms · Artificial vision · Intelligent traffic light · Pattern · FPGA

1 Introduction

Automatic vehicle traffic control has been of interest for many years; since this is a rather complicated problem and that every day takes on greater importance in everyday life, methods have been investigated for the implementation of automatic traffic monitoring systems, the same ones that aim to improve the traffic problems we currently have [1].

Over the years and the rapid advance in the development of new technologies, artificial vision techniques, mainly the recognition of intelligent patterns and systems such

E. F. Méndez—System Department. University of the Andes UNIANDES. Electromechanical Engineer. Master in Industrial Automation and Control Systems.
G. Mafla—Electronics, Control and Industrial Networks Engineer. Master in Industrial Automation and Control Systems.
J. Ortiz—Electronics, Control and Industrial Networks Engineer. Master's student in Industrial Automation and Control Systems.

Á. Rocha et al. (Eds.): WorldCIST 2020, AISC 1159, pp. 703–712, 2020.
https://doi.org/10.1007/978-3-030-45688-7_69

as fuzzy logic, neural networks and genetic algorithms, have become a topic of current affairs and of great application for the benefit of society [2], they have an enormous technical and scientific value for the innumerable fields in which it can be applied, including automatic control of vehicular traffic.

The FPGAs, are widely used in artificial vision techniques, such as vehicular traffic monitoring applications, these allow handling very high volumes of images and treating these images to detect objects, in most cases in real time [3].

That is why the design and implementation of a prototype of intelligent traffic lights through FPGAs, has the purpose of minimizing vehicular congestion.

In this article, the main aspects of the algorithms for pattern recognition are detailed in the first instance: color matching algorithm, cross correlation algorithm and optical character recognition algorithm (OCR), the design and implementation of the prototype for traffic lights intelligent and an experimental evaluation is developed from which it is concluded which is the most suitable pattern recognition algorithm for the implementation of intelligent traffic lights by means of FPGA. Finally, practical tests are carried out on the prototype implemented to verify its efficiency.

2 Methodology

The study was based on three algorithms: color matching algorithm, cross correlation algorithm and optical character recognition algorithm (OCR).

The application it was made in LabView 2012, for which a program has been developed for each proposed algorithm; this application aims to recognize the search pattern.

The scheme designed for the development of each of the applications is the one shown in Fig. 1.

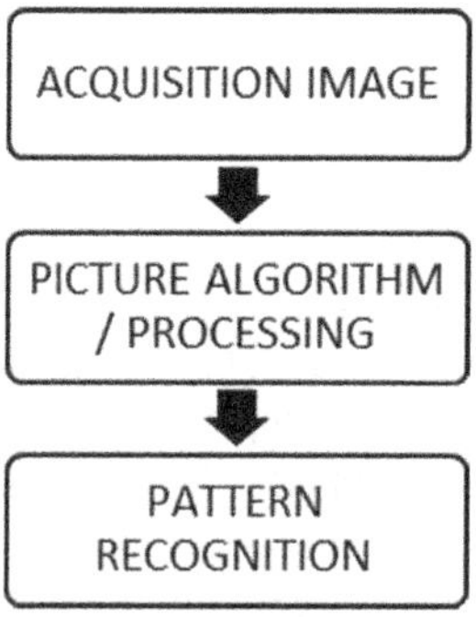

Fig. 1. Application scheme

Acquisition of the image: At this stage, the image to be processed is obtained by means of a camera.
Algorithm/Image processing: At this stage, the algorithm processes the acquired image for later analysis [4].

Pattern recognition: At this stage the quantitative and qualitative analysis of the image is performed for pattern recognition, this can be at the pixel level, borders, color, character, etc. [5].

2.1 Software Design

In the design of the software the graphic interface and the HMI for the control of the electronics of the system have been developed, it was created in LabView. As a first step we open the camera; Then we select the Vision Acquisition module as can be seen in Fig. 2, its function is to acquire the image, and with the IMAQ create module we select the type of image [4].

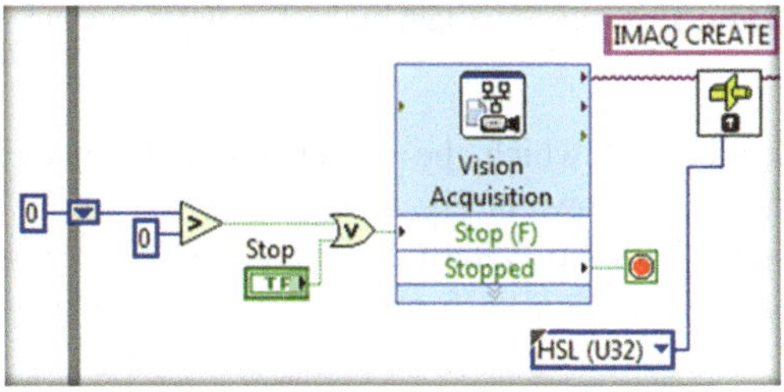

Fig. 2. Vision Acquisition module

Once the image is centered, we segment our regions of interest by time (image centered on the recognized pattern), each track will be segmented every 200 ms, for which we first assign the cutting coordinates in the image, this process can be observed in Fig. 3.

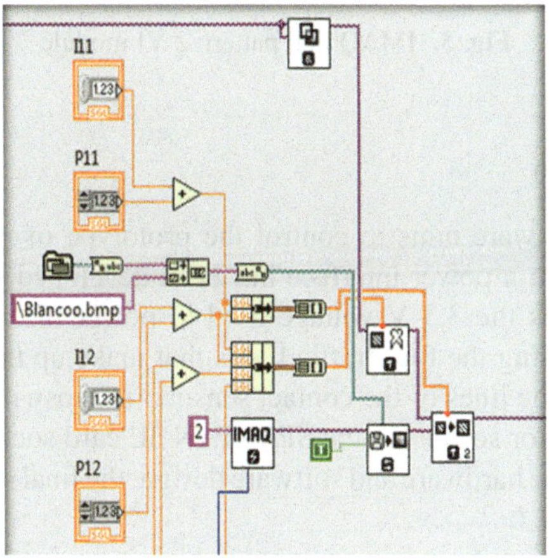

Fig. 3. Image segmentation process of the pattern recognized by the camera and performed in labview

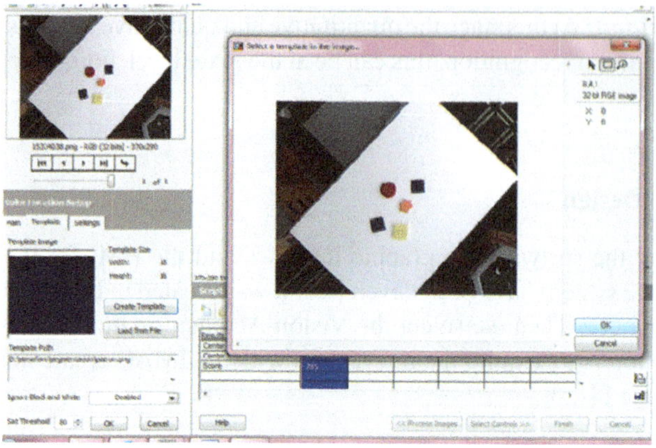

Fig. 4. Vision Assistant module - select template

The pattern is loaded, for which the Select Template Image tool of the Vision Assistant module is used, as shown in Fig. 4 [5].

Finally, using the IMAQ Find Pattern 2 VI module that can be seen in Fig. 5, the previously defined pattern search is performed [6].

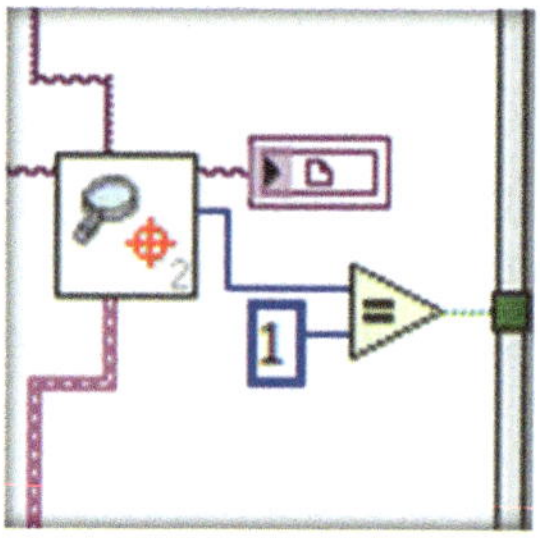

Fig. 5. IMAQ find pattern 2 VI module

2.2 Hardware Design

The design of the hardware aims to control the prototype of intelligent traffic lights implemented, for which a power interface has been developed, this interface has four plates, the first controls the 3.3 V voltage used to power the FPGA; the second has the function of controlling the four traffic lights that make up the track; the third is an interface for ordering the lines of the contact sensors (microswitch); Finally, the fourth board is a male connector socket for the SPARTAN 3E card socket [7].

After performing the hardware and software design, the final scheme of the prototype is the one shown in Fig. 6.

As can be seen, the image is first acquired, which is processed in the computer and subsequently the data acquired from the image is delivered to the application of the FPGA, whereby the traffic lights are controlled.

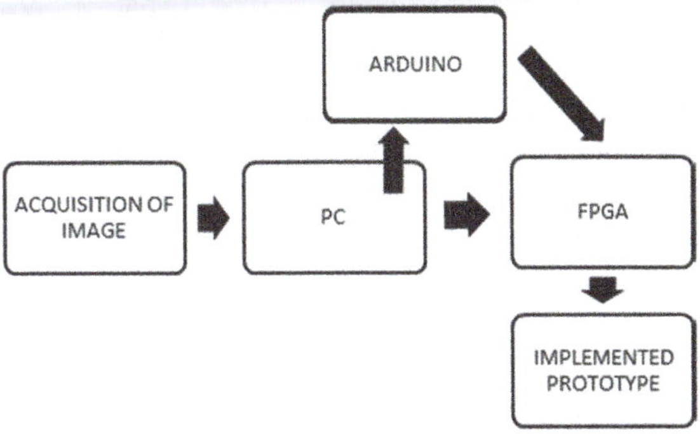

Fig. 6. Operating scheme

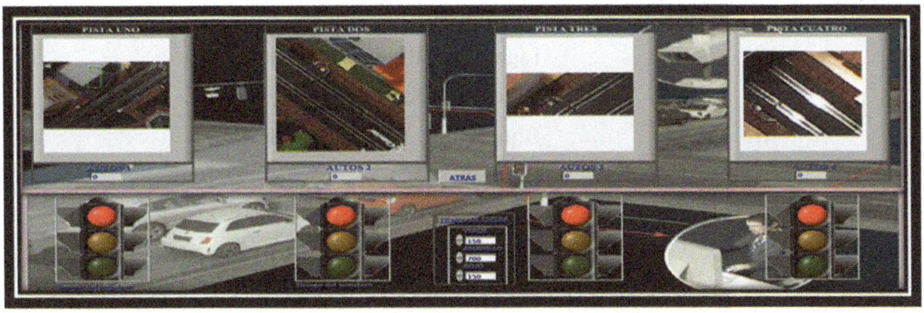

Fig. 7. Systems comparison chart

3 Tests and Results

The present system through a camera and the developed algorithm detects the cars present in each of the tracks and counts them, the information obtained is transmitted to the FPGA, it analyzes the data and makes decisions to control the traffic lights, which indicate the course that traffic must take.

After the development of the applications, a series of tests have been carried out that generate a set of data, which were used to determine the most appropriate algorithm, taking into account two variables that are considered as the most relevant, accuracy and response time.

3.1 Color Matching Algorithm

Color is an effective descriptor that facilitates the identification of objects and simplifies the extraction of the same within a scene. The color comparison is a process that requires access to the color of each pixel that makes up the image. The color matching algorithm aims to recognize patterns based on color, said pattern is previously established by the user. The tests for the color-matching algorithm were performed in a range of one to five patterns; they were increasing or decreasing randomly for each case (1, 4, 2, 5, 3). The color chosen for the search pattern was blue. The colors used for the tests can be seen in Table 1.

One of the first tests carried out was to choose the color of the pattern to be used, it was tested with a series of colors and based on the trial and error method, the dark blue color was selected as the pattern, because it has no similarity to any Another color present on stage.

Table 1. Color combinations

No. test	No. patterns	Colors
1	1	Yellow, Red, **Blue**, Orange, Light Blue
2	3	Orange, **Blue**, Yellow, **Blue, Blue**
3	5	**Blue**, **Blue**, **Blue**, **Blue**, **Blue**
4	4	Light Blue, **Blue**, **Blue**, **Blue**, **Blue**
5	2	**Blue**, Orange, Red, Orange, **Blue**
6	5	**Blue**, **Blue**, **Blue**, **Blue**, **Blue**
7	3	**Blue**, Yellow, **Blue**, Light Blue, **Blue**
8	1	Orange, **Blue**, Red, Light Blue, Yellow
9	4	**Blue**, **Blue**, Light Blue, **Blue**, **Blue**
10	2	Yellow, **Blue**, Red, **Blue**, Orange

3.2 Cross Correlation Algorithm

The cross-correlation of images aims to automatically locate a study point within an image. The correlation also called "matching" explains the process of automatic identification of homologous points in digital images. This is one of the most common methods for recognizing patterns in an image, and because it requires a greater number of processes, its execution time is high. As in the previous case, the tests were performed with a minimum of one pattern and a maximum of five patterns, they were increasing or decreasing randomly for each case (1, 4, 2, 5, 3). The form chosen for the search pattern was the square. Table 2 shows the figures used to perform the respective tests.

3.3 Optical Character Algorithm (OCR)

Optical character recognition is a process that aims to recognize the textual part of a digitized image. This process automatically identifies symbols or characters in a given image. The digitized image is the input of the OCR, resulting in a text file, it can be used for any application that needs it. The optical character recognition works with images in a gray scale, this facilitates the recognition of the characters, guaranteeing that they will be recognized in their entirety, and then be able to create the text file. To recognize these characters, a comparison is made of the patterns or templates that will contain all

Table 2. Figure combinations

No. Test	No. Patterns	Figures
1	1	
2	3	
3	5	
4	4	
5	2	
6	5	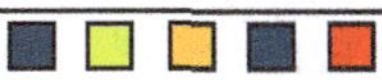
7	3	
8	1	
9	4	
10	2	

possible characters with each of the characters that make up the image. For the tests carried out with this algorithm, a database was previously created, in which the search characters were stored. These characters were C, E, G, R. Table 3 shows the information collected by this algorithm [9, 10].

3.4 Algorithm Selection

Table 4 show the results obtained in the analysis of variance based on each parameter.

Through the results based on the variance, it was possible to verify that the study of pattern recognition algorithms has allowed to select the color matching algorithm as the most suitable for the implementation of an intelligent traffic light by means of FPGAs, to minimize vehicular congestion.

3.5 Prototype Efficiency

After selecting the appropriate algorithm, we proceed to check the efficiency of the intelligent traffic lights prototype implemented [8].

To verify that the implemented traffic light prototype is intelligent, the experimental inductive method was used, for which 5 samples were taken randomly, test one without carriages on the track is shown in Fig. 8.

In test 2 a car is detected on track three, as can be seen in Fig. 9; when there is presence of cars the traffic light is activated.

Through the results obtained, it was possible to verify that the implemented traffic light prototype is intelligent, if the algorithm does not detect the presence of cars the traffic lights are deactivated, this means that they remain in the red light state, while if the algorithm detects a auto in any of the tracks the corresponding traffic light is activated.

Table 3. Data sampling

No. test	No. characters	Information collected	No. accesses
1	CEGR	CERE	3
2	CEAB	C???	1
3	1CEC	?C?C	2
4	REGR	CE	1
5	CECR	CC	2
6	F2E4	?E?	1
7	ZRGE	??CE	1
8	123R	???R	1
9	DGAJ	G?	1
10	QNJG	??C	0

Table 4. Analysis of variance for precision and time

		Precision		Time	
Algorithms	Samples	Average	Variance	Average	Variance
Colors match	10	100	0	3	0,444444444
Crossed correlation	10	45,666	935,3269156	377,3	123,1222222
OCR	10	32,5	423,6111111	311,4	89,82222222

By analyzing the results obtained from each system as shown in Table 5, it was found that the intelligent system through evaluation parameters (number of cars and waiting time) has minimized vehicular congestion. This system, using the color matching algorithm, acts intelligently, significantly reducing waiting times.

Through the results obtained a comparative table is made in which the advantages of the intelligent system are described over the traditional one, as shown in Fig. 7.

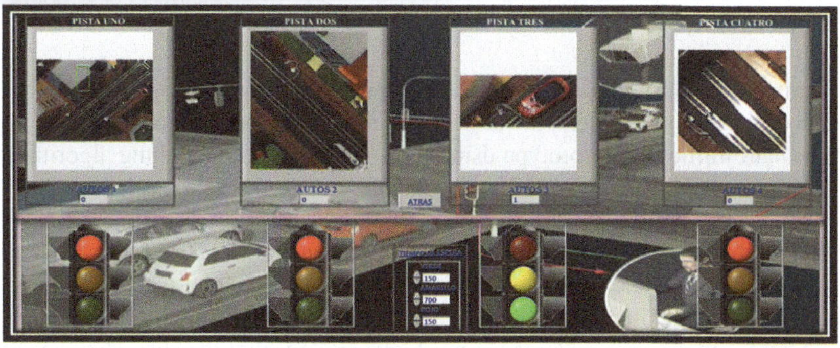

Fig. 8. Traffic lights deactivated

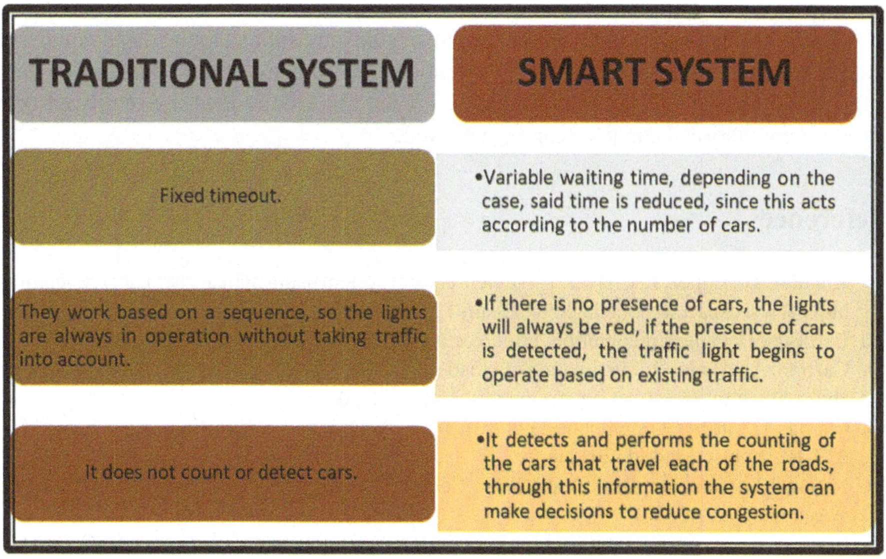

Fig. 9. Traffic light activation 3

Table 5. Data means

	No. cars	Waiting time (ms)
Traditional system average	15	60
Smart system average	15	18,75

4 Conclusions

When conducting the comparative study of the proposed pattern recognition algorithms, through evaluation parameters, it was obtained that the appropriate algorithm to implement a smart traffic light prototype using FPGAs is the color matching algorithm that has an accuracy of 100% and a response time of 3 ms.

When examining the results obtained from each system, it was possible to verify that the intelligent system, through evaluation parameters (number of cars and waiting time), minimized vehicle congestion. This system, using the color matching algorithm, acts intelligently, significantly reducing waiting times.

When recognizing the patterns we must take into account the characteristics of the camera and the light intensity, because they are predominant factors that directly affect the process, which if they do not meet the necessary requirements can give erroneous or null results.

The algorithms selected for the comparative study are subject to the disadvantage of computational complexity, making computation time in conventional machines a very important point. Therefore, the complexity of the algorithms must be improved or a greater computational capacity can be achieved.

References

1. Andrade, G., López, J., Chávez, P.: Vehicle control system using optical character recognition. http://www.dspace.espol.edu.ec/bitstream/123456789/1458/1/2973.pdf
2. Carrasco, J.: Pattern recognition. http://ccc.inaoep.mx/~ariel/recpat.pdf
3. Cazorla, A., Alados, L.: Estimation of cloud cover in sky images using the KNN classification algorithm. http://www.aet.org.es/congresos/xi/ten76.pdf
4. García, A.: Artificial Intelligence. Fundamentals, Practice and Applications, 1st edn, pp. 171–233. RC Books, Madrid (2012)
5. National Instruments: Basic LabVIEW: Course Manual, pp. 45–62. National Instruments, Washington DC (2006)
6. Nilsson, N.: Artificial Intelligence. A New Synthesis, 1st edn, pp. 33–62. McGraw-Hill, Madrid (2001). 75–99
7. Roncancio, H., Cifuentes, H.: Labview tutorial. http://perso.wanadoo.es/jovilve/tutoriales/016tutorlabview.pdf
8. Smith, S.: The Scientist and Engineer's Guide to Digital Signal Processing, 2nd edn, p. 254. Technical Publishing, San Diego (1999)
9. Sánchez, C., Sandonís, V.: Optical character recognition (OCR). http://www.it.uc3m.es/jvillena/irc/practicas/08-09/09.pdf
10. Tasiguano, C.: Development of vehicle license plate recognition algorithms. Thesis Ing. Electronics and Control. National Polytechnic School, Quito, Ecuador (2011)

Software and Systems Modeling

A Petri Net-Based Model of Self-adaptive Systems and Its (Semi-)Automated Support

Lorenzo Capra(✉)

Dipartimento di Informatica, Università degli Studi di Milano, Milan, Italy
capra@di.unimi.it

Abstract. Classical Petri Nets are not suitable for describing systems with an adaptable layout. A model based on Symmetric Nets (SN) has been recently introduced. It is composed of a SN emulating the behaviour of any base-level Place/Transition net (encoded as a marking), and an API for basic net-transformation primitives. In this paper, we discuss about the automation of the modelling process and the issues related to analysis complexity, proposing a more effective formalization of the base-level. The discussion is accompanied by a simple running example.

Keywords: Evolving systems · Petri nets · Emulation

1 Introduction

Classical Petri nets (PNs) are a central model for distributed systems, but are not designed to describe self-adaptation capabilities. For that purpose, a number of hybrid PN extensions have been proposed over the last decade, whose enhanced expressivity is not adequately supported by analysis techniques/tools.

A formal model for evolving systems based on the Symmetric Nets (SN) formalism (in origin, Well-formed Nets) [7] has been recently introduced [5]. The model, partly inspired by the "nets within nets" paradigm [10,12], builds on a "meta-level" SN which emulates any "base-level" Place/Transition (P/T) system with inhibitor arcs (known to be Turing-complete), encoded into the meta-net's state. An API for a basic set of *net-transformations* (SN subnets) can be profitably used to specify adaptation procedures operating on the emulated system. This approach is conceptually uniform and simple, differently from other proposals with similar objectives. SNs are *Colored* Petri Nets [9] with a structured syntax implicitly representing system symmetries, which are exploited to reduce the model's analysis complexity. The SN formalism, and its native stochastic extension (SSN), have the great advantage of being supported by a well engineered tool, GreatSPN [1]. Any P/T system can be emulated. Moreover, the design of adaptation procedures is unaware of emulator's inside.

As discussed in this paper, the entire modelling process may be automated, even if at present single steps are: the emulator's initial state is derived from

Á. Rocha et al. (Eds.): WorldCIST 2020, AISC 1159, pp. 715–725, 2020.
https://doi.org/10.1007/978-3-030-45688-7_70

a P/T net in PNML format, whereas the ALGEBRA module of GREATSPN is used to link the user-defined adaptation procedures to the emulator. Analysis complexity issues are also addressed. The benefits of a formalization of the base-level through SN (automatically unfolded into P/T nets) are discussed, A self-healing manufacturing system is used as a running example.

Related Work. The emulator-based approach has been introduced in [4] using Spec-inscribed nets [9]. Their higher data abstraction results in much more compact model than with SN. It has been implemented as an extensible library [3], at present, with reduced analysis capability. A survey on approaches combining higher-order tokens and the features of object-orientation can be found in [12]. The representative of this class of formalisms, Reference Nets [2], are supported by tool `Renew`. The formalism introduced in [8] extends Algebraic Higher-Order (AHO) nets with the main concepts of graph transformation systems. Common drawbacks of models based on higher-order languages/tokens are the lack of a clear semantics, the use of hybrid formalisms (hard to manage by non experts), consequently, a limited support in terms of analysis techniques. For a general survey on formalisms for self-adaptive systems we refer to [13].

2 The Formalisms

P/T nets [11] are finite bipartite graphs (see Fig. 1) whose nodes are partitioned into *places* and *transitions*, corresponding to sets P and T. Places and transitions are linked by weighted *input/output/inhibitor arcs* (the latter ending with a small circle). Each arc family (I, O, H) is described by a multiset on $P \times T$. A state, or *marking*, m is a multiset on P. A transition $t \in T$ is *enabled* in m iff $\forall p, I(p,t) \leq m(p) \wedge (H(p,t) = 0 \vee H(p,t) > m(p))$. When enabled, t may *fire* leading to m', where $m'(p) = m(p) + O(p,t) - I(p,t)$ (we write $m[t > m'$). A P/T system is a P/T net with an initial state m_0. Its *reachability graph* is a multi-graph whose nodes are the markings reachable from m_0, and such that there is an edge $m \xrightarrow{t} m'$ iff $m[t > m'$.

SN [7] have the same structure as P/T nets. As in any high-level PN formalism, nodes are associated with *domains*, defined as Cartesian products of finite *color classes*, denoted by capital letters, e.g., C. A class C may be partitioned into *static subclasses* $\{\mathrm{C}_j\}$: colours in a class represent entities of the same type, but only those within the same static subclass are guaranteed to *behave similarly*. A color class may also be *circularly ordered*.

A place's color domain, $cd(p)$, defines the type of tokens the place may hold. A place's *marking*, $M(p)$, is a multiset defined on $cd(p)$. A SN marking is the family of place markings. SN transitions are parametrized: the *instances* of t are elements of the transition's domain, $cd(t)$, implicitly defined by the variables annotating the arcs incident to t. A variable is denoted by a (possibly) indexed small letter which refers to a color class, e.g., c_2. A transition instance, also denoted (t, b), is a binding of t's variables with colors of proper type. A *guard* may be used to restrict $cd(t)$: it is defined in terms of *basic predicates*, which allow one to compare variables of the same type or test whether a color bound to a variable belongs to a given static subclass.

Input/output/inhibitor arcs are annotated by *functions* ($W^-[p,t]$, $W^+[p,t]$ and $W^h[p,t]$, respectively) $cd(t) \rightarrow Bag[cd(p)]$, expressed as linear combinations of *tuples* $\langle f_1, \ldots, f_k \rangle$ of *class functions*. A class-C function f_i, $cd(t) \rightarrow Bag[\mathrm{C}]$, is in turn a linear combination of functions in the set: $\{c_j, c_j{+}{+}, \mathrm{C}_q, All\}$. The symbol c_j (variable, or *projection*) maps a tuple in $cd(t)$ to the jth occurrence of color C; ++ gets the successor $\mathrm{mod}_{|\mathrm{C}|}$, if the class is ordered; constants C_q and *All* map to $\sum_{x \in \mathrm{C}_q} x$ and $\sum_{x \in \mathrm{C}} x$, respectively. We have ($b \in cd(t)$): $\langle f_1, \ldots, f_k \rangle(b) = f_1(b) \times \ldots f_k(b)$, where $\times$ is the multiset Cartesian product.

There are two different kinds of transitions: those representing *observable* (time-consuming) events, with a rectangle shape, and those representing *invisible* (logical) activities, drawn as tiny black bars. The latter take priority over the former. "Black" transitions may in turn have different priorities.

(t, b) *has concession* in marking M if $\forall p \quad W^-[p,t](b) \leq M(p) \wedge \forall c \in cd(p)$ $W^h[p,t](b)(c) = 0 \vee W^h[p,t](b)(c) > M(p)(c)$. An instance (t, b) with concession in M is *enabled* if no higher priority transition instance has. In this case it may fire, leading to M', where $\forall p \; M'(p) = M(p) - W^-[p,t](b) + W^+[p,t](b)$ (we write $M[(t,b) > M')$. If σ is a sequence of transition instances, $M[\sigma > M'$ means that M' is reachable from M through σ. M is said *vanishing* if some black transition is enabled, *tangible* otherwise. Assuming that the initial marking M_0 is tangible, and there are no vanishing loops, we may build the tangible reachability graph, whose nodes are tangible and there is an edge $M_i \xrightarrow{b} M_j$ iff $M_i[b\sigma > M_j$, where b is an observable instance whereas σ is a (possibly empty) vanishing sequence.

3 Running Example

A simple manufacturing system (MS) with self-healing capabilities is used throughout the paper. The MS has two symmetric production lines working pieces which are loaded two at a time, and evenly distributed to the lines. Pairs of refined pieces are assembled into the final artefact. Either line is periodically subject to failures, and repaired. The system's nominal behaviour, and the fault occurrence, are shown in Fig. 1: whenever either place broken_i is marked the attached line is blocked and the MS, without any action, would enter a deadlock.

A first adaptation scenario takes place upon a failure: at the end of it the MS layout looks like Fig. 2. During adaptation the MS continues working using the available line: the faulty line is detached, and the behaviour of both the loader and the assembler changes accordingly. The presence of pending pieces on the faulty line is a critical issue. Once adapted, the loader puts two row pieces at a time on the available line, whereas the assembler takes pairs of worked pieces from that line. The second adaptation scenario brings the MS back to its nominal behaviour as soon as the faulty component has been repaired. Once again, without stopping the system. Despite its simplicity, trying to describe the MS and its adaptation with PN is almost impossible. Our approach follows a clear separation of concerns and consists of representing the adaptation *procedures* as apart concurrent components (SN sub-nets), which monitor the current state/topology of the base-level system, possibly rearranging it. The base-level

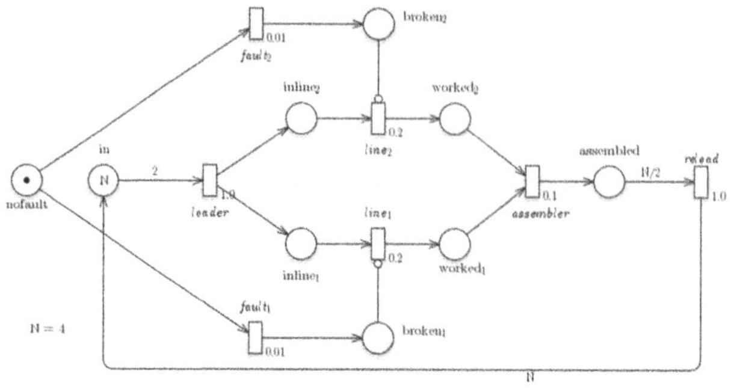

Fig. 1. MS nominal behaviour.

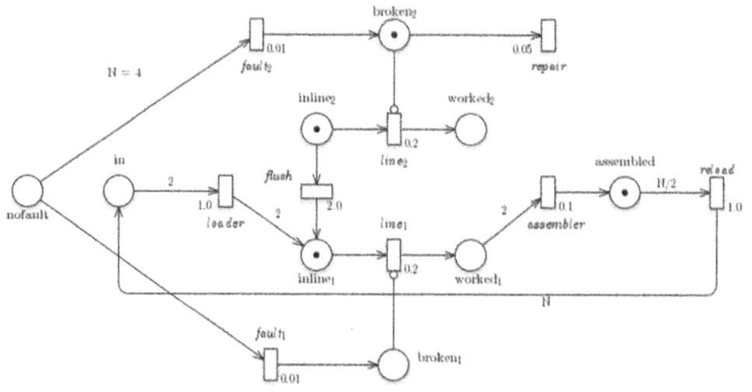

Fig. 2. Self-adaptation upon a fault on a line.

dynamics is emulated by a meta-level SN encoding it as a marking. Adaptation is implemented by read/write primitives (SN sub-nets) which safely operate on the system's encoding, through a simple API. One can thus easily represent disconnection/reconnection of a faulty/repaired line, migration of raw pieces from one line to the other, failure repair (through a new transition), and so forth.

The SN model in Fig. 3 represents in a compact, parametric way a more complex system formed by N copies of the MS in Fig. 1, which synchronize at the end of a production cycle. Through the unfolding module of GreatSPN it may be automatically translated into a P/T net. The advantages of such an approach will be discussed later.

4 The Modelling Framework

The building-block of the modelling approach is a quite complex SN, specified in [5], referred to as *emulator*[1]. The emulator reproduces the *interleaving* semantics of any P/T system encoded as a marking. Despite its apparent complexity, it allows for significant achievements in modelling dynamic systems.

[1] All the SN sources (GreatSPN .PNPRO files) are publicly available at https://github.com/SELab-unimi/sn-based-emulator.

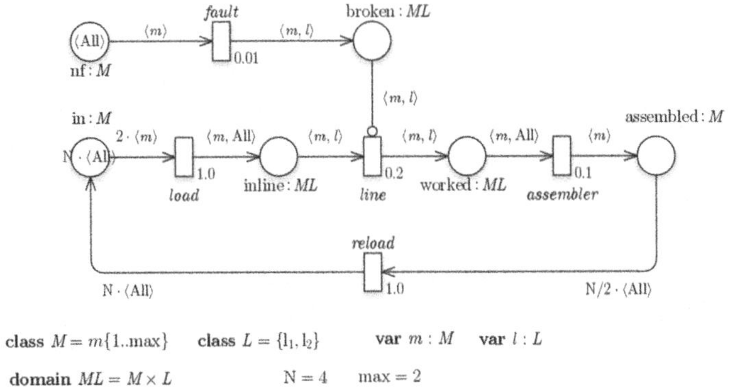

Fig. 3. A symmetric MS model

4.1 The Emulator

The model's annotations build on basic classes P and T, whose elements represent the nodes of a P/T net: letting $\mathcal{N} = (P, T, I, O, H, m_0)$ be a P/T system, $\mathrm{P} \supseteq P$, $\mathrm{T} \supseteq T$, i.e., classes P and T must be large enough to cover all possible evolutions. The emulator's places IN, OUT, H (with domain P×T) and MARK ($cd(\mathrm{MARK}) = \mathrm{P}$) encode the (current) structure and marking of $\mathcal{N}$, respectively. The initial marking is: $\forall \mathrm{pl}_i \in P$, $\forall \mathrm{tr}_j \in T$

$$M_0(\mathrm{IN})[\langle \mathrm{pl}_i, \mathrm{tr}_j \rangle)] = I(\mathrm{pl}_i, \mathrm{tr}_j) \ldots M_0(\mathrm{MARK})[\langle \mathrm{pl}_i \rangle] = m_0(\mathrm{pl}_i).$$

The emulator's behaviour is cyclic. A firing instance of the *only* observable transition, ($\mathrm{PT\_fire}, t = \mathrm{tr}_k$) matches the firing of a P/T transition, tr_k: it triggers a sequence of invisible transition instances reproducing the *atomic* state change caused by tr_k. This sequence is divided in three parts: first the marking of place MARK is updated, according to the P/T firing rule; two places encoding the set of transitions enabled before the firing of tr_k (enabList) and those to check upon it (checkList) are efficiently updated, taking into account the *structural conflicts* and *causal connection*; the transitions "to be checked" are tested for enabling.

Any reachable *tangible* marking matches a reachable marking of the encoded P/T system: the emulator's *tangible* reachability graph and the reachability graph of the encoded P/T system are *isomorphic*.

Lemma $m_i[\mathrm{tr}_k > m_j$ if and only if $M_i[\beta \cdot \sigma > M_j$, where M_i and M_j are tangible markings such that $M_i(\mathrm{MARK})$ and $M_j(\mathrm{MARK})$ encode m_i and m_j, respectively, $\beta := (\mathrm{PT\_fire}, t = \mathrm{tr}_k)$, and σ is a vanishing path.

4.2 The Adaptation API

Figure 4a isolates the set of emulator's places encoding the P/T system. This set represents the emulator's *evolutionary interface*. A basic but complete set of net-*read/write primitives* form the *evolutionary API*, which may be used to safely

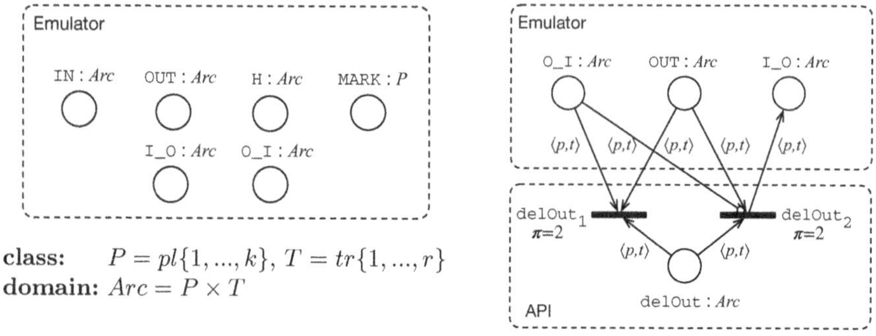

(a) Places encoding the base-level.

(b) The delOut primitive.

Fig. 4. Emulator section and the **delOut** API primitive example.

modify the emulated system. Each primitive is linked to the emulator's evolutionary interface: one can get information about the marking and the structure of the P/T system; add/remove nodes; set up the weight of arcs; change the marking, and so forth. A primitive is defined by a SN subnet which reads and/or consistently modifies the P/T system's encoding, (possibly) through invisible firing sequences. The evolutionary API is similar to the reflection API of most modern programming languages.

Figure 4b shows a simple example of primitive (the delOut operation), which decreases the weight of a base-level output arc. When a token $\langle pr, tr\rangle$ is put into place delOut (holding the input of the primitive), one of the two mutually exclusive transitions delOut_1, delOut_2 may become enabled. Its firing removes the token $\langle pr, tr\rangle$ from the OUT place and updates the marking of I_O and O_I accordingly. The priorities of transitions composing a primitive sub-net are relative: when bringing all together, the greatest priority in a primitive-net is set lower than the lowest priority in the emulator.

Other primitives are more complex, due to additional consistency checks they perform. As an example, the primitive which decreases the weight of an input arc (the argument) has to check whether the linked transition is currently either in enabList or in checkList; if not, it has to be added to checkList.

4.3 Self-adaptation Procedures

The component driving the system evolution is made up of a set of procedures, implementing a *feedback control loops*. A procedure is described by a SN, which may contain both observable and logical transitions, which is transparently linked to the emulator through the evolutionary API. Figure 5 shows the procedure managing a fault occurrence. This procedure is triggered whenever place Broken of the MS model is marked. A challenging point is that the MS execution keeps going while changes are being carried out. Even though single transformations are atomic, i.e, theoretically consistent, their sequence may bring the overall system into logically inconsistent states.

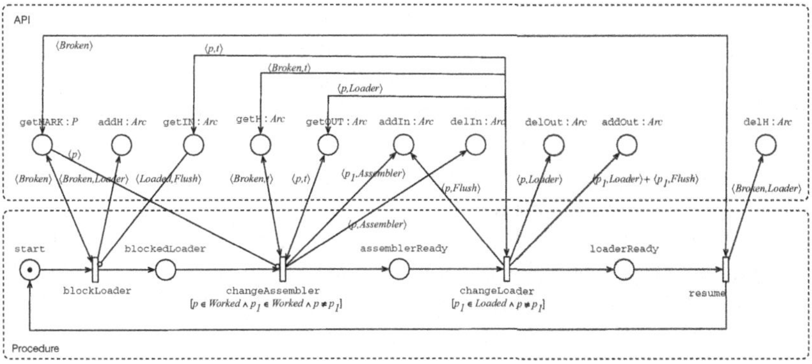

Fig. 5. The *fault managing* procedure.

A fault occurrence is checked by the blockLoader transition through the getMARK primitive. When blockLoader fires, it temporarily suspends the loader by linking it to place Broken with an inhibitor arc (addH primitive). Then changeAssembler modifies the arcs surrounding assembler, as shown in Fig. 2, through the addIN and delIN primitives. In a similar way, the procedure changes the loader behaviour to avoid loading of raw pieces into the faulty line (through addOut and delOut primitives). A new transition is inserted through which residual row pieces on the faulty line eventually move to the working one. Loading is resumed at the end of the procedure by removing the temporary inhibitor arc between Broken and Loader. The procedure which brings the system back to its default layout (after the faulty line has been repaired) is not described due to lack of space. The tricky point there is that the system must enter a safe state before the reconfiguration can take place. The specification of this procedure is available on the online repository mentioned before.

The emulator, the evolutionary API, and the adaptation procedures are connected using a simple place *superposition*. The composition process can be automatically performed using the ALGEBRA package of GREATSPN.

5 Complexity Issues

Table 1 reports some experiments conduced on the running example by using the `GreatSPN`'s ordinary and symbolic reachability-graph builders. The whole model, composed of the emulator (encoding the P/T system in Fig. 1) and the two adaptation procedures, has been analysed for number of worked pieces per cycle (N) 2 through 32. Color class T has been partitioned into subclasses holding elements of the same type. Two variants of the MS model are considered: the symmetric one (denoted SMS) described in Fig. 1, and the asymmetric one (MS), in which only one of the two lines is periodically off. The size of the RG and SRG are listed in terms of tangible and vanishing states. Building times are reported too. Execution times are reported, varying from dozens ms to a few hours. We observe that the SRG takes much more time than the RG.

The high number of black transitions in the emulator leads to an explosion of vanishing markings, as it is evident from Table 1. The state-space builder implemented in GreatSPN, in fact, has some drawbacks. In particular, it does not use any on-the-fly reduction technique of vanishing markings, which are eliminated (to get the tangible graph) after the whole state-space is built. Moreover, simultaneously enabled immediate transition instances are fired in an interleaved way (causing a combinatorial explosion of vanishing states), even when their are independent. While the first limitation may be faced only by reimplementing the state-space builder, the interleaving of immediate transitions instances has been tackled by using two orthogonal workarounds.

Ordering/Partitioning of Basic Classes – A major source of inefficiency is the enabling test of P/T transitions, which is performed by non-deterministically selecting the next transition to test. The transition interleaving due to non-determinism notably lowers if color class T is *ordered*. This potentially drops the number of vanishing paths from $n!$ to n, where $n = |\mathrm{T}|$. Just to give an idea, for $N = 32$ the number of vanishing markings (SMS model) lowers to around one million, and the RG building time to 300 s. Although ordering T is effective, it prevents from exploiting symmetries in stochastic SN (SSN): there T may need to be partitioned so that P/T transitions with the same firing rate belong the same subclass. But ordering a partitioned class causes its complete split.

An alternative solution (in SSN) is to exploit the partition of class T induced by time specification, to reduce the aforementioned non-determinism. The SN transition which takes the next P/T transition to check for enabling is split into a number of mutually exclusive copies with different priorities, so that elements belonging to different subclasses are considered in an arbitrary order. The gain, in terms of interleaving reduction, depends on the size of the biggest subclass (the smaller, the better). As for the MS example the achieved reduction is slightly lower than the one got with the ordering of T.

Structural Techniques – In [6] a calculus is defined[2] for computing *symbolic structural relations* between SN nodes. Structural relations between SN transitions are maps $cd(t) \to 2^{cd(t')}$. For example $SC(t,t')$ (the asymmetric Structural Conflict) maps an instance (t,b) to the set of instances $\{(t',b')\}$ that can disable (t,b) by withdrawing color-tuples from some input place of t or adding color-tuples into some inhibitor place of t. Such relations are *syntactically* expressed by using a simple extension of SN arc functions' grammar. By computing SC, CC, and the transitive closure, it is possible to check whether two transitions t and t' are *structurally independent*, meaning that there is no instance of t conflicting (either directly, or indirectly) with any instance of t', and vice versa. Independent SN transitions have been assigned different priorities, to reduce interleaving. It is worth noting that it is also possible to decrease the interleaving of instances of

[2] Its implementation is available at http://www.di.unito.it/~depierro/SNex/.

Table 1. Reachability graph size/building time.

Model	N	$\|RG\|$ (TM/VM)	Time (s)	$\|SRG\|$ (TM/VM)	Time (s)
MS	2	55/3484	0.15	55/3484	3.40
	4	184/13906	2.31	184/13906	10.06
	8	985/91586	8.04	985/91586	64.11
	16	7964/886842	77.11	7964/886842	622.91
	32	95568/11108025	1544	95568/11108025	$\approx$3.5 h
SMS	2	92/6708	1.71	48/3437	5.36
	4	276/23268	2.79	142/11986	10.06
	8	1289/131761	10.12	662/66663	91.66
	16	9103/1114027	99.05	4634/561461	816.71
	32	109236/13945378	1978	55448/7078872	$\approx$4.8 h

the same transition, which is a major concern in the SN emulator. Structural analysis has been also used to validate the emulator-based model.

Symmetry Exploitation – By setting an initial *symbolic* marking it is possible to build a *quotient-graph*, called symbolic reachability graph (SRG), which retains all the information of the ordinary RG. SRG nodes are *syntactical* equivalence classes of ordinary colored markings, where m, m' are equivalent if and only if m' is obtained from m through a permutation on basic classes preserving the partition into subclasses and the circular ordering. In stochastic SN a lumped CTMC is derived from the SRG, which can be solved instead of the original one. A symbolic marking (SM) is defined in terms of *dynamic subclasses*. Each dynamic subclass refers to a static subclass, or to a basic class, and has a cardinality. Dynamic subclasses represent *parametric partitions* of color (sub-)classes. Dynamic subclasses of an ordered class are ordered too. A simple way to set up an initial symbolic marking in the emulator is to replace in the ordinary initial marking colors $\{\mathrm{pl}_i\}$ and $\{\mathrm{tr}_i\}$ with (cardinality one) dynamic subclasses $\{zp_i\}$ and $\{ztr_i\}$, respectively. The resulting SRG nodes (SMs) represent classes of isomorphic *marked P/T nets*. Checking graph isomorphism is a demanding task and in our model corresponds to bring a SM into a canonical form [7]. Table 1 shows that in the case of symmetric MS the SRG size, as expected, is more or less the half of the ordinary RG. On the other side, there is an evidence that building the SRG is much more time consuming than building the ordinary RG. We believe that a major source of inefficiency, in the current implementation, is that the cardinality of an SM has to be computed, and this is done by explicitly enumerating the possible permutations represented by the SM.

In order to alleviate this problem, it is convenient to further refine the partitions of classes T and P induced by time specification, so that each subclass contains nodes which are known a priori as permutable. A way to automatically derive this information is starting from a base-level formalization in terms of SN, like that described in Fig. 3, which represents a parametric version of the

MS system. The idea is that all and only the P/T nodes which are the instances of an unfolded SN node are gathered in the same subclass. Very unlikely two base-level nodes of the same type corresponding to different SN nodes can be permuted in the graph structure. On the other side, we may assume that (possibly after a preliminary split) all instances of a SSN transition have the same firing rate.

Analysing systems composed of a large number of symmetric modules, on the other hand, is unfeasible without exploiting such symmetries. Consider, e.g, that a system composed of 4 MS like that in Fig. 1, each working 4 pieces per cycle, results in several dozens millions ordinary states, against just a few thousands symbolic ones.

6 Conclusion and Future Work

We have discussed full automation and complexity issues of a SN-based model for self-adaptive system supported by GreatSPN tool. A self-healing MS has been used as a running example. Orthogonal techniques based on ordering/partitioning of basic color classes, computation of structural relations, and symmetries have been proposed, discussed and compared.

References

1. Baarir, S., Beccuti, M., Cerotti, D., De Pierro, M., Donatelli, S., Franceschinis, G.: The GreatSPN tool: recent enhancements. SIGMETRICS Perform. Eval. Rev. **36**(4), 4–9 (2009). https://doi.org/10.1145/1530873.1530876
2. Cabac, L., Duvigneau, M., Moldt, D., Rölke, H.: Modeling dynamic architectures using nets-within-nets. In: Ciardo, G., Darondeau, P. (eds.) Applications and theory of Petri Nets 2005, pp. 148–167. Springer, Heidelberg (2005)
3. Camilli, M., Capra, L., Bellettini, C.: PNemu: an extensible modeling library for adaptable distributed systems. In: Donatelli, S., Haar, S. (eds.) Application and Theory of Petri Nets and Concurrency, pp. 80–90. Springer International Publishing, Cham (2019)
4. Capra, L.: A pure SPEC-inscribed PN model for reconfigurable systems. In: 2016 13th International Workshop on Discrete Event Systems (WODES), May 2016, pp. 459–465 (2016). https://doi.org/10.1109/WODES.2016.7497888
5. Capra, L., Camilli, M.: Towards evolving Petri Nets: a symmetric nets-based framework. IFAC PapersOnLine **51**(7), 480–485 (2018). https://doi.org/10.1016/j.ifacol.2018.06.343. 14th IFAC Workshop on Discrete Event Systems WODES 2018
6. Capra, L., De Pierro, M., Franceschinis, G.: Computing structural properties of symmetric nets. In: Campos, J., Haverkort, B.R. (eds.) Quantitative Evaluation of Systems, pp. 125–140. Springer International Publishing, Cham (2015)
7. Chiola, G., Dutheillet, C., Franceschinis, G., Haddad, S.: Stochastic well-formed coloured nets for symmetric modelling applications. IEEE Trans. Comput. **42**(11), 1343–1360 (1993)
8. Hoffmann, K., Ehrig, H., Mossakowski, T.: High-level nets with nets and rules as tokens. In: Proceedings of the 26th International Conference on Applications and Theory of Petri Nets, ICATPN 2005, pp. 268–288, Springer-Verlag, Heidelberg (2005)

9. Jensen, K., Rozenberg, G. (eds.): High-level Petri Nets: Theory and Application. Springer-Verlag, London (1991)
10. Lakos, C.: Object Oriented Modelling with Object Petri Net, pp. 1–37. Springer, Heidelberg (2001). https://doi.org/10.1007/3-540-45397-0_1
11. Reisig, W.: Petri Nets: An Introduction. Springer-Verlag, New York (1985)
12. Valk, R.: Object Petri Nets. In: Desel, J., Reisig, W., Rozenberg, G. (eds.) Lectures on Concurrency and Petri Nets. LNCS, vol. 3098, pp. 819–848. Springer, Heidelberg (2004)
13. Weyns, D., Iftikhar, M.U., de la Iglesia, D.G., Ahmad, T.: A survey of formal methods in self-adaptive systems. In: Proceedings of the Fifth International C* Conference on Computer Science and Software Engineering, C3S2E 2012, pp. 67–79. ACM, New York (2012). https://doi.org/10.1145/2347583.2347592

An Exploratory Study on the Simulation of Stochastic Epidemic Models

Carlos Balsa[1], Isabel Lopes[2,3](✉), José Rufino[1], and Teresa Guarda[3,4,5]

[1] Research Centre in Digitalization and Intelligent Robotics (CeDRI), Instituto Politécnico de Bragança, Campus de Santa Apolónia, 5300-253 Bragança, Portugal
{balsa,rufino}@ipb.pt

[2] Applied Management Research Unit (UNIAG), Instituto Politécnico de Bragança, Campus de Santa Apolónia, 5300-253 Bragança, Portugal
isalopes@ipb.pt

[3] Centro ALGORITMI, Escola de Engenharia - Universidade do Minho, Campus Azurém, 4800-058 Guimarães, Portugal
tguarda@gmail.com

[4] Universidad Estatal Península de Santa Elena – UPSE, La Libertad, Ecuador

[5] Universidad de las Fuerzas Armadas-ESPE, Sangolqui, Quito, Ecuador

Abstract. A small number of people infected with a contagious disease in a large community can lead to the rapid spread of the disease by many of the people in that community, leading to an epidemic. Mathematical models of epidemics allow estimating several impacts on the population, such as the total and maximum number of people infected, as well as the duration and the moment of greatest impact of the epidemic. This information is of great use for the definition of public health policies.

This work is concerned with the simulation of the spread of infectious diseases in small to medium communities by applying the Monte Carlo method to a Susceptibles-Infectives-Recovered (SIR) stochastic epidemic model. To minimize the computational effort involved, a simple parallelization approach was adopted and deployed in a small HPC cluster.

The simulations conducted show that an epidemic outbreak can occur even if the initial number of infected people is small, and that this probability decreases significantly with the vaccination of a population subset.

Keywords: Infectious diseases · Epidemic models · Stochastic models · Numerical simulations · Parallel computing

1 Introduction

An epidemic is the rapid spread of an infectious disease within a population, producing a large number of infected individuals in a short period of time. These may ultimately die, or become permanently incapacitated. Infectious diseases that can be responsible for an epidemic outbreak include well known diseases like HIV [12], Smallpox [5], SARS [9] and (H1N1) influenza [11], among others.

Á. Rocha et al. (Eds.): WorldCIST 2020, AISC 1159, pp. 726–736, 2020.
https://doi.org/10.1007/978-3-030-45688-7_71

Mathematical modeling of an epidemic, or epidemic modeling, aims at understanding the spreading of an infectious disease and to predict its future course. This information is of vital importance to the implementation of strategies to control the spread of an epidemic. For instance, mathematical models allow to estimate several effects of an epidemic, like the total number of infected people or the duration of the epidemic, and how and when to use prevention methods such as education, immunization or isolation, just to name the most common.

There are two main broad classes of epidemic mathematical models: *deterministic* and *stochastic*. Typically, deterministic models are suitable to model epidemics within large communities. The "epidemic process" in these models is governed by a system of differential equations and the evolution of the process is deterministic in the sense that no randomness is allowed. The community is considered homogeneous and it is assumed that individuals mix uniformly with each other. Hence, these models do not incorporate any arbitrariness (see [7]).

Stochastic models, in turn, have been successfully applied to small-size communities. Since the spread of an infectious disease is a random process, that takes place locally through the close contact with infectious individuals, stochastic models are more realistic in nature (see [4]). Stochastic models can be simulated by using Monte Carlo based methods. Essentially, this consists in the repetition of a random behavior of the population, a large number of times. For small-size populations, the simulation times are moderate; but, when applied to larger communities, this procedure has a high computational cost.

Epidemic mathematical models may also be a applied in other contexts. In Ecology, for instance, such models are useful for the simulation of the effects of the biological control of certain natural plagues (see [3]). As a result, the computational implementation of these epidemic mathematical methods is of great importance for the understanding of the various simulated real processes.

The focus of this paper is the implementation of several epidemic stochastic models and its application in small to medium communities, thus gaining insights for a future application to large-size populations. In order to contain the simulations time, their execution was accelerated by means of a simple parallelization approach aimed at taking advantage of a small commodity HPC cluster.

The rest of this paper is structured as follows: Sect. 2 introduces a simple deterministic epidemic model and defines a stochastic counterpart; Sect. 3 describes the computational strategy used to conduct the simulation of the stochastic model and presents the results of its execution; Sect. 4 provides final considerations on the work of this paper and lays out its future directions.

2 Deterministic vs Stochastic Epidemic Modeling

A typical approach when modeling the spread of an infectious disease in a community is to formulate a compartment model. A simple compartment model is to consider that, at each point in time, each individual of the population belongs to one, and only one, of the following compartments: S – an individual is Susceptible to catch the disease; I – an individual is Infective, meaning he has got the disease and is able to infect others; R – an individual has Recovered from infection and can no longer be infected again. This Susceptibles-Infectives-Recovered compartments model is simply referred to as the SIR model. Thus, in the SIR model, an individual can only move from compartment S to compartment I, and then to compartment R. As the name hints, the compartments of the SIR model are closely related to three different stages of the disease.

2.1 Deterministic Model

Let us consider a population, or community, of size $n \in \mathbb{N}$. Let $s(t)$, $i(t)$ and $r(t)$ denote, respectively, the fraction of the community that belongs to the compartments S, I and R, at a given instant $t \geq 0$, in time. Assuming that the population is closed, meaning that there are no births, deaths and immigration or emigration during the study period, then $s(t) + i(t) + r(t) = 1$, for any t.

Assuming that the population is homogeneous, individuals mix uniformly, and s, i and r are differentiable functions, then the variation of the fraction of the population in compartments S, I and R is given by the following system of differential equations:

$$\begin{cases} \frac{ds}{dt} = -\beta s i \\ \frac{di}{dt} = \beta s i - \gamma i \\ \frac{dr}{dt} = \gamma i \end{cases} \tag{1}$$

In the previous system, β is the rate of contact between susceptible and infectious individuals, and $1/\gamma$ is the infectious period. These differential equations, along with the initial values $s(0) = 1 - \epsilon$, $i(0) = \epsilon$ and $r(0) = 0$, fully define the deterministic SIR model. The parameter $\epsilon > 0$ represents the initial fraction of infectives and, although positive, it is assumed to be small.

The computational simulation of the deterministic SIR epidemic model is trivial. The problem with the initial values presented above can be rapidly solved by a numerical method like the four-order Runge-Kutta; also, the computational time does not depend on the dimension n of the population.

The values of β and γ are of paramount importance in this model. The ratio

$$R_0 = \frac{\beta}{\gamma} \tag{2}$$

can be interpreted as the average number of new infections caused by an infectious individual. This ratio is also referred to as the *basic reproduction number*. When $R_0 > 1$ an epidemic outbreak takes place, infecting a substantial part of the population, and when $R_0 < 1$ there is no major epidemic outbreak (see [7]).

This simple deterministic model also enables the deduction of a balance equation determining the fraction $z = r(\infty)$ of the population that is infected throughout the epidemic period (see [4]):

$$1 - z = (1 - \epsilon)\, e^{-R_0 z}. \tag{3}$$

In an homogeneous community where individuals mix uniformly with each other, the size of an epidemic outbreak is determined by the *basic reproduction number* (R_0) and by the infectious period ($1/\gamma$), and there is no incertitude or randomness in the final number of infected individuals. However, in some cases, this contradicts reality: for instance, the introduction of a small number of infected in a community may not necessarily have as a consequence a large outbreak (even if $R_0 > 1$); such situation can happen, for example, if the infectious isolate themselves from the rest of the community or, by chance, if their contacts are restricted to immune individuals. This motivates the formulation of models that are stochastic in nature.

2.2 Stochastic Model

A simple stochastic epidemic model is the stochastic counterpart of the SIR deterministic model. Like before, a closed homogeneous and uniformly mixing community of n individuals is assumed. However, $S(t) = n \times s(t)$, $I(t) = n \times i(t)$ and $R(t) = n \times r(t)$ now denote the number of susceptible, infectious and recovered individuals, respectively, at time t.

A brief description of the temporal dynamics of the SIR stochastic model follows, based on [4]. If at time $t = 0$ there is a number of m infectious, then $S(0) = n - m$, $I(0) = m$ and $R(0) = 0$. Infectious individuals have contact (adequate to the transmission of the disease) with other individuals randomly in time, according to a Poisson process with intensity β. Each contact is made with an individual randomly selected from the community. Any susceptible that receives such contact immediately becomes infective and starts spreading the disease according to the same rules. The infectious periods are independent and exponentially distributed with mean $1/\gamma$.

In a more realistic scenario, individuals tend to have different contact types with others. Hence, an improvement to this model is to consider a population with a non-uniform mixing, where an individual can have different average contact rates with different groups of individuals. In fact, most individuals usually have a few others with whom they mix at a much higher rate. In a *household epidemic model* individuals are grouped into households and the contact rate between pairs of individuals of the same household (β_H) is independent of the contact rates between pairs of individual of different household (β_G) [1,2].

There are also *stochastic lattice models* that represent spatial locations by means of a grid structure where each cell (pixel) is in a state and evolves according to stochastic transition rules [10]. A general survey of the main mathematical methods used in the epidemiology can be find in Brauer and co-workers [6].

3 Simulation Experiments

The basic SIR stochastic model introduced in the previous section allows to predict, without performing any simulations, the probability of a minor or a large epidemic outbreak. Also, in the case of a major outbreak, it allows to predict the final number of infected [4]. However, the model does not provision for the prediction of the outbreak duration, or the time at which the number of simultaneously infected reaches its peak. To obtain this information it is necessary to apply the Monte Carlo method a large number of times, getting the probability distribution for each of these variables. Once a large number of simulation repetitions is necessary in order to stabilize the distribution of frequencies, the application of the Monte Carlo method may take a lot of time and demand considerable computational effort. Furthermore, these requirements will scale up as the size of the population grows, calling for efficient computational methods.

3.1 Basic Algorithm

Algorithm 1 is the basic algorithm of the SIR stochastic model enriched with the Monte Carlo method, as proposed in this paper. Its inputs (line 1) are: the initial number m of infected, a parameter β for the Poisson distribution (rate of contact between susceptible and infectious), a parameter γ for the exponential distribution (reciprocal of the infectious period) and a vaccination rate υ.

Algorithm 1: Stochastic SIR epidemic model.

```
 1. inputs: m, β, γ and υ
 2. for j = 1...N_sim
 3.    choose randomly m individuals and the corresponding infectious period
 4.    while pop(i) > 0 for some i
 5.       for i = 1...n
 6.          if pop(i) > 0 then
 7.             pop(i) = pop(i) − 1
 8.             if pop(i) − 1 = 0 then pop(i) = −1 end if
 9.          endif
10.          generate the number of close contacts of the individual i
11.          if pop(i) = 0 and one of the contacted individuals is infectious then
12.             pop(i) = p
13.          endif
14.          if pop(i) = 0 and random value < υ then pop(i) = −2 end if
15.       end for
16.    end while
17. end for
```

The algorithm repeats (line 2) the simulation a number of times N_{sim}. In each simulation, the population is represented by a vector *pop* of n cells. Each cell *pop*(i) stores the current status of the i'th individual (with $i = 1, ..., n$) as follows: 0 for susceptible, > 0 for infected, -1 for recovered and -2 for vaccinated.

In the first day ($t = 0$) of a simulation (line 3), an initial number $i_0 = m$ of infected individuals are randomly selected; for each one of these individuals their specific cell in the population vector is set to p, where p is the infection period given by an exponential distribution with expected value $1/\gamma$; for the remaining individuals, that are susceptible, their cell value is set to 0.

In the next days ($t > 0$) the model dynamics repeat (while loop in line 4). At each day, the vector cell of the infected individuals decreases one unit (line 7); furthermore (line 8), when it reaches 0 (again susceptible) is set to -1 (recovered).

Next, for every person still susceptible, the number of close contacts with others is generated by a Poisson distribution with parameter β (line 10). These other persons are randomly selected. If there is an infected among them (line 11), the susceptible becomes infected and its cell value is set to p, where p is randomly generated by an exponential distribution with parameter γ (line 12).

Finally, if the person is still susceptible, a random number between 0 and 1 is drawn; if this number is smaller than the vaccination rate v, the person becomes vaccinated; otherwise, the person stays susceptible (line 14). The simulation stops when there are no more infectious individuals (condition in line 4 is false).

In order to get a probability distribution of the target variables, the number of simulations (N_{sim}) should be large (more simulations lead to better probability distributions). The variables targeted by the study presented in this paper are: the total number of infected individuals, the duration of the epidemic outbreak, the maximum number of simultaneous infected individuals, and the day in which this maximum happens (that is, the day corresponding to the epidemic peak).

3.2 Parallelization Approach

Depending on the particular combination of the input parameters, the sequential execution of Algorithm 1 may take a lot of time. Clearly, with large populations (large values of n), or a high number of simulations (large values of N_{sim}), the execution time will increase. But other factors, like the vaccination rate (v), have also a decisive influence: the lower this rate, the higher will be the propagation of the disease, thus delaying the reaching of the algorithm stop condition.

Thus, in order to be able to simulate a wide range of scenarios, while generating accurate probability distributions, in a reasonable amount of time, a simple parallelization approach was applied to Algorithm 1: the initial range of simulations ($1 \ldots N_{sim}$) was fully split into mutually exclusive sub-ranges, and these were assigned to independent processors; this was possible because each simulation is completely independent of the others, allowing for a SPMD (single program, multiple data) approach; depending on the relative computing power of the processors, the partition of the initial range may be homogeneous or heterogeneous; either way, after all processors exhaust their sub-range, the results of each simulation are joined to produce the final consolidated results; these are histograms concerning the four variables targeted by this work (see above).

3.3 Implementation Details

The computational environment used in this research is a small Linux cluster (32 CPU-cores overall), based on 8 homogeneous nodes (each with one Intel i7-4790K 4.0 GHz quad-core CPU) and with a single MATLAB installation network-shared by all nodes. The implementation of Algorithm 1, and of its parallelization strategy, were both made in order to take advantage of such environment.

Algorithm 1 was implemented in MATLAB [8], but with the lower and upper limits of the main loop (line 2) as input parameters, thus restraining the loop to a sub-range of simulations. That lower limit seeds the MATLAB random number generator, ensuring a different strain of random numbers per sub-range. Each instance of the algorithm saves the results of its sub-range of simulations in CSV files. These are later consolidated by a BASH script, which further explores MATLAB to generate histograms of the target variables and related graphics.

Another BASH script generates the sub-ranges, accordingly with the number of CPU-cores of the cluster intended to be used; for each sub-range, a specific job is submitted to the job manager of the cluster; when released, each job executes the MATLAB implementation of Algorithm 1, properly parametrized.

3.4 Simulations Results

Figures 1 and 2 exhibit the results of a simulation of Algorithm 1, with $N_{sim} = 10000$, for the spread of the influenza in a small city (with $n = 20000$ individuals) and in a medium city (with $n = 100000$ individuals), considering that a unique infectious individual ($m = 1$) is introduced in the community. The results of Fig. 1 concern a vaccination rate of $\upsilon = 0$, and for Fig. 2 the vaccination rate assumed was $\upsilon = 0.33$. For influenza, the basic reproduction number is $R_0 \approx 1.5$ and the mean of the infectious period is $1/\gamma \approx 4$ days [7].

In both figures, the left column shows results for a small city, and the right column presents the same kind of results for a medium city. Each column presents, top to bottom, the empirical distribution of the i) total number of infected individuals during the epidemic outbreak, ii) the duration of the epidemic, iii) the maximum number of infected simultaneously, and iv) the higher incidence day.

For Fig. 1, with regard to the total number of infected individuals, it can be observed, in both communities ($n = 20000$ and $n = 100000$), that the histogram is bimodal. Moreover, the simulation data reveals that a large number of simulations results in a small outbreak, and another portion results in a major outbreak. The proportion of simulations that results in a small outbreak is approximately 0.67 (in this case, a very small number of individuals are infected); conversely the proportion of simulations that results in a major outbreak is 0.33 (in which case, a large number of individuals become infected). In a major outbreak, the total number of infected individuals has a distribution approximately normal, with mean $Z = 11656$ if $n = 20000$, and with $Z = 58281$ if $n = 100000$.

All the results obtained from the heuristics are in agreement with the theoretical values provided for the general SIR stochastic model (see [4]), namely the total number of infected individual that can be obtained from the solution of Eq. (3). The epidemic has a duration close to 90 days if $n = 20000$, and close to 120 days if $n = 100000$. The maximum number of infected simultaneously is near 2200 individuals if $n = 20000$, and near 10500 if $n = 100000$. The higher incidence day is close to 45 if $n = 20000$, and near 55 if $n = 100000$.

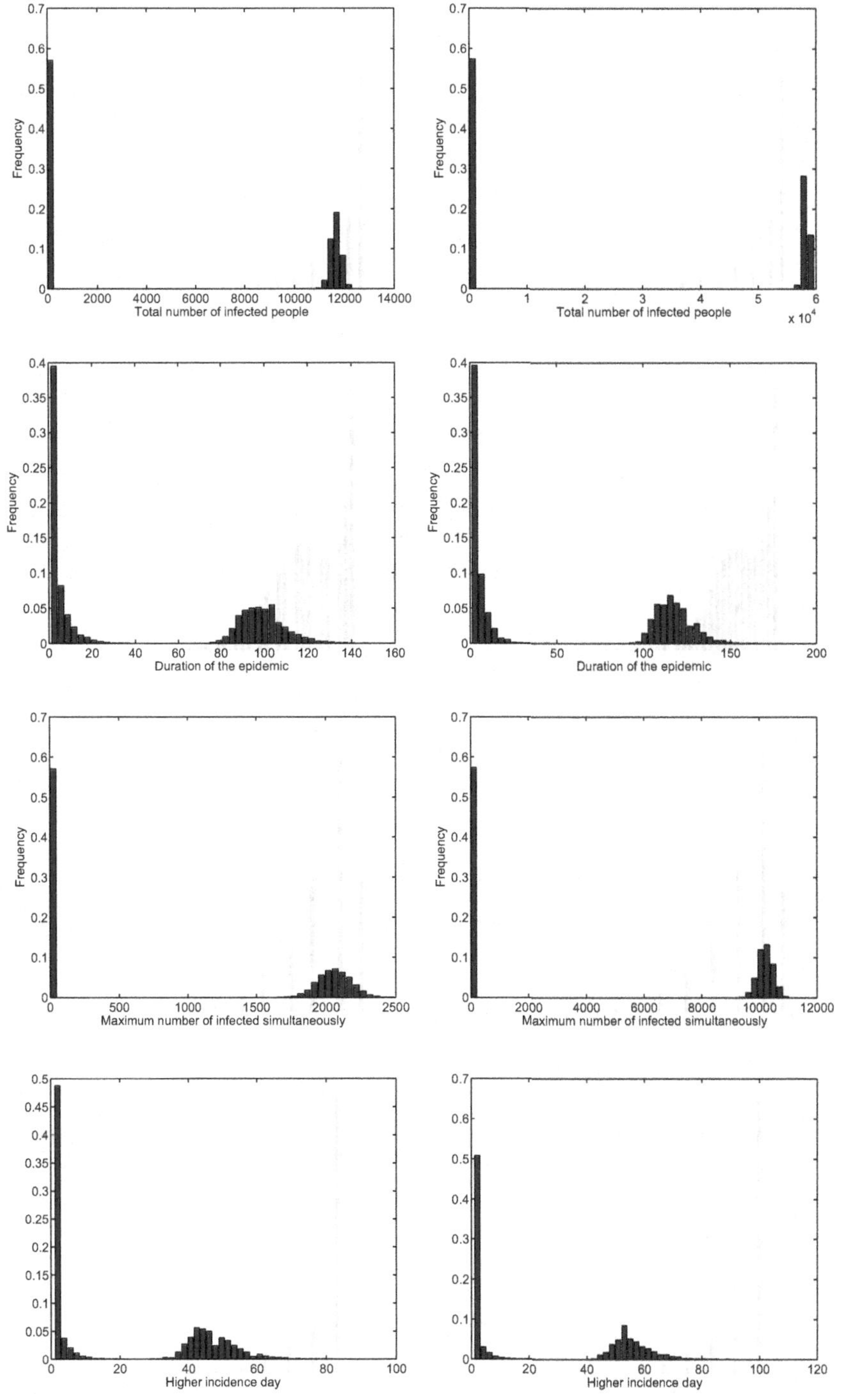

Fig. 1. Empirical distribution of the main epidemic parameters from 10000 simulation of the SIR stochastic epidemic model with $R_0 = 1.5$ and $\upsilon = 0$ in a community with 20000 (left) and 100000 (right) individuals, and one initial infectious ($m = 1$).

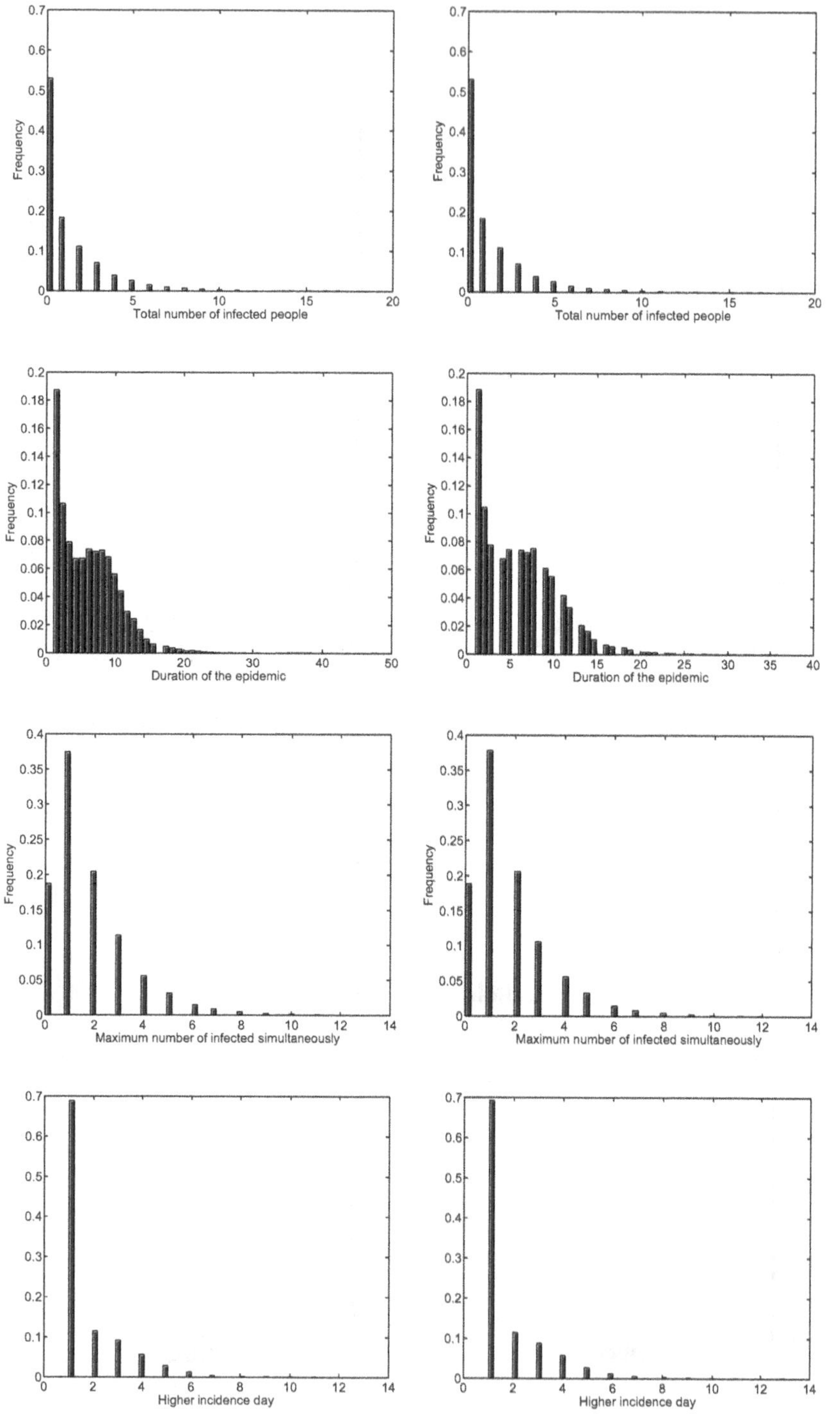

Fig. 2. Empirical distribution of the main epidemic parameters from 10000 simulation of the SIR stochastic epidemic model with $R_0 = 1.5$ and $v = 0.33$ in a community with 20000 (left) and 100000 (right) individuals, and one initial infectious ($m = 1$).

For the simulations ran with vaccination rate $v = 0.33$, the results presented in Fig. 2 show that the possibility of a major outbreak disappears. The proportion of simulations that results in a major outbreak is 0. The results obtained in the two communities are very similar. The mean of the total number of infected individuals is near $Z = 1$, the infectious disease period has a duration close to 4 days, the maximum number of infected simultaneously is 1 individual and the higher incidence day is 1 for both communities ($n = 20000$ and $n = 100000$).

All simulations were run using the full set of 32 CPU-cores of the test-bed cluster, working in parallel. With $v = 0.00$ (Fig. 1), the computations took ≈ 3 h 37 min for $n = 20000$, plus 17 h 05 min for $n = 100000$; with $v = 0.33$ (Fig. 2) the cluster worked ≈ 17 min for $n = 20000$, plus 1 h 13 min for $n = 100000$. As expected, the simulation times with $v = 0.33$ are lower than with $v = 0.00$ (though much lower than initially expected). Moreover, for both values of v, the simulations with $n = 100000$ are ≈5 times longer than with $n = 20000$, hinting at a linear increase of the computation time, in direct proportion to the size of the population. Thus, as the population grows, it is necessary to have more processors working on the problem in order to keep overall simulation times under reasonable bounds. Conversely, with more processors, the simulation times should decrease for a certain specific size of the population. These predictions are to be confirmed in future work, using an HPC cluster with much more processors.

4 Conclusions

Stochastic methods make possible to realistically simulate the spread of contagious diseases in a given community. Through the usage of these methods it is possible to conclude that a single infected person can cause the infection of a large number of people. In both populations tested (small and medium size), a 33% probability of an epidemic outbreak was observed. This risk is reversed if 33% of the population is vaccinated against the infectious disease.

The application of stochastic methods requires a large number of simulations which, in the case of large populations, will demand considerable computational resources. However, since the simulations are independent of each other, these methods are "embarrassingly parallel", easily suitable for computational parallelization. This option was already exploited in this work, where only small to medium size populations were considered, and will be of further use in future work, with larger populations and a larger number of simulations.

The results achieved in this work thus show that the proposed SIR stochastic algorithm can be used in small to medium scale problems, and pave the way for its application to large-scale problems. However, besides considering larger populations, future work will also tackle using the same algorithm in more realistic stochastic models, like the endemic and the household epidemic models.

The endemic model enables to simulate the entrances and exits that occur in the community. The household models enable to simulate the non-uniform mixing of the population. These models may be used in order to take into account the spatial distribution or the social structure of the community. The simulation of the dynamics of propagation of an infectious disease in a community may also be useful in spatial transmission models [13]. In these kinds of models the community can be considered as a unit of the spatial model as, for instance, a patch, a group or a node of the network.

Acknowledgement. UNIAG, R&D unit funded by the FCT – Portuguese Foundation for the Development of Science and Technology, Ministry of Science, Technology and Higher Education. Project n.° UIDB/04752/2020.

References

1. Ball, F., Britton, T., House, T., Isham, V., Mollison, D., Pellis, L., Tomba, G.S.: Seven challenges for metapopulation models of epidemics, including households models. Epidemics **10**, 63–67 (2015)
2. Ball, F., Mollison, D., Scalia-Tomba, G.: Epidemics with two levels of mixing. Ann. Appl. Probab. 46–89 (1997)
3. Balsa, C., Citerici, M., Lopes, I.M., Rufino, J.: Numerical simulation of the biological control of the chestnut gall wasp with T. sinensis. In: ICAIW 2019, vol. 2486, pp. 230–243 (2019)
4. Britton, T.: Stochastic epidemic models: a survey. Math. Biosci. **225**(1), 24–35 (2010)
5. Dasaklis, T.K., Rachaniotis, N., Pappis, C.: Emergency supply chain management for controlling a smallpox outbreak: the case for regional mass vaccination. Int. J. Syst. Sci.: Oper. Logist. **4**, 27–40 (2017)
6. Brauer, F., van den Driessche, P., Wu, J. (eds.): Mathematical Epidemiology. Springer, Heidelberg (2008)
7. Hethcote, H.W.: The mathematics of infectious diseases. SIAM Rev. **42**(4), 599–653 (2000)
8. Toolbox MATLAB Inc.: MATLAB and Statistics Toolbox Release (2012b)
9. Meyers, L.A., Pourbohloul, B., Newman, M.E., Skowronski, D.M., Brunham, R.C.: Network theory and SARS: predicting outbreak diversity. J. Theor. Biol. **232**, 71–81 (2005)
10. Lee, J.S.: Stochastic simulation and spatial statistics of large datasets using parallel computing. Ph.D. thesis, Western University. Electronic Thesis and Dissertation Repository. 1652 (2013). https://ir.lib.uwo.ca/etd/1652
11. Mao, L., Bian, L.: Spatial-temporal transmission of influenza and its health risks in an urbanized area. Comput. Environ. Urban Syst. **34**(3), 204–215 (2010)
12. May, R.M., Anderson, R.M.: Transmission dynamics of HIV infection. Nature **326**, 137–142 (1987). https://doi.org/10.1038/326137a0
13. Riley, S.: Large-scale spatial-transmission models of infectious disease. Science **316**(5829), 1298–1301 (2007)

Ankle Modeling and Simulation in the Context of Sport Activities

Nicoleta Negru, Monica Leba(✉), and Laura Marica

University of Petrosani, str. Universitatii, Petrosani, Romania
monicaleba@yahoo.com

Abstract. This paper presents a general framework for the use of modeling and simulation software combined with the 3D printer industry to be used in the sports world in the event of accidents being able to quickly create orthoses. It is shown that with the help of modeling and simulation software one can model and simulate the ankle movement so that later an orthosis can be made as faithful as it can be used. It is presented the fact that practicing in inappropriate conditions of sport presents many health risks. Among the most common mistakes were: incorrect heating of the muscles, improper hydration, intense workouts and execution in exercise rounds without the guidance of a trainer. Starting from these aspects, the research presented in this paper began. The evolution of the prostheses is presented historically and then a classification is made according to the fundamental characteristics from a functional point of view. It also presents the medical conditions in which it is necessary to use a prosthesis. There is presented a modeling and simulation performed in Matlab of the ankle movement so that later it will serve as a starting point to be able to perform a more accurate modeling of a prosthesis with the help of specialized software. To achieve the Matlab model it is considered that the ankle is similar to a 5-foot Stewart platform. To develop the mathematical driving model for this, the Denavit-Hartenberg model was developed for one of the legs. Taking into account in the first instance the medical conditions that must be fulfilled for the realization of a prosthesis, then the mathematical models and the simulations that can be realized with the help of specialized software can be realized prostheses as accurately as possible with the help of the new technologies of 3D printing.

Keywords: 3D modeling · 3D printing · Ankle · Orthoses

1 Introduction

The multitude and variety of sports branches, the increased number of performance athletes, the extremely high demands imposed on them determine a wide and varied range of conditions whose treatment must be so oriented, in order to obtain the complete and rapid recovery of the functional capacity of the injured segment.

The main objective of sports traumatology is not limited to the treatment of traumatic conditions in a reversible stage, but especially to the prevention of accidents by diagnosing and applying treatment in the preclinical stage.

Á. Rocha et al. (Eds.): WorldCIST 2020, AISC 1159, pp. 737–745, 2020.
https://doi.org/10.1007/978-3-030-45688-7_72

Knowing the causes that can lead to the appearance of traumas in the practice of athletes who have a great impotence not only for the correct diagnosis, but especially for finding measures to prevent their occurrence [1]. Medical physics is a branch that aims to maintain and rehabilitate human health through the use of modern technology. Medical physics is a modern and complex interdisciplinary field, which uses specific knowledge of physics, medicine, mathematics and computer science. Practically medical physics means the physics applied in medicine. The future of rehabilitation robotics is promising. We have developed a seed of technologies, from which we can see amazing, much more efficient results. Orthotics is a medical field that deals with the manufacture and application of orthoses. An orthosis is a device applied to a part of the human body to correct a deformity and improve a function or to lessen the symptoms of a disease.

2 History of Classical Prosthetics

The first hip arthroplasty was performed in 1947 by Robert Judet at Garches Hospital in Paris. In this case an acrylic was used to make the prosthesis. However, the history of knee arthroplasty began in 1860, when German surgeon Themistocles Gluck implanted the first hinge prosthesis, made of ivory. The field of prosthesis of the knee joint really began to develop when the Walldius hinge prosthesis was introduced in 1951. It was initially made of acrylic and later, in 1958, cobalt-chrome. Unfortunately, this type of prosthesis has not been successful for a long time. In 1968, he spoke of total knee arthroplasty, performed by Mr. R. Merryweather, using a total prosthesis introduced by Waldins, of Stockholm [2]. This prosthesis allowed a flexion-extension movement of −5 to 90°, being recommended for patients of advanced age, who did not lead an active life and suffered from osteoarthritis or rheumatoid arthritis. Mr. Merryweather implanted this type of prosthesis in 21 cases. In only two of them, complications appeared. Since then, joint replacement has become successful orthopedic treatments. In the world, between 500,000 and 1 million hip prostheses are implanted annually, and between 250,000 and 500,000 knee prostheses [3].

3 Overview of Orthoses and Medical Conditions

Orthosis is an externally applied system that supports or assists the skeletal neuromuscular system.

Orthoses have different shapes, are made of different materials and have the purpose of improving the locomotor function of the foot [4].

There are three categories of braces:

1) Rigid - designed to control the locomotive function of the leg, made of plastics or carbon fiber, used to eliminate foot pain and to control leg joints.
2) Soft - helps to maintain balance, eliminates pressure from painful points, made of soft and elastic materials.
3) Semi-rigid - used by athletes, allow dynamic improvement of balance during running are constructed of layers of soft materials reinforced with rigid materials [4].

Ankle injuries are the most common of all locomotor trauma locations, with sprain taking first place, and fractures second in that category.

Their high frequency is explained by the fact that at this level there are multiple movements controlled by the mobility needs of the foot exposed to different traumatic factors that act under different conditions [5].

Ankle injuries are the most common injuries suffered during physical activity, representing 15% to 30% of the total injuries, 85% of which are ankle sprains. It is estimated that one day there are 23,000 ankle sprains, equivalent to a sprain of 10,000 people per day. A person who suffers a sprain or "cracks" his ankle, the pain remains the same, the recovery time is different, depending on the severity of the accident. Normally, he will be able to walk after one or two weeks, if he is experiencing a low degree of pain, and in six or eight weeks he will be able to walk without crashing [6].

More difficult physical activities and sports will most certainly resume in about two or three months. When facing an ankle sprain, it should consider reducing inflammation and limiting the internal bleeding that can occur. For the purpose of faster recovery, we advise you to use creams and ointments for this condition, to be applied locally, two to three times a day and to use anti-inflammatory drugs.

Also, cold compresses should be applied, and when it is in a horizontal position, place a pillow under the injured foot, in order not to force the ankle and to improve circulation. A useful exercise would be to sit in a chair, foot to foot, and with the thumb of the affected foot, to draw letters from A to Z in the air. In order to perform this exercise correctly, it is important to maintain the more rigid the finger, so that the movement is achieved, for the most part, with the help of the ankle.

Practicing in inappropriate conditions of the sport poses a lot of health risks. Incorrect heating of the muscles, improper hydration, too demanding workouts and random execution of exercises without the guidance of a trainer are just a few aspects that you can ignore when entering the room and that can compromise not only the results, but also the immediate health, as follows:

- Muscle pain, stretching or contracting

Most injuries that occur as a result of improper sports are located in the muscles. These are of several types: stretches, contracts, and in very serious cases, muscle breaks:

- the contracts are characterized by muscular pain generally caused by the omission of the execution of the specific stretches from the end of the exercises - they are prolonged involuntary contractions of one or more groups of muscles;
- the stretches are different from the contracts and are caused by the overloading of the muscles in the absence of a pre-heating exercise;
- ruptures are the most serious and painful muscular disorders that can occur as a result of the wrong sport; as the name implies, it is characterized by the actual rupture of the overloaded muscle in the absence of a proper preheating [6].

- Dislocations

Dislocations are disorders of the locomotor system that result from trauma caused by overloading the joints, especially at the knees, ankles or wrists. The dislocations also occur due to insufficient heating of the body, but also to wearing improper shoes or incorrect use of the room equipment. It manifests through severe pain, local swelling and the inability to temporarily use the affected limbs [6].

- Sprains

Sprains are other frequent accidents that you may suffer from inadequate practice of gym workouts. They are localized, as are the dislocations, especially in the area of the ankles, wrists, knees or feet, and are characterized by stretching, bending or forced twisting of the articular ligaments. They are caused by sudden movements and are manifested by local pain, edema and temporary inability to move the affected areas. Mild sprains, where no ligament tearing or tearing has taken place, can be easily treated at home [6].

- Bruising

Regardless of whether you do high-performance or daily-maintenance sports, the bruises (bruises) are almost endless on the body. There are visible traces of subcutaneous haemorrhages, which are formed after applying pressure or a stroke. In general, improper use of the room equipment can lead to the formation of subcutaneous lesions in the form of bruises, which are harmless to health but extremely unsightly [4].

Whether you suffer from sprains, dislocations, stretches or muscle contractions or have chosen an unsightly bruise on the skin, diclofenac gel is the first help you need to quickly alleviate the unpleasant pains and symptoms of these accidents. Being applied locally, the anti-inflammatory with a high concentration of the active substance is absorbed more easily into the skin and relieves pain faster than treatments with oral administration [7]. Ankle orthoses develop in close collaboration with doctors and physiotherapists. As a result, they are extremely functional. All products are remarkable for their increased comfort and design and meet the requirements of each individual for orthotics and physiotherapy [8].

The posterior nocturnal orthosis is used when the patient is identified with neuromuscular problems, with disorders of the sciatica nerve, following injuries of the ligaments, soft tissues and tendons, following the conservative treatment of the injuries of the forearm and metatarsus, as well as of the tarsus. postoperative and conservative treatment of distal fibula (peroneal) fractures and stable forearm, metatarsal and ankle fractures. The orthosis contains velcro straps that ensure superior ankle stability and is perfect to ensure the transition after using a walker, the rear atelia (orthosis) for the night was designed with comfort in mind. Application and use is simple, orthosis reduces plantar flexion, making it ideal for patients with reduced ankle mobility, the essential elements are the three-dimensional adapted cushions on the support plate and the bands for flexion adjustment that control the level of tension, favoring healing.

Adaptable to the edema variable during the recovery process, the sole profile is specially designed to promote stability and a natural ride, the lining (foam) and the locking straps allow the quick and comfortable adjustment [9].

4 3D Printers in the Prosthetic Industry

3D printing is a process and, at the same time, a technology that allows the printing of an object in the three-dimensional version. Unlike the traditional processing techniques that were based on the elimination of materials by various methods (cutting, drilling), 3D printing is realized through an additive process. 3D printing has many advantages such as: the ability to produce a prototype that can be tested and quickly remodeled, to print very small or very large objects and to use sensitive materials (objects that by the traditional method of production would have taken a longer time). Essentially, a 3D printer is a type of industrial robot capable of printing under computer control. There are several 3D printing technologies: FDM - Thermoplastic Extrusion Modeling, SLS - Selective Laser Sintering, SLA - Stereolithography. The preference for one technology or another is dictated by the specificity of each field of use (medical field, jewelry industry, architecture, educational field, industrial design, etc.) [10].

In areas where high precision is needed, such as the medical field or the jewelry industry, very high resolution (SLA) printers are used. For these types of applications, printers generally use different types of resins as a raw material. Domain size (locally) in the North-West region of Romania 3D printing has been growing in recent years, the number of companies printing using this technique increasing from 2 in 2015 to 10 in 2018 [11]. Also, the large number of software specialists in the region, the large number of multinational companies and Start-ups in the field, good collaboration between research centers and companies, as well as the presence in the region of companies that represent final clients for 3D printers make this sub-domain one with potential for development in the region. A 3D printer is a device, first invented in the 1980s, that allows the creation of physical objects composed of either a single material or a variety of materials such as plastic, metal, glass, ceramic, resin, - geometry dimensional - after a 3D virtual modeling sketch. In other words, a 3D printer is an industrial robot capable of creating physical objects under computer control. 3D printing of an object is done through the process of superposition with added layer, until the object has been completely created as defined digitally. Each such layer can be viewed as a horizontal section of the object, more precisely a 2D slice, all layers being gradually joined together to form the final shape of the object. All current 3D printers use this layer overlap process, as well as several types of available technologies, the difference between them being the way the layers are created and merged. Some of them are based on melting or increasing the malleability of the material they work on, others on different processes, including the use of laser beams or ultraviolet radiation on materials receptive to them [12].

5 OpenSim Modeling and Simulation of Ankle Movement

Different simulations of human body movements can be made using OpenSim. Figure 1 shows such a simulation of a part of the human body, the part with the human ankle [13].

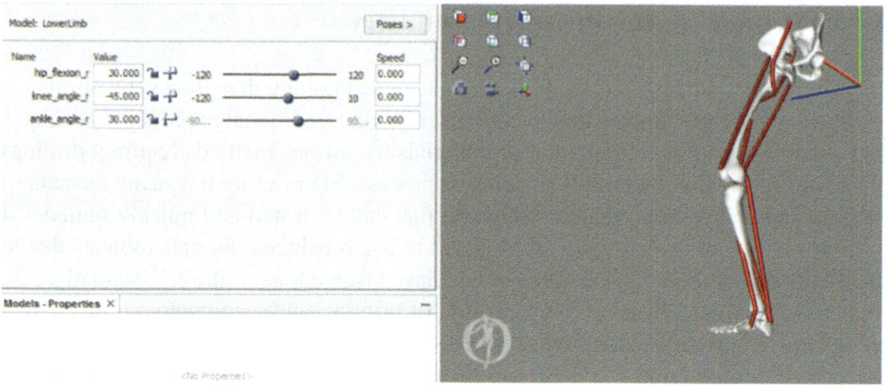

Fig. 1. OpenSim modeling and simulation of ankle movement

6 MatLab Modeling and Simulation of Ankle Movement

As has been considered in other specialized works, we consider in this work also the ankle as a 5-foot Stewart platform of the type shown in Fig. 2.

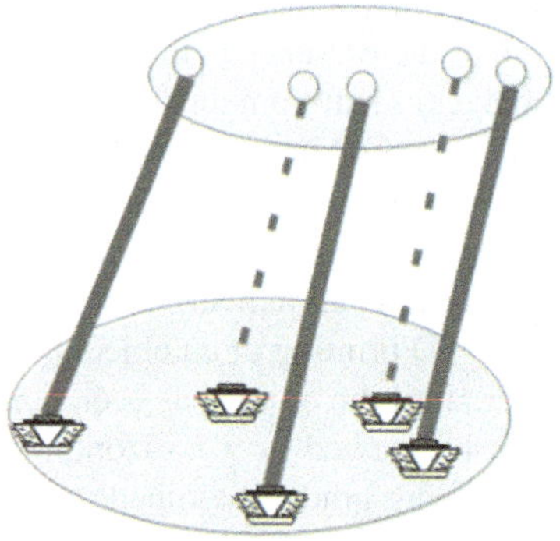

Fig. 2. Stewart platform

To achieve the mathematical driving model for this we have developed the Denavit-Hartenberg (DH) model for one of the legs [14]. We represented it simplified as in Fig. 2, consisting of 4 rotating couplings. In this figure you can also see the coordinate axes in each couple. I considered the first two couplings to rotate after the axes Z_0 and X_0, having the distance between them zero.

Then at a distance D_2, equal to the foot length within the Stewart platform, we considered the following 2 rotating couplings around the X_0 and Z_0 axes as well as the zero length between them.

Table 1 shows the DH coordinates for the foot in Fig. 2, and Eq. (1) represents the general DH matrix [14].

Table 1. Denavit-Hartenberg coordinates

Element	Coordinates			
	a_i	α_i	d_i	θi
1	0	$\sigma/2$	0	$\theta_1 + \sigma/2$
2	d_2	0	0	θ_2
3	0	$-\sigma/2$	0	$\theta_3 + \sigma$
4	0	0	0	θ_4

$$T_{i-1,i} = \begin{bmatrix} \cos\theta_i & -sin\theta_i * \cos\theta_i & sin\theta_i * \sin\alpha_i & \alpha_i * \cos\alpha_i \\ \sin\theta_i & \cos\theta_i * cos\alpha_i & -\cos\theta_i * sin\alpha_i & \alpha_i * \sin\alpha_i \\ 0 & sin\alpha\alpha_i & \cos\alpha_i & d_i \\ 0 & 0 & 0 & 1 \end{bmatrix} \quad (1)$$

From this we write T_{01}, T_{12}, T_{23}, T_{34} and multiply them successively by obtaining the matrix T_{04} from Eq. 2 [15].

$$T_{04} = \begin{bmatrix} s_1 * c_{23} * c_4 - c_1 * s_4 & -s_1 * c_{23} * s_4 - c_1 * c_4 & -s_1 * s_{23} & -d_1 * s_1 * c_2 \\ -c_1 * c_{23} * c_4 - s_1 * s_4 & c_1 * c_{23} * s_4 - s_1 * s_4 & c_1 * s_{23} & d_2 * c_1 * c_2 \\ -s_{23} * c_4 & s_{23} * s_4 & -c_{23} & d_2 * s_2 \\ 0 & 0 & 0 & 1 \end{bmatrix} \quad (2)$$

The last column in Eq. (2) represents the position of the end of the foot, and the left matrix up to the 3 × 3 dimension gives us its orientation. In Fig. 3 we have the MatLab-Simulink model for the leg analyzed above, together with the result of the graphical simulation represented in Fig. 4 [14].

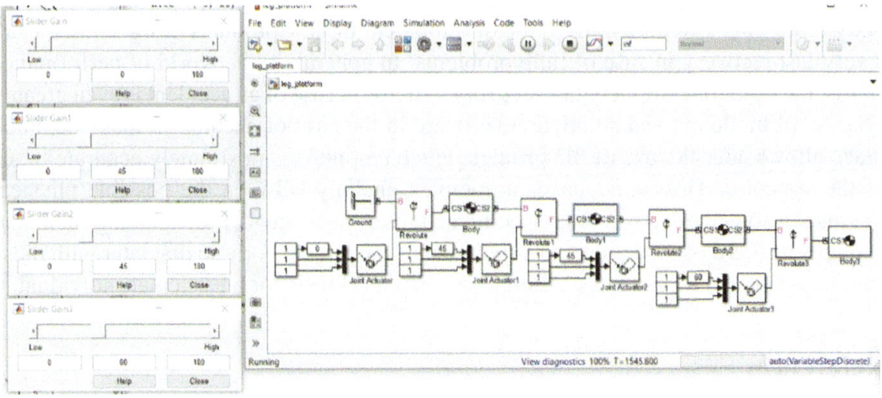

Fig. 3. MatlabSimulink model

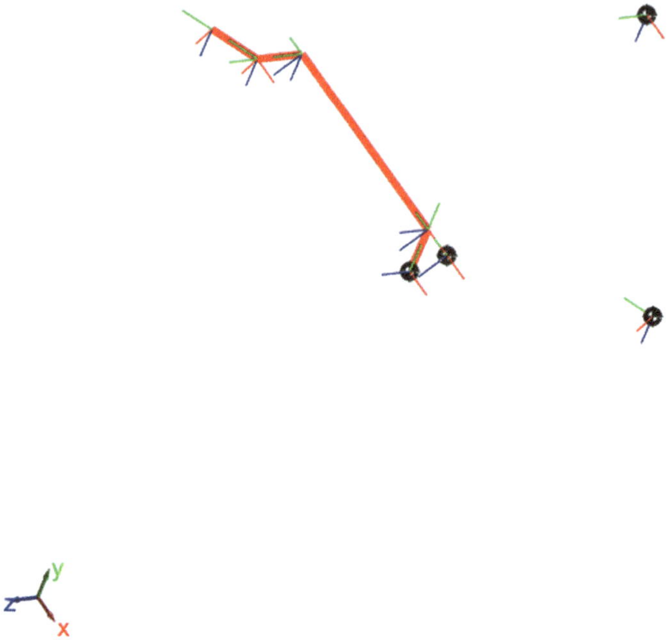

Fig. 4. Grafical simulation of MatlabSimulink model

So you can see how with the help of new technologies in 3D printing and with the help of modeling and simulation software, incredible results can be obtained that can be materialized through orthoses specially designed for each individual [16].

7 Conclusion

The use of new technologies for modeling and simulating the movements of the human body represents a step forward in the development of new products designed to ensure a very fast recovery in case of limb problems. In general, in the world of performance sports, the need for extremely fast recovery systems in case of an accident is even greater. The use of modeling and simulation software in the case of needing to make an ankle brace allows later the use of 3D printers which can make an extremely accurate brace as the structure. This work makes a complex analogy between the possible physical traumas of some athletes in their specific activities, new ways of recognizing them and explores the field of modeling and simulation of the human body so that later with these simulations it can be done, for example, a custom orthosis for a particular individual.

References

1. Wu, K.W.: Foot Orthoses: Principle and Clinical Applications, p. 97. Baltimore, Wiliams and Wikins (1990)

2. Schuster, R.: A history of orthopedics in podiatry. J. Am. Podiatr. Med. Assoc. **64**(5), 332–345 (1974)
3. Dagnall, J.C.: History of foot supports. Br. J. Chiropody **32**, 5–7 (1967)
4. D'Ambrosia, R.D.: Orthotic devices in running injuries. Clin. Sports Med. **4**, 611–618 (1985)
5. Root, M.L.: Development of the functional orthoses. Clin. Podiatr. Med. Surg. **11**(2), 183–210 (1994)
6. Hubbard, T.J., Wikstrom, E.A.: Ankle sprain: pathophysiology, predisposing factors, and management strategies. Open Access J. Sports Med. **1**, 115–122 (2010)
7. Van Gheluwe, B., Kirby, K.A.: Foot biomechanics and podiatry: research meets the clinical world. Footwear Sci. **1**, 79–80 (2009)
8. Bennett, Wilson A.: Limb prosthetics –1970. Artif. Limbs **14**(1), 1–52 (1970)
9. Page, P.: Current concepts in muscle stretching for exercise and rehabilitation. Int. J. Sports Phys. Ther. **7**(1), 109–119 (2012)
10. Naàji, A.: Application of Computer Graphics in Biomechanics. Vasile Goldis Universitatea de Vest, Arad (2003)
11. Radu, C.: Contribuţii la structurarea optimă a tehnologiilor de prototipare rapidă în vederea realizării elementelor specifice de protezare. Universitatea Transilvania din Brasov, Romania (2005)
12. Watt, A.: 3D Computer Graphics. Addison Wesley, Anglia (1995)
13. Risteiu, M., Leba, M., Arad, A.: Exoskeleton for improving quality of life for low mobility persons. Qual. Access Success **20**, 341–346 (2019). Supplament1
14. Risteiu, M.N., Rosca, S.D., Leba, M.: 3D modelling and simulation of human upper limb. In: IOP Conference Series: Materials Science and Engineering, vol. 572 (2019)
15. Rosca, S.D., Leba, M.: Using brain-computer-interface for robot arm control. In: MATEC Web of Conferences, vol. 121, p. 08006 (2017)
16. Negru, N., Leba, M., Rosca, S., Marica, L., Ionica, A.: A new approach on 3D scanning-printing technologies with medical applications. In: IOP Conference Series: Materials Science and Engineering, vol. 572, p. 012049. IOP Publishing (2019). https://doi.org/10.1088/1757-899x/572/1/012049

Autonomous Electric ATV Using IPM Based Inverter Control and Deep Learning

Cosmin Rus, Nelu Mija, and Monica Leba(✉)

University of Petrosani, Petrosani street Universitatii, Petrosani, Romania
monicaleba@yahoo.com

Abstract. This work focuses on the design, development and use of an electric ATV (all terrain vehicle) starting from the conversion of such a thermal vehicle. It was proposed and realized a revolutionary concept of driving this vehicle in the first phase in order to reduce the consumption of electricity then and in order to add the autonomous driving component. Modeling and simulation of the control part with inverter were performed. A deep learning algorithm has been developed that allows this autonomous vehicle to be driven both indoors and out by using images taken and then processed in real time.

Keywords: Self-driving-car · Electric ATV · Smart autonomous driving · Intelligent power inverter

1 Introduction

Historically, we can say that the development of the principles of self-guided vehicles began somewhere around 1950, as shown by research from the Frauenhofer Institute in Stuttgart/Germany. Notable are the three major categories relevant to development over the years, the automotive industry, application areas and control technology. Thus, in 1950 the automation of manual tractors in America was realized. In 1975, the first automated guided vehicles could be moved vertically and the use of these types of vehicles was fully integrated into production. In 1989 vehicles with sensor systems were built to be able to run at high speeds and in 1995 the first low-cost automatic guided vehicle was built [1, 2]. Regarding the control technology, we observe a continuous development such that in 1954 the optical guidance was used, in 1966 the inductive guidance, in 1968 the semiconductor control (TTL technology). Starting with 1976, the automatically guided vehicles have been controlled with the help of microprocessors and since 1982 the control is used with the help of a computer or even more recently with the help of embedded devices [3].

Autonomous vehicles with industrial uses represent an advantage in the flexibility of handling raw materials in production, as well as in the logistics field. In this sense, the development of these systems can continuously reduce the costs of production according to studies, based on manufacturing concepts such as "just-in-time" or "flexible manufacturing" and the current global context of production development based on continuous

Á. Rocha et al. (Eds.): WorldCIST 2020, AISC 1159, pp. 746–755, 2020.
https://doi.org/10.1007/978-3-030-45688-7_73

adaptation. The costs of handling the material represent a significant proportion in the production costs [4]. In this sense, the thematic approach becomes important due to the flexibility, characteristic of the production system. It is important to approach the interaction between machines, material handling and the computer system. The level of automation allows communication between the central control system and autonomous vehicles or between autonomous units. The criteria for evaluating the current transport architectures and the new technical realization possibilities provide an overview of the future research directions. Regardless of the scope, the tendency is to automate processes by introducing specific vehicles. Based on the evolution of the last fifty years of the vehicles used in the automotive industry, it can be stated that the technical and informational news of the respective times have been successfully accumulated and developed in the area of industrial space transfer. With the emergence of high-performance sensors, vehicles have also been developed with other functions, increasing in this way flexibility first and foremost. The products receive through integration artificial intelligence an increasing degree of adaptability, a fact confirmed also by the top on the direction of research and implementation of autonomous vehicles worldwide [4].

LoRa (Long Range) is a protocol of radio transmissions through which networks of intelligent objects are formed. The network consists of a star-of-stars topology, with gateways serving as transparent bridges, which transmit messages between sensors and the central server. Gateways connect to the network through traditional IP connections, and sensor devices use single-hop wireless communication to one or more gateways. The structure is similar to a mobile network, but instead of having a single interconnected network, LoRa allows the implementation of several independent networks over the same infrastructure. Thus, the LoRa infrastructure allows 4 gateways to cover a very large area. The entire LoRa communication protocol has low power consumption [5].

The Internet of Vehicles (IoV) 20 concept makes it possible to communicate dynamically between the vehicle and all other components of an intelligent traffic system, with several types of communication specific to the system: V2V communication (vehicle to vehicle), V2D communication (road vehicle), communication V2O (human vehicle), V2S communication (sensor vehicle). This system integrates a common database that allows the collection and access of more types of information related to vehicles, roads and complementary elements (e.g. smart phones). In addition, it allows the processing and dissemination of information securely to other similar information platforms [6].

2 The Autonomous Driving Technology

Driverless cars are built with a series of sensors, computer chips and artificial intelligence software programs that allow them to run on the road, detect other obstacles around them and drive safely on the road. Man does not have to do anything but sit in his chair and relax. This vehicle is capable of performing partially (semi-autonomous car) or fully (autonomous car) running functions, adapting to traffic, infrastructure and weather conditions. These machines use a wide range of sensors, cameras, radars and laser radars (Lidar) to build a true representation of the environment. In their advanced versions, autonomous cars can communicate with each other and receive information from the infrastructure elements (traffic lights, indicators, etc.). The major advantages of autonomous cars are the major potential for limiting accidents, improving traffic flow, improving quality of life, limiting pollution and helping people with reduced mobility [6].

Google is the world leader in driverless vehicle technology. Google was the first company to drive driverless cars on public streets. Their original prototype was a Toyota Prius and proved to be successful. The center of this technology revolves around a device called the Velodyne 64 Laser (Velodyne 64-beam Laser). This device is, in principle, a laser beam search device that is mounted on the roof of vehicles. It scans the environment and automatically produces detailed maps of the environment in 3D. There are 64 invisible laser beams that are distributed around the environment. The purpose of these laser beams is to detect any person, objects or other obstacles that may be around. After generating the 3D maps, the computer will measure the laser beams together with the map details. Based on this combination, different data models are produced that allow the vehicle to drive safely on the streets. Autonomous cars will be able to detect other vehicles or people who may be in front of them. Once detected, the car will brake automatically respecting the braking distance. You do not have to worry, these objects are detected for a long time, so braking will be calm and gradual. If someone suddenly throws himself in front of the car, then braking will be more aggressive [7].

3 Implementation

The research regarding the modeling, design and construction of an electric vehicle had as a starting point the purchase of an ATV type vehicle (All-Terrain Vehicle) having a classic thermal type engine (250 cc cubic petrol engine). The thermal engine part and the control and control part of this engine were removed and the functional platform of this vehicle was prepared in order to be able to mount an electric motor. The motor chosen is a three-phase asynchronous motor of 2,2 kW, 1420 rotations/ minute and cosφ 0.82, protection class IP 44.

A smart system was created from an electric ATV with a radio communication system with low power consumption (LoRa) that will allow the retrieval of data from different sensors placed on the vehicle in order to monitor the environmental quality of a city (air, noise, dust, humidity, temperature differences) and with the functionality of recovering a part of the consumed electricity during acceleration and movement in order to increase the autonomy. A patent application has also been filed for this development proposal having the number A/00201/28-03-2019. The invention described in the patent has as objective the creation of a smart electric vehicle with a LoRa communication system that allows the retrieval of data from different sensors located on the vehicle in order to monitor the environmental quality of a city (air, noise, dust, humidity), temperature differences) and with the functionality of recovering a part of the consumed electricity and during the acceleration and the movement in order to increase the autonomy. The current state of research in the field of electric vehicles presents the use for the propulsion of DC motors without brushes. Electric vehicles manufactured at the present time have a major disadvantage regarding the low autonomy, not having a function to recover part of the electricity consumed during acceleration and displacement but only during the braking period. The term LoRa (or LoRa technology) refers to a category of Long-Range radio communications with low power consumption. Unlike traditional digital radio technologies, LoRa technologies are capable of communicating data at distances of several kilometers or even tens of kilometers with extraordinary applicability in wireless (wireless) sensor networks, Internet of Things (IoT) and network creation of smart devices.

Functional prototype of the electric ATV is presented in Fig. 1.

Fig. 1. Functional prototype of the electric ATV

The main disadvantages of today's smart data transmission solutions other than LoRa are related to high energy consumption and the fact that data transmission protocols are complicated and difficult to troubleshoot. From the documentation, it can be seen that in Romania, LoRa technology is still in the pioneering stage and is not widespread in practical use. The invention presented in the patent aims to achieve a smart vehicle powered by one or more DC motors without brushes with permanent magnets which also benefit from an innovative recovery function. energy consumed and as a communication system uses a technical solution with low energy consumption and large area to cover.

In addition to all the above mentioned as an achievement regarding the development of a vehicle's automatic steering system, it is worth noting the development of a steering system that uses as a central element a control element like a Raspberry PI4 that takes over a camera. video data that are then used to create a route in order to avoid obstacles. It is used as an auxiliary control element a speed sensor which is mounted on the closed loop principle so that if the camera notices an obstacle the speed is reduced and if it is desired to maintain the same constant speed the same principle is used.

4 Modeling and Simulation

4.1 Modeling and Simulation of Motor Drive Control

Software-oriented intelligent power inverter is developed according to the block diagram of Fig. 2. It consists of hardware and software. The hardware consists of power electronics that is an intelligent power module (IPM). This is the core of the inverter in order to supply the induction motor with controlled voltage (U) and frequency (f) AC power. A second part of the intelligent power inverter hardware is the real-time data acquisition card (DAQ) connected to a PC, which is the third part. This system has two feedback connections, first for logic control and the second for analogical control through the transducers block. The transducers block can be replaced sometimes by software. The software is a flexible set of programs embedded on the IPM, DAQ and CPU.

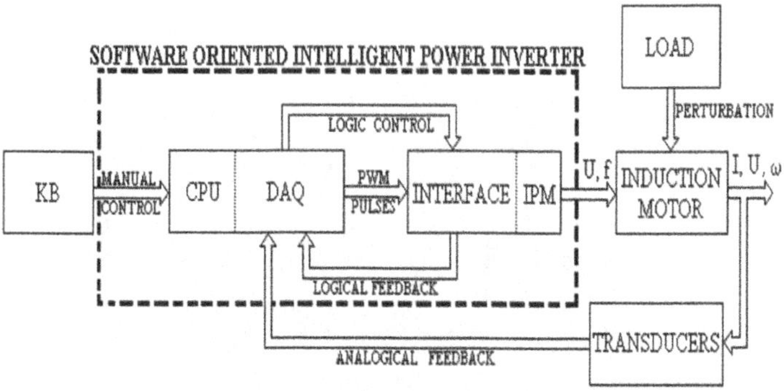

Fig. 2. Intelligent power inverter block diagram

This inverter can be used in many different applications, like: intelligent motion control of electrical drives, power conditioning in electrical networks and also for research. On the other hand, this allows large possibilities of advanced control algorithms implementation, hardware in-the-loop (HIL) simulation, a friendly graphic support and virtual measurement.

The intelligent inverter is based on maximum reusable resources (software) and minimum limited resources (hardware) and represents one low-cost future-solution for control of electrical drives. The concept of Software Oriented Intelligent Inverter consists of hierarchical control architecture, an embedded distributed intelligence and replacement of the hardware elements, where it is possible, by software [8].

The concept consists of three main principles: hierarchical control structure; embedded intelligence in each part and software-oriented design. Hierarchical control structure is based on three levels. First level is the execution level and consists of intelligent power module (IPM) and its programmable logic. The main task of this level is to generate controlled three-phased voltage and frequency. Another function of it is to supervise the power supply in order to detect over-current, under-voltage and heat stress malfunctions. This level informs DAQ regarding the normal or malfunction (ALM) conditions.

Second level is the intermediate (tactical) level and is represented by the data acquisition card (DAQ) and the embedded software. The main task of this level is to generate the control pulses for the IPM. Another function of it is to receive, convert and preprocess the analogical signals from the transducers. This level informs PC regarding the normal or malfunction conditions and sends the preprocessed digital signals. Third level is the superior (strategically) level and is represented by the PC and the real-time software. The main task of this level is to implement the mathematical model of either sine-delta pulse width modulation or space vector modulation, according to the appropriate control strategy and output to the DAQ the power parameters, as frequency (f), modulation index (R), modulation ratio (p). Another function of it is to receive and process the digital data from the DAQ level. This level informs the user about the system general state using the virtual measurement on the graphical user interface. Intelligent Power Module (IPM) has embedded intelligence in order to protect the power electronics devices (Tj) for over-current, under-voltage and over heat. All of these inform the DAQ through ALM signal. The embedded programmable logic is implemented in drivers and Over-heat protection blocks. In Fig. 3 is shown an equivalent diagram for an IPM of 1200 V and 100 [8].

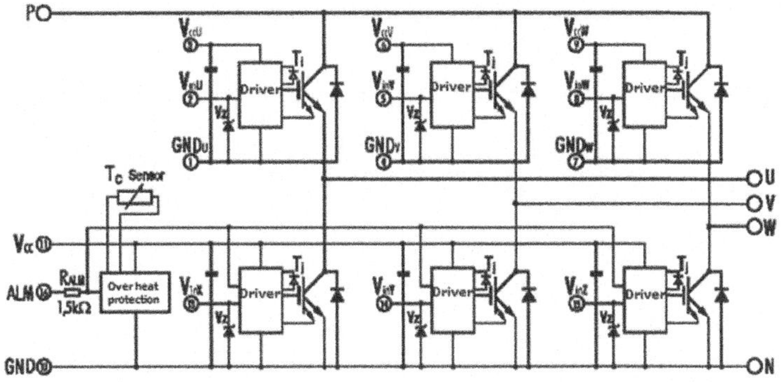

Fig. 3. Intelligent Power Module (IPM) with embedded programmable logic

Data Acquisition Card (DAQ) is programmable and has its appropriate own BIOS. This receives the control strategy and the necessary parameters (f, R, p for PWM or f, V, for SVM) from the PC, determines the six pulses and sends them to the IPM. PC has flexible software that solves the mathematical model of the control algorithm in order to determine the DAQ necessary parameters. Software-oriented design represents a system with strictly necessary hardware and embedded software. In these systems, the software replaces a great part of hardware components, like logical gates and analogical blocks, including some microcontrollers and DSP. By this reliable and real low-cost solution, the hardware becomes very simple, flexible and easy to use and debug [8].

In this case the hardware consists only of the power electronics (IPM), data acquisition card (DAQ), transducers, if any, and necessary interfaces.

To design the control software, it is crucial to develop a mathematical support of PWM pulses generator. Using the notations [9]: $R = \frac{A_{\sim}}{A_{\Delta}}$, the modulation index and $p = f_p/f_1$, modulation ratio, we have to solve a nonlinear equation, representing the sine-delta intersection angles α_{jk}^{i} (Fig. 4), as follows:

$$R \cdot \sin\left(\alpha + \frac{2 \cdot (i-1) \cdot \pi}{3}\right) = (-1)^{k-1} \cdot \frac{2 \cdot p}{\pi} \cdot \left(\alpha + \frac{2 \cdot (i-1) \cdot \pi}{3}\right) + \frac{(k-1) \cdot \pi}{p} \quad (1)$$

Parameters R and p are variable, i is the phase number and k are the pulse number. This nonlinear equation has not a solution represented by a mathematical function. Based on Fig. 5, representing the $k^{\underline{th}}$ PWM pulse [10], results:

$$\delta^{i}_{jk} = \delta_0 \cdot \left[1 + R \cdot \sin\left(\alpha^{i}_{k} + (-1)^{j} \cdot \delta_0\right)\right] \tag{2}$$

where:

$$\alpha^{i}_{k} = \frac{2 \cdot (i-1) \cdot \pi}{3} + (4 \cdot k - 1) \cdot \frac{\pi}{2 \cdot p}, \; \delta_0 = \frac{\pi}{2 \cdot p} \tag{3}$$

This allows a very good approximation of the solution, as follows:

$$\begin{aligned} \alpha^{i}_{jk} = {} & \frac{2 \cdot (i-1) \cdot \pi}{3} + \frac{(4 \cdot k - 1) \cdot \pi}{2 \cdot p} + (-1)^{j} \cdot \frac{\pi}{2 \cdot p} \\ & \cdot \left[1 + R \cdot \sin\left(\frac{2 \cdot (i-1) \cdot \pi}{3} + \frac{(4 \cdot k - 1) \cdot \pi}{2 \cdot p} + (-1)^{j} \cdot \frac{\pi}{2 \cdot p}\right)\right] \end{aligned} \tag{4}$$

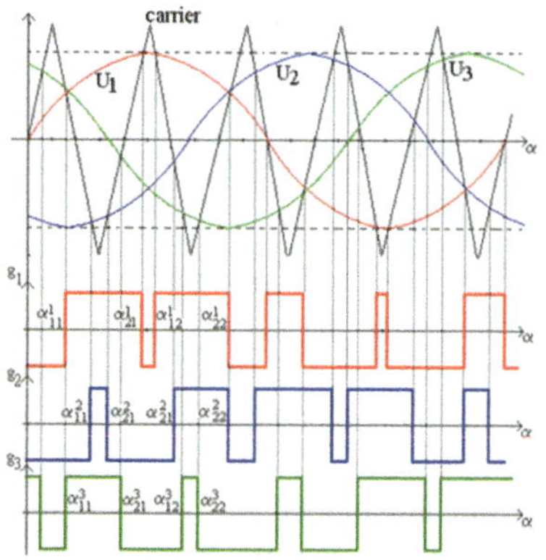

Fig. 4. Three-phased PWM pulses and α^{i}_{jk} angles

Having a good approximation of the solutions, it is possible to determine and generate exclusively by software all the control PWM pulses. Now there will be compared the exact solutions of Eq. (1) with the approximate ones obtained by the relations (4). All of the results are determined by simulation. The results show that the errors are less than 1.8%, which proves that the relations (4) can be used further.

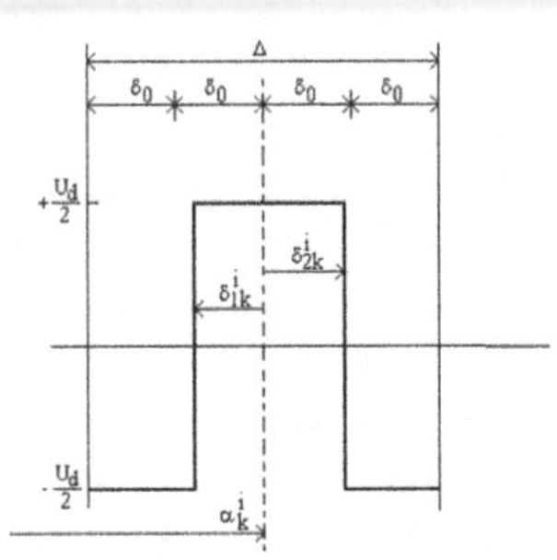

Fig. 5. The kth PWM pulse

A model of three phases IGBT inverter and its control driver was designed. The simulation model consists of two subsystems, called PWM pulses and Power inverter. The first one determines the angles by the formula (4) determined above and the second subsystem is the model of a real three phased IPM inverter. The simulation results are very close to the theoretical ones (Fig. 6).

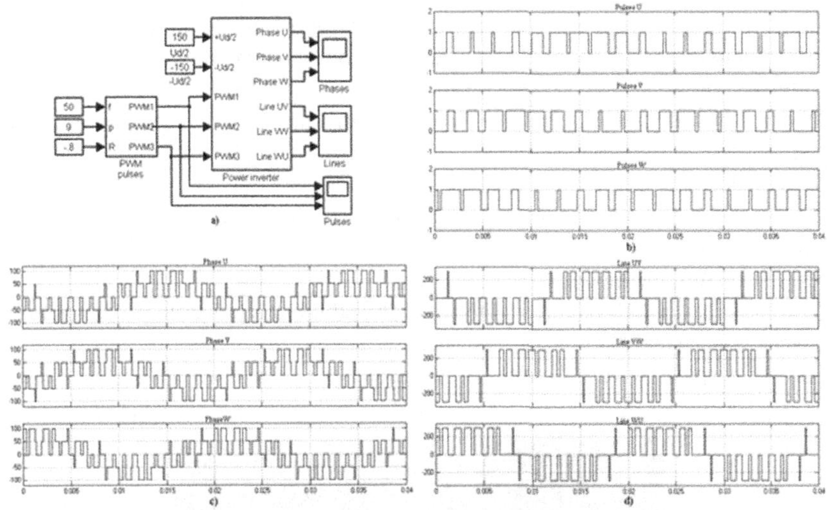

Fig. 6. Inverter simulation: a) Simulation model; b) Control pulses; c) Phase voltages; d) Line voltages

In order to implement a software-oriented intelligent power inverter it was necessary to design the hardware part of the inverter and make it as state-of-the-art in the System Control and Computer Engineering research laboratory of University of Petrosani. On this hardware there was implemented a software controller based on the theoretical support developed and presented in paragraph 4. The hardware of the intelligent power inverter has the internal structure and consists of the Intelligent Power Module (IPM) and its driver as main element of the inverter. This is supplied with DC voltage, which is connected to the AC power by means of KDCP power switch. The output of IPM, as a controlled three-phase AC voltage is routed to Out for induction machine, by means of inductances filter and KINV power switch and fusses. On the other hand, the main AC power comes from Input through other switches, by main switch MS via transducers block. All this hardware equipment is connected to an external PC by means of an interface. The interface has multi-I/O analog and digital features and is connected through 4 busses. First bus supplies the pulses to the IPM inverter's driver. Second bus is for logic control input/output of all the power switches of the inverter. Third is for analog inputs from the transducers block. Fourth is the input/output links to the DAQ card of the PC [8].

4.2 Environmental Identification for Autonomous Car - Deep Learning

Currently the most used deep learning methodologies for autonomous driving are Convolutional Neural Networks (CNN), Recurrent Neural Networks (RNN) and Deep Reinforcement Learning (DRL). Recent contributions in computer vision and deep learning technologies can improve vision detection and classification tasks easily [11].

Considering we have a number of possible scenarios that the autonomous system will encounter, the volume of data collected will be very large. The collected data are labeled according to various parameters such as: road type, car actions, weather conditions [12].

For training of autonomous car, we use a convolutional neural network (CNN) with the following steps:

Step 1: Load the training, validation and testing datasets using Python and NumPy library.

Step 2: Design and test the model architecture.

Design and implement a deep learning model that learns to recognize road objects, road types, weather conditions.

Step 3: Testing the model on new images.

Download at least 10 pictures from the web and use the model to predict action the of autonomous system [11].

For this purpose, we use samples of 20 images per second (FPS), Fig. 7.

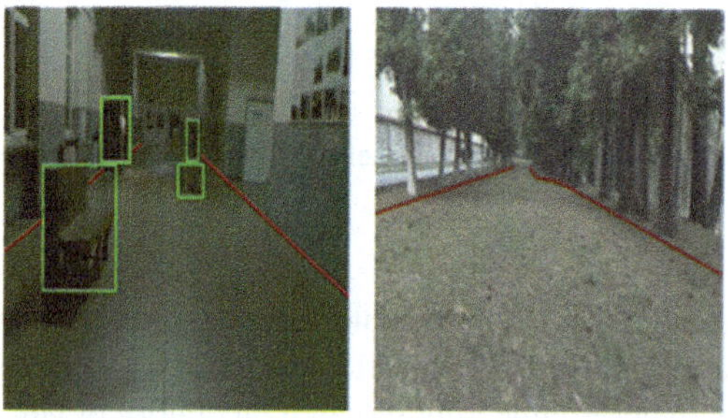

Fig. 7. Different road types and driving conditions for autonomous car

The entire control algorithm is based on a Raspberry Pi4 module that benefits from a camcorder. With the help of the camera, the images will be taken which will then be processed according to the deep learning algorithms. The result of the image processing is an input for the inverter that controls the engine. The engine drives the transmission part to which a control sequence based on a speed sensor is applied [11].

5 Conclusions

Making a smart electric ATV starting from a classic motorized functional platform allowed the use of new technologies and concepts meant to develop new areas of deployment of resources specific to the IoT in the context of its integration with Deep Learning branch. Research conducted on the concepts of autonomous control of some vehicles has revealed that a very cheap system can be made using only a Raspberry type control board and a proprietary video camera. With this system and using specific algorithms deep learning was able to mark a route that a vehicle can use for traveling and within which it can avoid potential obstacles. From a mechanical point of view, a very cheap vehicle prototype was obtained using only recycled materials. There have been modeling and simulation of party control with an inverter of an electric motor, the motor used in the ATV built.

References

1. Tokuyama, H.: Intelligent transportation systems in Japan Public Road, 60(2), Berghout (1996)
2. Bossom, L., Chevreuil, M., Burkert, A., Franco, G., Gaillet, J.F., Pencole, B., Schulz, H.J.: Transport Telematics System Architecture. Constraint analysis, mitigation stategies and recommendations, Bruxelles (1999)
3. Eynan, A., Rosenblatt, J.: An interleaving policy in automated storage/retrieval systems. Int. J. Prod. Res. **31**(1), 1–18 (2007). https://doi.org/10.1080/00207549308956710
4. Piecha, J.: Register and Data Process in Transport Telematics Systems. Monograph, Silesian University of Technology, Gliwice (2003)
5. Adelantado, F., Vilajosana, X., Tuset, P., Martinez, B., Melià-Seguí, J., Watteyne, T.: Understanding the limits of LoRaWAN. IEEE Commun. Mag. **55**(9), 34–40 (2017)
6. Rus, C., Negru, N., Patrascoiu, N.: Low-cost system to acquire environmental parameters in urban areas in the context of iot. J. Environ. Protect. Ecol. **20**(3), 1451–1461 (2019)
7. Poczter, S.L., Jankovic, L.M.: The Google car: driving toward a better future? J. Bus. Case Stud. **10**(1), 7–14 (2014)
8. Pop, E., Pop, M., Leba, M.: The hierarchical control of asynchronous drive system. In: Proceedings of INTERPARTNER 1999, pp. 240–245. Yalta (1999)
9. Meyer, M.: Leistungselektronik, Einführung, Grundlagen, Überblick. Springer, Berlin (1990)
10. Salam, Z.: Power Electronics and Drives. FKE-UTM, Skudai (2001)
11. Norris, D.J.: Beginning artificial intelligence with the Raspberry Pi. https://doi.org/10.1007/978-1-4842-2743-5. Apress, Barrington, New Hampshire, USA (2017)
12. Patrascoiu, N., Rus, C.: Traffic signal controller implemented through a virtual instrument. In: 15th International Conference on Engineering of Modern Electric Systems (EMES), pp. 29–32 (2019)

Applying Agile Software Engineering to Develop Computational Thinking in Children

Ângelo Magno de Jesus[1(✉)] and Ismar Frango Silveira[2]

[1] Cruzeiro do Sul University/Federal Institute of Minas Gerais, Av. Professor Mário Werneck, 2590, Buritis Belo Horizonte, MG 30575-180, Brazil
angelo.jesus@ifmg.edu.br

[2] Cruzeiro do Sul University, Galvão Bueno, 868 - Liberdade, São Paulo, SP 01506-000, Brazil
ismar.silveira@cruzeirodosul.edu.br

Abstract. The Computational Thinking (CT) mindset can support basic education by providing tools and strategies for students to solve problems in many subjects. In this context, system design is an important CT practice because it allows students to automate their solutions. This paper proposes a conceptual framework based on Agile Software Engineering practices to teach students digital game development and consequently improve CT skills. The maker culture was also taken into account in the proposed approach. The framework was applied to middle school students. Data were collected and analyzed by Rubric guided interviews and project evaluation. The results pointed out that the approach can be able to actively engage students in different aspects of software development.

Keywords: Agile Software Engineering · Maker culture · Computational Thinking

1 Introduction

Systems Engineering is an important practice of Computational Thinking (CT) [1]. CT is a problem-solving approach in which a computer can be used to automate the solution. According to [1] it is a fundamental skill for everyone everywhere. Therefore, CT can assist education at different levels including in K12. In the sense of elementary education, systems can be designed as digital games, animations, robots controlling and other elements that may bring the children's interests. Agile Software Engineering values and practices can also add a lot to the teaching of CT and System Design. The focus of these practices on collaboration between team members and on the working software makes them a promising way to apply CT in the classroom. Also in this context, Maker Culture can be easily integrated with Agile Software Engineering. Maker Culture is about the concept of "*do-it-yourself*" however with a technological bias [2]. Through this learning approach, students build and share their own technological artifacts.

Á. Rocha et al. (Eds.): WorldCIST 2020, AISC 1159, pp. 756–765, 2020.
https://doi.org/10.1007/978-3-030-45688-7_74

This paper describes a conceptual framework for applying Agile Software Engineering practices to teach children about CT, especially in the systems development perspective. As systems, digital games - which are software whose main requirement is to entertain its users [3] - were chosen. In [4] the application of the proposed framework for collaborative problem-solving in the use of digital games focused on programming learning was described. In [5] the authors describe the collaborative aspects of the proposed framework. This paper describes the Agile Software Engineering processes and maker culture practices involved in the design process. Also, new results achieved are presented and discussed. Digital Games development was chosen as the focus of the approach, due to the great attraction that it exerts on learners [2], besides allowing to work several concepts that are the state of the art of Computer Science [3]. The framework was applied to middle school students. The students' developed games were analyzed. A rubric guided interview was also conducted to collect students' awareness about the proposed development process.

2 Theoretical Background

2.1 Computational Thinking and System Design

The CT presents a set of skills that may be demanded in the information society. This skill set can be powerful tools for education as it has a lot to add to children's analytical abilities [1]. CT is thinking to solve problems, automate systems, or transform data by constructing models and representations [6]. Therefore, CT is closely related to systems engineering. According to [1] CT evolves - among other things - engineering thinking once computational thinkers build systems that interact with the real world. In this sense, [7] presents a model in which students modify an existing program to create their creations. The authors use applications with digital games, robotics and simulation systems.

2.2 Agile Software Engineering

Agile Software Engineering represents the core principles of this study. Agile Methods include collaborative working and problem solving for information system development - which is directly related to Computational Thinking and education. Agile Method values and practices are described in the Manifesto for Agile Software Development [9]. Some of the values and practices of Agile Methods that are directly related to Computational Thinking Education are described following.

- Simplicity: These methods focus on a leaner system development lifecycle (process) [10]. The team also opts to simplify the system design to get quick feedback on results [11]. This simplicity makes it easy to apply these methods as students and teachers need not worry about the complexity of a project process. As just learners, students also need not worry about overly elaborate system design. According to [9], simplicity is essential.

- Collective Work: In the Extreme Programming (XP) method, all developers are equally responsible for all programming parts of the system. This means that the production of source code of the system is collective [11]. Also, pair programming is a common XP practice. [9] describe this principle as individuals and interactions over processes and tools. Collaborative learning also assumes that all students are responsible for the end products of class activities.
- Working Software: the agile manifesto says working software over comprehensive documentation [9]. Generally, in the "traditional" software engineering process it is common for the entire system to be specified by a series of detailed documents. This documentation process can spend time and often be useless as there may be requirements mistakes. In this way, agile methods try to get only the necessary documentation to develop the system [11]. This suggests that students can directly practice programming while creating a useful artifact without taking too much time with preparatory processes.
- Motivation Based: [10] state that agile methods focus on building projects based on people's motivation and pride. [11] argues that the quality of system design and programming is directly related to the ability of developers to remain alert, creative and willing to solve problems. Therefore, it is important to maintain a sustainable pace (without excesses). Educational environments must also focus on students' motivation to learn, collaborate and solve problems.

Agile processes employ incremental iterative development. In agile projects, the software is produced incrementally, following a spiral life cycle [11]. The system grows gradually in repeating short stages. In each repetition, new elements and features are added to the system. In this life cycle, developers continually perform the steps of (1) Problem analysis; (2) Design and Prototyping; (3) Implementation and testing; and (4) Delivery. Due to its practicality, agile processes can be adapted to various activities and projects, including teaching-learning activities related to Computational Thinking. [8] propose a teaching framework based on a mapping of the agile software development process to the CT development process. Each iteration includes the following activities: Storyboard for the single scenes and feasibility analysis; Drawing and programming; Check conformance with the requirements; and Connect the newly developed scene with the existing ones. Our approach diverges from [8] in emphasizing pair programming and testing, active use of kanban, use of the reflection meeting among other aspects.

2.3 Maker Culture

The Maker Culture also can be used as a teaching strategy. The Maker generation is an extension, with more technological and technical applications, of the *Do-It-Your-Self* culture. This culture is based on the idea that anyone can build; fix; modify or manufacture the most diverse types of objects and projects [2]. The Maker universe is quite wide and can include the making of robots, software, electronic objects, apps, art artifacts, toys, carpentry objects among many others. Values of the Maker Movement Manifesto as Make; Share; Learn; Play; Support and Participate [12] represent great support to education.

3 Method

3.1 The Framework

The proposed framework is composed of the processes described in Fig. 1. Each step is described below.

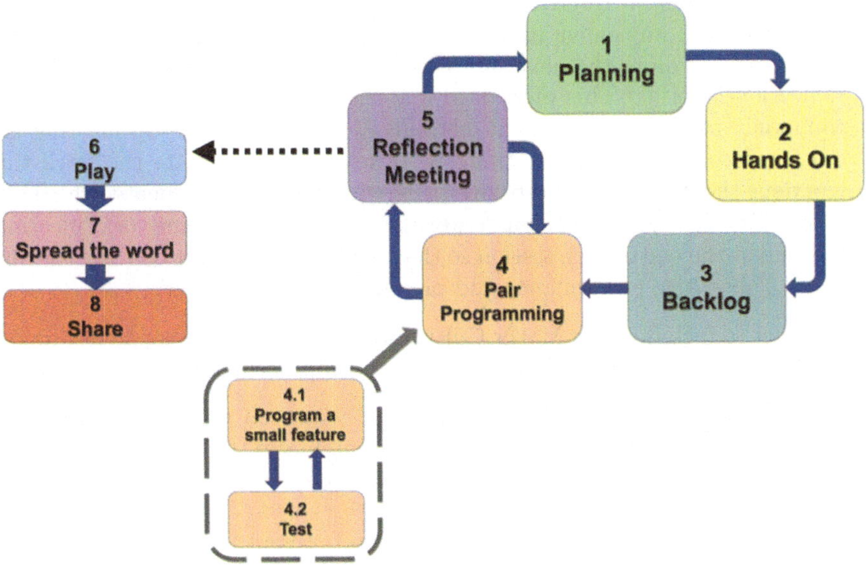

Fig. 1. Proposed framework - life cycle

1. Planning: At this stage, the students come together to define the game that will be developed. The teacher should assist students to perform a brainstorming to generate ideas for the game.
2. Hands on: Based on the previous step, the teacher should lead a workshop with the team and address key development concepts that could be applied to the game implementation process.
3. Backlog: At this stage, the student team should begin the process of raising all subtasks and building a product backlog. As described by [11], developers should use cards to know what features the customer wants. In our approach, the functionalities are written on post-its and fixed in a Kanban board which will be described later.
4. Pair Programming: the team starts programming the game. Students must choose one of the Backlog tasks and begin its implementation. Only one computer is used for implementation. The student who takes over the keyboard will assume the role of driver and the others take on the navigators role and should assist with the test and implementation. The teacher or the students themselves should check the time that each one will take over the keyboard.

This way, learners switch roles frequently. It must be emphasized that digital game development is an extremely complex task and can be a challenge even for experienced programmers. In this way, the teacher can provide a development guide containing many tips for implementing the tasks. This guide will be held by the navigators to engage members in the process more effectively. As can be seen from Fig. 1, this stage is divided into programming and testing substeps. The testing step is essential for the team to try out a simplified model of Test-Driven Development. In this way, students should test each small implementation performed. The subtask must be well tested before moving on to the next subtask.

5. Reflection Meeting: This procedure was inspired by the Stand Up Meetings commonly used in the agile XP method. As [11], the Stand-Up meeting is a quick meeting held every morning by developers to evaluate the work that was done the day before and to prioritize what will be implemented the day it starts. Due to the educational context of the proposed framework, meetings are held at the end of each session so that students can reflect on their learning and collaborative performance and prepare for the next class.

In order to include more Maker Culture values into the development process, three stages can be applied when the project is approaching the final iterations:

6. Play: When the project has advanced, the team can turn on a multimedia projector and have fun playing with their teammates the game they designed themselves. The main idea is to increase student engagement and a sense of pride.
7. Spread the word: The team can publicize and present the game developed to others at their own school or other schools. They can also present the game at science and technology fairs and events.
8. Share: In presentations, the team can ask the audience to interact with the game and give their opinion about it. Students may also share the game's source code on websites.

The proposed framework also recommends using the following materials to perform the activities:

- Timer: The purpose of the timer is to enable students to set the time to take turns in driver and navigator roles.
- "Board" and post-its: in agile engineering, developers use a board containing three columns: (1) uninitiated cards, (2) cards in progress, and (3) finished cards [11]. This mural is known as Kanban. A board (or even a paper) and post-its (to write the subtasks) should be used by the learners to build the Kanban.
- Makey-Makey: Maker Culture usually deals with tangible technological objects. The use of Makey-Makey can add to the project in this regard. Makey-Makey is a device that allows students to design their own controller with everyday materials like playdough or graphite pencils [13]. When the project is sufficiently evolved, students can build their own controls to interact with their games.

3.2 Game Development Workshop

The activities, which were called the game development workshop, were applied to middle school students from public institutions. Four teams were formed and each team consisted of 2 to 4 members. The workshops started with 14 students and 11 of them went to the end. Each class spent about an hour and took place weekly occurring from May/19 to October/19. There were around 12 classes per team. It should be noted that many students who participated had no access to computers at home. Students joined as volunteers and the workshops didn't assign any value or advantage to school grades. Due to the availability of the participants, some teams didn't remain with the same initial composition. Some member switches were held during the workshop. The Scratch development environment was chosen since its features support beginners in programming.

In the final stages of the workshop, projects were shared on the Scratch project base (https://scratch.mit.edu/explore/projects/all). Some team members presented their projects at a regional Science and Technology event that took place at the Federal Institute of MG, Brazil. During the presentations, some spectators were invited to interact with the games. Also, in the final classes students designed their own controls for the games using Makey-Makey. The workshop facilitators took a moment to let the students have fun playing their own games through their new control - Fig. 2(a).

Fig. 2. (a) Students playing the developed game. (b) A example of a developed game.

3.3 Data Collection and Analysis

Data collection and analysis were performed by two means: application of a rubric guided interview and analysis of the developed games. The rubric was employed so that students could self-evaluate their performance during the framework activities. The objective was to get students' awareness of their participation in the activities and to understand if the framework accomplished its objective of involving students in practices of design, collaboration, programming and development processes. The rubric was applied with the authors' support to clarify for students any doubts that might arise. Each aspect addressed in the rubric was an interview question that invited the participant to give their opinion. Students were also invited to make a general assessment of the activity as a whole. Therefore, during the guided interviews, students were encouraged to talk freely about their experience in the different aspects of the activity. This was done so that elements that were not covered under the rubric could be addressed. By starting the interview, the interviewer made it clear to the student that he/she had successfully completed the course and that there would be no harm in answering the questions honestly. The interviewer sought to make explicit that what was being evaluated was the teaching strategy and not the student himself. It should be noted that the activities were performed as extra-class practices and did not assign any value to the school grades. Students participated as volunteers, which may have corroborated fair answers. The rubric, shown in Table 1, was adapted from [14] and [15].

The analysis of the games developed by the students aimed to verify the quality of the products - this procedure was called project evaluation. Based on some requirements, the metric of product quality was defined. Therefore, a list of basic features that games should address was set. Features have been classified as completed, partially implemented, and not implemented.

4 Results

The rubrics were applied to each student individually. To make the data easier to understand, a score for each of the progression levels was assigned: Beginning (−2); Developing (−1); Proficient (+1); and Exceptional (+2). Thus, for each team and each aspect (design, programming, etc.) the average grade given by its members was calculated. A negative value obtained as a result could indicate that the approach had little student involvement. Otherwise, it may indicate that

Table 1. Applied rubric example

Category	Beginning	Developing	Proficient	Exceptional
Project design	- No clear purpose of project or organization - Does not provide a way for other people to interact with the game	- Has some sense of purpose and structure - Includes way for player to interact with the game, may need to be clearer or fit game's purpose better	- Has clear purpose, makes sense, has structure - Includes way for player to interact with game and, clear instructions	- Has multiple layers or, complex design - User interface fits content well, is complex; instructions are well-written and integrated into design
Process	- Student did not get involved in, design process - Did not use project time well and did not meet deadlines	- Student tried out the design process - Used project time well sometimes and met some deadlines	- Student used design process (stated problem, came up with ideas, chose solution, built and tested, presented results) - Used project time constructively, met deadlines	- Student made significant use of the design process - Used project time constructively, finished early or added additional elements
Programming	...	...	...	...
Collaboration	...	...	...	...

Table 2. Rubric result

Team	Project design	Programming	Process	Collaboration
1	1.5	1.5	1.25	2
2	0.3	1	1.3	1.3
3	1.5	1.5	1	1.5
4	1.5	1	1	1.5
m	1.2	1.25	1.14	1.6

the students were involved in the project which contributes to the acceptance of the framework. Table 2 illustrates the results obtained. The m item indicates the overall average taking into account all grades. The results shows that the proposed framework has positively involved learners in all evaluated aspects of the project.

In order to perform the analysis of the projects, a set of 8 features that represent product quality was defined. As all teams chose action games to implement, the following parameters were set: (F1) character movement; (F2) perform an attack; (F3) be defeated; (F4) make enemies sequences; (F5) defeat enemies; (F6) score; (F7) additional movements; (F8) add one more challenge. Note that the F1-F5 requirements represent basic elements that make the game playable. F6-F8 features, although important for fun, were considered additional elements. A feature is considered partially developed if it has bugs or the source code is incomplete. Figure 2(b) illustrates one of the developed games. Table 3 shows that most projects did not meet all of the defined parameters. On the other hand, all games were payable, which means all reached the minimum requirements (F1-F5).

Table 3. Project features: C - complete; P - partially developed; N - not developed; Cp - the percentage of C; Pp - the percentage of P; and Np - the percentage of N.

Team	F1	F2	F3	F4	F5	F6	F7	F8	Cp	Pp	Np
1	*C*	*C*	*C*	*C*	*C*	*C*	*N*	*P*	75%	12.5%	12.5%
2	*C*	*C*	*C*	*C*	*C*	*N*	*P*	*N*	62.5%	12.5%	25%
3	*C*	*C*	*C*	*C*	*C*	*C*	*C*	*C*	100%	0%	0%
4	*C*	*C*	*C*	*C*	*C*	*N*	*N*	*N*	62.5%	0%	37.5%

Overall, during the interviews students expressed interest in the workshop and highlighted their willingness to learn about game development. Also, other interesting questions came to light. Some of the highlighted subjects were described below - the names are fictitious to preserve the privacy of participants:

– Karen - showed interest in doing a second part of the course. Speech: "*... it was nice, it was good to participate, I want you to do another one*".

- Carlos - Discovered about the career in Game Development. Speech: *"... helped (...) to learn that a person can graduate and create games"*.
- Eliot (as Thomas) - Has shown interest in learning game development beyond the Scratch environment. Speech: *"I just wanted... if we will have next year to start over, I wanted to... on that program you showed once, you know? that one the game gets more developed, right?!"* (he was talking about the Unity Game Engine).
- Bruna - thinks that she can make a simple game by herself. Speech: *"... in the classes I came, I learned a lot, I learned to do a lot of things, I think I could make a simple game on my own"*.
- Thomas - concluded that the source code tests allowed the game to be presented to the public at the Science and Technology event.

Therefore, the students' engagement shown by interviews and the rubric was reinforced by the result of the analysis of the developed features that showed, in which the games reached between 60% and 100% of completed features.

5 Conclusion

This paper presented an Agile Software Engineering based conceptual framework for developing CT in students. The focus of the approach was on the engineering thinking dimension of the CT. Then, the approach involved the use of an agile life cycle and included some Maker Culture values. Data were collected through a workshop offered to middle school students. The results pointed out that the learners felt involved in the aspects of the project: design, programming, process and collaboration. The interviews revealed that the development of digital games was a significant motivational factor. The interviews also reveal some interesting viewpoints that some students reported, such as the discovery of a game developer career and the importance of tests for a public presentation of the software. An analysis of the implemented games showed that students were able to build minimally playable products.

As future work, the researchers intend to adapt and apply the framework for high school and college students.

References

1. Wing, J.M.: Computational thinking. Commun. ACM **49**, 33–35 (2006)
2. Tonéis, C.T.: Os Games na Sala de Aula: Games na Educação, ou a Gamificação na Educação. Bookess, São Paulo (2017)
3. Feijó, B., da Silva, F.S.C., Clua, E.: Introdução à Ciência da omputação com Jogos: aprendendo a programar com entretenimento. Elsevier, Rio de Janeiro (2010)
4. Jesus, Â.M., Silveira, I.F.: A collaborative game-based learning framework to improve computational thinking skills. Trans. Edutain. (2019, in press)
5. Jesus, Â.M., Silveira, I.F.: A cooperative learning strategy to computational thinking development. REnCiMa **10**, 192–211 (2019). https://doi.org/10.26843/rencima.v10i4.2387

6. Hu, C.: Computational thinking – what it might mean and what we might do about it. In.: Annual Conference on Innovation and Technology in Computer Science Education, pp. 223–227. ACM, Darmstadt (2011)
7. Lee, I., et al.: Computational thinking for youth in practice. ACM Inroads **2**, 32–37 (2011). https://doi.org/10.1145/1929887.1929902
8. Fronza, I., Ioini, N., Corral, L.: Teaching computational thinking using agile software engineering methods: a framework for middle schools. ACM Trans. Comput. Educ. **17**, 1–28 (2017). https://doi.org/10.1145/3055258
9. Beck, K., et al.: Manifesto for agile software development. https://agilemanifesto.org/
10. Pham, A., Pham, P.-V.: Scrum em Ação: gerenciamento e desenvolvimento ágil de projetos de software. Novatec, São Paulo (2011)
11. Teles, V. M.: Extreme Programming: Aprenda como encantar seus usuários desenvolvendo software com agilidade e alta qualidade. Novatec, São Paulo (2004)
12. Hatch, M.: The Maker Movement Manifesto: Rules for Innovation in the New World of Crafters, Hackers, and Tinkerers. McGraw-Hill, New York (2013)
13. Makey Makey. https://makeymakey.com/
14. Ross-Kleinmann, J.: Computational Thinking Rubrics. ScratchEd (2013). http://scratched.gse.harvard.edu/resources/computational-thinking-rubrics.html
15. ReadWriteThink: Cooperative Learning Rubric (2003). http://www.readwritethink.org/files/resources/lesson_images/lesson95/coop_rubric.pdf

The Influence of Personality on Software Quality – A Systematic Literature Review

Erica Weilemann(✉) and Philipp Brune

Neu-Ulm University of Applied Sciences, Neu-Ulm, Germany
{erica.weilemann,philipp.brune}@hs-neu-ulm.de
https://www.hs-neu-ulm.de/en/home/

Abstract. *Context:* Software quality is an important factor for reducing costs or increasing sales of a software developing company. Humans are involved in the whole software development process. This suggests that human factors have an influence on the result of the development process. *Aim:* We aimed to understand which distinct human factors or personality traits have an influence on which distinct software quality attributes. *Method:* For this purpose, we conducted a Systematic Literature Review following the guidelines of Kitchenham and Charters. *Results:* Existing studies show a link between human factors or personality types and software quality in general. There were only two studies which mentioned an influence on distinct software quality aspects: Extraversion influences decomposition, modularization, testability, functionality, re-usability and programming style. Personality in general influences memory consumption and runtime. *Conclusions:* We expected to find more such links but research conducted until now is far from being sufficient to provide a strong basis for making clear statements on the link between the personality of a software engineer and distinct software quality attributes.

Keywords: Software quality · Personality · Systematic literature review

1 Background

For quite some time now, the claim has raised for taking into account human factors in information systems (IS) research. Early studies demanded a shift of perspective away from "human has to conform to computers" to "computer has to serve the human needs" [15]. In the early 80s, Shneiderman presented four approaches to "include human factors during system design" [5]. Him too, he emphasized the need to take into account human factors during the system or software development process. Hanenberg mentions "an urgent need to consider human factors when reasoning about programming language and software engineering artifacts" [37]. Lately, there rose a request for communities which discuss and investigate the influence of human and social factors on software development problems and software engineering approaches [20].

Á. Rocha et al. (Eds.): WorldCIST 2020, AISC 1159, pp. 766–777, 2020.
https://doi.org/10.1007/978-3-030-45688-7_75

The interest in human factors in the field of software engineering is not new. There is a huge amount of literature in this field and researchers have already investigated, which software engineering topics have been addressed and which human factors were taken into account. Cruz et al. e.g. investigated "the methods used, topics addressed, personality tests applied, and the main findings produced in the research about personality in software engineering" [9]. The aim of Pirzadeh was to "identify and characterize human factors influencing the software development process from development lifecycle and software management perspectives" with the help of the existing literature [32]. And Maier investigated "personality traits on distinct hierarchical levels in three fields of information systems research" within a very restricted source of literature [7].

We found even more literature reviews on this topic. But they did not satisfy the DARE criteria (Database of Abstracts of Reviews of Effects[1]) given in [16] and because of this will not be listed. All these literature reviews covered a broad area: software engineering in general. Our specific interest was the impact of human factors or personality on the outcome of a software engineering process – the produced software. The listed literature reviews did not answer this question.

Precisely software quality is a very substantial factor for the success of a software producing company. Customer satisfaction can lead to subsequent orders which leads to increasing incomes for the company. On the other hand, a company could reduce costs. If we consider software maintainability e.g., "maintenance makes 40%–80% of the software costs" [13]. An improvement of software maintainability subsequently leads to a reduction of maintenance costs. If human factors affect software quality aspects like maintainability, performance, and others, software producing companies could save money by assigning the right tasks to the right people. In addition, the satisfaction and motivation of employees could be increased because they are more pleased with the results they produce.

2 Review Question

Our specific interest is the influence of the personality of a software engineer on software quality attributes. We do not restrict the term software engineer to a single role in the software engineering process. A software engineer could be a programmer as well as a software architect, a designer, an analyst or a tester. Also the term "personality" is used in a broad meaning in this study and includes not only different personality type theories but also diverse human factors resulting from other theories. Some of those factors are e.g. locus of control, value belief, motivation, and self-efficacy. The purpose of this approach was to receive a comprehensive idea of the influence of human factors on software quality attributes.

The research question is:

RQ: Which attributes of software quality are affected by the personality of a software engineer?

[1] http://www.crd.york.ac.uk/CRDWeb/AboutPage.asp.

This question was structured according to the recommendations of Petticrew and Roberts [31] using the **P**opulation **I**ntervention **C**omparison **O**utcomes **C**ontext (PICOC) criteria. The population are software engineers, not restricted to companies or universities. The intervention is the personality or personality type of the software engineer. The personality type is also the comparison. Outcomes are software quality attributes. The context of the research questions is not limited.

3 Review Methods

3.1 Data Sources and Search Strategies

Data Sources. As data sources, we chose three digital libraries as well as results of "manual" search. The three digital libraries were taken from Turner [27]. They are IEEE Explore[2], ACM Portal[3], and ScienceDirect[4]. Search was conducted in this sequence. So we looked for duplicates in this sequence, too. The question type is no question with a primary technical focus, so Kitchenham's question types [23] do not match. The Australian NHMR Guidelines [30] are more applicable. Type five matches: "Identifying whether a condition can be predicted". With the results, a prediction of software quality attributes should be possible by knowing the personality (type) of the software engineer. The question is meaningful and important to practitioners as well as researchers, because it will be possible to predict software quality attributes through personality type testing and it enables an optimization of forming a software development team with the purpose to increase software quality. For each question, according to Kitchenham and Charters [23] we identified population, intervention, comparison, outcomes, and context. We took the resulting phrases as basis for searching synonyms. In our search terms, we used the operators OR and AND as well as the wild card *.

Another source were bibliographies of articles which were results in the first phase. For one article, we had to contact the author because it was not accessible online.

Search Strings. Because of individual characteristics of the digital libraries, the search terms were customized to each digital library. The resulting search terms are the following:

- For IEEE Explore and ACM Portal: *((("information system" OR comput* OR IT OR IS OR program* OR system) AND (engineer OR professional* OR personnel OR people OR practitioner* OR producer OR role)) AND (personalit* OR "cognitive style" OR personality trait OR "human factor")).*

[2] http://ieeexplore.ieee.org.
[3] http://dl.acm.org.
[4] http://www.sciencedirect.com.

– For ScienceDirect: we conducted two searches in ScienceDirect. The first search string was: *("information system" OR comput* OR "IT" OR "IS" OR program* OR system) AND (engineer OR professional* OR personnel OR people OR practitioner* OR producer OR role) AND (personalit* OR "cognitive style" OR personality trait OR "human factor") AND NOT (education OR "HCI" OR "human computer interaction" OR agent OR avatar OR robot* OR game*) AND LIMIT-TO(topics, "social psychology, individual difference, personality trait, personality,emotional intelligence") [All Sources (Computer Science, Psychology, Social Sciences)].* From these results, no article was directly linked to the topic. So we conducted a second search with another search string. We held this string very simple. The second search string was: *"software quality" AND "personality" [All Sources (Computer Science, Psychology, Social Sciences)].*

We did not limit the results by the year of publication. In ACM Portal we conducted the search in "[any field]", not limited to titles. In IEEE we searched in "Full Text & Metadata", in ScienceDirect we searched in "All Fields".

3.2 Study Selection

Including and Excluding Criteria. As we were looking for the impact of the personality of a software engineer on software quality, all articles which describe the personality of a software engineer and software product quality were included. The search strings indicate that we obtained a huge amount of articles and thus had to apply excluding factors.

In one searching engine, it was possible to exclude some topics already in the search. But where this was not possible, we had to narrow down the amount of not directly linked studies. For this purpose we defined excluding criteria. Those were *human-computer-interaction; technical issues (reuse, automation, architecture; nothing to do with software engineering; school concepts; aviation, space flight systems, nuclear power plants; engineering tools; biometry; educational concepts; learning systems; knowledge management; health care; other engineering disciplines (e.g. mechanics); gameplay; gamification; biological/pharmaceutical issues; machine learning; learning styles; intelligent systems; virtual humans; virtual agents; patterns; CAD; automated knowledge acquisition; virtual reality; neural networks; music; cognitive engineering; automotive; car IT; indices/table of contents of conference proceedings; decision support systems; trust in virtual teams; tools for distributed software teams; robots; language is not English or German.*

Study Selection Process. The study selection process comprised three phases. In the first phase we scanned the titles for relevance. If titles showed no relevance for the research question, the study was excluded. If it was not clear, the study was not excluded. In the second phase we scanned the abstracts of the remaining papers and checked if the paper was relevant for our research question. If not, the study was excluded. In the third phase – the final selection phase – we read the complete paper and decided whether its content is relevant or not. The number of the papers in each phase can be found in Table 1.

3.3 Quality Assurance of the Studies

For each study 22 quality criteria were checked. Those were questions which could be answered with yes or no. Only in one case, a choice between yes, moderately, and no was possible. These criteria were e.g. "Does the study report clear, unambiguous findings based on evidence and argument?", "Were the basic data adequately described?", "Do the researchers explain the consequences of any problems with the validity/reliability of their measures?" and "Is the paper well/appropriately referenced?". We defined the criteria following Beecham et al. [36]. Our quality measure for each paper was the rate of positively answered questions. 0% means very poor quality, 100% means very high quality/complies with all our quality requirements. Papers with a quality level less than 50% were excluded from the study.

3.4 Data Extraction and Synthesis

For each study, several information were selected in an Excel datasheet. Those concerned some general information on the paper like title, author, paper id[5], year, and country of the study. In addition, they concerned the type of study, data collection methods, number of study subjects, type of subjects in the study (students or practitioners/experts), role of the subjects in the software development process, type of personality type test, the software development process model, whether the development process was agile, the team size, the programming language, and whether the findings are relevant for the research question or not (is the work directly linked to the research question or partly or not at all). Also we captured if group attributes were taken into account and which ones, if and which cognitive model or style was taken into account, which software quality attributes were taken into account, and if motivation was taken into account. We collected the key findings for each paper in the table as well as the research questions.

3.5 Threats to Validity

Investigator Bias: This systematic literature review was conducted by an individual researcher. Therefore, the probability of validity threats is higher than if several researchers would have conducted the systematic literature review together. Because of this, we implemented several (preventive) measures. During all phases, the individual researcher consulted a second researcher who has

[5] First two letters of the first author's surname + year of publishing + first four letters of the title, e.g.: M. Petticrew and H. Roberts. Systematic reviews in the social sciences: A practical guide. John Wiley & Sons, 2008 ⇒ paper id: Pe2008Syst .

gained experiences with systematic literature reviews. Interim results and next steps were discussed. In addition, in the data extraction phase, the individual researcher called in a separate group of researchers. The purpose and results are explained in *Threats to Data Extraction.*

Publication Bias: Since this systematic literature review did not intend to find any positive effects of treatments, this bias is low in this survey. The intention of this systematic literature review was of exploratory nature – investigate the link between personality and software quality. In addition, the search was not restricted to journal articles only. The individual researcher took into account several different sources of articles, e.g. proceedings, books, and gray literature.

Threats to the Identification of Primary Studies: To assure a high coverage of primary study inclusion, the researcher chose to use ACM Digital Library, IEEEXplore, and ScienceDirect as the main sources for automated search. These digital libraries are well known and among the most common databases for research in this area. In addition, the researcher conducted a parallel manual search. Sources were e.g. the bibliography of articles, and discussions with other researchers. Furthermore, the author customized the search strings for the different digital libraries to ensure as many relevant results as possible.

Threats to Data Extraction: As data extraction is one of the most important phases for the presentation of the results, the researcher focused special attention on the validation of the method. For this purpose, the researcher asked other researchers from the same faculty to extract data from randomly chosen articles. 5% of all articles remaining after phase three were chosen randomly and sent to the researchers. They received an Excel sheet of the same form the individual researcher used to extract data. They also received the Excel forms to asses the quality of the papers. The individual researcher explained in an e-mail the most important variables and how to handle the Excel sheet. With the help of the returned answers the individual researcher calculated the percent agreement and Krippendorff's Alpha [24] for the variables paper id, title, year, country, style of study, subjects of the study, number of the subjects, type of personality type test, software quality attribute, weight of the findings for the research question, and key findings. The percent agreement of all variables was 79.8%. This is a very high agreement. Krippendorff's Alpha for the distinct variables showed moderate to substantial agreement [25].

4 Results

At the end of our study, we did a lot of paperwork. The dimensions of this work are shown in Table 1 – this table shows the number of search results after each phase of searching and selecting.

During the analysis of the results we detected that some authors mixed the concepts of performance and software quality. Very often, software quality was one of the measures for performance. Thus, in the following we will use the terms "performance" and "software quality" with equal relevance. In many papers

Table 1. Number of search results after the three phases, including manual search

Digital library	Number of search results	Number of duplicates	Number of relevant articles after phase 1 + 2	Number of relevant articles after phase 3
IEEE Explore	14.797	0	278	52
ACM Digital Library	5.045	109	176	40
ScienceDirect	758	0	50	11
Manual Search	27	0	27	6
Total:	**20.627**	**109**	**531**	**109**

where the term software quality was used there was no precise definition of what was exactly meant by software quality, i.e., which quality attributes were considered. Or it was not told, which measures were taken into account to determine the respective quality attribute. In many cases, the only metric that was given was "grade point average", but the determinants which lead to the grade were not given. In some cases, software quality was substituted by personal performance, which often was measured by grade point average or external assessment of a superior.

The search results that are related to RQ can be split into two groups: (1) Personality traits that are important for a high team performance and (2) personality traits that contribute to improve software quality.

Results of (1) are omitted at this point due to page number limitations and can be requested from the authors.

Search results for (2) can be split into several topics which have an influence on software quality or personal performance. Some papers show the direct impact of distinct personality traits on performance or software quality attributes. Other papers argue for experience as a main influencing factor of performance. Some papers only speak of personality in general as an influencing factor on performance. One paper mentions the influence of individual differences in design considerations on memory consumption and runtime [33]. Design considerations are creative work and thus influenced by personality [22]. Another group of papers links cognitive abilities and software quality. Self-efficacy and values have been found to be reliable predictors for performance and competence. There was one group of papers that we could not cluster. These papers state in general the influence of human factors or skills on software quality and performance.

We also found papers which rejected a link between personality traits and software quality or performance. In his investigations, Gallivan could not find a significant link between "Openness to Experience" and job performance [26]. Darcy and Ma argue that "little research done to date has identified the personality profile for an excellent programmer" [11]. Papers covering these topics can be found in Table 2.

Table 2. Summary of influencing factors on software quality or performance. (t.b.c.)

Distinct personality traits and their impact on performance or software quality attributes			
This	*has impact on*	*reports who*	*group*
Extraversion	Exploratory testing	[40]	Distinct personality traits and their impact on performance or sofware quality attributes
Extraversion	Decomposition, modularization, testability, functionality, reusability, programming style	[2]	
Extraversion, Conscientiousness	High performance	[38]	
Extraversion, Openness to Experience	Open Source development	[34]	
Openness to Experience	Pair programming effectiveness	[39]	
Intuitive and Thinking	Code review	[10]	
Experience	Issue lead time/Bug fixing time prediction	[4]	Experience
Experience	Interpretation of user's needs, requirements elicitation	[21]	
Experience, reflection, motivation, personal characteristics (thoroughness, conscientiousness, patience or persistency, accurateness, creativeness), domain knowledge, specialized technical skills	High performing tester	[18]	
Self-efficacy	In major grade point average of computing students	[6]	Self-efficacy and values
Personality (high selfesteem, high selfefficacy, high locus of control, low neuroticism), cognitive ability, value belief	Object Oriented programming performance	[8]	
Values	High competence	[14]	
Short term memory performance	Location of high frequencies of errors; quality of a software system	[35]	Cognitive abilities
Cognitive weight of a program	Software complexity and physical size of program	[19]	
Student's perception of understanding of the module	Programming performance	[42]	Miscellaneous
Human factors, individuality	Design cost and quality including design time, design efficiency, and designer error rate	[1]	
Communication, domain knowledge	Requirements elicitation	[28, 43]	
Skilled people	High quality software	[29]	
Openness to Experience	High job performance	[12]	**No link** between personality (traits) and programming performance
Personality	Excellent programmer	Not enough research done to prove link; [11]	

5 Discussion

Many researchers investigated the influence of human factors or personality on performance and even software quality in general. Although it is a very interesting and important issue, the question of how the personality of a software engineer influences distinct software quality attributes has so far not yet been studied to a satisfying level. We could detect only two studies which explored this link AND named the investigated software quality attributes and measures. The authors of the other studies did not mention any reason for this imprecise information.

In addition, some studies used the Myers-Briggs-Type-Indicator whose reliability and validity has been questioned [17]. Most studies used student groups as study subjects. This makes it difficult to generalize the results to professionals, since personality changes with major life events like the first job [3,41].

6 Conclusions

This systematic literature review shows that there are connections between the personality of a software engineer and performance or software quality in general. But research done so far is not sufficient to provide a strong basis for making clear statements on the link between the personality of a software engineer and distinct software quality attributes. With this literature review we have shown the necessity of further studies in this area. Studies should include larger groups and study subjects should be experts. It is also important to chose a personality type test whose reliability and validity has been proven.

Further studies should be conducted among experts from different branches of industry. Studies could concentrate on particular roles of people in a software engineering process, e.g. architects, programmers, or testers. As mentioned above, a valid and reliable personality type test should be used for the investigations. Additionally, studies could concentrate on one aspect of software product quality and should clearly define the metrics that were used to measure the distinct aspect of software product quality. Having all these data, researchers could investigate all kinds of correlations between personality type and different aspects of software product quality.

Therefore, our next step is to concentrate on the link between the personality type of a programmer and the maintainability of the code, she or he produced.

Acknowledgments. The present work as part of the EVELIN project was funded by the German Federal Ministry of Education and Research (Bundesministerium für Bildung und Forschung) under grant number 01PL12022E. The authors are responsible for the content of this publication.

References

1. Aas, E., Sundsbo, I.: Harnessing the human factor for quality designs. IEEE Circ. Devices Mag. **11**(3), 24–28 (1995)
2. Acuña, S.T., Gómez, M., Juristo, N.: How do personality, team processes and task characteristics relate to job satisfaction and software quality? Inf. Softw. Technol. **51**(3), 627–639 (2009). http://www.sciencedirect.com/science/article/pii/S0950584908001080
3. Ahlers, A.: Die persönlichkeitsentwicklung ist niemals fertig. Zeit online (2018). https://www.zeit.de/wissen/2018-05/psychologie-persoenlichkeit-entwicklung-alter-forschung/seite-2
4. Anh, N.D., Cruzes, D.S., Conradi, R., Ayala, C.: Empirical validation of human factors in predicting issue lead time in open source projects. In: Proceedings of the 7th International Conference on Predictive Models in Software Engineering, pp. 1–10. ACM (2011)
5. Shneiderman, B.: Putting the human factor into systems development. In: Proceedings of the Eighteenth Annual Computer Personnel Research Conference, pp. 1–13. ACM (1981)
6. Belanger, F., Lewis, T., Kasper, G.M., Smith, W.J., Harrington, K.V.: Are computing students different? An analysis of coping strategies and emotional intelligence. IEEE Trans. Educ. **50**(3), 188–196 (2007)
7. Maier, C.: Personality within Information Systems research: a literature analysis. In: Proceedings of the ECIS 2012, p. Paper 101 (2012). http://aisel.aisnet.org/ecis2012/101
8. Cegielski, C.G., Hall, D.J.: What makes a good programmer? Commun. ACM **49**(10), 73–75 (2006)
9. Cruz, S., da Silva, F., Monteiro, C., Santos, C., dos Santos, M.: Personality in software engineering: preliminary findings from a systematic literature review. In: Proceedings of the 15th Annual Conference on Evaluation and Assessment in Software Engineering, pp. 1–10 (2011)
10. Da Cunha, A.D., Greathead, D.: Does personality matter?: an analysis of code-review ability. Commun. ACM **50**(5), 109–112 (2007)
11. Darcy, D., Ma, M.: Exploring individual characteristics and programming performance: implications for programmer selection. In: Proceedings of the 38th Annual Hawaii International Conference on System Sciences, p. 314a (2005)
12. Gallivan, M.J.: Examining it professionals' adaptation to technological change: the influence of gender and personal attributes. SIGMIS Database **35**(3), 28–49 (2004)
13. Glass, R.: Frequently forgotten fundamental facts about software engineering. IEEE Softw. **18**(3), 112–111 (2001)
14. Gonçalves, D., Britz, J.: Screening candidate systems engineers: exploratory results. In: Proceedings of the CSIR 3rd Biennial Conference 2010, p. 14 (2010)
15. Hoplin, H.P.: Trends in computing careers: human factors in MIS. In: Proceedings of the Twenty-Second Annual Computer Personnel Research Conference on Computer Personnel Research Conference, pp. 143–150. ACM (1986)
16. Higgins, J.P.T., Green, S.: Cochrane Handbook for Systematic Reviews of Interventions. Cochrane Book Series. Wiley-Blackwell, Chichester (2008)
17. Hunsley, J., Lee, C.M., Wood, J.M.: Controversial and questionable assessment techniques. In: Lilienfeld, S.O., Lynn, S.J., Lohr, J.M. (eds.) Science and Pseudoscience in Clinical Psychology, pp. 39–76. Guilford Press, New York (2003)

18. Iivonen, J., Mäntylä, M.V., Itkonen, J.: Characteristics of high performing testers: a case study. In: Proceedings of the 2010 ACM-IEEE International Symposium on Empirical Software Engineering and Measurement, p. Article No. 60. ACM (2010)
19. Shao, J., Wang, Y.: A new measure of software complexity based on cognitive weights. In: Proceedings of the Canadian Conference on Electrical and Computer Engineering, pp. 1333–1338 (2003)
20. John, M., Maurer, F., Tessem, B.: Human and social factors of software engineering. In: Proceedings of the 27th International Conference on Software Engineering, p. 686 (2005)
21. Kaiser, K.M.: The relationship of cognitive style to the derivation of information requirements. Newsl. ACM SIGCPR Comput. Pers. **10**(2), 2–12 (1985). http://delivery.acm.org/10.1145/1040000/1036375/p2-kaiser.pdf?ip=194.95.21.161&id=1036375&acc=ACTIVE%20SERVICE&key=2BA2C432AB83DA15%2EA95C2088598DDEA0%2E4D4702B0C3E38B35%2E4D4702B0C3E38B35&__acm__=1566808795_60284fe8dd40b051341a904aea6858bd
22. King, L.A., Walker, L.M., Broyles, S.J.: Creativity and the five-factor model. J. Res. Pers. **30**(2), 189–203 (1996)
23. Kitchenham, B., Charters, S.: Guidelines for performing systematic literature reviews in software engineering (2007)
24. Krippendorff, K.: Computing Krippendorff's alpha reliability. Departmental Papers (ASC), p. 43 (2007)
25. Landis, J.R., Koch, G.G.: The measurement of observer agreement for categorical data. Biometrics **33**(1), 159–174 (1977)
26. Gallivan, M.J.: Examining it professionals' adaptation to technological change: the influence of gender and personal attributes. SIGMIS Database **35**(3), 28–49 (2004)
27. Turner, M.: Digital libraries and search engines for software engineering research: an overview (2010). https://community.dur.ac.uk/ebse/resources/notes/tools/SearchEngineIndex_v5.pdf
28. Marnewick, A., Pretorius, J.H., Pretorius, L.: A perspective on human factors contributing to quality requirements: a cross-case analysis. In: Proceedings of the IEEE International Conference on Industrial Engineering and Engineering Management, pp. 389–393 (2011)
29. Mizuno, Y.: Software quality improvement. Computer **16**(3), 66–72 (1983)
30. NHMRC: How to review the evidence: systematic identification and review of the scientific literature. National Health and Medical Research Council, Canberra and Australia (2000)
31. Petticrew, M., Roberts, H.: Systematic Reviews in the Social Sciences: A Practical Guide. Blackwell Pub., Malden (2006)
32. Pirzadeh, L.: Human factors in software development: a systematic literature review (2010)
33. Prechelt, L., et al.: Comparing Java vs. C/C++ efficiency differences to interpersonal differences. Commun. ACM **42**(10), 109–112 (1999)
34. Rigby, P.C., Hassan, A.E.: What can OSS mailing lists tell us? A preliminary psychometric text analysis of the apache developer mailing list. In: Proceedings of the Fourth International Workshop on Mining Software Repositories, p. 23 (2007)
35. Rilling, J., Klemola, T.: Identifying comprehension bottlenecks using program slicing and cognitive complexity metrics. In: Proceedings of the 11th IEEE International Workshop on Program Comprehension, pp. 115–124 (2003)
36. Beecham, S., Baddoo, N., Hall, T., Sharp, H.: Protocol for a systematic literature review of motivation in software engineering, September 2006. http://uhra.herts.ac.uk/handle/2299/992

37. Hanenberg, S.: Faith, hope, and love: an essay on software science's neglect of human factors. SIGPLAN Not. **45**(10), 933–946 (2010)
38. Salleh, N., Mendes, E., Grundy, J., St. Burch, G.J.: An empirical study of the effects of personality in pair programming using the five-factor model. In: Proceedings of the 3rd International Symposium on Empirical Software Engineering and Measurement, pp. 214–225 (2009)
39. Salleh, N., Mendes, E., Grundy, J., St. Burch, G.J.: An empirical study of the effects of conscientiousness in pair programming using the five-factor personality model. In: Kramer, J., Bishop, J., Devanbu, P., Uchitel, S. (eds.) Software Engineering, pp. 577–586. ACM Press, New York (2010). https://doi.org/10.1145/1806799.1806883
40. Shoaib, L., Nadeem, A., Akbar, A.: An empirical evaluation of the influence of human personality on exploratory software testing. In: Proceedings of the IEEE 13th International Multitopic Conference, pp. 1–6 (2009)
41. Specht, J., Egloff, B., Schmukle, S.C.: Stability and change of personality across the life course: the impact of age and major life events on mean-level and rank-order stability of the big five. J. Pers. Soc. Psychol. **101**(4), 862–882 (2011)
42. Bergin, S., Reilly, R.: Programming: factors that influence success. SIGCSE Bull. **37**(1), 411–415 (2005)
43. Yanyan, Z., Renzuo, X.: The basic research of human factor analysis based on knowledge in software engineering. In: Proceedings of the International Conference on Computer Science and Software Engineering, pp. 1302–1305 (2008)

Laying the Foundation for Design System Ontology

Jan Slifka(✉) and Robert Pergl

Faculty of Information Technology, Czech Technical University in Prague, Prague 6, Prague 16000, Czech Republic
{jan.slifka,robert.pergl}@fit.cvut.cz

Abstract. There is a growing need for more client applications for different platforms while maintaining a consistent appearance. Managing this usually requires a lot of tedious labour work. In this paper, we explored what should be included in the design system based on the real-world needs, how to represent and formalise it using semantic web technologies to achieve evolvability and interoperability, and how to convert it into code automatically leveraging the Normalised System theory. Our solution is already a foundation for the ontology representing the design system and working prototype of the code generator using the ontology.

Keywords: Design systems · Client applications · Software engineering · Applied ontology · Evolvability

1 Introduction

Client applications are an essential part of the most software because they provide interfaces with which the users interact with. Nowadays, with a growing number of different mobile devices and platforms, there is a need for more client applications. There is a common need to have a consistent appearance across various platforms to achieve consistent user experience.

Maintaining the consistent visual presentation across all the platforms is a time-consuming task. Graphics designers create the design in a graphical tool, and it is then converted into code by the developers for each platform. It is not a one-off thing. The design evolves, and all the changes need to be addressed and applied in each platform, manually by developers.

The goal of this paper is to explore the concept of design systems and the real-world needs, use the technologies of semantic web to build a platform-agnostic representation of the design system and apply the principles and methodologies of Normalised Systems theory to create so-called expanders from the platform-agnostic representation into the actual platform-specific code.

The main benefits of our approach should be easier evolvability and maintainability of the design systems as well as reusability of design systems across different applications and platforms.

Á. Rocha et al. (Eds.): WorldCIST 2020, AISC 1159, pp. 778–787, 2020.
https://doi.org/10.1007/978-3-030-45688-7_76

In Sect. 2, we describe the concept and technologies and how we want to use them together. Then in Sect. 3, we introduce how the design system ontology should work in the ecosystem of related tools. We explain how we built and test the ontology in Sect. 4 and discuss future development in Sect. 5.

2 Methodology

We want to use the concept of Design Systems to represent the application appearance, formalise it using standard technologies of the semantic web that are suited for data interchange. Then, we want to apply the Normalised System theory to create so-called expanders to generate the actual code from the formal representation. The following subsections briefly describe these concepts and technologies we want to base our solution on.

2.1 Design Systems

There is a growing need in the industry for a way of how to represent a graphical user interface style and behaviour. These systems are usually referred to as *design systems*. Different companies and designers choose different approaches how to create, represent and maintain them [6,7,14]. These approaches usually include maintaining the design systems in graphical tools that designers use, such as Sketch [15] or Adobe XD [1]. The typical workflow, when converting these into code, includes manual labour work (sometimes with the help of ad-hoc plug-ins).

There is Atomic Design [5] which theoretically describes one of the possible approaches to design systems. The basic building block is called an atom. It represents a single element, such as an input field. Atoms are composed into molecules and organisms. Then, there are templates and eventually, final pages. Authors also created a tool called Pattern Lab [11] which can generate code. However, there is no formal definition of the design system. It focuses on web and offers no platform-agnostic representation that could be used on other platforms.

Another approach took the Open Design System [3] initiative, which tries to make the design systems semantic. Authors chose a method they called Nuclear Design, taking the inspiration in nuclear chemistry. The design system consists of sub-atomic particles, atoms, isotopes, molecules, mixtures and matter. While the idea looks promising, it is not finished, and there was no activity recently, and there are no published papers or books on the topic.

The idea of design systems is excellent and helpful, especially to keep consistency when creating multiple client applications on more platforms. The problem with current solutions is that they are either very tight to specific technologies on both ends or useful in theory only. The lack of interoperability or applicability weakens the powerful concept.

2.2 RDF, XSD and OWL

RDF (Resource Description Framework) [8] is a model for data interchange. It is used to describe facts using triples *subject – predicate – object* that forms a graph. We can use it to define relationships between different individuals. In addition, it defines a predicate called *rdf:type* which is used to describe that subject is of a certain type. XSD (XML Schema Definition) [16] is commonly used together with RDF for describing data types (e.g., integer or double).

The structure (vocabulary and allowed relationship) can be described using OWL (Web Ontology Language) [10]. OWL ontologies are used to formally define meaning, providing classes, properties, individuals and data values. It adds more semantic on top of RDF while the ontologies themselves are exchanged as RDF documents.

We believe that these technologies (which are nowadays used for, e.g., semantic web, open data or bioinformatics) are ideal for representing our Design System Ontology. We can use OWL to formally define classes, individuals, relationships and properties to describe the domain of design systems systematically. Moreover, the format is technology agnostic and machine-actionable [4] by default.

2.3 Normalised System Theory

Normalised Systems (NS) theory [9] provides general guidelines on how to build an evolvable system leveraging design concepts from software engineering, e.g., separation of concerns or data version transparency. At the same time, it is based on solid mathematical proofs.

Application of NS theory for building software includes using so-called *expanders*. These NS expanders generate the information systems that conform to the NS theory and can be further customised [12].

We want to follow the principles of NS and build the expanders for generating the code from the Design System Ontology that can be immediately used in the real application.

3 Approach

When building a client application, we can think of different dimensions that it consists of – platform, business logic and design system. These dimensions or concerns are not always clearly separated but tightly coupled instead.

However, if we achieved this separation of concerns, we would be able to change one dimension without affecting the others. The goal of the Design System Ontology and the proposed ecosystem is to be able to achieve that for the design system dimension. That means it should be seamless to replace one design system with another, without the need to touch the platform-specific code or the code responsible for the business logic. Equally, it should be possible to reuse the same design system for more application without extra work (Fig. 1).

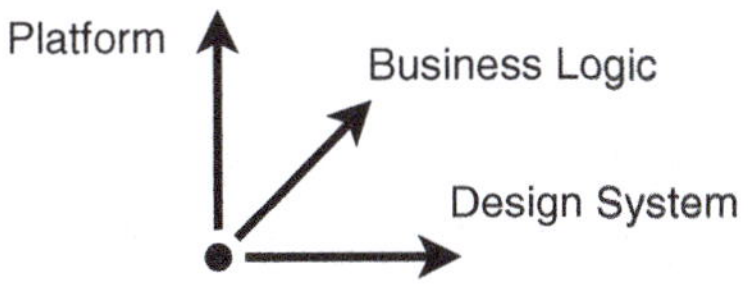

Fig. 1. Different dimensions of the client application

The idea is to create an ontology that would work as an interoperability point between the tools used by graphic designers and generated code. There could be a converter for each design tool that would transform the design system created within it into the Design System Ontology (specifically, an RDF document).

Then, for each target technology, an expander would be used to generate the desired code automatically. However, the conversion from DSO cannot cover every possible use case, and some manual tweaks will be necessary. When we change the code, we cannot simply regenerate it when DSO is updated because we would lose the custom changes. Luckily, we can use the harvesting mechanism from NST. The expanders generate NST compliant code with well-defined places for customisations. When generating the result code, the expander does it in three steps:

1. Harvest customisations from the defined locations in the previous version of the code
2. Generate a new code from the latest version of DSO
3. Insert harvested customisations into the newly generated code

Having the Design System Ontology as a middle point instead of converting a source from the design tool directly to the target code allows combining more entry points and output technologies, leaving us with the number of possible combinations equal to *(# of design system inputs) * (# of expanders)*. The whole workflow is illustrated in Fig. 2.

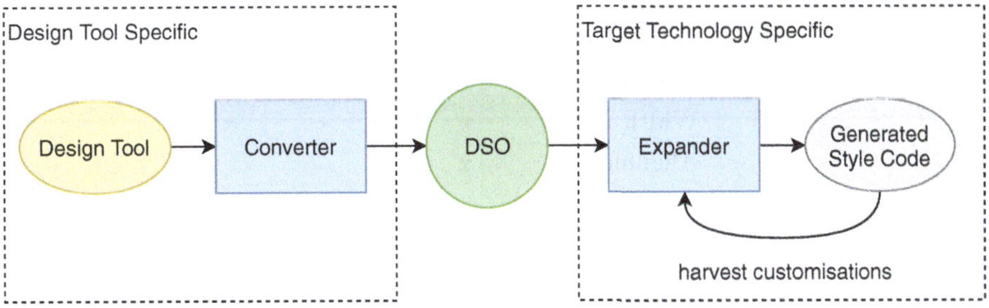

Fig. 2. Design System Ontology and its workflow

4 Results

First of all, we wanted to analyse what is possible in the design tools and target technologies. As a reference sample for design tools, we chose Sketch and Adobe XD. We went through the different features that can be set – e.g., various types of fills, shadows or text properties.

We selected the web to be the target platform for this prototype. On the web, we use Cascading Style Sheets (CSS) [17] to define styles for elements. Therefore, we wanted to analyse the intersection of what is possible in the design tools and what is achievable in CSS. That would serve as a basis for what we should include in the Design System Ontology.

The first version should focus on the definition of individual elements and support their styling. In the future, we want to go further and extend the ontology to support composing elements together and building components with different layout options.

4.1 Appearance Features

By analysing the appearance features that are available in the design tools, we found out that there are four categories – fills, borders, shadows and blurs. We took those and compared them to CSS features. Note that certain things are achievable in CSS using some hacks (e.g., using shadow for borders) or are not consistently compatible in all browsers (e.g., blurs). We decided to leave those out and use the features only for the purpose they were designed for. The detailed comparison can be found in Table 1.

Table 1. Appearance features comparison

Feature		Sketch	Adobe XD	CSS
Fill	Solid Colour	x	x	x
	Linear Gradient	x	x	x
	Radial Gradient	x	x	x
	Angular Gradient	x		
Border	Colour	x	x	x
	Width	x	x	x
	Alignment	x	x	
	Joins/Ends	x	x	
	Dash/Gap	x	x	
	Style			x
Shadow	Inner	x		x
	Outer	x	x	x
Blur	*Various options*	x	x	(x)

These appearance features can be applied to any element in the design system. There are some differences in cardinality. It is possible to configure multiple instances of each appearance class to each element in Sketch. However, this is not possible in Adobe XD nor CSS. Therefore, we decided not to support these and allow only one instance of fill, border, inner shadow and outer shadow for each element in the Design System Ontology.

4.2 Primitives

Based on the analysis of the features it turned out, we needed to define some primitives to represent values with units (e.g., *10px*) or colour. We could use the simple data types (such as a string) to encode everything, but we wanted to give the representation more semantics. Also, these primitives can be referenced in more places. For example, we can define a primary colour and use it for a colour fill of one element and border colour of another.

Table 2. Text features comparison

Sketch	Adobe XD	CSS
Type Face	Font	font-family
Weight	Weight	font-weight
Size	Font Size	font-size
Character Spacing	Character Spacing	letter-spacing
Line Height	Line Spacing	line-height
Color	-	color
Text Alignment	Text Alignment	text-align
Vertical Alignment	-	-
Text Alignment	Text Alignment	text-align
Decoration	Decoration	text-decoration
Transform	Transform	text-transform
Paragraph Spacing	Paragraph Spacing	-
-	Superscript/Subscript	vertical-align + font-size

4.3 Text

Besides appearance, we also wanted to configure various text properties for each element. Table 2 shows how different text properties are called in graphical tools and CSS. Most of the features are available across all compared, usually with a similar name.

4.4 Putting It Together

In previous sections, we analysed what is achievable in the design tools and the target technology of choice. Based on that, we created a hierarchy of classes that defines the Design System Ontology. A preview of the ontology classes and their relationships can be seen in Fig. 3 (primitives and data properties are skipped for brevity). We defined various classes representing styling features (such as fill or border) that can be applied to elements (such as button or input) and primitives, and we defined the relationships between these classes. We can use the ontology to describe a basic design system formally.

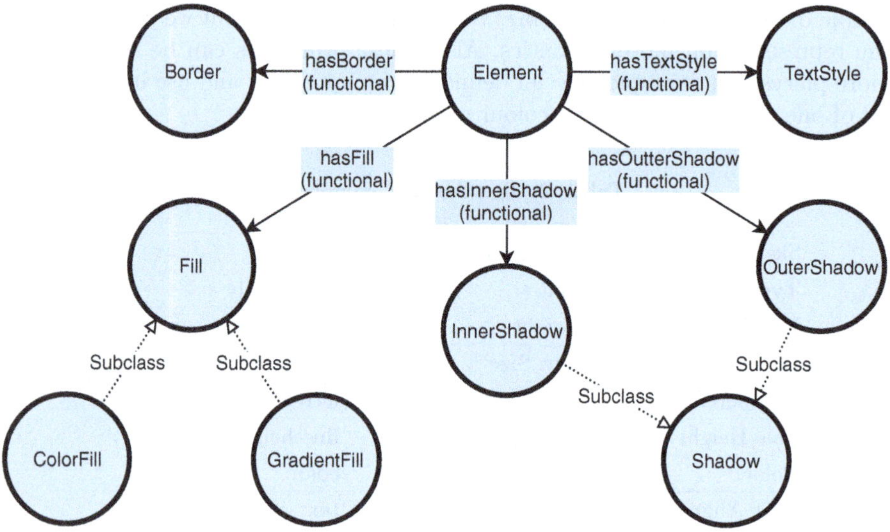

Fig. 3. Diagram showing different classes in the Design System Ontology

4.5 Applying the Ontology

The next step was the application of the ontology to test it is useful. We used Python [13] programming language to write an expander prototype which automatically converts DSO into platform-specific code, web for this example. The input for the expander is the RDF representation of DSO with the individuals representing various entities of the ontology. Figure 4 shows a visualisation of a part of the design system.

As for the expander output, we decided not to use CSS directly but to use SCSS [2] language instead. It is CSS preprocessor which allows us to use some features not available in CSS standard. The most important for us are *properties* and *mixins*.

Properties allow us to define specific values that can be used across the stylesheets. For example, every `Color` class can be represented as an SCSS property. We can use it instead of the actual value in the customisations. When the colour changes in the ontology, the SCSS code is regenerated, the customisations are injected, and everything works again. On the other hand, if we used the values, we would have to manually change every occurrence of the colour in the custom code.

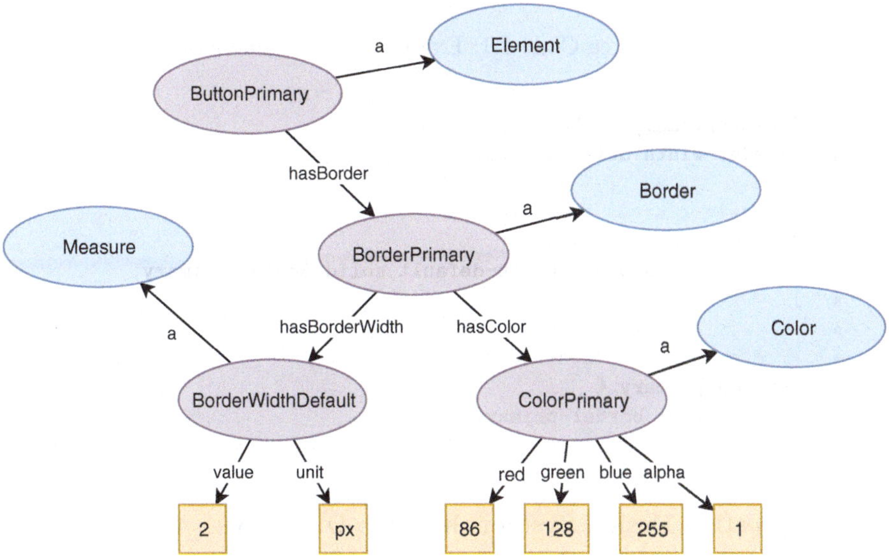

Fig. 4. Part of the design system represented in the Design System Ontology

Mixins are groups of CSS rules that can be included in other elements. We can use mixins to represent the style classes we have (fill, border, etc.), which will more precisely reflect the structure of the ontology and allows more flexible the reuse in the custom code.

This way, we created an expander following the principles of NST. Source Code 1.1 is the output SCSS code for the example in Fig. 4. The expander can be used to expand any design system represented in DSO into code automatically.

We could create more expanders that would take DSO as input and generate the code for different platforms. The main advantage is that we create the expander only once and then we can reuse it for many applications.

5 Future Outlook

In this paper, we prepared a foundation for the design system ontology that can already be used, and we built a prototype of an expander to test how it works. There are some next steps we want to undergo in the future.

5.1 Extending the Ontology

This is the first proposal of the Design System Ontology which covers only the styling of individual elements for now. We want to go further. We want to extend the ontology to also include more complicated components, layout options and other higher-level structures.

Source Code 1.1: Example of the generated SCSS code

```scss
// Primitives are expanded into properties
$color-primary: rgba(86, 128, 255, 1);
$border-width-default: 2px;

// Style classes are expanded into mixins
@mixin border-primary {
    border: $border-width-default solid $color-primary;
}

// Elements are expanded into CSS classes
.button-primary {
    @include border-primary;
}
```

5.2 Converting Design Systems into the Ontology

We focused on the ontology and prototype of the expander to get the code. There is the other side too – graphic tools and the converter to the ontology. We want to build these tools in the future to cover the whole life-cycle of the design system from the graphic tool to the application code.

6 Conclusion

In this paper, we designed the foundation of the Design System Ontology based on real-world needs. While the ontology now covers only basic elements and styles, it is ready to be further extended. The benefit of having the design system represented as a platform-agnostic ontology is that we have a single source of truth which is also machine-actionable. It improves evolvability. We only need to change the design system definition in one place from which it can be automatically applied to all the target platforms.

For that, we implemented an expander according to the Normalised System Theory that transforms the DSO representation fo the design system into actual usable code. Expanding the code is not a one-way process. The generated output can be further modified manually if necessary, these changes are harvested before regenerating the new version and injected back into the new version of the code. This way, we achieved better flexibility for features that are not yet or hard to cover by the ontology.

Our contribution here is a foundation of the ontology that can represent the design systems as well as the demonstration of how such ontology can be used to achieve better evolvability of design systems using the expanders based on the Normalised Systems theory.

Acknowledgements. This research was supported by the grant of Czech Technical University in Prague No. SGS17/211/OHK3/3T/18.

References

1. Adobe Inc.: Adobe XD. https://www.adobe.com/products/xd.html
2. Catlin, H., Weizenbaum, N., Eppstein, C., Anne, J., et al.: Sass. https://sass-lang.com
3. Dash, S., Murugesan, G.: OpenDesignSystem (2018). http://opendesignsystem.org/
4. DDI Alliance: Machine-actionable (definition) (2018). https://www.ddialliance.org/taxonomy/term/198. Acceessed 21 Sept 2019
5. Frost, B.: Atomic Design (2016). ISBN 978-0-9982966-0-9
6. Hacq, A.: Atomic design: how to design systems of components (2017). https://uxdesign.cc/atomic-design-how-to-design-systems-of-components-ab41f24f260e. Accessed 22 Aug 2019
7. Kopf, B.: Figma & Atomic Design Systems (2018). https://medium.com/@benkopf20/figma-atomic-design-systems-324a903b1215. Accessed 24 Aug 2019
8. Lassila, O., Swick, R.R., et al.: Resource description framework (RDF) model and syntax specification (1998)
9. Mannaert, H., Verelst, J., Bruyn, P.D.: Normalized systems theory from foundations for evolvable software toward a general theory for evolvable design (2016)
10. McGuinness, D.L., Van Harmelen, F., et al.: OWL web ontology language overview. W3C Recommendation **10**(10), 2004 (2004)
11. Muenzenmeyer, B., Lovely, E., Frost, B.: Pattern Lab. https://patternlab.io
12. Oorts, G., Huysmans, P., De Bruyn, P., Mannaert, H., Verelst, J., Oost, A.: Building evolvable software using normalized systems theory: a case study. In: 2014 47th Hawaii International Conference on System Sciences, pp. 4760–4769. IEEE, Waikoloa, HI (2014). https://doi.org/10.1109/HICSS.2014.585. http://ieeexplore.ieee.org/document/6759187/
13. Python Software Foundation: Python language reference. https://www.python.org
14. Saarinen, K.: Building a Visual Language (2018). https://airbnb.design/building-a-visual-language/. Accessed 22 Aug 2019
15. Sketch B.V.: Sketch. https://www.sketch.com
16. Thompson, H.S., Mendelsohn, N., Beech, D., Maloney, M.: W3C XML schema definition language (XSD) 1.1 part 1: Structures. The World Wide Web Consortium (W3C), W3C Working Draft Dec **3** (2009)
17. World Wide Web Consortium (W3C): CSS. https://www.w3.org/Style/CSS/specs.en.html

A New Swarm Algorithm Based on Orcas Intelligence for Solving Maze Problems

Habiba Drias[1(✉)], Yassine Drias[1,2], and Ilyes Khennak[1]

[1] LRIA Laboratory, USTHB, Algiers, Algeria
hdrias@usthb.dz
[2] Algiers University, Algiers, Algeria

Abstract. In this paper, a new swarm intelligence algorithm based on orcas behaviors is proposed for problem solving. The algorithm called Orcas Algorithm (OA) consists in simulating the orcas life style and in particular their social organization, their echolocation practice and their hunting techniques. The experimental study we conducted tested the designed algorithm as well as recent state-of-the-art evolutionary algorithms for comparison purposes. The experiments were performed using a public dataset describing mazes with four level of complexity. The overall obtained results clearly show the superiority of OA algorithm over the others. This finding opens the way to other problems to solve to benefit from OA robustness.

Keywords: Swarm intelligence · Orcas · Social organization · Echolocation · Hunting techniques · Orcas Algorithm · Maze

1 Introduction and Motivation

Swarm intelligence [1] has contributed immensely to artificial intelligence over the last three decades. Algorithms such as ACO (Ant Colony Optimization) [2] and BSO (Bee Swarm Optimization) [3] appeared first in the beginning of the advent of this research axis. Lately, birds and fishes behaviors were explored for designing another type of swarm algorithms. Then we have known a series of continuous evolutionary approaches such as PSO (Particle Swarm Optimization) [4], BA (Bat Algorithm) [5], FFA (Firefly Algorithm) [6], that were conceived basically for numeric functions. Recently, animal herding has attracted researchers for designing interesting algorithms such as EHO (Elephant Herding Optimization) [7], ESWSA (Elephant Swarm Water Search Algorithm) [8] and WOA (Whale Optimization Algorithm) [9]. In this work, we are interested in another kind of swarm intelligence that integrates several interesting comportments observed at once in a mammal. These features are found in Orcas also called killer whales. Orcas are seen as a highly social, organized and intelligent species, with an agility that depends essentially on their brain, as studies show that orcas have the second largest brain among marine mammals [10]. Their lifestyle aroused our curiosity and prompted us to first deepen our knowledge on

Á. Rocha et al. (Eds.): WorldCIST 2020, AISC 1159, pp. 788–797, 2020.
https://doi.org/10.1007/978-3-030-45688-7_77

this mammal species and then develop an intelligent algorithm based on their wise behaviors such as their social organization [10], their echolocation practice [11] for prey detection and their hunting skills [12]. The proposed algorithm is called OA (Orcas Algorithm) and is developed by simulating these different comportments. It is then applied to the problem of finding the optimal path in labyrinths, in order to evaluate its performance compared to recent algorithms of the literature like EHO, BA and WOA. In fact, OA encompasses at once all the intelligent features found in these algorithms and hence, it is more general than all of them.

This paper is organized around five main sections. The next one is a brief presentation of the life style of orcas with the aim to extract the most important characteristics to simulate later. The third describes the step-by-step design of our algorithm. The fourth section focuses on the application of OA to maze problem and exposes the experiments carried out on a public dataset with an analysis of the obtained results as well as a comparative study with state-of-the-art algorithms. The last section provides a conclusion on the most important achieved results.

2 Major Features of Orcas Behavior

The major identified aspects in orcas are social behavior evolving in groups, the phenomenon of echolocation and the hunting techniques. The social organization consists of a community of resident orcas, which is a particular kind of orcas that we considered in this work because of their attractive behaviors. The group is formed of clans, where each clan includes a set of pods, containing each of them a fixed number of orcas or individuals, which are led by a matriarch. Females take their independence as soon as they have a descent, while keeping the family ties.

For orcas, echolocation is a sensory ability for better navigation and more effective hunting. Clicks used during echolocation last less than a millisecond. At this moment, the sound crosses the water and bounces off the fish or any other kind of interest, thus bringing vibrations back to the orca with valuable information that will give them precise details about the prey: its size, its proximity, the depth of water and the presence of other predators.

The phenomenon that seems the most interesting in orcas style of life after that of echolocation is hunting. Since the beginning of their existence, orcas have been developing hunting techniques, which are transmitted over time through learning. Scientists have identified at least six kinds of them, namely the Wave Wash for attacking seals, the Karate Chop for sharks, the Carousel for small fish, the Pod Pin for narwhals, the Blowhole Block for larger cetaceans such as humpback whales and the D-Day for sea lions.

3 OA: The Orcas Algorithm

Just like PSO, OA is also based on the evolution of moves of individuals within the group. The space of the solutions of a problem to be solved corresponds to the set of positions that the artificial orcas can reach. The best position of the prey represents the optimal solution. A population of artificial orcas follows the same social structure of a community of resident orcas, as described in their social organization. It consists of clans that stand out from each other (strong neighborhood), where each clan contains a number of pods. The pods in a clan are defined by a weak neighborhood containing close solutions. The OA algorithm is designed to include a series of operating actions in a local region followed by an exploration phase to diversify search of other regions. In the former step, echolocation phenomenon followed by hunting acts is simulated whereas in the second one, the leaving of the group by females reaching independence is modeled. Concretely, in order to reach the optimal solution, the best current candidate solution is determined at each iteration of the algorithm. Once it is found, the other agents update their positions to it. This social interaction is a kind of stigmergy that is the central key in developing self organization of the group. Technically, the target is detected by Formulas (1), (2) and (3).

$$f_i(t) = f_{min} + (f_{max} - f_{min})\beta \tag{1}$$

$$v_i(t) = v_i(t-1) + (x_i(t-1) - x^*)f_i(t) \tag{2}$$

$$x_i^{'} = x_i(t-1) + v_i(t) \tag{3}$$

where:

- $f_i(t)$ is the frequency of the individual i at time t, which varies between two constants f_{min} and f_{max}.
- β is a random value in the range $[0, 1]$ drawn from a uniform distribution.
- $v_i(t)$, $x_i^{'}$, x^* represent respectively the velocity of the individual i at time t, the position of the new target and the best solution obtained from the previous iterations.

These equations simulate one unit of the echolocation phenomenon. It will be repeated for a while until finding the optimal solution. If the latter is not reached then the result of the third formula ($x_i^{'}$) will be used for hunting simulation. In this study, we focus on the so-called Carousel hunting technique. The method is quite similar to that of the noted bubble net feeding in humpback whales. Orcas swim around the prey on a spiral path while shrinking the circle they create. Two approaches are defined to describe this hunting behavior: the encircling shrinking mechanism and the updating of the spiral position. A spiral equation is set up between these two positions to simulate the helical movement of orcas. Then we assume there is a probability of choosing between the encircling narrowing mechanism and the spiral model to update the position of the orcas, which gives rise to Eq. (4).

$$x_i(t) = \begin{cases} x_i^{'} - A.d & if\, p < \alpha \\ x_i^{'} + d'e^{bl}cos2l & if\, p >= \alpha \end{cases} \tag{4}$$

where:

- $x_i(t)$ is the position of the individual i at time t.
- p is a random number in the interval $[0, 1]$ and α an empirical parameter between 0 and 1.
- $A = 2a.r - a$
- a is a constant that decreases linearly from 2 to 0 and r is a random value.
- d and d' represent respectively the distance between the position of the orca at time $t-1$ and the position of the new target, according to Formulas (5) and (6)
- b is a constant that defines the shape of the logarithmic spiral.
- l is a random number taken from the interval $[-1, 1]$.

$$d = |c.x_i' - x_i(t-1)| \quad with\ c = 2r \tag{5}$$

$$d' = |x_i' - x_i(t-1)| \tag{6}$$

This hunting action is iterated for a while until finding the optimal solution, otherwise the skip in other search region is operated. Inspired by the behavior of resident orcas females leaving the family group, the exploration phase consists in replacing at each generation, the worst individual $x_{worst,pj}$ of pod j belonging to clan i, by a random individual according to Eq. (7).

$$x_{worst,pj} = (x_{minli} + (x_{maxli} - x_{minli})rand) * b + (x_{ming} + (x_{maxg} - x_{ming})rand) * (1 - b) \tag{7}$$

where:

- x_{minli} and x_{maxli} are respectively the lower and upper bounds of the local search space of clan i.
- x_{ming} and x_{maxg} are respectively the lower and upper bounds of the population.
- b is a Boolean value drawn randomly.
- rand ϵ $[0, 1]$ is a chosen random number of a uniform distribution.

The Orcas Algorithm OA is outlined in Algorithm 1.

4 Experiments

The public dataset [13] on which we tested the OA algorithm represents a set of labyrinths, developed at the National University of Colombia in 2018. In order to evaluate the performance of our algorithm on several types of labyrinths, we used the similarly connected maze problems (SCMP) [14] and especially SCMP1 with a connectivity equal to 0, SCMP2 with a connectivity equal to 0.3, SCMP3 with a connectivity equal to 0.6 and SCMP4 with a connectivity equal to 1. Each type contains 10 labyrinths and 12 starting positions, which gives 120 instances for each type and thus a total number of instances equal to 480 instances.

The different algorithms tested have been implemented in Java programming language and for the experiments, we used a machine with a Processor Intel Core i5, 2.3 GHz, a Memory (RAM) with 8GO and Windows 10 system.

For parameters tuning, we varied each parameter as shown in Table 1, which adds up to a number of tests equal to 288 scenarios. For reasons of simplification, we initialized the speed of each individual with the value 0, while the distance between the pods of the same clan was set to 3. The size of the population is obtained by fixing the number of individuals in each pod at 5 and by varying the other two parameters that is, the number of clans and the number of pods in a clan as shown in Table 2.

Table 1. Tested empirical parameters for OA

Population size	Initial solution size	b	α	f_{min}	f_{max}
20	5	0	0,25	0	1
30	15	1	0,50	2	3
60	30	5	0,75	–	–
100	45	–	–	–	–

Table 2. Distribution of the number of clans and pods per clan for OA

Population size	Number of clans	Number of pods
20	2	2
30	2	3
60	3	4
100	4	5

4.1 Optimal Solutions and Maze Complexity

The fitness function is computed as the Manhattan distance between the exit cell and the last cell reached by the agent after executing its path on the instance. Figure 1 shows that the size of the population plays a crucial role in the search for the optimal solution. We also notice despite a small exception the latter is in continuous decrease with the growth of the population size for the 4 types of labyrinths. However, in Fig. 2, we observe the runtime in continuous slow increase with the growth of the size of the population. Both figures reveal that the complexity of the labyrinth increases the size of the optimal solution as well as the execution time.

Algorithm 1. OA (Orca Algorithm)

Input : dataset, empirical parameters : $\alpha, f_{max}, f_{min}, b, tmax$...
output: optimal or approximate solution
begin initialize the population according to **the orcas social organization**;
  calculate the fitness of each individual;
  determine the matriarch of each pod and each clan;
  t :=1 ;
  while ($t < tmax$) and (optimal solution not found) **do**
  for n times **do** *(* echolocation step for prey search*)*
   for each clan ci **do**
    for each pod pi **in** ci **do**
     for each individual i **in** pi **do**
      apply Equations (1), (2) and (3);
      update the matriarch;
  if (optimal solution not found)**then**
  for m times **do***(* hunting round *)*
   for each clan ci **do**
    for each pod pi **in** ci **do**
     for each individual i **in** pi **do**
      apply Equation (4);
      update the matriarch;
  if (optimal solution not found) **then** *(* change region for exploration*)*
   evaluate the individuals of pod pi;
   generate a new individual using Equation (7);
   if (the worst individual is worse than the new individual)
      then replace it with the new individual;
   t :=t+1 ;
end

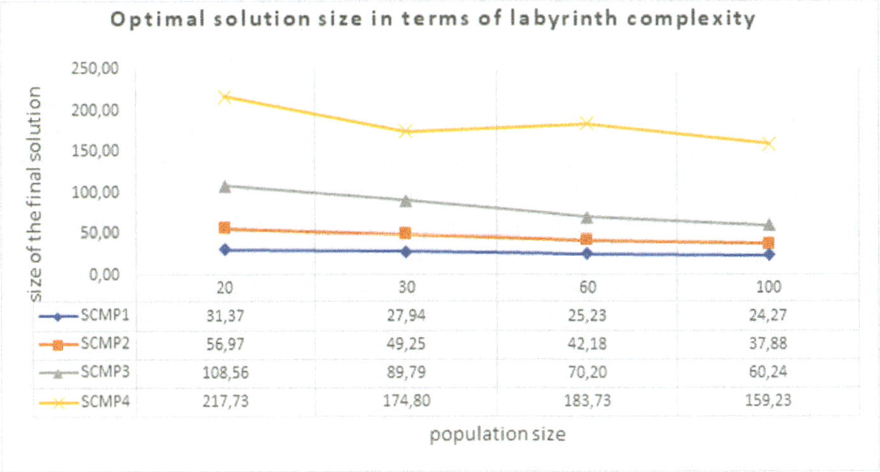

	20	30	60	100
SCMP1	31,37	27,94	25,23	24,27
SCMP2	56,97	49,25	42,18	37,88
SCMP3	108,56	89,79	70,20	60,24
SCMP4	217,73	174,80	183,73	159,23

Fig. 1. Optimal solution size computed by OA relatively to the labyrinth type complexity.

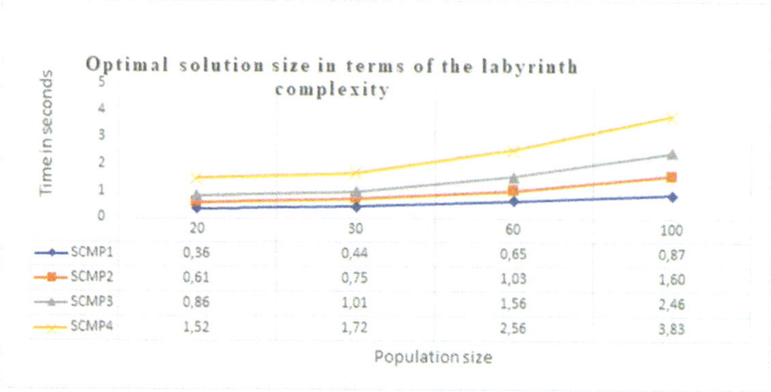

Fig. 2. OA runtime in terms of labyrinth complexity and population.

4.2 Comparison with Recent State-of-the-Art Algorithms

We selected WOA, EHO and BA as recent state-of-the-arts algorithms to experiment for comparison purposes because they have proven their ability to solve hard problems. Several applications of these algorithms were performed in various domains. For the empirical parameters, we used the same parameters values as OA so that we can compare the results obtained later. The concept of pod does not exist for these algorithms, we adjusted approximately the parameters values to reach equivalent population and clans sizes.

Analysis of the Results. Figure 3 shows the optimal results obtained by the algorithms WOA, EHO, BA and OA according to the type of labyrinths and the size of the population. In order to have consistent results, we focused on average sizes of solutions with a success rate of 100%. This allow us to have a better comparison between the different algorithms later, because a low success rate is found only with simple instances whose solution size is short, which may distort the comparison. We observe a low success rate with WOA for the four types of labyrinths. As with a very low complexity maze (SCMP1), it achieves a success rate that does not exceed 61% and 32% with labyrinths of very high complexity (SCMP4). We note that the best results for this algorithm are performed with a population size equal to 60 individuals. However, we note better results with OA, BA and EHO, where the success rate for very low complexity labyrinths (SCMP1) is 100%. Meanwhile, EHO finds difficulties with other types of labyrinths, as it does not exceed 98% for SCMP2, 91% for SCMP3 and 86% for SCMP4, while OA and BA were capable to reach a rate of 100% for other types but with a very large size of solution.

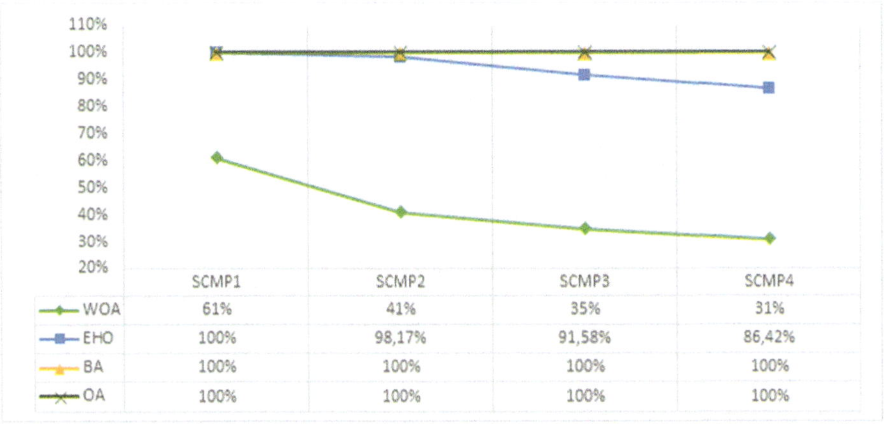

Fig. 3. Best success rates of the 4 algorithms for the 4 types of labyrinths.

4.3 Comparison Between the Best Success Rates

In this part, we undertake the comparison between the optimal results obtained by each algorithm. Note that an average solution size of a given type of maze is considered only in the case where the success rate of the latter is 100%.

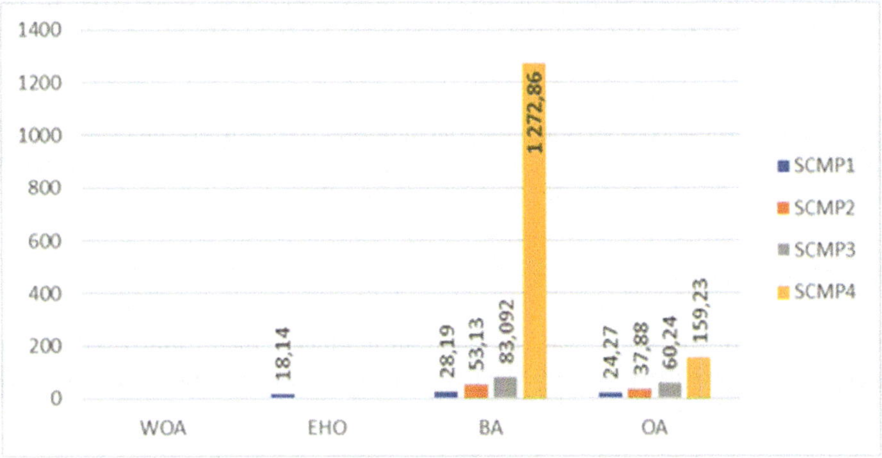

Fig. 4. Comparing the solution size for the 4 algorithms.

Figures 4 and 5 respectively show the average solution size and the execution time of the best results obtained by WOA, EHO, BA and OA algorithms. On the average size, we notice in the first place that it is null for WOA for the 4 types of labyrinths. This is explained by the fact that it does not reach a success

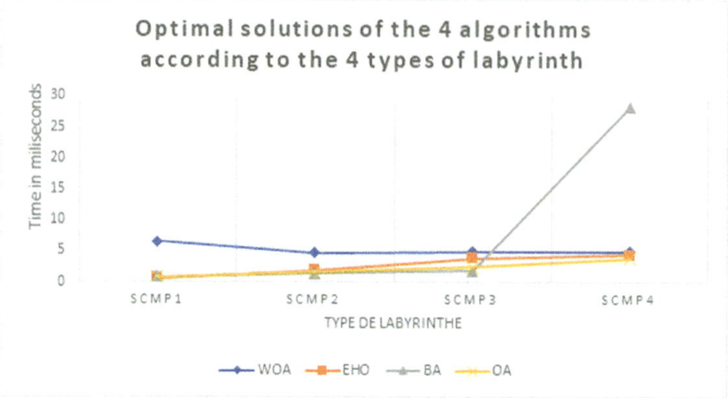

Fig. 5. Best runtime of the 4 algorithms according to the 4 types of mazes.

rate of 100%. However, we notice that EHO is able to solve only labyrinths of very low complexity but giving a similar result to that of OA which happens to be the best in these tests. Second, we remark that BA and OA solve all types of labyrinths. It is clear that among both of them, OA is the best because it does not exceed 160 moves unlike BA, which exceeds 1200 moves for the labyrinth of high complexity. For the execution time, we observe that WOA takes more time in the execution of SCMP1, SCMP2 compared to the other algorithms, followed by WOA, EHO and OA compared to BA in the execution of SCMP3. In the execution of SCMP4, WOA, EHO and OA are the fastest. However, BA records a delay of 28 ms relatively to the others. Note that WOA terminates rapidly because it is far from being able to reach a 100% success rate.

Following these comparisons, we conclude that WOA is not efficient for our problem and that EHO can solve only simple labyrinths. However, BA and OA are the most effective but only OA is the most efficient.

5 Conclusion

The purpose of this study was to design an intelligent algorithm based on the behavior of orcas expressed by their aptitude of social organization and their echolocation and hunting techniques ability.

The experimentation of the proposed OA algorithm and those of the related works, namely EHO, BA and WOA was performed for the maze problem on public datasets of different complexities. As a result, we established a performance ranking in terms of effectiveness and efficiency in solving complex labyrinths in decreasing order between these algorithms as follows: OA, BA, EHO, and WOA.

As future work, it remains interesting to extend the carousel hunting technique simulated in the OA algorithm by the other mentioned strategies.

References

1. Kennedy, J., Eberhart, R.C.: Swarm Intelligence. Morgan Kaufmann Publishers Inc., San Francisco ©(2001). ISBN 1-55860-595-9
2. Dorigo, M., Stützle, T.: Ant Colony Optimization. MIT Press, Cambridge (2004). ISBN 0-262-04219-3
3. Drias, H., Sadeg, S., Yahi, S.: Cooperative bees swarm for solving the maximum weighted satisfiability problem. LNCS, pp. 318–325. Springer (2005)
4. Zambrano-Bigiarini, M., Clerc, M., Rojas, R.: Standard particle swarm optimisation 2011 at CEC-2013: a baseline for future PSO improvements. In: 2013 IEEE Congress on Evolutionary Computation (CEC), pp. 2337–2344. https://doi.org/10.1109/CEC.2013.6557848
5. Yang, X.S.: A new metaheuristic bat-inspired algorithm. In: Gonzalez, J.R., et al. (eds.) NICSO 2010, vol. 284, pp. 65–74. Springer, Heidelberg (2010)
6. Yang, X.S., He, X.: Firefly algorithm: recent advances and applications. Int. J. Swarm Intelligence **1**(1), 36–50 (2013). https://doi.org/10.1504/IJSI.2013.055801
7. Wang, G.G., Dos Santos Coelho, L., Gao, X.Z., Deb, S.: A new metaheuristic optimisation algorithm motivated by elephant herding behaviour. Int. J. Bio-Inspired Comput. **8**(6), 394 (2016)
8. Mandal, S.: Elephant swarm water search algorithm for global optimization. Sādhanā **43**, 2 (2018). https://doi.org/10.1007/s12046-017-0780-z
9. Mirjalili, S., Lewis, A.: The whale optimization algorithm. Adv. Eng. Softw. **95**, 51–67 (2016)
10. Orca Social Organization: OrcaLab, 24 February 2019
11. All About Killer Whales. Communication and Echolocation — SeaWorld Parks and Entertainment. https://seaworld.org/animals/all-about/killer-whale/communication/. Accessed 30 Jan 2019
12. Killer Whale Hunting Strategies, 24 November 2014. http://www.pbs.org/wnet/nature/killer-whales-killer-weapon-brain/11352/. Accessed 02 Jan 2019
13. Welcome to Maze Benchmark. Maze Benchmark for Evolutionary Algorithms. https://mazebenchmark.github.io/. Accessed 15 Apr 2019
14. Bagnall, A.J., Zatuchna, Z.V.: On the classification of maze problems. In: Bull, L., Kovacs, T. (eds.) Foundations of Learning Classifier Systems, vol. 183, pp. 305–316. Springer, Heidelberg (2005)

Author Index

Á. Rocha et al. (Eds.): WorldCIST 2020, AISC 1159, pp. 799–802, 2020.
https://doi.org/10.1007/978-3-030-45688-7